Applied Mathematics for the Management, Life, and Social Sciences

Applied Mathematics for the Management, Life, and Social Sciences

Stanley I. Grossman
University of Montana

Wadsworth Publishing Company
Belmont, California
A Division of Wadsworth, Inc.

Mathematics Editor: Jim Harrison
Production: Del Mar Associates
Copy Editors: Linda L. Thompson, Rebecca Smith
Technical Illustrators: Pat Rogondino, Pam Posey
Cover Design: John Odam
Signing Representative: Jane Moulton

Cover Painting: Ben Nicholson, *October 1962*, The Tate
Gallery, London.

Printed in the United States of America

1 2 3 4 5 6 7 8 9 10—89 88 87 86 85

ISBN 0-534-04239-2

Library of Congress Cataloging in Publication Data

Grossman, Stanley I.
    Applied mathematics for the management, life, and
social sciences.

    Includes index.
    1. Mathematics—1961-   . I. Title.
QA37.2.G76   1985   510   84-26994
ISBN 0-534-04239-2

To Kerstin, Erik, and Aaron

CONTENTS

13 INTRODUCTION TO MULTIVARIABLE CALCULUS 734

APPENDIX Review of Three Topics in Algebra A-1

TABLES

ANSWERS TO ODD-NUMBERED PROBLEMS A-35

INDEX A-87

Preface

Students in business and the social and biological sciences require a variety of mathematical tools for their studies. *Applied Mathematics for the Management, Life, and Social Sciences* provides many of these tools. Approximately two-thirds of the book (Chapters 1–8) covers material commonly gathered under the topic *finite mathematics* and can be covered in one semester or two quarters. The remaining five chapters provide an introduction to calculus. These last chapters can be covered in one quarter or one semester.

Applied Mathematics for the Management, Life, and Social Sciences is an outgrowth of two courses I have taught many times at the University of Montana: Mathematics for Business and Mathematics for the Biological Sciences. The prerequisite for students in these courses is intermediate algebra, and that is the prerequisite for this text. However, for those students who need to review some frequently used topics in algebra, I have provided an appendix that includes detailed discussions of polynomials, quadratic equations, and rational expressions. This supplements the preliminary material in Chapter 1.

Some of the features of *Applied Mathematics for the Management, Life, and Social Sciences* are:

Examples. As a student, I learned this material from seeing examples and doing exercises. There are 668 examples—many more than are commonly found in texts at this level. The examples include all the necessary steps so that students can see clearly how to get from "A to B." In many instances, explanations are highlighted in color to make steps easier to follow.

Exercises. The text includes more than 3700 exercises—both drill and applied problems. More difficult problems are marked with an asterisk (*), and a few especially difficult ones are marked with two (**). In my opinion, exercises provide the most important learning tool in any undergraduate mathematics textbook. I stress to my students that no matter how well they think they understand my lectures or the textbook, they do not really know the material until they have worked problems. There is a vast difference between understanding someone else's solution and solving a new problem by yourself. Learning mathematics without doing problems is about as easy as learning to ski without going to the slopes.

Chapter Review Exercises. At the end of each chapter, I have provided a collection of review exercises. Any student able to do these exercises can feel confident that he or she understands the material in the chapter.

Applications. Although every similar text contains applied problems as illustrations, they are usually easily stated applications with fairly simple solutions. In this text, a great number of longer applied problems are provided throughout. I have also included large, section-long treatments of applications. (See Sections 2.6, 3.5, and 7.3.) Moreover, whenever possible I have used real data. For example, in the section on input-output analysis (see page 138), I use Leontief's original data to model the 1958 U.S. economy. As another example, in my discussion of statistical graphing techniques, I use data as it might come to a statistician (see page 314), rather than contrived data that do not really exemplify the statistical techniques being taught.

The calculus material (Chapters 9 through 13) is written to show calculus as a tool, rather than as a collection of abstract theorems. Applications with realistic data are included early and often. The following is a partial list of interesting applications:

The Dow Jones Averages (Problem 9.1.57 on page 480)

Introduction to the derivative by considering the velocity of a falling rock (Example 10.4.2 on page 538)

Velocity and marginal cost—first considered as an application of the derivative in Section 10.5 and then as an application of the indefinite integral or antiderivative in Section 12.2

Optimal fleet size for a car leasing company (Example 11.3.8 on page 623)

Exponential growth and decay (Section 11.6)

Finding the effective interest rate when a boat is purchased in installments (Example 11.7.6 on page 658)

GNP and the national debt (Example 12.3.12 on page 678)

The Cobb-Douglas Production Function (Examples 13.1.4 13.2.5, and 13.4.5 on pages 735, 746, and 771)

Business Mathematics. Several features are intended especially for business students, and I would like to mention three of them here. In Sections 4.2, 4.3, 6.2, 8.3, and 11.4, I have included problems taken from the uniform CPA exams. Chapter 8 contains a fairly complete treatment of business mathematics. The last section of the chapter covers the mathematics of bonds—a topic usually omitted from texts of this type. Section 11.4 contains an extensive discussion of marginal analysis.

Use of the Hand Calculator. Virtually all college and university students own or have access to a hand calculator. Problems that twenty years ago were computational monstrosities have become fairly easy with the aid of a calculator. It would be anachronistic to ignore this tool in any mathematics text—especially one that claims to be "applied." For that reason I have used the calculator in many examples and have suggested its use in a number of exercises. Examples and exercises that require the use of a calculator are marked with the symbol [⊞] .

In addition, there are two sections in the book that deal exclusively with numerical techniques and require the use of a calculator or computer: Section 11.7 discusses Newton's method in detail; the trapezoidal rule for numerical integration is covered in Section 12.10.

Answers and Other Aids. The answers to most odd-numbered problems appear at the back of the book. In addition, Tom Cromer, Mark Eastman, and Greg St. George have prepared a student's manual that contains worked-out solutions to all odd-numbered problems and an instructor's manual that contains solutions to all even-numbered problems.

Numbering of Examples, Problems, and Equations. Numbering in the book is fairly standard. Within each section, examples, problems, and equations are numbered consecutively, starting with 1. An example, problem, or equation outside the section in which it appears is referenced by chapter, section, and number. For example, Example 2 in Section 3.4 is called, simply, Example 2 in that section, but outside the section it is referred to as Example 3.4.2. Often, to make it easier to find an example, problem, or equation that appears earlier in the text, I have indicated the number of the page on which the item appears.

Chapter Interdependence. The following chart indicates interdependence—that is, which later chapters depend on the student's having mastered earlier material. Note that Chapter 13 may follow either Chapter 11 or Chapter 12.

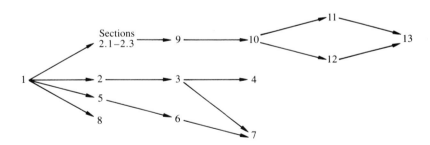

ACKNOWLEDGMENTS

I have used a variety of sources in the preparation of this text. I am especially grateful to the following publishers for permission to use valuable material:

Harper & Row, Publishers, Inc., for permission to use the example on page 650 and Problem 1 on page 651 in *Mathematical Analysis: Business and Economics Applications,* 4th ed., by Jean E. Weber, copyright © 1982.

The American Institute of Certified Public Accountants, Inc., for permission to reprint material from *Uniform CPA Examination Questions and Unofficial Answers,* copyright © 1975, 1976, 1977, 1978, 1979, 1980.

The American Association for the Advancement of Science, publishers of *Science '85,* for permission to excerpt material from an article by William Foster Allman that appeared in the July/August 1981 issue of *Science '81,* pages 36–37.

Dow Jones & Company, Inc., for permission to reprint their table of the Dow Jones Averages from April 15 to July 15, 1983.

CBS College Publishing for permission to reprint Problems 11, 12, 13, and 14 on pages 68–69 of *Managerial Economics,* 3rd ed. (1979), by James L. Pappas and Eugene F. Brigham.

John Wiley & Sons, Inc., for permission to adapt material from *Games and Decisions*, by Luce and Raiffa (1958), pages 64–65.

John Wiley & Sons, Inc. for permission to use Figure 5.5 on page 109 of *Elementary Statistics, an Applied Approach* (1978) by Neil R. Ullman.

Academic Press, Inc., for permission to excerpt material from pages 6–10 of *Applied Linear Programming for the Socioeconomic and Environmental Sciences* (1978) by Michael R. Greenberg.

Academic Press, Inc., for permission to use material from my text *Calculus,* 3rd ed. (1984), in Chapters 9–13 of this book.

Kent Publishing Company, a division of Wadsworth, Inc., 20 Providence Street, Boston, MA 02116, for permission to use Problems 12 and 13 (including Table P9-13) on page 473 of *Quantitative Models for Management* (1981) by K. Roscoe Davis and Patrick G. McKeown.

Ellis Horwood Limited, Market Cross House, Cooper Street, Chichester, England, for permission to use the following problems from their excellent text *Mathematical Models in the Social, Management and Life Sciences* (1980) by D. N. Burgher and A. D. Wood: Problems 12, 13, 14, 19, 20, and 21 on pages 59–60, Case Study 2 on page 73, Case Study 3 on page 75, Problems 6, 7, 8, 9, 10, 11, 14, and 16 on pages 78–79, and Case Study 1 on pages 88–89.

The British Museum in London for permission to reproduce the photo of the Rhind papyrus that appears on page 695 of this text.

West Publishing Company, for permission to use Problems 14 and 15 on pages 66–67, Problems 22 and 24 on pages 69–70, Problems 15 and 16 on page 127, and Problem 24 on page 130 of *Introduction to Management Science,* 2nd ed., by David R. Anderson, Dennis J. Sweeney, and Thomas A. Williams, copyright © 1979; and Problems 9, 10, 13, and 14 on pages 656–657 of *Introduction to Management Science,* 3rd ed., copyright © 1982.

Wadsworth Publishing Company for permission to use a variety of problems and examples from *Applied Finite Mathematics* (1977) by Robert Brown and Brenda Brown.

Finally, I owe a considerable debt to Professor James E. Turner of McGill University, my coauthor of *Mathematics for the Biological Sciences,* originally published by Macmillan in 1974. Some of the examples and exercises in the present text first appeared in that earlier work, and others were provided to me privately by Professor Turner.

Most of us really don't know what a book is like until it's been used in class and we get comments on how it works. That is why later editions of texts are sometimes better learning and teaching tools than earlier editions. However, although a first edition can never be a third edition, it can come close if we get and respond to enough advance help and comments from experienced teachers. The following reviewers have made invaluable contributions to the reliability and teachability of this text: Martin Billik, San Jose State University; Duane Blumberg, University of Southern Louisiana; Gerald L. Bradley, Claremont McKenna College; James Crenshaw, Southern Illinois University; Bruce Edwards, University of Florida; Garret J. Etgen, University of Houston; Harry D. Eylar, California State University, Long Beach; Jean Ferris, University of Colorado; Gary Grimes, Mount Hood Community College; Kenneth Hart, Ricks College; Anthony Lepre, Lehigh County Community College; Laurence Maher, North Texas State University; James Phillips, Cameron University; Richard Porter, Northeastern University; Ken Rager, Metropolitan State College; Eric Robinson, Ithaca College; Derald Rothmann, Moorhead State University; Robert Russell, West Valley College; Wes Sanders, Sam Houston State University; Michael Spinelli, Virginia Commonwealth University; Ann Watkins, Los Angeles Pierce College; and Al White, St. Bonaventure University.

Students depend on correct answers to problems as an important way of verifying their progress. Tom Cromer at the University of Alabama, Huntsville, Mark Eastman at the University of California, Santa Cruz, and Greg. St. George at the University of Montana provided answers to all problems. Equally important, they made many useful suggestions for the improvement of the problem sets that, I am confident, have greatly enhanced the usefulness of this text. I am grateful to all three of them.

Finally, I am especially grateful to Richard Jones, former Mathematics Editor at Wadsworth, and Jim Harrison, the present Mathematics Editor. Both provided the guidance, insight, and sometimes not-so-gentle prodding that helped me both to complete this book and to retain my sanity. They are the best in the business.

Stanley I. Grossman

1 SETS AND ALGEBRA

1.1 Sets

The idea of a *set* is a very general concept that comes up in virtually every area being studied. A **set** is a *well-defined* collection of distinct objects. The objects that make up the set are called **elements**, or **members**, of the set. For example, the pages of this book comprise a set. An element of the set is a particular page of the book. Other examples of sets are the set of raw materials used by one manufacturer, the set of banks in Atlanta, Georgia, the set of guests in a given resort hotel, and the set of species that became extinct between 1900 and 1980.

If you look at the examples given above, you might conclude that there are many different kinds of sets, which is true. But not every collection of objects makes up a set. For example, the collection of "large numbers" does not form a set because the collection is not well-defined; that is, there is no universal agreement about what constitutes a large number. If you must drive to work every day, then 50 miles is a large number of miles. On the other hand, distances of a few million miles are small to an astronomer. Similarly, the collections of good athletes, beautiful paintings, and delicious foods do not define sets because there is no general agreement on what is a good athlete, a beautiful painting, or a delicious food.

We will deal with sets in many sections of this book. The purpose of this section is to provide some useful facts about them.

Notation. We will denote sets by capital letters such as A, B, C, S, and T. We will use two different ways to denote sets; with each, we use brackets.

Example 1 The set S containing the numbers 2, 4, 6, 8 can be denoted by

$$S = \{2, 4, 6, 8\} \tag{1}$$

or

$$S = \{x : 0 < x < 10 \text{ and } x \text{ is an even integer}\}. \tag{2}$$

The second expression is read "S is the set of all x such that x is between 0 and 10 and x is an even integer." In (1), we simply listed the members of the set. In (2), we described properties satisfied by every member of that set. You should convince yourself that (1) and (2) describe the same set.

Example 2 The set P of primary colors can be denoted by

$$P = \{\text{red, yellow, blue}\}$$

or

$$P = \{x : x \text{ is a primary color}\}.$$

Example 3 The following are sets written in bracket notation.

$$A = \{x : x \text{ is a state in the United States}\}$$

$$B = \{x : x \text{ is a raw material used to produce an automobile}\}$$

$$C = \{x : x \text{ is a patient in a certain hospital}\}$$

If x is an element of a set S, we write

$$x \in S.$$

For example, in Example 3 Minneapolis $\in A$, iron $\in B$, and Peter Jones $\in C$ if Peter Jones is a patient in the hospital.
  If x is not an element of S, we write

$$x \notin S.$$

There are many facts that we will use about sets.

The Empty Set
This is the set with no elements and is denoted by ϕ.

$$\boxed{\phi \text{ denotes the empty set.}} \tag{3}$$

Equality
Two sets A and B are equal if they contain the same elements. We write $A = B$.

Example 4 Let $A = \{\text{California, Oregon, Washington, Hawaii, Alaska}\}$ and $B = \{x : x \text{ is a state in the United States bordered by the Pacific Ocean}\}$. Then $A = B$.

Subset

We say that A is a **subset** of B if every element of A is an element of B. Also, A is a **proper subset** of B if A is a subset of B but $A \neq B$.

$$A \subseteq B \text{ denotes } A \text{ is a subset of } B. \tag{4}$$

$$A \subset B \text{ denotes } A \text{ is a proper subset of } B. \tag{5}$$

Example 5 Let $A = \{$Washington, Oregon, California$\}$ and B be the set in Example 4. Then $A \subset B$.

Example 6 List the subsets of $A = \{x, y, z\}$.

Solution A subset can have 0, 1, 2, or 3 elements. Thus the subsets are

ϕ	No elements
$\{x\}, \{y\}, \{z\}$	One element
$\{x, y\}, \{x, z\}, \{y, z\}$	Two elements
$A = \{x, y, z\}$	Three elements

Note that A has eight subsets.

Universal Set

In a discussion of particular sets, the **universal set** is a set containing the elements of all other sets under discussion;

$$\Omega \text{ denotes the universal set.} \tag{6}$$

In Examples 4 and 5, Ω could be the set containing all states in the United States. If we are discussing sets of numbers, then Ω will usually be taken to be the set of all real numbers (see Section 1.2).

Union

The **union** of set A and set B is the set composed of the elements of A, together with the elements of B.

$$A \cup B \text{ denotes the union of } A \text{ and } B. \tag{7}$$

We have

$$A \cup B = \{x : x \in A \text{ or } x \in B \text{ or both}\}. \tag{8}$$

Example 7   Let $A = \{1, 2, 4, 7, 9, 13\}$ and $B = \{2, 5, 6, 7, 10\}$. Then $A \cup B$ is the set of numbers in A or B or both; that is,

$$A \cup B = \{1, 2, 4, 5, 6, 7, 9, 10, 13\}.$$

Example 8   Let $A = \{\text{male smokers in the United States}\}$ and $B = \{\text{fathers in the United States}\}$. Then $A \cup B$ is the set of all males in the population who are either smokers or fathers (or both).

Intersection

The **intersection** of set A and set B is the set of elements contained in *both* A and B.

$$A \cap B \text{ denotes the intersection of } A \text{ and } B. \tag{9}$$

We have

$$A \cap B = \{x : x \in A \text{ and } x \in B\}. \tag{10}$$

Example 9   If $A = \{1, 2, 4, 7, 9, 13\}$ and $B = \{2, 5, 6, 7, 10\}$, as in Example 7, then $A \cap B$ is the set of numbers in both A and B; that is,

$$A \cap B = \{2, 7\}.$$

Example 10   If A and B are the sets described in Example 8, then $A \cap B$ is the set of males in the population who are both fathers and smoke.

Venn Diagram

A **Venn diagram** gives a convenient way to depict the set operations we have discussed. We illustrate some Venn diagrams in Figure 1.

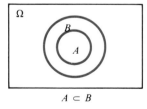

$A \subset B$

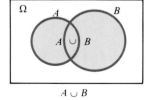

$A \cup B$

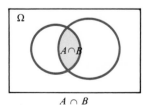

$A \cap B$

Figure 1

Complement

The **complement** of a set A is the set consisting of all elements of the universal set Ω that are not in A.

$$A^c \text{ denotes the complement of } A. \tag{11}$$

We have

$$A^c = \{x : x \in \Omega \text{ and } x \notin A\}. \tag{12}$$

Example 11 Let $A = \{x : x \text{ is a state east of the Mississippi River}\}$. Then $A^c = \{x : x \text{ is a state west of the Mississippi River}\}$. It is understood that the universal set Ω consists of the states of the United States.

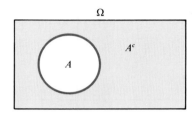

Figure 2

The complement of a set is depicted by a Venn diagram, as in Figure 2. Note that (see Problems 15 and 16)

$$A \cup A^c = \Omega \tag{13}$$

and

$$A \cap A^c = \phi. \tag{14}$$

Difference of Two Sets If A and B are two sets, then $A - B$, called the **difference** between A and B, is the set of elements in A that are not in B. We have

$$A - B = \{x : x \in A \text{ and } x \notin B\}. \tag{15}$$

Example 12 Let $A = \{\text{California, New York, Florida, Ohio}\}$ and $B = \{\text{Montana, Ohio, Florida, New Mexico, Minnesota}\}$. Then

$$A - B = \{\text{California, New York}\}$$

and

$$B - A = \{\text{Montana, New Mexico, Minnesota}\}.$$

The difference of two sets is shown in Figure 3.

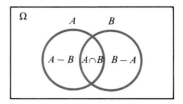

Figure 3

From Figure 3 it is evident that

$$A \cup B = (A - B) \cup (A \cap B) \cup (B - A). \tag{16}$$

Example 13 In a population, define M to be the set of males and S to be the set of stockbrokers. Then $M^c = \{$females$\}$, $S^c = \{$people who are not stockbrokers$\}$, $M - S = \{$males who are not stockbrokers$\}$, $S - M = \{$female stockbrokers$\}$, $M \cup S = \{$people who are males or stockbrokers or both$\}$, and $M \cap S = \{$male stockbrokers$\}$. It is easy to see that

$$M \cup S = (M - S) \cup (M \cap S) \cup (S - M).$$

Disjoint Sets Two sets A and B are said to be **disjoint** if they have no elements in common; that is

$$A \text{ and } B \text{ are disjoint if } A \cap B = \phi. \tag{17}$$

Example 14 For any set A, A and A^c are disjoint.

Example 15 Let $A = \{$fathers who smoke$\}$ and $B = \{$mothers who bowl$\}$. Then, since no one is simultaneously a father and a mother, $A \cap B = \phi$.

We will be dealing with sets and the set operations of union, intersection, and complement in many parts of this book. We close this section with one further example.

Example 16 Among a group of 170 business students, 70 are taking at least one course in accounting, 95 are taking economics, and 80 are taking mathematics. Suppose that 30 students are taking both mathematics and accounting, 35 are taking both mathematics and economics, and 15 are taking both accounting and economics. Suppose further that 5 students are taking all three subjects. How many students are taking exactly two of the three subjects?

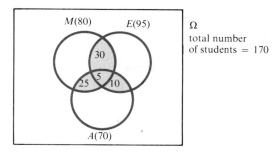

Figure 4

Solution This problem is most easily solved by using a Venn diagram. By adding the numbers in the shaded regions of Figure 4, we find that 25 + 30 + 10 = 65 students are studying exactly two of the three subjects. To obtain these numbers note, for example, that $M \cap E$ contains 35 students and $(M \cap E) \cap A$ contains 5 students. Thus there are 30 students in $M \cap E$ who are not in A. In this problem, note that the total number of students, as well as the total number taking each subject, are irrelevant.

PROBLEMS 1.1 In Problems 1–10, determine whether the given collection is a set.

1. The collection of naturally occurring chemical elements

2. The collection of small businesses in the United States

3. The collection of mountain ranges in the world

4. The collection of possible outcomes when two dice are thrown

5. The collection of good poker hands

6. The collection of U.S. tax laws

7. The collection of all students at your school who have an average of 3.0 (**B**) or better

8. The collection of good students in your school

9. The collection of monkeys who hate bananas

10. The collection of popular songs

11. Let $A = \{u, v, w, z\}$. Which of the following statements are true?

(a) $v \in A$ (b) $w \notin A$ (c) $\{u, w\} \subset A$

(d) $\{u, v, x\} \subset A$ (e) $A \subset A$ (f) $A \subseteq A$

(g) $\{w, z\} \subseteq A$ (h) $\phi \in A$ (i) $\phi \subset A$

12. Let $A = \{u, v, w, x\}$. List all the subsets of A.

13. In a large company, let Ω denote the set of its employees and let

$$A = \{x : x \text{ is a female employee}\},$$

$$B = \{x : x \text{ is an executive of the company}\},$$

$$D = \{x : x \text{ has a vested interest in the company's pension plan}\}.$$

Describe, in words, the following sets.

(a) $A \cup B$ (b) A^c (c) B^c

(d) $A \cap D$ (e) $B - D$ (f) $D - A$

(g) $A \cup B \cup D$ (h) $B \cap D^c$ (i) $A \cup (B \cap D)$

(j) $(A \cap B) \cap D$ (k) $(A \cap B) \cup D$ (l) $A \cap (B \cup D)$

14.   In Problem 13, show that $A \cup B = (A - B) \cup (A \cap B) \cup (B - A)$.

15.   For any set A, show that $A \cup A^c = \Omega$. [*Hint*: Show that if $x \in \Omega$, then either $x \in A$ or $x \in A^c$.]

16.   For any set A, show that $A \cap A^c = \phi$.

17.   (a) In Example 16, how many students are taking mathematics only? (b) Economics only? (c) How many are taking none of the three subjects?

18.   A company produces 100 products. Of these 63 require steel, 39 require aluminum, and 72 require copper. In addition, 44 require both steel and copper, 25 both aluminum and copper, and 20 both steel and aluminum. Finally, 13 require all three materials.

(a)   Draw a Venn diagram to illustrate this situation.

(b)   How many products require no steel, aluminum, or copper?

(c)   How many require exactly one of these materials?

(d)   How many require exactly two materials?

19.   In a study of blood groups, 10,000 people were tested. Of those, 5500 have antigen A, 2500 have antigen B, and 3000 have neither antigen. Define A, B, and O to be these three sets of people.

(a)   Draw a Venn diagram that illustrates these tests.

(b)   Describe in words the sets $A \cup B$, $A \cap B$, $A \cap O$, $(A \cup B) \cap O$, A^c and $A^c \cap B$.

20.   A public relations firm interviewed 1000 families and asked about the number of color television sets they owned and their annual income. The results are given in the table.

Color TV sets owned	Income			
	Less than $5000	$5001 to $10,000	$10,001 to $20,000	More than $20,000
None	73	26	12	2
One	109	84	118	58
Two	48	63	144	162
Three or more	2	8	26	6

Let A represent a family earning between $10,001 and $20,000; B, a family with two color TV sets; C, a family earning under $10,001; D, a family with at least one color TV set; and E, a family earning more than $20,000. Describe, in words, the following sets.

(a) $A \cup B \cup C$ (b) $C \cap D$ (c) $C - E$

(d) $E - B$ (e) $A \cap E$ (f) $B \cap D$

(g) $(D - B) \cup A$ (h) C^c (i) $(A \cap B)^c$

21. In Problem 20, determine the number of elements in each set.

22. In Problem 20, determine which of the given sets, if any, are proper subsets of other given sets.

23. Explain the difference between the sets ϕ and $\{\phi\}$. [*Hint*: Explain why $\{\phi\}$ is *not* the empty set.]

24. Using a Venn diagram, show that $(A \cup B)^c = A^c \cap B^c$.

25. Using a Venn diagram, show that $(A \cap B)^c = A^c \cup B^c$.

26. How many subsets has a set containing four elements?

27. Using a Venn diagram, show that $A \cup (B \cap C) = (A \cup B) \cap (A \cup C)$.

28. Let A and B be sets each containing a finite number of elements. Let $n(A)$ denote the number of elements of A. Using a Venn diagram, show that

$$n(A - B) = n(A) - n(A \cap B).$$

29. Use the result of Problem 28 and the formula (16) to show that $n(A \cup B) = n(A) + n(B) - n(A \cap B)$.

1.2 Sets of Real Numbers

The sets that we shall encounter most often in this book are sets of numbers. To desribe these sets, we begin with the **real number line** (or **real line**), which is a line that extends infinitely in both directions and contains a point labeled *zero* and called the **origin** (see Figure 1).

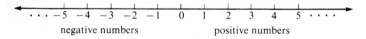

Figure 1

Points on the line are called **real numbers**, and the set of real numbers is denoted by $\mathbb{R}$. Unless otherwise stated, when dealing with sets of numbers the universal set will be understood to be $\mathbb{R}$. Numbers to the right of zero are called **positive** numbers while those to the left of zero are called **negative** numbers. There are certain sets of numbers that have special names. We list these sets below.

The Natural Numbers, Denoted by *N* These are the counting numbers:

$$N = \{1, 2, 3, 4, 5, \ldots\}. \tag{1}$$

The Integers, Denoted by *I* This is the set consisting of the natural numbers, their negatives, and zero:

$$I = \{0, \pm 1, \pm 2, \pm 3, \pm 4, \ldots\}. \tag{2}$$

Note that $N \subset I$.

The Rational Numbers, Denoted by Q   This is the set consisting of numbers that can be written as the quotient of two integers:

$$Q = \left\{ x : x = \frac{m}{n}, \text{ where } m \in I \text{ and } n \in I, n \neq 0 \right\}. \qquad (3)$$

Example 1   Let m be an integer. Then $m = m/1$, so that m is also a rational number. Thus $I \subset Q$.

Example 2   The following are rational numbers.

(a) $\dfrac{3}{4}$ (b) $\dfrac{-1}{2}$ (c) $\dfrac{125}{6125}$ (d) $\dfrac{6}{8}$ (e) $-\dfrac{21,785}{20,195,784}$

Note.   A rational number can be written in many ways. For example,

$$\frac{1}{2} = \frac{2}{4} = \frac{3}{6} = \frac{-7}{-14} = \frac{2001}{4002}.$$

Example 3   Every terminating decimal represents a rational number. For example,

$$0.17 = \frac{17}{100} \quad \text{and} \quad -0.0235 = \frac{-235}{10,000}.$$

It can be shown that every repeating decimal represents a rational number.

Example 4   Express $0.232323\ldots$ as a rational number.

Solution   Let $x = 0.232323\ldots$. Then $100x = 23.232323\ldots$. We write these expressions, one above the other, and then subtract.

$$100x = 23.232323\ldots$$
$$\underline{x = 0.232323\ldots}$$
$$99x = 23$$

and $x = \frac{23}{99}$. Note that if you divide $\frac{23}{99}$ on a hand calculator displaying ten digits, you will obtain 0.2323232323.

The relationship between $\mathbb{R}$, N, I, and Q is shown in the Venn diagram in Figure 2.

The preceding examples might give you the idea that almost all numbers are rational. This is not the case. It is true, although difficult to prove, that if we removed all the rational numbers from the real line we would have essentially

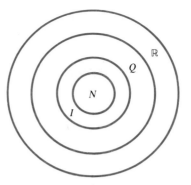

Figure 2

the same number of numbers with which we started. Numbers that are not rational are called **irrational** numbers. The number $\sqrt{2}$ is an irrational number. We show why this is true in Problem 29. To ten significant figures, $\sqrt{2} =$ 1.414213562.

Another well-known irrational number is π, which is defined to be the ratio of the circumference to the diameter of a circle (see Figure 3). To ten significant figures $\pi = 3.141592654$.

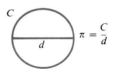

Figure 3

One kind of commonly occurring set of real numbers is called an *interval*.

Open Interval   An **open interval** is a set O of the form

$$O = \{x : a < x < b, \text{ where } a \text{ and } b \text{ are real numbers}\}. \qquad (4)$$

We write

$$O = (a, b).$$

Closed interval   A **closed interval** is a set C of the form

$$C = \{x : a \leq x \leq b, \text{ where } a \text{ and } b \text{ are real numbers}\}. \qquad (5)$$

We write

$$C = [a, b].$$

In both types of intervals the points a and b are called **endpoints** of the interval. Open and closed intervals are shown in Figure 4.

(a) open interval; a, b not included   (b) closed interval; a, b included

Figure 4

Note.   An open interval does *not* contain its endpoints, while a closed interval does contain its endpoints.

Half-Open (or Half-Closed) Interval   A half-open interval H is an interval of real numbers containing one, but not both, of its endpoints. We have either

$$H = \{x : a < x \le b\} \quad \text{or} \quad H = \{x : a \le x < b\} \tag{6}$$

and H is denoted by $(a, b]$ or $[a, b)$.

Example 5   The following are examples of intervals.

(a) $(3, 5)$ is open.
(b) $[-2, 4.1]$ is closed.
(c) $(1.2, 6]$ is half-open.
(d) $[-17, \sqrt{2})$ is half-open.

Example 6   Let $A = (1, 7)$, $B = [2, 3)$, and $C = [5, 8]$. Compute (a) $A \cup C$, (b) $A \cap C$, (c) B^c, (d) $B - A$, and (e) $(C - A) \cup B$.

Solution
(a) $A \cup C = \{x : x \in (1, 7) \text{ or } x \in [5, 8]\} = (1, 8]$.
(b) $A \cap C = \{x : x \in (1, 7) \text{ and } x \in [5, 8]\} = [5, 7)$ (since $5 \in A$ and $5 \in C$, but $7 \notin A$).
(c) $B^c = \{x : x \notin [2, 3)\} = \mathbb{R} - [2, 3) = \{x : x < 2 \text{ or } x \ge 3\}$.
(d) $B - A = \{x : x \in [2, 3) \text{ and } x \notin (1, 7)\} = \phi$ since $B \subset A$.
(e) $C - A = \{x : x \in [5, 8] \text{ and } x \notin (1, 7)\} = [7, 8]$. Thus $(C - A) \cup B = \{x : x \in [2, 3) \text{ or } x \in [7, 8]\} = [2, 3) \cup [7, 8]$.

We close this section by describing several kinds of **infinite** intervals. To do this we need to introduce the symbol ∞, **infinity**, which is not a number. The expression $x < \infty$ indicates that x is a real number.

Using this symbol, we can write the set of real numbers as

$$\mathbb{R} = \{x : -\infty < x < \infty\} = (-\infty, \infty). \tag{7}$$

Thus $(-\infty, \infty)$ is another kind of open interval. Using the infinity symbol, we can write other sets of real numbers as intervals: for example,

$$(a, \infty) = \{x : x > a\} \tag{8}$$

and

$$(-\infty, b) = \{x : x < b\}. \tag{9}$$

The intervals given in (8) and (9) are open intervals.
Similarly,

$$[a, \infty) = \{x : x \geq a\} \tag{10}$$

and

$$(-\infty, b] = \{x : x \leq b\} \tag{11}$$

are half-open intervals.

Example 7   If $b > a$, the interval $[a, b]$ can be written as

$$[a, b] = \{x : x \geq a \quad \text{and} \quad x \leq b\} = [a, \infty) \cap (-\infty, b].$$

For instance,

$$[-2, 5] = \{x : x \geq -2 \quad \text{and} \quad x \leq 5\} = [-2, \infty) \cap (-\infty, 5].$$

PROBLEMS 1.2   In Problems 1–8, let $A = \{1, 2, 4\}$, $B = \{-3, 0, 2, 4, 7\}$, and $C = \{-8, -3, 1, 5\}$. Compute each of the following.

1. $A \cup B$	**2.** $A \cap C$	**3.** $C - B$
4. $A - C$	**5.** $A \cup B \cup C$	**6.** $(A \cap B) \cap C$
7. $C - A^c$	**8.** B^c	

In Problems 9–14, write the repeating decimal as a rational number.

9. $0.22222\ldots$	**10.** $0.1313131313\ldots$
11. $0.000010101010\ldots$	**12.** $-6.25111111\ldots$
13. $0.205205205\ldots$	**14.** $1234.56565656\ldots$

In Problems 15–28, compute with $A = [1, 5]$, $B = [-7, 2)$, $C = (-1, 8)$, and $D = (5, 8]$.

15. $A \cup B$

16. $C - A$

17. $A - C$

18. $D \cup B$

19. $D \cap B$

20. $A \cup B \cup C \cup D$

21. $(A \cap B) \cap (C \cap D)$

22. $D - (A \cap B)$

23. $C - (B - A)$

24. $D - [C - (B - A)]$

25. $D - C$

26. B^c

27. $(D - B)^c$

28. $(A \cap C)^c$

*29. Show that $\sqrt{2}$ is irrational by assuming it is rational and obtaining a contradiction. Do this by using the following steps:

   (a) Write $\sqrt{2}$ as a rational number m/n, where m and n have no common factors.

   (b) Square the expression $\sqrt{2} = m/n$.

   (c) Show that m^2 must be even.

   (d) Using (c), show that m is even.

   (e) Using (b) and (d) show that n is even.

   (f) Show why (d) and (e) provide a contradiction of the assumption in (a).

*30. Using the procedure in Problem 29 show that $\sqrt{3}$ is irrational.

31. Write the open interval (a, b) as the intersection of two infinite intervals.

In Problems 32–37, write each interval as the intersection of two infinite intervals.

32. $[2, 3]$

33. $(-3, 7)$

34. $[-1, 6)$

35. $(-1, 6]$

36. $[4, 83)$

37. $(4, 83]$

1.3 Solving Linear Equations and Inequalities in One Variable

One of the central ideas in algebra is that of solving an equation for an unknown variable. If x stands for an unknown quantity and is the only unknown quantity, then an equation involving x is called an **equation in one variable**.

Example 1 The following are equations in one variable.

(a) $2x = 4$

(b) $x - 7 = 5x + 2$

(c) $x^2 + 4x + 3 = 0$

(d) $\dfrac{x}{2} = \dfrac{5}{x}$

Linear Equation

If an equation in one variable can be written in the form

$$ax + b = c \qquad (1)$$

where a, b, and c are real numbers and $a \neq 0$, then the equation is called **linear**.

Example 2   In Example 1 the equations in (a) and (b) are linear, while the other two equations are not. In (a), $2x = 4$ means that $2x - 4 = 0$, which is in form (1) with $a = 2$, $b = -4$, and $c = 0$. Likewise, in (b), $x - 7 = 5x + 2$ means that $4x + 2 = -7$, with $a = 4$, $b = 2$, and $c = -7$.

In order to solve linear equations, we rely on two rules, which follow from the basic laws of arithmetic.

Let a, b, and c denote real numbers.

ADDITION RULE

$$\text{If } a = b, \text{ then } a + c = b + c. \qquad (2)$$

MULTIPLICATION RULE

$$\text{If } a = b, \text{ then } ac = bc. \qquad (3)$$

From these two rules, we can deduce one other. Suppose that $a = b$. If $c \neq 0$, then $1/c$ is a real number and, from the multiplication rule,

$$a \cdot \frac{1}{c} = b \cdot \frac{1}{c}$$

or

DIVISION RULE

$$\text{If } a = b \text{ and } c \neq 0, \text{ then } \frac{a}{c} = \frac{b}{c}. \qquad (4)$$

We now use these rules to solve linear equations.

Example 3 Solve the equation $2x = 4$.

Solution If $2x = 4$, then, by the division rule,

$$\frac{2x}{2} = \frac{4}{2} \quad \text{or} \quad x = 2.$$

Example 4 Solve the equation $x - 7 = 5x + 2$.

Solution

$$x - 7 - x = 5x + 2 - x \qquad \text{Addition rule}$$

or

$$-7 = 4x + 2 \qquad x - x = 0 \text{ and } 5x - x = 4x$$
$$-7 - 2 = 4x + 2 - 2 \qquad \text{Addition rule}$$

or

$$-9 = 4x$$
$$x = -\tfrac{9}{4} \qquad \text{Division rule}$$

Before going further, some comments are in order. The last two examples were very easy—so easy, in fact, that there is some risk that an important point may be lost. It is unlikely that you see anything new in these two examples. The important point is that these equations can be solved because we used established rules. Without rules it would be very difficult to solve any mathematical equation. Throughout much of this text we will be developing rules for solving a very wide variety of problems. In fact, the three rules we used in this section can be generalized in a very natural way to solve equations involving matrices—a new and important topic that we will encounter in Chapter 3.

A frequently asked question is "Why bother to study this stuff at all?" Linear equations arise in virtually every area of science, social science, economics and business. We provide one example from business.

BREAK-EVEN ANALYSIS

Example 5 The manager of a shoelace company has determined that the company operates according to the following conditions.

(a) It makes only shoelaces, and sells them for 50¢ per pair.
(b) Ten people are employed, and they are paid 10¢ for each pair of shoelaces they make. Note that they are paid *only* for what they produce.
(c) Raw materials cost 24¢ for each pair of shoelaces made.
(d) The costs for equipment, plant, insurance, manager's salary, and fringe benefits (called *fixed costs*) amount to $2000 per month.

The manager of this shoelace factory is interested in how many pairs of shoe-laces have to be made and sold to break even. This means that the manager has to determine **total revenue** and **total cost** at different levels of production. The information given above is sufficient to formulate equations for the total revenue and the total cost, and from these equations we can write an equation to determine the **profit** at different levels of production. From this equation we can determine when the profit is zero; that is, when the factory breaks even. Let's discuss this problem in more detail.

The **total revenue function** indicates how much money is brought into the firm each month by the sale of its product. The total revenue for a month is the amount taken in before any expenses are paid. If q pairs of shoelaces are sold in a month, the total revenue for the shoelace factory is $0.50q$. Thus the total revenue function is

$$R = 0.50q \quad \text{(in dollars).}$$

We can also write the equation for the **total cost function**, which measures the number of dollars the firm must pay to produce and sell its product. The total cost is composed of two parts, fixed costs and variable costs. **Fixed costs** are those which remain constant regardless of the number of units produced. They include such costs as depreciation or rent on buildings, interest on investments, and so on. The fixed costs for the shoelace factory are given to be $2000 per month. Thus the cost for 1 month will always be at least $2000, even if no shoelaces are produced.

Variable costs are those directly related to the production of a commodity. The variable costs for producing q pairs of shoelaces is $0.10q$ for labor and $0.24q$ for raw materials. The total variable cost is $0.34q$. The total cost function is the sum of the fixed and variable costs. We have

$$\underset{\substack{\text{Fixed} \\ \text{costs}}}{\downarrow} \quad \underset{\substack{\text{Variable} \\ \text{costs}}}{\downarrow}$$

$$C = 2000 + 0.34q \quad \text{(in dollars).}$$

Example 6 Find the total revenue and total cost if the shoelace factory produces and sells the following amounts.

(a) 10,000 pairs of shoelaces in a month.
(b) 15,000 pairs of shoelaces in a month.

Solution

(a) $R = 0.50q$. Thus, when $q = 10,000$,

$$R = (0.50)(10,000) = \$5000.$$

Also

$$\overset{q = 10,000 \text{ here}}{C = 2000 + (0.34)(10,000)} = \$5400$$

(b)   Now $q = 15{,}000$, so

$$R = (0.50)(15{,}000) = \$7500$$

and

$$C = 2000 + (0.34)(15{,}000) = \$7100$$

Clearly the profit or loss for a month can be found by subtracting the total cost from the total revenue. In the previous example we see that producing and selling 10,000 pairs of shoelaces results in a loss of \$400 (\$5000 − \$5400), while producing and selling 15,000 pairs results in a profit of \$7500 − \$7100 = \$400. So the firm will "break even" at some point between 10,000 and 15,000 pairs of shoelaces.

If q units of a commodity are produced and sold, the total profit is found by subtracting the total cost from the total revenue.

> Total profit (or loss) = total revenue − total cost

We can define the **total profit function** as follows:

> $$P = R - C.$$

Example 7   In Example 5, at what level of production will the firm break even?

Solution   At the **break-even point**, there is no profit or loss. That is, $P = 0$.

$$R = 0.50q \quad \text{and} \quad C = 2000 + 0.34q$$

so

$$
\begin{aligned}
P = R - C &= (0.50q) - (2000 + 0.34q) \\
&= 0.50q - 0.34q - 2000 \\
&= 0.16q - 2000 \overset{\checkmark}{=} 0. \qquad \text{We set } P = 0.
\end{aligned}
$$

Then

$$0.16q = 2000$$

and

$$q = \frac{2000}{0.16} = 12{,}500.$$

That is, the break-even point is 12,500 pairs of shoelaces. If $q = 12{,}500$, then

$$R = 0.50(12{,}500) = 6250$$

and
$$C = 2000 + 0.34(12{,}500) = 2000 + 4250 = 6250.$$

Since
$$P = 0.16q - 2000,$$

we see that

if $q > 12{,}500$, the company makes a profit

and

if $q < 12{,}500$, the company loses money.

For example, if $q = 15{,}000$, then
$$P = 0.16(15{,}000) - 2000 = \$400$$

and if $q = 10{,}000$, then
$$P = 0.16(10{,}000) - 2000 = -\$400 \quad \text{(that is, a loss of \$400).}$$

LINEAR INEQUALITIES

We now turn to the subject of linear inequalities. **A linear inequality in one variable** is an inequality in one of the following forms.

$$ax + b < c \tag{5}$$
$$ax + b \le c \tag{6}$$
$$ax + b > c \tag{7}$$
$$ax + b \ge c \tag{8}$$

Here a, b, and c are real numbers and $a \ne 0$.

We have three additional rules for dealing with inequalities; let a, b, and c be real numbers.

ADDITION RULE FOR INEQUALITIES

If $a < b$, then $a + c < b + c$. $\tag{9}$

RULE FOR MULTIPLICATION BY A POSITIVE NUMBER

If $a < b$ and $c > 0$, then $ac < bc$. $\tag{10}$

**RULE FOR MULTIPLICATION
BY A NEGATIVE NUMBER**

If $a < b$ and $c < 0$, then $ac > bc$.   (11)

Rule (11) can be restated: *Multiplying both sides of an inequality by a negative number reverses the direction of the inequality.*

Example 8   Since $2 < 3$, we see by (9) that $2 + c < 3 + c$ for every real number c. For example, if $c = 4$, we have $6 = 2 + 4 < 3 + 4 = 7$, or $6 < 7$, and if $c = -5$, we have

$$2 - 5 < 3 - 5, \quad \text{or} \quad -3 < -2.$$

We remind you that $a < b$ means that a is to the *left* of b on the real number line.

Example 9   Since $2 < 3$, we see by (10) that $2c < 3c$ for any *positive* real number c. For example, if $c = 4$, then we see that $8 = 2 \cdot 4 < 3 \cdot 4 = 12$.

Example 10   Since $2 < 3$, we find from (11) that $2c > 3c$ for every *negative* real number c. For example, if $c = -5$, we see that $-10 = 2(-5) > 3(-5) = -15$. This is true because -10 is to the right of -15 on the number line (see Figure 1).

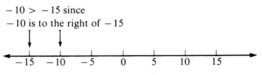

Figure 1

We can solve linear inequalities of the forms of (5), (6), (7), or (8) by using the three rules just discussed. By the **solution set** of an inequality, we mean the set of numbers that satisfy the inequality.

Example 11   Solve the inequality $2x < 4$.

Solution   Since $\frac{1}{2} > 0$, we have $\frac{1}{2}(2x) < \frac{1}{2}(4)$, or $x < 2$. Thus the solution set can be written as $\{x : x < 2\} = (-\infty, 2)$. The solution set is sketched in Figure 2.

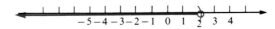

Figure 2

Example 12   Solve the inequality $-3x + 5 \geq 12$.

Solution

$$-3x + 5 \geq 12$$

$$-3x \geq 7 \qquad \text{We added } -5 \text{ to both sides.}$$

$$x \leq -\tfrac{7}{3}. \qquad \text{We multiplied by } -\tfrac{1}{3}$$
$$\text{(which reverses the inequality).}$$

Thus the solution set is $(-\infty, -\tfrac{7}{3}]$, as shown in Figure 3.

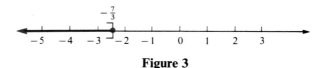

Figure 3

Example 13   Solve the inequalities $-3 < \dfrac{7 - 2x}{3} \leq 4$.

Solution   We need to find the set of numbers for which the inequalities are satisfied.

$$-9 < 7 - 2x < 12 \qquad \text{We multiplied by 3.}$$

$$-16 < \;\; -2x \;\; \leq 5 \qquad \text{We added } -7.$$

$$8 > x \geq -\tfrac{5}{2}, \quad \text{or} \quad x \in [-\tfrac{5}{2}, 8). \qquad \text{We multiplied by } -\tfrac{1}{2}$$
$$\text{(which reverses the inequalities).}$$

Thus the solution set is the half-open interval $[-\tfrac{5}{2}, 8)$, as sketched in Figure 4. Note that each of the steps in the computation served to simplify the term containing x.

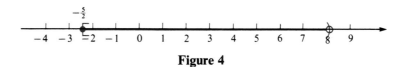

Figure 4

PROBLEMS 1.3   In Problems 1–13, solve the linear equation for the one unknown.

1. $3x = 6$ **2.** $3x = -6$

3. $4y = 5$ **4.** $-5y = 2$

5. $2x + 3 = 0$ **6.** $2x + 3 = -2$

7. $5x - 2 = 8$ **8.** $-5z + 2 = 7$

9. $4 - 7y = 3$ **10.** $-2 - 3z = -8$

11. $a + bx = c, \quad b \neq 0.$

12. $-a - bx = c, \quad b \neq 0.$

13. $a - bx = -c, \quad b \neq 0$

In Problems 14–40, find the solution set of the given inequality (or inequalities).

14. $x - 2 < 5$

15. $x - 2 \leq 5$

16. $x - 2 > 5$

17. $x - 2 \geq 5$

18. $-x + 2 \leq 3$

19. $-x + 4 > -5$

20. $2x - 7 \leq 2$

21. $-2x + 4 > 8$

22. $-2x + 4 \geq -8$

23. $-3x + 2 < -3$

24. $3x - 2 > 3$

25. $-7x + 4 \leq 10$

26. $1 \leq x + 2 \leq 4$

27. $-2 < x - 1 \leq 3$

28. $-3 \leq -x + 1 < 5$

29. $1 \leq 2x + 2 \leq 4$

30. $1 \leq 2x + 2 < 4$

31. $-1 < 2x - 2 < 4$

32. $-1 \leq 2x + 5 < 7$

33. $1 \leq 7x - 6 \leq 4$

34. $-4 \leq -3x + 5 < 8$

35. $2 < 3x + 4 \leq 7$

36. $-4 < \dfrac{2x - 4}{3} \leq 7$

37. $2 \geq \dfrac{4 - 2x}{5} > -4$

38. $\dfrac{1}{x} > 3$

39. $\dfrac{4}{3x - 2} \leq -2$

⋆40. $\dfrac{1}{x - 2} > \dfrac{2}{x + 3}$. [*Hint*: Keep track of whether you are multiplying by a positive or negative number.]

41. Solve the inequalities $a \leq \dfrac{bx + c}{d} < e$, with b and $d > 0$.

42. Suppose the price of a commodity is 40¢ each. If fixed costs are $200 and the variable costs amount to 20¢ per item, find each of the following.

(a) Total revenue function.

(b) Total cost function.

(c) Total profit function.

(d) Break-even point.

43. A product has a fixed cost of $1650 and a variable cost of $35 for each item produced during a given month.

(a) Write the equation that represents total cost.

(b) What will it cost to produce 215 items during the month?

44. The product of Problem 43 is sold for $85 per item.

(a) Write the equation representing the revenue function.

(b) What will be the revenue from sales of 50 items?

(c) What is the profit function for this product?

(d) What is the profit on 50 items?

(e) How many items must be sold in the month to avoid losing money?

1.4 Exponents and Roots

In this section we review the basic algebraic notions of taking powers and roots. Let n be a positive integer and let a be a real number. Then the expression a^n, read "a to the nth power," is defined by

$$a^n = \underbrace{a \cdot a \cdot a \cdots a}_{n \text{ times}}. \tag{1}$$

That is, a^n is the product of n terms, each of which is equal to a. In this setting, the number n is called an **exponent**.

Example 1 $6^2 = 6 \cdot 6 = 36$. This is read "6 squared."

Example 2 $\left(\dfrac{1}{2}\right)^3 = \dfrac{1}{2} \cdot \dfrac{1}{2} \cdot \dfrac{1}{2} = \dfrac{1}{8}$. This is read as "$\frac{1}{2}$ cubed."

Example 3 $(-2)^5 = (-2)(-2)(-2)(-2)(-2) = -32$. This is read as "$-2$ to the fifth power."

Example 4 $5^1 = 5$ since there is now only one term in product (1).

If $n = 0$, we define

$$a^0 = 1 \text{ for any real number } a \neq 0.$$

We will soon show why this makes sense.

Example 5 $8^0 = 1$, $(-7)^0 = 1$, and $(1.235)^0 = 1$.

We now define negative exponents. Let n be a positive integer and suppose that $a \neq 0$. Then

$$a^{-n} = \frac{1}{a^n}. \tag{2}$$

Example 6 $5^{-2} = \dfrac{1}{5^2} = \dfrac{1}{25} = 0.04$. This is read "5 to the minus 2."

Example 7 $\left(\dfrac{1}{2}\right)^{-3} = \dfrac{1}{\left(\frac{1}{2}\right)^3} = \dfrac{1}{\frac{1}{8}} = 8$

Example 8 $\left(\dfrac{5}{7}\right)^{-1} = \dfrac{1}{\frac{5}{7}} = \dfrac{7}{5}$

We now know how to compute a^m, where m is an integer. To extend this definition to noninteger exponents, we need to define the nth root of a number.

Let a be a real number. Suppose that there is a real number b such that $b^n = a$. Then b is called an nth root of a. This is denoted by $a^{1/n}$. We have

$$(a^{1/n})^n = \underbrace{a^{1/n}a^{1/n}\cdots a^{1/n}}_{n \text{ times}} = a. \tag{3}$$

Notation.   If $n = 2$, we call $a^{1/2}$ the **square root of a** and write

$$a^{1/2} = \sqrt{a}. \tag{4}$$

Similarly, we call $a^{1/3}$ the **cube root of a** and write

$$a^{1/3} = \sqrt[3]{a}. \tag{5}$$

Higher-order fractional powers are also given names; thus $a^{1/4}$ is called the **fourth root of a**, $a^{1/5}$ is called the **fifth root of a**, and so on. Finally, $a^{1/n}$ is called the **nth root of a**.

Example 9 $8^{1/3} = 2$ because $2^3 = 8$.

Example 10 $9^{1/2} = \pm 3$ because $3^2 = 9$ and $(-3)^2 = (-3)(-3) = 9$.

Example 11† $5^{1/2} = \pm 2.236067977$ to ten significant figures.

As the last two examples suggest, *every positive real number has two square roots, one positive and one negative*. However, if $x > 0$, then $\sqrt{x}$ will henceforth denote the *positive* square root of x. The negative square root of x will be denoted by $-\sqrt{x}$.

No negative number has a real square root. This follows from the fact that the square of every real number is nonnegative. Thus, for example, there is no real number b which satisfies $b^2 = -1$.

---

† The symbol ▦ indicates that a calculator was used in the computations.

Example 12 $(-1)^{1/3} = -1$ since $(-1)^3 = -1$.

Example 13 $(-32)^{1/5} = -2$ since $(-2)^5 = -32$.

Suppose that n is a positive odd integer. Then

$$a^n \begin{cases} \text{is positive if } a > 0, \\ \text{is negative if } a < 0. \end{cases}$$

This suggests the following fact, illustrated in Examples 12 and 13. *If n is odd, then every real number a has exactly one nth root.*

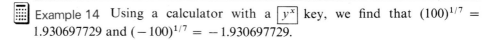 Example 14 Using a calculator with a $\boxed{y^x}$ key, we find that $(100)^{1/7} = 1.930697729$ and $(-100)^{1/7} = -1.930697729$.

Example 15 $(16)^{1/4} = \pm 2$ since $(2^4) = 16$ and $(-2)^4 = 16$.

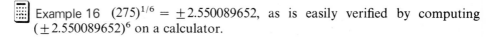 Example 16 $(275)^{1/6} = \pm 2.550089652$, as is easily verified by computing $(\pm 2.550089652)^6$ on a calculator.

The last two examples, together with the fact about square roots stated earlier, suggests that *if n is even and a > 0, then a has exactly two nth roots. If a < 0, then a has no nth root.* However, as with square roots, if $x > 0$, then $x^{1/n}$ will denote the positive nth root of x. The negative nth root of x will be denoted by $-x^{1/n}$.

We now extend our definitions to the case of rational exponents. Let $r = m/n$ be a rational number given in lowest terms. Then, for every real number a,

$$\boxed{a^r = a^{m/n} = (a^{1/n})^m.} \tag{6}$$

Example 17 $4^{3/2} = (4^{1/2})^3 = 2^3 = 8$

Example 18 $(-8)^{2/3} = (-8^{1/3})^2 = (-2)^2 = 4$

Example 19 $(32)^{-7/5} = (32^{1/5})^{-7} = 2^{-7} = \dfrac{1}{2^7} = \dfrac{1}{128}$

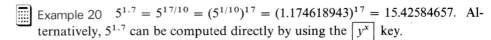 Example 20 $5^{1.7} = 5^{17/10} = (5^{1/10})^{17} = (1.174618943)^{17} = 15.42584657$. Alternatively, $5^{1.7}$ can be computed directly by using the $\boxed{y^x}$ key.

Example 21 $(-20{,}759)^{11/15} = [(-20{,}759)^{1/15}]^{11} = (-1.940053748)^{11}$
$= -1465.383784.$

Example 22 $(-2)^{3/2}$ does not exist because $(-2)^{1/2}$ does not exist.

We now summarize some properties of exponents and roots.

Let a and b be nonzero real numbers, let r and s be rational numbers, and assume that a^r and a^s are defined.

1. $a^0 = 1$	**2.** $a^1 = a$
3. $a^r a^s = a^{r+s}$	**4.** $\dfrac{a^r}{a^s} = a^{r-s}$
5. $(a^r)^s = (a^s)^r = a^{rs}$	**6.** $(ab)^r = a^r b^r$
7. $\left(\dfrac{a}{b}\right)^r = \dfrac{a^r}{b^r}$	**8.** $a^{-1} = \dfrac{1}{a}$

We will not prove these facts (see Problems 90, 91, and 92).

Note. If $a \neq 0$ and n is a positive integer, then from (3),

$$a^0 = a^{n-n} = \frac{a^n}{a^n} = 1.$$

This is why we define a^0 to be 1.

The following examples show how the facts given above can be used to simplify exponential expressions.

Example 23 $4^2 4^3 = 4^{2+3} = 4^5$. We can verify this by noting that $4^2 = 16$, $4^3 = 64$, and $4^2 4^3 = 16 \cdot 64 = 1024 = 4^5$.

Example 24 $\dfrac{3^7}{3^4} = 3^{7-4} = 3^3 = 27$. Again, this can be verified by computing $3^7 = 2187$, $3^4 = 81$, and $\dfrac{2187}{81} = 27 = 3^3$. In this example it was much easier to simplify the exponents before doing any computation.

Example 25 $\dfrac{50^9}{25^9} = \left(\dfrac{50}{25}\right)^9 = 2^9 = 512$. Here direct computations are horrendous. For example, $50^9 = 1,953,125,000,000,000$.

Example 26 $(2^3)^4 = 2^{3 \cdot 4} = 2^{12} = 4096$.

Example 27 $\dfrac{5^{-2} \cdot 5^4}{5^8 \cdot 5^{-5}} = \dfrac{5^{-2+4}}{5^{8-5}} = \dfrac{5^2}{5^3} = 5^{2-3} = 5^{-1} = \dfrac{1}{5}$.

Quantities that are initially bigger than 1 increase very rapidly as the exponent increases and those that are between 0 and 1 decrease very rapidly. To illustrate this point, we retell an old fable. A brave soldier in the service of a king won many battles. The king offered him his choice of reward. The soldier, being mathematically inclined, asked for what seemed to the king to be a small gift. He requested that one grain of gold be given him on the first day of the month, two grains the second day, four the third day, eight the fourth day, and so on until the 30-day month was over. That is, he asked that the number of grains be doubled each day.

The king granted this wish with pleasure, thinking the soldier a fool. But not for long. On the tenth day the man received $2^9 = 512$ grains of gold and on the 20th, $2^{19} = 524,288$ grains. The soldier, alas, never lived until the 30th day, since before then he would have depleted the king's (and the world's) supply of gold. (On the 30th day he would have been owed $2^{29} = 536,870,912$ grains.) The king found it more expedient to do away with the soldier. You thought that all fairy tales had happy endings?

PROBLEMS 1.4   In Problems 1–76 compute the indicated value(s), if it exists. Recall that the symbol ▦ indicates that a calculator is needed.

1. 8^3
2. $(-8)^3$
3. $8^{1/3}$
4. 8^{-3}

5. $(-8)^{-3}$
6. $8^{-1/3}$
7. $(-8)^{-1/3}$
8. 2^{10}

9. 10^2
10. 2^{-10}
11. 10^{-2}
12. $(-10)^2$

13. $(-2)^{10}$
14. $(-10)^{-2}$
15. $(-2)^{-10}$
16. 3^{-1}

17. $(-1)^{-1}$
18. $(\frac{1}{3})^{-1}$
19. $(-1)^{15}$
20. $(-1)^{22}$

21. $(\frac{3}{5})^{-1}$
22. $(\frac{5}{3})^{-1}$
23. $(\frac{81}{26})^{-1}$
24. $(1.04)^{-1}$

25. $\left(\dfrac{a}{b}\right)^{-1}$, $a, b \neq 0$
26. $(\frac{1}{4})^3$

27. $(\frac{1}{3})^4$
28. $(\frac{1}{4})^{1/2}$

29. $(\frac{1}{4})^{-1/2}$
30. $(0.03)^2$

31. 6^{-2}
32. $(27)^{2/3}$

33. $(-27)^{2/3}$
34. $(-27)^{3/2}$

▦ 35. $27^{3/2}$
36. $100^{1/2}$

37. $100^{-1/2}$
38. $1000^{1/3}$

39. $1000^{-1/3}$
40. $1000^{2/3}$

41. $1000^{-2/3}$
42. $128^{1/7}$

43. $128^{-3/7}$
44. $128^{5/7}$

45. $128^{8/7}$
46. $128^{-10/7}$

47. $(-128)^{-2/7}$
48. $(-128)^{6/7}$

49. $(-128)^{-9/7}$
50. $64^{1/2}$

51. $(-64)^{1/2}$
52. $64^{1/6}$

53. $64^{-1/6}$

54. $64^{5/6}$

55. $64^{-5/6}$

56. $(-64)^{6/5}$

57. $(-64)^{3/5}$

58. $10^{1/10}$

59. $(\frac{1}{10})^{10}$

60. $(-6)^{-3/7}$

61. $(14)^{3.2}$

62. $(3.1415)^{2.718}$

63. $(2.718)^{3.1415}$

64. $\dfrac{4^7}{4^5}$

65. $\dfrac{4^5}{4^7}$

66. $4^5 \cdot 4^2$

67. $\dfrac{2^5 2^{-3}}{2^8}$

68. $\dfrac{2^3 2^{-4}}{2^4 2^{-3}}$

69. $\dfrac{5^3 5^4}{5^2 5^5}$

70. $\dfrac{8^{-2}}{8^{-3}}$

71. $\dfrac{8^{-3}}{8^{-2}}$

72. $\dfrac{1.6^5}{0.8^5}$

73. $\dfrac{(3.15)^3}{(9.45)^3}$

74. $\dfrac{4^{1/2}4^{3/2}}{4^4 4^{-5/2}}$

75. $\dfrac{(-64)^{1/3}(-64)^{-2/3}}{(-64)^{4/3}(-64)^{2/3}}$

76. $\dfrac{7^{1/5}7^{-9/5}}{7^{12/5}7^{-14/5}}$

*77. Let $r = m/n$ be a rational number. Show that if m is odd, n is even, and $a > 0$, then there are exactly two distinct values for $a^{m/n}$.

78. Find two distinct values for $16^{3/4}$.

79. In Problem 77, explain why $a^{m/n}$ does not exist if $a < 0$.

80. If n is odd, show that there is exactly one value for $a^{m/n}$ for every real number a.

81. Show that if $a \neq 0$, then $(1/a)^{-1/n} = a^{1/n}$.

82. If $-1 < a < 1$, show that a^n decreases as n increases.

83. Find the smallest value of n such that $(\frac{1}{2})^n < 0.0001$.

84. Find the smallest value of n such that $(0.9)^n < 0.00000001$.

85. Find the smallest value of n such that $(1.2)^n > 1000$.

86. Find the smallest value of n such that $(1.001)^n > 1,000,000,000$.

87. Which is bigger: 4^5 or 5^4?

88. Which is bigger: 50^{51} or 51^{50}?

89. Guess which is bigger: 1000^{1001} or 1001^{1000}.

*90. Let $r = m/n$ and $s = p/q$, where m, n, p, and q are integers.
(a) Write $r + s$ as a single quotient of integers.
(b) Using (6), show that $a^{r+s} = a^r a^s$ for any real number a for which all expressions are defined.

*91. Show that $a^{r-s} = a^r/a^s$, where r and s are as in Problem 90.

*92. Show that $(a^r)^s = a^{rs}$, where r and s are as in Problem 90.

93. It has been estimated that the number of grains of sand on the beach in Coney Island is about 10^{20}.† After how many days would the soldier in the fable retold in this section be due to receive this number of grains of gold?

1.5 Absolute Value and Inequalities

Many inequalities can be written in terms of the *absolute value* of a real number. Look at Figure 1. The **absolute value** of a number a is defined as the distance from

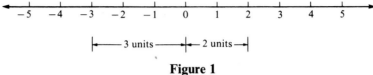

Figure 1

that number to zero and is written $|a|$. Since 2 is 2 units from zero, $|2| = 2$. The number -3 is 3 units from zero, so $|-3| = 3$. Alternatively, we may define

$$|a| = a \qquad \text{if } a \geq 0 \tag{1}$$

and

$$|a| = -a \qquad \text{if } a < 0. \tag{2}$$

The absolute value of a number is a nonnegative number. Note that, for example, $|5| = 5$ and $|-5| = -(-5) = 5$; thus numbers which are negatives of one another have the same absolute value. Another way to calculate absolute value is to observe that

$$|x| = \sqrt{x^2} \tag{3}$$

where, of course, the positive square root is taken. Therefore $|-3| = \sqrt{(-3)^2} = \sqrt{9} = 3$ and $|3| = \sqrt{(3)^2} = \sqrt{9} = 3$.

For all real numbers a and b, the following facts can be proven.

$$|-a| = |a| \tag{4}$$

$$|ab| = |a||b| \tag{5}$$

$$|a + b| \leq |a| + |b| \qquad \text{Triangle inequality} \tag{6}$$

---

† See James R. Newman, *The World of Mathematics*, 4 vols. (New York: Simon and Schuster, 1956) 3: 2007.

Property (4) is obviously true. A proof of property (5) is outlined in Problem 32 and a proof of the triangle inequality is suggested in Problem 36.

Example 1 $3 = |-2 + 5| \leq |-2| + |5| = 2 + 5 = 7$, which shows that the triangle inequality holds in this case.

In solving inequalities using absolute values, the following property will be very useful.

$$|x| < a \text{ is equivalent to } -a < x < a; \qquad (7)$$

that is, $\{x : |x| < a\} = \{x : -a < x < a\}$ (see Figure 2), which indicates that for any x in the open interval $(-a, a)$, the distance between x and zero is less than a.

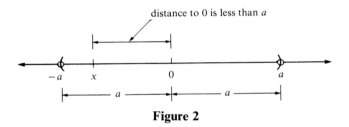

Figure 2

Example 2 Solve the inequality $|x| \leq 3$.

Solution $|x| \leq 3$ implies that the distance between x and zero is less than or equal to 3. Thus $-3 \leq x \leq 3$, and the solution set is the closed interval $[-3, 3]$. See Figure 3.

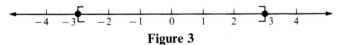

Figure 3

Example 3 Solve the inequality $|x - 4| < 5$.

Solution The distance between $x - 4$ and zero is less than 5, so that $-5 < x - 4 < 5$. Adding 4 to each term, we see that $-1 < x < 9$. Thus the solution set is the open interval $(-1, 9)$. See Figure 4. Another way to think of this set is as the set of all x such that x is within 5 units of 4.

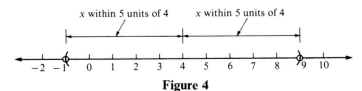

Figure 4

Another useful rule is

$$|x| > a \text{ is equivalent to } x > a \text{ or } x < -a. \tag{8}$$

This follows from (7).

Example 4 Solve the inequality $|x + 2| \geq 8$.

Solution The distance between $x + 2$ and 0 is greater than or equal to 8, so that either $x + 2 \geq 8$ or $x + 2 \leq -8$. Hence either $x \geq 6$ or $x \leq -10$. The solution set is $(-\infty, -10] \cup [6, \infty)$. See Figure 5.

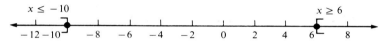

Figure 5

Example 5 Solve the inequality $|3x + 4| < 2$.

Solution We have $-2 < 3x + 4 < 2$. We subtract 4: $-6 < 3x < -2$. Then we divide by 3: $-2 < x < -\frac{2}{3}$, or $x \in (-2, -\frac{2}{3})$.

Example 6 Solve the inequality $|5 - 3x| \geq 1$.

Solution We have either $5 - 3x \geq 1$ or $5 - 3x \leq -1$. In the first case $-3x \geq -4$, which implies that $x \leq \frac{4}{3}$. In the second case $-3x \leq -6$, so that $x \geq 2$. The solution set is, therefore, $(-\infty, \frac{4}{3}] \cup [2, \infty)$.

PROBLEMS 1.5 In Problems 1–7, solve for x.

1. $x = |4 - 5|$ **2.** $x = |-6 - (-2)|$

3. $x = |2| - |-3|$ **4.** $x = ||2| - |-3||$

5. $|x| = 2, x > 0$ **6.** $|x| = 2, x < 0$

7. $|x| = 0$

In Problems 8–31, find the solution set of each inequality and sketch it on a number line.

8. $|x| \leq 4$ **9.** $|x| \geq 5$

10. $|x| \leq 0$ **11.** $|x| \geq 0$

12. $|x| < 3$ **13.** $|x| > -1$

14. $|x| \leq -1$ **15.** $|x - 2| < 1$

16. $|x + 3| \leq 4$

17. $|x + 3| \geq 4$

18. $|x + 6| > 3$

19. $|2x + 4| < 3$

20. $|-x + 2| < 3$

21. $|5 - x| \geq 1$

22. $|2 - x| \geq 0$

23. $|-3x - 4| > 2$

24. $|3x + 4| > 2$

*25. $|6 - 4x| \geq |x - 2|$

26. $\left|\dfrac{8 - 3x}{2}\right| \leq 3$

27. $\left|\dfrac{3x + 17}{4}\right| > 9$

28. $|ax + b| < c, a > 0, c > 0$

29. $|ax + b| \geq c, a < 0, c > 0$

*30. $x \leq |x|$

*31. $|2x| > |5 - 2x|$

32. Show that $|xy| = |x||y|$. [*Hint*: Deal with each of four cases separately: (1) $x \geq 0$, $y \geq 0$, (2) $x \geq 0$, $y < 0$, and so on.]

33. Show that if $x \geq 0$ and $y \geq 0$, then $|x + y| = |x| + |y|$.

34. If $x > 0$ and $y < 0$, show that $|x + y| < |x| + |y|$.

35. If $x < 0$ and $y < 0$, show that $|x + y| = |x| + |y|$.

36. Using Problems 33–35, prove the triangle inequality (6).

*37. Show that $||x| - |y|| \leq |x - y|$. [*Hint*: Write $x = (x - y) + y$ and apply the triangle inequality.]

38. Solve each inequality and graph its solution set.

    (a) $|2 - x| + |2 + x| \leq 10$

    (b) $|2 - x| + |2 + x| > 6$

    (c) $|2 - x| + |2 + x| \leq 4$

    (d) $|2 - x| + |2 + x| \leq 3.99$

39. Use absolute-value bars to translate each of the following statements into a single inequality.

    (a) $x \in (-4, 10)$

    (b) $x \notin (-3, 3)$

    (c) $x \notin [5, 11]$

    (d) $x \in (-\infty, 2] \cup [9, \infty)$

    (e) $x \in (-93, 4) \cap (-10, 50)$

40. Write single inequalities which are satisfied in each case.

    (a) All numbers x that are closer to 5 than to 0.

    (b) All numbers y that are closer to -2 than to 2.

41. Show that

$$\frac{s + t + |s - t|}{2}$$

equals the maximum of $\{s, t\}$.

42. Show that

$$\frac{s + t - |s - t|}{2}$$

equals the minimum of $\{s, t\}$.

43. (a) Show that

$$|A - B| \leq |A - W| + |W - B|$$

for all real numbers A, B, and W.

(b) Describe those situations in which the preceding less-than-or-equal statement is actually an equality.

44. For what choices of s is

$$3.72s > 4.06s?$$

*45. (a) Suppose that a and b are positive; show that

$$\sqrt{ab} \leq \frac{a + b}{2}.$$

[*Hint*: Use the fact that $(x - y)^2 \geq 0$ for all real numbers x and y.]

(b) Use the inequality of part (a) to prove that among all rectangles with an area of 225 cm^2, the one with shortest perimeter is a square.

(c) Use the inequality of part (a) to prove that among all rectangles with perimeter of 300 cm, the one with largest area is a square.

Review Exercises for Chapter 1

In Exercises 1–4, determine whether each given collection is a set.

1. The collection of corporations in the United States with net earnings over $1 million a year.

2. The collection of large corporations in the United States.

3. The collection of powerful nations.

4. The collection of member nations in the United Nations.

5. In a large corporation, let

$$M = \{\text{male workers}\},$$
$$F = \{\text{female workers}\},$$
$$E = \{\text{executives}\},$$
$$C = \{\text{clerical workers}\}.$$

Describe each set in words.

(a) $M \cup F$ (b) $M \cap F$ (c) $F \cap E$

(d) $M \cup C$ (e) F^c (f) E^c

(g) $E - F$ (h) $M - C$ (i) $(M \cap E) \cup C$

(j) $(C - F)^c$ (k) $C \cap (F \cup E)$ (l) $(M \cup E) - C$

(m) $(F \cup C \cup E)^c$.

6. A clothing manufacturer produces 260 items. Of these, 83 require rayon, 103 require wool, and 177 require cotton. In addition, 28 require rayon and wool, 50 require wool and cotton, and 35 require cotton and rayon. Finally, 10 items require all three materials.

(a) Draw a Venn diagram to illustrate this situation.

(b) How many items require no rayon, cotton, or wool?

(c) How many require exactly one of these materials?

(d) How many require exactly two?

In Exercises 7–18, use $A = [2, 7]$, $B = [-8, 4)$, $C = (-2, 9)$, and $D = (6, 12]$. Find each set.

7. $B \cup C$

8. $A \cap D$

9. B^c

10. $A - C$

11. $D - B$

12. $C - (B - A)$

13. $D - [C - (B - A)]$

14. $(C - B)^c$

15. $(A \cup C)^c$

16. $[D - (A \cap B)]^c$

17. $A \cap B \cap C$

18. $(A \cup B) \cap D$

19. Write $0.134134134\ldots$ as a rational number.

In Exercises 20–24, solve for the unknown quantity.

20. $2x = 7$

21. $-4x = 5 - x$

22. $3y - 4 = 6$

23. $8 - 5z = -4z + 3$.

24. $6w - 3 = 4 + 8w$

In Exercises 25–41, find the solution set of the given inequality (or inequalities).

25. $x - 3 < 4$

26. $x + 7 \geq 6$

27. $2x + 4 \leq 9$

28. $-7x + 3 > 8$

29. $-4 \leq x - 2 < 3$

30. $2 \leq 3x + 5 \leq 6$

31. $-2 < 4x - 7 \leq 6$

32. $|x| < 3$

33. $|x - 4| > 5$

34. $|x + \frac{1}{2}| \leq \frac{3}{4}$

35. $|2x - 5| < 3$

36. $|3x + 7| > -1$

37. $|4x - 9| < -2$

38. $|4x - 9| \geq 5$

39. $|4 - 3x| < 2$

40. $\left| \dfrac{6 - 3x}{4} \right| \leq 7$

41. $\left| \dfrac{5x - 8}{3} \right| > 6$

In Exercises 42–66, compute the indicated value(s), if it exists.

42. 5^3

43. $9^{1/2}$

44. $(-9)^{1/2}$

45. $9^{-1/2}$

46. $(-3)^2$

47. 2^{-3}

48. 6^0

49. 1.6^1

50. $(-1)^{20}$

51. $(-1)^{21}$

52. $(\frac{2}{3})^2$

53. $(\frac{2}{3})^{-3}$

54. $16^{1/4}$

55. $16^{-1/4}$

56. $-16^{1/4}$

57. $(\frac{1}{16})^{1/4}$

58. $(\frac{1}{16})^{-1/4}$

59. $20^{1/3}$

60. $(-20)^{-1/3}$

61. $(\frac{7}{3})^{-1}$

62. $(-\frac{3}{5})^{-1}$

63. $(2.035)^{1.426}$

64. $(85.2)^{0.85}$

65. $\dfrac{9^{1/2} \cdot 9^{-3/2}}{9^2 \cdot 9^{-5/2}}$

66. $\dfrac{10^{5/8} 10^{-7/8}}{10^{3/4} 10^{11/8}}$

67. Find the smallest value of n such that $(1.01)^n > 5000$.

2 LINES AND SYSTEMS OF LINEAR EQUATIONS

2.1 The Cartesian Coordinate System

In this section we describe the most common way of representing points in a plane: the **Cartesian coordinate system**.[†] To form the Cartesian coordinate system, we draw two mutually perpendicular number lines as in Figure 1:

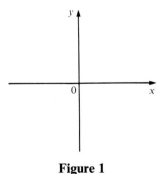

Figure 1

one horizontal line and one vertical line. The horizontal line is called the **x-axis** and the vertical line is called the **y-axis**. The point at which the lines meet is called the **origin** and is labeled 0.

_____

[†] This system is named after the great French mathematician and philosopher René Descartes (1596–1650), who is considered to be the inventor of analytic geometry. He is known for the statement "cogito ergo sum," "I think, therefore I am," which played a central role in his philosophical writings.

To every point in the plane we assign an **ordered pair** of numbers. The first element in the pair is called the **x-coordinate**, and the second element of the pair is called the **y-coordinate**.

The x-coordinate measures the number of units from the point to the y-axis. Points to the right of the y-axis have positive x-coordinates, while those to the left have negative x-coordinates.

The y-coordinate measures the number of units from the point to the x-axis. Points above the x-axis have a positive y-coordinate, while those below have a negative y-coordinate. Figure 2 shows a typical point (a, b), where $a > 0$ and $b > 0$.

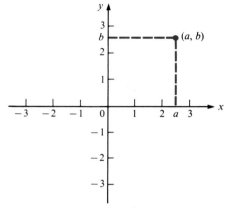

Figure 2

Two ordered pairs, or points, are **equal** if their first elements are equal and their second elements are equal. Note that $(1, 0)$ and $(1, 1)$ are *different* since their second elements are different. Note too that $(1, 2)$ and $(2, 1)$ are different points.

In Figure 3, several different points are depicted. Note again that $(1, 2) \neq (2, 1)$. as they represent different points in the plane.

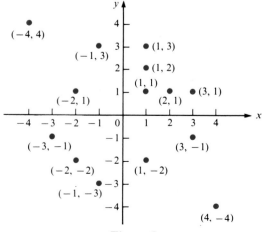

Figure 3

When points in the plane are represented by the Cartesian coordinate system, the plane is called the **Cartesian plane**, or the *xy*-plane, and is denoted by $\mathbb{R}^2$. We have

$$\mathbb{R}^2 = \{(x, y) : x \in \mathbb{R} \text{ and } y \in \mathbb{R}\}.$$

A glance at Figure 4 indicates that the *x*- and *y*-axes divide the *xy*-plane into four regions. These regions are called **quadrants** and are denoted as in the figure.

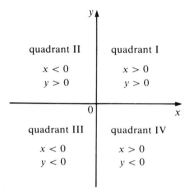

Figure 4

Example 1

(a) $(1, 3)$ is in the first quadrant since $1 > 0$ and $3 > 0$.
(b) $(-4, -7)$ is in the third quadrant since $-4 < 0$ and $-7 < 0$.
(c) $(-2, 5)$ is in the second quadrant.
(d) $(7, -3)$ is in the fourth quadrant.

Let (x_1, y_1) and (x_2, y_2) be two points in the *xy*-plane (see Figure 5). Then, using the Pythagorean theorem, it can be shown that the distance, d, between them is given by

$$d = \sqrt{(x_1 - x_2)^2 + (y_1 - y_2)^2}. \tag{1}$$

Example 2 Find the distance between the points $(2, 5)$ and $(-3, 7)$.

Solution Let $(x_1, y_1) = (2, 5)$ and $(x_2, y_2) = (-3, 7)$, so that from (1),

$$d = \sqrt{(2 - (-3))^2 + (5 - 7)^2} = \sqrt{5^2 + (-2)^2} = \sqrt{29}.$$

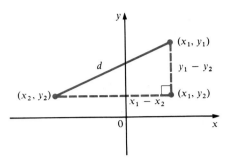

Figure 5

PROBLEMS 2.1   In Problems 1–10, sketch each point in the xy-plane. If the point is not on the x- or y-axis, determine the quadrant in which it lies.

1. $(3, -2)$ **2.** $(4, 3)$ **3.** $(2, 0)$ **4.** $(0, -5)$ **5.** $(-4, -1)$

6. $(-2, 3)$ **7.** $(\frac{1}{2}, \frac{1}{3})$ **8.** $(\frac{1}{3}, -\frac{3}{2})$ **9.** $(0, \frac{3}{4})$ **10.** $(-\frac{2}{3}, -\frac{7}{3})$

In Problems 11–17, find the distance between the given points.

11. $(1, 3), (4, 7)$ **12.** $(-7, 2), (4, 3)$

13. $(8, -1), (-2, 0)$ **14.** $(\frac{1}{2}, \frac{1}{3}), (\frac{1}{3}, \frac{1}{2})$

15. $(-3, -7), (-1, -2)$ **16.** $(a, b), (b, a)$

17. $(a, b), (0, 0)$.

2.2 Linear Functions and Their Graphs

In Example 1.3.5 on page 17, we showed that if a manufacturer of shoelaces has fixed costs of $2000 and additional costs of 34¢ for each pair of shoelaces produced, then the total cost function is given by

$$C = 2000 + 0.34q \quad \text{(in dollars)}, \tag{1}$$

where q is the number of pairs (quantity) produced. Equation (1) is really a rule that says "you tell me what q is and I'll tell you what C is." For example, if $q = 1000$, then

$$C = 2000 + 0.34(1000) = \$2340.$$

This rule is an example of a *linear function*.

Linear Function   A **linear function** is a rule that takes the form

$$ax + by = c, \tag{2}$$

where a and b are not both equal to zero. Often, linear functions are written in the form

$$y = mx + b. \qquad (3)$$

In this case x is called the **independent variable** and y is called the **dependent variable**.

Note. The equation (2) is also called a **linear equation in two variables** if neither a nor b is zero.

Example 1 Let $y = 3x - 5$. Compute y for each value of x.

(a) $x = 2$ (b) $x = -7$ (c) $x = 0$

Solution

(a) $y = 3 \cdot 2 - 5 = 6 - 5 = 1$
(b) $y = 3(-7) - 5 = -21 - 5 = -26$
(c) $y = 3 \cdot 0 - 5 = 0 - 5 = -5$

Example 2 Let $2x + 5y = 3$.

(a) Find y if $x = 4$. (b) Find x if $y = -6$.

Solution

(a) Substituting $x = 4$ in the equation $2x + 5y = 3$, we have

$2 \cdot 4 + 5y = 3$

$8 + 5y = 3$ We multiplied through.

$5y = 3 - 8 = -5$ We subtracted 8 from both sides.

$y = -1$ We divided by 5.

(b) Substituting $y = -6$ in $2x + 5y = 3$, we obtain

$2x + 5(-6) = 3$

$2x - 30 = 3$ We multiplied through.

$2x = 33$ We added 30 to both sides.

$x = \dfrac{33}{2}$ We divided by 2.

Graph of a Linear Function The **graph** of the linear function $ax + by = c$ is the set of points in the xy-plane whose coordinates satisfy the equation.

Example 3   Sketch the graph of the equation $y = 2x + 1$.

Solution   Table 1 shows some points on the graph.

TABLE 1

x	y = 2x + 1	Corresponding point
−5	−9	(−5, −9)
−4	−7	(−4, −7)
−3	−5	(−3, −5)
−2	−3	(−2, −3)
−1	−1	(−1, −1)
0	1	(0, 1)
1	3	(1, 3)
2	5	(2, 5)
3	7	(3, 7)
4	9	(4, 9)
5	11	(5, 11)

We plot these 11 points in Figure 1. It appears that the points lie on a straight line.

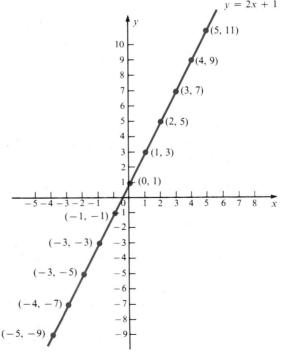

Figure 1

The following fact generalizes the last example.

The graph of a linear function is a straight line. (4)

We will not prove this fact, but we will use it to graph a number of linear functions. *Since a straight line is determined by two points, we can graph a linear function by finding two points and then drawing the line that passes through them.*

Example 4   Sketch the graph of $y = -2x + 3$.

Solution   When $x = 0$, $y = 3$, and when $y = 0$, $2x = 3$, or $x = \frac{3}{2}$. Thus two points on the line are $(0, 3)$ and $(\frac{3}{2}, 0)$. Using these points, we obtain the graph in Figure 2.

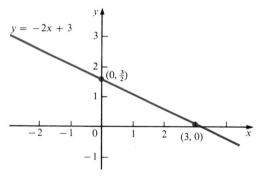

Figure 2

In Example 4 we found the points where the line crossed the x- and y-axes.

x-intercept and y-intercept   The **x-intercept** of a linear function is the x-coordinate of the point where its graph crosses the x-axis. The **y-intercept** is the y-coordinate of the point where its graph crosses the y-axis.

Note.   The x-intercept can be found by setting $y = 0$ in (2) or (3). The y-intercept can be found by setting $x = 0$ in (2) or (3). In Example 4 we found that the x-intercept is 3 and the y-intercept is $\frac{3}{2}$.

Example 5   Sketch the graph of $3x + 2y = 6$.

Solution   Setting $y = 0$, we have

$$3x + 2 \cdot 0 = 6$$
$$3x = 6$$
$$x = 2.$$

Setting $x = 0$, we have

$$3 \cdot 0 + 2y = 6$$

$$0 + 2y = 6$$

$$y = 3.$$

Thus the x-intercept is 2 and the y-intercept is 3. Using these values, we obtain the line in Figure 3.

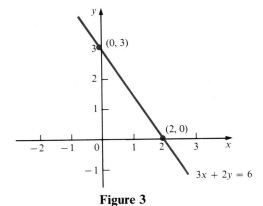

Figure 3

Example 6   Graph $x = 3$.

Solution   Every point that has the x-coordinate 3 is on this graph. Some of these points are $(3, 0)$, $(3, -2)$, $(3, 5)$, and $(3, 25)$. The x-intercept is 3. There is no y-intercept, since a point on the y-axis has x-coordinate 0, not 3. The line is *vertical*, as shown in Figure 4.

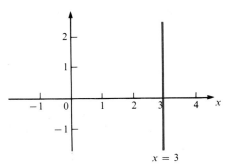

Figure 4

In general,

> the graph of $x = a$ is a vertical line for any constant a.

Example 7   Sketch the graph of $y = -2$.

Solution   Every point on this graph has y-coordinate -2. The y-intercept is -2, and there is no x-intercept since at an x-intercept, $y = 0$. The graph is *horizontal* and is shown in Figure 5.

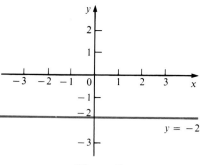

Figure 5

In general,

> the graph of $y = a$ is a horizontal line for any constant a.

Example 8   Sketch the graph of the cost function

$$C = 2000 + 0.34q$$

Solution   We note that this formula takes the form of (3). The only difference is that the variables are q and C instead of x and y. When $q = 0$, $C = 2000$. This is the C-intercept. When $q = 1000$, $C = 2340$. Thus two points on the graph are $(0, 2000)$ and $(1000, 2340)$. This is all we need to sketch the graph in Figure 6.

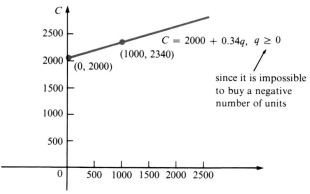

Figure 6

PROBLEMS 2.2   In Problems 1–12, a linear function is given. Find the y-value that corresponds to the given x-value.

1. $y = 3x;\quad x = 2$

2. $y = 2x + 5;\quad x = -1$

3. $y = -x + 2;\quad x = 4$

4. $y = -7x + 5;\quad x = 0$

5. $y = -\dfrac{x}{5} + \dfrac{2}{5};\quad x = 3$

6. $y = \dfrac{-x - 7}{12};\quad x = -2$

7. $2x + 4y = 8;\quad x = 6$

8. $-x + 4y = 10;\quad x = -2$

9. $3x - 5y = 15;\quad x = 3$

10. $\frac{1}{2}x + \frac{1}{3}y = 1;\quad x = 4$

11. $8x + 9y = 10;\quad x = -6$

12. $-4x - 3y = 1;\quad x = -1$

In Problems 13–28, a linear function is given. (a) Find the x- and y-intercepts (if any) of its graph. (b) Draw the graph.

13. $y = x$

14. $y = -x$

15. $x + y = 1$

16. $x - y = 3$

17. $y = 3x - 4$

18. $y = -2x + 7$

19. $y = \dfrac{x}{2} + 3$

20. $x - 2y = 8$

21. $2x - y = 8$

22. $2x + y = 8$

23. $x + 2y = 8$

24. $-x - 2y = 8$

25. $-2x - y = 8$

26. $4x + 5y = 20$

27. $-3x + 4y = 12$

28. $3x - 4y = 12.$

29. Sketch the revenue function $R = 0.5q$ (see Example 1.3.5 on page 17).

30. The price of a commodity is 40¢ each. Answer each question if fixed costs are $200 and the variable costs amount to 20¢ per item.

 (a) Find and sketch the total revenue function.

 (b) Find and sketch the total cost function.

31. The Thunder Power Company has the following rate schedule for electricity consumers.

   $6 for the first 30 kilowatt (kWh) or less each month.

   7¢ for each kWh over 30.

 (a) Find a linear function which gives monthly cost as a function of number of kWh of electricity used. Assume that at least 30 kWh will be used each month.

 (b) Graph the function.

 (c) What is the cost of 75 kWh in 1 month?

32. Some people paying federal income tax in the United States do not use tax tables. In this case they must file a tax computation schedule (Schedule TC). Tax rates for single taxpayers in 1980 who filed Schedule TC are given on the next page.

 (a) Find a linear function that gives federal income tax due in 1980 as a function of income for single taxpayers earning between $12,900 and $15,000 a year.

 (b) Graph this function.

 (c) Determine the tax due on an income of $14,000.

From 1980 U.S. Federal Tax Tables

If the amount on Schedule TC, Part I, line 3, is:		Enter on Schedule TC, Part I, line 4:	
Not over $2,300		-0-	

Over—	But not over—		of the amount over—
$2,300	$3,400	14%	$2,300
$3,400	$4,400	$154 + 16%	$3,400
$4,400	$6,500	$314 + 18%	$4,400
$6,500	$8,500	$692 + 19%	$6,500
$8,500	$10,800	$1,072 + 21%	$8,500
$10,800	$12,900	$1,555 + 24%	$10,800
$12,900	$15,000	$2,059 + 26%	$12,900
$15,000	$18,200	$2,605 + 30%	$15,000
$18,200	$23,500	$3,565 + 34%	$18,200
$23,500	$28,800	$5,367 + 39%	$23,500
$28,800	$34,100	$7,434 + 44%	$28,800
$34,100	$41,500	$9,766 + 49%	$34,100
$41,500	$55,300	$13,392 + 55%	$41,500
$55,300	$81,800	$20,982 + 63%	$55,300
$81,800	$108,300	$37,677 + 68%	$81,800
$108,300	—	$55,697 + 70%	$108,300

33. (a) Find a linear function that gives tax due in 1980 as a function of income for single taxpayers earning between $28,800 and $34,100 a year.

(b) Graph this function.

(c) Determine the tax due on an income of $32,750.

34. (a) Find a linear function that gives tax due in 1980 as a function of income for single taxpayers earning over $108,300.

(b) Graph this function.

(c) What is the tax due on an income of $500,000?

2.3 The Slope and Equation of a Line

In the last section we saw that the graph of a linear equation in two variables is a straight line. It is also true that

> the x- and y-coordinates of the points on a line satisfy an equation of the form $ax + by = c$. (1)

In this section we will show how the equation of a line can be found. We first discuss the *slope* of a line, which is a measure of the relative rate of change of the *x*- and *y*-coordinates of points on the line as we move along the line.

Slope

Let L denote a nonvertical line and let (x_1, y_1) and (x_2, y_2) be two points on the line. Then the **slope** of the line, denoted by m, is given by

SLOPE OF A LINE

$$m = \text{slope of } L = \frac{y_2 - y_1}{x_2 - x_1} = \frac{\Delta y}{\Delta x} \tag{2}$$

Here Δy and Δx denote the changes in y and x, respectively (see Figure 1).[†] If L is vertical, then the slope is *undefined*.[‡]

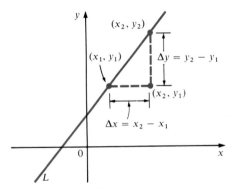

Figure 1

Remark.   It is not difficult to show, using similar triangles, that the ratio $(y_2 - y_1)/(x_2 - x_1)$ is the same no matter which two points (x_1, y_1) and (x_2, y_2) are chosen on a given line. Thus the slope of a line is well defined.

Suppose $m > 0$ and $x_2 > x_1$. Then $x_2 - x_1 > 0$ and, from (2),

$$y_2 - y_1 = \overset{>0}{\underset{\downarrow}{m}}(\overbrace{x_2 - x_1}^{>0}) > 0$$

so that $y_2 > y_1$. That is, if m is positive, then y increases as x increases. Analogously, if $m < 0$ and if $x_2 > x_1$, then

$$y_2 - y_1 = \overset{<0}{\underset{\downarrow}{m}}(\overbrace{x_2 - x_1}^{>0}) < 0.$$

[†] Δ is the capital Greek letter "delta."
[‡] In some books a line parallel to the *y*-axis is said to have an *infinite slope*.

If *m* is negative, then *y* decreases as *x* increases. This means that the following are true.

> **1.** If $m > 0$, the graph of the line will rise as we move from left to right along the *x*-axis.
>
> **2.** If $m < 0$, the graph of the line will fall as we move from left to right along the *x*-axis.

These facts are illustrated in Figure 2.

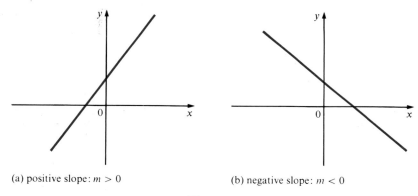

(a) positive slope: $m > 0$ (b) negative slope: $m < 0$

Figure 2

We will discuss two cases separately. In Figure 3(a) we have drawn the line $y = a$, which is horizontal. Here, as *x* changes, *y* does not change at all (since *y* is equal to the constant *a*). Therefore $\Delta y/\Delta x = 0/\Delta x = 0$.

> Horizontal lines have a slope of zero.

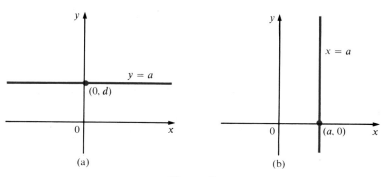

(a)   (b)

Figure 3

In Figure 3(b) we have drawn the line $x = a$, which is vertical. Here, when y changes, x does not change at all. In this case the slope is undefined.

The slope of a vertical line is undefined.

Example 1   Find the slopes of the lines containing the given pairs of points. Then sketch these lines.

(a) $(2, 3), (-1, 4)$ (c) $(2, 6), (-1, 6)$
(b) $(1, -3), (4, 0)$ (d) $(3, 1), (3, 5)$

Solution

(a) $m = \dfrac{\Delta y}{\Delta x} = \dfrac{4 - 3}{-1 - 2} = \dfrac{1}{-3} = -\dfrac{1}{3}$

(b) $m = \dfrac{\Delta y}{\Delta x} = \dfrac{0 - (-3)}{4 - 1} = \dfrac{3}{3} = 1$

(c) $m = \dfrac{6 - 6}{-1 - 2} = \dfrac{0}{-3} = 0$. That is, as the x-coordinate changes, the y-coordinate does not vary. This line is horizontal.

(d)   Here the slope is undefined since the line is vertical. (The x-coordinate of both points have the constant value 3.) The lines are shown in Figure 4.

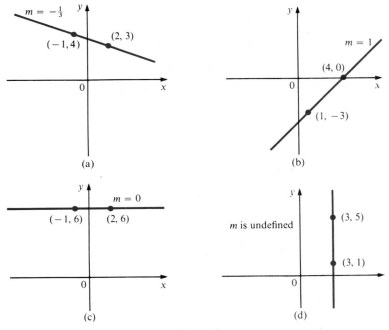

Figure 4

We now state two useful facts. Let L_1 and L_2 be two lines not parallel to the coordinate axes with slopes m_1 and m_2, respectively.

L_1 is parallel to L_2 if and only if[†] $m_1 = m_2$. $\hspace{2cm}$ (3)

L_1 is perpendicular[‡] to L_2 if and only if $m_1 = -\dfrac{1}{m_2}$. $\hspace{1cm}$ (4)

We can rephrase these statements as follows.

The slopes of parallel lines are the same. $\hspace{2cm}$ (5)

The slopes of perpendicular lines are negative reciprocals of one another. $\hspace{0.3cm}$ (6)

Example 2 The line joining the points $(1, -1)$ and $(2, 1)$ is parallel to the line joining the points $(0, 4)$ and $(-2, 0)$ because the slope of each line is 2.

Example 3 Let the line L_1 contain the two points $(2, -6)$ and $(1, 4)$. Find the slope of a line L_2 that is perpendicular to L_1.

Solution The slope of L_1 is $m_1 = [4 - (-6)]/(1 - 2) = -10$. Thus

$$m_2 = \frac{-1}{-10} = \frac{1}{10}.$$

An **equation** of a line is an equation in the variables x and y satisfied by the coordinates of every point on the line. As with a linear function, the **graph** of an equation in x and y is the set of all points in the xy-plane whose coordinates satisfy the equation. If we know two points on a line, then we can find an equation of the line.

Point-Slope Equation of a Line If a line is vertical, then it has the equation $x = a$. If x is not vertical, then it has slope given by (2):

$$m = \frac{y_2 - y_1}{x_2 - x_1},$$

or

$$y_2 - y_1 = m(x_2 - x_1). \qquad \text{We multiplied both sides by } x_2 - x_1. \qquad (7)$$

[†] The words *if and only if* mean that each of the two statements implies the other. For example, (3) states that if L_1 is parallel to L_2, then $m_1 = m_2$ and, if $m_1 = m_2$, then L_1 is parallel to L_2.
[‡] Two lines are perpendicular if they meet at right angles.

Let (x_1, y_1) be a point on a line with slope m. If (x, y) is any other point on the line, then—from (7)—its coordinates must satisfy the following equation:

POINT-SLOPE EQUATION OF A LINE

$$y - y_1 = m(x - x_1) \tag{8}$$

Equation (8) is called the **point-slope equation** of a line.

Example 4   Find a point-slope equation of the line passing through the points $(-1, -2)$ and $(2, 5)$.

Solution   We first compute

$$m = \frac{5 - (-2)}{2 - (-1)} = \frac{7}{3}$$

Thus, if we choose $(x_1, y_1) = (2, 5)$, a point-slope equation of the line is

$$y - 5 = \tfrac{7}{3}(x - 2).$$

Choosing $(x_1, y_1) = (-1, -2)$, we obtain another (equivalent) point-slope equation of the line:

$$y - (-2) = \tfrac{7}{3}(x - (-1))$$

or

$$y + 2 = \tfrac{7}{3}(x + 1).$$

Both equations yield the same graph, which is shown in Figure 5.

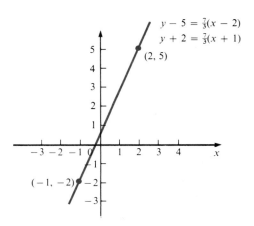

Figure 5

As the last example shows, there are many equivalent point-slope equations of a line. In fact, there are an infinite number of them—one for each point on the line. A more commonly used equation of a line is given below.

Slope-Intercept Equation of a Line   Let m be the slope and b be the y-intercept of a line. Then the **slope-intercept equation** of the line is the equation

---

SLOPE-INTERCEPT EQUATION OF A LINE

$$y = mx + b \qquad (9)$$

Example 5   Find the slope-intercept equation of the line passing through $(-1, -2)$ and $(2, 5)$.

Solution   In Example 4 we found the equation

$$y - 5 = \tfrac{7}{3}(x - 2).$$

Then

$$y - 5 = \tfrac{7}{3}x - \tfrac{14}{3} \qquad \text{We multiplied through.}$$

$$y = \tfrac{7}{3}x - \tfrac{14}{3} + 5 \qquad \text{We added 5 to both sides.}$$

$$y = \tfrac{7}{3}x + \tfrac{1}{3}. \qquad \text{We observed that } 5 = \tfrac{15}{3} \text{ and } -\tfrac{14}{3} + \tfrac{15}{3} = \tfrac{1}{3}.$$

Note that when $x = 0$, $y = \tfrac{1}{3}$, so $\tfrac{1}{3}$ is the y-intercept. Thus the last equation is the slope-intercept equation of the line.

Standard Equation of a Line   A **standard equation** of a line is an equation of the form

---

STANDARD EQUATION OF A LINE

$$ax + by = c \qquad (10)$$

---

where a and b are not both equal to zero.

Example 6   Find a standard equation of the line passing through $(-1, -2)$ and $(2, 5)$.

Solution   In Example 5 we found that

$$y = \tfrac{7}{3}x + \tfrac{1}{3}.$$

Then

$$3y = 7x + 1 \qquad \text{We multiplied through by 3.}$$

$$3y - 7x = 1 \qquad \text{We subtracted 7x from both sides.}$$

A standard equation of the line is $-7x + 3y = 1$. Another standard equation is $7x - 3y = -1$.

Example 7   Find the slope-intercept equation of the line passing through the point (2, 3) and parallel to the line whose equation is $y = -3x + 5$.

Solution   Parallel lines have the same slope. The slope of the line $y = -3x + 5$ is -3 since the line is given in slope-intercept form. Then, using (8), a point-slope equation of the line is

$y - 3 = -3(x - 2)$

$y - 3 = -3x + 6$ We multiplied through.

$\quad y = -3x + 9.$ We added 3 to both sides to obtain the slope-intercept equation.

Both lines are sketched in Figure 6.

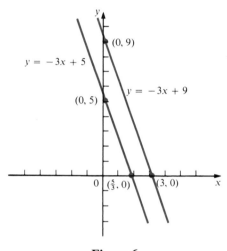

Figure 6

Example 8   Find the slope-intercept equation of the line passing through $(-1, 3)$ and perpendicular to the line $2x + 3y = 4$.

Solution   From $2x + 3y = 4$, we obtain

$$3y = -2x + 4$$

or

$$y = -\tfrac{2}{3}x + \tfrac{4}{3} \qquad \text{We divided by 3.}$$

This means that the slope of the line $2x + 3y = 4$ is $-\tfrac{2}{3}$, so that the line we seek has the slope

$$m = \frac{-1}{-\tfrac{2}{3}} = \frac{3}{2}.$$

We use (8) to find the slope-intercept equation of the line passing through $(-1, 3)$ with slope $\frac{3}{2}$.

$$y - 3 = \tfrac{3}{2}(x + 1)$$

$2(y - 3) = 3(x + 1)$ We multiplied by 2.

$2y - 6 = 3x + 3$ We multiplied through.

$2y = 3x + 9$ We added 6 to both sides.

$y = \tfrac{3}{2}x + \tfrac{9}{2}$ We divided by 2 to obtain the slope-intercept equation.

The two lines are sketched in Figure 7.

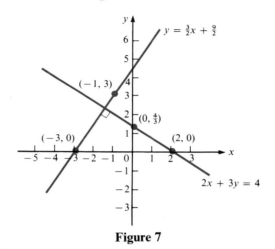

Figure 7

Table 2 summarizes properties of straight lines.

TABLE 2

Equation	Description of Line
$x = a$	*Vertical line.* x-intercept is a. No y-intercept. No slope.
$y = a$	*Horizontal line.* No x-intercept. y-intercept is a. Slope $= 0$.
$y - y_1 = m(x - x_1)$	*Point-slope form* of line with slope m passing through the point (x_1, y_1)
$y = mx + b$	*Slope-intercept form* of line with slope m and y-intercept b. x-intercept $= -b/m$ if $m \neq 0$.
$ax + by = c$	*Standard form.* Slope is $-a/b$ if $b \neq 0$. x-intercept is c/a if $a \neq 0$ and y-intercept is c/b if $b \neq 0$.

There is another type of problem we will encounter.

Example 9 Find the point of intersection of the lines $2x + 3y = 7$ and $-x + y = 4$, if one exists.

Solution The point of intersection, which we will label (a, b), must satisfy both equations. For the first equation we have

$$y = -\tfrac{2}{3}x + \tfrac{7}{3}$$

and for the second,

$$y = x + 4.$$

The lines have different slopes $(-\tfrac{2}{3}$ and $1)$ and are therefore not parallel, so they do have a point of intersection. At (a, b),

$$b = -\tfrac{2}{3}a + \tfrac{7}{3} = a + 4 \quad \text{or} \quad \tfrac{5}{3}a = \tfrac{7}{3} - 4 = -\tfrac{5}{3},$$

and $a = -1$. Then $b = a + 4 = 3$, and the point of intersection is $(-1, 3)$. The two lines are given in Figure 8.

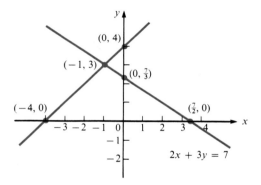

Figure 8

Example 10 **Break-Even Analysis.** In Example 1.3.5 on page 17, we discussed the case of a shoelace manufacturer whose total revenue function was given by

$$R = 0.50q$$

and whose total cost function was

$$C = 2000 + 0.34q.$$

Profit, P, is given by

$$P = R - C.$$

In that example we found the **break-even point**, or the level of production at which profit is zero. If $P = 0$, then $R = C$. Since the graphs of R and C are straight lines, we need to find the point of intersection of the lines to find the value q for which $R = C$. Setting $R = C$, we have

$$0.50q = 2000 + 0.34q$$

or, as we found in Example 1.3.7, $q = 12,500$. The functions R and C are sketched in Figure 9.

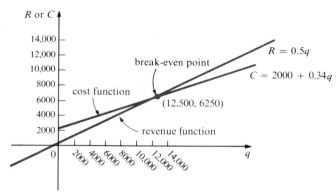

Figure 9

PROBLEMS 2.3

In Problems 1–11, find the slope of the line passing through the two given points. Sketch the lines.

1. $(1, 6), (2, 4)$

2. $(-3, 4), (7, 9)$

3. $(-1, -2), (-3, -4)$

4. $(4, 0), (0, 4)$

5. $(-6, 5), (7, -2)$

6. $(1, 7), (-4, 7)$

7. $(2, -3), (5, -3)$

8. $(-2, 4), (-2, 6)$

9. $(0, a), (a, 0), a \neq 0$

10. $(a, b), (b, a), ab \neq 0$

11. $(a, b), (c, d), a \neq c$

In Problems 12–20, two pairs of points are given. Determine whether the two lines containing these pairs of points are parallel, perpendicular, or neither.

12. $(1, 8), (2, 9);$ $(1, 2), (0, 1)$

13. $(3, -1), (2, 4);$ $(2, 0), (5, 7)$

14. $(0, 2), (-2, 0);$ $(0, 3), (3, 0)$

15. $(0, 5), (2, -1);$ $(0, 0), (-1, 3)$

16. $(5, 2), (1, 7);$ $(2, 5), (7, 1)$

17. $(1, -2), (2, 4);$ $(4, 1), (-8, 2)$

18. $(3, 2), (5, -2);$ $(0, 6), (-5, 6)$

19. $(3, 1), (3, 7);$ $(2, 4), (-1, 4)$

20. $(4, 3), (4, 1);$ $(-2, 4), (-2, 0)$

21. Suppose $a > 0$ and $a + h > 0$. Show that the straight line through $(a, \sqrt{a})$ and $(a + h, \sqrt{a + h})$ has slope $1/(\sqrt{a + h} + \sqrt{a})$. [*Hint:* $(\sqrt{B} - \sqrt{A})(\sqrt{B} + \sqrt{A}) = B - A$ if $A, B \geq 0$.]

In Problems 22–33, find the slope-intercept form, a standard form, and a point-slope form of the equation of the straight line when either two points on the line or a point and the slope of the line are given. Sketch the graph of the line in the xy-plane.

22. $(1, 2), (3, 6)$

23. $(-2, 3), (4, -1)$

24. $(3, 7), m = \frac{1}{2}$

25. $(4, -7), m = 0$

26. $(-3, -7)$, m undefined

27. $(3, -\frac{1}{2})$, $(\frac{1}{3}, 0)$

28. $(-2, -4)$, $(3, 7)$

29. $(5, -1)$, $(8, 2)$

30. $(7, -3)$, $m = -\frac{4}{3}$

31. $(-5, 1)$, $m = \frac{3}{7}$

32. (a, b), (c, d)

33. (a, b), $m = c$.

34. Find the slope-intercept equation of the line parallel to the line $2x + 5y = 6$ and passing through the point $(-1, 1)$.

35. Find the slope-intercept equation of the line parallel to the line $5x - 7y = 3$ and passing through the point $(2, 5)$.

36. Find a standard equation of the line perpendicular to the line $x + 3y = 7$ and passing through the point $(0, 1)$.

37. Find the slope-intercept equation of the line perpendicular to the line $2x - \frac{3}{2}y = 7$ and passing through the point $(-1, 4)$.

38. Find a standard equation of the line perpendicular to the line $ax + by = c$ and passing through the point (α, β). Assume that $a \neq 0$ and $b \neq 0$.

In Problems 39–44, find the point of intersection (if there is one) of the two lines.

39. $x - y = 7$; $\quad 2x + 3y = 1$

40. $y - 2x = 4$; $\quad 4x - 2y = 6$

41. $4x - 6y = 7$; $\quad 6x - 9y = 12$

42. $4x - 6y = 10$; $\quad 6x - 9y = 15$

43. $3x + y = 4$; $\quad y - 5x = 2$

44. $3x + 4y = 5$; $\quad 6x - 7y = 8$

45. A commodity sells for 40¢. Fixed costs are $200 and the variable cost is 20¢ per item.

    (a) Find the total cost and total revenue functions.

    (b) Sketch the functions and determine the break-even point.

46. Answer the questions of Problem 45 if a product sells for $85 per item, fixed costs are $1650, and the variable cost is $35 per item.

47. In Problem 2.2.31 on page 44, we gave the following rates for users of electric power sold by the Thunder Power Company:

$6 for the first 30 kWh or less each month.

7¢ for each kWh over 30.

New rates are to go into effect:

$4.50 for the first 20 kWh or less each month.

8¢ for each kWh over 20.

Evidently, a consumer who uses less than 20 kWh each month will save money with the new rate. What is the break-even point? That is, what is the monthly level of usage (in kilowatt hours) for which the bill will be the same under both the new and the old rates?

48. Refer to the tax table given in Problem 2.2.32 on page 45. What is the slope of the graph of the function that gives federal income tax in 1980 as a function of income for single taxpayers earning between $18,200 and $23,500 a year?

49. Answer the question of Problem 48 for single taxpayers earning between $55,300 and $81,800 a year.

50. The price of a gallon of gasoline is a linear function of the octane rating. If 85-octane gas costs $1.43 and 90-octane gas costs 1.49\frac{1}{2}$, what is the price of 95-octane gas?

2.4 Systems of Two Linear Equations in Two Unknowns

Consider the following system of two linear equations in two unknowns:

$$a_{11}x + a_{12}y = b_1$$
$$a_{21}x + a_{22}y = b_2$$
(1)

where $a_{11}, a_{12}, a_{21}, a_{22}, b_1$, and b_2 are given numbers. Each of these equations is the equation of a straight line in the xy-plane.

The slope of the first line is $-a_{11}/a_{12}$; the slope of the second line is $-a_{21}/a_{22}$ (if $a_{12} \neq 0$ and $a_{22} \neq 0$). A **solution** to system (1) is a pair of numbers, denoted (x, y), that satisfies (1). The questions that naturally arise are whether (1) has any solutions and if so, how many? We will answer these questions after looking at some examples. In these examples we will make use of two important facts from elementary algebra.[†]

ADDITION OF EQUALS RULE

If $a = b$ and $c = d$, then $a + c = b + d$.
(2)

MULTIPLICATION RULE

If $a = b$ and c is any real number, then $ca = cb$.
(3)

The first rule states that if we add two equations together, we obtain a third, valid equation. The second rule states that if we multiply both sides of an equation by a constant, we obtain a second, valid equation.

Example 1   Consider the system

$$x - y = 7$$
$$x + y = 5$$
(4)

From (2), we may add these equations together to obtain

$$2x = 12$$
$$x = 6. \qquad \text{We divided by 2.}$$

[†] Similar facts called the *addition rule* and *multiplication rule*, respectively, were discussed in Section 1.3.

Then, from the second equation,

$$6 + y = 5$$
$$y = 5 - 6 = -1$$

Thus the pair $(6, -1)$ satisfies system (4) and the way we found the solution shows that it is the only pair of numbers to do so. That is, system (2) has a **unique solution**. In problems where there is a unique solution, it is easy to check the answer.

Check.

$$x - y = 6 - (-1) = 7$$
$$x + y = 6 + (-1) = 5$$

Example 2   Consider the system

$$x - \ y = 7$$
$$2x - 2y = 14 \tag{5}$$

It is apparent that these two equations are **equivalent**. That is, they are equations of the same straight line. To see this multiply the first by 2. (This is permitted by (3)). Then $x - y = 7$, or $y = x - 7$. Thus the pair $(x, x - 7)$ is a solution to system (5) for any real number x; that is, system (5) has an **infinite number of solutions**. For example, the following pairs are solutions: $(7, 0)$, $(0, -7)$, $(8, 1)$, $(1, -6)$, $(3, -4)$, and $(-2, -9)$.

Example 3   Consider the system

$$x - \ y = 7$$
$$2x - 2y = 13 \tag{6}$$

Multiplying the first equation by 2 (which, again, is permitted by (3)) gives us $2x - 2y = 14$. This contradicts the second equation, since $2x - 2y$ cannot be equal to both 13 and 14 at the same time. Thus system (6) has *no* solution. In this case, the system is said to be **inconsistent**.

It is easy to explain, geometrically, what is going on in the preceding examples. First we repeat that the equations in system (1) are both equations of straight lines. A solution to (1) is a point (x, y) that lies on both lines. If the two lines are not parallel, then they intersect at a single point. If they are parallel, then either they never intersect (no points in common) or they are the same line (infinite number of points in common). In Example 1 the lines have slopes of 1 and -1, respectively. Thus they are not parallel. They have the single point $(6, -1)$ in common. In Example 2 the lines are parallel (slope of 1) and coincident. In Example 3 the lines are parallel and distinct. These relationships are all illustrated in Figure 1.

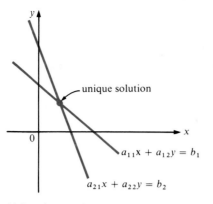

(a) lines intersecting at one point

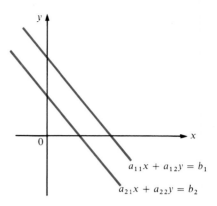

(b) parallel lines

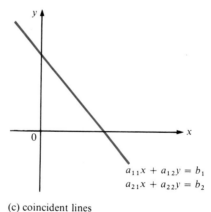

(c) coincident lines

Figure 1

Summarizing, we find that

---

a system of two linear equations in two unknowns has no
solution, exactly one solution, or an infinite number of
solutions.

In the next section we see that this fact holds for a system of m linear equations
in n unknowns where $m > 1$ and $n > 1$ are integers.

Example 4   The Sunrise Porcelain Company manufactures ceramic cups and
saucers. For each cup or saucer, a worker measures a fixed amount of material
and puts it into a forming machine, from which it is automatically glazed and

dried. On the average, a worker needs 3 minutes to get the process started for a cup and 2 minutes for a saucer. The material for a cup costs 25¢ and the material for a saucer costs 20¢. If $44 is allocated daily for production of cups and saucers, how many of each can be manufactured in an 8-hour work day if a worker is working every minute and exactly $44 is spent on materials?

Solution Let x denote the number of cups and y the number of saucers produced in an 8-hour day. Then, since there are 480 minutes in 8 hours, we obtain the following equations for x and y.

$$3x + 2y = 480 \qquad \text{Time or labor equation}$$

$$0.25x + 0.20y = 44 \qquad \text{Cost equation}$$

Multiplying the cost equation by 10, we obtain

$$2.5x + 2y = 440.$$

Subtracting this from the labor equation, we have

$$0.5x = 40, \quad \text{or} \quad x = 80.$$

We then have

From the labor equation

$$2y = 480 - 3x = 480 - 3(80) = 480 - 240 = 240$$

or

$$y = 120.$$

Thus the solution is 80 cups and 120 saucers can be manufactured in an 8-hour day. The labor and cost equations are sketched in Figure 2.

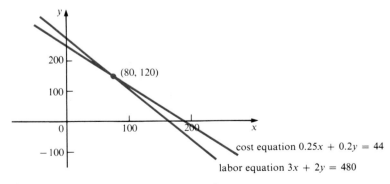

Figure 2

Check.

$$3x + 2y = 3 \cdot 80 + 2 \cdot 120 = 240 + 240 = 480$$

$$0.25x + 0.20y = 0.25(80) + 0.2(120) = 20 + 24 = 44$$

Example 5   Answer the question of Example 4 if the materials for a cup and saucer cost 15¢ and 10¢, respectively, and $24 is spent in an 8-hour day.

Solution   The cost equation is now

$$0.15x + 0.10y = 24$$

or, multiplying by 20,

$$3x + 2y = 480. \tag{7}$$

This is identical to the labor equation. This means that we have only one equation in the unknowns x and y, and our problem has an infinite number of solutions. Realistically, what does this mean? Remember that, although this was not stated explicitly, we must have $x \geq 0$ and $y \geq 0$ (the Sunrise Company cannot produce a negative number of cups or saucers). Thus the smallest value of x is zero (no cups produced). Then equation (7) reads

$$3 \cdot 0 + 2y = 480, \quad \text{or} \quad y = 240.$$

Thus no cups and 240 saucers is one solution. At the other extreme, we may have $y = 0$. Then equation (7) becomes

$$3x + 2 \cdot 0 = 480, \quad \text{or} \quad x = 160$$

and 160 cups and no saucers is another solution.
    Solving (7) for y in terms of x, we find that

$$2y = 480 - 3x, \quad \text{or} \quad y = 240 - \tfrac{3}{2}x.$$

Thus the infinite set of solutions to our problem can be written as

$$(x, y) = (x, 240 - \tfrac{3}{2}x), \quad \text{where } 0 \leq x \leq 160.$$

For example, if $x = 50$, then $y = 240 - \tfrac{3}{2} \cdot 50 = 240 - 75 = 165$. The labor and cost equations for this problem are graphed in Figure 3.

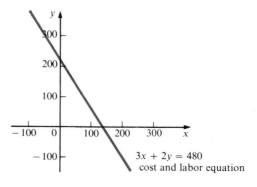

$3x + 2y = 480$
cost and labor equation

Figure 3

Example 6   Answer the question of Example 5 if $25 is spent in an 8-hour day.

Solution   The cost equation is now

$$0.15x + 0.10y = 25$$

or, again multiplying by 20,

$$3x + 2y = 500.$$

Subtracting the labor equation (7), we obtain

$$0 = 20$$

and the system of equations is inconsistent. This does not mean that the Sunrise Company is unable to manufacture cups and saucers. It simply means that the data in the problem are inconsistent. That is, there is no way to work exactly 8 hours spending 3 minutes and 15¢ for each cup, 2 minutes and 10¢ for each saucer, and a total of $25 for materials. Either the cost or labor equation must be changed in order to obtain an answer. The labor and cost equations for the problem are shown in Figure 4.

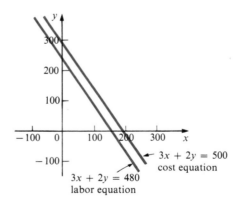

Figure 4

PROBLEMS 2.4   In Problems 1–12, find all solutions (if any) to the given systems.

1. $x - 3y = 4$
    $-4x + 2y = 6$

2. $2x - y = -3$
    $5x + 7y = 4$

3. $2x - 8y = 5$
    $-3x + 12y = 8$

4. $2x - 8y = 6$
    $-3x + 12y = -9$

5. $6x + y = 3$
    $-4x - y = 8$

6. $3x + y = 0$
    $2x - 3y = 0$

7. $4x - 6y = 0$
$-2x + 3y = 0$

8. $5x + 2y = 3$
$2x + 5y = 3$

9. $2x + 3y = 4$
$3x + 4y = 5$

10. $ax + by = c$
$ax - by = c$

11. $ax + by = c$
$bx + ay = c$

12. $ax - by = c$
$bx + ay = d$

13. Find conditions on a and b such that the system in Problem 10 has a unique solution.

14. Find conditions on a, b, and c such that the system in Problem 11 has an infinite number of solutions.

15. Find conditions on a, b, c, and d such that the system in Problem 12 has no solutions.

In Problems 16–21, find the point of intersection (if there is one) of the two lines.

16. $x - y = 7$; $2x + 3y = 1$

17. $y - 2x = 4$; $4x - 2y = 6$

18. $4x - 6y = 7$; $6x - 9y = 12$

19. $4x - 6y = 10$; $6x - 9y = 15$

20. $3x + y = 4$; $y - 5x = 2$

21. $3x + 4y = 5$; $6x - 7y = 8$

22. A zoo keeps birds (two-legged) and beasts (four-legged). If the zoo contains 60 heads and 200 feet, how many birds and how many beasts live there?

23. A mutual fund has two investment plans. In plan A, 80% of one's money is invested in blue-chip stocks and 20% is invested in riskier "glamour" stocks. In plan B, 40% is invested in blue-chip stocks and 60% is invested in glamour stocks. If the firm invests a total of $3 million in blue-chip stocks and $1 million in glamour stocks, how much money has been put in each of plans A and B?

24. The Atlas Tool Company manufactures pliers and scissors. Each pair of pliers contains 2 units of steel and 4 units of aluminum. Each pair of scissors requires 1 unit of steel and 3 units of aluminum. How many pairs of pliers and scissors can be made from 140 units of steel and 290 units of aluminum?

25. Answer the question in Problem 24 if each pair of scissors requires 2 units of aluminum and all other information is unchanged.

26. Answer the question in Problem 25 if only 280 units of aluminum are available and all other information is unchanged.

27. Ryland Farms in northwestern Indiana grows soybeans and corn on its 500 acres of land. During the planting season, 1200 hours of planting time will be available. Each acre of soybeans requires 2 hours, while each acre of corn requires 6 hours. If all the land and hours are to be utilized, how many acres of each crop should be planted?

28. Spina Food Supplies, Inc. manufactures frozen pizzas. Art Spina, President of Spina Food Supplies, personally supervises the production of both types of frozen pizzas produced by the company: Spina's regular and Spina's super deluxe. He currently has 150 lb of dough mix and 800 oz of topping mix available. Each regular pizza uses 1 lb of dough mix and 4 oz of topping, whereas each super deluxe uses 1 lb of dough and 8 oz of topping mix. How many of each type of pizza should he make in order to use all his dough and topping mix?

2.5 *m* Equations in *n* Unknowns: Gauss–Jordan Elimination

In this section we describe a method for finding all solutions (if any) to a system of *m* linear equations in *n* unknowns. In doing so we shall see that, as in the case of two equations in two unknowns, such a system has no solutions, exactly one solution, or an infinite number of solutions. Before launching into the general method, let us look at some simple examples.

Example 1   Solve the system

$$2x + 8y + 6z = 20$$
$$4x + 2y - 2z = -2 \qquad (1)$$
$$3x - y + z = 11$$

Solution   Here we seek three numbers *x*, *y*, and *z*, such that the three equations in (1) are satisfied. Our method of solution will be to simplify the equations so that solutions can be readily identified. We begin by dividing the first equation by 2. This gives us

$$x + 4y + 3z = 10$$
$$4x + 2y - 2z = -2 \qquad (2)$$
$$3x - y + z = 11$$

As we saw in the last section (addition of equals rule), adding two equations together leads to a third, valid equation. We may use this new equation in place of either of the two equations used to obtain it. We begin simplifying system (2) by multiplying both sides of the first equation in (2) by -4 and adding this new equation to the second equation. This gives us

$$\begin{array}{ll} -4x - 16y - 12z = -40 & \text{We multiplied the first equation by } -4. \\ \underline{4x + 2y - 2z = -2} & \text{This is the second equation.} \\ -14y - 14z = -42 & \end{array}$$

The equation $-14y - 14z = -42$ is our new second equation and the system is now

$$x + 4y + 3z = 10$$
$$-14y - 14z = -42 \qquad (3)$$
$$3x - y + z = 11$$

It is important to note that any solution to (1) is also a solution to (3), and vice versa. This is because of the two rules of Section 2.4. In this case we say that systems (1) and (3) are **equivalent**. We then multiply the first equation by -3 and add it to the third equation.

$$x + 4y + 3z = 10$$
$$- 14y - 14z = -42$$
$$- 13y - 8z = -19$$

Again, we note that systems (2) and (3) have the same solutions. Note that in the system above, the variable x has been eliminated from the second and third equations. Next we divide the second equation by -14.

$$x + 4y + 3z = 10$$
$$y + z = 3$$
$$- 13y - 8z = -19$$

We multiply the second equation by -4 and add it to the first and then multiply the second equation by 13 and add it to the third.

$$x - z = -2$$
$$y + z = 3$$
$$5z = 20$$

We divide the third equation by 5.

$$x - z = -2$$
$$y + z = 3$$
$$z = 4$$

Finally, we add the third equation to the first and then multiply the third equation by -1 and add it to the second to obtain the system

$$x = 2$$
$$y = -1 \qquad (4)$$
$$z = 4.$$

In each step we obtained a system equivalent to system (1). This means that the solution(s) to (1) are the same as the solution(s) to (4). System (4) obviously has the unique solution $x = 2$, $y = -1$ and $z = 4$, so this is the unique solution to the original system (1). We write this solution in the form $(2, -1, 4)$. The method we have used here is called *Gauss-Jordan elimination.*[†]

[†] Named after the great German mathematician Karl Friedrich Gauss (1777–1855) and the French mathematician Camille Jordan (1838–1922).

Check.

$$2x + 8y + 6z = 2(2) + 8(-1) + 6(4) = 4 - 8 + 24 = 20$$

$$4x + 2y - 2z = 4(2) + 2(-1) - 2(4) = 8 - 2 - 8 \ = -2$$

$$3x - y + z = 3(2) - (-1) + 4 = 6 + 1 + 4 = 11$$

Before going on to another example, let us summarize what we have done in this example.

1. We divided to make the coefficient of x in the first equation equal to 1.

2. We "eliminated" the x-terms in the second and third equations. That is, we made the coefficients of these terms equal to zero by multiplying the first equation by appropriate numbers and then adding it to the second and third equations, respectively.

3. We divided to make the coefficient of the y-term in the second equation equal to 1 and then proceeded to use the second equation to eliminate the y-terms in the first and third equations.

4. We divided to make the coefficient of the z-term in the third equation equal to 1 and then used the third equation to eliminate the z-terms in the first and second equations.

At every step, we obtained systems that were equivalent—that is, each system had the same set of solutions as the one that preceded it. This follows from the two rules in Section 2.4.

Before solving other systems of equations, we introduce notation that makes it easier to write down each step in our procedure. A **matrix** is a rectangular array of numbers. For example, the coefficients of the variables x, y, and z in system (1) can be written as the entries of a matrix A, called the **coefficient matrix** of the system:

$$A = \begin{pmatrix} 2 & 8 & 6 \\ 4 & 2 & -2 \\ 3 & -1 & 1 \end{pmatrix} \qquad \text{This is the coefficient matrix of our system.}$$

We study properties of matrices in Chapter 3. We introduce them here for convenience of notation. Using matrix notation, system (1) can be represented as the **augmented matrix**

$$\begin{pmatrix} 2 & 8 & 6 & | & 20 \\ 4 & 2 & -2 & | & -2 \\ 3 & -1 & 1 & | & 11 \end{pmatrix} \qquad (5)$$

For example, the first row in the augmented matrix (5) is read $2x + 8y + 6z = 20$. Note that each row of the augmented matrix corresponds to one of the equations in the system.

If we use this form, the solution to Example 1 looks like this.

This becomes a 1.

$$\begin{pmatrix} 2 & 8 & 6 & | & 20 \\ 4 & 2 & -2 & | & -2 \\ 3 & -1 & 1 & | & 11 \end{pmatrix} \rightarrow \begin{pmatrix} 1 & 4 & 3 & | & 10 \\ 4 & 2 & -2 & | & -2 \\ 3 & -1 & 1 & | & 11 \end{pmatrix}$$

These become 0.

This becomes 1.

$$\rightarrow \begin{pmatrix} 1 & 4 & 3 & | & 10 \\ 0 & -14 & -14 & | & -42 \\ 3 & -1 & 1 & | & 11 \end{pmatrix} \rightarrow \begin{pmatrix} 1 & 4 & 3 & | & 10 \\ 0 & -14 & -14 & | & -42 \\ 0 & -13 & -8 & | & -19 \end{pmatrix}$$

$$\rightarrow \begin{pmatrix} 1 & 4 & 3 & | & 10 \\ 0 & 1 & 1 & | & 3 \\ 0 & -13 & -8 & | & -19 \end{pmatrix} \rightarrow \begin{pmatrix} 1 & 0 & -1 & | & -2 \\ 0 & 1 & 1 & | & 3 \\ 0 & 0 & 5 & | & 20 \end{pmatrix}$$

This becomes 1.

These become 0.

$$\rightarrow \begin{pmatrix} 1 & 0 & -1 & | & -2 \\ 0 & 1 & 1 & | & 3 \\ 0 & 0 & 1 & | & 4 \end{pmatrix} \rightarrow \begin{pmatrix} 1 & 0 & 0 & | & 2 \\ 0 & 1 & 0 & | & -1 \\ 0 & 0 & 1 & | & 4 \end{pmatrix}$$

Again we can easily see the solution $x = 2$, $y = -1$, $z = 4$.

Example 2 Solve the system

$$2x + 8y + 6z = 20$$
$$4x + 2y - 2z = -2$$
$$-6x + 4y + 10z = 24$$

Solution We proceed as in Example 1, first writing the system as an augmented matrix.

$$\begin{pmatrix} 2 & 8 & 6 & | & 20 \\ 4 & 2 & -2 & | & -2 \\ -6 & 4 & 10 & | & 24 \end{pmatrix}$$

$$\begin{pmatrix} 1 & 4 & 3 & | & 10 \\ 4 & 2 & -2 & | & -2 \\ -6 & 4 & 10 & | & 24 \end{pmatrix}$$ Divide the first row by 2.

$$\begin{pmatrix} 1 & 4 & 3 & | & 10 \\ 0 & -14 & -14 & | & -42 \\ 0 & 28 & 28 & | & 84 \end{pmatrix}$$ Multiply the first row by −4 and add it to the second and then multiply the first by 6 and add it to the third.

$$\begin{pmatrix} 1 & 4 & 3 & | & 10 \\ 0 & 1 & 1 & | & 3 \\ 0 & 28 & 28 & | & 84 \end{pmatrix}$$ Divide the second row by −14.

$$\begin{pmatrix} 1 & 0 & -1 & | & -2 \\ 0 & 1 & 1 & | & 3 \\ 0 & 0 & 0 & | & 0 \end{pmatrix}$$ Multiply the second row by −4 and add it to the first; then multiply the second row by −28 and add it to the third.

This is equivalent to the system of equations

$$x \quad - z = -2$$
$$y + z = 3$$
$$0 = 0$$

This is as far as we can go. There are now only two nonzero equations in the three unknowns x, y, and z and there are an infinite number of solutions. To see this, let z be chosen. Then $y = 3 - z$ and $x = -2 + z$. This will be a solution for any number z. We write these solutions in the form $(-2 + z, 3 - z, z)$. For example, if $z = 0$ we obtain the solution $(-2, 3, 0)$. For $z = 10$ we obtain the solution $(8, -7, 10)$.

ELEMENTARY ROW OPERATIONS

We now introduce some terminology. We have seen that multiplying (or dividing) the sides of an equation by a nonzero number gives us a new, valid equation. Moreover, adding a multiple of one equation to another equation in a system gives us another valid equation. Finally, if we interchange two equations in a system of equations, we obtain an equivalent system. These three operations, when applied to the rows of the augmented matrix representation of a system of equations, are called **elementary row operations**.

To sum up, the following three elementary row operations can be applied to the augmented matrix representation of a system of equations.

> **1.** Replace a row with a nonzero multiple of that row.
> **2.** Replace a row with the sum of the row and a multiple of some other row.
> **3.** Interchange two rows.

The process of applying elementary row operations to simplify an augmented matrix is called **row reduction**.

Notation.

1. $M_i(c)$ stands for "replace the ith row by the ith row *multiplied* by c."
2. $A_{i,j}(c)$ stands for "replace the jth row with the sum of the jth row and the ith row multiplied by c."
3. $P_{i,j}$ stands for "interchange (permute) rows i and j."
4. $A \rightarrow B$ indicates that the augmented matrices A and B are equivalent; that is, the systems they represent have the same solution.

In Example 1 we saw that by using the elementary row operations (1) and (2) several times we could obtain a system in which the solutions to the system were given explicitly. In the examples that follow we shall use our new notation to indicate the steps we are performing.

Example 3   Solve the system

$$2x + 8y + 6z = 20$$
$$4x + 2y - 2z = -2 \tag{6}$$
$$-6x + 4y + 10z = 30.$$

Solution   We use the augmented-matrix form and proceed exactly as in Example 2 to obtain, successively, the following systems. (Note how, in each step, we use either elementary row operation 1 or 2.)

$$\begin{pmatrix} 2 & 8 & 6 & | & 20 \\ 4 & 2 & -2 & | & -2 \\ -6 & 4 & 10 & | & 30 \end{pmatrix} \xrightarrow{M_1(\frac{1}{2})} \begin{pmatrix} 1 & 4 & 3 & | & 10 \\ 4 & 2 & -2 & | & -2 \\ -6 & 4 & 10 & | & 30 \end{pmatrix}$$

$$\xrightarrow[A_{1,3}(6)]{A_{1,2}(-4)} \begin{pmatrix} 1 & 4 & 3 & | & 10 \\ 0 & -14 & -14 & | & -42 \\ 0 & 28 & 28 & | & 90 \end{pmatrix}$$

$$\xrightarrow{M_2(-\frac{1}{14})} \begin{pmatrix} 1 & 4 & 3 & | & 10 \\ 0 & 1 & 1 & | & 3 \\ 0 & 28 & 28 & | & 90 \end{pmatrix}$$

$$\xrightarrow[A_{2,3}(-28)]{A_{2,1}(-4)} \begin{pmatrix} 1 & 0 & -1 & | & -2 \\ 0 & 1 & 1 & | & 3 \\ 0 & 0 & 0 & | & 6 \end{pmatrix}$$

$$\xrightarrow{M_3(\frac{1}{6})} \begin{pmatrix} 1 & 0 & -1 & | & -2 \\ 0 & 1 & 1 & | & 3 \\ 0 & 0 & 0 & | & 1 \end{pmatrix}$$

The last equation now reads $0x + 0y + 0z = 1$, which is impossible since $0 \neq 1$. Thus system (6) has *no* solution. As in the system of two unknowns, we say that the system is **inconsistent**.

Let us take another look at these three examples. In Example 1, the original coefficient matrix was

$$A_1 = \begin{pmatrix} 2 & 8 & 6 \\ 4 & 2 & -2 \\ 3 & -1 & 1 \end{pmatrix}.$$

In the process of row reduction A_1 was "reduced" to the matrix

$$R_1 = \begin{pmatrix} 1 & 0 & 0 \\ 0 & 1 & 0 \\ 0 & 0 & 1 \end{pmatrix}.$$

In Example 2 we started with

$$A_2 = \begin{pmatrix} 2 & 8 & 6 \\ 4 & 2 & -2 \\ -6 & 4 & 10 \end{pmatrix}$$

and ended up with

$$R_2 = \begin{pmatrix} 1 & 0 & -1 \\ 0 & 1 & 1 \\ 0 & 0 & 0 \end{pmatrix}.$$

In Example 3 we began with

$$A_3 = \begin{pmatrix} 2 & 8 & 6 \\ 4 & 2 & -2 \\ -6 & 4 & 10 \end{pmatrix}$$

and again ended up with

$$R_3 = \begin{pmatrix} 1 & 0 & -1 \\ 0 & 1 & 1 \\ 0 & 0 & 0 \end{pmatrix}.$$

The matrices R_1, R_2, and R_3 are called the *reduced row-echelon forms* of the matrices A_1, A_2, and A_3, respectively. In general, we have the following definition.

Reduced Row-Echelon Form   A matrix is in **reduced row-echelon form** if the following four conditions hold:

1. All rows (if any) consisting entirely of zeros appear at the bottom of the matrix.

2. The first (starting from the left) nonzero number in any row not consisting entirely of zeros is 1.

3. If two successive rows do not consist entirely of zeros, then the first 1 in the lower row occurs farther to the right than the first 1 in the higher row.

4. Any column containing the first 1 in a row has zeros everywhere else.

Example 4   The following matrices are in reduced row-echelon form.

$$\begin{pmatrix} 1 & 0 & 0 \\ 0 & 1 & 0 \\ 0 & 0 & 1 \end{pmatrix} \qquad \begin{pmatrix} 1 & 0 & 0 & 0 \\ 0 & 1 & 0 & 0 \\ 0 & 0 & 0 & 1 \end{pmatrix} \qquad \begin{pmatrix} 1 & 0 & 0 & 5 \\ 0 & 0 & 1 & 2 \end{pmatrix}$$

$$\begin{pmatrix} 1 & 0 \\ 0 & 1 \end{pmatrix} \qquad \begin{pmatrix} 1 & 0 & 2 & 5 \\ 0 & 1 & 3 & 6 \\ 0 & 0 & 0 & 0 \end{pmatrix}$$

Example 5 The following matrices are *not* in reduced row-echelon form

(a) $\begin{pmatrix} 1 & 0 & 0 \\ 0 & 0 & 0 \\ 0 & 1 & 0 \end{pmatrix}$ Condition (1) is violated.

(b) $\begin{pmatrix} 1 & 0 & 0 \\ 0 & 2 & 0 \\ 0 & 0 & 1 \end{pmatrix}$ Condition (2) is violated.

(c) $\begin{pmatrix} 1 & 0 & 0 \\ 0 & 0 & 1 \\ 0 & 1 & 0 \end{pmatrix}$ Condition (3) is violated.

(d) $\begin{pmatrix} 1 & 0 & 3 \\ 0 & 1 & 0 \\ 0 & 0 & 1 \end{pmatrix}$ Condition (4) is violated.

As we saw in Examples 1, 2, and 3, there is a strong connection between the reduced row-echelon form of a matrix and the existence of a unique solution to the system. In Example 1, the reduced row-echelon form of the *coefficient matrix* (that is, the first three columns of the augmented matrix) had a 1 in each row and there was a unique solution. In Examples 2 and 3, the reduced row-echelon form of the coefficient matrix had a row of zeros and the system had either no solution or an infinite number of solutions. This turns out always to be true in any system with the same number of equations as unknowns.

GAUSS-JORDAN ELIMINATION

In general, the process of solving a system of equations by reducing the co-efficient matrix to its reduced row-echelon form is called **Gauss-Jordan elimination.**

The general $m \times n$ system of m linear equations in n unknowns is given by

$$\begin{array}{c} a_{11}x_1 + a_{12}x_2 + a_{13}x_3 + \cdots + a_{1n}x_n = b_1 \\ a_{21}x_1 + a_{22}x_2 + a_{23}x_3 + \cdots + a_{2n}x_n = b_2 \\ a_{31}x_1 + a_{32}x_2 + a_{33}x_3 + \cdots + a_{3n}x_n = b_3 \\ \vdots \qquad \vdots \qquad \vdots \qquad \vdots \qquad \vdots \qquad \vdots \\ a_{m1}x_1 + a_{m2}x_2 + a_{m3}x_3 + \cdots + a_{mn}x_n = b_m \end{array} \tag{7}$$

In system (7) all the a's and b's are given real numbers. The problem is to find all sets of n numbers, denoted by $(x_1, x_2, x_3, \ldots, x_n)$, that satisfy each of the m equations in (7). The number a_{ij} is the coefficient of the variable x_j in the ith equation.

We solve system (7) by writing the system as an augmented matrix and row-reducing the matrix to its reduced row echelon form.

Note.   To make things simpler, we will limit ourselves to equations with up to four unknowns and will generally use the letters x, y, z, and w to denote the variables.

Example 6   Solve the system

$$x + 3y - 5z + \ w = 4$$
$$2x + 5y - 2z + 4w = 6$$

Solution   We write this system as an augmented matrix and row-reduce.

$$\begin{pmatrix} 1 & 3 & -5 & 1 & | & 4 \\ 2 & 5 & -2 & 4 & | & 6 \end{pmatrix}$$

$$\xrightarrow{A_{1,2}(-2)} \begin{pmatrix} 1 & 3 & -5 & 1 & | & 4 \\ 0 & -1 & 8 & 2 & | & -2 \end{pmatrix}$$
Remember: $A_{1,2}(-2)$ means that we multiply the first row by -2 and add it to the second row.

$$\xrightarrow{M_2(-1)} \begin{pmatrix} 1 & 3 & -5 & 1 & | & 4 \\ 0 & 1 & -8 & -2 & | & 2 \end{pmatrix}$$
$M_2(-1)$ means that we multiply the second row by -1.

$$\xrightarrow{A_{2,1}(-3)} \begin{pmatrix} 1 & 0 & 19 & 7 & | & -2 \\ 0 & 1 & -8 & -2 & | & 2 \end{pmatrix}$$

This is as far as we can go. The coefficient matrix is in reduced row-echelon form. There are evidently an infinite number of solutions. The variables z and w can be chosen arbitrarily. Then $y = 2 + 8z + 2w$ and $x = -2 - 19z - 7w$. All solutions are, therefore, represented by $(-2 - 19z - 7w, 2 + 8z + 2w, z, w)$. For example, if $z = 1$ and $w = 2$, we obtain the solution $(-35, 14, 1, 2)$.

As you will see if you do a lot of system solving, the computations can become very messy. It is a good rule of thumb to use a calculator whenever the fractions become unpleasant.

Example 7   Solve the system
$$x + 2y + 2w = 5$$
$$z = 3. \tag{8}$$

Solution   There are now two equations in four unknowns. The augmented matrix is

$$\begin{pmatrix} 1 & 2 & 0 & 2 & | & 5 \\ 0 & 0 & 1 & 0 & | & 3 \end{pmatrix}$$

Note that this matrix is already in reduced row-echelon form. The equations can be written in the form

$$x = 5 - 2y - 2w$$
$$z = 3. \tag{9}$$

TABLE 1

Machine type	Available time (in machine hours per week)
Milling machines	1950
Lathes	1490
Grinders	2160

Evidently, we may choose any values for y and w and then determine x from (9). For example, $y = 1$ and $w = 2$ leads to $x = 5 - 2 \cdot 1 - 2 \cdot 2 = 5 - 2 - 4 = -1$, so one solution to system (8) is $(-1, 1, 3, 2)$. Evidently, there are an infinite number of solutions, which may be written as

$$(5 - 2y - 2w, y, 3, w).$$

This indicates that:

1. y and w are arbitrary;

2. $z = 3$;

3. $x = 5 - 2y - 2w$.

Example 8   A manufacturing firm has discontinued production of a certain unprofitable product line, creating considerable excess production capacity. Management is planning to devote this excess capacity to three products, which we call products 1, 2, and 3. The available capacity on the machines used to produce these products is summarized in Table 1. The number of machine-hours required for each unit of the respective products is given in Table 2. How many units of each product should be manufactured in order to use all the available production capacity?

Solution   Let x, y, and z denote the number of units produced each week of each of the three products. Since each unit of product 1 requires 0.2 hours on a milling machine, the number of hours needed each week on the milling machines to produce x units is $0.2x$. Similarly, $0.5y$ and $0.3z$ represent the weekly requirements (in hours) on the milling machines to produce y units of product 2 and z

TABLE 2   Productivity (in Machine Hours per Unit)

Machine type	Product 1	Product 2	Product 3
Milling machines	0.2	0.5	0.3
Lathes	0.3	0.4	0.1
Grinders	0.1	0.6	0.4

units of product 3, respectively. Since 1950 hours are available on milling machines each week, we have (assuming that all capacity is to be used)

$$0.2x + 0.5y + 0.3z = 1950 \qquad \text{Milling machine equation.}$$

The equations for utilizing all the capacity of the other two machine types are obtained in a like manner.

$$0.3x + 0.4y + 0.1z = 1490 \qquad \text{Lathe equation}$$

$$0.1x + 0.6y + 0.4z = 2160 \qquad \text{Grinder equation}$$

This is a system of three equations in three unknowns. To simplify matters algebraically, we first multiply each equation by 10 to eliminate the decimals. Then we row-reduce in the usual way.

$$\begin{pmatrix} 2 & 5 & 3 & | & 19{,}500 \\ 3 & 4 & 1 & | & 14{,}900 \\ 1 & 6 & 4 & | & 21{,}600 \end{pmatrix} \xrightarrow{M_1(\frac{1}{2})} \begin{pmatrix} 1 & \frac{5}{2} & \frac{3}{2} & | & 9750 \\ 3 & 4 & 1 & | & 14{,}900 \\ 1 & 6 & 4 & | & 21{,}600 \end{pmatrix}$$

$$\xrightarrow[A_{1,3}(-1)]{A_{1,2}(-3)} \begin{pmatrix} 1 & \frac{5}{2} & \frac{3}{2} & | & 9750 \\ 0 & -\frac{7}{2} & -\frac{7}{2} & | & -14{,}350 \\ 0 & \frac{7}{2} & \frac{5}{2} & | & 11{,}850 \end{pmatrix}$$

$$\xrightarrow{M_2(-\frac{2}{7})} \begin{pmatrix} 1 & \frac{5}{2} & \frac{3}{2} & | & 9750 \\ 0 & 1 & 1 & | & 4100 \\ 0 & \frac{7}{2} & \frac{5}{2} & | & 11{,}850 \end{pmatrix}$$

$$\xrightarrow[A_{2,3}(-\frac{7}{2})]{A_{2,1}(-\frac{5}{2})} \begin{pmatrix} 1 & 0 & -1 & | & -500 \\ 0 & 1 & 1 & | & 4100 \\ 0 & 0 & -1 & | & -2500 \end{pmatrix}$$

$$\xrightarrow{M_3(-1)} \begin{pmatrix} 1 & 0 & -1 & | & -500 \\ 0 & 1 & 1 & | & 4100 \\ 0 & 0 & 1 & | & 2500 \end{pmatrix}$$

$$\xrightarrow[A_{3,2}(-1)]{A_{3,1}(1)} \begin{pmatrix} 1 & 0 & 0 & | & 2000 \\ 0 & 1 & 0 & | & 1600 \\ 0 & 0 & 1 & | & 2500 \end{pmatrix}$$

Thus

$$x = 2000 \text{ units of product 1}$$

$$y = 1600 \text{ units of product 2}$$

$$z = 2500 \text{ units of product 3}$$

must be produced in order to ensure full capacity.

Check.

$0.2x + 0.5y + 0.3z = 0.2(2000) + 0.5(1600) + 0.3(2500)$
$$= 400 + 800 + 750 = 1950$$

$0.3x + 0.4y + 0.1z = 0.3(2000) + 0.4(1600) + 0.1(2500)$
$$= 600 + 640 + 250 = 1490$$

$0.1x + 0.6y + 0.4z = 0.1(2000) + 0.6(1600) + 0.4(2500)$
$$= 200 + 960 + 1000 = 2160$$

Example 9   Three species of bacteria coexist in a test tube and feed on three foods. Suppose that a bacterium of the ith species consumes on the average an amount a_{ij} of the jth food per day. Suppose that $a_{11} = 1$, $a_{12} = 1$, $a_{13} = 1$, $a_{21} = 1$, $a_{22} = 2$, $a_{23} = 3$, $a_{31} = 1$, $a_{32} = 3$, and $a_{33} = 5$. Suppose further that there are 15,000 units of the first food supplied daily to the test tube, 30,000 units of the second food, and 45,000 units of the third food. Assuming that all food is consumed, what are the populations of the three species that can coexist in this environment?

Solution   Let x, y, and z be the populations of the three species that can be supported by the given foods. Using the information supplied above, we see that, for example, species 1 consumes $a_{12}x = x$ units of food 2; species 2 consumes $a_{22}y = 2y$ units of food 2; and species 3 consumes $a_{32}z = 3z$ units of food 2. Hence $x + 2y + 3z =$ total supply of food 2 $= 30,000$. Doing a similar calculation for each of the other two foods, we obtain the following system

$$x + \ y + \ z = 15,000$$
$$x + 2y + 3z = 30,000 \qquad (10)$$
$$x + 3y + 5z = 45,000$$

Upon solving, we obtain

$$\begin{pmatrix} 1 & 1 & 1 & | & 15,000 \\ 1 & 2 & 3 & | & 30,000 \\ 1 & 3 & 5 & | & 45,000 \end{pmatrix} \xrightarrow{\substack{A_{1,2}(-1) \\ A_{1,3}(-1)}} \begin{pmatrix} 1 & 1 & 1 & | & 15,000 \\ 0 & 1 & 2 & | & 15,000 \\ 0 & 2 & 4 & | & 30,000 \end{pmatrix}$$

$$\xrightarrow{\substack{A_{2,1}(-1) \\ A_{2,3}(-2)}} \begin{pmatrix} 1 & 0 & -1 & | & 0 \\ 0 & 1 & 2 & | & 15,000 \\ 0 & 0 & 0 & | & 0 \end{pmatrix}$$

Thus if z is chosen arbitrarily, we have an infinite number of solutions given by $(z, 15,000 - 2z, z)$. Of course x, y, and z are nonnegative so $15,000 - 2z \geq 0$, which implies that $0 \leq z \leq 7500$. The total population that can coexist is 15,000 (from the first equation in system (10), and $0 \leq x = z \leq 7500$ and $y = 15,000 - 2z$. If $z = 5000$, for example, then $x = y = z = 5000$. If $z = 2000$, then $x = z = 2000$, and $y = 11,000$.

PROBLEMS 2.5 In Problems 1–20, use Gauss-Jordan elimination to find all solutions, if any, to the given systems.

1.
$$x - 2y + 3z = 11$$
$$4x + y - z = 4$$
$$2x - v + 3z = 10$$

2.
$$-2x + y + 6z = 18$$
$$5x + 8z = -16$$
$$3x + 2y - 10z = -3$$

3.
$$3x + 6y - 6z = 9$$
$$2x - 5y + 4z = 6$$
$$-x + 16y + 14z = -3$$

4.
$$3x + 6y - 6z = 9$$
$$2x - 5y + 4z = 6$$
$$5x + 28y - 26z = -8$$

5.
$$x + y - z = 7$$
$$4x - y + 5z = 4$$
$$2x + 2y - 3z = 0$$

6.
$$x + y - z = 7$$
$$4x - y + 5z = 4$$
$$6x + y + 3z = 18$$

7.
$$x + y - z = 7$$
$$4x - y + 5z = 4$$
$$6x + y + 3z = 20$$

8.
$$x - 2y + 3z = 0$$
$$4x + y - z = 0$$
$$2x - y + 3z = 0$$

9.
$$x + y - z = 0$$
$$4x - y + 5z = 0$$
$$6x + y + 3z = 0$$

10.
$$2y + 5z = 6$$
$$x - 2z = 4$$
$$2x + 4y = -2$$

11.
$$x + 2y - z = 4$$
$$3x + 4y - 2z = .7$$

12.
$$x + 2y - 4z = 4$$
$$-2x - 4y + 8z = -8$$

13.
$$x + 2y - 4z = 4$$
$$-2x - 4y + 8z = -9$$

14.
$$x + 2y - z + w = 7$$
$$3x + 6y - 3z + 3w = 21$$

15.
$$2x + 6y - 4z + 2w = 4$$
$$x - z + w = 5$$
$$-3x + 2y - 2z = -2$$

16.
$$x - 2y + z + w = 2$$
$$3x + z - 2w = -8$$
$$y - z - w = 1$$
$$-x + 6y - 2z = 7$$

17.
$$x - 2y + z + w = 2$$
$$3x + 2z - 2w = -8$$
$$4y - z - w = 1$$
$$5x + 3z - w = -3$$

18.
$$x - 2y + z + w = 2$$
$$3x + 2z - 2w = -8$$
$$4y - z - w = 1$$
$$5x + 3z - w = 0$$

19.
$$x + y = 4$$
$$2x - y = 7$$
$$3x + 2y = 8$$

20.
$$x + y = 4$$
$$2x - 3y = 7$$
$$3x - 2y = 11$$

In Problems 21–29, determine whether the given matrix is in reduced row-echelon form

21. $\begin{pmatrix} 1 & 1 & 0 \\ 0 & 1 & 1 \\ 0 & 0 & 1 \end{pmatrix}$

22. $\begin{pmatrix} 2 & 0 & 0 \\ 0 & 1 & 0 \\ 0 & 0 & -1 \end{pmatrix}$

23. $\begin{pmatrix} 1 & 0 & 1 & 0 \\ 0 & 1 & 1 & 0 \\ 0 & 0 & 0 & 0 \end{pmatrix}$

24. $\begin{pmatrix} 1 & 0 & 0 & 0 \\ 0 & 0 & 1 & 0 \\ 0 & 0 & 0 & 1 \end{pmatrix}$

25. $\begin{pmatrix} 0 & 1 & 0 & 0 \\ 1 & 0 & 0 & 0 \\ 0 & 0 & 0 & 0 \end{pmatrix}$

26. $\begin{pmatrix} 1 & 0 & 1 & 2 \\ 0 & 1 & 3 & 4 \end{pmatrix}$

27. $\begin{pmatrix} 1 & 0 \\ 0 & 1 \\ 0 & 0 \end{pmatrix}$ **28.** $\begin{pmatrix} 1 & 0 & 0 \\ 0 & 0 & 0 \\ 0 & 0 & 1 \end{pmatrix}$ **29.** $\begin{pmatrix} 1 & 0 & 0 & 4 \\ 0 & 1 & 0 & 5 \\ 0 & 1 & 1 & 6 \end{pmatrix}$

In Problems 30–36, use the elementary row operations to reduce the given matrices to reduced row-echelon form.

30. $\begin{pmatrix} 1 & 1 \\ 2 & 3 \end{pmatrix}$ **31.** $\begin{pmatrix} -1 & 6 \\ 4 & 2 \end{pmatrix}$ **32.** $\begin{pmatrix} 1 & -1 & 1 \\ 2 & 4 & 3 \\ 5 & 6 & -2 \end{pmatrix}$

33. $\begin{pmatrix} 2 & -4 & 8 \\ 3 & 5 & 8 \\ -6 & 0 & 4 \end{pmatrix}$ **34.** $\begin{pmatrix} 2 & -4 & -2 \\ 3 & 1 & 6 \end{pmatrix}$ **35.** $\begin{pmatrix} 2 & -7 \\ 3 & 5 \\ 4 & -3 \end{pmatrix}$

36. $\begin{pmatrix} 1 & 1 & 1 & 1 \\ 2 & 2 & 2 & 2 \\ 3 & 3 & 3 & 3 \\ 4 & 4 & 4 & 4 \end{pmatrix}$

37. In Example 8, how many units of each product should be manufactured in order to use all the available production capacity for the data in Tables 3 and 4?

TABLE 3

Machine type	Available time (in machine hours per week)
Milling machines	1281
Lathes	942
Grinders	1185

TABLE 4 Productivity (in Machine Hours per Unit)

Machine type	Product 1	Product 2	Product 3
Milling machines	0.2	0.5	0.4
Lathes	0.1	0.4	0.3
Grinders	0.3	0.3	0.5

38. The Robinson Farm in Illinois has 1000 acres to be planted with soybeans, corn, and wheat. During the planting season Mrs. Robinson has 3700 labor-hours of planting time available to her. Each acre of soybeans requires 2 labor-hours, each acre of corn requires 6 labor-hours and each acre of wheat requires 6 labor-hours. Seed to plant an acre of soybeans costs $12; seed for an acre of corn costs $20 and seed for an acre of wheat costs $8. Mrs. Robinson has $12,600 on hand to pay for seed. How should the 1000 acres be planted in order to use all the available land, labor, and seed money?

39. A traveler just returned from Europe spent $30 a day for housing in England, $20 a day in France and $20 a day in Spain. For food the traveler spent $20 a day in England, $30 a day in France, and $20 a day in Spain. The traveler spent $10 a day in each country for incidental expenses. The traveler's records of the trip indicate a total of $340 spent for housing, $320 for food, and $140 for incidental expenses while traveling in these countries. Calculate the number of days the traveler spent in each of the countries or show that the records must be incorrect, because the amounts spent are incompatible with each other.

40. An intelligence agent knows that 60 aircraft, consisting of fighter planes and bombers, are stationed at a certain secret airfield. The agent wishes to determine how many of the 60 are fighter planes and how many are bombers. There is a type of rocket carried by both sorts of planes; the fighter carries six of these rockets, the bomber only two. The agent learns that 250 rockets are required to arm every plane at this airfield. Furthermore, the agent overhears a remark that there are twice as many fighter planes as bombers at the base (that is, the number of fighter planes minus twice the number of bombers equals zero). Calculate the number of fighter planes and bombers at the airfield or show that the agent's information must be incorrect, because it is inconsistent.

41. An investor remarks to a stockbroker that all her stock holdings are in three companies, Eastern Airlines, Hilton Hotels, and McDonald's, and that 2 days ago the value of her stocks went down $350 but yesterday the value increased by $600. The broker recalls that 2 days ago the price of Eastern Airlines stock dropped by $1 a share, Hilton Hotels dropped $1.50, but the price of McDonald's stock rose by $0.50. The broker also remembers that yesterday the price of Eastern Airlines stock rose $1.50, there was a further drop of $0.50 a share in Hilton Hotels stock, and McDonald's stock rose $1.

Show that the broker does not have enough information to calculate the number of shares the investor owns of each company's stock, but that when the investor says that she owns 200 shares of McDonald's stock, the broker can calculate the number of shares of Eastern Airlines and Hilton Hotels.

42. In Example 9, assume that there are 20,000 units of the first food, 30,000 units of the second, and 40,000 units of the third supplied daily to the test tube. Assuming that all three foods are consumed, what populations of the three species can coexist in the environment? Are these populations unique?

43. Suppose that an experiment has five possible outcomes with probabilities $p_1, p_2, p_3,$ $p_4,$ and p_5. If $p_1 = p_2 + p_3, p_3 + p_4 = 2p_2, p_2 + p_3 + p_4 = p_5,$ and $p_1 + p_2 = p_5,$ determine the probabilities of the five outcomes. [*Hint*: In any experiment, the sum of the probabilities of the outcomes is 1.]

44. The activities of a grazing animal can be classified roughly into three categories: (1) grazing, (2) moving (to new grazing areas or to avoid predators), and (3) resting. The net energy gain (above maintenance requirements) from grazing is 200 calories per hour. The net energy losses in moving and resting are 150 and 50 calories per hour, respectively.

(a) How should the day be divided among the three activities so that the energy gains during grazing exactly compensate for energy losses during moving and resting?

(b) Is this division of the day unique?

45. Suppose that the grazing animal of Problem 44 must rest for at least 6 hours every day. How should the day be divided?

46. Consider the system

$$2x - y + 3z = a$$
$$3x + y - 5z = b$$
$$-5x - 5y + 21z = c$$

Show that the system is inconsistent if $c \neq 2a - 3b$.

47. Consider the system

$$2x + 3y - z = a$$
$$x - y + 3z = b$$
$$3x + 7y - 5z = c$$

Find conditions on a, b, and c such that the system is consistent.

 48. Solve the following system using a hand calculator and carrying 5 decimal places of accuracy.

$$2y - z - 4w = 2$$
$$x - y + 5z + 2w = -4$$
$$3x + 3y - 7z - w = 4$$
$$-x - 2y + 3z = -7$$

 49. Follow the directions of Problem 48 for the system

$$3.8x + 1.6y + 0.9z = 3.72$$
$$-0.7x + 5.4y + 1.6z = 3.16$$
$$1.5x + 1.1y - 3.2z = 43.78$$

50. The system of equations in (7) is called **homogeneous** if the numbers $b_1, b_2, \ldots, b_m$ are all equal to zero. Explain why every homogeneous system of equations is consistent.

In Problems 51–63, find all solutions to the homogeneous systems.

51. $2x - y = 0$
  $3x + 4y = 0$

52. $x - 5y = 0$
  $-x + 5y = 0$

53. $x + y - z = 0$
  $2x - 4y + 3z = 0$
  $3x + 7y - z = 0$

54. $x + y - z = 0$
  $2x - 4y + 3z = 0$
  $-x - 7y + 6z = 0$

55. $x + y - z = 0$
  $2x - 4y + 3z = 0$
  $-5x + 13y - 10z = 0$

56. $2x + 3y - z = 0$
  $6x - 5y + 7z = 0$

57. $4x - y = 0$
  $7x + 3y = 0$
  $-8x + 6y = 0$

58. $x - y + 7z - w = 0$
  $2x + 3y - 8z + w = 0$

59. $x - 2y + z + w = 0$
  $3x + 2z - 2w = 0$
  $4y - z - w = 0$
  $5x + 3z - w = 0$

60. $-2x + 7w = 0$
  $x + 2y - z + 4w = 0$
  $3x - z + 5w = 0$
  $4x + 2y + 3z = 0$

61.
$$2x - y = 0$$
$$3x + 5y = 0$$
$$7x - 3y = 0$$
$$-2x + 3y = 0$$

62.
$$x - 3y = 0$$
$$-2x + 6y = 0$$
$$4x - 12y = 0$$

63.
$$x + y - z = 0$$
$$4x - y + 5z = 0$$
$$-2x + y - 2z = 0$$
$$3x + 2y - 6z = 0$$

64. Show that the homogeneous system

$$a_{11}x + a_{12}y = 0$$
$$a_{21}x + a_{22}y = 0$$

has an infinite number of solutions if and only if $a_{11}a_{22} - a_{12}a_{21} = 0$. Explain this result geometrically.

*65. Show that a homogeneous system of m equations in n unknowns has an infinite number of solutions if $n > m$.

*66. Consider the system

$$2x - 3y + 5z = 0$$
$$-x + 7y - z = 0$$
$$4x + 11y + kz = 0$$

For what value of k will the system have nonzero solutions?

**67. Consider the homogeneous system of three equations in three unknowns:

$$a_{11}x + a_{12}y + a_{13}z = 0$$
$$a_{21}x + a_{22}y + a_{23}z = 0$$
$$a_{31}x + a_{32}y + a_{33}z = 0$$

Find conditions on the coefficients a_{ij} such that the zero solution is the only solution.

2.6 A Model of Traffic Flow (Optional)

In this section we show how solving a system of linear equations can help solve a practical problem involving traffic flow. To make things more concrete, we begin with a map showing a small section of downtown Atlanta, Georgia, as in Figure 1.

On the map we have indicated the traffic flow in and out of each street; the units are vehicles per hour (vph). Since traffic flow varies greatly during the day, we shall assume that the numbers given represent average traffic flow at the peak period, which is approximately 4 P.M. to 5:30 P.M.

Suppose now that a political group is planning a demonstration on Merritts Avenue between Courtland Street and Peachtree Street for 5 P.M. Wednesday. The Atlanta police can, to a certain extent, control the traffic flow by resetting traffic lights, stationing police officers at certain key intersections, or closing a critical street to all vehicular traffic. If traffic is slowed on Merritts Avenue, it will increase on adjacent streets. The question becomes one of minimizing

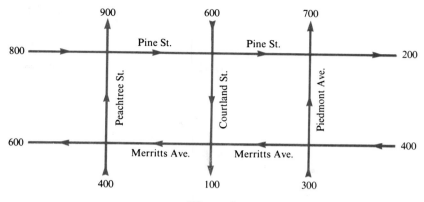

Figure 1

traffic flow on Merritts Avenue (between Courtland and Peachtree) without causing traffic tie-ups on the other streets.

To solve our minimization problem, we add some labels to our map, as in Figure 2. Here we have labeled the six junctions A through F and have denoted the traffic flow between adjacent junctions by the variables x_1 through x_7. The problem now is to minimize x_4, subject to the constraints imposed by the problem.

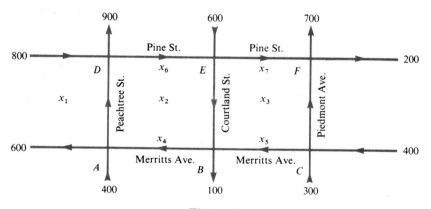

Figure 2

To find these constraints, let's look, for example, at the junction at B. The traffic flowing into B is, from the map, $x_2 + x_5$. The traffic flowing out is $x_4 + 100$. Assuming that no traffic is backed up at B, the "in" traffic must equal the "out" traffic. Thus we obtain the equation

$$x_2 + x_5 = x_4 + 100$$

or

$$x_2 - x_4 + x_5 = 100.$$

Doing this analysis at each junction, we obtain the following system of six equations in seven unknowns:

$$
\begin{array}{llll}
at\ A & x_1 & - x_4 & = -200 \\[4pt]
at\ B & x_2 & - x_4 + x_5 & = 100 \\[4pt]
at\ C & x_3 & + x_5 & = 700 \\[4pt]
at\ D & x_1 & - x_6 & = 100 \\[4pt]
at\ E & x_2 & - x_6 + x_7 & = 600 \\[4pt]
at\ F & x_3 & + x_7 & = 900.
\end{array}
$$

We write this system in augmented matrix form and solve it by row reduction.

$$
\left(\begin{array}{ccccccc|c}
1 & 0 & 0 & -1 & 0 & 0 & 0 & -200 \\
0 & 1 & 0 & -1 & 1 & 0 & 0 & 100 \\
0 & 0 & 1 & 0 & 1 & 0 & 0 & 700 \\
1 & 0 & 0 & 0 & 0 & -1 & 0 & 100 \\
0 & 1 & 0 & 0 & 0 & -1 & 1 & 600 \\
0 & 0 & 1 & 0 & 0 & 0 & 1 & 900
\end{array}\right)
$$

Remember:
$A_{1,4}(-1)$ means that we multiply the first row by -1 and add it to the fourth row.

$$
\xrightarrow{A_{1,4}(-1)}
\left(\begin{array}{ccccccc|c}
1 & 0 & 0 & -1 & 0 & 0 & 0 & -200 \\
0 & 1 & 0 & -1 & 1 & 0 & 0 & 100 \\
0 & 0 & 1 & 0 & 1 & 0 & 0 & 700 \\
0 & 0 & 0 & 1 & 0 & -1 & 0 & 300 \\
0 & 1 & 0 & 0 & 0 & -1 & 1 & 600 \\
0 & 0 & 1 & 0 & 0 & 0 & 1 & 900
\end{array}\right)
$$

$$
\xrightarrow{A_{2,5}(-1)}
\left(\begin{array}{ccccccc|c}
1 & 0 & 0 & -1 & 0 & 0 & 0 & -200 \\
0 & 1 & 0 & -1 & 1 & 0 & 0 & 100 \\
0 & 0 & 1 & 0 & 1 & 0 & 0 & 700 \\
0 & 0 & 0 & 1 & 0 & -1 & 0 & 300 \\
0 & 0 & 0 & 1 & -1 & -1 & 1 & 500 \\
0 & 0 & 1 & 0 & 0 & 0 & 1 & 900
\end{array}\right)
$$

$$
\xrightarrow{A_{3,7}(-1)}
\left(\begin{array}{ccccccc|c}
1 & 0 & 0 & -1 & 0 & 0 & 0 & -200 \\
0 & 1 & 0 & -1 & 1 & 0 & 0 & 100 \\
0 & 0 & 1 & 0 & 1 & 0 & 0 & 700 \\
0 & 0 & 0 & 1 & 0 & -1 & 0 & 300 \\
0 & 0 & 0 & 1 & -1 & -1 & 1 & 500 \\
0 & 0 & 0 & 0 & -1 & 0 & 1 & 200
\end{array}\right)
$$

$$\begin{array}{c} A_{4,1}(1) \\ A_{4,2}(1) \\ A_{4,5}(-1) \\ \longrightarrow \end{array} \left(\begin{array}{ccccccc|c} 1 & 0 & 0 & 0 & 0 & -1 & 0 & 100 \\ 0 & 1 & 0 & 0 & 1 & -1 & 0 & 400 \\ 0 & 0 & 1 & 0 & 1 & 0 & 0 & 700 \\ 0 & 0 & 0 & 1 & 0 & -1 & 0 & 300 \\ 0 & 0 & 0 & 0 & -1 & 0 & 1 & 200 \\ 0 & 0 & 0 & 0 & -1 & 0 & 1 & 200 \end{array} \right)$$

$$\begin{array}{c} M_5(-1) \\ \longrightarrow \end{array} \left(\begin{array}{ccccccc|c} 1 & 0 & 0 & 0 & 0 & -1 & 0 & 100 \\ 0 & 1 & 0 & 0 & 1 & -1 & 0 & 400 \\ 0 & 0 & 1 & 0 & 1 & 0 & 0 & 700 \\ 0 & 0 & 0 & 1 & 0 & -1 & 0 & 300 \\ 0 & 0 & 0 & 0 & 1 & 0 & -1 & -200 \\ 0 & 0 & 0 & 0 & -1 & 0 & 1 & 200 \end{array} \right)$$

$$\begin{array}{c} A_{5,2}(-1) \\ A_{5,3}(-1) \\ A_{5,6}(1) \\ \longrightarrow \end{array} \left(\begin{array}{ccccccc|c} 1 & 0 & 0 & 0 & 0 & -1 & 0 & 100 \\ 0 & 1 & 0 & 0 & 0 & -1 & 1 & 600 \\ 0 & 0 & 1 & 0 & 0 & 0 & 1 & 900 \\ 0 & 0 & 0 & 1 & 0 & -1 & 0 & 300 \\ 0 & 0 & 0 & 0 & 1 & 0 & -1 & -200 \\ 0 & 0 & 0 & 0 & 0 & 0 & 0 & 0 \end{array} \right)$$

This is as far as we can go. Evidently, there are an infinite number of solutions. Using the last matrix above, we can write the variables in terms of x_6 and x_7:

$$x_1 = x_6 + 100$$
$$x_2 = x_6 - x_7 + 600$$
$$x_3 = -x_7 + 900$$
$$x_4 = x_6 + 300$$
$$x_5 = x_7 - 200.$$

Since x_6 must be nonnegative (otherwise traffic would be moving backwards on a one-way street), we must have

$$x_4 \geq 300.$$

Thus to minimize the traffic flow on Merritts Avenue between Courtland and Peachtree (without backing up traffic), the Atlanta Police must allow for a flow of 300 vph there and close off traffic on Pine Street between Peachtree and Courtland (because to get $x_4 = 300$ we must have $x_6 = 0$). Finally, with $x_6 = 0$, we have

$$x_1 = 100$$
$$x_2 = -x_7 + 600$$
$$x_3 = -x_7 + 900$$
$$x_4 = 300$$
$$x_5 = x_7 - 200.$$

From the second equation we must have $x_7 \leq 600$. From the last equation, we see that $x_7 \geq 200$. Thus we obtain the final solution to our problem. To obtain minimum traffic flow at x_4, we must have

$$x_1 = \qquad 100$$
$$0 \leq x_2 \leq 400 \qquad \text{Since } 200 \leq x_7 \leq 600$$
$$300 \leq x_3 \leq 700$$
$$x_4 = 300$$
$$0 \leq x_5 \leq 400$$
$$x_6 = 0$$
$$200 \leq x_7 \leq 600.$$

PROBLEMS 2.6 On the following maps, minimize the traffic flow at the street indicated with asterisks (****).

1.

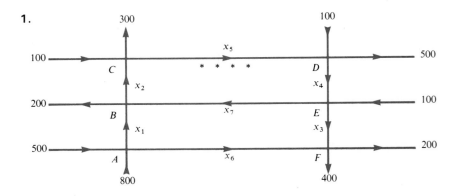

2.

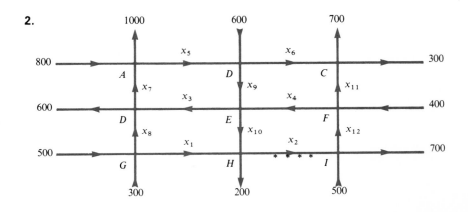

Review Exercises for Chapter 2

1. Determine the quadrant in which each point lies and draw the point in the xy-plane.

(a) $(4, -3)$

(b) $(-2, -6)$

(c) $(1, 1.7)$

(d) $(-5, 2)$

2. Find the distance between each pair of points.

(a) $(1, 5), (-3, 2)$

(b) $(-6, 1), (-11, -4)$

In Exercises 3–7 a linear function is given.

(a) Find the y-value that corresponds to the given x-value.

(b) Find the x- and y-intercepts (if any) of its graph.

(c) Sketch the graph.

3. $y = -2x; x = 5$

4. $x + y = 2; x = -3$

5. $y = 7x - 4; x = 2$

6. $3x + 4y = 12; x = -6$

7. $y = 4; x = 1$

In Exercises 8–13, find the slope-intercept equation of a straight line when either two points on it or its slope and one point are given. Also, find a standard form and a point-slope form of the equation of the line.

8. $(2, 5), (-1, 3)$

9. $(-2, 4), m = 3$

10. $(3, -1), (1, -3)$

11. $(-1, 4), m = 2$

12. $(1, 4), (1, 7)$

13. $(3, -8), (-8, -8).$

14. Find the equation of the line parallel to the line $2x - 5y = 6$ and passing through the point $(4, -2)$.

15. Find the equation of the line perpendicular to the line $3x + 2y = 7$ and passing through the point $(-2, 3)$.

In Exercises 16–30, find all solutions (if any) to the given systems.

16.
$$3x + 6y = 9$$
$$-2x + 3y = 4$$

17.
$$3x + 6y = 9$$
$$2x + 4y = 6$$

18.
$$3x - 6y = 9$$
$$-2x + 4y = 6$$

19.
$$x + y + z = 2$$
$$2x - y + 2z = 4$$
$$-3x + 2y + 3z = 8$$

20.
$$x + y + z = 0$$
$$2x - y + 2z = 0$$
$$-3x + 2y + 3z = 0$$

21.
$$x + y + z = 2$$
$$2x - y + 2z = 4$$
$$-x + 4y + z = 2$$

22.
$$x + y + z = 2$$
$$2x - y + 2z = 4$$
$$-x + 4y + z = 3$$

23.
$$x + y + z = 0$$
$$2x - y + 2z = 0$$
$$-x + 4y + z = 0$$

24.
$$2x + y - 3z = 0$$
$$4x - y + z = 0$$

25.
$$x + y = 0$$
$$2x + y = 0$$
$$3x + y = 0$$

26. $x + y = 1$
         $2x + y = 3$
         $3x + y = 4$

27. $x + y + z + w = 4$
          $2x - 3y - z + 4w = 7$
        $-2x + 4y + z - 2w = 1$
          $5x - y + 2z + w = -1$

28. $x + y + z + w = 0$
         $2x - 3y - z + 4w = 0$
       $-2x + 4y + z - 2w = 0$
         $5x - y + 2z + w = 0$

29. $x + y + z + w = 0$
         $2x - 3y - z + 4w = 0$
       $-2x + 4y + z - 2w = 0$

30. $x + \quad z = 0$
        $y - \quad w = 4$

In Exercises 31–35, determine whether the given matrix is in reduced row-echelon form.

31. $\begin{pmatrix} 1 & 0 & 0 & 0 \\ 0 & 1 & 0 & 3 \\ 0 & 0 & 1 & 3 \end{pmatrix}$

32. $\begin{pmatrix} 1 & 8 & 1 & 0 \\ 0 & 1 & 5 & -7 \\ 0 & 0 & 1 & 4 \end{pmatrix}$

33. $\begin{pmatrix} 1 & 0 \\ 0 & 3 \\ 0 & 0 \end{pmatrix}$

34. $\begin{pmatrix} 1 & 0 & 2 & 0 \\ 0 & 1 & 3 & 0 \end{pmatrix}$

35. $\begin{pmatrix} 1 & 1 & 1 & 1 \\ 0 & 1 & 1 & 1 \end{pmatrix}$

In Exercises 36 and 37, reduce the matrices to reduced row-echelon form.

36. $\begin{pmatrix} 2 & 8 & -2 \\ 1 & 0 & -6 \end{pmatrix}$

37. $\begin{pmatrix} 1 & -1 & 2 & 4 \\ -1 & 2 & 0 & 3 \\ 2 & 3 & -1 & 1 \end{pmatrix}$

3 MATRICES AND MATRIX OPERATIONS

In Section 2.5 we defined a *matrix* and used augmented matrices to simplify (or at least make less cumbersome) the procedure for solving a linear system of equations. In this chapter we again define a matrix, discuss some elementary matrix operations, and show how matrices can arise in practical situations.

3.1 Matrices

An $m \times n$ **matrix** A is a rectangular array of mn numbers arranged in a definite order in m rows and n columns.[†]

$$A = \begin{pmatrix} a_{11} & a_{12} & \cdots & a_{1j} & \cdots & a_{1n} \\ a_{21} & a_{22} & \cdots & a_{2j} & \cdots & a_{2n} \\ \vdots & \vdots & & \vdots & & \vdots \\ a_{i1} & a_{i2} & \cdots & a_{ij} & \cdots & a_{in} \\ \vdots & \vdots & & \vdots & & \vdots \\ a_{m1} & a_{m2} & \cdots & a_{mj} & \cdots & a_{mn} \end{pmatrix}$$

The number a_{ij} appearing in the ith row and jth column of A is called the i,j **component** of A. For convenience, the matrix A is written $A = (a_{ij})$. Usually, matrices will be denoted by capital letters.

If A is an $m \times n$ matrix with $m = n$, then A is called a **square** matrix. An $m \times n$ matrix with all components equal to zero is called the $m \times n$ **zero matrix**

[†] *Historical note:* The term *matrix* was first used in 1850 by the British mathematician James Joseph Sylvester (1814–1897) to distinguish matrices from determinants (which we will not discuss in this text). In fact, matrix was intended to mean "mother of determinants."

Example 1 The following are $m \times n$ matrices for various values of m and n.

(a) $A = \begin{pmatrix} 1 & 3 \\ 4 & 2 \end{pmatrix}$, 2×2 (square)

(b) $A = \begin{pmatrix} -1 & 3 \\ 4 & 0 \\ 1 & -2 \end{pmatrix}$, 3×2

(c) $\begin{pmatrix} -1 & 4 & 1 \\ 3 & 0 & 2 \end{pmatrix}$, 2×3

(d) $\begin{pmatrix} 1 & 6 & -2 \\ 3 & 1 & 4 \\ 2 & -6 & 5 \end{pmatrix}$, 3×3 (square)

(e) $\begin{pmatrix} 0 & 0 & 0 & 0 \\ 0 & 0 & 0 & 0 \end{pmatrix}$, 2×4 zero matrix

Example 2 Find the 1,2, 3,1, and 2,2 components of

$$A = \begin{pmatrix} 1 & 6 & 4 \\ 2 & -3 & 5 \\ 7 & 4 & 0 \end{pmatrix}$$

Solution The 1,2 component is the number in the first row and the second column. We have shaded the first row and the second column, and it is evident that the 1,2 component is 6.

From the shaded matrices below we see that the 3,1 component is 7 and the 2,2 component is -3.

Size and Equality of Matrices An $m \times n$ matrix is said to have the **size** $m \times n$. Two matrices $A = (a_{ij})$ and $B = (b_{ij})$ are **equal** if (1) they have the same size, and (2) corresponding components are equal.

Example 3 Are the following matrices equal?

(a) $\begin{pmatrix} 4 & 1 & 5 \\ 2 & -3 & 0 \end{pmatrix}$ and $\begin{pmatrix} 1+3 & 1 & 2+3 \\ 1+1 & 1-4 & 6-6 \end{pmatrix}$

(b) $\begin{pmatrix} -2 & 0 \\ 1 & 3 \end{pmatrix}$ and $\begin{pmatrix} 0 & -2 \\ 1 & 3 \end{pmatrix}$

(c) $\begin{pmatrix} 1 & 0 \\ 0 & 1 \end{pmatrix}$ and $\begin{pmatrix} 1 & 0 & 0 \\ 0 & 1 & 0 \end{pmatrix}$

Solution

(a) Yes; both matrices are 2×3, and $1 + 3 = 4$, $2 + 3 = 5$, $1 + 1 = 2$, $1 - 4 = -3$, and $6 - 6 = 0$.

(b) No; $-2 \neq 0$, so the matrices are unequal because, for example, the 1,1 components are unequal.

(c) No; the first matrix is 2×2 and the second matrix is 2×3, so they do not have the same size.

ROW AND COLUMN VECTORS

A matrix with one row and n columns is called an **n-component row vector** or, more simply, a **row vector**. A matrix with n rows and one column is called an **n-component column vector**, or **column vector**. Row and column vectors will usually be denoted by boldface, lowercase letters, such as **a**, **b**, **p**, **q**, **x**, or **y**.

Example 4   The following are row vectors.

(a) $(2 \quad 5)$

(b) $(-6 \quad 0 \quad 4)$

(c) $(5 \quad 0 \quad -2 \quad 3)$

(d) $(0 \quad 0 \quad 0)$, zero vector

Example 5   The following are column vectors.

(a) $\begin{pmatrix} 2 \\ 5 \end{pmatrix}$ (b) $\begin{pmatrix} -6 \\ 0 \\ 4 \end{pmatrix}$

(c) $\begin{pmatrix} 5 \\ 0 \\ -2 \\ 3 \end{pmatrix}$ (d) $\begin{pmatrix} 0 \\ 0 \\ 0 \end{pmatrix}$, zero vector

Example 6   Suppose that the buyer for a manufacturing plant must order different quantities of steel, aluminum, oil, and paper. The buyer can keep track of the quantities to be ordered with a single column (or row) vector. The vector $\begin{pmatrix} 10 \\ 30 \\ 15 \\ 60 \end{pmatrix}$ indicates that 10 units of steel, 30 units of aluminum, 15 units of oil, and 60 units of paper would be ordered.

Example 7 In Example 6 we saw how the vector $\begin{pmatrix} 10 \\ 30 \\ 15 \\ 60 \end{pmatrix}$ could represent order

quantities for four different products used by one manufacturer. Suppose that there were five different plants. Then the 4×5 matrix

$$
\begin{array}{cc}
& \text{Plants} \\
Q = \begin{array}{c} \\ \\ \\ \\ \end{array}
\begin{array}{ccccc}
1 & 2 & 3 & 4 & 5 \\
\begin{pmatrix} 10 & 20 & 15 & 16 & 25 \\ 30 & 10 & 20 & 25 & 22 \\ 15 & 22 & 18 & 20 & 13 \\ 60 & 40 & 50 & 35 & 45 \end{pmatrix}
\end{array}
\begin{array}{l} \text{steel} \\ \text{aluminum} \\ \text{oil} \\ \text{paper} \end{array}
\end{array}
\quad \underline{\text{Products}}
$$

could represent the orders for the four products in each of the five plants. We can see, for example, that plant 4 orders 25 units of aluminum, while plant 2 orders 40 units of paper.

Example 8 Five laboratory animals are fed on three different foods. If c_{ij} is defined to be the daily consumption of the ith food by the jth animal, then

$$
C = (c_{ij}) = \begin{pmatrix} c_{11} & c_{12} & c_{13} & c_{14} & c_{15} \\ c_{21} & c_{22} & c_{23} & c_{24} & c_{25} \\ c_{31} & c_{32} & c_{33} & c_{34} & c_{35} \end{pmatrix}
$$

is a 3×5 matrix that records all daily consumption. This is a convenient way to keep records.

Example 9 The table gives the distances in miles between the cities listed.

	Boston	New York	Chicago	Denver	San Francisco
Boston	0	208	980	2025	3186
New York	208	0	850	1833	3049
Chicago	980	850	0	1038	2299
Denver	2025	1833	1038	0	1270
San Francisco	3186	3049	2299	1270	0

(a) Write these data as a matrix.

(b) What is the fourth row?

(c) What is the second column?

(d) What is the 4,2 component?

Solution

(a)   The matrix is given by

$$A = \begin{array}{c} \\ \\ \\ \\ \\ \end{array} \begin{pmatrix} \overset{\text{Boston}}{0} & \overset{\text{New York}}{208} & \overset{\text{Chicago}}{980} & \overset{\text{Denver}}{2025} & \overset{\text{San Francisco}}{3186} \\ 208 & 0 & 850 & 1833 & 3049 \\ 980 & 850 & 0 & 1038 & 2299 \\ 2025 & 1833 & 1038 & 0 & 1270 \\ 3186 & 3049 & 2299 & 1270 & 0 \end{pmatrix} \begin{array}{l} \text{Boston} \\ \text{New York} \\ \text{Chicago} \\ \text{Denver} \\ \text{San Francisco} \end{array}$$

Note.   The labels are optional. They make the matrix easier to read.

(b)   The fourth row is the row vector

$$(2025 \quad 1833 \quad 1038 \quad 0 \quad 1270).$$

(c)   The second column is the column vector

$$\begin{pmatrix} 208 \\ 0 \\ 850 \\ 1833 \\ 3049 \end{pmatrix}.$$

(d)   The 4,2 component is 1883. This is the distance from Denver to New York.

Having defined matrices and seen how they could arise, we now turn to the question of adding matrices and multiplying them by real numbers. For historical reasons, real numbers encountered when dealing with matrices are called **scalars.**[†]

MULTIPLICATION OF A MATRIX BY A SCALAR

Let $A = (a_{ij})$ be an $m \times n$ matrix and let c be a scalar. Then the $m \times n$ matrix cA is given by

$$cA = (ca_{ij}) = \begin{pmatrix} ca_{11} & ca_{12} & \cdots & ca_{1n} \\ ca_{21} & ca_{22} & \cdots & ca_{2n} \\ \vdots & \vdots & & \vdots \\ ca_{m1} & ca_{m2} & \cdots & ca_{mn} \end{pmatrix}$$

[†] The study of vectors and matrices essentially began with the work of the Irish mathematician, Sir William Rowan Hamilton (1805–1865). Hamilton used the word *scalar* to denote a number that could take on "all values contained on the one *scale* of progression of numbers from negative to positive infinity" in a paper published in *Philosophy Magazine* in 1844.

In other words, $cA = (ca_{ij})$ is the matrix obtained by multiplying each component of A by c.

Example 10   Let $A = \begin{pmatrix} 1 & -3 & 4 & 2 \\ 3 & 1 & 4 & 6 \\ -2 & 3 & 5 & 7 \end{pmatrix}$. Then $2A = \begin{pmatrix} 2 & -6 & 8 & 4 \\ 6 & 2 & 8 & 12 \\ -4 & 6 & 10 & 14 \end{pmatrix}$,

$-3A = \begin{pmatrix} -3 & 9 & -12 & -6 \\ -9 & -3 & -12 & -18 \\ 6 & -9 & -15 & -21 \end{pmatrix}$, and $0A = \begin{pmatrix} 0 & 0 & 0 & 0 \\ 0 & 0 & 0 & 0 \\ 0 & 0 & 0 & 0 \end{pmatrix}$.

Example 11   The sales of four products by a national retail chain in three different months is given in Table 1.

TABLE 1   Sales (in Thousands of Units)

Month	Product			
	I	II	III	IV
April 1982	6	19	14	46
May 1982	8	28	12	40
June 1982	4	26	17	55

We can represent these data in a 3×4 matrix S.

$$S = \begin{pmatrix} 6 & 19 & 14 & 46 \\ 8 & 28 & 12 & 40 \\ 4 & 26 & 17 & 55 \end{pmatrix}$$

Suppose now that in 1983, sales in all months and in all categories increased by 50%. Increasing by 50% is the same as multiplying by $1\frac{1}{2}$. Thus the new matrix representing 1983 sales of the four products in the 3 months is $1.5S$.

$$1.5S = \begin{matrix} & & & \text{Product} & \\ & \text{I} & \text{II} & \text{III} & \text{IV} \\ & \begin{pmatrix} 9 & 28.5 & 21 & 69 \\ 12 & 42 & 18 & 60 \\ 6 & 39 & 25.5 & 82.5 \end{pmatrix} & \begin{matrix} \text{April 1983} \\ \text{May 1983} \\ \text{June 1983} \end{matrix} \end{matrix}$$

For example, we see that while May 1982 sales of product III were 12,000 units (since these numbers represent thousands of units), sales of product III in May 1983 amounted to 18,000 units.

We now turn to the addition of matrices.

ADDITION OF MATRICES

Let $A = (a_{ij})$ and $B = (b_{ij})$ be two $m \times n$ matrices. Then the sum of A and B is the $m \times n$ matrix $A + B$ given by

$$A + B = (a_{ij} + b_{ij}) = \begin{pmatrix} a_{11} + b_{11} & a_{12} + b_{12} & \cdots & a_{1n} + b_{1n} \\ a_{21} + b_{21} & a_{22} + b_{22} & \cdots & a_{2n} + b_{2n} \\ \vdots & \vdots & & \vdots \\ a_{m1} + b_{m1} & a_{m2} + b_{m2} & \cdots & a_{mn} + b_{mn} \end{pmatrix}$$

That is, $A + B$ is the $m \times n$ matrix obtained by adding the corresponding components of A and B.

Warning. The sum of two matrices is defined only when both matrices have the same size. Thus, for example, it is not possible to add the matrices

$$\begin{pmatrix} 1 & 2 & 3 \\ 4 & 5 & 6 \end{pmatrix} \quad \text{and} \quad \begin{pmatrix} -1 & 0 \\ 2 & -5 \\ 4 & 7 \end{pmatrix}.$$

Example 12

$$\begin{pmatrix} 2 & 4 & -6 & 7 \\ 1 & 3 & 2 & 1 \\ -4 & 3 & -5 & 5 \end{pmatrix} + \begin{pmatrix} 0 & 1 & 6 & -2 \\ 2 & 3 & 4 & 3 \\ -2 & 1 & 4 & 4 \end{pmatrix} = \begin{pmatrix} 2 & 5 & 0 & 5 \\ 3 & 6 & 6 & 4 \\ -6 & 4 & -1 & 9 \end{pmatrix}$$

Example 13 Let $A = \begin{pmatrix} 1 & 2 & 4 \\ -7 & 3 & -2 \end{pmatrix}$ and $B = \begin{pmatrix} 4 & 0 & 5 \\ 1 & -3 & 6 \end{pmatrix}$.
Calculate $-2A + 3B$.

Solution

$$-2A + 3B = (-2)\begin{pmatrix} 1 & 2 & 4 \\ -7 & 3 & -2 \end{pmatrix} + (3)\begin{pmatrix} 4 & 0 & 5 \\ 1 & -3 & 6 \end{pmatrix}$$
$$= \begin{pmatrix} -2 & -4 & -8 \\ 14 & -6 & 4 \end{pmatrix} + \begin{pmatrix} 12 & 0 & 15 \\ 3 & -9 & 18 \end{pmatrix} = \begin{pmatrix} 10 & -4 & 7 \\ 17 & -15 & 22 \end{pmatrix}$$

Example 14 The retail chain of Example 11 has two stores in southern California. Sales of each store (in hundreds of units) in three different months are given in Table 2. As in Example 11, we can represent the sales in each store by a 3×4 matrix.

$$A = \begin{pmatrix} 7 & 12 & 2 & 28 \\ 5 & 14 & 8 & 17 \\ 6 & 9 & 5 & 33 \end{pmatrix}, \quad B = \begin{pmatrix} 6 & 21 & 8 & 41 \\ 10 & 19 & 14 & 33 \\ 2 & 26 & 5 & 28 \end{pmatrix}.$$

TABLE 2

Store A Sales (in Hundreds of Units)					Store B Sales (in Hundreds of Units)				
	Product					**Product**			
Month	I	II	III	IV	Month	I	II	III	IV
April 1982	7	12	2	28	April 1982	6	21	8	41
May 1982	5	14	8	17	May 1982	10	19	14	33
June 1982	6	9	5	33	June 1982	2	26	5	28

Then the matrix $A + B$ represents total sales for the two stores for each product in each month.

$$A + B = \begin{pmatrix} 13 & 33 & 10 & 69 \\ 15 & 33 & 22 & 50 \\ 8 & 35 & 10 & 61 \end{pmatrix}$$

Thus, for example, we find that 2200 units of product III were sold in May 1982 in stores A and B combined.

PROBLEMS 3.1   In Problems 1–14, determine the size of the given matrix and indicate which matrices are square.

1. $\begin{pmatrix} 1 & 2 \\ 3 & 4 \end{pmatrix}$

2. $\begin{pmatrix} -1 & 2 \\ 3 & 1 \\ 1 & 6 \end{pmatrix}$

3. $\begin{pmatrix} 0 & 0 \\ 0 & 0 \end{pmatrix}$

4. $\begin{pmatrix} 0 & 0 & 0 \\ 0 & 0 & 0 \end{pmatrix}$

5. $\begin{pmatrix} 0 & 0 \\ 0 & 0 \\ 0 & 0 \end{pmatrix}$

6. $\begin{pmatrix} 1 & 0 & 0 \\ 0 & 1 & 0 \\ 0 & 0 & 1 \end{pmatrix}$

7. $\begin{pmatrix} 1 & 3 & 2 & 4 \\ 2 & 1 & 0 & 6 \end{pmatrix}$

8. $(1 \quad 0 \quad 2)$

9. $\begin{pmatrix} 1 \\ 0 \\ 2 \end{pmatrix}$

10. $\begin{pmatrix} 1 & 3 \\ 0 & 6 \\ 2 & 2 \\ 4 & 9 \end{pmatrix}$

11. $\begin{pmatrix} 3 & -6 & 2 \\ 1 & 7 & 2 \\ -1 & 4 & 6 \end{pmatrix}$

12. $\begin{pmatrix} a & b \\ c & d \end{pmatrix}$

13. $(a \quad b \quad c \quad d)$

14. $\begin{pmatrix} a \\ b \\ c \\ d \end{pmatrix}$

In Problems 15–19, determine whether the two matrices are equal.

15. $\begin{pmatrix} 1 & 0 \\ 0 & 1 \end{pmatrix}$ and $\begin{pmatrix} 0 & 1 \\ 1 & 0 \end{pmatrix}$

16. $\begin{pmatrix} 0 & 3 \\ 1 & 0 \end{pmatrix}$ and $\begin{pmatrix} 2-2 & 1+2 \\ 3-2 & -2+2 \end{pmatrix}$

17. $\begin{pmatrix} 0 & 0 & 0 \\ 0 & 0 & 0 \end{pmatrix}$ and $\begin{pmatrix} 0 & 0 \\ 0 & 0 \\ 0 & 0 \end{pmatrix}$

18. $\begin{pmatrix} 0 & 1 & 0 \\ 0 & 0 & 1 \\ 1 & 0 & 0 \end{pmatrix}$ and $\begin{pmatrix} 0 & 0 & 1 \\ 1 & 0 & 0 \\ 0 & 1 & 0 \end{pmatrix}$

19. $\begin{pmatrix} 1-2 & 3 & 1 \\ 0 & 4-5 & 3 \\ 2 & 6 & 2+3 \end{pmatrix}$ and $\begin{pmatrix} -1 & 1+2 & 1 \\ 5-5 & -1 & 5-2 \\ 1+1 & 6 & 5 \end{pmatrix}$

20. Let

$$A = \begin{pmatrix} 1 & 6 & -2 & 3 \\ 4 & 0 & 2 & 6 \\ -1 & 4 & 3 & 1 \end{pmatrix}$$

(a) Write the first row.

(b) Write the 1,3 component.

(c) Write the 3,2 component.

(d) Write the 2,4 component.

In Problems 21–30, perform the indicated computation, if possible.

21. $3\begin{pmatrix} 2 & 4 \\ 7 & -2 \end{pmatrix}$

22. $\begin{pmatrix} 3 & 1 \\ -1 & 4 \end{pmatrix} + \begin{pmatrix} 6 & -2 \\ 3 & 0 \end{pmatrix}$

23. $-\begin{pmatrix} 1 & 2 & 5 \\ 7 & -2 & 0 \end{pmatrix}$

24. $2\begin{pmatrix} 2 & -1 & 1 \\ 1 & 5 & 6 \end{pmatrix} - 5\begin{pmatrix} 1 & 3 & 2 \\ -1 & 1 & 4 \end{pmatrix}$

25. $\begin{pmatrix} 6 & 1 & 2 \\ 3 & -1 & 4 \end{pmatrix} + \begin{pmatrix} 1 & 6 \\ 2 & 4 \\ 3 & 2 \end{pmatrix}$

26. $\begin{pmatrix} 3 & 1 & 4 \\ 1 & 0 & 5 \\ -6 & 7 & 2 \end{pmatrix} + \begin{pmatrix} 7 & -2 & 3 \\ 5 & 0 & 6 \\ 0 & 1 & 2 \end{pmatrix}$

27. $5\begin{pmatrix} 2 & 1 & 6 \\ 0 & 1 & 4 \\ -3 & 2 & 4 \end{pmatrix} + 3\begin{pmatrix} -2 & 1 & 5 \\ 6 & 2 & 1 \\ 0 & 5 & 3 \end{pmatrix}$

28. $\begin{pmatrix} a & b \\ c & d \end{pmatrix} - \begin{pmatrix} e & f \\ g & h \end{pmatrix}$

29. $\begin{pmatrix} 5 & 6 & 2 \\ 3 & 4 & 1 \\ 0 & -7 & 2 \end{pmatrix} + \begin{pmatrix} 0 & 0 & 0 \\ 0 & 0 & 0 \\ 0 & 0 & 0 \end{pmatrix}$

30. $\begin{pmatrix} 1 & 6 \\ 2 & 3 \\ 4 & 7 \end{pmatrix} - \begin{pmatrix} 0 & 0 \\ 0 & 0 \\ 0 & 0 \end{pmatrix}$

In Problems 31–42, perform the indicated computation with $A = \begin{pmatrix} 1 & 3 \\ 2 & 5 \\ -1 & 2 \end{pmatrix}$,

$B = \begin{pmatrix} -2 & 0 \\ 1 & 4 \\ -7 & 5 \end{pmatrix}$, and $C = \begin{pmatrix} -1 & 1 \\ 4 & 6 \\ -7 & 3 \end{pmatrix}$.

31. $3A$

32. $A + B$

33. $A - C$

34. $2C - 5A$

35. $0B$ (0 is the scalar zero.)

36. $-7A + 3B$

37. $A + B + C$

38. $C - A - B$

39. $2A - 3B + 4C$

40. $7C - B + 2A$

41. Find a matrix D such that $2A + B - D$ is the 3×2 zero matrix.

42. Find a matrix E such that $A + 2B - 3C + E$ is the 3×2 zero matrix.

In Problems 43–50, perform the indicated computation with

$$A = \begin{pmatrix} 1 & -1 & 2 \\ 3 & 4 & 5 \\ 0 & 1 & -1 \end{pmatrix}, B = \begin{pmatrix} 0 & 2 & 1 \\ 3 & 0 & 5 \\ 7 & -6 & 0 \end{pmatrix}, \text{ and } C = \begin{pmatrix} 0 & 0 & 2 \\ 3 & 1 & 0 \\ 0 & -2 & 4 \end{pmatrix}.$$

43. $A - 2B$

44. $3A - C$

45. $A + B + C$

46. $2A - B + 2C$

47. $C - A - B$

48. $4C - 2B + 3A$

49. Find a matrix D such that $A + B + C + D$ is the 3×3 zero matrix.

50. Find a matrix E such that $3C - 2B + 8A - 4E$ is the 3×3 zero matrix.

51. Let $A = (a_{ij})$ be an $m \times n$ matrix and let O denote the $m \times n$ zero matrix. Show that $0A = O$ and $O + A = A$. Similarly, show that $1A = A$.

52. Let $A = (a_{ij})$ and $B = (b_{ij})$ be $m \times n$ matrices. Compute $A + B$ and $B + A$ and show that they are equal.

53. If k is a scalar and A and B are as in Problem 52, compute $k(A + B)$ and $kA + kB$ and show that they are equal.

54. If $A = (a_{ij})$, $B = (b_{ij})$, and $C = (c_{ij})$ are $m \times n$ matrices, compute $(A + B) + C$ and $A + (B + C)$ and show that they are equal.

55. Consider the "graph" joining the four points in the figure. Construct a 4×4 matrix having the property that $a_{ij} = 0$ if point i is not connected (joined by a line) to point j and $a_{ij} = 1$ if point i is connected to point j.

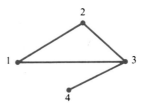

56. Follow the directions of Problem 55 (this time constructing a 5×5 matrix) for the accompanying graph.

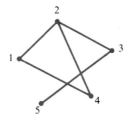

57. Consider the consumption matrix (see Example 8).

$$C = \begin{pmatrix} 1 & 0.5 & 3 & 8 & 0.25 \\ 1.5 & 2 & 6 & 6 & 0.3 \\ 2 & 1.5 & 4 & 9 & 0.6 \end{pmatrix}$$

(a) How many units of the third food are fed to the second animal each day?

(b) How many units of the second food are fed to the third animal each day?

(c) Find a matrix showing the amounts of each food fed to each animal in a 1-week period.

58. Referring to Example 11, if sales increase 25% between 1982 and 1983, find a matrix that represents total sales (in thousands of units) of the four products in the months April, May, and June of 1983.

59. Answer the question of Problem 58 if sales *decrease* 40% in 1983.

60. Referring to Example 14, if sales of store A increase 30% in 1983 while sales in store B decrease 20%, find a matrix that represents combined sales of the four products in April, May, and June of 1983.

61. A polling organization gives the results of asking the preferences of a sample of 1089 voters in a coming election for governor in the table.

	Democratic candidate	Republican candidate	Other candidates	Undecided
Democrats	415	91	6	77
Republicans	65	281	4	63
Independents	31	19	8	29

(a) Write these data as a matrix.

(b) What numbers form the fourth column?

(c) What numbers form the second row?

(d) What number is the 2,4 component?

(e) What number is the 4,2 component?

62. The cost (in cents) of ingredients for a recipe of each of two types of desserts is listed in the table.

	Brownies	Cookies
Sugar	25	8
Butter	13	20
Eggs	0	7
Flour	3	6
Vanilla	3	3
Chocolate	28	0

(a) If a bakery makes 50 recipes of brownies, how much money will be spent for chocolate?

(b) If a bakery makes 30 recipes of brownies and 40 of cookies, how much will it spend for butter?

(c) If a bakery makes 50 recipes of brownies and 20 of cookies, how much will it spend for sugar?

(d) Write the matrix corresponding to the table.

(e) What size is the matrix?

(f) What numbers form the sixth row of the matrix?

(g) What is the 3,1 component?

3.2 Matrix Products

In this section we see how two matrices can be multiplied together. Quite obviously, we could define the product of two $m \times n$ matrices $A = (a_{ij})$ and $B = (b_{ij})$ to be the $m \times n$ matrix whose ijth component is $a_{ij} b_{ij}$. However, for just about all the important applications involving matrices, another kind of product is needed. Let us try to see why this is the case.

Example 1 Suppose that a manufacturer produces four items. The demand for the items is given by the demand vector $\mathbf{d} = (30 \quad 20 \quad 40 \quad 10)$ (a 1×4 matrix). The price per unit that the manufacturer receives for the items is given by the

price vector $\mathbf{p} = \begin{pmatrix} \$20 \\ \$15 \\ \$18 \\ \$40 \end{pmatrix}$ (a 4×1 matrix). If the demand is met, how much money

will the manufacturer receive?

Solution Demand for the first item is 30 and the manufacturer receives \$20 for each of the first item sold. Thus $(30)(20) = \$600$ is received from the sales of the first item. By continuing this reasoning, we see that the total amount of money received is $(30)(20) + (20)(15) + (40)(18) + (10)(40) = 600 + 300 + 720 + 400 = \$2020$.

In the last example we multiplied a row vector by a column vector and obtained a scalar. In general, we have the following definition.

Dot Product Let $\mathbf{p} = (p_1 \quad p_2 \quad \cdots \quad p_n)$ be an n-component row vector and $q = \begin{pmatrix} q_1 \\ q_2 \\ \vdots \\ q_n \end{pmatrix}$ be an

n-component column vector. Then the **dot product** (or **scalar product**) of $\mathbf{p}$ and $\mathbf{q}$, denoted by $\mathbf{p} \cdot \mathbf{q}$, is given by

$$\boxed{\mathbf{p} \cdot \mathbf{q} = p_1 q_1 + p_2 q_2 + \cdots + p_n q_n} \tag{1}$$

We can write this as

$$(p_1 p_2 \cdots p_n) \begin{pmatrix} q_1 \\ q_2 \\ \vdots \\ q_n \end{pmatrix} = p_1 q_1 + p_2 q_2 + \cdots + p_n q_n \qquad \text{This is a real number.}$$

(2)

Warning. When taking the dot product of **p** and **q**, it is necessary that **p** and **q** have the same number of components.

Example 2   Let $\mathbf{a} = (2 \quad -3 \quad 4 \quad -6)$ and $\mathbf{b} = \begin{pmatrix} 1 \\ 2 \\ 0 \\ 3 \end{pmatrix}$. Compute $\mathbf{a} \cdot \mathbf{b}$.

Solution   Here $\mathbf{a} \cdot \mathbf{b} = (2)(1) + (-3)(2) + (4)(0) + (-6)(3) = 2 - 6 + 0 - 18 = -22$.

We now define the product of two matrices.

Matrix Product   Let $A = (a_{ij})$ be an $m \times n$ matrix and let $B = (b_{ij})$ be an $n \times p$ matrix. Then the **product** of A and B is an $m \times p$ matrix $C = (c_{ij})$, where

$$c_{ij} = (i\text{th row of } A) \cdot (j\text{th column of } B) \qquad (3)$$

That is, the ijth element of AB is the dot product of the ith row of A and the jth column of B. If we write this out, we obtain

$$c_{ij} = a_{i1}b_{1j} + a_{i2}b_{2j} + \cdots + a_{in}b_{nj} \qquad (4)$$

Warning. Two matrices can be multiplied together only if the number of columns of the first matrix is equal to the number of rows of the second. Otherwise the vectors that are the ith row of A and the jth column of B will not have the same number of components, and the dot product in equation (3) will not be defined. To illustrate this, we write the matrices A and B:

$$
\begin{array}{c}
i\text{th row of } A \rightarrow
\end{array}
\begin{pmatrix}
a_{11} & a_{12} & \cdots & a_{1n} \\
a_{21} & a_{22} & \cdots & a_{2n} \\
\vdots & \vdots & & \vdots \\
a_{i1} & a_{i2} & \cdots & a_{in} \\
\vdots & \vdots & & \vdots \\
a_{m1} & a_{m2} & \cdots & a_{mn}
\end{pmatrix}
\overset{\begin{array}{c}j\text{th column}\\ \text{of } B \\ \downarrow\end{array}}{
\begin{pmatrix}
b_{11} & b_{12} & \cdots & b_{1j} & \cdots & b_{1p} \\
b_{21} & b_{22} & \cdots & b_{2j} & \cdots & b_{2p} \\
\vdots & \vdots & & \vdots & & \vdots \\
b_{n1} & b_{n2} & \cdots & b_{nj} & \cdots & b_{np}
\end{pmatrix}}.
$$

The row and column vectors shaded above must have the same number of components.

Example 3 If $A = \begin{pmatrix} 1 & 3 \\ -2 & 4 \end{pmatrix}$ and $B = \begin{pmatrix} 3 & -2 \\ 5 & 6 \end{pmatrix}$, calculate AB and BA.

Solution A is a 2×2 matrix and B is a 2×2 matrix, so $C = AB = (2 \times 2) \times (2 \times 2)$ is also a 2×2 matrix. If $C = (c_{ij})$, what is c_{11}? We know that

$$c_{11} = (\text{1st row of } A) \cdot (\text{1st column of } B).$$

Rewriting the matrices, we have

$$
\text{1st row of } A \rightarrow
\begin{pmatrix} 1 & 3 \\ -2 & 4 \end{pmatrix}
\overset{\begin{array}{c}\text{1st column}\\ \text{of } B \\ \downarrow\end{array}}{
\begin{pmatrix} 3 & -2 \\ 5 & 6 \end{pmatrix}}
$$

Thus

$$c_{11} = (1 \quad 3)\begin{pmatrix} 3 \\ 5 \end{pmatrix} = 3 + 15 = 18.$$

Similarly, to compute c_{12} we have

$$
\text{1st row of } A \rightarrow
\begin{pmatrix} 1 & 3 \\ -2 & 4 \end{pmatrix}
\overset{\begin{array}{c}\text{2nd column}\\ \text{of } B \\ \downarrow\end{array}}{
\begin{pmatrix} 3 & -2 \\ 5 & 6 \end{pmatrix}}
$$

and

$$c_{12} = (1 \quad 3)\begin{pmatrix} -2 \\ 6 \end{pmatrix} = -2 + 18 = 16.$$

Continuing, we find that $c_{21} = (-2 \quad 4)\begin{pmatrix} 3 \\ 5 \end{pmatrix} = -6 + 20 = 14$; and $c_{22} = (-2 \quad 4)\begin{pmatrix} -2 \\ 6 \end{pmatrix} = 4 + 24 = 28$. Thus $C = AB = \begin{pmatrix} 18 & 16 \\ 14 & 28 \end{pmatrix}$. Similarly, leaving out the intermediate steps, we see that

$$C' = BA = \begin{pmatrix} 3 & -2 \\ 5 & 6 \end{pmatrix}\begin{pmatrix} 1 & 3 \\ -2 & 4 \end{pmatrix} = \begin{pmatrix} 3+4 & 9-8 \\ 5-12 & 15+24 \end{pmatrix} = \begin{pmatrix} 7 & 1 \\ -7 & 39 \end{pmatrix}$$

Remark. Example 3 illustrates an important fact: *Matrix products do not, in general, commute*; that is, $AB \neq BA$ in general. It sometimes happens that $AB = BA$, but this will be the exception, not the rule. In fact, as the next example illustrates, it may occur that AB is defined, while BA is not. Thus we must be careful of *order* when multiplying two matrices together.

Example 4   Let $A = \begin{pmatrix} 2 & 0 & -3 \\ 4 & 1 & 5 \end{pmatrix}$ and $B = \begin{pmatrix} 7 & -1 & 4 & 7 \\ 2 & 5 & 0 & -4 \\ -3 & 1 & 2 & 3 \end{pmatrix}$. Calculate AB.

Solution   We have

$$\begin{array}{cc} A & B \\ (2 \times 3) \times & (3 \times 4) \end{array}$$

These are equal.

so that $C = AB$ is a 2×4 matrix. Let $C = (c_{ij})$. Then to compute c_{11}, we have

1st column of B

1st row of A → $\begin{pmatrix} 2 & 0 & -3 \\ 4 & 1 & 5 \end{pmatrix}\begin{pmatrix} 7 & -1 & 4 & 7 \\ 2 & 5 & 0 & -4 \\ -3 & 1 & 2 & 3 \end{pmatrix}$,

or

$$c_{11} = (2 \quad 0 \quad -3)\begin{pmatrix} 7 \\ 2 \\ -3 \end{pmatrix} = 23.$$

Similarly, to compute c_{12} we write

2nd column of B

1st row of A → $\begin{pmatrix} 2 & 0 & -3 \\ 4 & 1 & 5 \end{pmatrix}\begin{pmatrix} 7 & -1 & 4 & 7 \\ 2 & 5 & 0 & -4 \\ -3 & 1 & 2 & 3 \end{pmatrix}$, so that

$$c_{12} = (2 \quad 0 \quad -3)\begin{pmatrix} -1 \\ 5 \\ 1 \end{pmatrix} = -5.$$

Continuing, we have

$$c_{13} = (2 \quad 0 \quad -3)\begin{pmatrix} 4 \\ 0 \\ 2 \end{pmatrix} = 2 \qquad c_{14} = (2 \quad 0 \quad -3)\begin{pmatrix} 7 \\ -4 \\ 3 \end{pmatrix} = 5$$

$$c_{21} = (4 \quad 1 \quad 5)\begin{pmatrix} 7 \\ 2 \\ -3 \end{pmatrix} = 15 \qquad c_{22} = (4 \quad 1 \quad 5)\begin{pmatrix} -1 \\ 5 \\ 1 \end{pmatrix} = 6$$

$$c_{23} = (4 \quad 1 \quad 5)\begin{pmatrix} 4 \\ 0 \\ 2 \end{pmatrix} = 26 \qquad c_{24} = (4 \quad 1 \quad 5)\begin{pmatrix} 7 \\ -4 \\ 3 \end{pmatrix} = 39$$

Hence $AB = \begin{pmatrix} 23 & -5 & 2 & 5 \\ 15 & 6 & 26 & 39 \end{pmatrix}$. This completes the problem. Note that the product BA is *not* defined since the number of columns of B (four) is not equal to the number of rows of A (two).

Example 5   In Example 3.1.11 on page 92, we obtained the following matrix representation of the number of units (in thousands) of four items sold by a large retail chain in three successive months.

$$\begin{array}{cccc} \text{I} & \text{II} & \text{III} & \text{IV} \end{array}$$
$$S = \begin{pmatrix} 6 & 19 & 14 & 46 \\ 8 & 28 & 12 & 40 \\ 4 & 26 & 17 & 55 \end{pmatrix} \begin{array}{l} \text{April} \\ \text{May} \\ \text{June} \end{array}$$

The gross (before-tax) profit made and the taxes paid (in hundreds of dollars) for each thousand units sold of each item are given in Table 1.

TABLE 1

Item	Profit (in Hundreds of Dollars)	Taxes (in Hundreds of Dollars)
I	2	1
II	3	2
III	5	3
IV	4	2

How much profit was made and how much tax was paid on sales of the four items for each of the months of April, May, and June?

Solution First, we write the information in Table 1 in matrix form to obtain what we may call the **profit-tax matrix**, P.

$$P = \begin{matrix} & \text{Profit} & \text{Taxes} & \\ & \begin{pmatrix} 2 & 1 \\ 3 & 2 \\ 5 & 3 \\ 4 & 2 \end{pmatrix} & & \begin{matrix} \text{I} \\ \text{II} \\ \text{III} \\ \text{IV} \end{matrix} \end{matrix}$$

Our problem now is really a problem in matrix multiplication. To see this, let us compute the profit in May, for example. In May there were 8 units of product I sold and the profit per unit was 2, so the total profit on sales in May of product I was $(8)(2) = 16$. (Actually, the profit was $(8000)(\$200) = \$1,600,000$, but—for simplicity—we will stick to the smaller numbers.) Similarly, the profit on product II in May was $(28)(3) = 54$. Thus we see that the total profit from sales of all four products in May is

$$(8)(2) + (28)(3) + (12)(5) + (40)(4) = 320.$$

But this is the dot product of the second (May) row of S and the first (profit) column of P. Analogously, we quickly find that the total tax paid in June is equal to the dot product of the third (June) row of S and the second (tax) column of P. Therefore we compute

$$SP = \begin{pmatrix} 6 & 19 & 14 & 46 \\ 8 & 28 & 12 & 40 \\ 4 & 16 & 17 & 55 \end{pmatrix} \begin{pmatrix} 2 & 1 \\ 3 & 2 \\ 5 & 3 \\ 4 & 2 \end{pmatrix} = \begin{matrix} \text{Profit} & \text{Taxes} & \\ \begin{pmatrix} 323 & 178 \\ 320 & 180 \\ 361 & 197 \end{pmatrix} & & \begin{matrix} \text{April} \\ \text{May} \\ \text{June} \end{matrix} \end{matrix}$$

The last matrix tells us at a glance that gross profits and taxes in May were $\$32,000,000$ and $\$18,000,000$ so that after-tax profit was $\$14,000,000$.

Example 6 **First- and Second-Order Contact to a Contagious Disease.** Suppose that three persons have contracted a contagious disease. A second group of six persons is questioned to determine who has been in contact with the three infected persons. A third group of seven persons is then questioned to determine contacts with any of the six persons in the second group. We define the 3×6 matrix $A = (a_{ij})$ by defining $a_{ij} = 1$ if the jth person in the second group has had contact with the ith person in the first group and $a_{ij} = 0$ otherwise. Similarly, we define the 6×7 matrix $B = (b_{ij})$ by defining $b_{ij} = 1$ if the jth person in the third group has had contact with the ith person in the second group and $b_{ij} = 0$ otherwise. These two matrices describe the *direct*, or *first-order*, *contacts* between the groups.

For example, we could have

$$A = \begin{pmatrix} 0 & 0 & 1 & 0 & 1 & 0 \\ 1 & 0 & 0 & 1 & 0 & 0 \\ 0 & 0 & 1 & 1 & 0 & 1 \end{pmatrix} \quad \text{and} \quad B = \begin{pmatrix} 0 & 0 & 1 & 0 & 0 & 1 & 0 \\ 0 & 0 & 1 & 1 & 0 & 0 & 0 \\ 1 & 0 & 0 & 0 & 0 & 1 & 1 \\ 0 & 0 & 1 & 1 & 0 & 0 & 0 \\ 0 & 1 & 0 & 1 & 0 & 0 & 0 \\ 1 & 0 & 0 & 0 & 0 & 1 & 0 \end{pmatrix}$$

In this case we have $a_{24} = 1$, which means that the fourth person in the second group has had contact with the second infected person. Analogously, $b_{33} = 0$, which means that the third person in the third group has not had contact with the third person in the second group.

We may be interested in studying the *indirect*, or *second-order, contacts* between the seven persons in the third group and the three infected people in the first group. The matrix product $C = AB$ describes these second-order contacts. The *ij*th component

$$c_{ij} = a_{i1}b_{1j} + a_{i2}b_{2j} + a_{i3}b_{3j} + a_{i4}b_{4j} + a_{i5}b_{5j} + a_{i6}b_{6j}$$

gives the number of second-order contacts between the *j*th person in the third group and the *i*th person in the infected group. With the given matrices A and B, we have

$$C = AB = \begin{pmatrix} 1 & 1 & 0 & 1 & 0 & 1 & 1 \\ 0 & 0 & 2 & 1 & 0 & 1 & 0 \\ 2 & 0 & 1 & 1 & 0 & 2 & 1 \end{pmatrix}$$

For example, the component $c_{23} = 2$ implies that there are two second-order contacts between the third person in the third group and the second contagious person. Note that the sixth person in the third group has had $1 + 1 + 2 = 4$ indirect contacts with the infected group. Only the fifth person has had no contacts.

The number 1 plays a special role in arithmetic. If x is any real number, then $1 \cdot x = x \cdot 1 = x$; that is, we can multiply any number on the right or the left by 1 and get the original number back. In matrix multiplication, a similar fact holds for square matrices.

Identity Matrix

The **$n \times n$ identity matrix** is the $n \times n$ matrix with 1s down the *main diagonal* and 0s everywhere else:

$$I_n = (b_{ij}) \quad \text{where} \quad b_{ij} = \begin{cases} 1 & \text{if } i = j \\ 0 & \text{if } i \neq j \end{cases} \qquad (5)$$

Remark. The **main diagonal** of a square matrix $A = (a_{ij})$ consists of the components a_{11}, a_{22}, a_{33}, and so on. The main diagonal in the 3×3 matrix A is circled:

$$A = \begin{pmatrix} 5 & -2 & 3 \\ 1 & 2 & 4 \\ 3 & 5 & -4 \end{pmatrix}.$$

Main diagonal

Example 7
$$I_3 = \begin{pmatrix} 1 & 0 & 0 \\ 0 & 1 & 0 \\ 0 & 0 & 1 \end{pmatrix} \quad \text{and} \quad I_5 = \begin{pmatrix} 1 & 0 & 0 & 0 & 0 \\ 0 & 1 & 0 & 0 & 0 \\ 0 & 0 & 1 & 0 & 0 \\ 0 & 0 & 0 & 1 & 0 \\ 0 & 0 & 0 & 0 & 1 \end{pmatrix}$$

It is not hard to show that I_n plays the role for $n \times n$ matrices that 1 plays for real numbers. Let A be an $n \times n$ matrix. Then the ijth component of $C = AI_n$ is, by (4),

$$c_{ij} = a_{i1}b_{1j} + a_{i2}b_{2j} + \cdots + a_{ij}b_{jj} + \cdots + a_{in}b_{nj} = a_{ij} \qquad (6)$$

In (6), the only nonzero term is $a_{ij}b_{jj} = a_{ij}$ (since $b_{jj} = 1$ and $b_{ij} = 0$ if $i \neq j$), so $c_{ij} = a_{ij}$. Thus $AI_n = A$. In a similar fashion we can show that $I_nA = A$. In summary, we have

$$AI_n = I_nA = A \qquad (7)$$

We will drop the subscript n since the size of the identity matrix we are using will always be obvious.

Example 8 We observe that

$$\begin{pmatrix} 2 & -3 & 1 \\ 4 & 0 & 6 \\ -2 & 4 & 5 \end{pmatrix}\begin{pmatrix} 1 & 0 & 0 \\ 0 & 1 & 0 \\ 0 & 0 & 1 \end{pmatrix} = \begin{pmatrix} 2 & -3 & 1 \\ 4 & 0 & 6 \\ -2 & 4 & 5 \end{pmatrix}$$

In some problems it is necessary to multiply three or more matrices together. This could pose a problem, however. Suppose we wanted to find the product ABC. Since we can multiply only two matrices at a time, we have two choices: We could multiply A by the product of B and C to obtain the matrix $A(BC)$, or we could multiply C by the product of A and B to obtain the matrix $(AB)C$ (remember, we cannot change the order). But if the matrices $A(BC)$ and $(AB)C$ were different, then the product ABC would not make sense. Fortunately, this cannot happen because of the following rule.

Associative Law for Matrix Multiplication   Let $A = (a_{ij})$ be an $n \times m$ matrix, $B = (b_{ij})$ an $m \times p$ matrix, and $C = (c_{ij})$ a $p \times q$ matrix. Then the **associative law**

$$\boxed{A(BC) = (AB)C} \tag{8}$$

holds and ABC, defined by either side of (8), is an $n \times q$ matrix. We will not prove this fact, but we can verify it in particular cases.

Example 9   Verify the associative law for $A = \begin{pmatrix} 1 & -3 \\ 0 & 2 \end{pmatrix}$, $B = \begin{pmatrix} 2 & -1 & 4 \\ 3 & 1 & 5 \end{pmatrix}$, and $C = \begin{pmatrix} 0 & -2 & 1 \\ 4 & 3 & 2 \\ -5 & 0 & 6 \end{pmatrix}$.

Solution   We first note that A is 2×2, B is 2×3, and C is 3×3. Hence all products used in the statement of the associative law are defined and the resulting product will be a 2×3 matrix. We then calculate

$$AB = \begin{pmatrix} 1 & -3 \\ 0 & 2 \end{pmatrix}\begin{pmatrix} 2 & -1 & 4 \\ 3 & 1 & 5 \end{pmatrix} = \begin{pmatrix} -7 & -4 & -11 \\ 6 & 2 & 10 \end{pmatrix}$$

$$(AB)C = \begin{pmatrix} -7 & -4 & -11 \\ 6 & 2 & 10 \end{pmatrix}\begin{pmatrix} 0 & -2 & 1 \\ 4 & 3 & 2 \\ -5 & 0 & 6 \end{pmatrix} = \begin{pmatrix} 39 & 2 & -81 \\ -42 & -6 & 70 \end{pmatrix}.$$

Similarly,

$$BC = \begin{pmatrix} 2 & -1 & 4 \\ 3 & 1 & 5 \end{pmatrix}\begin{pmatrix} 0 & -2 & 1 \\ 4 & 3 & 2 \\ -5 & 0 & 6 \end{pmatrix} = \begin{pmatrix} -24 & -7 & 24 \\ -21 & -3 & 35 \end{pmatrix}$$

$$A(BC) = \begin{pmatrix} 1 & -3 \\ 0 & 2 \end{pmatrix}\begin{pmatrix} -24 & -7 & 24 \\ -21 & -3 & 35 \end{pmatrix} = \begin{pmatrix} 39 & 2 & -81 \\ -42 & -6 & 70 \end{pmatrix}.$$

Thus $(AB)C = A(BC)$.

From now on we shall write the product of three matrices simply as ABC. We can do this because $(AB)C = A(BC)$; we get the same answer no matter how the multiplication is carried out (provided that we do not commute any of the matrices).

The associative law can be extended to longer products. For example, if AB, BC, and CD are defined, then

$$ABCD = (AB)(CD) = A(BC)D = [(AB)C]D = A[B(CD)]. \qquad (9)$$

PROBLEMS 3.2 In Problems 1–6, compute the dot product of the two vectors.

1. $(2 \quad 3), \begin{pmatrix} 4 \\ -2 \end{pmatrix}$

2. $(3 \quad -7), \begin{pmatrix} -4 \\ 0 \end{pmatrix}$

3. $(1 \quad 2 \quad 3), \begin{pmatrix} 4 \\ 5 \\ 6 \end{pmatrix}$

4. $(-3 \quad 1 \quad 7), \begin{pmatrix} -2 \\ 4 \\ 2 \end{pmatrix}$

5. $(2 \quad -3 \quad 1 \quad 4), \begin{pmatrix} 3 \\ 0 \\ 2 \\ 6 \end{pmatrix}$

6. $(-1 \quad 8 \quad 4 \quad 1), \begin{pmatrix} 5 \\ 0 \\ -2 \\ 4 \end{pmatrix}$

7. Let $\mathbf{a}$ be an n-vector. Show that $\mathbf{a} \cdot \mathbf{a} \geq 0$.

8. In Example 1, suppose that the demand vector is $\mathbf{d} = (25 \quad 45 \quad 20 \quad 30)$ and the price vector is $\mathbf{p} = \begin{pmatrix} \$12 \\ \$25 \\ \$16 \\ \$20 \end{pmatrix}$. If the manufacturer meets the demand, how much money will be received?

9. In Problem 8, let $\mathbf{c} = \begin{pmatrix} \$8 \\ \$15 \\ \$12 \\ \$14 \end{pmatrix}$ denote the costs of producing one unit of each of the four items. If the manufacturer meets the demand, what will be the cost of production?

10. In Examples 8 and 9, let $\mathbf{f}$ denote the profit vector—that is, $\mathbf{f}$ is a four-component column vector showing the profit earned by selling one unit of each of the four items.

  (a) Write $\mathbf{f}$ in terms of $\mathbf{p}$ and $\mathbf{c}$.

  (d) Write the total profit in terms of $\mathbf{f}$ and $\mathbf{d}$.

  (c) Write the total profit in terms of $\mathbf{p}$, $\mathbf{c}$, and $\mathbf{d}$.

  (d) Using the data of Problems 8 and 9, compute the total profit.

In Problems 11–25, perform the indicated computation.

11. $\begin{pmatrix} 2 & 3 \\ -1 & 2 \end{pmatrix} \begin{pmatrix} 4 & 1 \\ 0 & 6 \end{pmatrix}$

12. $\begin{pmatrix} 3 & -2 \\ 1 & 4 \end{pmatrix} \begin{pmatrix} -5 & 6 \\ 1 & 3 \end{pmatrix}$

13. $\begin{pmatrix} 1 & -1 \\ 1 & 1 \end{pmatrix} \begin{pmatrix} -1 & 0 \\ 2 & 3 \end{pmatrix}$

14. $\begin{pmatrix} -5 & 6 \\ 1 & 3 \end{pmatrix} \begin{pmatrix} 3 & -2 \\ 1 & 4 \end{pmatrix}$

15. $\begin{pmatrix} -4 & 5 & 1 \\ 0 & 4 & 2 \end{pmatrix} \begin{pmatrix} 3 & -1 & 1 \\ 5 & 6 & 4 \\ 0 & 1 & 2 \end{pmatrix}$

16. $\begin{pmatrix} 7 & 1 & 4 \\ 2 & -3 & 5 \end{pmatrix} \begin{pmatrix} 1 & 6 \\ 0 & 4 \\ -2 & 3 \end{pmatrix}$

17. $\begin{pmatrix} 1 & 6 \\ 0 & 4 \\ -2 & 3 \end{pmatrix} \begin{pmatrix} 7 & 1 & 4 \\ 2 & -3 & 5 \end{pmatrix}$

18. $\begin{pmatrix} 1 & 4 & -2 \\ 3 & 0 & 4 \end{pmatrix} \begin{pmatrix} 0 & 1 \\ 2 & 3 \end{pmatrix}$

19. $\begin{pmatrix} 1 & 4 & 6 \\ -2 & 3 & 5 \\ 1 & 0 & 4 \end{pmatrix} \begin{pmatrix} 2 & -3 & 5 \\ 1 & 0 & 6 \\ 2 & 3 & 1 \end{pmatrix}$

20. $\begin{pmatrix} 2 & -3 & 5 \\ 1 & 0 & 6 \\ 2 & 3 & 1 \end{pmatrix} \begin{pmatrix} 1 & 4 & 6 \\ -2 & 3 & 5 \\ 1 & 0 & 4 \end{pmatrix}$

21. $(1 \quad 4 \quad 0 \quad 2) \begin{pmatrix} 3 & -6 \\ 2 & 4 \\ 1 & 0 \\ -2 & 3 \end{pmatrix}$

22. $\begin{pmatrix} 3 & 2 & 1 & -2 \\ -6 & 4 & 0 & 3 \end{pmatrix} \begin{pmatrix} 1 \\ 4 \\ 0 \\ 2 \end{pmatrix}$

23. $\begin{pmatrix} 3 & -2 & 1 \\ 4 & 0 & 6 \\ 5 & 1 & 9 \end{pmatrix} \begin{pmatrix} 1 & 0 & 0 \\ 0 & 1 & 0 \\ 0 & 0 & 1 \end{pmatrix}$

24. $\begin{pmatrix} 1 & 0 & 0 \\ 0 & 1 & 0 \\ 0 & 0 & 1 \end{pmatrix} \begin{pmatrix} 3 & -2 & 1 \\ 4 & 0 & 6 \\ 5 & 1 & 9 \end{pmatrix}$

25. $\begin{pmatrix} a & b & c \\ d & e & f \\ g & h & j \end{pmatrix} \begin{pmatrix} 1 & 0 & 0 \\ 0 & 1 & 0 \\ 0 & 0 & 1 \end{pmatrix}$, where $a, b, c, d, e, f,$ g, h, j are real numbers.

26. Find a matrix $A = \begin{pmatrix} a & b \\ c & d \end{pmatrix}$ such that $A\begin{pmatrix} 2 & 3 \\ 1 & 2 \end{pmatrix} = \begin{pmatrix} 1 & 0 \\ 0 & 1 \end{pmatrix}$.

*27. Let $a_{11}, a_{12}, a_{21},$ and a_{22} be given real numbers such that $a_{11}a_{22} - a_{12}a_{21} \neq 0.$ Find numbers $b_{11}, b_{12}, b_{21},$ and b_{22} such that $\begin{pmatrix} a_{11} & a_{12} \\ a_{21} & a_{22} \end{pmatrix} \begin{pmatrix} b_{11} & b_{12} \\ b_{21} & b_{22} \end{pmatrix} = \begin{pmatrix} 1 & 0 \\ 0 & 1 \end{pmatrix}.$

28. Verify the associative law for multiplication for the matrices $A = \begin{pmatrix} 2 & -1 & 4 \\ 1 & 0 & 6 \end{pmatrix}$, $B = \begin{pmatrix} 1 & 0 & 1 \\ 2 & -1 & 2 \\ 3 & -2 & 0 \end{pmatrix}$, and $C = \begin{pmatrix} 1 & 6 \\ -2 & 4 \\ 0 & 5 \end{pmatrix}$.

29. As in Example 6, suppose that two persons have contracted a contagious disease. These persons have contacts with a second group who in turn have contacts with a third group. Let $A = \begin{pmatrix} 1 & 1 & 0 & 0 & 1 \\ 0 & 1 & 1 & 1 & 0 \end{pmatrix}$ represent the contacts between the contagious group and the members of group 2, and let

$$B = \begin{pmatrix} 1 & 1 & 0 & 1 \\ 0 & 1 & 0 & 1 \\ 0 & 0 & 1 & 0 \\ 0 & 0 & 1 & 1 \\ 0 & 1 & 1 & 0 \end{pmatrix}$$

represent the contacts between groups 2 and 3.

(a) How many people are in group 2?

(b) How many are in group 3?

(c) Find the matrix of second-order contacts between groups 1 and 3.

30. Answer the questions of Problem 29 for $A = \begin{pmatrix} 1 & 1 & 1 & 1 & 1 & 0 \\ 0 & 0 & 1 & 1 & 0 & 1 \end{pmatrix}$ and

$$B = \begin{pmatrix} 0 & 0 & 1 & 0 & 0 \\ 0 & 1 & 1 & 0 & 1 \\ 1 & 0 & 1 & 0 & 1 \\ 0 & 1 & 1 & 1 & 0 \\ 0 & 0 & 0 & 0 & 0 \\ 1 & 0 & 1 & 1 & 0 \end{pmatrix}.$$

31. An investor plans to buy 100 shares of telephone stock, 200 shares of oil stock, 400 shares of automobile stock, and 100 shares of airline stock. The telephone stock is selling for $46 a share, the oil stock for $34 a share, the automobile stock for $15 a share, and the airline stock for $10 a share.

(a) Express the numbers of shares as a row vector.

(b) Express the prices of the stocks as a column vector.

(c) Use matrix multiplication to compute the total cost of the investor's purchases.

32. A manufacturer of custom-designed jewelry has orders for two rings, three pairs of earrings, five pins, and one necklace. The manufacturer estimates that it takes 1 hour of labor to make a ring, $1\frac{1}{2}$ hours to make a pair of earrings, $\frac{1}{2}$ hour for each pin, and 2 hours to make a necklace.

(a) Express the manufacturer's orders as a row vector.

(b) Express the labor-hour requirements for the various types of jewelry as a column vector.

(c) Use matrix multiplication to calculate the total number of hours of labor it will require to complete all the orders.

33. A company pays its executives a salary and gives them shares of its stock as an annual bonus. Last year, the president of the company received $80,000 and 50 shares of stock, each of the three vice-presidents were paid $45,000 and 20 shares of stock, and the treasurer was paid $40,000 and 10 shares of stock.

(a) Express the payments to the executives in money and stock by means of a 2×3 matrix.

(b) Express the number of executives of each rank by means of a column vector.

(c) Use matrix multiplication to calculate the total amount of money and the total number of shares of stock the company paid these executives last year.

34. A tourist returns from a European trip with the following foreign currency: 1000 Austrian schillings. 20 British pounds, 100 French francs, 5000 Italian lire, and 50 German marks. In American money, a schilling was worth $0.055, the pound $1.80, the franc $0.20, the lira $0.001, and the mark $0.40.

(a) Express the quantity of each currency by means of a row vector.

(b) Express the value of each currency in American money by means of column vector.

(c) Use matrix multiplication to compute how much the tourist's foreign currency was worth in American money.

35. A family consists of two adults, one teenager, and three young children. Each adult consumes $\frac{1}{5}$ loaf of bread, no milk, $\frac{1}{10}$ pound of coffee, and $\frac{1}{8}$ pound of cheese in an average day. The teenager eats $\frac{2}{5}$ loaf of bread, drinks 1 quart of milk but no coffee, and eats $\frac{1}{8}$ pound of cheese. Each child eats $\frac{1}{5}$ loaf of bread, drinks $\frac{1}{2}$ quart of milk and no coffee, and eats $\frac{1}{16}$ pound of cheese.

(a)   Express the daily consumption of bread, milk, coffee, and cheese by the various types of family members using a matrix.

(b)   Express the number of family members of the various types by means of a column vector.

(c)   Use matrix multiplication to calculate the total amount of bread, milk, coffee, and cheese consumed by this family in an average day.

36.   Sales, unit gross profits, and unit taxes for sales of a large corporation are given in the table below.

	Product				Profit (in hundreds of dollars)	Taxes (in hundreds of dollars)
Month	I	II	III	Item		
January	4	2	20	I	3.5	1.5
February	6	1	9	II	2.75	2
March	5	3	12	III	1.5	0.6
April	8	2.5	20			

Find a matrix that shows total profits and taxes in each of the 4 months.

37.   Let A be a square matrix. Then A^2 is defined simply as AA, A^3 is defined as A^2A, and so on. Calculate $\begin{pmatrix} 2 & -1 \\ 4 & 6 \end{pmatrix}^2$.

38.   (a)   Calculate A^2, where $A = \begin{pmatrix} 1 & -2 & 4 \\ 2 & 0 & 3 \\ 1 & 1 & 5 \end{pmatrix}$.

(b)   Calculate A^3, where $A = \begin{pmatrix} -1 & 2 \\ 3 & 4 \end{pmatrix}$.

39.   Calculate A^2, A^3, A^4, and A^5, where

$$A = \begin{pmatrix} 0 & 1 & 0 & 0 \\ 0 & 0 & 1 & 0 \\ 0 & 0 & 0 & 1 \\ 0 & 0 & 0 & 0 \end{pmatrix}.$$

40.   Calculate A^2, A^3, A^4, and A^5, where

$$A = \begin{pmatrix} 0 & 1 & 0 & 0 & 0 \\ 0 & 0 & 1 & 0 & 0 \\ 0 & 0 & 0 & 1 & 0 \\ 0 & 0 & 0 & 0 & 1 \\ 0 & 0 & 0 & 0 & 0 \end{pmatrix}.$$

41. An $n \times n$ matrix A has the property that its matrix product with any $n \times n$ matrix is the zero matrix. Prove that A is the zero matrix.

42. A **probability matrix** is a square matrix with two properties: (1) every component is nonnegative (≥ 0) and (2) the sum of the elements in each row is 1. The following are probability matrices.

$$P = \begin{pmatrix} \frac{1}{4} & \frac{1}{4} & \frac{1}{2} \\ 0 & 1 & 0 \\ \frac{1}{3} & \frac{1}{3} & \frac{1}{3} \end{pmatrix} \quad \text{and} \quad Q = \begin{pmatrix} 1 & 0 & 0 \\ \frac{1}{4} & \frac{1}{3} & \frac{5}{12} \\ 0 & 0 & 1 \end{pmatrix}$$

Show that PQ is a probability matrix.

* **43.** Let P be a probability matrix. Show that P^2 is a probability matrix.

**   44.** Let P and Q be probability matrices of the same size. Prove that PQ is a probability matrix.

45. A round-robin tennis tournament can be organized in the following way. Each of the n players plays all the others, and the results are recorded in an $n \times n$ matrix R as follows:

$$R_{ij} = \begin{cases} 1 & \text{if the } i\text{th player beats the } j\text{th player,} \\ 0 & \text{if the } i\text{th player loses to the } j\text{th player,} \\ 0 & \text{if } i = j. \end{cases}$$

The ith player is then assigned the score $S_i = R_{i1} + R_{i2} + \cdots + R_{in} + \frac{1}{2}(R_{i1}^2 + R_{i2}^2 + \cdots + R_{in}^2)$ where R_{ij}^2 is the ijth component of R^2.

(a) In a tournament with four players,

$$R = \begin{pmatrix} 0 & 1 & 0 & 0 \\ 0 & 0 & 1 & 1 \\ 1 & 0 & 0 & 0 \\ 1 & 0 & 1 & 0 \end{pmatrix}.$$

Rank the players according to their scores.

(b) Interpret the meaning of the scores.

46. Let O be the $m \times n$ zero matrix and let A be an $n \times p$ matrix. Show that $OA = O_1$ where O_1 is the $m \times p$ zero matrix.

47. The **distributive law for matrix multiplication** states that if A is an $m \times n$ matrix and B and C are $n \times p$ matrices, then

$$A(B + C) = AB + AC.$$

Verify the distributive law for the matrices

$$A = \begin{pmatrix} 1 & 2 & 4 \\ 3 & -1 & 0 \end{pmatrix}, \quad B = \begin{pmatrix} 2 & 7 \\ -1 & 4 \\ 6 & 0 \end{pmatrix}, \quad C = \begin{pmatrix} -1 & 2 \\ 3 & 7 \\ 4 & 1 \end{pmatrix}.$$

48. Show that the $n \times n$ identity matrix is unique. [*Hint:* Suppose that for every $n \times n$ matrix A, $IA = AI = A$ and $JA = AJ = A$. Show that $I = J$.]

3.3 Matrices and Systems of Linear Equations

In Section 2.5 on page 71, we discussed the following systems of m equations in n unknowns:

$$
\begin{aligned}
a_{11}x_1 + a_{12}x_2 + \cdots + a_{1n}x_n &= b_1 \\
a_{21}x_1 + a_{22}x_2 + \cdots + a_{2n}x_n &= b_2 \\
\vdots \qquad \vdots \qquad\qquad \vdots \qquad \vdots \\
a_{m1}x_1 + a_{m2}x_2 + \cdots + a_{mn}x_n &= b_m.
\end{aligned}
\tag{1}
$$

We define the matrix

$$
A = \begin{pmatrix}
a_{11} & a_{12} & \cdots & a_{1n} \\
a_{21} & a_{22} & \cdots & a_{2n} \\
\vdots & \vdots & & \vdots \\
a_{m1} & a_{m2} & \cdots & a_{mn}
\end{pmatrix},
$$

the vector $\mathbf{x} = \begin{pmatrix} x_1 \\ x_2 \\ \vdots \\ x_n \end{pmatrix}$, and the vector $\mathbf{b} = \begin{pmatrix} b_1 \\ b_2 \\ \vdots \\ b_m \end{pmatrix}$. Since A is an $m \times n$ matrix and $\mathbf{x}$ is an $n \times 1$ matrix, the matrix product $A\mathbf{x}$ is an $m \times 1$ matrix. It is not difficult to see that system (1) can be written as

$$
\boxed{A\mathbf{x} = \mathbf{b}}
\tag{2}
$$

Example 1   Consider the system

$$
\begin{aligned}
2x + 8y + 6z &= 20 \\
4x + 2y - 2z &= -2 \\
3x - \;\; y + \;\; z &= 11
\end{aligned}
\tag{3}
$$

(see Example 2.5.1 on page 64). This can be written in the form $A\mathbf{x} = \mathbf{b}$, with

$$
A = \begin{pmatrix} 2 & 8 & 6 \\ 4 & 2 & -2 \\ 3 & -1 & 1 \end{pmatrix}, \qquad
\mathbf{x} = \begin{pmatrix} x \\ y \\ z \end{pmatrix}, \qquad
\mathbf{b} = \begin{pmatrix} 20 \\ -2 \\ 11 \end{pmatrix}.
$$

It is obviously easier to write out system (1) in the form $A\mathbf{x} = \mathbf{b}$. There are many other advantages, too. In Section 3.4 we will see how a square system can be solved almost at once if we know a matrix called the *inverse* of A. Even without that, as we saw in Chapter 2, computations are much easier to make by using an augmented matrix. Let us repeat the computations of Example 2.5.1 starting with the augmented matrix:

$$\begin{pmatrix} 2 & 8 & 6 & | & 20 \\ 4 & 2 & -2 & | & -2 \\ 3 & -1 & 1 & | & 11 \end{pmatrix} \xrightarrow{M_1(\frac{1}{2})} \begin{pmatrix} 1 & 4 & 3 & | & 10 \\ 4 & 2 & -2 & | & -2 \\ 3 & -1 & 1 & | & 11 \end{pmatrix}$$

Recall that $M_1(\frac{1}{2})$ means multiply the first row by $\frac{1}{2}$ and $A_{1,2}(-4)$ means that the first row is multiplied by -4 and added to the second row.

$$\xrightarrow{A_{1,2}(-4)} \begin{pmatrix} 1 & 4 & 3 & | & 10 \\ 0 & -14 & -14 & | & -42 \\ 3 & -1 & 1 & | & 11 \end{pmatrix}$$

$$\xrightarrow{A_{1,3}(-3)} \begin{pmatrix} 1 & 4 & 3 & | & 10 \\ 0 & -14 & -14 & | & -42 \\ 0 & -13 & -8 & | & -19 \end{pmatrix}$$

$$\xrightarrow{M_2(-\frac{1}{14})} \begin{pmatrix} 1 & 4 & 3 & | & 10 \\ 0 & 1 & 1 & | & 3 \\ 0 & -13 & -8 & | & -19 \end{pmatrix}$$

$$\xrightarrow[A_{2,3}(13)]{A_{2,1}(-4)} \begin{pmatrix} 1 & 0 & -1 & | & -2 \\ 0 & 1 & 1 & | & 3 \\ 0 & 0 & 5 & | & 20 \end{pmatrix}$$

$$\xrightarrow{M_3(\frac{1}{5})} \begin{pmatrix} 1 & 0 & -1 & | & -2 \\ 0 & 1 & 1 & | & 3 \\ 0 & 0 & 1 & | & 4 \end{pmatrix}$$

$$\xrightarrow[A_{3,2}(-1)]{A_{3,1}(1)} \begin{pmatrix} 1 & 0 & 0 & | & 2 \\ 0 & 1 & 0 & | & -1 \\ 0 & 0 & 1 & | & 5 \end{pmatrix}$$

The last augmented matrix tells us that $x = 2$, $y = -1$, and $z = 5$, as we already knew.

In this last example it is important to note that the last system of equations can be written as

$$I\mathbf{x} = \mathbf{s} \tag{4}$$

where $I = \begin{pmatrix} 1 & 0 & 0 \\ 0 & 1 & 0 \\ 0 & 0 & 1 \end{pmatrix}$ and $\mathbf{s}$ is the solution vector $\begin{pmatrix} 2 \\ -1 \\ 5 \end{pmatrix}$. We make use of this fact in Section 3.4.

Example 2 Write the following system in the form $A\mathbf{x} = \mathbf{b}$.

$$3x + 2y - 6z = 4$$
$$x + y \qquad = 7$$

Solution   The coefficient matrix of this system is given by

$$A = \begin{pmatrix} 3 & 2 & -6 \\ 1 & 1 & 0 \end{pmatrix}.$$

Let

$$\mathbf{x} = \begin{pmatrix} x \\ y \\ z \end{pmatrix} \quad \text{and} \quad \mathbf{b} = \begin{pmatrix} 4 \\ 7 \end{pmatrix}.$$

Then

$$A\mathbf{x} = \begin{pmatrix} 3 & 2 & -6 \\ 1 & 1 & 0 \end{pmatrix} \begin{pmatrix} x \\ y \\ z \end{pmatrix} = \begin{pmatrix} 3x + 2y - 6z \\ x + y \end{pmatrix} = \begin{pmatrix} 4 \\ 7 \end{pmatrix}.$$

The system can be written $A\mathbf{x} = \mathbf{b}$ with A, $\mathbf{x}$, and $\mathbf{b}$ as given above.

Example 3   Write the system represented by the augmented matrix

$$\begin{pmatrix} 1 & 4 & -2 & | & 3 \\ 4 & 6 & 1 & | & 7 \\ 2 & 0 & 3 & | & -2 \end{pmatrix}.$$

Solution   As we have seen, the coefficient matrix of the system is the matrix to the left of the vertical bar; that is,

$$A = \begin{pmatrix} 1 & 4 & -2 \\ 4 & 6 & 1 \\ 2 & 0 & 3 \end{pmatrix}.$$

Then, with $\mathbf{x} = \begin{pmatrix} x \\ y \\ z \end{pmatrix}$ and $\mathbf{b} = \begin{pmatrix} 3 \\ 7 \\ -2 \end{pmatrix}$, we have

$$A\mathbf{x} = \mathbf{b}$$

or

$$\begin{pmatrix} 1 & 4 & -2 \\ 4 & 6 & 1 \\ 2 & 0 & 3 \end{pmatrix} \begin{pmatrix} x \\ y \\ z \end{pmatrix} = \begin{pmatrix} 3 \\ 7 \\ -2 \end{pmatrix}.$$

But

$$A\mathbf{x} = \begin{pmatrix} 1 & 4 & 2 \\ 4 & 6 & 1 \\ 2 & 0 & 3 \end{pmatrix} \begin{pmatrix} x \\ y \\ z \end{pmatrix} = \begin{pmatrix} x + 4y - 2z \\ 4x + 6y + z \\ 2x + 3z \end{pmatrix}$$

Thus the system is

$$\begin{aligned} x + 4y - 2z &= 3 \\ 4x + 6y + z &= 7 \\ 2x + 3z &= -2 \end{aligned}$$

PROBLEMS 3.3   In Problems 1–6, write the given system in the form $A\mathbf{x} = \mathbf{b}$.

1. $2x - y = 3$
   $4x + 5y = 7$

2. $x - y + 3z = 11$
   $4x + y - z = -4$
   $2x - y + 3z = 10$

3. $3x + 6y - 7z = 0$
   $2x - y + 3z = 1$

4. $4x - y + z - w = -7$
   $3x + y - 5z + 6w = 8$
   $2x - y + z = 9$

5. $y - z = 7$
   $x + z = 2$
   $3x + 2y = -5$

6. $2x + 3y - z = 0$
   $-4x + 2y + z = 0$
   $7x + 3y - 9z = 0$

In Problems 7–15, write the system of equations represented by the given augmented matrix.

7. $\begin{pmatrix} 1 & 1 & -1 & | & 7 \\ 4 & -1 & 5 & | & 4 \\ 6 & 1 & 3 & | & 20 \end{pmatrix}$

8. $\begin{pmatrix} 0 & 1 & | & 2 \\ 1 & 0 & | & 3 \end{pmatrix}$

9. $\begin{pmatrix} 2 & 0 & 1 & | & 2 \\ -3 & 4 & 0 & | & 3 \\ 0 & 5 & 6 & | & 5 \end{pmatrix}$

10. $\begin{pmatrix} 2 & 3 & 1 & | & 2 \\ 0 & 4 & 1 & | & 3 \\ 0 & 0 & 0 & | & 0 \end{pmatrix}$

11. $\begin{pmatrix} 1 & 0 & 0 & 0 & | & 2 \\ 0 & 1 & 0 & 0 & | & 3 \\ 0 & 0 & 1 & 0 & | & -5 \\ 0 & 0 & 0 & 1 & | & 6 \end{pmatrix}$

12. $\begin{pmatrix} 2 & 3 & 1 & | & 0 \\ 4 & -1 & 5 & | & 0 \\ 3 & 6 & -7 & | & 0 \end{pmatrix}$

13. $\begin{pmatrix} 6 & 2 & 1 & | & 2 \\ -2 & 3 & 1 & | & 4 \\ 0 & 0 & 0 & | & 2 \end{pmatrix}$

14. $\begin{pmatrix} 3 & 1 & 5 & | & 6 \\ 2 & 3 & 2 & | & 4 \end{pmatrix}$

15. $\begin{pmatrix} 7 & 2 & | & 1 \\ 3 & 1 & | & 2 \\ 6 & 9 & | & 3 \end{pmatrix}$

16. Solve the system represented by the augmented matrix of Problem 9.

17. Solve the system represented by $\begin{pmatrix} 1 & 2 & -4 & | & 4 \\ -2 & -4 & 8 & | & -8 \end{pmatrix}$.

18. Solve the system represented by $\begin{pmatrix} 1 & 2 & -4 & | & 4 \\ -2 & -4 & 8 & | & -9 \end{pmatrix}$.

19. Solve the homogeneous system represented by $\begin{pmatrix} 1 & -2 & 3 & | & 0 \\ 4 & 1 & -1 & | & 0 \\ 2 & -1 & 3 & | & 0 \end{pmatrix}$.

20. Solve the homogeneous[†] system represented by $\begin{pmatrix} 1 & 1 & -1 & | & 0 \\ 4 & -1 & 5 & | & 0 \\ 6 & 1 & 3 & | & 0 \end{pmatrix}$.

[†] See Problem 2.5.50 on page 79.

21. Solve the system represented by the augmented matrix

$$\left(\begin{array}{cccc|c} 1 & 3 & -2 & 1 & 3 \\ 2 & -6 & 4 & -1 & 2 \\ 4 & 12 & -8 & 2 & 4 \\ -3 & 0 & 6 & -2 & -8 \end{array}\right).$$

22. Solve the homogeneous system represented by the augmented matrix

$$\left(\begin{array}{ccccc|c} 1 & 2 & -3 & 5 & 4 & 0 \\ -2 & 4 & 7 & -3 & 5 & 0 \\ -4 & 0 & 13 & -13 & -3 & 0 \end{array}\right).$$

23. Three chemicals are combined to form three grades of fertilizer. A unit of grade I fertilizer requires 10 kg of chemical A, 30 of B, and 60 of C. A unit of grade II requires 20 kg of A, 30 of B, and 50 of C. A unit of grade III requires 50 kg of A and 50 of C. If 1600 kg of A, 1200 of B, and 3200 of C are available, how many units of the three grades should be produced to use all available supplies? [*Hint:* To solve this problem, first write the resulting system in the form $A\mathbf{x} = \mathbf{b}$.]

3.4 The Inverse of a Square Matrix

In this section we define a kind of matrix central to matrix theory. We begin with a simple example. Let $A = \begin{pmatrix} 2 & 5 \\ 1 & 3 \end{pmatrix}$ and $B = \begin{pmatrix} 3 & -5 \\ -1 & 2 \end{pmatrix}$. Then an easy computation shows that $AB = BA = I$, where $I = \begin{pmatrix} 1 & 0 \\ 0 & 1 \end{pmatrix}$ is the 2×2 identity matrix. The matrix B is called the *inverse* of A and is written A^{-1}. In general, we have the following.

The Inverse of a Matrix   Let A and B be square, $n \times n$ matrices. Suppose that

$$AB = BA = I \tag{1}$$

Then B is called the **inverse** of A and is written as A^{-1}. We have

$$\boxed{AA^{-1} = A^{-1}A = I} \tag{2}$$

If A has an inverse, then A is said to be **invertible**.

Remark 1. From this definition it immediately follows that $(A^{-1})^{-1} = A$ if A is invertible.

Remark 2. This definition does *not* state that every square matrix has an inverse. In fact there are many square matrices that have no inverse. (See, for instance, Example 2.)

In the computation done above, we see that

$$\begin{pmatrix} 2 & 5 \\ 1 & 3 \end{pmatrix}^{-1} = \begin{pmatrix} 3 & -5 \\ -1 & 2 \end{pmatrix} \quad \text{and} \quad \begin{pmatrix} 3 & -5 \\ -1 & 2 \end{pmatrix}^{-1} = \begin{pmatrix} 2 & 5 \\ 1 & 3 \end{pmatrix}.$$

Consider the system

$$A\mathbf{x} = \mathbf{b}$$

and suppose that A is invertible. Then

$$A^{-1}A\mathbf{x} = A^{-1}\mathbf{b} \qquad \text{We multiplied on the left by } A^{-1}.$$

$$I\mathbf{x} = A^{-1}\mathbf{b} \qquad A^{-1}A = I$$

$$\mathbf{x} = A^{-1}\mathbf{b} \qquad I\mathbf{x} = \mathbf{x}$$

That is,

> If A is invertible, the system $A\mathbf{x} = \mathbf{b}$ has the unique solution $\mathbf{x} = A^{-1}\mathbf{b}$.

This is one of the reasons we study matrix inverses.

There are three basic questions that come to mind once we have defined the inverse of a matrix.

Question 1.   Can a matrix have more than one inverse?

Question 2.   What matrices do have inverses?

Question 3.   If a matrix has an inverse, how can we compute it?

We answer all three questions in this section. The first one is the easiest. Suppose that B and C are two inverses for A. We can show that $B = C$. By equation (1) we have $AB = BA = I$ and $AC = CA = I$. Then $B(AC) = BI = B$ and $(BA)C = IC = C$. But $B(AC) = (BA)C$ by the associative law of matrix multiplication. Hence $B = C$, and this means that A can have, at most, one inverse.

The other two questions are more difficult to answer. Rather than starting by giving you what seem to be a set of arbitrary rules, we first look at what happens in the 2×2 case.

Example 1   Let $A = \begin{pmatrix} 2 & -3 \\ -4 & 5 \end{pmatrix}$. Compute A^{-1} if it exists.

Solution   Suppose that A^{-1} exists. We write $A^{-1} = \begin{pmatrix} x & y \\ z & w \end{pmatrix}$ and use the fact that $AA^{-1} = I$. Then

$$AA^{-1} = \begin{pmatrix} 2 & -3 \\ -4 & 5 \end{pmatrix}\begin{pmatrix} x & y \\ z & w \end{pmatrix} = \begin{pmatrix} 2x - 3z & 2y - 3w \\ -4x + 5z & -4y + 5w \end{pmatrix} = \begin{pmatrix} 1 & 0 \\ 0 & 1 \end{pmatrix}$$

The last two matrices can be equal only if each of their corresponding components are equal. This means that

$$2x \quad\quad - 3z \quad\quad = 1 \tag{3}$$

$$2y \quad\quad - 3w = 0 \tag{4}$$

$$-4x \quad\quad + 5z \quad\quad = 0 \tag{5}$$

$$- 4y \quad\quad + 5w = 1 \tag{6}$$

This is a system of four equations in four unknowns. Note that there are two equations involving x and z only (equations (3) and (5)) and two equations involving y and w only (equations (4) and (6)). We write these two systems in augmented matrix form.

$$\begin{pmatrix} 2 & -3 & | & 1 \\ -4 & 5 & | & 0 \end{pmatrix} \tag{7}$$

$$\begin{pmatrix} 2 & -3 & | & 0 \\ -4 & 5 & | & 1 \end{pmatrix}. \tag{8}$$

Now, we know from Section 2.5 that if system (7) (in the variables x and z) has a unique solution, then Gauss-Jordan elimination of (7) will result in

$$\begin{pmatrix} 1 & 0 & | & x \\ 0 & 1 & | & z \end{pmatrix}$$

where (x, z) is the unique pair of numbers that satisfies $2x - 3y = 1$ and $-4x + 5z = 0$. Similarly, row reduction of (8) will result in

$$\begin{pmatrix} 1 & 0 & | & y \\ 0 & 1 & | & w \end{pmatrix}$$

where (y, w) is the unique pair of numbers that satisfies $2y - 3w = 0$ and $-4y + 5w = 1$.

Since the coefficient matrices in (7) and (8) are the same, we can perform the row reductions on the two augmented matrices simultaneously, by considering the new augmented matrix

$$\begin{pmatrix} 2 & -3 & | & 1 & 0 \\ -4 & 5 & | & 0 & 1 \end{pmatrix}. \tag{9}$$

If A^{-1} is invertible, then the system defined by (3), (4), (5), and (6) has a unique solution and, by what we said above, Gauss-Jordan elimination will result in

$$\begin{pmatrix} 1 & 0 & | & x & y \\ 0 & 1 & | & z & w \end{pmatrix}.$$

We now carry out the computation, noting that the matrix on the left in (9) is A and the matrix on the right in (9) is I:

$$\begin{pmatrix} 2 & -3 & | & 1 & 0 \\ -4 & 5 & | & 0 & 1 \end{pmatrix} \xrightarrow{M_1(\frac{1}{2})} \begin{pmatrix} 1 & -\frac{3}{2} & | & \frac{1}{2} & 0 \\ -4 & 5 & | & 0 & 1 \end{pmatrix}$$

$$\xrightarrow{A_{1,2}(4)} \begin{pmatrix} 1 & -\frac{3}{2} & | & \frac{1}{2} & 0 \\ 0 & -1 & | & 2 & 1 \end{pmatrix}$$

$$\xrightarrow{M_2(-1)} \begin{pmatrix} 1 & -\frac{3}{2} & | & \frac{1}{2} & 0 \\ 0 & 1 & | & -2 & -1 \end{pmatrix}$$

$$\xrightarrow{A_{2,1}(\frac{3}{2})} \begin{pmatrix} 1 & 0 & | & -\frac{5}{2} & -\frac{3}{2} \\ 0 & 1 & | & -2 & -1 \end{pmatrix}.$$

Thus $x = -\frac{5}{2}$, $y = -\frac{3}{2}$, $z = -2$, $w = -1$, and $A^{-1} = \begin{pmatrix} -\frac{5}{2} & -\frac{3}{2} \\ -2 & -1 \end{pmatrix}$. We still must check our answer. We have

$$AA^{-1} = \begin{pmatrix} 2 & -3 \\ -4 & 5 \end{pmatrix}\begin{pmatrix} -\frac{5}{2} & -\frac{3}{2} \\ -2 & -1 \end{pmatrix} = \begin{pmatrix} 1 & 0 \\ 0 & 1 \end{pmatrix}$$

and

$$A^{-1}A = \begin{pmatrix} -\frac{5}{2} & -\frac{3}{2} \\ -2 & -1 \end{pmatrix}\begin{pmatrix} 2 & -3 \\ -4 & 5 \end{pmatrix} = \begin{pmatrix} 1 & 0 \\ 0 & 1 \end{pmatrix}.$$

Thus A is invertible and $A^{-1} = \begin{pmatrix} -\frac{5}{2} & -\frac{3}{2} \\ -2 & -1 \end{pmatrix}.$

Example 2 Let $A = \begin{pmatrix} 1 & 2 \\ -2 & -4 \end{pmatrix}$. Calculate A^{-1} if it exists.

Solution If $A^{-1} = \begin{pmatrix} x & y \\ z & w \end{pmatrix}$ exists, then

$$AA^{-1} = \begin{pmatrix} 1 & 2 \\ -2 & -4 \end{pmatrix}\begin{pmatrix} x & y \\ z & w \end{pmatrix} = \begin{pmatrix} x + 2z & y + 2w \\ -2x - 4z & -2y - 4w \end{pmatrix} = \begin{pmatrix} 1 & 0 \\ 0 & 1 \end{pmatrix}.$$

This leads to the system

$$\begin{aligned} x \qquad + 2z \qquad &= 1 \\ y \qquad + 2w &= 0 \\ -2x \qquad - 4z \qquad &= 0 \\ -2y \qquad - 4w &= 1. \end{aligned} \tag{10}$$

Using the same reasoning as in Example 1, we can write this system in the augmented matrix form $(A|I)$ and row-reduce.

$$\begin{pmatrix} 1 & 2 & | & 1 & 0 \\ -2 & -4 & | & 0 & 1 \end{pmatrix} \xrightarrow{A_{1,2}(2)} \begin{pmatrix} 1 & 2 & | & 1 & 0 \\ 0 & 0 & | & 2 & 1 \end{pmatrix}$$

This is as far as we can go. The last line reads $0 = 2$ or $0 = 1$, depending on which of the two systems of equations (in x and z or in y and w) is being solved. Thus system (10) is inconsistent and A is not invertible.

The last two examples illustrate a procedure that always works when you are trying to find the inverse of a matrix.

PROCEDURE FOR COMPUTING THE INVERSE OF A SQUARE MATRIX *A*

Step 1.   Write the augmented matrix $(A|I)$.

Step 2.   Use row reduction to reduce the matrix A to its reduced row echelon form.

Step 3.   Decide if A is invertible.

(a)   If A can be reduced to the identity matrix I, then A^{-1} will be the matrix to the right of the vertical bar.

(b)   If the row reduction of A leads to a row of zeros to the left of the vertical bar, then A is not invertible.

Remark.   We can rephrase (a) and (b) as follows.

A square matrix A is invertible if and only if its reduced row echelon form is the identity matrix.

Example 3   Let $A = \begin{pmatrix} a & b \\ c & d \end{pmatrix}$. Compute A^{-1} if it exists.

Solution We assume that $a \neq 0$ (if $a = 0$, see Problem 26). Then, using the procedure outlined above, we have

$$\begin{pmatrix} a & b & | & 1 & 0 \\ c & d & | & 0 & 1 \end{pmatrix} \xrightarrow{M_1(1/a)} \begin{pmatrix} 1 & \dfrac{b}{a} & | & \dfrac{1}{a} & 0 \\ c & d & | & 0 & 1 \end{pmatrix}$$

$$\xrightarrow{A_{1,2}(-c)} \begin{pmatrix} 1 & \dfrac{b}{a} & | & \dfrac{1}{a} & 0 \\ 0 & d - \dfrac{bc}{a} & | & -\dfrac{c}{a} & 1 \end{pmatrix}$$

$$= \begin{pmatrix} 1 & \dfrac{b}{a} & | & \dfrac{1}{a} & 0 \\ 0 & \dfrac{ad - bc}{a} & | & -\dfrac{c}{a} & 1 \end{pmatrix} \qquad d - \dfrac{bc}{a} = \dfrac{ad}{a} - \dfrac{bc}{a} = \dfrac{ad - bc}{a}$$

Now, before we go any further, we must consider two cases.

Case I. $ad - bc = 0$.

The second row of A has been reduced to a row of zeros and A is not invertible.

Case II. $ad = bc \neq 0$.

We can continue.

$$\xrightarrow{M_2(a/(ad-bc))} \begin{pmatrix} 1 & \dfrac{b}{a} & | & \dfrac{1}{a} & 0 \\ 0 & 1 & | & \dfrac{-c}{ad-bc} & \dfrac{a}{ad-bc} \end{pmatrix}$$

$$\xrightarrow{A_{2,1}(-b/a)} \begin{pmatrix} 1 & 0 & | & \dfrac{d}{ad-bc} & \dfrac{-b}{ad-bc} \\ 0 & 1 & | & \dfrac{-c}{ad-bc} & \dfrac{a}{ad-bc} \end{pmatrix}$$

In the last step we computed

$$\dfrac{-c}{ad-bc}\left(\dfrac{-b}{a}\right) + \dfrac{1}{a} = \dfrac{bc}{a(ad-bc)} + \dfrac{1}{a} = \dfrac{bc}{a(ad-bc)} + \dfrac{ad-bc}{a(ad-bc)}$$

$$= \dfrac{bc + (ad-bc)}{a(ad-bc)} = \dfrac{ad}{a(ad-bc)} = \dfrac{d}{ad-bc}.$$

Thus (in the case $ad - bc \neq 0$), we have found that

$$A^{-1} = \begin{pmatrix} \dfrac{d}{ad-bc} & \dfrac{-b}{ad-bc} \\[2mm] \dfrac{-c}{ad-bc} & \dfrac{a}{ad-bc} \end{pmatrix} = \frac{1}{ad-bc}\begin{pmatrix} d & -b \\ -c & a \end{pmatrix}.$$

Check.

$$A^{-1}A = \frac{1}{ad-bc}\begin{pmatrix} d & -b \\ -c & a \end{pmatrix}\begin{pmatrix} a & b \\ c & d \end{pmatrix}$$

$$= \frac{1}{ad-bc}\begin{pmatrix} ad-bc & 0 \\ 0 & ad-bc \end{pmatrix} = \begin{pmatrix} 1 & 0 \\ 0 & 1 \end{pmatrix}.$$

You should also verify that $AA^{-1} = I$.

We summarize the result of the last example.

Let $A = \begin{pmatrix} a & b \\ c & d \end{pmatrix}$.

1. A is invertible if and only if $ad - bc \neq 0$.

2. If $ad - bc \neq 0$, then

$$A^{-1} = \frac{1}{ad-bc}\begin{pmatrix} d & -b \\ -c & a \end{pmatrix}.$$

(11)

Remark.   The quantity $ad - bc$ is called the **determinant** of A and is abbreviated det A. We will not discuss determinants any further in this text.

Example 4   Let $A = \begin{pmatrix} 6 & -7 \\ 2 & 1 \end{pmatrix}$. Compute A^{-1} if it exists.

Solution $ad - bc = (6)(1) - (-7)(2) = 6 + 14 = 20 \neq 0$. Thus A^{-1} exists and

$$A^{-1} = \frac{1}{20}\begin{pmatrix} 1 & 7 \\ -2 & 6 \end{pmatrix}.$$

Check.

$$\frac{1}{20}\begin{pmatrix} 1 & 7 \\ -2 & 6 \end{pmatrix}\begin{pmatrix} 6 & -7 \\ 2 & 1 \end{pmatrix} = \frac{1}{20}\begin{pmatrix} 20 & 0 \\ 0 & 20 \end{pmatrix} = \begin{pmatrix} 1 & 0 \\ 0 & 1 \end{pmatrix}.$$

Similarly,

$$\begin{pmatrix} 6 & -7 \\ 2 & 1 \end{pmatrix}\begin{pmatrix} \frac{1}{20} & \frac{7}{20} \\ -\frac{2}{20} & \frac{6}{20} \end{pmatrix} = \begin{pmatrix} 1 & 0 \\ 0 & 1 \end{pmatrix}.$$

Example 5 Let $A = \begin{pmatrix} 2 & 8 & 6 \\ 4 & 2 & -2 \\ 3 & -1 & 1 \end{pmatrix}$ (see Example 3.3.1 on page 112). Calculate A^{-1} if it exists.

Solution We first put I next to A in an augmented matrix form,

$$\begin{pmatrix} 2 & 8 & 6 & | & 1 & 0 & 0 \\ 4 & 2 & -2 & | & 0 & 1 & 0 \\ 3 & -1 & 1 & | & 0 & 0 & 1 \end{pmatrix},$$

and then carry out the row reduction.

$$\xrightarrow{M_1(\frac{1}{2})} \begin{pmatrix} 1 & 4 & 3 & | & \frac{1}{2} & 0 & 0 \\ 4 & 2 & -2 & | & 0 & 1 & 0 \\ 3 & -1 & 1 & | & 0 & 0 & 1 \end{pmatrix}$$

$$\xrightarrow[A_{1,3}(-3)]{A_{1,2}(-4)} \begin{pmatrix} 1 & 4 & 3 & | & \frac{1}{2} & 0 & 0 \\ 0 & -14 & -14 & | & -2 & 1 & 0 \\ 0 & -13 & -8 & | & -\frac{3}{2} & 0 & 1 \end{pmatrix}$$

$$\xrightarrow{M_2(-\frac{1}{14})} \begin{pmatrix} 1 & 4 & 3 & | & \frac{1}{2} & 0 & 0 \\ 0 & 1 & 1 & | & \frac{2}{14} & -\frac{1}{14} & 0 \\ 0 & -13 & -8 & | & -\frac{3}{2} & 0 & 1 \end{pmatrix}$$

$$\xrightarrow[A_{2,3}(13)]{A_{2,1}(-4)} \begin{pmatrix} 1 & 0 & -1 & | & -\frac{1}{14} & \frac{4}{14} & 0 \\ 0 & 1 & 1 & | & \frac{2}{14} & -\frac{1}{14} & 0 \\ 0 & 0 & 5 & | & \frac{5}{14} & -\frac{13}{14} & 1 \end{pmatrix}$$

$$\xrightarrow{M_3(\frac{1}{5})} \begin{pmatrix} 1 & 0 & -1 & | & -\frac{1}{14} & \frac{4}{14} & 0 \\ 0 & 1 & 1 & | & \frac{2}{14} & -\frac{1}{14} & 0 \\ 0 & 0 & 1 & | & \frac{1}{14} & -\frac{13}{70} & \frac{1}{5} \end{pmatrix}$$

$$\xrightarrow[A_{3,2}(-1)]{A_{3,1}(1)} \begin{pmatrix} 1 & 0 & 0 & | & 0 & \frac{1}{10} & \frac{1}{5} \\ 0 & 1 & 0 & | & \frac{1}{14} & \frac{8}{70} & -\frac{1}{5} \\ 0 & 0 & 1 & | & \frac{1}{14} & -\frac{13}{70} & \frac{1}{5} \end{pmatrix}$$

Since A has now been reduced to I, we have

$$A^{-1} = \begin{pmatrix} 0 & \frac{1}{10} & \frac{1}{5} \\ \frac{1}{14} & \frac{8}{70} & -\frac{1}{5} \\ \frac{1}{14} & -\frac{13}{70} & \frac{1}{5} \end{pmatrix} = \frac{1}{70}\begin{pmatrix} 0 & 7 & 14 \\ 5 & 8 & -14 \\ 5 & -13 & 14 \end{pmatrix}$$

Check.

$$A^{-1}A = \frac{1}{70}\begin{pmatrix} 0 & 7 & 14 \\ 5 & 8 & -14 \\ 5 & -13 & 14 \end{pmatrix}\begin{pmatrix} 2 & 8 & 6 \\ 4 & 2 & -2 \\ 3 & -1 & 1 \end{pmatrix} = \frac{1}{70}\begin{pmatrix} 70 & 0 & 0 \\ 0 & 70 & 0 \\ 0 & 0 & 70 \end{pmatrix} = I.$$

We can also verify that $AA^{-1} = I$.

Warning. It is easy to make numerical errors in computing A^{-1}. Therefore it is essential to check the computations by verifying that $A^{-1}A = I$.

Example 6 Let $A = \begin{pmatrix} 2 & 4 & 3 \\ 0 & 1 & -1 \\ 3 & 5 & 7 \end{pmatrix}$. Calculate A^{-1} if it exists.

Solution We proceed as in Example 5 to obtain, successively, the following augmented matrices.

$$\begin{pmatrix} 2 & 4 & 3 & | & 1 & 0 & 0 \\ 0 & 1 & -1 & | & 0 & 1 & 0 \\ 3 & 5 & 7 & | & 0 & 0 & 1 \end{pmatrix} \xrightarrow{M_1(\frac{1}{2})} \begin{pmatrix} 1 & 2 & \frac{3}{2} & | & \frac{1}{2} & 0 & 0 \\ 0 & 1 & -1 & | & 0 & 1 & 0 \\ 3 & 5 & 7 & | & 0 & 0 & 1 \end{pmatrix}$$

$$\xrightarrow{A_{1,3}(-3)} \begin{pmatrix} 1 & 2 & \frac{3}{2} & | & \frac{1}{2} & 0 & 0 \\ 0 & 1 & -1 & | & 0 & 1 & 0 \\ 0 & -1 & \frac{5}{2} & | & -\frac{3}{2} & 0 & 1 \end{pmatrix}$$

$$\xrightarrow[A_{2,3}(1)]{A_{2,1}(-2)} \begin{pmatrix} 1 & 0 & \frac{7}{2} & | & \frac{1}{2} & -2 & 0 \\ 0 & 1 & -1 & | & 0 & 1 & 0 \\ 0 & 0 & \frac{3}{2} & | & -\frac{3}{2} & 1 & 1 \end{pmatrix}$$

$$\xrightarrow{M_3(\frac{2}{3})} \begin{pmatrix} 1 & 0 & \frac{7}{2} & | & \frac{1}{2} & -2 & 0 \\ 0 & 1 & -1 & | & 0 & 1 & 0 \\ 0 & 0 & 1 & | & -1 & \frac{2}{3} & \frac{2}{3} \end{pmatrix}$$

$$\xrightarrow[A_{3,2}(1)]{A_{3,1}(-\frac{7}{2})} \begin{pmatrix} 1 & 0 & 0 & | & 4 & -\frac{13}{3} & -\frac{7}{3} \\ 0 & 1 & 0 & | & -1 & \frac{5}{3} & \frac{2}{3} \\ 0 & 0 & 1 & | & -1 & \frac{2}{3} & \frac{2}{3} \end{pmatrix}$$

Thus

$$A^{-1} = \begin{pmatrix} 4 & -\frac{13}{3} & -\frac{7}{3} \\ -1 & \frac{5}{3} & \frac{2}{3} \\ -1 & \frac{2}{3} & \frac{2}{3} \end{pmatrix}$$

Check.

$$A^{-1}A = \begin{pmatrix} 4 & -\frac{13}{3} & -\frac{7}{3} \\ -1 & \frac{5}{3} & \frac{2}{3} \\ -1 & \frac{2}{3} & \frac{2}{3} \end{pmatrix} \begin{pmatrix} 2 & 4 & 3 \\ 0 & 1 & -1 \\ 3 & 5 & 7 \end{pmatrix} = \begin{pmatrix} 1 & 0 & 0 \\ 0 & 1 & 0 \\ 0 & 0 & 1 \end{pmatrix}$$

Example 7   Let $A = \begin{pmatrix} 1 & -3 & 4 \\ 2 & -5 & 7 \\ 0 & -1 & 1 \end{pmatrix}$. Calculate A^{-1} if it exists.

Solution   Proceeding as before we obtain, successively,

$$\begin{pmatrix} 1 & -3 & 4 & | & 1 & 0 & 0 \\ 2 & -5 & 7 & | & 0 & 1 & 0 \\ 0 & -1 & 1 & | & 0 & 0 & 1 \end{pmatrix} \xrightarrow{A_{1,2}(-2)} \begin{pmatrix} 1 & -3 & 4 & | & 1 & 0 & 0 \\ 0 & 1 & -1 & | & -2 & 1 & 0 \\ 0 & -1 & 1 & | & 0 & 0 & 1 \end{pmatrix}$$

$$\xrightarrow[A_{2,3}(1)]{A_{2,1}(3)} \begin{pmatrix} 1 & 0 & 1 & | & -5 & 3 & 0 \\ 0 & 1 & -1 & | & -2 & 1 & 0 \\ 0 & 0 & 0 & | & -2 & 1 & 1 \end{pmatrix}.$$

This is as far as we can go. The matrix A *cannot* be reduced to the identity matrix and we can conclude that A is *not* invertible.

There is another way to see that the matrix of Example 7 is not invertible. Let **b** be any 3-vector and consider the system $A\mathbf{x} = \mathbf{b}$. If we tried to solve this by Gauss-Jordan elimination, we would end up with an equation that reads either $0 = c \neq 0$ or $0 = c$—that is, the system either has no solution or it has an infinite number of solutions. The one possibility ruled out is the case in which the system has a unique solution. But if A^{-1} existed, then there would be a unique solution given by $\mathbf{x} = A^{-1}\mathbf{b}$. We are left to conclude again that

> if in the row reduction of A we end up with a row of zeros, then A is *not* invertible.   (12)

We have seen that if A^{-1} exists, then the unique solution to the system $A\mathbf{x} = \mathbf{b}$ is given by $\mathbf{x} = A^{-1}\mathbf{b}$. We now exploit this fact.

Example 8   Solve the system

$$2x + 4y + 3z = 6$$
$$y - z = -4$$
$$3x + 5y + 7z = 7.$$

Solution   This system can be written as $A\mathbf{x} = \mathbf{b}$, where $A = \begin{pmatrix} 2 & 4 & 3 \\ 0 & 1 & -1 \\ 3 & 5 & 7 \end{pmatrix}$ and

$\mathbf{b} = \begin{pmatrix} 6 \\ -4 \\ 7 \end{pmatrix}$. In Example 6 we found that A^{-1} exists and

$$A^{-1} = \begin{pmatrix} 4 & -\frac{13}{3} & -\frac{7}{3} \\ -1 & \frac{5}{3} & \frac{2}{3} \\ -1 & \frac{2}{3} & \frac{2}{3} \end{pmatrix}$$

Thus the unique solution is given by

$$\mathbf{x} = \begin{pmatrix} x \\ y \\ z \end{pmatrix} = A^{-1}\mathbf{b} = \begin{pmatrix} 4 & -\frac{13}{3} & -\frac{7}{3} \\ -1 & \frac{5}{3} & \frac{2}{3} \\ -1 & \frac{2}{3} & \frac{2}{3} \end{pmatrix}\begin{pmatrix} 6 \\ -4 \\ 7 \end{pmatrix} = \begin{pmatrix} 25 \\ -8 \\ -4 \end{pmatrix}$$

Example 9   A farmer feeds cattle on a mixture of three standard feeds, which we will call type A, type B, and type C. Suppose that a standard unit of type A feed supplies a steer with 10% of the calories, 10% of the protein, and 5% of the carbohydrates it needs each day. Similarly, type B supplies 10% of the calories and 5% of the protein but no carbohydrates, and type C has 5% of the calories, 5% of the protein, and 10% of the carbohydrates. How many units of each type of feed should the farmer give a steer each day so that it gets 100% of the amount of calories, protein, and carbohydrates it requires?

Solution   Let x, y, and z be the number of units of foods A, B, and C, respectively, fed to a steer each day. Since each unit of food A supplies 10% of the calories required, $10x$ represents the percentage of the daily calorie requirement supplied by x units of food A. Analogously, $10y$ and $5z$ represent the percentages of the daily calorie requirement supplied by foods B and C, respectively. Since we require 100% of the daily calorie requirement, we obtain

$$10x + 10y + 5z = 100 \qquad \text{Calorie equation}$$

Similarly, considering the daily requirements of proteins and carbohydrates, we obtain

$$10x + 5y + 5z = 100 \qquad \text{Protein equation}$$

and

$$5x + 10z = 100 \qquad \text{Carbohydrate equation}$$

This system can be written in the form $A\mathbf{x} = \mathbf{b}$, where

$$A = \begin{pmatrix} 10 & 10 & 5 \\ 10 & 5 & 5 \\ 5 & 0 & 10 \end{pmatrix}, \qquad \mathbf{x} = \begin{pmatrix} x \\ y \\ z \end{pmatrix}, \quad \text{and} \quad \mathbf{b} = \begin{pmatrix} 100 \\ 100 \\ 100 \end{pmatrix}.$$

We solve the system by computing A^{-1}. We write

$$\left(\begin{array}{ccc|ccc} 10 & 10 & 5 & 1 & 0 & 0 \\ 10 & 5 & 5 & 0 & 1 & 0 \\ 5 & 0 & 10 & 0 & 0 & 1 \end{array}\right) \xrightarrow{M_1(\frac{1}{10})} \left(\begin{array}{ccc|ccc} 1 & 1 & \frac{1}{2} & \frac{1}{10} & 0 & 0 \\ 10 & 5 & 5 & 0 & 1 & 0 \\ 5 & 0 & 10 & 0 & 0 & 1 \end{array}\right)$$

$$\xrightarrow[A_{1,3}(-5)]{A_{1,2}(-10)} \left(\begin{array}{ccc|ccc} 1 & 1 & \frac{1}{2} & \frac{1}{10} & 0 & 0 \\ 0 & -5 & 0 & -1 & 1 & 0 \\ 0 & -5 & \frac{15}{2} & -\frac{1}{2} & 0 & 1 \end{array}\right)$$

$$\xrightarrow{M_2(-\frac{1}{5})} \left(\begin{array}{ccc|ccc} 1 & 1 & \frac{1}{2} & \frac{1}{10} & 0 & 0 \\ 0 & 1 & 0 & \frac{1}{5} & -\frac{1}{5} & 0 \\ 0 & -5 & \frac{15}{2} & -\frac{1}{2} & 0 & 1 \end{array}\right)$$

$$\xrightarrow[A_{2,3}(5)]{A_{2,1}(-1)} \left(\begin{array}{ccc|ccc} 1 & 0 & \frac{1}{2} & -\frac{1}{10} & \frac{1}{5} & 0 \\ 0 & 1 & 0 & \frac{1}{5} & -\frac{1}{5} & 0 \\ 0 & 0 & \frac{15}{2} & \frac{1}{2} & -1 & 1 \end{array}\right)$$

$$\xrightarrow{M_3(\frac{2}{15})} \left(\begin{array}{ccc|ccc} 1 & 0 & \frac{1}{2} & -\frac{1}{10} & \frac{1}{5} & 0 \\ 0 & 1 & 0 & \frac{1}{5} & -\frac{1}{5} & 0 \\ 0 & 0 & 1 & \frac{1}{15} & -\frac{2}{15} & \frac{2}{15} \end{array}\right)$$

$$\xrightarrow{A_{3,1}(-\frac{1}{2})} \left(\begin{array}{ccc|ccc} 1 & 0 & 0 & -\frac{2}{15} & \frac{4}{15} & -\frac{1}{15} \\ 0 & 1 & 0 & \frac{1}{5} & -\frac{1}{5} & 0 \\ 0 & 0 & 1 & \frac{1}{15} & -\frac{2}{15} & \frac{2}{15} \end{array}\right).$$

Thus

$$A^{-1} = \left(\begin{array}{ccc} -\frac{2}{15} & \frac{4}{15} & -\frac{1}{15} \\ \frac{1}{5} & -\frac{1}{5} & 0 \\ \frac{1}{15} & -\frac{2}{15} & \frac{2}{15} \end{array}\right) = \frac{1}{15}\left(\begin{array}{ccc} -2 & 4 & -1 \\ 3 & -3 & 0 \\ 1 & -2 & 2 \end{array}\right)$$

Check.

$$A^{-1}A = \frac{1}{15}\left(\begin{array}{ccc} -2 & 4 & -1 \\ 3 & -3 & 0 \\ 1 & -2 & 2 \end{array}\right)\left(\begin{array}{ccc} 10 & 10 & 5 \\ 10 & 5 & 5 \\ 5 & 0 & 10 \end{array}\right)$$

$$= \frac{1}{15}\left(\begin{array}{ccc} 15 & 0 & 0 \\ 0 & 15 & 0 \\ 0 & 0 & 15 \end{array}\right) = \left(\begin{array}{ccc} 1 & 0 & 0 \\ 0 & 1 & 0 \\ 0 & 0 & 1 \end{array}\right).$$

To complete the problem, we compute

$$\mathbf{x} = \left(\begin{array}{c} x \\ y \\ z \end{array}\right) = A^{-1}\mathbf{b} = \frac{1}{15}\left(\begin{array}{ccc} -2 & 4 & -1 \\ 3 & -3 & 0 \\ 1 & -2 & 2 \end{array}\right)\left(\begin{array}{c} 100 \\ 100 \\ 100 \end{array}\right) = \frac{1}{15}\left(\begin{array}{c} 100 \\ 0 \\ 100 \end{array}\right) = \left(\begin{array}{c} \frac{100}{15} \\ 0 \\ \frac{100}{15} \end{array}\right).$$

That is, a mixture of $\frac{100}{15} = 6\frac{2}{3}$ units of food A, no food B and $6\frac{2}{3}$ units of food C will provide exactly 100% of the daily requirements of calories, proteins, and carbohydrates. Also, since A^{-1} exists, the solution is unique. This means that *no* other combinations of the three foods will provide exactly 100% of the three daily requirements.

Example 10   In Example 9, how many units of each of the three foods should be supplied to a steer each day to meet the following requirements: 90% of the calories, 70% of the proteins, and 80% of the carbohydrates?

Solution   Now we must solve $A\mathbf{x} = \mathbf{b}$, where A and $\mathbf{x}$ are as before and $\mathbf{b} = \begin{pmatrix} 90 \\ 70 \\ 80 \end{pmatrix}$.

$$\mathbf{x} = A^{-1}\mathbf{b} = \frac{1}{15}\begin{pmatrix} -2 & 4 & -1 \\ 3 & -3 & 0 \\ 1 & -2 & 2 \end{pmatrix}\begin{pmatrix} 90 \\ 70 \\ 80 \end{pmatrix} = \frac{1}{15}\begin{pmatrix} 20 \\ 60 \\ 110 \end{pmatrix} = \begin{pmatrix} \frac{20}{15} \\ 4 \\ \frac{110}{15} \end{pmatrix}$$

Thus the required diet is $\frac{20}{15} = \frac{4}{3}$ units of food A, 4 units of food B, and $\frac{110}{15} = \frac{22}{3} = 7\frac{1}{3}$ units of food C. Note that this answer was obtained with relatively little work once A^{-1} was known.

PROBLEMS 3.4   In Problems 1–15, determine whether the given matrix is invertible. If it is, calculate the inverse.

1. $\begin{pmatrix} 2 & 1 \\ 3 & 2 \end{pmatrix}$

2. $\begin{pmatrix} -1 & 6 \\ 2 & -12 \end{pmatrix}$

3. $\begin{pmatrix} 0 & 1 \\ 1 & 0 \end{pmatrix}$

4. $\begin{pmatrix} 1 & 1 \\ 3 & 3 \end{pmatrix}$

5. $\begin{pmatrix} a & a \\ b & b \end{pmatrix}$

6. $\begin{pmatrix} 1 & 1 & 1 \\ 0 & 2 & 3 \\ 5 & 5 & 1 \end{pmatrix}$

7. $\begin{pmatrix} 3 & 2 & 1 \\ 0 & 2 & 2 \\ 0 & 0 & -1 \end{pmatrix}$

8. $\begin{pmatrix} 1 & 1 & 1 \\ 0 & 1 & 1 \\ 0 & 0 & 1 \end{pmatrix}$

9. $\begin{pmatrix} 1 & 6 & 2 \\ -2 & 3 & 5 \\ 7 & 12 & -4 \end{pmatrix}$

10. $\begin{pmatrix} 3 & 1 & 0 \\ 1 & -1 & 2 \\ 1 & 1 & 1 \end{pmatrix}$

11. $\begin{pmatrix} 2 & -1 & 4 \\ -1 & 0 & 5 \\ 19 & -7 & 3 \end{pmatrix}$

12. $\begin{pmatrix} 1 & 2 & 3 \\ 1 & 1 & 2 \\ 0 & 1 & 2 \end{pmatrix}$

13. $\begin{pmatrix} 1 & 1 & 1 & 1 \\ 1 & 2 & -1 & 2 \\ 1 & -1 & 2 & 1 \\ 1 & 3 & 3 & 2 \end{pmatrix}$

14. $\begin{pmatrix} 1 & 0 & 2 & 3 \\ -1 & 1 & 0 & 4 \\ 2 & 1 & -1 & 3 \\ -1 & 0 & 5 & 7 \end{pmatrix}$

15. $$\begin{pmatrix} 1 & -3 & 0 & -2 \\ 3 & -12 & -2 & -6 \\ -2 & 10 & 2 & 5 \\ -1 & 6 & 1 & 3 \end{pmatrix}$$

16. Show that if A and B are invertible matrices, then AB is invertible and $(AB)^{-1} = B^{-1}A^{-1}$.

17. Show that the matrix $\begin{pmatrix} 3 & 4 \\ -2 & -3 \end{pmatrix}$ is equal to its own inverse.

18. Let $A = \begin{pmatrix} 0 & b \\ c & d \end{pmatrix}$.

    (a) Show that A^{-1} exists if and only if $bc \neq 0$.

    (b) If $bc \neq 0$, show that $A^{-1} = -\dfrac{1}{bc} \begin{pmatrix} d & -b \\ -c & 0 \end{pmatrix}$.

In Problems 19–28, solve by computing the inverse of an appropriate matrix.

19. Three chemicals are combined to form three grades of fertilizer. A unit of grade I fertilizer requires 10 kg of chemical A, 30 of B, and 60 of C. A unit of grade II requires 20 kg of A, 30 of B, and 50 of C. A unit of grade III requires 50 kg of A and 50 of C. If 1600 kg of A, 1200 of B, and 3200 of C are available, how many units of the three grades should be produced to use all available supplies?

20. Three species of squirrels have been introduced to an island, with a total initial population of 2000. After 10 years, species I has doubled its population and species II has increased by 50%. Species III becomes extinct. If the population increase in species I equals the increase in species II and if the total population has increased by 500, determine the initial populations of the three species. [*Hint:* Using all the information in the problem, write as a system $A\mathbf{x} = \mathbf{b}$, where $\mathbf{x}$ is the vector of initial populations.]

21. The Robinson Farm in Illinois has 1000 acres to be planted with soybeans, corn, and wheat. During the planting season, there are 3700 labor-hours of planting time available to Mrs. Robinson. Each acre of soybeans requires 2 labor-hours, each acre of corn requires 6 labor-hours and each acre of wheat requires 6 labor-hours. Seed to plant an acre of soybeans costs $12, seed for an acre of corn costs $20, and seed for an acre of wheat costs $8. Mrs. Robinson has $12,600 on hand to pay for seed. How should the 1000 acres be planted in order to use all the available land, labor, and seed money?

22. A witch's magic cupboard contains 10 oz of ground four-leaf clovers and 14 oz of powdered mandrake root. The cupboard will replenish itself automatically provided she uses up exactly all her supplies. A batch of love potion requires $3\frac{1}{13}$ oz of ground four-leaf clovers and $2\frac{2}{13}$ oz of powdered mandrake root. One recipe of a well-known (to witches) cure for the common cold requires $5\frac{5}{13}$ oz of four-leaf clovers and $10\frac{10}{13}$ oz of mandrake root. How much of the love potion and the cold remedy should the witch make in order to use up the supply in the cupboard exactly?

23. A factory for the construction of quality furniture has two divisions: a machine shop where the parts of the furniture are fabricated, and an assembly and finishing division where the parts are put together into the finished product. Suppose there are 12 employees in the machine shop and 20 in the assembly and finishing division and that each employee works an 8-hour day. Suppose further that the factory produces only two products: chairs and tables. A chair requires $\frac{384}{17}$ hours of machine shop time and $\frac{480}{17}$ hours of assembly and finishing time. A table requires $\frac{240}{17}$ hours of machine shop time and $\frac{640}{17}$ hours of assembly and finishing time. Assuming that there is an unlimited demand for these products and that the manufacturer wishes to keep all employees busy, how many chairs and how many tables can this factory produce each day?

24. An ice cream shop sells only ice cream sodas and milk shakes. It puts 1 oz of syrup and 4 oz of ice cream in an ice cream soda and 1 oz of syrup and 3 oz of ice cream in a milk shake. If the store used 4 gal of ice cream and 5 qt of syrup in one given day, how many ice cream sodas and how many milk shakes did it sell on that day? [*Hint:* 1 qt = 32 oz; 1 gal = 128 oz.]

25. A farmer feeds his cattle a mixture of two types of feed. One standard unit of type A feed supplies a steer with 10% of its minimum daily requirement of protein and 15% of its requirement of carbohydrates. Type B feed contains 12% of the requirement of protein and 8% of the requirement of carbohydrates in a standard unit. If the farmer wishes to feed his cattle exactly 100% of their minimum daily requirement of protein and carbohydrates, how many units of each type of feed should he give a steer each day?

26. Answer the question of Problem 25 if the farmer wishes to satisfy 90% of the daily requirement of protein and 110% of the daily requirement of carbohydrates.

27. A large corporation pays its vice-presidents a salary of $100,000 a year, 100 shares of stock, and an entertainment allowance of $20,000. A division manager receives $70,000 in salary, 50 shares of stock, and $5000 for official entertainment. The assistant manager of a division receives $40,000 in salary, but neither stock nor entertainment allowance. If the corporation pays out $1,600,000 in salaries, 1000 shares of stock, and $150,000 in expense allowances to its vice-presidents, division managers, and assistant division managers in a year, how many vice-presidents, division managers, and assistant division managers does the company have?

28. An automobile service station employs mechanics and station attendants. Each works 8 hours a day. An attendant pumps gas, while mechanics are expected to spend $\frac{3}{4}$ of their time repairing automobiles and $\frac{1}{4}$ of their time pumping gas. Suppose it takes $\frac{1}{10}$ hour to service an automobile that comes in for gas. If the service station owner wants to be able to sell gas to 320 cars a day and have 24 hours of mechanics' time available for repair work, how many attendants and how many mechanics should be hired?

In Problems 29–35, compute the reduced row echelon form of the given matrix and use it to determine directly whether the given matrix is invertible.

29. The matrix of Problem 1.   **30.** The matrix of Problem 4.

31. The matrix of Problem 7.   **32.** The matrix of Problem 9.

33. The matrix of Problem 11.   **34.** The matrix of Problem 13.

35. The matrix of Problem 14.

36. Calculate the inverse of $A = \begin{pmatrix} 2 & 0 & 0 \\ 0 & 3 & 0 \\ 0 & 0 & 4 \end{pmatrix}$.

37. A square matrix $A = (a_{ij})$ is called **diagonal** if all its elements off the main diagonal are zero. That is, $a_{ij} = 0$ if $i \neq j$. (The matrix of Problem 36 is diagonal.) Show that a diagonal matrix is invertible if and only if each of its diagonal components is nonzero.

38. Let

$$A = \begin{pmatrix} a_{11} & 0 & \cdots & 0 \\ 0 & a_{22} & \cdots & 0 \\ \vdots & \vdots & & \vdots \\ 0 & 0 & \cdots & a_{nn} \end{pmatrix}$$

be a diagonal matrix such that each of its diagonal components is nonzero. Calculate A^{-1}.

39. Calculate the inverse of $A = \begin{pmatrix} 2 & 1 & -1 \\ 0 & 3 & 4 \\ 0 & 0 & 5 \end{pmatrix}$.

40. Show that the matrix $A = \begin{pmatrix} 1 & 0 & 0 \\ -2 & 0 & 0 \\ 4 & 6 & 1 \end{pmatrix}$ is not invertible.

*41. A square matrix is called **upper (lower) triangular** if all its elements below (above) the main diagonal are zero. (The matrix of Problem 39 is upper triangular and the matrix of Problem 40 is lower triangular.) Show that an upper or lower triangular matrix is invertible if and only if each of its diagonal elements is nonzero.

**42. Show that the inverse of an invertible upper triangular matrix is upper triangular. [*Hint:* First prove the result for a 3×3 matrix.]

3.5 Input-Output Analysis

Macroeconomics is the branch of economics that deals with the broad and general aspects of an economy as in, for example, the relationships among the income, investments, and expenditures of a country as a whole. Many tools have been developed for dealing with problems in macroeconomics. We shall discuss one of the most important of these tools in this section.

To introduce our model we suppose that the United States Congress voted a large decrease in expenditures for highway construction. If there were no increase in other funding, then we would expect a reduction in income and employment. On the other hand, suppose that the government increased its military spending by an amount equal to the decrease in highway construction spending. What would be the change, if any, in income and employment?

The answer is complicated by the fact that highway construction projects and the military use money in different ways. So, while there might be an increase in income and employment among workers in industries such as aircraft and ship building, these might not offset the losses and unemployment in the construction industry (at least in the short run). The problem is that in the U.S. economy, there are many goods being produced and services being performed that are highly interrelated. The effects of increases or cutbacks in one industry are often felt in many other industries as well.

A model for analyzing these effects was developed by the American economist Wassily W. Leontief in 1936.[†] This model (or procedure) is called **input-output analysis**. Before describing this model in detail, we give a simple example.

Example 1 Consider a very simplified model of an economy in which two items are produced: automobiles (including trucks) and steel. Each year there is an **external demand** of 360,000 tons of steel and 110,000 automobiles. The word

---

[†] This model was used in Leontief's pioneering paper "Qualitative Input and Output Relations in the Economic System of the United States," *Review of Economic Statistics* 18(1936):105-125. An updated version of this model appears in Leontief's book *Input-Output Analysis* (New York: Oxford University Press, 1966). Leontief won the Nobel prize in economics in 1973 for his development of input-output analysis.

external here means that the demand comes from outside the ceonomy. For example, if this were a model of a part of the United States' economy, then the external demand could come from other countries (so that the steel and automobiles would be exported), from other industries in the United States, and from private individuals.

However, the external demand is not the only demand on the two industries. It takes steel to make cars. It also takes cars to make cars, since automobile manufacturing plants require cars and trucks to transport component materials and employees. Similarly, the steel industry requires steel (for appropriate machinery) and automobiles (for product and worker transport) in its operations. Thus each of the two industries in the system places demands on itself and the other industry. These demands are called **internal demands**.

In our simplified model we assume that the steel industry requires $\frac{1}{4}$ ton of steel and $\frac{1}{12}$ of an automobile (or truck) to produce 1 ton of steel (that is, one car or truck is used in the production of 12 tons of steel). Also, the automobile industry requires $\frac{1}{2}$ ton of steel and $\frac{1}{9}$ of an automobile to produce one car. The question posed by Leontief's input-output model is then: How many tons of steel and how many automobiles must be produced each year so that the supply of each is equal to the total demand?

Solution   We let x and y denote the number of tons of steel and the number of automobiles, respectively, in a given year. This is the supply. If, for example, it takes $\frac{1}{4}$ ton of steel to produce one ton of steel, then it takes $\frac{1}{4}x$ tons of steel to produce x tons of steel. Similarly, it takes $\frac{1}{2}y$ tons of steel to produce y automobiles. Thus the total internal demand on the steel industry is $\frac{1}{4}x + \frac{1}{2}y$, and the total demand (adding in the external demand) is $\frac{1}{4}x + \frac{1}{2}y + 360{,}000$. Similarly, the total demand on the automobile industry is $\frac{1}{12}x + \frac{1}{9}y + 110{,}000$. Setting supply equal to demand, we obtain the system

$$x = \tfrac{1}{4}x + \tfrac{1}{2}y + 360{,}000$$
$$y = \tfrac{1}{12}x + \tfrac{1}{9}y + 110{,}000. \tag{1}$$

Since $x - \frac{1}{4}x = \frac{3}{4}x$ and $y - \frac{1}{9}y = \frac{8}{9}y$, we can rewrite system (1) as

$$\tfrac{3}{4}x - \tfrac{1}{2}y = 360{,}000$$
$$-\tfrac{1}{12}x + \tfrac{8}{9}y = 110{,}000. \tag{2}$$

We solve system (2) by row reduction.

$$\begin{pmatrix} \frac{3}{4} & -\frac{1}{2} & 360{,}000 \\ -\frac{1}{12} & \frac{8}{9} & 110{,}000 \end{pmatrix} \xrightarrow{M_1(\frac{4}{3})} \begin{pmatrix} 1 & -\frac{2}{3} & 480{,}000 \\ -\frac{1}{12} & \frac{8}{9} & 110{,}000 \end{pmatrix}$$

$$\xrightarrow{A_{1,2}(\frac{1}{12})} \begin{pmatrix} 1 & -\frac{2}{3} & 480{,}000 \\ 0 & \frac{5}{6} & 150{,}000 \end{pmatrix}$$

$$\xrightarrow{M_2(\frac{6}{5})} \begin{pmatrix} 1 & -\frac{2}{3} & 480{,}000 \\ 0 & 1 & 180{,}000 \end{pmatrix}$$

$$\xrightarrow{A_{2,1}(\frac{2}{3})} \begin{pmatrix} 1 & 0 & 600{,}000 \\ 0 & 1 & 180{,}000 \end{pmatrix}.$$

Thus, in order that supply exactly equal demand, 600,000 tons of steel and 180,000 automobiles (or trucks) must be produced.

We now describe the general Leontief input-output model. Suppose an economic system has n industries. Again, there are two kinds of demands on each industry. First, there is the *external* demand from outside the system. If the system is a country, for example, then the external demand could be from another country. Second, there is the demand placed on one industry by another industry in the same system. As we have discussed, in the United States there is a demand on the output of the steel industry by the automobile industry, for example.

Let e_i represent the external demand placed on the ith industry. Let a_{ij} represent the internal demand placed on the ith industry by the jth industry. More precisely, a_{ij} represents the number of units of the output of industry i needed to produce 1 unit of the output of industry j. Let x_i represent the output of industry i. Now we assume that the output of each industry is equal to its demand (that is, there is no overproduction). The total demand is equal to the sum of the internal and external demands. To calculate the internal demand on industry 2, for example, we note that $a_{21}x_1$ is the demand on industry 2 made by industry 1. Thus the total internal demand on industry 2 is $a_{21}x_1 + a_{22}x_2 + \cdots + a_{2n}x_n$.

We are led to the following system of equations obtained by equating the total demand with the output of each industry.

$$
\begin{aligned}
a_{11}x_1 + a_{12}x_2 + \cdots + a_{1n}x_n + e_1 &= x_1 \\
a_{21}x_1 + a_{22}x_2 + \cdots + a_{2n}x_n + e_2 &= x_2 \\
\vdots \qquad \vdots \qquad\qquad \vdots \qquad \vdots \qquad \vdots & \\
a_{n1}x_1 + a_{n2}x_2 + \cdots + a_{nn}x_n + e_n &= x_n
\end{aligned}
\tag{3}
$$

Or, rewriting (3),

$$
\begin{aligned}
(1 - a_{11})x_1 - \quad a_{12}x_2 - \cdots - \quad a_{1n}x_n &= e_1 \\
- a_{21}x_1 + (1 - a_{22})x_2 - \cdots - \quad a_{2n}x_n &= e_2 \\
\vdots \qquad\qquad \vdots \qquad\qquad \vdots \qquad \vdots & \\
- a_{n1}x_1 - \quad a_{n2}x_2 - \cdots + (1 - a_{nn})x_n &= e_n.
\end{aligned}
\tag{4}
$$

System (4) of n equations in n unknowns is very important in economic analysis.

It is often convenient to write the numbers a_{ij} in a matrix A, called the **technology matrix**. We have

$$
A = \begin{pmatrix}
a_{11} & a_{12} & \cdots & a_{1n} \\
a_{21} & a_{22} & \cdots & a_{2n} \\
\vdots & \vdots & & \vdots \\
a_{n1} & a_{n2} & \cdots & a_{nn}
\end{pmatrix}.
\tag{5}
$$

Note that the technology matrix is a square matrix.

Example 2   In an economic system with three industries, suppose that the external demands are, respectively, 10, 25, and 20. Suppose that $a_{11} = 0.2$, $a_{12} = 0.5$, $a_{13} = 0.15$, $a_{21} = 0.4$, $a_{22} = 0.1$, $a_{23} = 0.3$, $a_{31} = 0.25$, $a_{32} = 0.5$, and $a_{33} = 0.15$. Find the output in each industry such that supply exactly equals demand.

Solution   Here $n = 3$, $1 - a_{11} = 0.8$, $1 - a_{22} = 0.9$, and $1 - a_{33} = 0.85$. Then system (4) is

$$0.8x_1 - 0.5x_2 - 0.15x_3 = 10$$
$$-0.4x_1 + 0.9x_2 - 0.3x_3 = 25$$
$$-0.25x_1 - 0.5x_2 + 0.85x_3 = 20$$

Solving this system by using a calculator, we obtain successively (using five-decimal-place accuracy and Gauss-Jordan elimination)

$$\begin{pmatrix} 0.8 & -0.5 & -0.15 & 10 \\ -0.4 & 0.9 & -0.3 & 25 \\ -0.25 & -0.5 & 0.85 & 20 \end{pmatrix}$$

This is the technology matrix

$$\xrightarrow{M_1(\frac{1}{0.8})} \begin{pmatrix} 1 & -0.625 & -0.1875 & 12.5 \\ -0.4 & 0.9 & -0.3 & 25 \\ -0.25 & -0.5 & 0.85 & 20 \end{pmatrix}$$

$$\xrightarrow[A_{1,3}(0.25)]{A_{1,2}(0.4)} \begin{pmatrix} 1 & -0.625 & -0.1875 & 12.5 \\ 0 & 0.65 & -0.375 & 30 \\ 0 & -0.65625 & 0.80313 & 23.125 \end{pmatrix}$$

$$\xrightarrow{M_2(\frac{1}{0.65})} \begin{pmatrix} 1 & -0.625 & -0.1875 & 12.5 \\ 0 & 1 & -0.57692 & 46.15385 \\ 0 & -0.65625 & 0.80313 & 23.125 \end{pmatrix}$$

$$\xrightarrow[A_{2,3}(0.65625)]{A_{2,1}(0.625)} \begin{pmatrix} 1 & 0 & -0.54808 & 41.34616 \\ 0 & 1 & -0.57692 & 46.15385 \\ 0 & 0 & 0.42453 & 53.41346 \end{pmatrix}$$

$$\xrightarrow{M_3(1/0.42453)} \begin{pmatrix} 1 & 0 & -0.54808 & 41.34616 \\ 0 & 1 & -0.57692 & 46.15385 \\ 0 & 0 & 1 & 125.81787 \end{pmatrix}$$

$$\xrightarrow[A_{3,2}(0.57692)]{A_{3,1}(0.54808)} \begin{pmatrix} 1 & 0 & 0 & 110.30442 \\ 0 & 1 & 0 & 118.74070 \\ 0 & 0 & 1 & 125.81787 \end{pmatrix}$$

We conclude that the outputs needed for supply to equal demand are, approximately, $x_1 = 110$, $x_2 = 119$, and $x_3 = 126$.

If $A = (a_{ij})$ is the technology matrix, then

$$I - A = \begin{pmatrix} 1 & 0 & 0 & \cdots & 0 \\ 0 & 1 & 0 & \cdots & 0 \\ 0 & 0 & 1 & \cdots & 0 \\ \vdots & \vdots & \vdots & & \vdots \\ 0 & 0 & 0 & \cdots & 1 \end{pmatrix} - \begin{pmatrix} a_{11} & a_{12} & a_{13} & \cdots & a_{1n} \\ a_{21} & a_{22} & a_{23} & \cdots & a_{2n} \\ a_{31} & a_{32} & a_{33} & \cdots & a_{3n} \\ \vdots & \vdots & \vdots & & \vdots \\ a_{n1} & a_{n2} & a_{n3} & \cdots & a_{nn} \end{pmatrix}$$

$$= \begin{pmatrix} 1 - a_{11} & a_{12} & a_{13} & \cdots & a_{1n} \\ a_{21} & 1 - a_{22} & a_{23} & \cdots & a_{2n} \\ a_{31} & a_{32} & 1 - a_{33} & \cdots & a_{3n} \\ \vdots & \vdots & \vdots & & \vdots \\ a_{n1} & a_{n2} & a_{n3} & \cdots & a_{nn} \end{pmatrix} \qquad (6)$$

Thus system (4) can be written

$$(I - A)\mathbf{x} = \mathbf{e}, \qquad (7)$$

where $\mathbf{e} = \begin{pmatrix} e_1 \\ e_2 \\ \vdots \\ e_n \end{pmatrix}$. The matrix $I - A$ in this model is called the **Leontief matrix**.

Assuming that the Leontief matrix is invertible, we can write the output vector $\mathbf{x}$ as

$$\mathbf{x} = (I - A)^{-1}\mathbf{e} \qquad (8)$$

There is an advantage to writing the output vector in the form of (8). The technology matrix A is the matrix of internal demands, which—over relatively long periods of time—remain fixed. However, the external demand vector $\mathbf{e}$ may change with some frequency. It is generally a long computation to find $(I - A)^{-1}$. But once we have done so, we can find the output vector $\mathbf{x}$ corresponding to any demand vector $\mathbf{e}$ by a simple matrix multiplication. Without computing $(I - A)^{-1}$, we could solve the problem only by using Gauss-Jordan elimination every time we changed the vector $\mathbf{e}$.

 Example 3   In an economic system with three industries, suppose that the technology matrix A is given by

$$A = \begin{pmatrix} 0.2 & 0.5 & 0.15 \\ 0.4 & 0.1 & 0.3 \\ 0.25 & 0.5 & 0.15 \end{pmatrix}$$

Find the total output corresponding to each external demand vector.

(a) $\mathbf{e} = \begin{pmatrix} 10 \\ 25 \\ 20 \end{pmatrix}$ (b) $\mathbf{e} = \begin{pmatrix} 15 \\ 20 \\ 40 \end{pmatrix}$ (c) $\mathbf{e} = \begin{pmatrix} 30 \\ 100 \\ 50 \end{pmatrix}$

Solution   The Leontief matrix is

$$I - A = \begin{pmatrix} 0.8 & -0.5 & -0.15 \\ -0.4 & 0.9 & -0.3 \\ -0.25 & -0.5 & 0.85 \end{pmatrix}$$

Part (a) was solved by Gauss-Jordan elimination in Example 2. We now solve all three problems essentially at once by computing $(I - A)^{-1}$. As in Example 2, we carry our computations to five decimal places (to the right of the decimal point).

$$\begin{pmatrix} 0.8 & -0.5 & -0.15 & | & 1 & 0 & 0 \\ -0.4 & 0.9 & 0.3 & | & 0 & 1 & 0 \\ 0.25 & -0.5 & 0.85 & | & 0 & 0 & 1 \end{pmatrix}$$

$$\xrightarrow{M_1(\frac{1}{0.8})} \begin{pmatrix} 1 & -0.625 & -0.1875 & | & 1.25 & 0 & 0 \\ -0.4 & 0.9 & -0.3 & | & 0 & 1 & 0 \\ -0.25 & -0.5 & 0.85 & | & 0 & 0 & 1 \end{pmatrix}$$

$$\begin{array}{c} A_{1,2}(0.4) \\ \xrightarrow{A_{1,3}(0.25)} \end{array} \begin{pmatrix} 1 & -0.625 & -0.1875 & | & 1.25 & 0 & 0 \\ 0 & 0.65 & -0.375 & | & 0.5 & 1 & 0 \\ 0 & -0.65625 & 0.80313 & | & 0.3125 & 0 & 1 \end{pmatrix}$$

$$\xrightarrow{M_2(\frac{1}{0.65})} \begin{pmatrix} 1 & -0.625 & -0.1875 & | & 1.25 & 0 & 0 \\ 0 & 1 & -0.57692 & | & 0.76923 & 1.53846 & 0 \\ 0 & -0.65625 & 0.80313 & | & 0.3125 & 0 & 1 \end{pmatrix}$$

$$\begin{array}{c} A_{2,1}(0.625) \\ \xrightarrow{A_{2,3}(0.65625)} \end{array} \begin{pmatrix} 1 & 0 & -0.54808 & | & 1.73077 & 0.96154 & 0 \\ 0 & 1 & -0.57692 & | & 0.76923 & 1.53846 & 0 \\ 0 & 0 & 0.42453 & | & 0.81731 & 1.00961 & 1 \end{pmatrix}$$

$$\xrightarrow{M_3(\frac{1}{0.42453})} \begin{pmatrix} 1 & 0 & -0.54808 & | & 1.73077 & 0.96154 & 0 \\ 0 & 1 & -0.57692 & | & 0.76923 & 1.53846 & 0 \\ 0 & 0 & 1 & | & 1.92521 & 2.37818 & 2.35555 \end{pmatrix}$$

$$\begin{array}{c} A_{3,1}(0.54808) \\ \xrightarrow{A_{3,2}(0.57692)} \end{array} \begin{pmatrix} 1 & 0 & 0 & | & 2.78594 & 2.26497 & 1.29103 \\ 0 & 1 & 0 & | & 1.87992 & 2.91048 & 1.35896 \\ 0 & 0 & 1 & | & 1.92521 & 2.37818 & 2.35555 \end{pmatrix}$$

Thus

$$(I - A)^{-1} = \begin{pmatrix} 2.78594 & 2.26497 & 1.29103 \\ 1.87992 & 2.91048 & 1.35896 \\ 1.92521 & 2.37818 & 2.35555 \end{pmatrix}$$

This must be checked!

$$\begin{pmatrix} 2.78594 & 2.26497 & 1.29103 \\ 1.87992 & 2.91048 & 1.35896 \\ 1.92521 & 2.37818 & 2.35555 \end{pmatrix} \begin{pmatrix} 0.8 & -0.5 & -0.15 \\ -0.4 & 0.9 & -0.3 \\ -0.25 & -0.5 & 0.85 \end{pmatrix}$$

$$= \begin{pmatrix} 1.00001 & 0 & -0.00001 \\ 0 & 0.99999 & -0.00002 \\ 0.00001 & -0.00002 & 0.99998 \end{pmatrix}$$

This verifies the correctness of our answer and points out the slight (in this case) inaccuracies caused by the round-off error.

Now we can solve our problems.

(a)
$$\begin{pmatrix} x_1 \\ x_2 \\ x_3 \end{pmatrix} = \begin{pmatrix} 2.78594 & 2.26497 & 1.29103 \\ 1.87992 & 2.91048 & 1.35896 \\ 1.92521 & 2.37818 & 2.35555 \end{pmatrix} \begin{pmatrix} 10 \\ 25 \\ 20 \end{pmatrix} = \begin{pmatrix} 110.30 \\ 118.74 \\ 125.82 \end{pmatrix}$$

and $x_1 \approx 110$, $x_2 \approx 119$, $x_3 \approx 126$. This is the answer we obtained in Example 2.

(b)
$$\begin{pmatrix} x_1 \\ x_2 \\ x_3 \end{pmatrix} = \begin{pmatrix} 2.78594 & 2.26497 & 1.29103 \\ 1.87992 & 2.91048 & 1.35896 \\ 1.92521 & 2.37818 & 2.35555 \end{pmatrix} \begin{pmatrix} 15 \\ 20 \\ 40 \end{pmatrix} = \begin{pmatrix} 138.73 \\ 140.77 \\ 170.66 \end{pmatrix}$$

Now $x_1 \approx 139$, $x_2 \approx 141$, and $x_3 \approx 171$.

(c)
$$\begin{pmatrix} x_1 \\ x_2 \\ x_3 \end{pmatrix} = \begin{pmatrix} 2.78594 & 2.26497 & 1.29103 \\ 1.87992 & 2.91048 & 1.35896 \\ 1.92521 & 2.37818 & 2.35555 \end{pmatrix} \begin{pmatrix} 30 \\ 100 \\ 50 \end{pmatrix} = \begin{pmatrix} 374.63 \\ 415.39 \\ 413.35 \end{pmatrix}$$

Thus $x_1 \approx 375$, $x_2 \approx 415$, and $x_3 \approx 413$.

Note that all these answers can be checked by inserting the computed values x_1, x_2, and x_3 into the original equation, $(I - A)\mathbf{x} = \mathbf{e}$.

Remark. It took us a little more work in part (a) to compute $(I - A)^{-1}$ than it took us in Example 2 to solve the system by row reduction. However, once we had $(I - A)^{-1}$, we were able to solve parts (b) and (c) with very little additional work.

 Example 4 Leontief used his model to analyze the 1958 American economy.[†] Leontief divided the economy into 81 sectors and grouped them into six families of related sectors. For simplicity, we treat each family of sectors as a single sector so we can treat the American economy as an economy with six industries.

[†] *Scientific American* (April, 1965): 26–27.

TABLE 1

Sector	Examples
Final nonmetal (FN)	Furniture, processed food
Final metal (FM)	Household appliances, motor vehicles
Basic metal (BM)	Machine-shop products, mining
Basic nonmetal (BN)	Agriculture, printing
Energy (E)	Petroleum, coal
Services (S)	Amusements, real estate

TABLE 2   Internal Demands in 1958 U.S. Economy

	FN	FM	BM	BN	E	S
FN	0.170	0.004	0	0.029	0	0.008
FM	0.003	0.295	0.018	0.002	0.004	0.016
BM	0.025	0.173	0.460	0.007	0.011	0.007
BN	0.348	0.037	0.021	0.403	0.011	0.048
E	0.007	0.001	0.039	0.025	0.358	0.025
S	0.120	0.074	0.104	0.123	0.173	0.234

These industries are listed in Table 1. The input-output table, Table 2, gives internal demands in 1958 based on Leontief's figures. The units in the table are millions of dollars. Thus, for example, the number 0.173 in the 6,5 position means that in order to produce $1 million worth of energy, it is necessary to provide $0.173 million = $173,000 worth of services. Similarly, the 0.037 in the 4,2 position means that in order to produce $1 million worth of final metal, it is necessary to expend $0.037 million = $37,000 on basic nonmetal products.

TABLE 3   External Demands on 1958 U.S. Economy (Millions of Dollars)

FN	$99,640
FM	$75,548
BM	$14,444
BN	$33,501
E	$23,527
S	$263,985

Finally, Leontief estimated the following demands on the 1958 American economy (in millions of dollars), as listed in Table 3. In order to run the American economy in 1958 and meet all external demands, how many units in each of the six sectors had to be produced?

Solution   The technology matrix is given by

$$A = \begin{pmatrix} 0.170 & 0.004 & 0 & 0.029 & 0 & 0.008 \\ 0.003 & 0.295 & 0.018 & 0.002 & 0.004 & 0.016 \\ 0.025 & 0.173 & 0.460 & 0.007 & 0.011 & 0.007 \\ 0.348 & 0.037 & 0.021 & 0.403 & 0.011 & 0.048 \\ 0.007 & 0.001 & 0.039 & 0.025 & 0.358 & 0.025 \\ 0.120 & 0.074 & 0.104 & 0.123 & 0.173 & 0.234 \end{pmatrix},$$

and

$$\mathbf{e} = \begin{pmatrix} 99,640 \\ 75,548 \\ 14,444 \\ 33,501 \\ 23,527 \\ 263,985 \end{pmatrix}.$$

To obtain the Leontief matrix, we subtract to obtain

$$I - A = \begin{pmatrix} 1 & 0 & 0 & 0 & 0 & 0 \\ 0 & 1 & 0 & 0 & 0 & 0 \\ 0 & 0 & 1 & 0 & 0 & 0 \\ 0 & 0 & 0 & 1 & 0 & 0 \\ 0 & 0 & 0 & 0 & 1 & 0 \\ 0 & 0 & 0 & 0 & 0 & 1 \end{pmatrix}$$

$$- \begin{pmatrix} 0.170 & 0.004 & 0 & 0.029 & 0 & 0.008 \\ 0.003 & 0.295 & 0.018 & 0.002 & 0.004 & 0.016 \\ 0.025 & 0.173 & 0.460 & 0.007 & 0.011 & 0.007 \\ 0.348 & 0.037 & 0.021 & 0.403 & 0.011 & 0.048 \\ 0.007 & 0.001 & 0.039 & 0.025 & 0.358 & 0.025 \\ 0.120 & 0.074 & 0.104 & 0.123 & 0.173 & 0.234 \end{pmatrix}$$

$$= \begin{pmatrix} 0.830 & -0.004 & 0 & -0.029 & 0 & -0.008 \\ -0.003 & 0.705 & -0.018 & -0.002 & -0.004 & -0.016 \\ -0.025 & -0.173 & 0.540 & -0.007 & -0.011 & -0.007 \\ -0.348 & -0.037 & -0.021 & 0.597 & -0.011 & -0.048 \\ -0.007 & -0.001 & -0.039 & -0.025 & 0.642 & -0.025 \\ -0.120 & -0.074 & -0.104 & -0.123 & -0.173 & 0.766 \end{pmatrix}$$

The computation of the inverse of a 6×6 matrix is a tedious affair. Carrying three decimal places on a calculator, we obtain the matrix below. Intermediate steps are omitted.

$$(I - A)^{-1} = \begin{pmatrix} 1.234 & 0.014 & 0.006 & 0.064 & 0.007 & 0.018 \\ 0.017 & 1.436 & 0.057 & 0.012 & 0.020 & 0.032 \\ 0.071 & 0.465 & 1.877 & 0.019 & 0.045 & 0.031 \\ 0.751 & 0.134 & 0.100 & 1.740 & 0.066 & 0.124 \\ 0.060 & 0.045 & 0.130 & 0.082 & 1.578 & 0.059 \\ 0.339 & 0.236 & 0.307 & 0.312 & 0.376 & 1.349 \end{pmatrix}$$

Therefore the "ideal" output vector is given by

$$\mathbf{x} = (I - A)^{-1}\mathbf{e} = \begin{pmatrix} 1.234 & 0.014 & 0.006 & 0.064 & 0.007 & 0.018 \\ 0.017 & 1.436 & 0.057 & 0.012 & 0.020 & 0.032 \\ 0.071 & 0.465 & 1.877 & 0.019 & 0.045 & 0.031 \\ 0.751 & 0.134 & 0.100 & 1.740 & 0.066 & 0.124 \\ 0.060 & 0.045 & 0.130 & 0.082 & 1.578 & 0.059 \\ 0.339 & 0.236 & 0.307 & 0.312 & 0.376 & 1.349 \end{pmatrix} \begin{pmatrix} 99,640 \\ 75,548 \\ 14,444 \\ 33,501 \\ 23,527 \\ 263,985 \end{pmatrix}$$

$$= \begin{pmatrix} 131,161 \\ 120,324 \\ 79,194 \\ 178,936 \\ 66,703 \\ 426,542 \end{pmatrix}$$

This means that it would require 131,161 units ($131,161 million worth) of final nonmetal products, 120,324 units of final metal products, 79,194 units of basic metal products, 178,936 units of basic nonmetal products, 66,703 units of energy and 426,542 service units to run the U.S. economy and meet the external demands in 1958.

PROBLEMS 3.5

1. In the Leontief input-output model, suppose that there are three industries. Suppose further that $e_1 = 10$, $e_2 = 15$, $e_3 = 30$, $a_{11} = \frac{1}{3}$, $a_{12} = \frac{1}{2}$, $a_{13} = \frac{1}{6}$, $a_{21} = \frac{1}{4}$, $a_{22} = \frac{1}{4}$, $a_{23} = \frac{1}{8}$, $a_{31} = \frac{1}{12}$, $a_{32} = \frac{1}{3}$, and $a_{33} = \frac{1}{6}$. Find the output of each industry such that supply exactly equals demand.

2. Answer the question of Problem 1 if $a_{11} = 0.1$, $a_{12} = 0$, $a_{13} = 0$, $a_{21} = 0.05$, $a_{22} = 0.01$, $a_{23} = 0.05$, $a_{31} = 0.1$, $a_{32} = 0.2$, $a_{33} = 0.1$, $e_1 = 10$, $e_2 = 25$, $e_3 = 15$.

3. An economist is called in to advise a country whose economy consists of three industries. The internal demands of the three industries are $a_{11} = 0.2$, $a_{12} = 0.4$, $a_{13} = 0.2$, $a_{21} = 0.7$, $a_{22} = 0.3$, $a_{23} = 0.8$, $a_{31} = 0.3$, $a_{32} = 0.4$, $a_{33} = 0.1$. The economist is appalled by these figures and suggests that the country is in very serious economic trouble. Why does she draw this conclusion?

4. Find the output vector **x** in the Leontief input-output model if $n = 3$, $\mathbf{e} = \begin{pmatrix} 30 \\ 20 \\ 40 \end{pmatrix}$, and

$$A = \begin{pmatrix} \frac{1}{5} & \frac{1}{5} & 0 \\ \frac{2}{5} & \frac{2}{5} & \frac{3}{5} \\ \frac{1}{5} & \frac{1}{10} & \frac{2}{5} \end{pmatrix}.$$

5. Find the output vector **x** in Problem 4 if $\mathbf{e} = \begin{pmatrix} 10 \\ 40 \\ 15 \end{pmatrix}$.

6. Find the output vector **x** in Problem 4 if $\mathbf{e} = \begin{pmatrix} 35 \\ 100 \\ 60 \end{pmatrix}$.

*7. Consider a very simple economy of three industries, A, B, and C, represented in the given table. Data are in millions of dollars of products. Find the technology and Leontief matrices corresponding to this input-output system.

Producer	User			External demand	Total output
	A	B	C		
A	90	150	225	75	540
B	135	150	300	15	600
C	270	200	300	130	900

[*Hint:* Consider the meaning of the coefficients a_{ij} in technology matrix A.]

8. In Problem 7, suppose that the external demand changes to 50 for A, 20 for B, and 60 for C. Compute the new output vector.

9. Answer the question in Problem 8 if the external demand is 80 for A, 100 for B, and 120 for C.

*10. In a situation similar to that of Problem 7, the economy is represented in the given table.

Producer	User			External demand	Total output
	A	B	C		
A	80	100	100	40	320
B	80	200	60	60	400
C	80	100	100	20	300

Determine the output vector for the economy if the external demand changes to 120 for A, 40 for B, and 10 for C.

11. Answer the question of Problem 10 if the external demand changes to 60 for A, 60 for B, and 60 for C.

12. A much simplified version of an input-output table for the 1958 Israeli economy divides that economy into three sectors—agriculture, manufacturing, and energy— with the following result.[†]

	Agriculture	Manufacturing	Energy
Agriculture	0.293	0	0
Manufacturing	0.014	0.207	0.017
Energy	0.044	0.010	0.216

(a) How many units of agricultural production are required to produce one unit of agricultural output?

(b) How many units of agricultural production are required to produce 200,000 units of agricultural output?

(c) How many units of agricultural product go into the production of 50,000 units of energy?

(d) How many units of energy go into the production of 50,000 units of agricultural products?

13. Continuing Problem 12, exports (in thousands of Israeli pounds) in 1958 were

Agriculture	138,213
Manufacturing	17,597
Energy	1,786

(a) Compute the technology and Leontief matrices.

(b) Determine the number of Israeli pounds worth of agricultural products, manufactured goods, and energy required to run this model of the Israeli economy and export the stated value of products.

14. The interdependence among the motor-vehicle industry and other basic industries in the 1958 American economy is described by the following input-output table for motor vehicles (V), steel (S), glass (G), and rubber and plastics (R).

	V	S	G	R
V	0.298	0.002	0	0
S	0.088	0.212	0	0.002
G	0.010	0	0.050	0.006
R	0.029	0.003	0.004	0.030

[†] Wassily Leontief, *Input-Output Economics* (New York: Oxford University Press, 1966), 54–57.

The external demand for these products in millions of dollars is

V	5444
S	3276
G	119
R	943

How many millions of dollars of each of the four industries was required to run the economy and satisfy outside demand?

15. An input-output analysis of the 1963 British economy [†] is simplified below in terms of four sectors: nonmetals (N), metals (M), energy (E), and services (S).

	N	M	E	S
N	0.184	0.101	0.355	0.059
M	0.062	0.199	0.075	0.031
E	0.029	0.023	0.150	0.015
S	0.104	0.112	0.075	0.076

The external demands (in millions of pounds) are

N	10,271
M	5,987
E	1,161
S	13,780

How many millions of pounds of the output of each sector were required to run the British economy in 1963 and satisfy external demand?

3.6 Cryptography (Optional)

Cryptography is the science of writing or deciphering codes. Although this subject is often associated with military endeavors, cryptography has emerged as an important area of business practice. Large corporations, who deal with huge amounts of computerized data, must constantly guard against what is termed *industrial espionage*, the theft of important information by competitors.

---

† L. S. Berman, "Development of Input-Output Statistics," ed. W. F. Grossling, *Input-Output in the United Kingdom*, Proc. 1968 Manchester Conf. (London: Frank Cass, 1970): 34–35.

Today, many extremely sophisticated techniques have been developed to ensure that large amounts of information can be transmitted in confidence. This has been accomplished after a great deal of highly sophisticated research carried out by modern cryptographers.

In this book we will not, of course, be able to describe the latest techniques for making and breaking codes. However, we will illustrate how some simple codes can be written and deciphered using some elementary matrix techniques.

Most of you are familiar with a CRYPTOGRAM, which is a puzzle that appears in a great number of daily newspapers. Typically, you are given a message like the following:

LOK   KPGIF   MTGQ   VPLVOKY   LOK   HCGU.

We can decipher this by using the "decoder" below:

A B C D E F G H I J K L M N O P Q R S T U V W X Y Z
P M V Q K S Z O T B W I U J C X E G Y L D S H A F N

Noting that L stands for T, O stands for H, and so on, we arrive at the message

THE EARLY BIRD CATCHES THE WORM.

A CRYPTOGRAM uses a very crude, easily deciphered kind of code. We now show how matrices can be employed to create a code that is far more difficult to break. We begin by assigning to each letter its ordered place in the alphabet. This gives us the following association.

A B C D E F G H I J K L M N O P Q R S T U V W X Y Z
1 2 3 4 5 6 7 8 9 10 11 12 13 14 15 16 17 18 19 20 21 22 23 24 25 26

$$(1)$$

Suppose we wish to encode the message

MATRICES ARE FRIENDLY.

We break the message into units of equal length. If we choose a length of two letters, we obtain

MA   TR   IC   ES   AR   EF   RI   EN   DL   YX.

$$(2)$$

The X at the end simply fills in the space. If we use our numerical code (1), we can write (2) as a set of 2-vectors

$$\begin{pmatrix}13\\1\end{pmatrix} \begin{pmatrix}20\\18\end{pmatrix} \begin{pmatrix}9\\3\end{pmatrix} \begin{pmatrix}5\\19\end{pmatrix} \begin{pmatrix}1\\18\end{pmatrix} \begin{pmatrix}5\\6\end{pmatrix} \begin{pmatrix}18\\9\end{pmatrix} \begin{pmatrix}5\\14\end{pmatrix} \begin{pmatrix}4\\12\end{pmatrix} \begin{pmatrix}25\\24\end{pmatrix}. \quad (3)$$

We choose an invertible, integer-valued, 2×2 matrix A, whose inverse is also integer-valued. One such matrix is

$$A = \begin{pmatrix} 1 & 3 \\ 1 & 4 \end{pmatrix}.$$

Continuing, we multiply each of the 2-vectors in (3) on the left by A. For example,

$$A\begin{pmatrix} 13 \\ 1 \end{pmatrix} = \begin{pmatrix} 1 & 3 \\ 1 & 4 \end{pmatrix}\begin{pmatrix} 13 \\ 1 \end{pmatrix} = \begin{pmatrix} 16 \\ 17 \end{pmatrix}.$$

We thereby obtain the new set of vectors

$$\begin{pmatrix} 16 \\ 17 \end{pmatrix} \begin{pmatrix} 74 \\ 92 \end{pmatrix} \begin{pmatrix} 18 \\ 21 \end{pmatrix} \begin{pmatrix} 62 \\ 81 \end{pmatrix} \begin{pmatrix} 55 \\ 73 \end{pmatrix} \begin{pmatrix} 23 \\ 29 \end{pmatrix} \begin{pmatrix} 45 \\ 54 \end{pmatrix} \begin{pmatrix} 47 \\ 61 \end{pmatrix} \begin{pmatrix} 40 \\ 52 \end{pmatrix} \begin{pmatrix} 97 \\ 121 \end{pmatrix}. \quad (4)$$

Finally, we write (4) as

16 17 74 92 18 21 62 81 55 73
23 29 45 54 47 61 40 52 97 121. (5)

This is our new, coded message. This would be very difficult to decode without knowing the matrix A. Knowing A, however, makes it relatively simple. We begin by rearranging the numbers in (5) into groups of 2-vectors. Since, for example,

$$\begin{pmatrix} 16 \\ 17 \end{pmatrix} = A\begin{pmatrix} 13 \\ 1 \end{pmatrix},$$

we have

$$\begin{pmatrix} 13 \\ 1 \end{pmatrix} = A^{-1}\begin{pmatrix} 16 \\ 17 \end{pmatrix}.$$

To verify this, we note that

$$A^{-1} = \begin{pmatrix} 4 & -3 \\ -1 & 1 \end{pmatrix} \quad \text{so that} \quad A^{-1}\begin{pmatrix} 16 \\ 17 \end{pmatrix} = \begin{pmatrix} 4 & -3 \\ -1 & 1 \end{pmatrix}\begin{pmatrix} 16 \\ 17 \end{pmatrix} = \begin{pmatrix} 13 \\ 1 \end{pmatrix} = \begin{pmatrix} M \\ A \end{pmatrix}.$$

Multiplying each of the vectors in (4) by A^{-1} will give us the vectors in (3), which can then be directly converted via (1) into the message (2). In this context, the matrix A is called the **encoding matrix**, and the matrix A^{-1} is called the **decoding matrix**.

As another example, we can make this even more difficult for the potential decoder by taking the information in larger chunks. If we choose units three letters in length, then (2) becomes

MAT   RIC   ESA   REF   RIE   NDL   YXX.   (6)

which, using (1), is translated into

$$\begin{pmatrix} 13 \\ 1 \\ 20 \end{pmatrix} \begin{pmatrix} 18 \\ 9 \\ 3 \end{pmatrix} \begin{pmatrix} 5 \\ 19 \\ 1 \end{pmatrix} \begin{pmatrix} 18 \\ 5 \\ 6 \end{pmatrix} \begin{pmatrix} 18 \\ 9 \\ 5 \end{pmatrix} \begin{pmatrix} 14 \\ 4 \\ 12 \end{pmatrix} \begin{pmatrix} 25 \\ 24 \\ 24 \end{pmatrix}. \quad (7)$$

To encode this message now requires an invertible, integer-valued 3×3 matrix with an integer-valued inverse. One such matrix is

$$A = \begin{pmatrix} 1 & 2 & 3 \\ 1 & 1 & 2 \\ 0 & 1 & 2 \end{pmatrix},$$

which has the inverse

$$A^{-1} = \begin{pmatrix} 0 & 1 & -1 \\ 2 & -2 & -1 \\ -1 & 1 & 1 \end{pmatrix}.$$

Then, for example,

$$A \begin{pmatrix} 13 \\ 1 \\ 20 \end{pmatrix} = \begin{pmatrix} 1 & 2 & 3 \\ 1 & 1 & 2 \\ 0 & 1 & 2 \end{pmatrix} \begin{pmatrix} 13 \\ 1 \\ 20 \end{pmatrix} = \begin{pmatrix} 75 \\ 54 \\ 41 \end{pmatrix}.$$

This gives us the new set of 3-vectors

$$\begin{pmatrix} 75 \\ 54 \\ 41 \end{pmatrix} \begin{pmatrix} 45 \\ 33 \\ 15 \end{pmatrix} \begin{pmatrix} 46 \\ 26 \\ 21 \end{pmatrix} \begin{pmatrix} 46 \\ 35 \\ 17 \end{pmatrix} \begin{pmatrix} 51 \\ 37 \\ 19 \end{pmatrix} \begin{pmatrix} 58 \\ 42 \\ 28 \end{pmatrix} \begin{pmatrix} 145 \\ 97 \\ 72 \end{pmatrix} \quad (8)$$

Finally, we write (8) as

$$75 \quad 54 \quad 41 \quad 45 \quad 33 \quad 15 \quad 46 \quad 26 \quad 21 \quad 46 \quad 35$$
$$17 \quad 51 \quad 37 \quad 19 \quad 58 \quad 42 \quad 28 \quad 145 \quad 94 \quad 72. \quad (9)$$

To decode (9) requires knowledge of A^{-1}. For example,

$$A^{-1} \begin{pmatrix} 75 \\ 54 \\ 41 \end{pmatrix} = \begin{pmatrix} 0 & 1 & -1 \\ 2 & -2 & -1 \\ -1 & 1 & 1 \end{pmatrix} \begin{pmatrix} 75 \\ 54 \\ 41 \end{pmatrix} = \begin{pmatrix} 13 \\ 1 \\ 20 \end{pmatrix} = \begin{pmatrix} M \\ A \\ T \end{pmatrix}.$$

PROBLEMS 3.6

1. Using the method of this section, encode the message

MOZART CONQUERS ALL

using the encoding matrix

$$A = \begin{pmatrix} 7 & 5 \\ 4 & 3 \end{pmatrix}.$$

2. Do Problem 1 using the encoding matrix $A = \begin{pmatrix} 1 & -1 & 0 \\ 4 & -2 & 3 \\ 2 & 1 & 5 \end{pmatrix}.$

3. Using the matrix of Problem 1, decode

132 \quad 78 \quad 140 \quad 81 \quad 148 \quad 85 \quad 77 \quad 46 \quad 175 \quad 103 \quad 171 \quad 100.

4. Using the matrix of Problem 2, decode

1 \quad 63 \quad 100 \quad 3 \quad 19 \quad 17 \quad 0 \quad 51 \quad 81 \quad 17 \quad 117 \quad 115 \quad 9 \quad 96 \quad 115 \quad -17 \quad 52 \quad 158.

Review Exercises for Chapter 3

In Exercises 1–8, perform the indicated computations.

1.
$$3\begin{pmatrix} -2 & 1 \\ 0 & 4 \\ 2 & 3 \end{pmatrix}$$

2.
$$\begin{pmatrix} 1 & 0 & 3 \\ 2 & -1 & 6 \end{pmatrix} + \begin{pmatrix} 2 & 0 & 4 \\ -2 & 5 & 8 \end{pmatrix}$$

3.
$$5\begin{pmatrix} 2 & 1 & 3 \\ -1 & 2 & 4 \\ -6 & 1 & 5 \end{pmatrix} - 3\begin{pmatrix} -2 & 1 & 4 \\ 5 & 0 & 7 \\ 2 & -1 & 3 \end{pmatrix}$$

4.
$$\begin{pmatrix} 2 & 3 \\ -1 & 4 \end{pmatrix}\begin{pmatrix} 5 & -1 \\ 2 & 7 \end{pmatrix}$$

5.
$$\begin{pmatrix} 2 & 3 & 1 & 5 \\ 0 & 6 & 2 & 4 \end{pmatrix}\begin{pmatrix} 5 & 7 & 1 \\ 2 & 0 & 3 \\ 1 & 0 & 0 \\ 0 & 5 & 6 \end{pmatrix}$$

6.
$$\begin{pmatrix} 2 & 3 & 5 \\ -1 & 6 & 4 \\ 1 & 0 & 6 \end{pmatrix}\begin{pmatrix} 0 & -1 & 2 \\ 3 & 1 & 2 \\ -7 & 3 & 5 \end{pmatrix}$$

7.
$$\begin{pmatrix} 1 & 0 & 3 & -1 & 5 \\ 2 & 1 & 6 & 2 & 5 \end{pmatrix}\begin{pmatrix} 7 & 1 \\ 2 & 3 \\ -1 & 0 \\ 5 & 6 \\ 2 & 3 \end{pmatrix}$$

8.
$$\begin{pmatrix} 1 & -1 & 2 \\ 3 & 5 & 6 \\ 2 & 4 & -1 \end{pmatrix}\begin{pmatrix} 2 \\ 1 \\ 3 \end{pmatrix}$$

9. Verify the associative law of matrix multiplication for the matrices

$$A = \begin{pmatrix} 2 & 3 & 1 \\ 0 & 4 & 6 \end{pmatrix}, \quad B = \begin{pmatrix} 1 & 0 & 2 \\ 0 & 3 & 3 \\ 5 & 1 & -1 \end{pmatrix}, \quad \text{and} \quad C = \begin{pmatrix} 5 & 6 \\ -1 & 2 \\ 0 & 1 \end{pmatrix}.$$

In Exercises 10–14, calculate the reduced row-echelon form and the inverse of the given matrix (if the inverse exists).

10.
$$\begin{pmatrix} 2 & 3 \\ -1 & 4 \end{pmatrix}$$

11.
$$\begin{pmatrix} -1 & 2 \\ 2 & -4 \end{pmatrix}$$

12.
$$\begin{pmatrix} 1 & 2 & 0 \\ 2 & 1 & -1 \\ 3 & 1 & 1 \end{pmatrix}$$

13.
$$\begin{pmatrix} -1 & 2 & 0 \\ 4 & 1 & -3 \\ 2 & 5 & -3 \end{pmatrix}$$

14.
$$\begin{pmatrix} 2 & 0 & 4 \\ -1 & 3 & 1 \\ 0 & 1 & 2 \end{pmatrix}$$

In Exercises 15–17, first write the system in the form $A\mathbf{x} = \mathbf{b}$, then calculate A^{-1}, and, finally, use matrix multiplication to obtain the solution vector.

15.
$$\begin{aligned} x - 3y &= 4 \\ 2x + 5y &= 7 \end{aligned}$$

16.
$$\begin{aligned} x + 2y &= 3 \\ 2x + y - z &= -1 \\ 3x + y + z &= 7 \end{aligned}$$

17.
$$\begin{aligned} 2x \quad\;\; + 4z &= 7 \\ -x + 3y + z &= -4 \\ y + 2z &= 5 \end{aligned}$$

 18. In the Leontief input-output model, suppose that there are three industries. Suppose further that $e_1 = 78$, $e_2 = 99$, $e_3 = 96$, $a_{11} = \frac{1}{2}$, $a_{12} = \frac{1}{4}$, $a_{13} = \frac{1}{4}$, $a_{21} = \frac{1}{4}$, $a_{22} = \frac{1}{8}$, $a_{23} = \frac{1}{3}$, $a_{31} = \frac{1}{8}$, $a_{32} = \frac{1}{6}$, and $a_{33} = \frac{1}{4}$. Find the output of each industry such that supply exactly equals demand.

19. A much-simplified version of Leontief's 42-sector analysis of the 1947 American economy divides the economy into just three sectors: agriculture, manufacturing, and the household (the sector of the economy which produces labor). It consists of the given input-output table.

	Agriculture	Manufacturing	Household
Agriculture	0.245	0.102	0.051
Manufacturing	0.099	0.291	0.279
Household	0.433	0.372	0.011

The external demands (in billions of dollars) are

Agriculture	2.88
Manufacturing	31.45
Household	30.91

(a) Find the technology and Leontief matrices corresponding to this model.

(b) Determine the output of each of the three sectors necessary to run the economy and meet external demand.

20. Encode the message

BUSINESS STUDENTS LOVE MATHEMATICS

using the encoding matrix

$$A = \begin{pmatrix} 5 & 6 \\ 4 & 5 \end{pmatrix}$$

21. Using the matrix of Exercise 20, decode

148 120 159 131 159 131 148 120 55 45 94 76 111 90 63 52

101 84 130 105 210 172 148 122 55 45.

4 LINEAR PROGRAMMING

4.1 Linear Inequalities in Two Variables

In Sections 2.2 and 2.3, we saw how to find the equation and the graph of a straight line. In this section we will show how to sketch linear inequalities in two variables. The techniques we develop in this section will be very useful when we discuss linear programming in the remaining sections of this chapter.

Before citing general rules, we give three examples.

Example 1 Sketch the set of points that satisfy the inequality $y > -2x + 3$.

Solution We begin by drawing the graph of the line $y = -2x + 3$ in Figure 1. Since the line extends infinitely far in both directions, we can think of this line (or any other straight line) as dividing the xy-plane into two **half-planes**. In Figure 1 we have labeled these half-planes as *upper half-plane* and *lower half-plane*. The set $L = \{(x, y): y = -2x + 3\}$ is the set of points on the line. We define two other sets by

$$A = \{(x, y): y > -2x + 3\} \quad \text{and} \quad B = \{(x, y): y < -2x + 3\}$$

Since for any pair (x, y) we have $y = -2x + 3$, $y > -2x + 3$, or $y < -2x + 3$, we see that every point in $\mathbb{R}^2$ is in exactly one of the sets L, A, or B; that is,

$$\mathbb{R}^2 = L \cup A \cup B$$

We can see that A is precisely the upper half-plane in Figure 1. To see why, look at Figure 2. Let (x^*, y^*) be in A. Then, by the definition of A, $y^* > -2x^* + 3$, so that the point (x^*, y^*) lies above the line $y = -2x + 3$. This follows because the y-coordinate of the point (x^*, y^*) is greater than (higher than) the

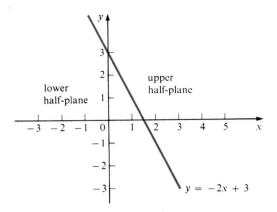

Figure 1

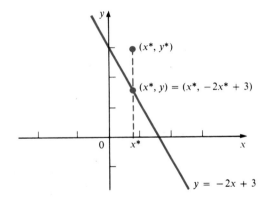

Figure 2

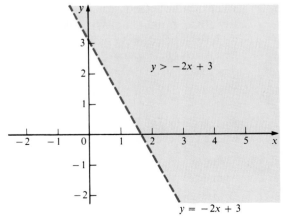

Figure 3

y-coordinate of the point $(x^*, -2x^* + 3)$, which is on the line. Thus the set of points that satisfy $y > -2x + 3$ is precisely the upper half-plane shaded in Figure 3. The dotted line in the figure indicates that points on the line do *not* satisfy the inequality.

Example 2   Sketch the set of points that satisfy the inequality $y \geq -2x + 3$.

Solution   The only difference between this set and the set of Example 1 is that points on the line $y = -2x + 3$ *are* now included. We indicate this by drawing a solid line, as in Figure 4.

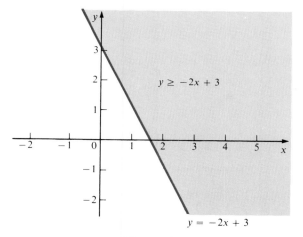

Figure 4

Example 3   Sketch the set of points that satisfy the inequality $y < -2x + 3$.

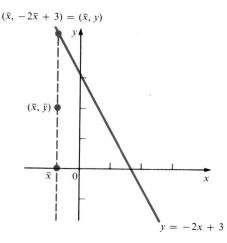

Figure 5

Solution   As in Example 1, let $B = \{(x, y) : y < -2x + 3\}$ and let $(\bar{x}, \bar{y})$ be in B. Then, as in Figure 5, $\bar{y} < -2\bar{x} + 3$, so that the point $(\bar{x}, \bar{y})$ lies *below* the line $y = -2x + 3$. Thus the set of points that satisfy $y < -2x + 3$ is the lower half-plane shown in Figure 6. Again, the dotted line indicates that points on the line are not included in the set.

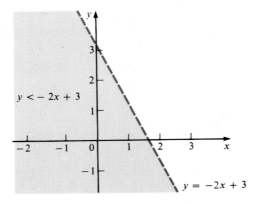

Figure 6

We now generalize these examples.

Linear Inequality in Two Variables   A **linear equality** in two variables is an inequality that can be written in one of the four forms

$$ax + by > c \tag{1}$$

$$ax + by \geq c \tag{2}$$

$$ax + by < c \tag{3}$$

$$ax + by \leq c \tag{4}$$

where a, b, and c are real numbers and a and b are not both equal to zero.

Remark.   Actually, there are only two distinct forms. For, if $ax + by < c$, then $-ax - by > -c$ (see equation (1.3.11) on page 20) and if $ax + by \leq c$, then $-ax - by \geq -c$.

There is a fairly easy method to use in graphing the set of points that satisfy one of these four inequalities. We illustrate this with an example.

Example 4   Sketch the set of points that satisfy $2x - 3y < 6$.

Solution   In Figure 7(a), we first sketch the line $2x - 3y = 6$. Since no point on the line satisfies the given inequality, we draw a dotted line. As in Examples 1, 2, and 3, the set of points we seek is one of the two half-planes into which the *xy*-plane has been divided by the line. Which one? The simplest way to tell is to select

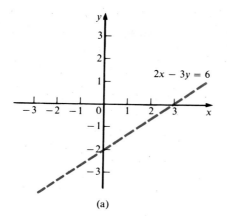

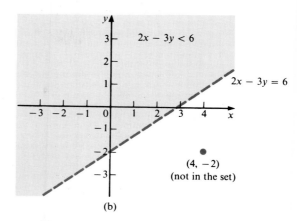

Figure 7

a **test point**, such as $(0, 0)$. Now $2 \cdot 0 - 3 \cdot 0 = 0 < 6$, so $(0, 0)$ is in the set $\{(x, y) : 2x - 3y < 6\}$. Thus the set we seek is the half-plane containing $(0, 0)$, as indicated in Figure 7(b). Why did we choose $(0, 0)$? Because it is the easiest point to check. But any test point can be used. For example, let's try the point $(4, -2)$. Then

$$2(4) - 3(-2) = 14 > 6.$$

Thus the half-plane containing $(4, -2)$ is *not* the half-plane we want. This leads us to the same graph as before.

We now state rules for sketching an inequality in one of the forms (1)–(4).

> To sketch the set of points satisfying a linear inequality in form (1), (2), (3) or (4):
>
> **1.** Draw the line $ax + by = c$. Use a dotted line if equality is not included in the inequality ((1) or (3)) and a solid line if it is ((2) or (4)).
>
> **2.** Pick any point in $\mathbb{R}^2$ not on the line and use it as a test point. If the coordinates of the test point satisfy the inequality, then the set sought is the half-plane containing the test point. Otherwise, it is the other half-plane.

Remark. We have now seen that the set of points that satisfy a linear inequality is a half-plane. If equality is excluded, then the half-plane is called an **open half-plane**. If equality is included, then it is called a **closed half-plane**. These definitions are similar to the definitions of open and closed intervals.

Example 5   Sketch the points that satisfy $4x + 2y \geq 4$.

Solution   We first sketch the graph of the line $4x + 2y = 4$, using a solid line since equality is included (see Figure 8(a)). Then, using (0, 0) as a test point, we

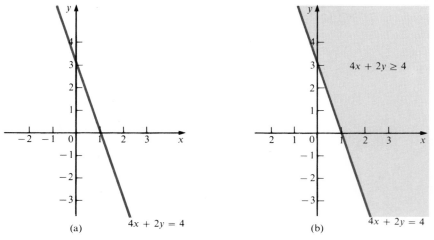

Figure 8

see that $4(0) + 2(0) = 0$, which is not greater than or equal 4, so that (0, 0) is not in the solution set. Thus our solution set is the closed half-plane sketched in Figure 8(b).

Example 6   Sketch the points in the plane whose x-coordinates satisfy $1 \leq x \leq 4$.

Solution   This set is really the intersection of two half-planes. The graph of the set $x \leq 4$ is the half-plane to the left of the line (and including the line) $x = 4$.

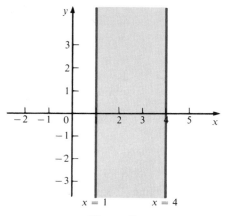

Figure 9

Similarly, the graph of $x \geq 1$ is the half-plane to the right of the line $x = 1$. Putting these together, we get the infinite *strip* sketched in Figure 9.

Example 7 Sketch the set of points that satisfy $-2 < x < 3$ and $0 < y \leq 5$.

Solution This set is the intersection of the four half-planes given by the inequalities $x > -2$, $x < 3$, $y > 0$, and $y \leq 5$. Three of these half-planes are open and the fourth is closed. The intersection is the rectangle in Figure 10.

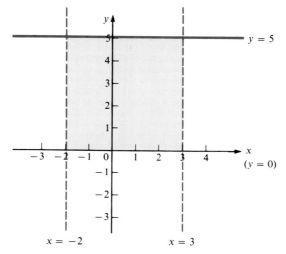

Figure 10

Example 8 Sketch the set of points that satisfy the inequalities $x + 2y \geq 2$ and $-2x + 3y < 6$.

Solution We begin by drawing, in Figure 11(a), the lines whose equations are given by $x + 2y = 2$ and $-2x + 3y = 6$. The coordinates $(0, 0)$ satisfy the

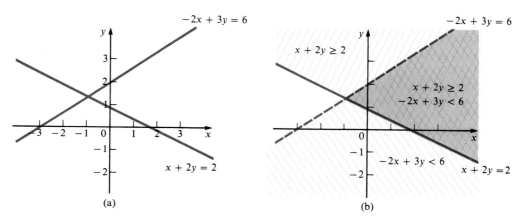

(a)

(b)

Figure 11

second inequality, but not the first. This means that the half-plane $\{(x, y):$ $-2x + 3y < 6\}$, which contains the point $(0, 0)$, is the set of points below the line $-2x + 3y = 6$, while the half-plane $\{(x, y): x + 2y \geq 2\}$, which does not contain the point $(0, 0)$, is the set of points on and above the line $x + 2y = 2$. Thus the set of points that satisfy both inequalities is the intersection of these two half-planes. This solution set is sketched in Figure 11(b).

Example 9 Sketch the set of points that satisfy the inequalities $x + y \leq 1$ and $2x + 2y \geq 6$.

Solution The two half-planes that are the solution sets of these two inequalities are sketched in Figure 12. Since these two sets are disjoint, their intersection is empty and, therefore, there is *no* point that satisfies both inequalities.

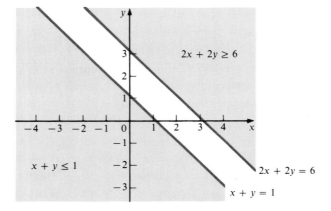

Figure 12

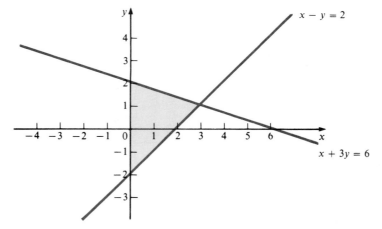

Figure 13

Example 10 Sketch the points that satisfy the inequalities

$$x + 3y \le 6$$
$$x - y \le 2$$
$$x \ge 0.$$

Solution The lines $x + 3y = 6$, $x - y = 2$, and $x = 0$ (the y-axis) are shown in Figure 13. The solution set is the shaded region bounded by these lines.

Example 11 Sketch the solution set of the inequalities

$$-x + y \le 1$$
$$x + 2y \le 6$$
$$2x + 3y \ge 3$$
$$-3x + 8y \ge 4.$$

Solution The lines $-x + y = 1$, $x + 2y = 6$, $2x + 3y = 3$, and $-3x + 8y = 4$ are shown in Figure 14. The solution set of the four inequalities is the shaded

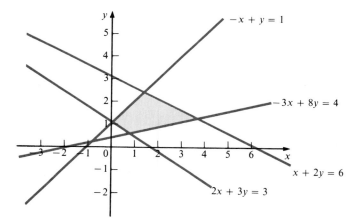

Figure 14

region in the figure. Note that in this case the solution set is a four-sided region in the first quadrant with four "corners." Solution sets like this will come up fairly frequently in the sections that follow.

PROBLEMS 4.1 In the following problems, sketch the set of points that satisfy the given inequalities.

1. $x \le 3$

2. $y < 2$

3. $y \ge -4$

4. $x \le \frac{3}{2}$

5. $y > \frac{2}{3}$

6. $x + y > 2$

7. $x + y \leq 2$ **8.** $2x - y < 4$ **9.** $2x - y \geq 4$

10. $y - 2x < 4$ **11.** $x - 2y > -4$ **12.** $3x - y < 3$

13. $y - 3x > -3$ **14.** $y - 3x \leq 3$ **15.** $y + 3x \geq 3$

16. $y \leq 2x - 5$ **17.** $y > 4x - 3$ **18.** $x < -2y + 7$

19. $y \leq 4x - 3$ **20.** $3x + 4y \leq 6$ **21.** $-3x + 4y > 6$

22. $3x - 4y \geq 6$ **23.** $-3x - 4y > 6$ **24.** $x - \dfrac{y}{2} > 4$

25. $\dfrac{x - y}{3} \leq 2$ **26.** $\dfrac{x}{2} - \dfrac{y}{3} \geq 1$ **27.** $\dfrac{x}{3} + \dfrac{y}{2} < -1$

28. $\dfrac{x}{3} - \dfrac{y}{5} \geq \dfrac{1}{2}$ **29.** $-3 \leq x < 0$ **30.** $1 < y \leq 6$

31. $0 \leq x \leq 2, 0 \leq y \leq 3$ **32.** $-1 \leq x < 4, -2 \leq y < 2$

33. $-\frac{1}{2} < x < 1, \frac{1}{2} \leq y < 2$ **34.** $|x| < 2, |y| < 3$

*35. $|x - 1| < 4, |y + 2| \leq 3$ **36.** $x + y \geq 1, 2x - 3y \leq 6$

37. $x + y \leq 1, 2x + 3y \geq 5$ **38.** $x - y \leq 2, 2y - 3x > 6$

39. $x + 2y \leq 2, 2x + 4y \geq 4$ **40.** $x + 2y < 2, 2x + 4y > 4$

41. $x + y \leq 2, 5x + 2y \geq 4$ **42.** $3x - 4y \leq 6, 2x + 3y > 3$

43. $-x + 2y \leq 4, 3x + 2y \leq 6$

44. $-x + 2y \leq 4, 3x + 2y \leq 6, x \geq 0, y \geq 0$

45. $x - y \leq 2, x + 3y \geq 6, x \geq 0, y \geq 0$

46. $2x + y \geq 1, x + 2y \geq 1, x + y \leq 3, x \geq 0, y \geq 0$

4.2 Linear Programming: Introduction

The problem of determining the maximum or minimum of a given function occurs in many applications of mathematics to business, economics, the biological sciences, and other disciplines. It is not difficult to think of examples of such problems. How can a businessman maximize profits or minimize costs? At what currency exchange rate will the balance of payments be most favorable? How can the food requirements of an animal be satisfied with a minimum expenditure of energy?

Maximization and minimization problems are often subject to constraints or limits on the variables. For example, a businessman is always limited by a finite supply of capital. Each of us could make virtually unlimited profits if we had unlimited sums to invest. A warehouse supervisor has limited space for storage. Biological variables may be constrained by physiological limits or by limits of resource availability. Some constraints are obvious by definition of the variables. A supermarket manager, for example, cannot order a negative number of pounds of tomatoes.

In this chapter, we consider the special problem of maximizing or minimizing linear functions of several variables subject to linear constraints. Instead of giving a general definition of the problem at this point, we begin with several simple examples.

Example 1 The Grant Furniture Company manufactures dining room tables and chairs. Each takes 20 board feet (bd ft) and 4 hours of labor. Each table requires 50 bd ft, but only 3 hours of labor. The manufacturer has 3300 bd ft of lumber available and a staff able to provide 380 hours of labor. Finally, the manufacturer has determined that there is a net profit of $3 for each chair sold and $6 for every table sold. For simplicity, we assume that needed materials (such as nails or varnish) are available in sufficient quantities. How many tables and chairs should the company manufacture in order to maximize its profit, assuming that each item manufactured is sold?

Solution The problem as stated seems difficult—there are lots of things going on. We begin simplifying the problem by putting all the information into a table.

TABLE 1   Data for the Grant Furniture Company

Raw material	Amount needed per unit		Total available
	Chair	Table	
Wood (board feet)	20	50	3300
Labor (hours)	4	3	380
Net unit profit (dollars)	3	6	

We now let x denote the number of chairs and y the number of tables produced by the company. Since it takes 20 bd ft of lumber to make one chair, it takes $20x$ bd ft of lumber to make x chairs. Similarly, it takes $50y$ bd ft of lumber to make y tables. Thus the data in the first line of Table 1 can be expressed algebraically by the linear inequality

$$20x + 50y \leq 3300. \qquad \text{Lumber inequality}$$

Analogously, the linear inequality representing the information in the second line of Table 1 is

$$4x + 3y \leq 380. \qquad \text{Labor inequality}$$

These two inequalities represent two of the **constraints** of this problem. They express, in mathematical terms, the obvious fact that raw materials and labor are finite (limited) quantities. There are two other constraints. Since the company cannot manufacture negative amounts of the two items, we must have

$$x \geq 0 \quad \text{and} \quad y \geq 0.$$

The profit P earned when x chairs and y tables are manufactured is given (from the third line of Table 1) by

$$P = 3x + 6y. \qquad \text{Profit equation}$$

Putting all this information together, we can state our problem in a form that we will soon recognize as a **standard linear programming problem**: Maximize

$$P = 3x + 6y \tag{1}$$

subject to the constraints

$$20x + 50y \le 3300 \tag{2}$$

$$4x + 3y \le 380 \tag{3}$$

$$x \ge 0, \ y \ge 0. \tag{4}$$

In this problem, the linear function[†] given by (1) is called the **objective function**. Any point in the constraint set is called a **feasible solution**. Our problem is to find the point (or points) in the constraint set at which the objective function is a maximum. Our first method for solving this problem will employ techniques of the last section. We begin to find our solution by graphing the **constraint set**, which is the solution set of the inequalities. This is done in Figure 1. Consider

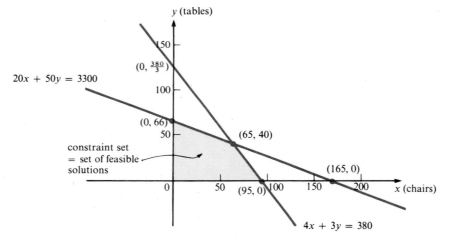

Figure 1

the lines $3x + 6y = C$ for different values of the constant C. Some of the lines are sketched in Figure 2. Each line $3x + 6y = C$ is called a **constant profit line** for this problem. To see why, consider the line $3x + 6y = 30$. For every point (x, y) lying both on this line and in the constraint set, the manufacturer makes a profit of $30. Some points are $(10, 0)$ (10 chairs and no tables), $(6, 2)$ (6 chairs and 2 tables), and $(0, 5)$ (no chairs and 6 tables). See Figure 3. From the manufacturer's point of view, these three points are equivalent because each leads to the same $30 profit.

Now, consider the constant profit line $3x + 6y = 60$. This line lies to the right of the line $3x + 6y = 30$ and is a "nicer" line for the manufacturer because every

---

[†] See Section 2.2.

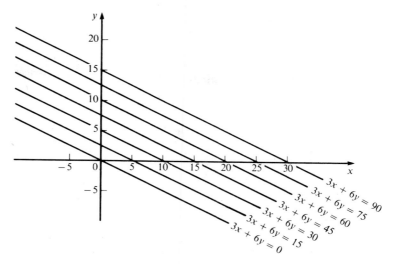

Figure 2

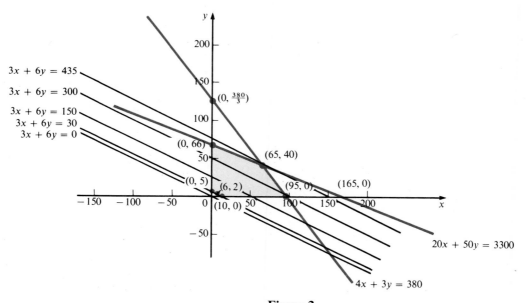

Figure 3

point on it and in the constraint set gives a profit of $60. Two such points are (20, 0) and (12, 4).

By now, the pattern may be getting clearer. All of the constant profit lines are parallel (each has slope $-\frac{1}{2}$) and the profit increases as we move to the right from one line to the next. Each new line (to the right) leads to a higher profit. Our method now is to move to the right as much as we can while still remaining in the constraint set. From Figure 3, we see that the "last" constant profit line is the line that intersects the constraint set at the single point (65, 40). This means that the largest profit is earned when 65 chairs and 40 tables are manufactured. This yields a profit of $3 \cdot 65 + 6 \cdot 40 = \$435$.

Note. It may seem at first glance that the company can make more profit by putting as much as possible into the more profitable tables. From Figure 1, we see that as many as 66 tables (the largest value of y in the constraint set is 66) can be manufactured, which yields a profit of $6 \cdot 66 = \$396$. On the other hand, the manufacture of 95 chairs and no tables gives a profit of \$195. Thus the company does indeed do better by making 65 chairs and 40 tables.

The method used in Example 1 to solve the linear programming problem is called the **graphical method**. This method illustrates what is going on, but it is very impractical to use for two reasons: First, it is necessary to use very precise drawings to obtain the solution, and second, it can be used only with problems involving two variables because graphs in three dimensions are unwieldy and graphs in more than three dimensions cannot be drawn.

We now introduce another method. To do so, we take another look at Example 1. The problem was to maximize

$$P = 3x + 6y \tag{5}$$

subject to the constraints

$$20x + 50y \leq 3300 \tag{6}$$

$$4x + 3y \leq 380 \tag{7}$$

$$x \geq 0 \tag{8}$$

$$y \geq 0. \tag{9}$$

In Figure 4, we again sketch the constraint set of this problem. The constraint set is the intersection of four sets: $S_1 = \{(x, y): 20x + 50y \leq 3300\}$, $S_2 = \{(x, y): 4x + 3y \leq 380\}$, $S_3 = \{(x, y): x \geq 0\}$, and $S_4 = \{(x, y): y \geq 0\}$. Each of these four sets is the solution set of a linear inequality, and each is bounded by a

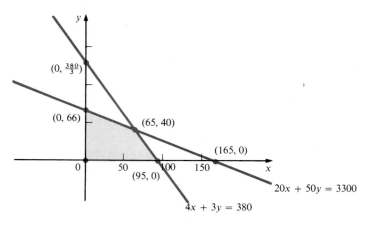

Figure 4

straight line. The intersection of any two of these lines is a point in the plane and, if the point is in the constraint set, then the point is called a **corner point** of the constraint set. In Table 2 we list all the possible and actual corner points. A point is a possible corner point if it is the intersection of two of the lines that determine the constraint set. It is an actual corner point if it is in the constraint set; that is, it is an actual corner point if it is a feasible solution.

TABLE 2

The two lines that determine the point	Possible corner point	Feasible solution? (Actual corner point?) (Is it in the constraint set?)
$20x + 50y = 3300$ $4x + 3y = 380$	$(65, 40)$	Yes.
$20x + 50y = 3300$ $x = 0$	$(0, 66)$	Yes.
$20x + 50y = 3300$ $y = 0$	$(165, 0)$	No. (Constraint (7) is violated since $4 \cdot 165 + 3 \cdot 0 = 660$, which is > 380.)
$4x + 3y = 380$ $x = 0$	$(0, \frac{380}{3})$	No. (Constraint (6) is violated since $20 \cdot 0 + 50 \cdot \frac{380}{3} = \frac{19,000}{3}$, which is > 3300.)
$4x + 3y = 380$ $y = 0$	$(95, 0)$	Yes.
$x = 0$ $y = 0$	$(0, 0)$	Yes.

The following statement is true.

> *The maximum and minimum values of the objective function of any linear programming problem always are taken at corner points.*

In our problem there are four actual corner points. So, in order to find a solution, we need only to evaluate the objective function at each corner point and choose the point that gives the maximum value. We do this in Table 3.

TABLE 3

Corner point	Value of objective function $P = 3x + 6y$	
(65, 40)	$3 \cdot 65 + 6 \cdot 40 = 435$	**Maximum value**
(0, 65)	$3 \cdot 0 + 6 \cdot 65 = 390$	
(95, 0)	$3 \cdot 95 + 6 \cdot 0 = 285$	
(0, 0)	$3 \cdot 0 + 6 \cdot 0 = 0$	

Thus we see, as we saw in Example 1, that the maximum profit of $435 is earned when 65 chairs and 40 tables are manufactured.

Example 2   In Example 1, suppose that the profits are $3 per chair and $10 per table and that all other data remain the same. How can the furniture company maximize its profits under these conditions?

Solution   The problem is to maximize

$$P = 3x + 10y$$

subject to

$$20x + 50y \leq 3300$$
$$4x + 3y \leq 380$$
$$x \geq 0, y \geq 0.$$

Now we have the same constraint set as in the example at the beginning of this section. In Table 4 we evaluate the objective function at each of the four corner

TABLE 4

Corner point	Value of objective function $P = 3x + 10y$
(65, 40)	$3 \cdot 65 + 10 \cdot 40 = 595$
(0, 65)	$3 \cdot 0 + 10 \cdot 65 = 650$
(95, 0)	$3 \cdot 95 + 10 \cdot 0 = 285$
(0, 0)	$3 \cdot 0 + 10 \cdot 0 = 0$

points obtained earlier. We find that the maximum profit of $650 is earned when no chairs and 65 tables are manufactured.

Example 3   A mountain lake in a national park is stocked each spring with two species of fish, S_1 and S_2. The average weight of the fish stocked is 4 lb for S_1 and 2 lb for S_2. Two foods, F_1 and F_2, are available in the lake. The average requirement of a fish of species S_1 is 1 unit of F_1 and 3 units of F_2 each day. The corresponding requirement of S_2 is 2 units of F_1 and 1 unit of F_2. If 500 units of F_1 and 900 units of F_2 are available daily, how should the lake be stocked to maximize the weight of fish supported by the lake?

Solution   Let x_1 and x_2 denote the numbers of fish of species S_1 and S_2 stocked in the lake. The total weight W of fish stocked is given by

$$W = 4x_1 + 2x_2. \tag{10}$$

The total consumption of food F_1 is $x_1 + 2x_2$, since each fish of the first species consumes one unit of F_1 and each fish of the second species consumes 2 units of F_1. Similarly, the total consumption of food F_2 is $3x_1 + x_2$. Since 500 units of F_1 and 900 units of F_2 are available, we have

$$x_1 + 2x_2 \le 500 \quad \text{and} \quad 3x_1 + x_2 \le 900. \tag{11}$$

Finally, we have the obvious constraints

$$x_1 \ge 0 \quad \text{and} \quad x_2 \ge 0 \tag{12}$$

since the lake cannot be stocked with a negative number of fish of either species.
    This is another typical problem of linear programming.
    Maximize

$$W = 4x_1 + 2x_2$$

subject to

$$x_1 + 2x_2 \le 500 \tag{13}$$

$$3x_1 + x_2 \le 900 \tag{14}$$

$$x_1 \ge 0 \tag{15}$$

$$x_2 \ge 0. \tag{16}$$

We will solve this by the corner-point method. To visualize what is going on, we sketch the constraint set. In Figure 5, the straight lines $x_1 + 2x_2 = 500$ and $3x_1 + x_2 = 900$ are shown in the x_1x_2-plane. The information needed to solve this problem is given in Table 5.

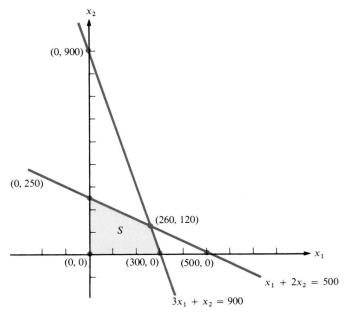

Figure 5

TABLE 5

The two lines that determine the point	Possible corner point	Feasible solution? (Actual corner point?)	Value of objective function $W = 4x_1 + 2x_2$ at corner point
$x_1 + 2x_2 = 500$ $3x_1 + x_2 = 900$	(260, 120)	Yes.	1280
$x_1 + 2x_2 = 500$ $x_1 \quad\;\; = 0$	(0, 250)	Yes.	500
$x_1 + 2x_2 = 500$ $x_2 \quad\;\; = 0$	(500, 0)	No. (Constraint (14) is violated.)	—
$3x_1 + x_2 = 900$ $x_1 \quad\;\; = 0$	(0, 900)	No. (Constraint (13) is violated.)	—
$3x_1 + x_2 = 900$ $x_2 \quad\;\; = 0$	(300, 0)	Yes.	1200
$x_1 = 0$ $x_2 = 0$	(0, 0)	Yes.	0

We find the maximum value of 1280 at $x_1 = 260$ and $x_2 = 120$. This means that the lake can support a maximum weight of 1280 lb if 260 fish of species S_1 and 120 fish of species S_2 are stocked.

Example 4 The water-supply manager for a midwest city must find a way to supply at least 10 million gal of potable (drinkable) water per day (mgd). The supply may be drawn from the local reservoir or from a pipeline to an adjacent town. The local reservoir has a daily yield of 5 mgd, which may not be exceeded. The pipeline can supply no more than 10 mgd because of its size. On the other hand, by contractual agreement it must pump out at least 6 mgd. Finally, reservoir water costs $300 for 1 million gal and pipeline water cost $500 for 1 million gal. How can the manager minimize daily water costs?

Solution   Let x denote the number of reservoir gallons and y denote the number of pipeline gallons (in millions of gallons) pumped per day. Then the problem is: Minimize

$$C = 300x + 500y$$

subject to

$x + y \geq 10$	To meet the city water requirements
$x \leq 5$	Reservoir capacity
$y \leq 10$	Pipeline capacity
$y \geq 6$	Pipeline contract
$x \geq 0$	
$y \geq 0$	

The constraint set for this problem is sketched in Figure 6. From Figure 6 we can see that there are four corner points (see Table 6).

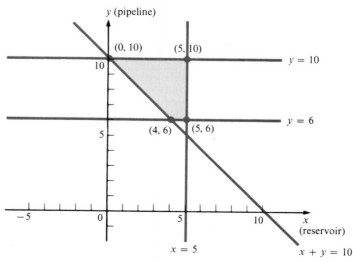

Figure 6

TABLE 6

Corner point	Value of objective function $C = 300x + 500y$ at corner point
(0, 10)	5000
(5, 10)	6500
(4, 6)	4200
(5, 6)	4500

The minimum value of the constraint function over the corner points is 4200 at (4, 6). That is, the manager should draw 4 million gal per day from the reservoir and 6 million gal per day from the pipeline at a daily cost of $4 \cdot 300 + 6 \cdot 500 =$ $4200.

In the examples we considered in this section, there were two variables (which we denoted by x and y or x_1 and x_2. The graphical method fails, as we have stated, if there are more than two variables. The corner-point method works with more than two variables, but the work required to compute the possible corner points can be tremendous. In the problems set we ask you to solve some linear programming problems involving three variables by the corner-point method. You will see how quickly computations can become cumbersome (see Problems 47–51).

In the next three sections we describe a much more efficient method for solving linear programming problems with more than two variables. We close this section by illustrating two of the difficulties that can arise when solving a linear programming problem.

Example 5 Solve the following linear programming problem:
Maximize

$$f = 2x + 3y$$

subject to

$$x + \ y \geq 5$$

$$6x + 2y \geq 12$$

$$x \geq 0, y \geq 0.$$

Solution The constraint set is sketched in Figure 7. It is clear that x and y can take arbitrarily large values and still remain in the constraint set. Therefore f can take on arbitrary large values and the problem has no solution. In this situation we say that the problem is **unbounded**.

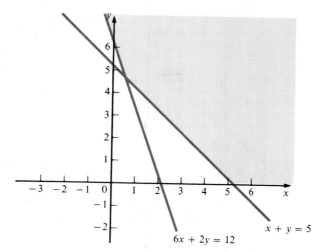

Figure 7

Example 6   Solve the problem.
Maximize

$$f = 2x + 3y$$

subject to

$$x + \ y \geq 5$$
$$2x + 3y \leq 6$$
$$x \geq 0, y \geq 0.$$

Solution   The linear inequalities are sketched in Figure 8. It is evident that the

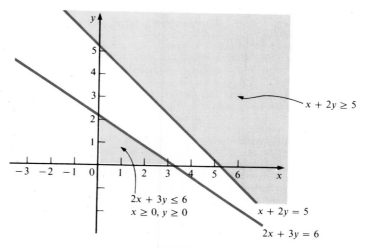

Figure 8

constraint set is empty. Thus there are no feasible solutions and we say that the problem is **infeasible**.

PROBLEMS 4.2

1. In Example 1, find the combination of tables and chairs that will maximize the manufacturer's profit if the profits are $5 per chair and $5 per table. Assume that all other data are unchanged.

2. Answer the question in Problem 1 if profits are $8 per chair and $2 per table.

3. Answer the question of Example 1 using the data in Table 7.

TABLE 7

Raw material	Amount needed per unit		Total available
	Chair	Table	
Wood (board feet)	30	40	11,400
Labor (hours)	4	6	1650
Net unit profit (dollars)	5	6	

4. Answer the question in Problem 3 if profits are $2 per chair and $8 per table and all other data are unchanged.

5. Answer the question in Problem 3 if profits are $8 per chair and $2 per table.

*6. In Example 1, assume that a table and a chair require the same amount of wood and labor. If the unit profit is $3 per chair and $4 per table, show that if lumber and labor are limited, then the manufacturer can always maximize profit by manufacturing tables only.

7. In Example 4, how can the water manager minimize costs if there is no lower limit to the number of gallons that must pass through the pipeline?

In Problems 8–20, solve the given linear programming problem using the graphical or the corner-point method. Find the values of x and y at which the given objective function is maximized or minimized.

8. Maximize

$$f = 3x + 4y$$

subject to

$$x + y \leq 4$$
$$2x + y \leq 5$$
$$x \geq 0, y \geq 0.$$

9. Maximize

$$f = 4x + 3y$$

subject to

$$x + y \leq 4$$
$$2x + y \leq 5$$
$$x \geq 0, y \geq 0.$$

10. Maximize

$$f = x + y$$

subject to

$$3x + 4y \leq 12$$
$$2x + y \leq 8$$
$$x \geq 0, y \geq 0.$$

11. Maximize

$$f = 2x + 3y$$

subject to

$$x + y \le 4$$
$$2x + 3y \le 10$$
$$4x + 2y \le 12$$
$$x \ge 0, y \ge 0.$$

12. Maximize

$$f = 3x + 5y$$

subject to

$$10x + y \le 10$$
$$x + 10y \le 10$$
$$2x + 3y \le 6$$
$$x \ge 0, y \ge 0.$$

13. Maximize

$$f = 5x + 3y$$

subject to

$$10x + y \le 10$$
$$x + 10y \le 10$$
$$2x + 3y \le 6$$
$$x \ge 0, y \ge 0.$$

14. Maximize

$$f = 12x + y$$

subject to

$$10x + y \le 10$$
$$x + 10y \le 10$$
$$2x + 3y \le 6$$
$$x \ge 0, y \ge 0$$

15. Maximize

$$f = x + 12y$$

subject to

$$10x + y \le 10$$
$$x + 10y \le 10$$
$$2x + 3y \le 6$$
$$x \ge 0, y \ge 0.$$

16. Minimize

$$g = 4x + 5y$$

subject to

$$x + 2y \ge 3$$
$$x + y \ge 4$$
$$x \ge 0, y \ge 0.$$

17. Minimize

$$g = 4x + 5y$$

subject to

$$x + 2y \ge 4$$
$$x + y \ge 3$$
$$x \ge 0, y \ge 0.$$

18. Minimize

$$g = 12x + 8y$$

subject to

$$3x + 2y \ge 1$$
$$4x + y \ge 1$$
$$x \ge 0, y \le 0.$$

19. Minimize

$$g = 3x + 7y$$

subject to

$$5x + y \ge 1$$
$$2x + 3y \ge 2$$
$$x \ge 0, y \ge 0.$$

20. Minimize

$$g = 3x + 2y$$

subject to

$$x + 2y \ge 1$$
$$2x + y \ge 2$$
$$5x + 4y \ge 10$$
$$x \ge 0, y \ge 0.$$

21. Determine the number of fish of species S_1 and S_2, with a total weight of 1200 lb, that can coexist in the lake of Example 3. Plot the corresponding points in the plane.

22. Suppose that 1000 units of F_1 and 1800 units of F_2 are available daily in Example 3. How should the lake be stocked to maximize the weight of fish supported by the lake?

23. As in Problem 22, how should the lake be stocked if 1000 units of F_1 and 1000 units of F_2 are available daily?

24. Suppose that two types of food are available in a lake in fixed daily amounts and that the daily requirements for these foods of average fish of two species are known. Formulate a general problem of stocking the lake in order to maximize the number of fish supported by the lake.

25. In Example 3, how should the lake be stocked in order to maximize the total number of fish supported by the lake?

26. In Problem 22, how should the lake be stocked to maximize the total number of fish supported by the lake?

27. In Problem 23, how should the lake be stocked to maximize the total number of fish supported by the lake? In this case, what is the total weight of fish stocked in the lake?

28. The Goody Goody Candy Company makes two kinds of gooey candy bars from caramel and chocolate. Each bar weighs 4 oz. Bar A has 3 oz of caramel and 1 oz of chocolate. Bar B has 2 oz of each. Bar A sells for 30¢ and bar B sells for 54¢. The company has stocked 90 lb of chocolate and 144 lb of caramel. How many units of each type of candy should be made in order to maximize the company's income?

29. Two foods contain carbohydrates and proteins only. Food I costs 50¢ per pound and is 90% carbohydrates (by weight). Food II costs $1 per pound and is 60% carbohydrates. What diet of these two foods provides at least 2 lb of carbohydrates and 1 lb of proteins at minimum cost? What is the cost per pound of this diet?

30. Spina Food Supplies, Inc. is a manufacturer of frozen pizzas. Art Spina, president of Spina Food Supplies, personally supervises the production of both types of frozen pizzas produced by the company: Spina's regular and Spina's super deluxe. Art makes a profit of $0.50 for each regular produced and $0.75 for each super deluxe. He currently has 150 lb of dough mix available and 800 oz of topping mix. Each regular pizza uses 1 lb of dough mix and 4 oz of topping, whereas each super deluxe uses 1 lb of dough and 8 oz of topping mix. Based upon past demand, Art knows that he can sell at most 75 super deluxe and 125 regular pizzas. How many regular and super deluxe pizzas should Art make in order to maximize profits?

31. A predator requires 10 units of food A, 12 units of food B, and 12 units of food C as its average daily consumption. These requirements are satisfied by feeding on two prey species. One prey of species I provides 5, 2, and 1 units of foods A, B, and C, respectively. An individual prey of species II provides 1, 2, and 4 units of A, B, and C, respectively. To capture and digest a prey of species I requires 3 units of energy, on the average. The corresponding energy requirement for species II is 2 units of energy. How many prey of each species should the predator capture to meet its food requirements with minimum expenditure of energy?

*32. (a) Sketch the constraint set for the following linear programming problem.
Maximize

$$f = 2x_1 + 3x_2$$

subject to

$$2x_1 + 5x_2 \le 10$$
$$3x_1 + 4x_2 \le 12$$
$$x_1 \ge 0, x_2 \ge 0.$$

(b) Sketch the constraint set for the following problem.
Minimize

$$g = 10y_1 + 12y_2$$

subject to

$$2y_1 + 3y_2 \ge 2$$
$$5y_1 + 4y_2 \ge 3$$
$$y_1 \ge 0, y_2 \ge 0.$$

(c) Show, using the graphical technique, that the maximum value of f in (a) is equal to the minimum value of g in (b). [*Note*: Parts (a) and (b) are called **dual** problems. We shall discuss dual problems in Section 4.5].

Problems 33–44 are taken from recent CPA exams.[†] The date of the exam from which each problem is taken is given in parentheses before the statement of the problem.

33. (November 1977) The Hale Company manufactures products A and B, each of which requires two processes, polishing and grinding. The contribution margin is $3 for product A and $4 for product B. The graph shows the maximum number of units of each product that may be processed in the two departments. Considering the constraints (restrictions) on processing, which combination of products A and B maximizes the total contribution margin?

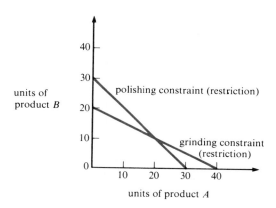

(a)   0 units of A and 20 units of B.   (c)   30 units of A and 0 units of B.

(b)   20 units of A and 10 units of B.   (d)   40 units of A and 0 units of B.

34. (November 1977) The Sanch Company plans to expand its sales force by opening several new branch offices. Sanch has $5,200,000 in capital available for new branch offices. Sanch will consider opening only two types of branches; 10-person branches (type A) and 5-person branches (type B). Expected initial cash outlays are $650,000 for a type A branch and $335,000 for a type B branch. Expected annual cash inflow, net of income taxes, is $46,000 for a type A branch and $18,000 for a type B branch. Sanch will hire no more than 100 employees for the new branch offices and will not open more than 10 branch offices. Linear programming will be used to help decide how many branch offices should be opened.

   In a system of equations for a linear programming model, which of the following equations would **not** represent a constraint (restriction)?

(a) $A + B \leq 10$ (c) $\$46{,}000A + \$18{,}000B \leq \$64{,}000$

(b) $10A + 5B \leq 100$ (d) $\$650{,}000A + \$335{,}000B \leq \$5{,}200{,}000$

35. (November 1975) Patsy, Inc., manufactures two products, X and Y. Each product must be processed in each of three departments: machining, assembling, and finishing. The hours needed to produce one unit of product per department and the maximum possible hours per department follow:

[†] Material from Uniform CPA Examination Questions and Unofficial Answers, copyright © 1975, 1976, 1977, 1978, 1979, 1980 by the American Institute of Certified Public Accountants, Inc., is reprinted with permission.

Department	Production hours per unit		Maximum capacity in hours
	X	Y	
Machining	2	1	420
Assembling	2	2	500
Finishing	2	3	600

Other restrictions are:

$$X \geq 50$$

$$Y \geq 50.$$

The objective function is to maximize profits where profit $= \$4X + \$2Y$. Given the objective and constraints, what is the most profitable number of units of X and Y, respectively, to manufacture?

(a) 150 and 100

(b) 165 and 90

(c) 170 and 80

(d) 200 and 50

36. (May 1978)   Hayes Company manufactures two models, standard and deluxe. Each product must be processed in each of two departments, grinding and finishing. The standard model requires 2 hours of grinding and 3 hours of finishing. The deluxe model requires 3 hours of grinding and 4 hours of finishing. The contribution margin is $3.50 for the standard model and $5.00 for the deluxe model. Hayes has four grinding machines and five finishing machines, which run 16 hours a day for 6 days a week. How would the restriction (constraint) for the finishing department be expressed?

(a) $3X + 4Y = 5$

(b) $3X + 4Y \leq 480$

(c) $\$3.50X + \$5.00Y = 5$

(d) $\$3.50(3)X + \$5.00(4)Y \leq 480$

Problems 37 and 38 are based on the following information.

Milligan Company manufactures two models, small and large. Each model is processed as follows.

	Machining department	Polishing department
Small (X)	2 hours	1 hour
Large (Y)	4 hours	3 hours

The available time for processing the two models is 100 hours a week in the machining department and 90 hours a week in the polishing department. The contribution margin expected is $5 for the small model and $7 for the large model.

37. (November 1978)   How would the objective function (maximization of total contribution margin) be expressed?

(a) $5X + 7Y$

(b) $5X + 7Y \leq 190$

(c) $5X(3) + 7Y(7) \leq 190$

(d) $12X + 10Y$

38. (November 1978)   How would the restriction (constraint) for the machining department be expressed?

(a) $2(5X) + 4(7Y) \leq 100$

(b) $2X + 4Y$

(c) $2X + 4Y \leq 100$

(d) $5X + 7Y \leq 100$

39. (November 1979)   Milford Company manufactures two models, medium and large. The contribution margin expected is $12 for the medium model and $20 for the large model. The medium model is processed 2 hours in the machining department and 4 hours in the polishing department. The large model is processed 3 hours in the machining department and 6 hours in the polishing department. How would the formula for determining the maximization of total contribution margin be expressed?

(a) $5X + 10Y$

(b) $6X + 9Y$

(c) $12X + 20Y$

(d) $12X(2 + 4) + 20Y(3 + 6)$

40. (November 1979)   A company manufactures two models, X and Y. Model X is processed 4 hours in the machining department and 2 hours in the polishing department. Model Y is processed 9 hours in the machining department and 6 hours in the polishing department. The available time for processing the two models is 200 hours a week in the machining department and 180 hours a week in the polishing department. The contribution margins expected are $10 for model X and $14 for model Y. How would the restriction (constraint) for the polishing department be expressed?

(a) $2X + 6Y \leq 180$

(b) $6X + 15Y \leq 180$

(c) $2(10X) + 6(14Y) \leq 180$

(d) $10X + 14Y \leq 180$

41. (May 1979)   The Pauley Company plans to expand its sales force by opening several new branch offices. Pauley has $10,400,000 in capital available for new branch offices. Pauley will consider opening only two types of branches; 20-person branches (type A) and 10-person branches (type B). Expected initial cash outlays are $1,300,000 for a type A branch and $670,000 for a type B branch. Expected annual cash inflow, net of income taxes, is $92,000 for a type A branch and $36,000 for a type B branch. Pauley will hire no more than 200 employees for the new branch offices and will not open more than 20 branch offices. Linear programming will be used to help decide how many branch offices should be opened.

In a system of equations for a linear programming model, which of the following equations would **not** represent a constraint (restriction)?

(a) $A + B \leq 20$

(b) $20A + 10B \leq 200$

(c) $\$92,000A + \$36,000B \leq \$128,000$

(d) $\$1,300,000A + \$670,000B \leq \$10,400,000$

Problems 42–44 are based on the following information.

The Random Company manufactures two products, Zeta and Beta. Each product must pass through two processing operations. All materials are introduced at the start of process 1. There are *no* work-in-process inventories. Random may produce either one product exclusively or various combinations of both products subject to the following constraints.

	Process 1	Process 2	Contribution margin per unit
Hours required to produce 1 unit of:			
Zeta	1 hour	1 hour	$4.00
Beta	2 hours	3 hours	5.25
Total capacity in hours per day	1000 hours	1275 hours	

A shortage of technical labor has limited Beta production to 400 units per day. There are *no* constraints on the production of Zeta other than the hour constraints in the above schedule. Assume that all relationships between capacity and production are linear and that all of the above data and relationships are deterministic rather than probabilistic.

42. (May 1975)   Given the objective to maximize total contribution margin, what is the production constraint for process 1?

(a) Zeta + Beta $\leq$ 1000

(b) Zeta + 2 Beta $\leq$ 1000

(c) Zeta + Beta $\geq$ 1000

(d) Zeta + 2 Beta $\geq$ 1000

*43. (May 1975)   Given the objective to maximize total contribution margin, what is the labor constraint for production of Beta?

(a) Beta $\leq$ 400

(b) Beta $\geq$ 400

(c) Beta $\leq$ 425

(d) Beta $\geq$ 425

44. (May 1975)   What is the objective function of the data presented?

   (a)   Zeta + 2 Beta = $9.25

   (b) $4.00 Zeta + 3($5.25) Beta = total contribution margin

   (c) $4.00 Zeta + $5.25 Beta = total contribution margin

   (d)   2($4.00) Zeta + 3($5.25) Beta = total contribution margin

45. Show that the following problems are unbounded.

   (a)   Maximize

$$f = x + 3y$$

   subject to

$$x + 2y \geq 3$$
$$4x - y \leq 6$$
$$x \geq 0, y \geq 0.$$

   (b)   Maximize

$$f = x_1 + x_2 + 2x_3$$

   subject to

$$x_1 + x_2 + x_3 \geq 2$$
$$x_1 - x_2 + x_3 \leq 8$$
$$x_1 \geq 0, x_2 \geq 0, x_3 \geq 0.$$

46. Show that the following problems are infeasible.

   (a)   Maximize

$$f = 2x + 7y$$

   subject to

$$2x + 5y \leq 8$$
$$4x + 6y \geq 11$$
$$x \geq 0, y \geq 0.$$

   (b)   Maximize

$$w = 4x_1 - x_2 + 9x_3$$

   subject to

$$2x_1 + 3x_2 + x_3 \leq 8$$
$$4x_1 + x_2 + 2x_3 \leq 6$$
$$8x_1 + 7x_2 + 4x_3 \geq 25$$
$$x_1 \geq 0, x_2 \geq 0, x_3 \geq 0.$$

We can also define a corner point in a linear programming problem with three variables. Suppose the constraint set is the set of 3-vectors whose coordinates satisfy a number of linear inequalities in three variables. A **possible corner point** is any solution to exactly *three* of the linear equations obtained by changing the inequalities to equations. A **feasible solution (actual corner point)** is a possible corner point that satisfies all the inequalities. It can be shown that the maximum or minimum of a linear function over the constraint set occurs at a corner point. If there are four variables instead of three, then a possible corner point is any solution to exactly four of the linear equations, and so on.

*47. Find all the corner points of the constraint set determined by the inequalities

$$x_1 + x_2 + x_3 \leq 15$$
$$2x_1 + x_2 + 2x_3 \leq 26$$
$$5x_1 + 2x_2 + 3x_3 \leq 43$$
$$x_1 \geq 0$$
$$x_2 \geq 0$$
$$x_3 \geq 0$$

[*Hint:* There are 20 ways to choose three equations from the six equations given above. Each solution to the system of three equations in three unknowns is a possible corner point. Each possible corner point must be tested to see if it is an actual corner point.]

*48. Find the maximum and minimum values of the function $f(x_1, x_2, x_3) = 3x_1 - 2x_2 + 2x_3$ subject to the constraints of Problem 47.

*49. A classic problem of linear programming is the *diet problem*. The goal is to determine the quantities of certain foods that meet certain nutritional needs at a minimum cost.

For simplicity, we limit ourselves to three foods, milk, beef, and eggs, and three vitamins, A, C, and D. The data for this problem are given in Table 8.

TABLE 8   Quantities of Vitamins in Milligrams (mg)

Vitamin	1 gal milk	1 lb beef	1 dozen eggs	Minimum daily requirements
A	1	1	10	1 mg
C	100	10	10	50 mg
D	10	100	10	10 mg
Cost	$2.00	$2.50	$0.80	

(a) Let x_1 denote the number of gallons of milk, x_2 the number of pounds of beef, and x_3 the number of dozen eggs consumed daily. Write the minimum problem of linear programming in three variables whose solution is the minimum cost. Your constraint set should have six inequalities.

(b) Find the 20 possible corner points of this constraint set.

(c) Find the feasible solutions. [*Hint*: There are nine of them].

(d) Evaluate the cost at each feasible solution.

(e) What is the minimum cost and how is it achieved?

*50. In a large hospital, surgical operations are classified into three categories according to their average times of 30 minutes, 1 hour, and 2 hours. The hospital receives a fee of $100, $150, or $200 for an operation in categories I, II, or III, respectively. If the hospital has eight operating rooms which are in use an average of 10 hours per day, how many operations of each type should the hospital schedule in order to (a) maximize its revenue and (b) maximize the total number of operations?

*51. A company producing canned mixed fruit has a stock of 10,000 lb of pears, 12,000 lb of peaches, and 8000 lb of cherries. The company produces three fruit mixtures, which it sells in 1-lb cans. The first mixture is half pears and half peaches and sells for 30¢. The second mixture has equal amounts of the three fruits and sells for 40¢. The third mixture is half peaches and half cherries and sells for 50¢. How many cans of each mixture should be produced to maximize the return?

4.3 Slack Variables

As we saw in the last section, linear programming problems involve a number of linear inequalities. In the method discussed in the next section for solving these problems, it is necessary to change inequalities to equations.

Example 1 Consider the linear inequality

$$2x_1 + 5x_2 \leq 40 \tag{1}$$

Inequality (1) states that either $2x_1 + 5x_2$ is less than 40 or it is equal to 40. If it is less than 40, then there is some "slack" in the inequality. Let us denote this slack by s_1. Then s_1 is the difference between 40 and the sum $2x_1 + 5x_2$; that is, by our definition of s_1,

$$s_1 = 40 - 2x_1 - 5x_2 \tag{2}$$

or, rewriting equation (2),

$$2x_1 + 5x_2 + s_1 = 40 \tag{3}$$

Notice that by introducing a new variable, we have transformed inequality (1) into equation (3). Note also that either $s_1 = 0$ (if $2x_1 + 5x_2 = 40$) or $s_1 > 0$ (if $2x_1 + 5x_2 < 40$) so that, in either case, we have

$$s \geq 0 \tag{4}$$

The variable s_1 is called a *slack variable*.

Slack Variable

Consider the linear inequality in n variables

$$a_{11}x_1 + a_{12}x_2 + \cdots + a_{1n}x_n \leq b_1 \tag{5}$$

Then the **slack variable** s_1 is defined by

$$s_1 = b_1 - a_{11}x_1 - a_{12}x_2 - \cdots - a_{1n}x_n$$

so that inequality (5) is equivalent to

$$\boxed{\begin{aligned} a_{11}x_1 + a_{12}x_2 + \cdots + a_{1n}x_n + s_1 &= b_1 \\ s_1 &\geq 0 \end{aligned}} \tag{6}$$

Example 2 Write the following system of linear inequalities as a system of linear equations by using slack variables:

$$3x_1 + 2x_2 + 5x_3 \leq 10$$
$$6x_1 + 8x_2 + 12x_3 \leq 250. \tag{7}$$

Solution Let

$$s_1 = 10 - 3x_1 - 2x_2 - 5x_3 \tag{8}$$

and

$$s_2 = 250 - 6x_1 - 8x_2 - 12x_3. \tag{9}$$

Then, because of (7), we find that $s_1 \geq 0$ and $s_2 \geq 0$. Finally, rewriting (8) and (9), we obtain the system

$$3x_1 + 2x_2 + 5x_3 + s_1 \qquad = 10$$
$$6x_1 + 8x_2 + 12x_3 \qquad + s_2 = 250$$
$$s_1 \geq 0, \ s_2 \geq 0$$

Consider the system of equations

$$x_1 + 2x_2 + 3x_3 + 5x_4 = 10 \tag{10}$$
$$2x_1 + 7x_2 + 12x_3 + x_4 = 44$$

Let us solve the system by row reduction.

Recall that $A_{1,2}(-2)$ means that we multiply the first row by -2 and add it to the second row. Also, $M_2(\frac{1}{3})$ means that we multiply the second row by $\frac{1}{3}$.

$$\begin{pmatrix} 1 & 2 & 3 & 5 & | & 10 \\ 2 & 7 & 12 & 1 & | & 44 \end{pmatrix} \xrightarrow{A_{1,2}(-2)} \begin{pmatrix} 1 & 2 & 3 & 5 & | & 10 \\ 0 & 3 & 6 & -9 & | & 24 \end{pmatrix}$$

$$\xrightarrow{M_2(\frac{1}{3})} \begin{pmatrix} 1 & 2 & 3 & 5 & | & 10 \\ 0 & 1 & 2 & -3 & | & 8 \end{pmatrix}$$

$$\xrightarrow{A_{2,1}(-2)} \begin{pmatrix} 1 & 0 & -1 & 11 & | & -6 \\ 0 & 1 & 2 & -3 & | & 8 \end{pmatrix}$$

This is as far as we can go. The equations now read (from the last augmented matrix)

$$x_1 \qquad - x_3 + 11x_4 = -6$$
$$x_2 + 2x_3 - 3x_4 = 8$$

We can write the infinite number of solutions to this system as

$$x_1 = -6 + x_3 - 11x_4 \tag{11}$$
$$x_2 = 8 - 2x_3 + 3x_4 \tag{12}$$

$$x_3, x_4 \text{ arbitrary.}$$

In this form we say that the variables x_1 and x_2 are **basic variables** and the variables x_3 and x_4 are **nonbasic variables**. That is, the solutions (11) and (12) to system (10) are given in such a way that the basic variables are written in terms of the nonbasic variables.

Note. One solution to (10) can be obtained immediately from (11) and (12) by setting the nonbasic variables x_3 and x_4 equal to zero; if $x_3 = x_4 = 0$, then $x_1 = -6$, $x_2 = 8$, and a solution is

$$(-6, 8, 0, 0).$$

Example 3 Represent the solution to system (10) with basic variables x_2 and x_3 and nonbasic variables x_1 and x_4.

Solution   The problem is to express x_2 and x_3 in terms of x_1 and x_4. From (11),

$$x_3 = 6 + x_1 + 11x_4. \qquad (13)$$

From (12),

$$x_2 = 8 - 2x_3 + 3x_4 \overset{\text{Using (13)}}{=} 8 - 2(6 + x_1 + 11x_4) + 3x_4$$
$$= 8 - 12 - 2x_1 - 22x_4 + 3x_4 = -4 - 2x_1 - 19x_4.$$

Thus the solutions to (10) can be written

$$\begin{aligned} x_2 &= -4 - 2x_1 - 19x_4 \\ x_3 &= 6 + x_1 + 11x_4 \\ &\quad x_1, x_4 \text{ arbitrary.} \end{aligned} \qquad (14)$$

We can obtain this answer in another way. We rewrite system (10) with x_2 and x_3 appearing first.

$$2x_2 + 3x_3 + x_1 + 5x_4 = 10$$
$$7x_2 + 12x_3 + 2x_1 + x_4 = 44$$

Then we row-reduce.

$$\begin{pmatrix} 2 & 3 & 1 & 5 & | & 10 \\ 7 & 12 & 2 & 1 & | & 44 \end{pmatrix} \xrightarrow{M_1(\frac{1}{2})} \begin{pmatrix} 1 & \frac{3}{2} & \frac{1}{2} & \frac{5}{2} & | & 5 \\ 7 & 12 & 2 & 1 & | & 44 \end{pmatrix}$$

$$\xrightarrow{A_{1,2}(-7)} \begin{pmatrix} 1 & \frac{3}{2} & \frac{1}{2} & \frac{5}{2} & | & 5 \\ 0 & \frac{3}{2} & -\frac{3}{2} & -\frac{33}{2} & | & 9 \end{pmatrix}$$

$$\xrightarrow{M_2(\frac{2}{3})} \begin{pmatrix} 1 & \frac{3}{2} & \frac{1}{2} & \frac{5}{2} & | & 5 \\ 0 & 1 & -1 & -11 & | & 6 \end{pmatrix}$$

$$\xrightarrow{A_{2,1}(-\frac{3}{2})} \begin{pmatrix} 1 & 0 & 2 & 19 & | & -4 \\ 0 & 1 & -1 & -11 & | & 6 \end{pmatrix}$$

The last system reads (remember x_2 and x_3 are written first)

$$x_2 + 2x_1 + 19x_4 = -4$$
$$x_3 - x_1 - 11x_4 = 6$$

or

$$x_2 = -4 - 2x_1 - 19x_4 \qquad (15)$$
$$x_3 = 6 + x_1 + 11x_4. \qquad (16)$$

This is system (14).

Note.   We can obtain another solution to system (10) by setting the new nonbasic variables equal to zero; if $x_1 = x_4 = 0$ then, from (15) and (16), $x_2 = -4$ and $x_3 = 6$. A solution to (10) is

$$(0, -4, 6, 0)$$

Basic and Nonbasic Variables   Suppose that a system of m equations in n unknowns, $n > m$, has an infinite number of solutions. Suppose also that $n - m$ of the variables x_{m+1}, $x_{m+2}, \ldots, x_n$ can be chosen arbitrarily and that the remaining variables $x_1, x_2, \ldots, x_m$ can be written in terms of $x_{m+1}, x_{m+2}, \ldots, x_n$. Then $x_1, x_2, \ldots, x_m$ are called **basic variables** and $x_{m+1}, x_{m+2}, \ldots, x_n$ are called **nonbasic variables**.

Example 4   Consider the system of linear inequalities

$$x_1 + 2x_2 + 4x_3 \le 10$$
$$2x_1 + 5x_2 + 6x_3 \le 30$$
$$x_1 + 3x_2 + 4x_3 \le 50.$$

(a)   Write this system as a system of linear equations by introducing slack variables.

(b)   Solve the resulting system with x_1, x_2, and x_3 as the basic variables and the slack variables as the nonbasic variables.

Solution

(a)   Defining the slack variables s_1, s_2, and s_3 as we have done earlier, we have

$$
\begin{aligned}
x_1 + 2x_2 + 4x_3 + s_1 \qquad\qquad &= 10 \\
2x_1 + 5x_2 + 6x_3 \qquad + s_2 \qquad &= 30 \\
x_1 + 3x_2 + 4x_3 \qquad\qquad + s_3 &= 50
\end{aligned}
\tag{17}
$$

$$s_1, s_2, s_3 \ge 0$$

(b)   We solve this system of three equations ($m = 3$) in six unknowns ($n = 6$) by row reduction.

$$
\begin{pmatrix}
1 & 2 & 4 & 1 & 0 & 0 & \vline & 10 \\
2 & 5 & 6 & 0 & 1 & 0 & \vline & 30 \\
1 & 3 & 4 & 0 & 0 & 1 & \vline & 50
\end{pmatrix}
\xrightarrow[A_{1,3}(-1)]{A_{1,2}(-2)}
\begin{pmatrix}
1 & 2 & 4 & 1 & 0 & 0 & \vline & 10 \\
0 & 1 & -2 & -2 & 1 & 0 & \vline & 10 \\
0 & 1 & 0 & -1 & 0 & 1 & \vline & 40
\end{pmatrix}
$$

$$
\xrightarrow[A_{2,3}(-1)]{A_{2,1}(-2)}
\begin{pmatrix}
1 & 0 & 8 & 5 & -2 & 0 & \vline & -10 \\
0 & 1 & -2 & -2 & 1 & 0 & \vline & 10 \\
0 & 0 & 2 & 1 & -1 & 1 & \vline & 30
\end{pmatrix}
$$

$$
\xrightarrow{M_3(\frac{1}{2})}
\begin{pmatrix}
1 & 0 & 8 & 5 & -2 & 0 & \vline & -10 \\
0 & 1 & -2 & -2 & 1 & 0 & \vline & 10 \\
0 & 0 & 1 & \frac{1}{2} & -\frac{1}{2} & \frac{1}{2} & \vline & 15
\end{pmatrix}
$$

$$
\xrightarrow[A_{3,2}(2)]{A_{3,1}(-8)}
\begin{pmatrix}
1 & 0 & 0 & 1 & 2 & -4 & \vline & -130 \\
0 & 1 & 0 & -1 & 0 & 1 & \vline & 40 \\
0 & 0 & 1 & \frac{1}{2} & -\frac{1}{2} & \frac{1}{2} & \vline & 15
\end{pmatrix}
$$

The coefficient matrix is now in reduced row-echelon form and the system can be written

$$x_1 \qquad + s_1 + 2s_2 - 4s_3 = -130$$
$$x_2 \quad - s_1 \qquad + s_3 = \qquad 40$$
$$x_3 + \tfrac{1}{2}s_1 - \tfrac{1}{2}s_2 + \tfrac{1}{2}s_3 = \qquad 15$$

or

$$x_1 = -130 - s_1 - 2s_2 + 4s_3$$
$$x_2 = \quad 40 + s_1 - s_3$$
$$x_3 = \quad 15 - \tfrac{1}{2}s_1 + \tfrac{1}{2}s_2 - \tfrac{1}{2}s_3.$$

As in Example 3, we can obtain a solution to system (17) by setting the nonbasic variables to zero; if $s_1 = s_2 = s_3 = 0$, then $x_1 = -130$, $x_2 = 40$, and $x_3 = 15$, and a solution to (17) is

$$(-130, 40, 15, 0, 0, 0).$$

In the next section we will describe a new and far more efficient method for solving linear programming problems. The first step of this method involves the introduction of slack variables, and subsequent steps involve changes in the basic and nonbasic variables of the problem.

PROBLEMS 4.3

In Problems 1–5, write the system of linear inequalities as a system of linear equations by introducing slack variables.

1. $x_1 + x_2 \le 3$
$2x_1 + x_2 \le 7$

2. $3x_1 + x_2 - x_3 \le 4$
$2x_1 + x_2 + x_3 \le 6$

3. $2x_1 + x_2 \le 10$
$3x_1 + 2x_2 \le 30$
$4x_1 + 7x_2 \le 20$

4. $3x_1 + x_2 + 2x_3 \le 15$
$2x_1 + 3x_2 + 7x_3 \le 12$
$4x_1 + 8x_2 + 5x_2 \le 8$

5. $7x_1 + x_2 + 3x_3 + x_4 \le 8$
$3x_1 + 2x_2 + 5x_3 + 12x_4 \le 12$
$2x_1 + 5x_2 + 8x_3 + 2x_4 \le 9$

6. Consider the system

$$x_1 + 2x_2 \le 5$$
$$3x_1 + 7x_2 \le 20.$$

(a) Write this as a system of linear equations by defining slack variables s_1 and s_2.

(b) Write all solutions to this linear system with x_1 and x_2 the basic variables and s_1 and s_2 the nonbasic variables.

7. In Problem 6(b), write the solutions with x_1 and s_1 basic and x_2 and s_2 nonbasic.

8. In Problem 6(b), write the solutions with x_2 and s_1 basic and x_1 and s_2 nonbasic.

9. Answer the questions of Problem 6 for the system

$$2x_1 + 5x_2 \leq 12$$
$$4x_1 + 9x_2 \leq 20$$

10. In Problem 9, write the solutions with x_1 and s_2 basic and x_2 and s_1 nonbasic.

In Problems 11–20, (a) write the system of linear inequalities as a system of linear equations by introducing slack variables. (b) Solve the system by writing the indicated basic variables in terms of the nonbasic variables.

11. $x_1 + 2x_2 + x_3 \leq 8$
 $2x_1 + 5x_2 + 5x_3 \leq 35$
 Basic: x_1, x_2
 Nonbasic: s_1, s_2, x_3

12. $x_1 + 2x_2 + x_3 \leq 8$
 $2x_1 + 5x_2 + 5x_3 \leq 35$
 Basic: x_1, s_1
 Nonbasic: x_2, x_3, s_2

13. $x_1 + 2x_2 + x_3 \leq 8$
 $2x_1 + 5x_2 + 5x_3 \leq 35$
 Basic: x_2, x_3
 Nonbasic: x_1, s_1, s_2

14. $x_1 + 2x_2 + x_3 \leq 8$
 $2x_1 + 5x_2 + 5x_3 \leq 35$
 Basic: s_1, s_2
 Nonbasic: x_1, x_2, x_3

15. $x_1 + 3x_2 \leq 5$
 $2x_1 + 7x_2 \leq 20$
 $3x_1 + 8x_2 \leq 40$
 Basic: x_1, x_2, s_1
 Nonbasic: s_2, s_3

16. $x_1 + 3x_2 \leq 5$
 $2x_1 + 7x_2 \leq 20$
 $3x_1 + 8x_2 \leq 40$
 Basic: x_2, s_2, s_3
 Nonbasic: x_1, s_1

17. $2x_1 + 4x_2 + 8x_3 \leq 12$
 $2x_1 + 5x_2 + 12x_3 \leq 25$
 $3x_1 + 6x_2 + 13x_3 \leq 60$
 Basic: s_1, s_2, s_3
 Nonbasic: x_1, x_2, x_3

18. $2x_1 + 4x_2 + 8x_3 \leq 12$
 $2x_1 + 5x_2 + 12x_3 \leq 25$
 $3x_1 + 6x_2 + 13x_3 \leq 60$
 Basic: x_1, x_2, x_3
 Nonbasic: s_1, s_2, s_3

19. $2x_1 + 4x_2 + 8x_3 \leq 12$
 $2x_1 + 5x_2 + 12x_3 \leq 25$
 $3x_1 + 6x_2 + 13x_3 \leq 60$
 Basic: x_1, s_1, s_3
 Nonbasic: x_2, x_3, s_2

20. $2x_1 + 4x_2 + 8x_3 \leq 12$
 $2x_1 + 5x_2 + 12x_3 \leq 25$
 $3x_1 + 6x_2 + 13x_3 \leq 60$
 Basic: x_1, x_3, s_1
 Nonbasic: x_2, s_2, s_3

21. (May 1976 CPA exam) In a system of equations for a linear programming model, what can be done to equalize an equality such as $3x + 2y \leq 15$?

 (a) Nothing.

 (b) Add a slack variable.

 (c) Add a tableau.

 (d) Multiply each element by -1.

4.4 The Simplex Method I: The Standard Maximizing Problem

The corner-point method discussed in Section 4.2 can be extremely tedious if the number of variables and constraints is large. In this section we describe a method for solving linear programming problems much more efficiently. The method is called the **simplex method**. It was developed in 1947 by the American mathematician George B. Dantzig. In 1976 President Gerald Ford awarded Dantzig the National Medal of Science, the nation's highest award in science. At the White House ceremony Dantzig was cited "for inventing linear programming and discovering methods that led to wide-scale scientific and technical applications to important problems in logistics, scheduling, and network optimization, and to the use of computers in making efficient use of mathematical theory."

To give you some notion of the idea behind the simplex method, look at the constraint set sketched in Figure 1.[†]

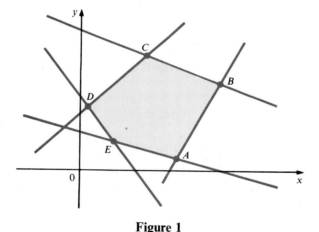

Figure 1

Suppose we wish to maximize an objective function f over this constraint set. We start by finding one corner point, say the one labeled A. Any line segment joining two corner points is called an **edge** of the constraint set (or simplex). Dantzig was able to prove the following important fact.

> If f does not take its maximum value at the corner point A,
> then there is an edge starting at A along which f increases.

[†] This (shaded) set in the plane is also called a **simplex**. Simplices exist in three or more dimensions, but we shall not attempt to sketch them.

This simply stated result gives the main idea behind the simplex method. We start at A, and if f does not take its maximum at A, then we find an edge of the constraint set along which f increases. This brings us to a new corner point. If f takes its maximum there, we are done. If not, we continue to follow another edge along which f increases to reach another corner point. Since there are a finite number of corner points, an optimal solution will be reached after a finite number of steps. We will never return to the same corner point twice, since the objective function increases at each step.

This method avoids the necessity of determining all corner points. Another great advantage of the simplex method is that its steps can be efficiently performed by a computer.

This all sounds easy, but a process for finding the "next" corner point and evaluating f at that point is not at all obvious. The simplex method is a method for obtaining this next corner point. It involves a number of technical steps. Before we describe these steps, we describe the type of problem that can be solved by the simplex method.

STANDARD MAXIMIZING PROBLEM OF LINEAR PROGRAMMING

Find the n-vector $\mathbf{x} = (x_1, x_2, \ldots, x_n)$ that maximizes the object function

$$f = c_1 x_1 + c_2 x_2 + \cdots + c_n x_n \qquad (1)$$

and satisfies the $m + n$ linear inequalities

$$
\begin{aligned}
a_{11} x_1 + a_{12} x_2 + \cdots + a_{1n} x_n &\leq b_1 \\
a_{21} x_1 + a_{22} x_2 + \cdots + a_{2n} x_n &\leq b_2 \\
\vdots \qquad \vdots \qquad\qquad \vdots \quad\ \ \vdots \\
a_{m1} x_1 + a_{m2} x_2 + \cdots + a_{mn} x_n &\leq b_m \\
\end{aligned}
\qquad (2)
$$

$$x_1 \geq 0,\ x_2 \geq 0, \ldots, x_n \geq 0.$$

Here $b_1 \geq 0,\ b_2 \geq 0, \ldots b_m \geq 0$

To solve the problem in (1) and (2), we introduce the slack variables $s_1, s_2, \ldots, s_m$. Then the problem becomes:
Maximize

$$f = c_1 x_1 + c_2 x_2 + \cdots + c_n x_n + 0 \cdot s_1 + 0 \cdot s_2 + \cdots + 0 \cdot s_m \qquad (3)$$

subject to

$$
\begin{aligned}
a_{11}x_1 + a_{12}x_2 + \cdots + a_{1n}x_n + s_1 \quad\quad\quad\quad &= b_1 \\
a_{21}x_1 + a_{22}x_2 + \cdots + a_{2n}x_n \quad\quad + s_2 \quad\quad &= b_2 \\
\vdots \quad\quad \vdots \quad\quad\quad \vdots \quad\quad\quad\quad\quad \vdots \quad\quad\quad & \quad (4) \\
a_{m1}x_1 + a_{m2}x_2 + \cdots + a_{mn}x_n \quad\quad\quad\quad + s_m &= b_m \\
x_1 \ge 0,\, x_2 \ge 0, \ldots, x_n \ge 0,\, s_1 \ge 0,\, s_2 \ge 0, \ldots, s_m &\ge 0.
\end{aligned}
$$

Initially, we can solve for $s_1, s_2, \ldots, s_m$ in terms of $x_1, x_2, \ldots, x_n$, so our basic variables are, initially, $s_1, s_2, \ldots, s_m$ and our nonbasic variables are $x_1, x_2, \ldots, x_n$. We make the assumption that all the b's are nonnegative; thus setting the nonbasic variables equal to zero in (4) gives us our initial corner point $(0, 0, \ldots, 0)$. If one or more of the b's is negative, then techniques exist for solving the problem, but we will not discuss these techniques here. Note that $f = 0$ at $(0, 0, \ldots, 0)$.

Remark. The only requirement for a corner point if the equality constraints are satisfied is that all variables have nonnegative values.

INITIAL SIMPLEX TABLEAU

A convenient way to write the information contained in this problem is to define the **initial simplex tableau**.

x_1	x_2	$\cdots$	x_n	s_1	s_2	$\cdots$	s_m		
a_{11}	a_{12}	$\cdots$	a_{1n}	1	0	$\cdots$	0	b_1	s_1
a_{21}	a_{22}	$\cdots$	a_{2n}	0	1	$\cdots$	0	b_2	s_2
$\vdots$	$\vdots$		$\vdots$	$\vdots$	$\vdots$		$\vdots$	$\vdots$	$\vdots$
a_{m1}	a_{m2}	$\cdots$	a_{mn}	0	0	$\cdots$	1	b_m	s_m
c_1	c_2	$\cdots$	c_n	0	0	$\cdots$	0	f	

These are, initially, the basic variables.

$$(5)$$

The variables on the right side indicate that $s_1, s_2, \ldots, s_n$ *are the basic variables*. All other entries conform exactly to the equations they replace. The first line, for example, reads

$$a_{11}x_1 + a_{12}x_2 + \cdots + a_{1n}x_n + s_1 + 0 \cdot s_2 + \cdots + 0 \cdot s_m = b_1.$$

Indicators

The numbers in the bottom row of the simplex tableau (except for the entry in the lower right-hand corner) are called **indicators** of the tableau.

The object of the simplex method is to perform elementary row operations on the rows of (5) so that the indicators all become nonpositive. As we will soon see, this will indicate that we have found the maximum value for f. To achieve this, we work with tableau (5) to eliminate, successively, all positive indicators.

THE SIMPLEX METHOD—CHOOSING A PIVOT

> **Step 1.** First, choose any column in (5) with a positive indicator (if there is more than one positive indicator, select any one) and consider all *positive* components in that column. Suppose the jth column is chosen.
>
> **Step 2.** Then define a **pivot** of this column to be a positive component a_{ij} in the jth column such that b_i/a_{ij} is a minimum for all such a_{ij}. If this minimum is not unique, choose any element in the jth column giving the minimum quotient.

Example 1   Find all pivots of the given initial simplex tableau.

These are the b_i's.

x_1	x_2	x_3	x_4	s_1	s_2		
2	−1	4	6	1	0	1	s_1
3	3	2	7	0	1	2	s_2
1	1	$\frac{1}{2}$	−2	0	0	f	

These are positive indicators.

Solution   First we note that there are three positive indicators, so there are at least three pivots. We find these one at a time.
   Column 1.   There are two positive components, so we form the ratios

$$\frac{b_1}{a_{11}} = \frac{1}{2} \quad \text{and} \quad \frac{b_2}{a_{21}} = \frac{2}{3}.$$

Since $\frac{1}{2} < \frac{2}{3}$, the pivot in the first column is 2.
   Column 2.   There is only one positive component in the second column— the number 3. This is the pivot.
   Column 3.   We form the ratios

$$\frac{b_1}{a_{13}} = \frac{1}{4} \quad \text{and} \quad \frac{b_2}{a_{23}} = \frac{2}{2} = 1.$$

Since $\frac{1}{4} < 1$, the pivot is 4.

The tableau is drawn below with the 3 pivots circled.

x_1	x_2	x_3	x_4	s_1	s_2	
②	-1	④	6	1	0	1	s_1
3	③	2	7	0	1	2	s_2
1	1	$\frac{1}{2}$	-2	0	0	f

Example 2 Find all pivots of the given initial simplex tableau.

x_1	x_2	x_3	x_4	x_5	s_1	s_2	s_3	s_4	These are the b_j's.
1	1	3	4	4	1	0	0	0	2	s_1
0	2	1	2	1	0	1	0	0	2	s_2
-1	5	0	1	2	0	0	1	0	1	s_3
1	-4	1	3	2	0	0	0	1	3	s_4
1	1	-1	0	2	0	0	0	0	f

These are the positive indicators.

Solution   There are positive indicators in columns 1, 2, and 5.
   Column 1.   There are two positive entries in this column. We form the ratios

$$\frac{b_1}{a_{11}} = \frac{2}{1} = 2 \quad \text{and} \quad \frac{b_4}{a_{41}} = \frac{3}{1} = 3.$$

Since $2 < 3$, the pivot is 1.
   Column 2.   There are three positive components in this column. We form the ratios

$$\frac{b_1}{a_{12}} = \frac{2}{1} = 2, \qquad \frac{b_2}{a_{22}} = \frac{2}{2} = 1, \qquad \frac{b_3}{a_{32}} = \frac{1}{5}.$$

Since $\frac{1}{5}$ is the smallest ratio, the pivot is 5.
   Column 5.   There are four positive components in this column. We form the ratios

$$\frac{b_1}{a_{15}} = \frac{2}{4} = \frac{1}{2}, \qquad \frac{b_2}{a_{25}} = \frac{2}{1} = 2, \qquad \frac{b_3}{a_{35}} = \frac{1}{2}, \qquad \frac{b_4}{a_{45}} = \frac{3}{2}.$$

There is a tie. The smallest ratio is $\frac{1}{2}$, and there are two ratios equal to this value. Therefore two pivots in column 5 are 4 and 2. We redraw the tableau with the pivots circled.

x_1	x_2	x_3	x_4	x_5	s_1	s_2	s_3	s_4		
①	1	3	4	④	1	0	0	0	2	s_1
0	2	1	2	1	0	1	0	0	2	s_2
−1	⑤	0	1	②	0	0	1	0	1	s_3
1	−4	1	3	2	0	0	0	1	3	s_4
1	1	−1	0	2	0	0	0	0	f	

THE SIMPLEX METHOD—PIVOTING

> **Step 3.** Once a pivot is chosen, use row reduction to make the pivot equal to 1 and make all other components in the pivot column equal to 0.
>
> **Step 4.** Replace the basic variable on the right-hand side of the pivot row by the nonbasic variable that heads the pivot column.

Remark. Carrying out Steps 3 and 4 is called **pivoting**.

Example 3 In Example 1, we found the circled pivot in the third column. Pivot on this component.

x_1	x_2	x_3	x_4	s_1	s_2		
2	−1	④	6	1	0	1	s_1
3	3	2	7	0	1	2	s_2
1	1	$\frac{1}{2}$	−2	0	0	f	

Solution

Step 3. We must divide the first row by 4 to make the coefficient in the first row and third column equal to 1. Then we use the first row to make zero the numbers 2 and $\frac{1}{2}$ in the third column. In carrying out these steps, we use the same notation we used in the row reduction of Chapter 2.

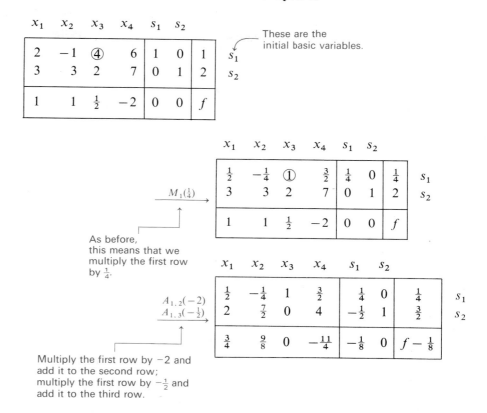

x_1	x_2	x_3	x_4	s_1	s_2	
2	-1	④	6	1	0	1	s_1
3	3	2	7	0	1	2	s_2
1	1	$\frac{1}{2}$	-2	0	0	f

These are the initial basic variables.

$M_1(\frac{1}{4})$

As before, this means that we multiply the first row by $\frac{1}{4}$.

x_1	x_2	x_3	x_4	s_1	s_2	
$\frac{1}{2}$	$-\frac{1}{4}$	①	$\frac{3}{2}$	$\frac{1}{4}$	0	$\frac{1}{4}$	s_1
3	3	2	7	0	1	2	s_2
1	1	$\frac{1}{2}$	-2	0	0	f

$A_{1,2}(-2)$
$A_{1,3}(-\frac{1}{2})$

Multiply the first row by -2 and add it to the second row; multiply the first row by $-\frac{1}{2}$ and add it to the third row.

x_1	x_2	x_3	x_4	s_1	s_2	
$\frac{1}{2}$	$-\frac{1}{4}$	1	$\frac{3}{2}$	$\frac{1}{4}$	0	$\frac{1}{4}$	s_1
2	$\frac{7}{2}$	0	4	$-\frac{1}{2}$	1	$\frac{3}{2}$	s_2
$\frac{3}{4}$	$\frac{9}{8}$	0	$-\frac{11}{4}$	$-\frac{1}{8}$	0	$f - \frac{1}{8}$

Step 4. Since we have pivoted on the first row and the third column, the basic variable in the first row, s_1, becomes nonbasic, and the nonbasic variable heading the third column, x_3, becomes basic. Thus, at the completion of the pivoting operation, the tableau looks like this:

x_1	x_2	x_3	x_4	s_1	s_2	
$\frac{1}{2}$	$-\frac{1}{4}$	1	$\frac{3}{2}$	$\frac{1}{4}$	0	$\frac{1}{4}$	x_3
2	$\frac{7}{2}$	0	4	$-\frac{1}{2}$	1	$\frac{3}{2}$	s_2
$\frac{3}{4}$	$\frac{9}{8}$	0	$-\frac{11}{4}$	$-\frac{1}{8}$	0	$f - \frac{1}{8}$

x_3 ← This is the new basic variable.

Note how x_3 is displayed on the right as our new basic variable. This means that we can write x_3 in terms of the nonbasic variables. To do so, we interpret the first row of the tableau:

$$\tfrac{1}{2}x_1 - \tfrac{1}{4}x_2 + x_3 + \tfrac{3}{2}x_4 + \tfrac{1}{4}s_1 = \tfrac{1}{4}$$

or

$$x_3 = \tfrac{1}{4} - \tfrac{1}{2}x_1 + \tfrac{1}{4}x_2 - \tfrac{3}{2}x_4 - \tfrac{1}{4}s_1.$$

Example 4   Pivot on the circled component in the given tableau.

x_1	x_2	x_2	x_4	x_5	s_1	s_2	s_3	s_4		
1	1	3	4	4	1	0	0	0	2	s_1
0	2	1	2	1	0	1	0	0	2	s_2
−1	⑤	0	1	2	0	0	1	0	1	s_3
1	−4	1	3	2	0	0	0	1	3	s_4
1	1	−1	0	2	0	0	0	0	f	

Solution   We perform the required row operations.

These are the initial basic variables.

x_1	x_2	x_3	x_4	x_5	s_1	s_2	s_3	s_4		
1	1	3	4	4	1	0	0	0	2	s_1
0	2	1	2	1	0	1	0	0	2	s_2
−1	⑤	0	1	2	0	0	1	0	1	s_3
1	−4	1	3	2	0	0	0	1	3	s_4
1	1	−1	0	2	0	0	0	0	f	

$\xrightarrow{M_3(\frac{1}{5})}$

x_1	x_2	x_3	x_4	x_5	s_1	s_2	s_3	s_4		
1	1	3	4	4	1	0	0	0	2	s_1
0	2	1	2	1	0	1	0	0	2	s_2
$-\tfrac{1}{5}$	1	0	$\tfrac{1}{5}$	$\tfrac{2}{5}$	0	0	$\tfrac{1}{5}$	0	$\tfrac{1}{5}$	s_3
1	−4	1	3	2	0	0	0	1	3	s_4
1	1	−1	0	2	0	0	0	0	f	

$\begin{array}{l}A_{3,1}(-1)\\A_{3,2}(-2)\\A_{3,4}(4)\\A_{3,5}(-1)\end{array}\longrightarrow$

x_1	x_2	x_3	x_4	x_5	s_1	s_2	s_3	s_4		
$\tfrac{6}{5}$	0	3	$\tfrac{19}{5}$	$\tfrac{18}{5}$	1	0	$-\tfrac{1}{5}$	0	$\tfrac{9}{5}$	s_1
$\tfrac{2}{5}$	0	1	$\tfrac{8}{5}$	$\tfrac{1}{5}$	0	1	$-\tfrac{2}{5}$	0	$\tfrac{8}{5}$	s_2
$-\tfrac{1}{5}$	1	0	$\tfrac{1}{5}$	$\tfrac{2}{5}$	0	0	$\tfrac{1}{5}$	0	$\tfrac{1}{5}$	s_3
$\tfrac{1}{5}$	0	1	$\tfrac{19}{5}$	$\tfrac{18}{5}$	0	0	$\tfrac{4}{5}$	1	$\tfrac{19}{5}$	s_4
$\tfrac{6}{5}$	0	−1	$-\tfrac{1}{5}$	$\tfrac{8}{5}$	0	0	$-\tfrac{1}{5}$	0	$f-\tfrac{1}{5}$	

Finally, the nonbasic variable x_2 heading the second column replaces the basic variable s_3 in the third row.

x_1	x_2	x_3	x_4	x_5	s_1	s_2	s_3	s_4		
$\frac{6}{5}$	0	3	$\frac{19}{5}$	$\frac{18}{5}$	1	0	$-\frac{1}{5}$	0	$\frac{9}{5}$	s_1
$\frac{2}{5}$	0	1	$\frac{8}{5}$	$\frac{1}{5}$	0	1	$-\frac{2}{5}$	0	$\frac{8}{5}$	s_2
$-\frac{1}{5}$	1	0	$\frac{1}{5}$	$\frac{2}{5}$	0	0	$\frac{1}{5}$	0	$\frac{1}{5}$	x_2 ← This is the new basic variable.
$\frac{1}{5}$	0	1	$\frac{19}{5}$	$\frac{18}{5}$	0	0	$\frac{4}{5}$	1	$\frac{19}{5}$	s_4
$\frac{6}{5}$	0	-1	$\frac{1}{5}$	$\frac{8}{5}$	0	0	$-\frac{1}{5}$	0	$f - \frac{1}{5}$	

> After each pivot operation the variables listed on the right side of the simplex tableau are the basic variables.

Why do we follow these four steps? It can be shown that two important things happen.

1. The numbers in the right-hand column give us a new corner point with all nonbasic variables taking the value zero.

2. At this new corner point the objective function has increased.

Example 5 In the last tableau of Example 3, we have $x_3 = \frac{1}{4}$ and $x_1 = x_2 = x_4 = 0$, since x_1, x_2, and x_4 are nonbasic variables. The last equation is

$$\tfrac{3}{4}x_1 + \tfrac{9}{8}x_2 - \tfrac{11}{4}x_4 - \tfrac{1}{8}s_1 = f - \tfrac{1}{8}$$

But $x_1 = x_2 = x_4 = s_1 = 0$, since these are all nonbasic variables; thus $f - \frac{1}{8} = 0$, or $f = \frac{1}{8}$. Thus f has increased from the value 0 (at $(0, 0, 0, 0)$) to the value $\frac{1}{8}$ at $(0, 0, \frac{1}{4}, 0)$.

Note that the value $s_2 = \frac{3}{2}$ (obtained by setting $x_1 = x_2 = x_4 = s_1 = 0$ in the tableau) means that there is a slack of $\frac{3}{2}$ in the second linear inequality in the original linear programming problem.

Example 6 In the last tableau of Example 4, we see that $x_2 = \frac{1}{5}$ and $x_1 = x_3 = x_4 = x_5 = 0$ (these are nonbasic variables). Here f has increased from the value 0 (at $(0, 0, 0, 0)$) to the value $\frac{1}{5}$ (at $(0, \frac{1}{5}, 0, 0, 0)$).

TERMINAL TABLEAU

The pivoting process outlined above is repeated until all positive indicators have been eliminated. We finally reach a **terminal tableau** when the indicators are all negative or zero.

Example 7 The following is a terminal tableau.

	x_1	x_2	x_3	x_4	s_1	s_2	s_3	s_4		
	0	1	0	1	$\frac{1}{3}$	1	2	0	2	x_4
	1	2	0	0	$\frac{1}{2}$	$\frac{1}{2}$	1	0	3	x_1
	0	0	1	0	0	1	3	0	1	x_3
	0	3	0	0	2	0	4	1	2	s_4
	0	−1	0	0	$-\frac{1}{3}$	$-\frac{1}{2}$	−1	0	$f - 20$	

All indicators are nonpositive.

Since all the indicators are nonpositive, we see that f has a maximum of 20 at the point

x_2 is nonbasic

$$(3, 0, 1, 2).$$

Why? We write out the last equation of the tableau:

$$0 \cdot x_1 - x_2 + 0 \cdot x_3 + 0 \cdot x_4 - \tfrac{1}{3}s_1 - \tfrac{1}{2}s_2 - s_3 + 0 \cdot s_4 = f - 20$$

or

$$-x_2 - \tfrac{1}{3}s_1 - \tfrac{1}{2}s_2 - s_3 = f - 20$$

or

$$f = 20 - x_2 - \tfrac{1}{3}s_1 - \tfrac{1}{2}s_2 - s_3$$

Since all variables are nonnegative, we see that

$$-x_2 - \tfrac{1}{3}s_1 - \tfrac{1}{2}s_2 - s_3 \le 0$$

Thus $f \le 20$. We observe that $f = 20$ when $x_2 = s_1 = s_2 = s_3 = 0$. The first four equations of the terminal tableau read

$$x_2 \quad + x_4 + \tfrac{1}{3}s_1 + \ s_2 + 2s_3 \quad\quad = 2$$
$$x_1 + 2x_2 \quad\quad\quad + \tfrac{1}{2}s_1 + \tfrac{1}{2}s_2 + \ s_3 \quad\quad = 3$$
$$x_3 \quad\quad\quad + \ s_2 + 3s_3 \quad\quad = 1$$
$$3x_2 \quad + \quad 2s_1 + \quad 4s_3 + s_4 = 2.$$

Since $x_2 = s_1 = s_2 = s_3 = 0$, these equations reduce to

$$x_4 = 2$$

$$x_1 = 3$$

$$x_3 = 1$$

$$s_4 = 2$$

Thus we see that f does indeed reach its maximum of 20 at $(3, 0, 1, 2)$.

We summarize the simplex method for solving a standard linear programming maximization problem.

SIMPLEX METHOD

1. Write the inequalities as equalities by introducing slack variables.

2. Initially, define the basic variables to be the slack variables.

3. Write all information in an initial simplex tableau.

4. Follow Steps 1 and 2 to choose a pivot in a column with a positive indicator. A positive indicator is a positive number in the bottom row.

5. Follow Steps 3 and 4 to pivot on the component chosen in 4.

6. Continue Steps 4 and 5 until a terminal tableau (no positive indicators) is reached.

7. Read the solution from the terminal tableau: If $f - M$ is in the lower, right-hand box, then the maximum value of f is M. The values of the basic variables are given in the right-hand column. All nonbasic variables are set equal to zero.

Example 8   Maximize

$$f = x_1 + x_2 + x_3$$

subject to

$$x_1 + 2x_2 + 3x_3 \le 1$$

$$2x_1 + x_2 + x_3 \le 2$$

$$x_1 \ge 0, \quad x_2 \ge 0, \quad x_3 \ge 0.$$

Solution   The initial simplex tableau is

x_1	x_2	x_3	s_1	s_2		
①	2	3	1	0	1	s_1
2	1	1	0	1	2	s_2
1	1	1	0	0	f	

These are the initial basic variables.

If we start with the first column (it has a positive indicator), we can either pivot on $a_{11} = 1$ or $a_{21} = 2$, since $\frac{1}{1} = \frac{2}{2} = 1$. Choosing a_{11} (since it already has the value 1), we pivot to obtain

	x_1	x_2	x_3	s_1	s_2		
$A_{1,2}(-2)$	1	2	3	1	0	1	x_1 ← This is the new
$A_{1,3}(-1)$	0	-3	-5	-2	1	0	s_2 basic variable.
	0	-1	-2	-1	0	$f - 1$	

Since all the indicators are nonpositive, this is a terminal tableau. We find that $x_1 = 1$, $x_2 = x_3 = 0$ (x_2 and x_3 are nonbasic variables), and $f = 1$ at the point $(1, 0, 0)$. This is our solution. Note that f increased from 0 (at $(0, 0, 0)$) to 1 (at $(1, 0, 0)$).

Example 9 Determine the maximum of

$$f(x_1, x_2, x_3) = 3x_1 - 2x_2 + 2x_3$$

subject to the constraints

$$x_1 + x_2 + x_3 \leq 15$$
$$2x_1 + x_2 + 2x_3 \leq 26$$
$$5x_1 + 2x_2 + 3x_3 \leq 43$$
$$x_1 \geq 0, \quad x_2 \geq 0, \quad x_3 \geq 0.$$

Solution We write the initial simplex tableau after introducing the slack variables s_1, s_2, and s_3.

x_1	x_2	x_3	s_1	s_2	s_3		
1	1	1	1	0	0	15	s_1
2	1	②	0	1	0	26	s_2
5	2	3	0	0	1	43	s_3
3	-2	2	0	0	0	f	

These are the initial basic variables.

There are two positive indicators. Choosing the third column, we find that

$$\tfrac{15}{1} = 15, \qquad \tfrac{26}{2} = 13, \qquad \tfrac{43}{3} = 14\tfrac{1}{3}$$

Since 13 is the smallest ratio, we pivot on the circled component.

	x_1	x_2	x_3	s_1	s_2	s_3		
	1	1	1	1	0	0	15	s_1
	1	$\frac{1}{2}$	1	0	$\frac{1}{2}$	0	13	s_2
$\xrightarrow{M_2(\frac{1}{2})}$	5	2	3	0	0	1	43	s_3
	3	-2	2	0	0	0	f	

	x_1	x_2	x_3	s_1	s_2	s_3		
$A_{2,1}(-1)$	0	$\frac{1}{2}$	0	1	$-\frac{1}{2}$	0	2	s_1
$A_{2,3}(-3)$	1	$\frac{1}{2}$	1	0	$\frac{1}{2}$	0	13	x_3 ← This is the new basic variable.
$\xrightarrow{A_{2,4}(-2)}$	②	$\frac{1}{2}$	0	0	$-\frac{3}{2}$	1	4	s_3
	1	-3	0	0	-1	0	$f - 26$	

Note that f takes the value 26 at $(0, 0, 13)$. Initially, f was 0 at $(0, 0, 0)$. There is one positive indicator remaining. We form the ratios

$$\frac{13}{1} = 13, \qquad \frac{4}{2} = 2$$

and so we pivot on the circled component.

	x_1	x_2	x_3	s_1	s_2	s_3		
	0	$\frac{1}{2}$	0	1	$-\frac{1}{2}$	0	2	s_1
	1	$\frac{1}{2}$	1	0	$\frac{1}{2}$	0	13	x_3
$\xrightarrow{M_3(\frac{1}{2})}$	1	$\frac{1}{4}$	0	0	$-\frac{3}{4}$	$\frac{1}{2}$	2	s_3
	1	-3	0	0	-1	0	$f - 26$	

	x_1	x_2	x_3	s_1	s_2	s_3		
	0	$\frac{1}{2}$	0	1	$-\frac{1}{2}$	0	2	s_1
$A_{3,2}(-1)$	0	$\frac{1}{4}$	1	0	$\frac{5}{4}$	$-\frac{1}{2}$	11	x_3
$\xrightarrow{A_{3,4}(-1)}$	1	$\frac{1}{4}$	0	0	$-\frac{3}{4}$	$\frac{1}{2}$	2	x_1 ← This is the new basic variable.
	0	$-\frac{13}{4}$	0	0	$-\frac{1}{4}$	$-\frac{1}{2}$	$f - 28$	

All the indicators are nonpositive, so this is a terminal tableau; since $f = 28 - \frac{13}{4}x_2 - \frac{1}{4}s_2 - \frac{1}{2}s_3$, we see that the maximum value of f is 28, obtained when $x_2 = s_2 = s_3 = 0$ and $x_1 = 2$, $x_3 = 11$, $s_1 = 2$. That is, f takes its maximum value of 28 at the corner point $(2, 0, 11)$.

Note. The result $s_1 = 2$ means that there is a slack of 2 in the first constraint of the problem at an optimal solution. We verify this by observing that at $(2, 0, 11)$,

$$x_1 + x_2 + x_3 = 2 + 0 + 11 = 13,$$

which is 2 less than 15.

Example 10 A manufacturing firm has discontinued production of a certain unprofitable product line. This has created considerable excess production capacity. Management is considering devoting this excess capacity to one or more of three products; call them products 1, 2, and 3. The available capacity on the company's machines that might limit output is summarized in Table 1.

TABLE 1

Machine type	Available time (in machine hours per week)
Milling machine	200
Lathe	100
Grinder	60

The number of machine-hours required for each unit of the respective products is given in Table 2.

TABLE 2 **Productivity (in Machine-Hours Per Unit)**

Machine type	Product 1	Product 2	Product 3
Milling machine	8	2	3
Lathe	4	3	0
Grinder	2	1	1

The sales department indicates that the sales potential for the three products exceeds the maximum production rate. The unit profit would be $20, $6, and $8 on products 1, 2, and 3, respectively.

How much of each product should the firm produce in order to maximize profits?

Solution Let x_1, x_2, and x_3 denote the number of units of products 1, 2, and 3, respectively. Then we obtain the following problem:
  Maximize

$$P = 20x_1 + 6x_2 + 8x_3 \qquad \text{Profit equation}$$

subject to

$$8x_1 + 2x_2 + 3x_3 \le 200 \qquad \text{Milling machine constraint}$$
$$4x_1 + 3x_2 \qquad \le 100 \qquad \text{Lathe constraint}$$
$$2x_1 + x_2 + x_3 \le 60 \qquad \text{Grinder constraint}$$
$$x_1 \ge 0, \quad x_2 \ge 0, \quad x_3 \ge 0.$$

After introducing three slack variables, we can write the following initial simplex tableau.

x_1	x_2	x_3	s_1	s_2	s_3		
8	2	3	1	0	0	200	s_1
4	3	0	0	1	0	100	s_2
2	1	①	0	0	1	60	s_3
20	6	8	0	0	0	P	

We can pivot in any of the first three columns. Since one goal is to minimize work, we look first and observe that if we pivot in column 3, the pivot is 1 (since $\frac{60}{1} < \frac{200}{3}$). We then obtain

x_1	x_2	x_3	s_1	s_2	s_3		
②	−1	0	1	0	−3	20	s_1
4	3	0	0	1	0	100	s_2
2	1	1	0	0	1	60	x_3
4	−2	0	0	0	−8	$P - 480$	

$A_{3,1}(-3)$
$A_{3,4}(-8)$

x_3 ⟵ This is the new basic variable.

There is one positive indicator (in the first column). Since $\frac{20}{2} = 10$ is smaller than $\frac{100}{4} = 25$ and $\frac{60}{2} = 30$, we pivot on the 2 to obtain, successively,

	x_1	x_2	x_3	s_1	s_2	s_3		
	1	$-\frac{1}{2}$	0	$\frac{1}{2}$	0	$-\frac{3}{2}$	10	s_1
	4	3	0	0	1	0	100	s_2
$\xrightarrow{M_1(\frac{1}{2})}$	2	1	1	0	0	1	60	x_3
	4	-2	0	0	0	-8	$P - 480$	

	x_1	x_2	x_3	s_1	s_2	s_3		
$A_{1,2}(-4)$	1	$-\frac{1}{2}$	0	$\frac{1}{2}$	0	$-\frac{3}{2}$	10	x_1 ← This is the new
$A_{1,3}(-2)$	0	5	0	-2	1	6	60	s_2 basic variable.
$\xrightarrow{A_{1,4}(-4)}$	0	2	1	-1	0	4	40	x_3
	0	0	0	-2	0	-2	$P - 520$	

The last tableau is a terminal tableau. We find that the maximum profit of $520 is achieved when $x_1 = 10$, $x_3 = 40$, and $x_2 = 0$ (this is a nonbasic variable). That is, by producing 10 units of product 1, 40 units of product 3, and no units of product 2, the firm can maximize its profits. As a check, note that

$$20x_1 + 6x_2 + 8x_3 = 20 \cdot 10 + 6 \cdot 0 + 8 \cdot 40 = 520.$$

Finally, the quantity $s_2 = 60$ means that there is a slack of 60 in the second constraint. As a second check, we observe that

$$4x_1 + 3x_2 = 4 \cdot 10 + 3 \cdot 0 = 40,$$

which is 60 less than 100.

PROBLEMS 4.4   In Problems 1–3, determine the pivots of the given initial simplex tableau.

1.

2	-1	2	1	0	1
-2	0	3	0	1	2
1	1	1	0	0	f

2.

1	1	1	1	0	1
-1	0	1	0	1	2
2	1	3	0	0	f

3.

1	2	3	1	0	0	5
2	3	1	0	1	0	3
3	1	2	0	0	1	1
2	-1	3	0	0	0	f

In Problems 4–9, write the initial simplex tableau for the given linear programming problem and circle all the possible pivots.

4. Maximize

$$f = 2x_1 + x_2$$

subject to

$$x_1 + x_2 \le 1$$
$$2x_1 + 5x_2 \le 2$$
$$x_1 \ge 0, x_2 \ge 0.$$

5. Maximize

$$f = x_1 - x_2$$

subject to

$$2x_1 + 3x_2 \le 7$$
$$5x_1 + 8x_2 \le 4$$
$$x_1 \ge 0, x_2 \ge 0.$$

6. Maximize

$$f = 4x_1 + 3x_2$$

subject to

$$x_1 + 2x_2 \le 5$$
$$3x_1 + 2x_2 \le 7$$
$$5x_1 + 3x_2 \le 14$$
$$x_1 \ge 0, x_2 \ge 0.$$

7. Maximize

$$f = 3x_1 + 2x_2 + 4x_3$$

subject to

$$x_1 + x_2 + x_3 \le 5$$
$$2x_1 + x_2 + 3x_3 \le 6$$
$$x_1 \ge 0, x_2 \ge 0, x_3 \ge 0.$$

8. Maximize

$$f = 2x_1 + x_2 + 3x_3$$

subject to

$$x_1 - x_2 - x_3 \le 5$$
$$-x_1 + x_2 + 2x_3 \le 6$$
$$2x_1 - x_2 + x_3 \le 7$$
$$x_1 \ge 0, x_2 \ge 0, x_3 \ge 0.$$

9. Maximize

$$f = x_1 + x_2 - 3x_3$$

subject to

$$x_1 + x_2 + x_3 \le 5$$
$$x_1 - 2x_2 + 2x_3 \le 6$$
$$2x_1 - x_2 + x_3 \le 4$$
$$x_1 \ge 0, x_2 \ge 0, x_3 \ge 0.$$

In Problems 10–14, find terminal tableaux for the initial simplex tableau.

10.

x_1	x_2	x_3	x_4	s_1	s_2		
2	-1	4	6	1	0	1	s_1
3	3	2	7	0	1	2	s_2
1	1	$\frac{1}{2}$	-2	0	0	f	

11.

x_1	x_2	x_3	s_1	s_2		
2	3	2	1	0	2	s_1
-1	0	4	0	1	2	s_2
1	2	0	0	0	f	

12. The tableau of Problem 1.

13. The tableau of Problem 2.

14. The tableau of Problem 3.

In Problems 15–26, solve by the simplex method.

15. Problem 5.

16. Problem 4.

17. Maximize

$$f = 4x_1 + 5x_2$$

subject to

$$2x_1 + 3x_2 \leq 6$$
$$3x_1 + 2x_2 \leq 5$$
$$x_1 \geq 0, x_2 \geq 0.$$

18. Maximize

$$f = x_1 + 2x_2 + x_3$$

subject to

$$x_1 + x_2 \qquad \leq 2$$
$$x_2 \qquad \leq 1$$
$$x_2 + 2x_3 \leq 3$$
$$x_1 \geq 0, x_2 \geq 0, x_3 \geq 0.$$

19. Problem 9.

20. Problem 8.

21. Maximize

$$f = 5x_1 + 8x_2$$

subject to

$$x_1 + x_2 \leq 3$$
$$x_1 + 2x_2 \leq 4$$
$$x_1 \qquad \leq \tfrac{5}{2}$$
$$x_2 \leq \tfrac{3}{2}$$
$$x_1 \geq 0, x_2 \geq 0.$$

22. Maximize

$$f = 5x_1 + x_2 + 3x_3$$

subject to

$$x_1 \qquad \leq 3$$
$$4x_2 + x_3 \leq 2$$
$$x_1 - x_2 \qquad \leq 0$$
$$x_3 \leq 1$$
$$x_1 \geq 0, x_2 \geq 0, x_3 \geq 0.$$

23. Maximize

$$f = x_1 + 2x_2 + 2x_3$$

subject to

$$x_1 + 3x_2 + 6x_3 \leq 12$$
$$3x_1 + 2x_2 + 4x_3 \leq 10$$
$$-x_1 + 2x_2 + x_3 \leq 5$$
$$x_1 \geq 0, x_2 \geq 0, x_3 \geq 0.$$

24. Maximize

$$f = x_1 + 2x_2 + 3x_3$$

subject to

$$x_1 + x_2 - x_3 \leq 1$$
$$x_1 - x_2 + x_3 \leq 2$$
$$-x_1 + x_2 + x_3 \leq 3$$
$$x_1 \geq 0, x_2 \geq 0, x_3 \geq 0.$$

25. Maximize

$$f = x_1 - x_2 + x_3$$

subject to

$$x_1 + x_2 + 2x_3 \leq 5$$
$$2x_1 + x_2 + x_3 \leq 7$$
$$2x_1 - x_2 + 3x_3 \leq 8$$
$$x_1 + 2x_2 + 5x_3 \leq 9$$
$$x_1 \geq 0, x_2 \geq 0, x_3 \geq 0.$$

26. Maximize

$$f = 5x_1 + 7x_2 + 15x_3 + 6x_4$$

subject to

$$x_1 + 2x_2 \qquad + x_4 \leq 1$$
$$x_1 + 3x_2 + x_3 \qquad \leq 2$$
$$x_1 + 4x_2 + 3x_3 + 2x_4 \leq 3$$
$$x_1 + \qquad 5x_3 + 3x_4 \leq 4$$
$$x_1 \geq 0, x_2 \geq 0, x_3 \geq 0, x_4 \geq 0.$$

27. Liva's Lumber, Inc., manufactures three types of plywood. The table summarizes the production hours per unit in each of three production operations and other data for the problem.

Plywood	Operations (hours)			Profit per unit
	I	II	III	
Grade A	2	2	4	$40
Grade B	5	5	2	$30
Grade X	10	3	2	$20
Maximum time available	900	400	600	

How many units of each grade of lumber should be produced?

28. Ye Olde Cording Winery in Peoria, Illinois, makes three kinds of authentic German wine: Heidelberg Sweet, Heidelberg Regular, and Deutschland Extra Dry. The raw materials, labor, and profit for a gallon of each of these wines is summarized in the table.

Wine	Grapes grade A (bushels)	Grapes grade B (bushels)	Sugar (pounds)	Labor (work-hours)	Profit per gallon
Heidelberg Sweet	1	1	2	2	$1.00
Heidelberg Regular	2	0	1	3	$1.20
Deutschland Extra Dry	0	2	0	1	$2.00

If the winery has 150 bushels of grade A grapes, 150 bushels of grade B grapes, 80 pounds of sugar, and 225 hours of labor available during the next week, what product mix of wines will maximize the company's profit?

(a) Solve by the simplex method.

(b) Interpret all slack variables.

(c) An increase in which resources could improve the company's profit?

29. Kirkman Brothers Ice Cream Parlors sell three different flavors of Dairy Sweet ice milk: chocolate, vanilla, and banana. Due to extremely hot weather and a high demand for its products, Kirkman has run short of its supply of ingredients: milk, sugar and cream. Hence Kirkman will not be able to fill all the orders received from its retail outlets, the ice cream parlors. Due to these circumstances, Kirkman has decided to make the best amounts of the three flavors given the constraints on supply of the basic ingredients. The company will then ration the ice milk to the retail outlets.

Kirkman has collected the following data on profitability of the various flavors, availability of supplies, and amounts required by each flavor.

		Usage per gallon		
Flavor	Profit per gallon	Milk (gallons)	Sugar (pounds)	Cream (gallons)
Chocolate	$1.00	0.45	0.50	0.10
Vanilla	$0.90	0.50	0.40	0.15
Banana	$0.95	0.40	0.40	0.20
Maximum available		200	150	60

Determine the optimal product mix for Kirkman Brothers. What additional resources could be used?

30. A politician plans to walk the length of her state to attract attention to her candidacy for office, get better acquainted with the problems of her state, and get a chance to talk with many voters. She will spend part of her time in fast walking, part in leisurely strolling, and part in stopping to talk to voters. To balance the time spent in achieving the various goals of her walk, she decides to walk as far as she can each hour while spending at least $\frac{1}{4}$ hour in conversation and limiting her time of fast walking to no more than the sum of talking time and strolling time combined. If her fast walking speed is 3 miles per hour, her strolling speed is 1 mile per hour, and she stands still while she is talking to voters, what fraction of each hour should she devote to each activity in order to go as far as possible in the hour?

31. An employee of an ice cream shop wishes to make the richest (measured in calories) ice cream soda for his friends that he can fit in a 12-oz glass. The ingredients of a soda are syrup, cream, soda water, and ice cream. In order to look and taste like a soda, the mixture must contain no more than 4 oz of ice cream, at least as much soda water as the total amount of syrup and cream combined, and no more than 1 oz more of syrup than of cream. If the syrup is 75 calories an ounce, cream is 50 calories an ounce, ice cream is 40 calories an ounce, and soda water contains no calories, how many ounces of each ingredient should the employee use?

*32. An investor has $10,000 from which she would like to make as much money as possible. She plans to invest some money in stocks, some in bonds, and put the rest in a savings account. The investor believes she can earn 8% on the money invested in stocks and 7% on bonds. The savings bank pays 5% interest. Since stocks are a rather risky investment, she decides to invest no more in stocks than half the amount invested in bonds, and no more in stocks than she puts in the bank. The investor also will keep at least $2000 in the bank in case she needs cash on short notice. How much money should she invest in stocks, how much in bonds, and how much should she put in the bank?

33. A jeweler makes rings, earrings, pins, and necklaces. He wishes to work no more than 40 hours a week. It takes him 2 hours to make a ring, 2 hours to make a pair of earrings, 1 hour to make a pin, and 4 hours to make a necklace. He estimates that he can sell no more than 10 rings, 10 pairs of earrings, 15 pins, and 3 necklaces in a week. The jeweler charges $50 for a ring, $80 for a pair of earrings, $25 for a pin, and $200 for a necklace. How many rings, earrings, pins, and necklaces should the jeweler make in order to make the largest possible gross earnings?

34. A company producing canned mixed fruit has a stock of 10,000 pounds of pears, 12,000 pounds of peaches, and 8000 pounds of cherries. The company produces three fruit mixtures, which it sells in 1-lb cans. The first mixture is half pears and half peaches and sells for 30¢. The second mixture has equal amounts of the three fruits and sells for 40¢. The third mixture is half peaches and half cherries and sells for 50¢. How many cans of each mixture should be produced to maximize the return?

35. In a large hospital, surgical operations are classified into three categories according to their average times of 30 minutes, 1 hour, and 2 hours. The hospital receives a fee of $100, $150, or $200 for an operation in categories I, II, or III, respectively. If the hospital has eight operating rooms in use an average of 10 hours per day, how many operations of each type should the hospital schedule in order to (a) maximize its revenue and (b) maximize the total number of operations?

Solve Problems 36–38 by the simplex method.

36. Problem 4.2.28 on page 172.

37. Problem 4.2.31 on page 172.

38. Problem 4.2.30 on page 172.

4.5 The Simplex Method II: The Dual Minimum Problem

There is a remarkable connection between the maximum and minimum problems of linear programming. To every maximum problem is associated a minimum problem, called the **dual** of the maximum problem. Conversely, the maximum problem is the dual of the associated minimum problem. This association is useful, since the solution of one problem is closely related to the solution of its dual problem.

Before we define the dual, we define the transpose of a matrix.

Transpose

Let $A = (a_{ij})$ be an $m \times n$ matrix. Then the **transpose** of A, written A^t, is the $n \times m$ matrix obtained by interchanging the rows and columns of A. Succinctly, we may write $A^t = (a_{ji})$. In other words,

$$\text{if } A = \begin{pmatrix} a_{11} & a_{12} & \cdots & a_{1n} \\ a_{21} & a_{22} & \cdots & a_{2n} \\ \vdots & \vdots & & \vdots \\ a_{m1} & a_{m2} & \cdots & a_{mn} \end{pmatrix} \quad \text{then} \quad A^t = \begin{pmatrix} a_{11} & a_{21} & \cdots & a_{m1} \\ a_{12} & a_{22} & \cdots & a_{m2} \\ \vdots & \vdots & & \vdots \\ a_{1n} & a_{2n} & \cdots & a_{mn} \end{pmatrix} \quad (1)$$

Simply put, the ith row of A is the ith column of A^t and the jth column of A is the jth row of A^t.

Example 1 Find the transposes of the matrices.

$$A = \begin{pmatrix} 2 & 3 \\ 1 & 4 \end{pmatrix} \qquad B = \begin{pmatrix} 2 & 3 & 1 \\ -1 & 4 & 6 \end{pmatrix} \qquad C = \begin{pmatrix} 1 & 2 & -6 \\ 2 & -3 & 4 \\ 0 & 1 & 2 \\ 2 & -1 & 5 \end{pmatrix}$$

Solution   Interchanging the rows and columns of each matrix, we obtain

$$A^t = \begin{pmatrix} 2 & 1 \\ 3 & 4 \end{pmatrix}, \quad B^t = \begin{pmatrix} 2 & -1 \\ 3 & 4 \\ 1 & 6 \end{pmatrix}, \quad C^t = \begin{pmatrix} 1 & 2 & 0 & 2 \\ 2 & -3 & 1 & -1 \\ -6 & 4 & 2 & 5 \end{pmatrix}.$$

Note, for example, that 4 is the component in row 2 and column 3 of C, while 4 is the component in row 3 and column 2 of C^t. That is, the 2, 3 component of C is the 3, 2 component of C^t.

We may now define the dual.

DUAL PROBLEM OF LINEAR PROGRAMMING

The following maximum and minimum problems are called **dual** problems.

1. Maximize

$$f = c_1 x_1 + c_2 x_2 + \cdots + c_n x_n \tag{2}$$

subject to

$$a_{11}x_1 + a_{12}x_2 + \cdots + a_{1n}x_n \le b_1$$
$$a_{21}x_1 + a_{22}x_2 + \cdots + a_{2n}x_n \le b_2$$
$$\vdots \qquad \vdots \qquad \qquad \vdots \qquad \vdots \tag{3}$$
$$a_{m1}x_1 + a_{m2}x_2 + \cdots + a_{mn}x_n \le b_m$$
$$x_1 \ge 0, \quad x_2 \ge 0, \ldots, x_n \ge 0.$$

2. Minimize

$$g = b_1 y_1 + b_2 y_2 + \cdots + b_m y_m \tag{4}$$

subject to

$$a_{11}y_1 + a_{21}y_2 + \cdots + a_{m1}y_m \ge c_1$$
$$a_{12}y_1 + a_{22}y_2 + \cdots + a_{m2}y_m \ge c_2$$
$$\vdots \qquad \vdots \qquad \qquad \vdots \qquad \vdots \tag{5}$$
$$a_{1n}y_1 + a_{2n}y_2 + \cdots + a_{mn}y_m \ge c_n$$
$$y_1 \ge 0, \quad y_2 \ge 0, \ldots, y_m \ge 0.$$

We observe the following.

1. If the maximum problem involves m inequalities in n variables, then its dual minimum problem involves n inequalities in m variables.

2. The matrix of coefficients of the inequalities in (5) is the transpose of the matrix of coefficients of the inequalities in (3).

3. The b_i's in (3) are the coefficients of the y_i's in (4).

4. The c_i's in (5) are the coefficients of the x_i's in (2).

Example 2 Find the dual minimum problem of the following maximum problem.

Maximize

$$f = 3x_1 + 2x_2 \tag{6}$$

subject to

$$x_1 + 2x_2 \leq 5$$
$$3x_1 + 4x_2 \leq 8$$
$$2x_1 + x_2 \leq 4 \tag{7}$$
$$x_1 \geq 0, \quad x_2 \geq 0.$$

Solution If we look at the dual problems (2), (3) and (4), and (5), we observe that the number of variables in the minimum problem is equal to the number of $\leq$ linear inequalities in the maximum problem. This number is 3. Also, since the coefficients of the y_i's in the expression for g are the b_i's in (3), we have:

Minimize

$$g = 5y_1 + 8y_2 + 4y_3. \tag{8}$$

The matrix of coefficients of the inequalities (7) is

$$\begin{pmatrix} 1 & 2 \\ 3 & 4 \\ 2 & 1 \end{pmatrix}$$

Its transpose is

$$\begin{pmatrix} 1 & 3 & 2 \\ 2 & 4 & 1 \end{pmatrix}$$

Thus, from (8) and observations 2 and 4, we have:

Minimize

$$g = 5y_1 + 8y_2 + 4y_3$$

subject to

$$y_1 + 3y_2 + 2y_3 \geq 3$$
$$2y_1 + 4y_2 + y_3 \geq 2$$
$$y_1 \geq 0, \quad y_2 \geq 0, \quad y_3 \geq 0.$$

We write the two problems side by side to illustrate how they are related.

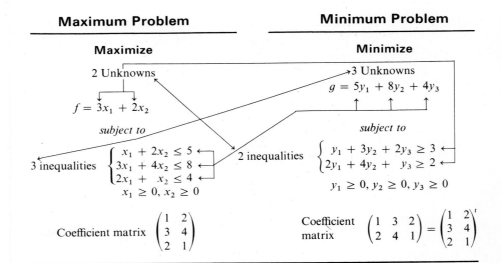

Maximum Problem	**Minimum Problem**

Maximize — 2 Unknowns

$$f = 3x_1 + 2x_2$$

subject to

3 inequalities
$$\begin{cases} x_1 + 2x_2 \le 5 \\ 3x_1 + 4x_2 \le 8 \\ 2x_1 + x_2 \le 4 \end{cases}$$
$$x_1 \ge 0, \, x_2 \ge 0$$

Coefficient matrix $\begin{pmatrix} 1 & 2 \\ 3 & 4 \\ 2 & 1 \end{pmatrix}$

Minimize — 3 Unknowns

$$g = 5y_1 + 8y_2 + 4y_3$$

subject to

2 inequalities
$$\begin{cases} y_1 + 3y_2 + 2y_3 \ge 3 \\ 2y_1 + 4y_2 + y_3 \ge 2 \end{cases}$$
$$y_1 \ge 0, \, y_2 \ge 0, \, y_3 \ge 0$$

Coefficient matrix $\begin{pmatrix} 1 & 3 & 2 \\ 2 & 4 & 1 \end{pmatrix} = \begin{pmatrix} 1 & 2 \\ 3 & 4 \\ 2 & 1 \end{pmatrix}^t$

Example 3   Determine the dual problem of:

Minimize

$$g = 3y_1 + 4y_2 + 6y_3$$

subject to

$$4y_1 + 7y_2 + y_3 \ge 3$$
$$y_1 + 3y_2 + 5y_3 \ge 7$$
$$2y_1 + y_2 + 4y_3 \ge 10$$
$$y_1 \ge 0, \quad y_2 \ge 0, \quad y_3 \ge 0.$$

Solution   The transpose of $\begin{pmatrix} 4 & 7 & 1 \\ 1 & 3 & 5 \\ 2 & 1 & 4 \end{pmatrix}$ is $\begin{pmatrix} 4 & 1 & 2 \\ 7 & 3 & 1 \\ 1 & 5 & 4 \end{pmatrix}$.

Thus the dual maximum problem is:

Maximize

$$f = 3x_1 + 7x_2 + 10x_3$$

subject to

$$4x_1 + x_2 + 2x_3 \leq 3$$
$$7x_1 + 3x_2 + x_3 \leq 4$$
$$x_1 + 5x_2 + 4x_3 \leq 6$$
$$x_1 \geq 0, x_2 \geq 0, x_3 \geq 0.$$

Why do we study the dual problem? We do so because of the following remarkable result that tells us that, in a certain sense, dual problems have the same solutions.

FUNDAMENTAL THEOREM OF LINEAR PROGRAMMING

Let f be the objective function of a linear programming maximum problem, and let g be the objective function of the corresponding dual minimum problem. Then the maximum problem for f has a solution if and only if the minimum problem for g has a solution. Furthermore, $(x_1, x_2, \ldots, x_n)$ and $(y_1, y_2, \ldots, y_m)$ are optimal solutions of the two problems if and only if f evaluated at $(x_1, x_2, \ldots, x_n) = g$ evaluated at $(y_1, y_2, \ldots, y_m)$.

We now show how the fundamental theorem of linear programming and the simplex method can be used to solve a **standard minimizing problem**. This is a minimum problem taking the form of (4) and (5).

SOLVING A STANDARD MINIMUM PROBLEM OF LINEAR PROGRAMMING

1. Write the dual maximum problem.

2. Solve this maximum problem by the simplex method.

3. The minimum value for g (in (4)) equals the maximum value of f (in (2)).

4. The values of $y_1, y_2, \ldots, y_m$ that minimize g are the negatives of the coefficients of $s_1, s_2, \ldots, s_m$ in the last row of the terminal tableau.

Example 4 Suppose that Steps 1 and 2 have been carried out, resulting in the following terminal tableau for the dual maximum problem.

x_1	x_2	s_1	s_2	s_3		
1	0	0	4	3	50	s_1
0	1	0	7	2	80	x_1
0	0	1	6	4	65	x_2
-4	-2	-8	-3	-5	$f - 125$	

The negatives of these numbers are the values of y_1, y_2, and y_3.

Then the solution to the original minimum problem is $g = 125$ when $y_1 = 8$, $y_2 = 3$, and $y_3 = 5$.

Example 5 Determine, by the simplex method, the minimum of the objective function $g = 3y_1 + 2y_2$ subject to

$$5y_1 + y_2 \geq 10$$
$$2y_1 + 2y_2 \geq 12$$
$$y_1 + 4y_2 \geq 12$$
$$y_1 \geq 0, \quad y_2 \geq 0.$$

Solution The dual maximum problem is:

Maximize

$$f = 10x_1 + 12x_2 + 12x_3$$

subject to

$$5x_1 + 2x_2 + x_3 \leq 3$$
$$x_1 + 2x_2 + 4x_3 \leq 2$$
$$x_1 \geq 0, \quad x_2 \geq 0, \quad x_3 \geq 0.$$

The initial simplex tableau for this problem is

x_1	x_2	x_3	s_1	s_2		
5	2	1	1	0	3	s_1
1	②	4	0	1	2	s_2
10	12	12	0	0	f	

We choose the circled component as a pivot to obtain

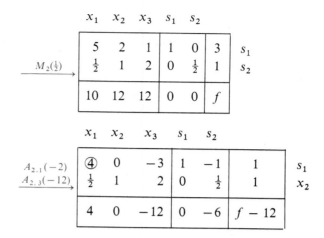

	x_1	x_2	x_3	s_1	s_2	
	5	2	1	1	0	3	s_1
$\xrightarrow{M_2(\frac{1}{2})}$	$\frac{1}{2}$	1	2	0	$\frac{1}{2}$	1	s_2
	10	12	12	0	0	f

	x_1	x_2	x_3	s_1	s_2	
$A_{2.1}(-2)$	④	0	-3	1	-1	1	s_1
$\xrightarrow{A_{2.3}(-12)}$	$\frac{1}{2}$	1	2	0	$\frac{1}{2}$	1	x_2
	4	0	-12	0	-6	$f-12$

Pivoting again on the circled component, we have

	x_1	x_2	x_3	s_1	s_2	
	1	0	$-\frac{3}{4}$	$\frac{1}{4}$	$-\frac{1}{4}$	$\frac{1}{4}$	s_1
$\xrightarrow{M_1(\frac{1}{4})}$	$\frac{1}{2}$	1	2	0	$\frac{1}{2}$	1	x_2
	4	0	-12	0	-6	$f-12$

	x_1	x_2	x_3	s_1	s_2	
$A_{1.2}(-\frac{1}{2})$	1	0	$-\frac{3}{4}$	$\frac{1}{4}$	$-\frac{1}{4}$	$\frac{1}{4}$	x_1
$\xrightarrow{A_{1.3}(-4)}$	0	1	$\frac{19}{8}$	$-\frac{1}{8}$	$\frac{5}{8}$	$\frac{7}{8}$	x_2
	0	0	-9	-1	-5	$f-13$

$$\uparrow \qquad \uparrow$$
$$-y_1 \qquad -y_2$$

We have reached a terminal tableau. The solution to the maximum problem is

A nonbasic variable

$$f = 13 \quad \text{at} \quad x_1 = \tfrac{1}{4},\, x_2 = \tfrac{7}{8},\, x_3 = 0.$$

The solution to the original problem is

$$g = 13 \quad \text{at} \quad y_1 = 1 \quad \text{and} \quad y_2 = 5.$$

Check. At (1, 5),

$$g = 3y_1 + 2y_2 = 3 \cdot 1 + 2 \cdot 5 = 13.$$

Example 6 A dog-food manufacturer advertises that one can of its all-meat product provides the minimum daily requirements for carbohydrates and proteins of the average 20-lb dog. The meats available for use are beef, horse-meat, and liver. An ounce of beef costs 1.5¢ and yields 0.5 oz of carbohydrate and 0.2 oz of protein. An ounce of horsemeat costs 1¢ and yields 0.6 oz of carbo-hydrate and 0.1 oz of protein. An ounce of liver costs 2.5¢ and yields 0.4 oz of carbohydrate and 0.3 oz of protein. The minimum daily requirements of the average 20-lb dog are estimated to be 6 ounces of carbohydrates and 3.1 ounces of protein. What combination of the three meats should the manufacturer choose to satisfy these requirements at minimum cost?

Solution Let y_1, y_2, and y_3 denote the number of ounces of beef, horsemeat, and liver, respectively, to be used in the product. Then the problem is:

Minimize

$$g = 1.5y_1 + y_2 + 2.5y_3$$

subject to

$$0.5y_1 + 0.6y_2 + 0.4y_3 \geq 6$$
$$0.2y_1 + 0.1y_2 + 0.3y_3 \geq 3.1$$
$$y_1 \geq 0, \quad y_2 \geq 0, \quad y_3 \geq 0.$$

The dual maximum problem is

Maximize

$$f = 6x_1 + 3.1x_2$$

subject to

$$0.5x_1 + 0.2x_2 \leq 1.5$$
$$0.6x_1 + 0.1x_2 \leq 1$$
$$0.4x_1 + 0.3x_2 \leq 2.5$$
$$x_1 \geq 0, \quad x_2 \geq 0.$$

The initial simplex tableau is

x_1	x_2	s_1	s_2	s_3		
0.5	(0.2)	1	0	0	1.5	s_1
0.6	0.1	0	1	0	1	s_2
0.4	0.3	0	0	1	2.5	s_3
6	3.1	0	0	0	f	

We can pivot in the first or second column. A few computations on a calculator show that if we pivot on the circled component, then all the indicators will be nonpositive after one pivoting operation.

	x_1	x_2	s_1	s_2	s_3		
	2.5	1	5	0	0	7.5	s_1
	0.6	0.1	0	1	0	1	s_2
	0.4	0.3	0	0	1	2.5	s_3
	6	3.1	0	0	0	f	

$\xrightarrow{\quad M_1(5) \quad}$

	x_1	x_2	s_1	s_2	s_3		
	2.5	1	5	0	0	7.5	x_2
	0.35	0	−0.5	1	0	0.25	s_2
	−0.35	0	−1.5	0	1	0.25	s_3
	−1.75	0	−15.5	0	0	$f - 23.25$	

$\xrightarrow[\ A_{1,4}(-3.1)\ ]{\begin{array}{c} A_{1,2}(-0.1) \\ A_{1,3}(-0.3) \end{array}}$

The last tableau is terminal, and the solution is given by $g = 23.25$ at $y_1 = 15.5$, $y_2 = 0$, and $y_3 = 0$. This means that the minimum cost for a can of dog food is $23\frac{1}{4}$¢. This is achieved by using $15\frac{1}{2}$ oz of beef and no horsemeat or liver.

PROBLEMS 4.5 In Problems 1–10, find the transpose of the given matrix.

1. $\begin{pmatrix} -1 & 4 \\ 6 & 5 \end{pmatrix}$

2. $\begin{pmatrix} 3 & 0 \\ 1 & 2 \end{pmatrix}$

3. $\begin{pmatrix} 2 & 3 \\ -1 & 2 \\ 1 & 4 \end{pmatrix}$

4. $\begin{pmatrix} 2 & -1 & 0 \\ 1 & 5 & 6 \end{pmatrix}$

5. $\begin{pmatrix} 1 & 2 & 3 \\ -1 & 0 & 4 \\ 1 & 5 & 5 \end{pmatrix}$

6. $\begin{pmatrix} 1 & 2 & 3 \\ 2 & 4 & -5 \\ 3 & -5 & 7 \end{pmatrix}$

7. $\begin{pmatrix} 1 & 0 & 1 & 0 \\ 0 & 1 & 0 & 1 \end{pmatrix}$

8. $\begin{pmatrix} 2 & -1 \\ 2 & 4 \\ 1 & 6 \\ 1 & 5 \end{pmatrix}$

9. $\begin{pmatrix} a & b & c \\ d & e & f \\ g & h & j \end{pmatrix}$

10. $\begin{pmatrix} 0 & 0 & 0 \\ 0 & 0 & 0 \end{pmatrix}$

In Problems 11–20, find the dual problem of the given linear programming problem.

11. Maximize

$$f = 2x_1 + 5x_2$$

subject to

$$x_1 + 2x_2 \leq 5$$
$$3x_1 + 2x_2 \leq 7$$
$$x_1 + x_2 \leq 1$$
$$x_1 \geq 0, x_2 \geq 0.$$

12. Maximize

$$f = 4x_1 + 3x_2$$

subject to

$$x_1 - x_2 \leq 5$$
$$3x_1 - 2x_2 \leq 6$$
$$x_1 \geq 0, x_2 \geq 0.$$

13. Minimize

$$g = 2y_1 + 3y_2$$

subject to

$$2y_1 + y_2 \geq 1$$
$$y_1 + 2y_2 \geq 1$$
$$y_1 \geq 0, y_2 \geq 0.$$

14. Minimize

$$g = 5y_1 + 3y_2$$

subject to

$$2y_1 + y_2 \geq 1$$
$$y_1 + 2y_2 \geq 1$$
$$y_1 + y_2 \geq 3$$
$$y_1 \geq 0, y_2 \geq 0.$$

15. Maximize

$$f = x_1 + x_2 + x_3$$

subject to

$$x_1 + x_2 + x_3 \leq 5$$
$$2x_1 + x_2 + 3x_3 \leq 6$$
$$x_1 \geq 0, x_2 \geq 0, x_3 \geq 0.$$

16. Maximize

$$f = 2x_1 + 8x_2 + 3x_3$$

subject to

$$x_1 - x_2 - x_3 \leq 5$$
$$-x_1 + x_2 + 2x_3 \leq 6$$
$$2x_1 - x_2 + x_3 \leq 7$$
$$x_1 \geq 0, x_2 \geq 0, x_3 \geq 0.$$

17. Minimize

$$g = 2y_1 + 5y_2 + 3y_3$$

subject to

$$y_1 + 2y_2 + y_3 \geq 13$$
$$4y_1 + y_2 + 2y_3 \geq 21$$
$$-3y_1 - y_2 + 4y_3 \geq 11$$
$$y_1 \geq 0, y_2 \geq 0, y_3 \geq 0.$$

18. Maximize

$$W = 4x_1 - x_2 + 9x_3$$

subject to

$$2x_1 + 3x_2 + x_3 \leq 8$$
$$4x_1 + x_2 + 2x_3 \leq 6$$
$$8x_1 + 7x_2 + 4x_3 \leq 25$$
$$x_1 \geq 0, x_2 \geq 0, x_3 \geq 0.$$

19. Maximize

$$f = x_1 + 2x_2 - x_3 + 5x_4$$

subject to

$$x_1 + 2x_2 + 3x_3 + 4x_4 \leq 12$$
$$x_1 \geq 0, x_2 \geq 0, x_3 \geq 0, x_4 \geq 0.$$

20. Minimize

$$g = 3y_1 + y_2 + 5y_3 + 12y_4$$

subject to

$$y_1 + y_2 + y_3 + y_4 \geq 10$$
$$2y_1 - y_2 + y_3 + 2y_4 \geq 14$$
$$5y_1 - 8y_2 - 3y_3 + 3y_4 \geq 5$$
$$2y_1 - y_2 - 5y_3 + 3y_4 \geq 0$$
$$y_1 \geq 0, y_2 \geq 0, y_3 \geq 0, y_4 \geq 0.$$

21. Solve Problem 13 by the simplex method.

22. Solve Problem 14 by the simplex method.

23. Solve Problem 17 by the simplex method.

24. Solve Problem 20 by the simplex method.

25. Two foods contain carbohydrates and proteins only. Food I costs 50¢ per pound and is 90% carbohydrates (by weight). Food II costs $1 per pound and is 60% carbohydrates. What diet of these two foods provides 2 lb of carbohydrates and 1 lb of proteins at minimum cost? What is the cost per pound of this diet?

26. Continuing Problem 25, suppose that a third food is available, which costs $2 per pound and is 30% carbohydrates and 70% protein. What diet of these three foods provides 2 lb of carbohydrates and 1 lb of proteins at minimum cost? What is the cost per pound of this diet?

*27. In the production of fertilizers, three chemicals are combined in different mixtures or grades and are sold in 100-lb units. Suppose that the three chemicals cost 20¢, 15¢, and 5¢ a pound, respectively. In all mixtures, there must be at least 20 lb of the first chemical, and the amount of the third chemical in a mixture must not be greater than the amount of the second chemical. How many pounds of each chemical should be put into a 100-lb bag of fertilizer in order to minimize the cost of a bag of fertilizer? [*Hint:* Because of the equality constraint, this can be written as a problem involving only two variables.]

28. A woman wishes to design a weekly exercise schedule, which will involve jogging, bicycling, and swimming. In order to vary the exercise, she plans to spend at least as much time bicycling as she devotes to jogging and swimming combined. Also, she wishes to swim at least 2 hours a week, because she enjoys that activity more than the others. If jogging consumes 600 calories an hour, bicycling uses 300 calories an hour, and swimming uses 300 calories an hour and if she wishes to burn up a total of at least 3000 calories a week through exercise, how many hours should she devote to each type of exercise if she wishes to reach this goal in the smallest number of hours?

Review Exercises for Chapter 4

In Exercises 1–12, sketch the set of points that satisfy the given inequality or inequalities.

1. $x \le 4$

2. $x - y < 4$

3. $x + y \ge -1$

4. $4x + 3y > 12$

5. $3x - 4y \le 12$

6. $y \le 2x + 1$

7. $2y < 6 - 3x$

8. $\dfrac{x}{2} - \dfrac{y}{3} \ge 1$

9. $|x| > 2, |y| < 1$

10. $0 \le x \le 3, -1 < y < 4$

11. $x + y \le 1, 3x - 2y \le 6$

12. $2x - 2y > 1, x + 2y \ge 3$

In Exercises 13–16, solve the given linear programming problem by the graphical method.

13. Maximize
$$f = 2x + 5y$$
subject to
$$2x + y \le 4$$
$$x + 3y \le 8$$
$$x \ge 0, y \ge 0.$$

14. Minimize
$$g = x + 2y$$
subject to
$$x + y \ge 3$$
$$2x + 3y \ge 6$$
$$x \ge 0, y \ge 0.$$

15. Minimize

$$g = 3x + 2y$$

subject to

$$x + 2y \geq 4$$
$$2x + 4y \geq 6$$
$$5x + \quad y \geq 10$$
$$x \geq 0, y \geq 0.$$

16. Maximize

$$f = 5x + 3y$$

subject to

$$x + 10y \leq 10$$
$$2x + \quad 5y \leq 5$$
$$3x + \quad 2y \leq 6$$
$$x \geq 0, y \geq 0.$$

In Exercises 17 and 18, solve the given linear programming by the corner-point method.

17. Maximize

$$f = 2x_1 + 3x_2$$

subject to

$$-x_1 + \quad x_2 \leq 5$$
$$2x_1 - 3x_2 \leq 6$$
$$x_1 \geq 0, x_2 \geq 0.$$

18. Minimize

$$f = 4x_1 + 5x_2$$

subject to

$$x_1 + 3x_2 \geq 3$$
$$3x_1 + \quad x_2 \geq 3$$
$$x_1 + \quad x_2 \leq 7$$
$$x_1 \geq 0, x_2 \geq 0.$$

19. Find the dual of Exercise 17.

20. Find the dual of Exercise 18.

In Exercises 21–24, find the transpose of the given matrix.

21. $\begin{pmatrix} 2 & 3 \\ 1 & 7 \end{pmatrix}$

22. $\begin{pmatrix} 1 & 4 \\ -2 & 3 \\ 0 & 4 \end{pmatrix}$

23. $\begin{pmatrix} 1 & 3 & 2 \\ 4 & 1 & -6 \\ 3 & 0 & 5 \end{pmatrix}$

24. $\begin{pmatrix} 2 & -1 & 0 \\ 1 & 4 & 2 \end{pmatrix}$

25. Write the initial simplex tableau for the problem:
Maximize

$$f = 2x_1 + x_2 + 4x_3$$

subject to

$$x_1 + 2x_2 + \quad x_3 \leq 13$$
$$3x_1 + 4x_2 - 2x_3 \leq 6$$
$$-4x_1 + 6x_2 + 3x_3 \leq 11$$
$$x_1 \geq 0, x_2 \geq 0, x_3 \geq 0.$$

26. What is the first pivot of the tableau you found in Exercise 25?

27. What is the dual of the problem in Exercise 25?

28. Find a terminal tableau for the problem in Exercise 25.

29. Find the solution to both the problem of Exercise 25 and its dual.

30. Find the dual of the problem:
   Minimize

$$g = 3y_1 - y_2 + 4y_3$$

   subject to

$$3y_1 + 2y_2 + y_3 \geq 4$$
$$-6y_1 + 8y_2 + 3y_3 \geq 12$$
$$9y_1 - 10y_2 + 5y_3 \geq 8$$
$$y_1 \geq 0, y_2 \geq 0, y_3 \geq 0.$$

31. Write the initial simplex tableau for the dual of the problem in Exercise 30.

32. Find the terminal tableau for the tableau of Exercise 30.

33. Determine the solution of the problem in Exercise 30.

In Exercises 34–39, solve by the simplex method.

34. Maximize

$$f = x_1 + x_2 + x_3$$

   subject to

$$x_1 + 2x_2 + x_3 \leq 5$$
$$-x_1 + x_2 + x_3 \leq 3$$
$$- x_2 + x_3 \leq 1$$
$$x_1 \geq 0, x_2 \geq 0, x_3 \geq 0.$$

35. Minimize

$$g = 4y_1 - 4y_2 + 4y_3$$

   subject to

$$y_1 - 2y_2 + y_3 \geq 2$$
$$y_1 + y_2 - y_3 \geq 5$$
$$y_1 \geq 0, y_2 \geq 0.$$

36. Maximize

$$f = 4x_1 + 2x_2 + 5x_3$$

   subject to

$$x_1 + 2x_2 + 3x_3 \leq 6$$
$$x_1 + x_2 + 2x_3 \leq 4$$
$$x_1 \geq 0, x_2 \geq 0, x_3 \geq 0.$$

37. Maximize

$$f = x_1 - x_2 + x_3$$

   subject to

$$3x_1 + x_2 + 2x_3 \leq 8$$
$$x_1 + 2x_2 + 4x_3 \leq 6$$
$$7x_1 + 4x_2 + 8x_3 \leq 25$$
$$x_1 \geq 0, x_2 \geq 0, x_3 \geq 0.$$

38. Minimize

$$g = 5y_1 - y_2 + y_3$$

   subject to

$$y_1 + y_2 + y_3 \geq 7$$
$$2y_1 + 3y_2 - 7y_3 \geq 9$$
$$-3y_1 + y_2 + 4y_3 \geq 2$$
$$y_1 \geq 0, y_2 \geq 0, y_3 \geq 0.$$

39. Minimize

$$g = 2y_1 + y_2 + 3y_2$$

   subject to

$$y_1 + y_2 + 3y_3 \geq 3$$
$$2y_1 + 3y_2 + y_3 \geq 6$$
$$y_1 \geq 0, y_2 \geq 0, y_3 \geq 0.$$

40. A baker plans to bake cakes and cookies. Each cake requires $\frac{5}{2}$ cups of flour and 2 cups of sugar, while a batch of cookies uses 1 cup of flour and $\frac{1}{2}$ cup of sugar. The baker wishes to use no more than 70 cups of flour and 50 cups of sugar in all. If she can sell each cake for $10 and each batch of cookies for $3, how many cakes and how many batches of cookies should she sell in order to make the greatest income?

41. The baker of Exercise 40 runs out of flour and sugar. The minimum reorder is 200 cups of flour and 120 cups of sugar. If each cup of flour costs 25¢ and each cup of sugar costs 40¢, how many cakes and how many cookies should she sell in order to minimize her total reorder costs?

42. Ryland Farms, in northwestern Indiana, grows soybeans and corn on its 500 acres of land. An acre of soybeans brings a $100 profit and an acre of corn brings a $200 profit. Because of government regulations, no more than 200 acres can be planted in soybeans. During the planting season 1200 hours of planting time will be available. Each acre of soybeans requires 2 hours, while each acre of corn requires 6 hours. How many acres of soybeans and how many acres of corn should be planted in order to maximize profits?

5 INTRODUCTION TO PROBABILITY

5.1 Introduction

Almost every human or natural event contains uncertainties that we attempt to analyze in terms of an intuitive concept of probability. Daily newspapers give "odds" on all sorts of sporting events and cite the "probability of precipitation" tomorrow. Insurance companies sell policies and charge premiums based on the probability that the insured will not have a fire, have an accident, suffer a theft, or die in a certain time period.

How likely is it that it will rain tomorrow? Will an insurance company make a profit if it charges a certain premium on its automobile liability policies? How likely is it that you will win money if you gamble in Las Vegas next weekend? In order to answer these questions, it is necessary to have at least a rudimentary knowledge of the theory probability and statistics.

Historically, probability developed from a mathematical analysis of gambling games. Games of chance have been played for more than 5000 years. An early game, similar to modern gambling with dice, involved astragali—small, roughly rectangular bones from the ankle of a mammal.[†] Each throw of such a bone had four possible outcomes. The four outcomes were not equally likely, due to the lack of symmetry of the bones. This must have made the intuitive calculation of a player's chance very difficult.

In this chapter we describe some of the techniques used in determining the probabilities of certain events. Before doing so, it is useful to understand the difference between probability and statistics. This is best illustrated with a simple example.

[†] This game is described in F. N. David's paper, "Dicing and Gaming," *Biometrika*, 42 (1955): 1–15.

Suppose I flip a coin and ask, "What is the probability that it comes out heads?" You can answer that question only if you assume that I am flipping a *fair* coin; that is, one for which a head and a tail are equally likely to occur. If you suspect that I am flipping the coin in a careful manner in order to affect the outcome, or if you suspect that the coin is weighted to make heads more likely than tails, you will be unable to answer the question. But if you assume that the coin is fair, you can reason as follows: There are two possible outcomes and each is equally likely. Thus, the outcome is a head approximately half the time, so the probability of a head is $\frac{1}{2}$, or 50%. At this point you do not know precisely what the word *probability* means. Nevertheless, you have obtained the correct answer. That is, you computed a probability based on known information. This is an example in probability theory.

Let us change the problem. Suppose I come to you with my coin and ask, "Is this a fair coin?" Without further information, your only reasonable response is, "How the heck do I know?" But, after that, you may wish to do an experiment. The most obvious thing to try is to flip the coin several times, 100 times for example. If you get 53 heads and 47 tails, it might be reasonable to assume that the coin is fair. On the other hand, if you get 92 heads and 8 tails, you should be suspicious. What you are trying to solve by this experiment is a problem in statistics.

In probability theory you are *given* information (the fairness of the coin in our example) and are asked to determine the probabilities that certain things occur. For example, if you know that a coin is fair, then it would be reasonable to ask you to compute the probability that if you flip the coin three times you obtain exactly two heads. Of course, this question is meaningless unless you know that the coin is fair or, if it is not, the way it is balanced. Determining the properties of the coin is a problem for the statistician.

Once we have made a very simplified distinction between probability and statistics, it should be stressed that the two subjects are very closely interrelated. In the example given above, a statistical test might be needed to determine (or at least be fairly certain) that the coin is fair. But once that is determined, techniques from probability theory can be used to answer other questions, such as the likelihood of getting two heads in three tosses.

In this text we have separated probability and statistics by placing probability theory in Chapter 5 and an introduction to statistics in Chapter 6. But you should never lose sight of the fact that, in practical situations, a knowledge of both is essential.

5.2 The Mathematics of Counting I: The Fundamental Principle of Counting

In our introduction we cited a very simple example in probability theory—finding the probability of obtaining a head when a fair coin is tossed. We obtained the answer by reasoning that as there are two possible outcomes (heads and tails) and each is equally likely, there is one chance out of two of obtaining a head; therefore the probability of obtaining a head is $\frac{1}{2}$.

Now let us ask a similar but more difficult question. In poker, a *flush* consists of 5 cards of the same suit (that is, 5 spades, 5 hearts, 5 diamonds, or 5 clubs). If you are dealt 5 cards from a standard deck of 52 cards, what is the probability of obtaining a flush? We shall defer the answer to this question until Section 5.3. However, let us see how we could approach the problem even before we define the word *probability*. We could reason as in the coin problem. Suppose that there are x ways to deal 5 cards from a deck of 52 and that there are y ways to obtain a flush. Then, assuming that all deals are equally likely (that is, no one is cheating), we could reason that there are y chances of obtaining a flush out of x possible hands—so the probability of obtaining a flush is y/x.

The reasoning here is the same as before, but now we are faced with the problem of computing x and y. That is, we must *count* the number of poker hands and the number of flushes. It turns out (see Examples 5.3.10 and 5.3.11) that $x = 2,598,960$ (the number of possible poker hands) and $y = 5148$ (the number of flushes) so that the probability of being dealt a flush is

$$\frac{5148}{2,598,960} \approx 0.00198 \approx \frac{1}{505}.$$

That is, a flush will be dealt, on the average, in 1 out of every 505 hands.

These examples illustrate that in order to solve problems in probability, it is necessary to be able to count. In this section and in Section 5.3 we discuss techniques for counting in many different situations.

Before stating our first general principle, we begin with two examples.

Example 1 A large hotel chain supplies its guests with toothpaste and soap. The chain has been offered substantial discounts on three brands of toothpaste and four brands of soap. In how many ways can it choose one brand of toothpaste and one brand of soap?

Solution Let us represent the brands of toothpaste by A, B, and C and the brands of soap by 1, 2, 3, and 4. The possibilities are listed below.

A1	B1	C1
A2	B2	C2
A3	B3	C3
A4	B4	C4

That is, there are 12 possibilities in all. We note that we could pair each of the 3 brands of toothpaste with one of 4 brands of soap. This tells us that there are $4 + 4 + 4$, or $3 \times 4 = 12$, possibilities.

Another way to see this is to use a **tree diagram**. In Figure 1 the first set of branches in the tree contains the brands of toothpaste and the second, or auxiliary, set of branches lists the brands of soap. From the figure we can see that there are $3 \times 4 = 12$ possibilities. In order to obtain the number of ways of pairing a brand of toothpaste with a brand of soap, we *multiplied* the number of each kind together.

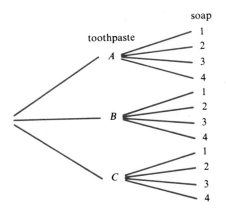

Figure 1   Tree Diagram of Toothpaste–Soap Possibilities

Example 2   A wealthy art collector has two paintings by Picasso, four by Matisse, and three by Van Gogh. She has agreed to lend one painting by each artist to a local art museum for an exhibition of 20th century art. In how many ways can this be done?

Solution   We again obtain the answer in two ways. First, we list the possibilities.

$$
\begin{array}{ll}
P_1 M_1 V_1 & P_2 M_1 V_1 \\
P_1 M_1 V_2 & P_2 M_1 V_2 \\
P_1 M_1 V_3 & P_2 M_1 V_3 \\
P_1 M_2 V_1 & P_2 M_2 V_1 \\
P_1 M_2 V_2 & P_2 M_2 V_2 \\
P_1 M_2 V_3 & P_2 M_2 V_3 \\
P_1 M_3 V_1 & P_2 M_3 V_1 \\
P_1 M_3 V_2 & P_2 M_3 V_2 \\
P_1 M_3 V_3 & P_2 M_3 V_3 \\
P_1 M_4 V_1 & P_2 M_4 V_1 \\
P_1 M_4 V_2 & P_2 M_4 V_2 \\
P_1 M_4 V_3 & P_2 M_4 V_3
\end{array}
$$

We see that there are 24 possibilities. Here P_1 stands for the first Picasso painting, M_3 the third Matisse painting, and so on. Note that 24 is the product of 2, 4, and 3.

Figure 2 shows the possibilities in a tree diagram. Note that there are 2 choices for the first branch (a Picasso), 4 choices for the second branch (a Matisse) and 3 choices for the third branch (a Van Gogh). There are $2 \times 4 \times 3 = 24$ choices in all.

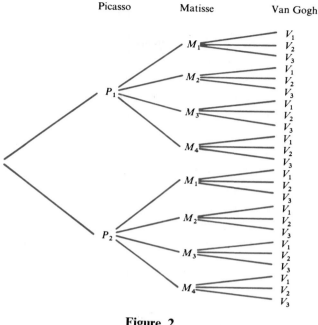

Figure 2

These two examples illustrate the **fundamental principle of counting**.

FUNDAMENTAL PRINCIPLE OF COUNTING

Two Activities. If activity 1 can be performed in n_1 ways and activity 2 can be performed in n_2 ways, then the two activities can be performed in $n_1 \cdot n_2$ ways.

Three Activities. If activities 1, 2, and 3 can be performed in n_1, n_2, and n_3 ways, respectively, then the three activities can be performed in $n_1 \cdot n_2 \cdot n_3$ ways.

k Activities. If activities 1, 2, 3, . . . , k can be performed in $n_1, n_2, n_3, \ldots, n_k$ ways, respectively, then the k activities can be performed in $n_1 \cdot n_2 \cdot n_3 \cdots n_k$ ways.

Note. In Example 1 there were two activities; $n_1 = 3$ and $n_2 = 4$, so $n_1 n_2 = 3 \cdot 4 = 12$. In Example 2 there were three activities; $n_1 = 2$, $n_2 = 4$, and $n_3 = 3$, so $n_1 n_2 n_3 = 2 \cdot 4 \cdot 3 = 24$.

Example 3   In a certain environment there are 14 species of fruit flies, 17 species of moths, and 13 species of mosquitoes. A biologist wishes to choose one species of each type for an experiment. In how many ways can this be done?

Solution   There are three activities with $n_1 = 14$, $n_2 = 17$, and $n_3 = 13$; by the fundamental principle of counting, there are $14 \cdot 17 \cdot 13 = 3094$ different ways of selecting one species of each type. In this example it would be very time-consuming to provide a list of the possibilities without the aid of a computer.

Example 4   A coin is flipped six times. What are the possible outcomes of heads and tails that can be obtained?

Solution   Here we are performing 6 identical activities, each of which has 2 possible outcomes—H or T. Thus, by the fundamental principle of counting, there are $2 \cdot 2 \cdot 2 \cdot 2 \cdot 2 \cdot 2 = 2^6 = 64$ possible outcomes. Four of them are *HHTHTT*, *THTHTH*, *HTTTHH* and *TTTTTT*.

Example 5   At a race track horses compete for first place (win), second place (place) and third place (show). If five horses are running, how many different outcomes are possible, assuming that there are no ties?

Solution   Any one of the 5 horses can win, so there are 5 ways to choose the winner. The winning horse cannot also come in second, so once the winner is chosen, there are 4 choices for second place. Similarly, once the first and second choices are determined, there are 3 possibilities for third place. Using the fundamental principle of counting, there are $5 \cdot 4 \cdot 3 = 60$ possible outcomes.

We can also use a tree diagram to illustrate the possibilities, as done in Figure 3. The first branch of the tree lists the possible winners, the second branch lists the possible place horses and the third branch lists the possible show horses. Note that there are 5 choices for the first branch and, once that is chosen, there are 4 choices for the second branch. Finally, once the first two branches are chosen, there are 3 choices for the third branch—leading to $5 \cdot 4 \cdot 3 = 60$ possible paths through the tree.

PROBLEMS 5.2   In Problems 1–12, a procedure or activity is described. In each problem determine the number of possible outcomes. Solve first by using a tree diagram and then by using the fundamental principle of counting.

1.   A coin is flipped twice. Assume that *HT* and *TH* are different.

2.   A coin is flipped three times.

3.   Two dice are thrown (a die is a cube whose sides are numbered 1 through 6). Assume, for example, that (2, 5) and (5, 2) are different.

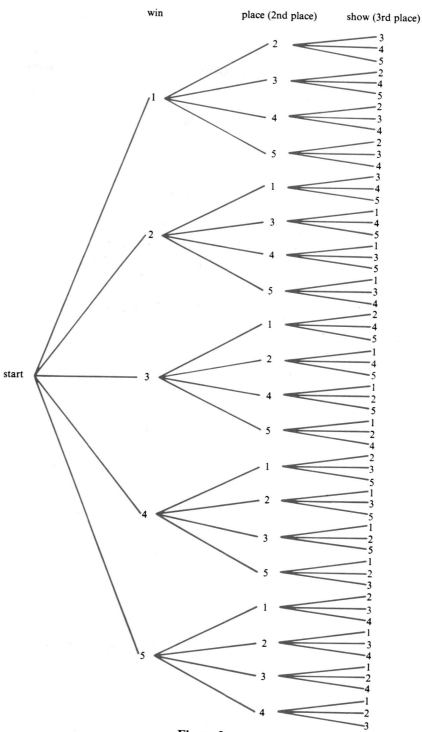

Figure 3

4. A student committee selects a chairperson and a vice-chairperson from among its five members (no student can be both).

5. A coin is flipped and a die is thrown. [*Hint*: Two possible outcomes are $H5$ and $T2$.]

6. A clinic stocks one of five brands of aspirin and one of four brands of antiseptic cream.

7. A businessman with a large fleet of automobiles must choose, from bids, a type of car from among three possibilities, a type of tire from between two brands, and a brand of gasoline from among four brands.

8. Four applicants take an aptitude test for a job and are ranked according to their performances on the test (assume that there are no ties).

9. Seven politicians compete in a municipal election in which the first place winner is elected mayor, the second place winner is elected city council chairperson and the third place winner is elected city clerk and recorder.

10. A family has two children (assume that GB and BG are different).

11. A family has three children.

12. A family has four children.

13. A biologist attempts to classify 46,200 species of insects by assigning to each species three letters from the alphabet. Will the classification be completed? If not, determine the number of letters that should be used.

 14. It is estimated that there are 2 million species of insects, 1 million species of plants, 20,000 species of fish, and 8700 species of birds. If one species from each of these four categories is to be chosen for a comparative study, in how many ways can this be done?

15. How many different "words" of three letters can be formed from the four letters A, U, G, and C if repetitions are allowed? (The words of the genetic code are formed from triplets or codons of four bases: adenine, uracil, guanine, and cytosine.)

16. In a primitive religion a professional diviner appeals to the god by reciting a verse. To choose the appropriate verse, 16 smooth palm nuts are grasped by the diviner between both hands. The diviner then attempts to grasp them all in his right hand. Since palm nuts are relatively large, they are difficult to grasp in one hand. If one or two nuts remain in the left hand, this number is recorded. This procedure is repeated eight times, producing a sequence of eight numbers, which determines the verse that the diviner recites. How many verses must the diviner know?

17. In a comparative study of digital watches, five characteristics are to be examined. If there are six recognizable differences in each of three characteristics and five recognizable differences in each of the remaining two characteristics, what is the maximum number of different brands of digital watches that could be distinguished by these five characteristics?

18. A license tag in a certain state consists of one letter followed by four numbers. How many possible license plates are there?

19. In Problem 18 how many license plates are there if the number zero cannot be used?

20. In Problem 18 how many license plates are there if no number can be used more than once?

21. Eight horses complete at a race track. Assuming no ties, how many possible outcomes are there (only first, second, and third place finishes are considered)?

*22. In Problem 21 how many outcomes are there if ties are allowed?

23. Four classes of students each choose one representative on a committee. If the classes contain 47, 51, 54, and 55 students, in how many ways can the representatives be chosen?

5.3 The Mathematics of Counting II: Permutations and Combinations

In problems involving probability, it is often necessary to count the number of ways in which a set of objects can be arranged. For example, let us count the number of ways the letters A, B, and C can be arranged; they are as follows.

$$ABC \qquad BAC \qquad CAB$$

$$ACB \qquad BCA \qquad CBA$$

Clearly, there are six possible orders in which the three letters can be written. Each such ordering, or rearrangement, is called a *permutation*.

Permutation of *n* Objects   In general, we have the following: A **permutation** of n objects is a rearrangement or ordering of the n objects.

Note.   When we use the word *permutation*, we imply that the order in which things are written or done is important.

In problems involving permutations, *order is important.*

Thus the permutations ABC and ACB are *different* permutations because the ordering in the two are different.

Example 1   How many permutations are there of the letters A, B, C, D, and E?

Solution   We could solve this problem by writing out all the permutations. This would take a lot of space and time and, moreover, would be unnecessary. Let us think of the 5 letters as "sitting" in chairs: There are 5 choices for the letter in the first chair. Suppose it is a B. Then there are 4 choices for the letter

in the second chair, 3 for the third chair and 2 for the fourth chair. Finally, there is 1 letter that remains to be placed in the fifth chair. Thus, by the fundamental principle of counting, there are $5 \cdot 4 \cdot 3 \cdot 2 \cdot 1 = 120$ permutations of the 5 letters.

Note.   In Example 1 we could have been talking about any set of 5 objects. Thus we have shown that *there are 120 permutations of 5 objects.*

Before continuing, it is useful to introduce the *factorial* notation.

Factorial.   The product of the first n positive integers is called $\boldsymbol{n}$ **factorial** and is denoted by $n!$ By convention, zero factorial is equal to 1. We have

$$
\begin{aligned}
0! &= 1 \\
1! &= 1 \\
2! &= 2 \cdot 1 = 2 \\
3! &= 3 \cdot 2 \cdot 1 = 6 \\
4! &= 4 \cdot 3 \cdot 2 \cdot 1 = 24 \\
&\vdots \\
n! &= n(n-1)(n-2)\cdots 3 \cdot 2 \cdot 1
\end{aligned}
\tag{1}
$$

Note that

$$
\begin{aligned}
4! &= 4 \cdot 3 \cdot 2 \cdot 1 = 4 \cdot 3! \\
5! &= 5 \cdot 4 \cdot 3 \cdot 2 \cdot 1 = 5 \cdot 4!
\end{aligned}
$$

and, in general

$$
n! = n \cdot (n-1)!
\tag{2}
$$

We can now restate the result of Example 1 in a more compact notation: *There are 5! permutations of 5 objects.*

We can extend this result. If, in Example 1, there were n objects instead of 5 letters, then we could place them in n chairs. There are n choices for the first chair, $n - 1$ choices for the second chair, and so on. Using the same reasoning as in Example 1, we conclude that

there are $n!$ permutations of n objects.

Example 2   Eight brands of refrigerator are ranked according to fixed criteria. Assuming no ties, how many rankings are possible?

Solution   Here the order in which the brands are ranked is important, and so we can rephrase the question as, "How many permutations are there of 8 objects?" The answer is $8! = 8 \cdot 7 \cdot 6 \cdot 5 \cdot 4 \cdot 3 \cdot 2 \cdot 1 = 40{,}320$.

Very often, we are not interested in all possible orderings of n objects but in the possible orders of some subset of the n objects. For example, in Example 5.2.5 on page 224 we found that, in a race with 5 horses, there are $5 \cdot 4 \cdot 3 = 60$ ways for 3 of the horses to finish first, second, and third. Let us look at this number more closely. We note that

$$60 = 5 \cdot 4 \cdot 3 = \frac{5 \cdot 4 \cdot 3 \cdot 2 \cdot 1}{2 \cdot 1} = \frac{5!}{2!} = \frac{5!}{(5-3)!} \tag{3}$$

We will return to expression (3) in a moment, but first we provide a useful definition.

Permutation of n Objects Taken k at a Time   A **permutation of n objects taken k at a time** (with $0 \leq k \leq n$) is any selection of k objects in a definite order from the n objects. We denote the number of permutations of n objects taken k at a time by $P_{n,k}$.

Remark.   In (3) we found that

$$P_{5,3} = 60 = \frac{5!}{(5-3)!}$$

Example 3   How may four-letter "words" (not necessarily English) can be formed from the letters of the word *STRANGE*?

Solution   First we note that we are asking the question, "How many permutations are there of 7 objects taken 4 at a time?" Some of these permutations are *STAR, RTAS, RATS, GEAR,* and *GRAE*. Clearly, the order in which the letters are chosen is important, which is why we have a permutation. Reasoning as before, we note that there are 7 ways to choose the first letter, 6 ways to choose the second letter, 5 ways to choose the third letter and 4 ways to choose the fourth letter. Then, using the fundamental principle of counting, we conclude that $7 \cdot 6 \cdot 5 \cdot 4 = 840$ four-letter words can be made from the letters of the word *STRANGE*.

Remark.   We showed in the last example that the number of permutations of 7 objects taken 4 at a time is given by

$$P_{7,4} = 840 = 7 \cdot 6 \cdot 5 \cdot 4 = \frac{7 \cdot 6 \cdot 5 \cdot 4 \cdot 3 \cdot 2 \cdot 1}{3 \cdot 2 \cdot 1}$$

$$= \frac{7!}{3!} = \frac{7!}{(7-4)!} \tag{4}$$

Before going further, let us derive a general formula for $P_{n,k}$. If we wish to choose k objects from n objects where order is important, then there are n ways to choose the first object, $n - 1$ ways to choose the second, $n - 2$ ways to choose the third, and so on; there are $n - (k - 1) = n - k + 1$ ways to choose the kth object. That is,

$$P_{n,k} = n(n-1)(n-2)(n-3)\cdots(n-k+1). \tag{5}$$

Now, multiplying and dividing both sides of (5) by $(n-k)!$, we obtain

$$P_{n,k} = \frac{n(n-1)(n-2)(n-3)\cdots(n-k+1)(n-k)!}{(n-k)!}$$

$$= \frac{n(n-1)(n-2)\cdots(n-k+1)(n-k)(n-k-1)(n-k-2)\cdots 3\cdot 2\cdot 1}{(n-k)!}$$

$$= \frac{n!}{(n-k)!}$$

Thus

$$\boxed{P_{n,k} = \frac{n!}{(n-k)!}} \tag{6}$$

Note. Formula (6) generalizes the specific formulas obtained in (3) and (4).

Example 4   Compute

(a) $P_{8,2}$

(b) $P_{10,4}$

Solution

(a) $P_{8,2} = 8!/(8-2)! = 8!/6!$. There are two ways to finish this computation. First, we can use brute force to compute $8! = 40{,}320$, $6! = 720$, and $40{,}320/720 = 56$. However, we can save time and needless computation by writing

$$\frac{8!}{6!} = \frac{8\cdot 7\cdot 6\cdot 5\cdot 4\cdot 3\cdot 2\cdot 1}{6\cdot 5\cdot 4\cdot 3\cdot 2\cdot 1} = 8\cdot 7 = 56.$$

(b) $P_{10,4} = \dfrac{10!}{(10-4)!} = \dfrac{10!}{6!} = \dfrac{10\cdot 9\cdot 8\cdot 7\cdot 6!}{6!} = 10\cdot 9\cdot 8\cdot 7 = 5040.$

The last example suggests the following useful rule.

RULE

When computing $P_{n,k}$, always cancel the terms in the denominator with corresponding terms in the numerator before doing any other computations.

Example 5 First, second, and third prizes are to be awarded in a competition among 20 persons. In how many ways can the prizes be distributed?

Solution Since order counts here (John getting first prize and Susan getting second prize is a different outcome from Susan first and John second), we recognize this as a permutation of 20 objects taken 3 at a time:

$$P_{20,3} = \frac{20!}{(20-3)!} = \frac{20!}{17!} = \frac{20 \cdot 19 \cdot 18 \cdot 17!}{17!} = 20 \cdot 19 \cdot 18 = 6840$$

There is another type of permutation problem we will encounter.

Example 6 How many distinct permutations are there of the letters of the word *FIDDLED*?

Solution There are 3*D*'s in *FIDDLED*. Let us label them D_1, D_2, and D_3. The following are all permutations of *FIDDLED*.

$$FILED_1D_2D_3$$
$$FILED_1D_3D_2$$
$$FILED_2D_1D_3$$
$$FILED_2D_3D_1$$
$$FILED_3D_1D_2$$
$$FILED_3D_2D_1$$

These six permutations are indistinguishable; they all appear as *FILEDDD*. We see that for every permutation of *FIDDLED*, there are a total of $6 = 3!$ permutations that contain the same letters in the same order. The permutations of *FIDDLED* appear in groups of 6. Therefore, to obtain the number of distinct permutations, we must divide the total number of permutations by $3! = 6$. That is,

$$\frac{\text{number of distinct}}{\text{permutations}} = \frac{\text{number of permutations}}{6} = \frac{7!}{6} = 7 \cdot 5! = 7 \cdot 120 = 840.$$

In the counting problems so far encountered, we were interested in the number of orderings of a set of objects. There is a type of counting problem in which the ordering of objects is not relevant. For example, we may wish to choose 25 people from a population of 1000 in order to conduct a survey. The order in which the people are chosen is not of interest. Instead, we are interested in the number of ways that the group of 25 people may be chosen. To see how this is computed, we begin with a simple example.

Example 7 From five people on a committee, three are to be chosen for a subcommittee. In how many ways can this be done?

Solution Let us, for simplicity, label the 5 people as A, B, C, D, and E. Then we can list the possible subcommittees.

$$
\begin{array}{ccccc}
ABC & ABE & ACE & BCD & BDE \\
ABD & ACD & ADE & BCE & CDE
\end{array} \tag{7}
$$

This is all! If you think that this list is too small, then look again. Since the order in which the subcommittee members are chosen is *not* relevant, we see, for example, that *ABC* (listed) and *CAB* (unlisted) represented the *same* subcommittee. Thus we can conclude that 10 subcommittees can be formed.

Let us solve the problem in a different way. We know that the number of permutations of 5 objects taken 3 at a time is $5!/2! = 5 \cdot 4 \cdot 3 = 60$. These 60 permutations include *ABC*, *ACB*, *BAC*, *BCA*, *CAB* and *CBA*. As permutations, these six are different. But they all represent the *same* subcommittee. Obviously, the list of 60 permutations contains many duplications if order does not matter. More precisely, each subcommittee listed in (7) contains 3 objects which can be permuted in $3! = 6$ ways. We see that for every subcommittee listed in (7), there are $3! = 6$ permutations of three of the letters *ABCDE*. Thus there are 6 times as many permutations as there are subcommittees; putting it another way, to obtain the number of subcommittees we must divide the number of permutations by 6. Hence

$$
\text{number of subcommittees} = \frac{P_{5,3}}{3!} = \frac{60}{6} = 10.
$$

Before continuing, we have an important definition. Assume that $0 \le k \le n$.

Combination of n Objects Taken k at a Time A **combination of n objects taken k at a time** is any selection of k of the n objects without regard to order. The symbol $\binom{n}{k}$ is used to denote the number of combinations of n objects taken k at a time.

Remark 1. In Example 7 we found that $\binom{5}{3} = 10$.

Remark 2. In some books the symbol $C_{n,k}$ is used to denote the number of combinations of n objects taken k at a time.

How many combinations of n objects taken k at a time are there? Reasoning as in Example 7, we observe that there are $P_{n,k} = n!/(n-k)!$ permutations of the n objects taken k at a time. But any set of k objects can be permuted in $k!$ ways. Therefore there are $k!$ permutations for every combination. Thus, in order to determine the number of combinations of n objects taken k at a time, we must divide the number of permutations of n objects taken k at a time by $k!$. That is,

$$\binom{n}{k} = \frac{P_{n,k}}{k!} = \frac{n!}{(n-k)!k!}. \tag{8}$$

Example 8 Compute

(a) $\binom{8}{3}$

(b) $\binom{10}{6}$

Solution As before, we cancel everything in the denominator before making further computations.

(a) $\binom{8}{3} = \frac{8!}{(8-3)!3!} = \frac{8!}{5!3!} = \frac{8 \cdot 7 \cdot 6 \cdot 5!}{5!3!} = \frac{8 \cdot 7 \cdot 6}{3 \cdot 2} = 8 \cdot 7 = 56$

(b) $\binom{10}{6} = \frac{10!}{(10-6)!6!} = \frac{10!}{4!6!} = \frac{10 \cdot 9 \cdot 8 \cdot 7}{4 \cdot 3 \cdot 2} = \frac{10 \cdot 9 \cdot 7}{3} = 10 \cdot 3 \cdot 7 = 210$

We emphasize a very useful fact.

> In problems involving combinations, order is *not* important.

Example 9 A common procedure in quality control is to take a sample of a manufactured product and test it for defects. From a collection of 12 electric razors, a manufacturer wishes to select 3 for extensive testing. In how many ways can this be done?

Solution Since the order in which the razors are chosen is not important, we are being asked to compute the number of combinations of 12 objects taken 3 at a time. The answer is, therefore,

$$\binom{12}{3} = \frac{12!}{9!3!} = \frac{12 \cdot 11 \cdot 10}{3 \cdot 2} = 2 \cdot 11 \cdot 10 = 220.$$

Example 10 A poker hand consists of 5 cards selected from a standard deck of 52 cards. How many poker hands are there?

Solution   We discussed this question briefly at the beginning of Section 5.2. Now we can give a complete answer by noting that since the order in which the cards are dealt is irrelevant, we must compute the number of combinations of 52 objects taken 5 at a time.

$$\frac{48}{4 \cdot 3} = 4 \text{ and } \frac{50}{5 \cdot 2} = 5$$

$$\binom{52}{5} = \frac{52!}{47!5!} = \frac{52 \cdot 51 \cdot 50 \cdot 49 \cdot 48}{5 \cdot 4 \cdot 3 \cdot 2} = 52 \cdot 51 \cdot 5 \cdot 49 \cdot 4 = 2{,}598{,}960.$$

Example 11   A flush in poker consists of five cards of the same suit. How many flushes are there?

Solution   In order to make a flush, we must do two things: First, we must choose a suit, and second, we must choose 5 cards in that suit. There are 4 ways to choose a suit (since there are 4 suits: spades, hearts, diamonds, and clubs). As each suit contains 13 cards, there are

$$\binom{13}{5} = \frac{13!}{8!5!} = \frac{13 \cdot 12 \cdot 11 \cdot 10 \cdot 9}{5 \cdot 4 \cdot 3 \cdot 2} = 13 \cdot 11 \cdot 9 = 1287$$

ways to choose 5 cards from 13 cards. Thus, from the fundamental principle of counting, there are $4 \cdot 1287 = 5148$ possible flushes in a standard deck of 52.[†]

Example 12   Six boys and 11 girls in a class are suspected to have an infectious disease. Blood samples are to be taken from 2 of the boys and 3 of the girls to test for the disease. In how many ways can this be done?

Solution   There are

$$\binom{6}{2} = \frac{6!}{4!2!} = 15$$

ways to choose the two boys and

$$\binom{11}{3} = \frac{11!}{8!3!} = 165$$

ways to choose the 3 girls. By the fundamental principle of counting, there are $15 \cdot 165 = 2475$ ways to choose 2 boys and 3 girls.

Note.   In the last example we solved the problem the way we did by assuming that the orders in which the boys and girls were chosen did not matter so that we were dealing with combinations.

[†] These include straight flushes, which are 5 cards of the same suit that are in order. Thus ♡4 ♡5 ♡6 ♡7 ♡8 is a straight flush.

To the Student. The topics with which many students have the most difficulty involve permutations and combinations. The formulas in this section can easily be memorized. It's learning when to use each formula that is difficult. Facility comes only with *practice*!

To help you with this difficulty, the problem set is divided into three parts. Part 1 includes only problems involving permutations. Part 2 contains only problems involving combinations. In Part 3, both types of problem are included.

In the real world, problems don't jump at you and say, "We are permutation problems." You must determine that for yourself. Remember, if you are unsure whether to use a formula for a permutation or a combination, ask yourself the question, "Does order count?" If the answer is yes, you have a permutation problem. If the answer is no, you are dealing with a problem involving combinations. Finally, in some problems no formula will be helpful. In those cases it is necessary to think hard and use common sense.

PROBLEMS 5.3   *Part 1   Permutation Problems*

1. How many permutations are there of seven objects?

2. How many ways can six children be lined up?

3. How many ways can the letters of the word *CATNIP* be rearranged?

4. How many four-letter words (not necessarily in English) can be made from the letters of the word *CATNIP*?

In Problems 5–18, compute the number of permutations.

5. $P_{5,3}$	6. $P_{6,1}$	7. $P_{10,3}$
8. $P_{9,5}$	9. $P_{12,8}$	10. $P_{13,12}$
11. $P_{8,7}$	12. $P_{4,0}$	13. $P_{n,0}$
14. $P_{6,6}$	15. $P_{n,n}$	16. $P_{n,n-1}$
17. $P_{n,n-2}$	18. $P_{n,1}$	

19. A scrabble player has seven distinct letters in front of her and wishes to play a four-letter word. If she chooses to test each possible four-letter permutation before playing, how many words must she test?

20. If the scrabble player of Problem 19 tests all six-letter words and it takes her 2 seconds to test each word, how long will it take?

21. An environmental group ranks the 50 members of the state legislature according to their actions on certain key issues. Assuming no ties, how many rankings are possible?

22. The group of Problem 21 finds the 12 legislators with the worst environmental records and accords them the title "the dirty dozen." In how many ways can a dirty-dozen list be compiled? Here the dozen are listed in order with the worst record listed first.

23. The environmentalists of Problem 21 determine the ten legislators with the best records and put their names on an environmental honor roll. In how many ways can the honor roll be determined? Here the members of the honor roll are listed in order with the best record listed first.

24. How many distinct permutations are there of the letters of the word *RABBIT*?

25. How many distinct permutations are there of the letters of the word *ERROR*?

*26.   How many distinct permutations are there of the letters of the word *BARBAROUS*?

Part 2   Combination Problems

In Problems 27–47, determine the number of combinations.

27. $\binom{5}{3}$ 28. $\binom{6}{1}$ 29. $\binom{5}{2}$ 30. $\binom{6}{5}$

31. $\binom{10}{3}$ 32. $\binom{10}{7}$ 33. $\binom{9}{5}$ 34. $\binom{9}{4}$

35. $\binom{12}{8}$ 36. $\binom{13}{1}$ 37. $\binom{13}{12}$ 38. $\binom{12}{4}$

39. $\binom{8}{7}$ 40. $\binom{4}{0}$ 41. $\binom{6}{6}$ 42. $\binom{n}{0}$

43. $\binom{n}{n}$ 44. $\binom{n}{n-1}$ 45. $\binom{n}{1}$ 46. $\binom{n}{2}$

47. $\binom{n}{n-2}$

48.   Show that $\binom{n}{k} = \binom{n}{n-k}$. Explain why this must be true.

49.   From a ten-person committee, how many different four-person subcommittees can be formed?

50.   An exam contains ten questions and a student must answer six of them. In how many ways can this be done?

 51.   A bridge hand contains 13 cards. From a standard deck of 52 cards, how many bridge hands can be dealt?

52.   The supreme court has nine members. In how many ways can a five-to-four decision be reached?

53.   In Problem 52, in how many ways can a six-to-three decision be reached?

54.   A salesman carries 14 brands of shirts, but can show only 6 brands at any one sales call. In how many ways can these 6 brands be chosen?

55.   A certain course covers ten topics in probability and eight topics in matrix theory. The final exam will have five questions with at most one from any topic. Three questions will be on probability and two will be on matrix theory. In how many ways can the topics examined be chosen?

56.   Six persons are to be chosen from a group of ten men and ten women.

   (a)   What is the number of ways that the six persons can be chosen?

   (b)   What is the number of ways that more men than women can be chosen?

*57.   How many poker hands contain a full house (three tens and two kings, for example)? [*Hint:* Do this problem in several steps. First, compute the number of ways of choosing a denomination to give you three of a kind. In how many ways can the three be chosen out of the four in the denomination (three tens out of four tens, for example)? Then determine the number of ways to choose a denomination giving a pair and compute the number of ways of, for instance, choosing two kings from among four kings. Finally, use the fundamental principle of counting.]

Part 3 Assorted Counting Problems

58. A ten-member committee chooses a chairperson, vice-chairperson, secretary and treasurer. In how many ways can this be done?

59. In how many ways can a four-person subcommittee be chosen from the committee of Problem 58?

60. In a genetics experiment, four white peas, seven red peas, and five pink peas are chosen for pollination from a sample of ten white, ten red, and ten pink peas. (The color of the peas refers to the color of their flowers.) In how many ways can this be done?

61. How many "words" can be formed from the symbols X and Y if each word must contain at least one X and if the maximum length of the words is three letters, with the order of the letters not being relevant? (The X- and Y-chromosomes determine sex. Normal females and males are XX and XY, but nondisjunction of the sex chromosomes may give rise to X-, XXX-, XXY-, and XYY-chromosomes.)

62. Five drugs are used in the treatment of a disease. It is believed that the sequence in which the drugs are used may have a considerable effect on the result of treatment. In how many different orders can the drugs be administered?

63. How many poker hands contain a straight (five cards in ascending order, such as 5 6 7 8 9 or 9 10 J Q K)?

*64. How many poker hands contain exactly three of a kind?

*65. How many poker hands contain two pairs?

*66. How many poker hands contain only one pair?

67. Suppose that there are five highways joining cities A and B and three highways joining cities B and C. How many different routes join cities A and C?

68. In Problem 67, in how many ways can the round trip from city A to city C be made

(a) without traveling the same route twice?

(b) without traveling the same highway twice?

69. How many distinct groups of five letters can be chosen from the letters, $A, B, C, D, E,$ $F, G,$ and H?

70. How many distinct five-letter "words" can be made from the letters $A, B, C, D, E, F,$ $G,$ and H?

71. Among 20 teachers in a mathematics department, students give prizes to the best teacher and the worst teacher. In how many ways can the prizes be awarded?

72. In the same math department, two faculty members are chosen to represent the department in a university committee. In how many ways can this be done?

73. On an accounting test there are 20 multiple-choice questions, each with four possible answers. If a student guesses on each question, how many possible sets of answers are there?

74. If, on the test of Problem 73, the student must first select eight questions and then guess, how many possible sets of answers are there?

75. Of 100 selected stocks on the New York Stock Exchange, 37 advanced and 51 declined. In how many different ways could this happen?

76. A committee of the House of Representatives has 9 Democrats and 7 Republicans. In how many ways can a subcommittee consisting of 3 Democrats and 2 Republicans be chosen?

77. In Problem 76, in how many ways can a majority (Democratic) chairman and vice-chairman and a minority (Republican) chairman and vice-chairman be chosen?

78. A football team contains (offensively) 7 backs, 11 linemen, 5 ends, and 3 quarterbacks. In how many ways can the coach select a starting team of 3 backs, 5 linemen, 2 ends, and 1 quarterback?

*79. Fifteen children must be placed on four teams containing four, three, five, and three members, respectively. In how many ways can this be done?

80. A laboratory cage contains eight white mice and six brown mice. Find the number of ways of choosing five mice from the cage if

 (a) they can be of either color.

 (b) three must be white and two must be brown.

 (c) they must be of the same color.

81. A committee contains 12 members. A minimum quorum for meetings of the committee consists of 8 members. In how many ways can a quorum occur?

*82. Three types of bacteria are cultured in nine test tubes. Three test tubes contain bacteria of the first type, four contain bacteria of the second type, and two contain bacteria of the third type. In how many distinct ways can the test tubes be arranged in a row in a test tube rack, assuming that we cannot distinguish among test tubes containing the same types of bacteria?

83. Mr. and Mrs. Smith and four other people are seated in six chairs (in a row). In how many seating arrangements will the Smiths end up sitting together?

84. The regions of the country are divided into area-code zones, and in each zone the telephone numbers start with a three-digit prefix. Each area code has a zero or a one as its second digit. Assume that each three-digit prefix contains no zero.

 (a) What is the maximum number of area codes?

 (b) What is the maximum number of distinct telephone numbers in a given area-code zone?

 (c) What is the maximum number of distinct telephone numbers in the country?

**85. In Problem 84 how many telephone numbers in a zone will have exactly one repeated digit?

*86. In Problem 84 how many telephone numbers in a zone will have at least one repeated digit?

87. In problem 84 how many telephone numbers in a zone start with the number 23?

*88. In Problem 84 how many telephone numbers in a zone contain the numbers 2 and 3?

*89. Five people, labeled P_1, P_2, P_3, P_4, and P_5, are to be tested for blood types A, B, AB, and O.

 (a) How many different distributions of blood types are possible among these five people?

 (b) In how many of these distributions is there at least one person of each blood type?

90. In the computer diagnosis of a certain disease, the computer is programmed to take into account two primary and six secondary symptoms. The disease is diagnosed positively if at least one of the primary symptoms and at least two of the secondary symptoms are present.

 (a) How many combinations of the symptoms can occur?

 (b) How many conbinations of the symptoms lead to a negative diagnosis?

 (c) How many combinations of the symptoms lead to a positive diagnosis?

91. For the purposes of this problem, two families are said to have the *same structure* if the number of children is the same and the sexes of the children taken in order of birth are the same. We restrict our attention to families of five children.

(a) How many different structures are possible for such families?

(b) In how many of these structures are both the oldest and the youngest children girls?

(c) In how many structures is the fifth child the second girl of the family?

5.4 Sample Spaces and Equiprobable Spaces

The common feature of every situation involving probabilities is an action, or occurrence, which can take place in several ways. It may rain tomorrow or it may not. You may collect on your insurance policy or you may not. You may win at roulette or you may not. We analyze these situations by comparing the likelihood of occurrences of the various possibilities.

The theory of probability is developed as a study of the outcomes of trials of an experiment. An **experiment** is a phenomenon to be observed according to a clearly defined procedure. It may be as simple as tossing a coin and observing the outcome or as complex as polling 1000 people from a large population to determine their preferences on a variety of social, economic, and political issues. A **trial** of an experiment is a single performance of the experiment.

Sample Space of an Experiment   The **sample space** S of an experiment is the *set* of all possible **outcomes** of one trial of the experiment. If the experiment has a finite number of outcomes, the sample space is said to be *finite*.

Example 1 Consider the experiment of tossing a coin once and observing the outcome. The sample space is $S = \{H, T\}$. The possible outcomes are H and T.

Example 2 Consider the experiment of tossing a coin twice and observing the outcome. Then

$$S = \{HH, HT, TH, TT\}.$$

The possible outcomes are HH, HT, TH, and TT. It is often useful to illustrate the possibilities using a tree diagram. This is done in Figure 1.

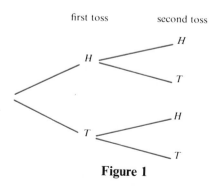

Figure 1

Example 3   A couple has three children. The sex and order of birth of the children is observed. Describe the sample space of this experiment.

Solution   The possibilities are illustrated in the tree diagram of Figure 2.

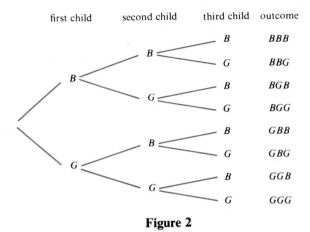

Figure 2

Clearly, the sample space is

$$S = \{BBB, BBG, BGB, BGG, GBB, GBG, GGB, GGG\}.$$

Example 4   A couple has three children and the number of girls is recorded. Describe the sample space of this experiment.

Solution   Here $S = \{0, 1, 2, 3\}$, since we have been asked only for the number of girls. We can also illustrate this in a tree diagram (Figure 3). Note that two of the possible outcomes can occur in more than one way.

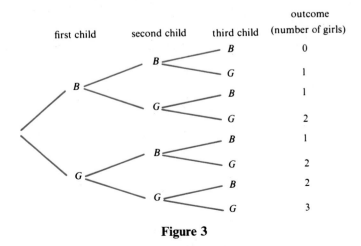

Figure 3

Remark.   Examples 3 and 4 illustrate the important point that the sample space of an experiment depends not only on what is being done but also on what question is asked.

Example 5   One thousand people are polled and asked whether they are for or against the peacetime draft. The number of people for the draft is recorded. The sample space is

$$S = \{0, 1, 2, \ldots, 1000\}.$$

Example 6   A fair coin is tossed until the first head appears. The number of tosses is recorded. Describe the sample space.

Solution   The coin must be tossed at least once. However, there is no upper limit to the number of tosses. It may be very unlikely that a tail comes up 100 or 1000 times in a row, but it is not impossible. Thus the sample space is

$$S = \{1, 2, 3, \ldots\}.$$

This is an example of an **infinite** sample space. We shall not deal with infinite sample spaces in this chapter.

Example 7   Two dice are tossed. The two numbers that appear are recorded. The sample space contains 36 pairs of numbers:

$$S = \{(1, 1), (1, 2), \ldots, (1, 6), (2, 1), \ldots, (2, 6), \ldots, (6, 1), \ldots, (6, 6)\}.$$

Example 8   Two dice are tossed and the sum of the numbers that turn up is recorded. The sample space is given by

$$S = \{2, 3, 4, 5, 6, 7, 8, 9, 10, 11, 12\}.$$

Remark.   We can illustrate the sample spaces of Examples 7 and 8 in a single tree diagram, as in Figure 4.

Example 9   Human blood type is classified according to the presence or absence of the antigens A, B, and Rh. If the antigen A or B is present, it is listed. If both are absent, the blood is said to be type O. If the Rh antigen is present, then the blood type is said to be *positive*. Otherwise it is called *negative*. If an experiment consists of choosing a single individual and determining his or her blood type, the sample space is given by

$$S = \{O^+, A^+, B^+, AB^+, O^-, A^-, B^-, AB^-\}.$$

Example 10   One hundred digital watches are tested for a given defect and the number of watches with the defect is recorded. The sample space consists of the numbers from 0 to 100. That is,

$$S = \{0, 1, 2, \ldots, 100\}.$$

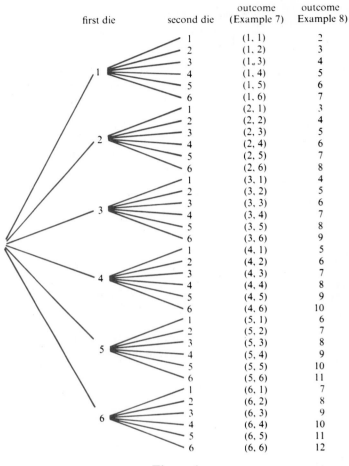

first die	second die	outcome (Example 7)	outcome Example 8
1	1	(1, 1)	2
	2	(1, 2)	3
	3	(1, 3)	4
	4	(1, 4)	5
	5	(1, 5)	6
	6	(1, 6)	7
2	1	(2, 1)	3
	2	(2, 2)	4
	3	(2, 3)	5
	4	(2, 4)	6
	5	(2, 5)	7
	6	(2, 6)	8
3	1	(3, 1)	4
	2	(3, 2)	5
	3	(3, 3)	6
	4	(3, 4)	7
	5	(3, 5)	8
	6	(3, 6)	9
4	1	(4, 1)	5
	2	(4, 2)	6
	3	(4, 3)	7
	4	(4, 4)	8
	5	(4, 5)	9
	6	(4, 6)	10
5	1	(5, 1)	6
	2	(5, 2)	7
	3	(5, 3)	8
	4	(5, 4)	9
	5	(5, 5)	10
	6	(5, 6)	11
6	1	(6, 1)	7
	2	(6, 2)	8
	3	(6, 3)	9
	4	(6, 4)	10
	5	(6, 5)	11
	6	(6, 6)	12

Figure 4

As these ten examples illustrate, sample spaces are all around us. In order to define the probability that something happens, we need first to define an event.

Event

An **event** E is a subset of the sample space S. That is, an event is a set of possible outcomes chosen from S. The event E is called a **simple event** if it contains exactly one of the outcomes. The empty set ϕ, which is a subset of S, is called the **impossible event,** while S itself is called the **certain event**.

Example 11   As in Example 2, a coin is tossed twice and the outcome is observed. What are the events?

Solution   The sample space is $\{HH, HT, TH, TT\}$. There are four simple events:

$$\{HH\}, \quad \{HT\}, \quad \{TH\}, \quad \{TT\}.$$

There are six events containing two outcomes:

$$\{HH, HT\}, \quad \{HH, TH\}, \quad \{HH, TT\}, \quad \{HT, TH\}, \quad \{HT, TT\}, \quad \{TH, TT\}.$$

| Head on first toss | Head on second toss | Both tosses the same | Both tosses different | Tail on second toss | Tail on first toss |

There are four events containing three outcomes:

$$\{HH, HT, TH\}, \quad \{HH, HT, TT\}, \quad \{HH, TH, TT\}, \quad \{HT, TH, TT\}.$$

At least one head   At least one tail

There are two other subsets:

$$\phi \qquad \text{The impossible event}$$

and

$$S = \{HH, HT, TH, TT\} \qquad \text{The certain event}$$

Counting, we find that our sample space contains 16 events.

Example 12   Two dice are tossed and (as in Example 8) the sum of the numbers that turn up is recorded. Describe each event.

(a)   The sum is odd.

(b)   The sum is greater than 8.

Solution   The sample space is given by

$$S = \{2, 3, 4, 5, 6, 7, 8, 9, 10, 11, 12\}$$

(a) $E = \{\text{sum is odd}\} = \{3, 5, 7, 9, 11\}$

(b) $E = \{\text{sum} > 8\} = \{9, 10, 11, 12\}$

Our next step is to define what we mean by the probability of an event in a sample space. We define this central concept in two stages. The easiest definition of probability is the intuitive definition already used in Sections 5.1 and 5.2.

Equiprobable Space   Suppose that in an experiment all the outcomes (simple events) in its sample space S are equally likely to occur. Then S is said to be an **equiprobable** (or **uniform**) **space**.

Probability of Events in a Finite Equiprobable Space   Suppose that S is a finite equiprobable space. The **probability** of an event E in S, written $P(E)$, is defined to be the number of outcomes in E divided by the number of outcomes in S. That is, if $n(E)$ denotes the number of outcomes in E,

$$P(E) = \frac{n(E)}{n(S)} = \frac{\text{number of outcomes in } E}{\text{number of outcomes in } S}. \tag{1}$$

With the definition of probability given in (1), we can compute an astonishingly large number of probabilities. To do so, we need to do two things. First, we must make sure that we are dealing with an equiprobable space. This is not always obvious (or correct). Second, we must, if necessary, use the counting techniques described in Sections 5.2 and 5.3.

Example 13   A fair coin is tossed. What is the probability that the result is a head?

Solution   The word *fair* means that H and T are equally likely. Thus with $E = \{H\}$ and $S = \{H, T\}$, we have

$$P(E) = \frac{n(E)}{n(S)} = \frac{1}{2}.$$

Example 14   Two fair dice are thrown. What is the probability that the sum of the numbers showing is 7?

Solution   Look at Examples 7 and 8. The experiment has at least two different sample spaces. Which one should we use? The answer is suggested in Figure 4. Since the dice are fair, every number is just as likely as every other number to turn up on any one die. That means that each of the 36 pairs in Example 7 is equally likely to occur. In Example 7, the event $E = \{$sum of 7$\}$ is the subset of S given by

$$E = \{(1, 6), (6, 1), (2, 5), (5, 2), (4, 3), (3, 4)\}.$$

Thus

$$P(\text{sum of 7}) = P(E) = \frac{n(E)}{n(S)} = \frac{6}{36} = \frac{1}{6}.$$

Warning.   If we make the mistake of using the sample space of Example 8, we obtain the wrong answer. Here $E = \{7\}$ contains only one outcome (that is, it is a simple event), suggesting that $P(E) = n(E)/n(S) = \frac{1}{11}$. The problem here is that the space $S = \{2, 3, \ldots, 12\}$ is *not* an equiprobable space because the outcomes in it are *not* equally likely. Certainly 2, which can occur in only one way $((1, 1))$ is less likely to occur than 7, which can occur in six ways.

A key to determining that a sample space is an equiprobable space is the use in the problem of the words *at random*.

> If people or objects are chosen at random, then each person or object is equally likely to be chosen.

Example 15 Suppose that in a group of ten persons, four are male. If two are chosen at random, what is the probability that (a) both are male, (b) both are female, and (c) one is male and one is female?

Solution Let A be the event that both are male, B the event that both are female, and C the event that there is one of each. The sample space S consists of pairs of people and, since the order in which the people are chosen is irrelevant,

S contains $\binom{10}{2} = 45$ outcomes.

(a) There are $\binom{4}{2} = 6$ ways to choose 2 males from among the 4 present, so

$$P(A) = \tfrac{6}{45} = \tfrac{2}{15}.$$

(b) There are $\binom{6}{2} = 15$ ways to choose 2 females, so

$$P(B) = \tfrac{15}{45} = \tfrac{1}{3}.$$

(c) There are 4 ways to choose a male and 6 ways to choose a female, so—by the fundamental principle of counting—there are $4 \cdot 6 = 24$ outcomes in C. Thus

$$P(C) = \tfrac{24}{45} = \tfrac{8}{15}.$$

Note. $P(A) + P(B) + P(C) = \tfrac{2}{15} + \tfrac{1}{3} + \tfrac{8}{15} = \tfrac{2}{15} + \tfrac{5}{15} + \tfrac{8}{15} = 1$. This follows from rules of probability given in the next section.

Example 16 What is the probability of obtaining a flush when a five-card poker hand is dealt?

Solution Before answering this question, we must first assume that all poker hands are equally likely to occur; that is, we assume that the poker game is honest and the cards are well shuffled. Then $E = \{flush\}$ and, from Example 5.3.11 on page 234, $n(E) = 5148$. Similarly, from Example 5.3.10 on page 233, $n(S) = 2{,}598{,}960$. Thus

$$P(E) = \frac{5{,}148}{2{,}598{,}960} = 0.0019807923 \approx \frac{1}{505}.$$

Example 17 Of 20 commercial vehicles which break down at the same time, 15 have been repaired within 3 days. Suppose that 5 vehicles were chosen at random from the 20. What is the probability that exactly 3 were repaired within 3 days?

Solution The event E whose probability we seek is given by

$$E = \{3 \text{ are repaired, 2 are not}\}.$$

To obtain $n(E)$, we must compute the number of ways of choosing 3 vehicles from the 15 repaired vehicles and 2 vehicles from the remaining group. Thus, from the fundamental principle of counting,

$$n(E) = \binom{15}{3}\binom{5}{2} = 455 \cdot 10 = 4550$$

$$n(S) = \binom{20}{5} = 15{,}504$$

so that

$$P(E) = \frac{n(E)}{n(S)} = \frac{4550}{15{,}504} \approx 0.29.$$

Remark. We could solve this problem because the words *at random* suggest that all combinations of five vehicles were equally likely to be chosen.

Example 18 What is the probability that in a family of three children there are (a) exactly two girls, (b) two or more girls?

Solution Before starting, we have to assume that boys and girls are equally likely.† With this assumption, the sample space of Example 3 is an equiprobable space and we can see, from the tree diagram in Figure 2 on page 240, that the event {2 girls} occurs in 3 ways and the event {3 girls} occurs in 1 way. Since the event {2 or more girls} is the event {2 girls or 3 girls} and since the sample space contains 8 outcomes, we have the following.

(a) $P(\text{exactly 2 girls}) = \frac{3}{8}$

(b) $P(\text{two or more girls}) = \dfrac{3+1}{8} = \dfrac{1}{2}$

PROBLEMS 5.4   In Examples 1–5, (a) describe the sample space of the experiment and (b) find the probability of an elementary event in the sample space.

1. Draw a card at random from a standard 52-card deck.

2. Choose at random an integer from 1 to 10.

3. Choose five persons at random from a group of ten.

4. Dial a seven-digit number at random.

5. Choose a chairperson and co-chairperson at random from a six-person committee.

6. The numbers 1 to 100 are written on slips of paper and placed in a bowl. After the bowl is thoroughly shaken, one of the slips of paper is drawn at random.

---

† Actually, in the United States over the past 20 or 30 years, approximately 51.2% of all births were boys and 48.8% were girls.

(a) What is the probability that the number drawn is greater than 75?

(b) What is the probability that the number drawn is divisible by 3?

(c) What is the probability that the number drawn is divisible by 15?

7. A professor assigns 12 different grades to the 12 students in his class. Because of a computer error, the grades are distributed at random on the transcripts of his students.

(a) What is the probability that every student receives his or her correct grade?

(b) What is the probability that at least one receives an incorrect grade?

(c) What is the probability that exactly 11 students receive their correct grades?

8. A cage contains six white mice and four brown mice. Consider the experiment of drawing three mice at random from the cage and observing the colors of those drawn.

(a) Describe the sample space of the experiment.

(b) Calculate the probabilities of the four possible distributions of color (three white, two white and one brown, and so on).

9. Six persons are chosen at random from a group of 20 men and 8 women.

(a) What is the probability that all 6 chosen are men?

(b) What is the probability that 3 men and 3 women will be selected?

10. A chimpanzee is placed at a toy typewriter with the letters A, B, C, D, and E. The chimpanzee types four keys at random.

(a) What is the probability that the word *BEAD* is typed?

(b) What is the probability that all typed letters are the same?

In Problems 11–15, a 5-card poker hand is dealt at random from a standard 52-card deck. Find the probability of the given hand.

*11. A full house.   **12.** A straight.   *13. 3 of a kind only.

*14. Two pairs only.   *15. One pair.

16. A bridge hand consists of 13 cards from a deck of 52 cards. If a bridge hand is dealt at random, what is the probability that all 13 cards will be of the same suit?

17. An accounting quiz contains five multiple-choice questions with four possible answers for each question. If a student guesses on all five questions, what is the probability that he will get a score of 100%?

18. A local union has eight members, two of whom are women. Two of the members are chosen by a lottery to represent the union on a bargaining council. Find the probability of each event.

(a) Both women are chosen.

(b) One man and one woman are chosen.

(c) Two men are chosen.

*19. Of a group of blood donors, three have type A blood, eight have type O blood, three have type B blood and two have type AB blood. If three people are chosen at random, what is the probability that exactly two have the same blood type?

*20. Suppose you flip a coin eight times in a row.

   (a) What is the probability of getting the outcome $HHTTHHTT$?

   (b) What is the probability of getting all heads?

   (c) What is the probability of getting at least one head?

   (d) What is the probability that the first toss is a tail?

21. A family is known to have four children.

   (a) What is the probability that all the children are of the same sex?

   (b) What is the probability that there is no girl in this family older than any boy in the family?

22. In a group of 12 people, 4 people are under 20 years of age, 5 are between ages 20 and 40, and 3 are over 40 years old. Six people are chosen at random from this group.

   (a) What is the sample space of this experiment? What is the probability of an elementary event?

   (b) What is the probability that the 3 people over 40 are chosen?

   (c) What is the probability that the 6 youngest people are chosen?

23. In a group of 15 people, 10 are right-handed, 4 are left-handed, and one is ambi-dextrous. Four people are to be chosen at random from this group.

   (a) What is the sample space of this experiment?

   (b) What is the probability of an elementary event in this experiment?

   (c) What is the probability that the 4 left-handed people are chosen?

24. What are the sample spaces of the following experiments? Determine the number of elementary events for each experiment and the probability of each elementary event.

   (a) Choose three integers at random from 1 to 100 without repetition.

   (b) Choose two blue objects and three red objects at random from a set consisting of five blue objects and four red objects.

25. A family has five children. Define A to be the event that the oldest two children are both girls and B to be the event that the youngest three children are all girls.

   (a) Define the events $A \cup B$, $A \cap B$, $A - B$ and $B - A$.

   (b) Determine the probabilities of the six events in this problem.

5.5 Finite Probability Spaces

Equiprobable spaces are not the only sample spaces that arise in practical situations. In fact, most sample spaces that occur in applications are *not* equiprobable spaces. For example, it may rain 20% of the days in a certain city in July in which there are cloudy days 30% of the time and clear days 50% of the time. Then the sample space of observed weather is $S = \{$rainy, cloudy, clear$\}$ which is not an equiprobable space. The probability of rain on a given, randomly chosen day is not $\frac{1}{3}$.

As another familiar example, consider the experiment of flipping a coin twice and counting the number of heads. The sample space is $S = \{0, 1, 2\}$. Since the four equally likely possible outcomes of flipping a coin twice are HH,

HT, TH, and TT, we know that one head (HT or TH) is twice as likely as two heads (HH), so the space $\{0, 1, 2\}$ is not an equiprobable space.

Evidently we need a more general definition of probability.

Probability Space   A **probability space** S is a sample space, together with a rule P that assigns to each subset of S a real number; P satisfies the following four axioms:

> **1.** $P(\phi) = 0$ (1)
>
> **2.** $P(S) = 1$ (2)
>
> **3.** $0 \leq P(E) \leq 1$ for any event $E \subset S$.   (3)
>
> **4.** If $A, B \subset S$ and $A \cap B = \phi$, then
>
> $$P(A \cup B) = P(A) + P(B).$$ (4)

The rule P is called a **probability function** and the number $P(E)$ is called the **probability** of the event E.

Some explanation of this definition is called for. The most difficult question is, "From where does the rule P come?" The answer is not always given by mathematical theory—except in an equiprobable space. In our weather example, we might assign the probabilities 0.2 to rain, 0.3 to cloudy, and 0.5 to clear weather. This would conform to our observations. But other numbers that added up to 1 would still not violate our definition, although they might not make much sense.

The four axioms are easily explained. Axiom 1 says that we assign a probability of zero to the impossible event. Axiom 2 says that the sample space (which we called the certain event) occurs with probability 1. Put another way, Axiom 2 says that if we do an experiment, then something in the sample space will occur 100% of the time. Axiom 3 says that any event is at least as likely to occur as the impossible event and not more likely to occur than the certain event.

Axiom 4 is best illustrated by a Venn diagram (see Figure 1). Here $A \cap B = \phi$ means that A and B are disjoint (see equation (1.1.17) on page 6). In this case we say that the events A and B are **mutually exclusive**. Think of probabilities as

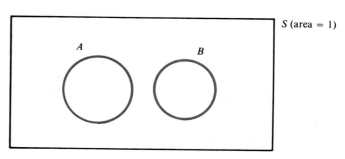

Figure 1

areas, with the certain event having an area of 1. Then $P(A)$ represents the area of the set A and $P(B)$ represents the area of set B. Clearly, if $A \cap B = \phi$, the area of the set $A \cup B$ is the sum of areas of A and B. That is $P(A \cup B) = P(A) + P(B)$. This is Axiom 4.

Example 1 Consider the experiment of tossing a fair coin three times in succession and observing the number of heads that appear.

(a) What is the probability of each of the simple events?

(b) What is the probability that all three tosses give the same outcome?

(c) If A is the event {fewer than two heads} and B is the event {two heads}, determine $P(A \cup B)$.

Solution The coin can be tossed three times in succession in eight ways: HHH, $HHT, HTH, HTT, THH, THT, TTH$, and TTT. Each of these ways is equally probable with probability $\frac{1}{8}$ (since the coin is fair). The experiment in which we are interested has the sample space $S = \{0, 1, 2, 3\}$. The event {0 heads} can occur in only one way (TTT) and therefore has probability $\frac{1}{8}$. The event {1 head} can occur in three ways and has probability $\frac{3}{8}$. We conclude that

(a) $P(0) = \frac{1}{8}, P(1) = \frac{3}{8}, P(2) = \frac{3}{8}$, and $P(3) = \frac{1}{8}$.

(b) All three tosses give the same outcome in two ways (HHH and TTT); therefore, the probability of this event is $\frac{2}{8} = \frac{1}{4}$.

(c) A occurs if no head or one head appears. This can occur in four ways, so $P(A) = \frac{4}{8} = \frac{1}{2}$. We have already seen that $P(B) = \frac{3}{8}$. Since A and B are mutually exclusive (fewer than 2 heads and exactly 2 heads cannot occur at the same time), we have, from Axiom 4, $P(A \cup B) = P(A) + P(B) = \frac{1}{2} + \frac{3}{8} = \frac{7}{8}$.

Any event E is the (disjoint) union of the simple events contained in E. The following generalizes Axiom 4.

5. The probability of any event E is the sum of the probabilities of the simple events contained in E. (5)

Since S is an event and $P(S) = 1$, we have the following from (5).

6. The probabilities of the simple events in a sample space always add up to 1. (6)

Rule (6) is often quite useful as a check of our computation of probabilities. After we have computed the probabilities of the simple events in a sample space, we can find their sum. If the sum is not 1, then we have made an error. In Example 1 we note that $\frac{1}{8} + \frac{3}{8} + \frac{3}{8} + \frac{1}{8} = 1$.

Example 2 A California aircraft manufacturer has determined the following probabilities for the number of orders for the coming year for its largest wide-body jet.[†]

Number of orders	Probability
0	0.05
1	0.10
2	0.25
3	0.20
4	0.20
5	0.10
6 or more	0.10
	1.00

Note that the sum of the probabilities is 1

Calculate the probabilities of the following events.

(a) {Fewer than 3 orders}.

(b) {5 or more orders}.

(c) {Between 2 and 5 orders}.

(d) {at most 1 order}.

(e) {An odd number of orders}.

Solution

(a) $P(<3) = P(0) + P(1) + P(2) = 0.05 + 0.1 + 0.25 = 0.4$

(b) $P(\geq 5) = P(5) + P(\geq 6) = 0.1 + 0.1 = 0.2$

(c) $P(2 \leq \text{number of orders} \leq 5) = P(2) + P(3) + P(4) + P(5)$
$$= 0.25 + 0.2 + 0.2 + 0.1 = 0.75$$

(d) $P(\leq 1) = P(0) + P(1) = 0.05 + 0.1 = 0.15$

(e) $P(\text{odd}) = P(1) + P(3) + P(5) = 0.1 + 0.2 + 0.1 = 0.4$

[†] As stated in the introduction, the determination of these probabilities is a problem in statistics and may be a very difficult problem indeed. Certainly the manufacturer must take several factors into account, including past order patterns, current increases or decreases in passenger-miles flown, the effect of competition—both foreign and domestic, the price of fuel, and so on.

Let E be an event in the sample space S. Then E^c, the complement of E, is called the **complementary event** of E. Since $E \cup E^c = S$ and $E \cap E^c = \phi$, we have, from Axiom 4, $P(E) + P(E^c) = P(S) = 1$, or

$$\textbf{7.}\ P(E^c) = 1 - P(E) \tag{7}$$

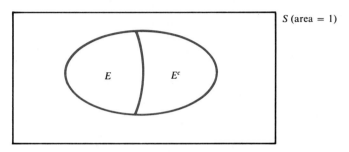

Figure 2

Formula (7) can be very useful in computing probabilities in certain cases, as Examples 3 and 4 show.

Example 3   In Example 2, what is the probability that the manufacturer will receive at least one order for its wide-body jet for the coming year?

Solution   Let E be the event {at least one order}. Then $P(E) = P(1) + P(2) + P(3) + P(4) + P(5) + P(\geq 6)$. We could find this sum, but there is an easier way to solve this problem. We note that $E^c = \{no\ sales\}$ and $P(E^c) = P(0) = 0.05$. Thus, from (7), $P(E) = 1 - 0.05 = 0.95$.

Example 4   A fair coin is flipped ten times in a row. What is the probability that not all the tosses will result in heads?

Solution   We seek $P(E)$, where E is the event $\{<10\ heads\}$. At this point we cannot compute this directly without writing out the $2^{10} = 1024$ sequences, a typical one of which is $HTHHHTTHTH$. However, $E^c = \{10\ heads\}$ and $P(E^c) = P(10\ heads) = P(HHHHHHHHHH) = \frac{1}{1024}$, since the sample space containing the 1024 possible outcomes is an equiprobable space. Thus

$$P(E) = 1 - P(E^c) = 1 - \tfrac{1}{1024} = \tfrac{1023}{1024}.$$

We now derive three more basic results about probability spaces. We assume you are familiar with the set theory material of Section 1.1. If you are not, reread that section.

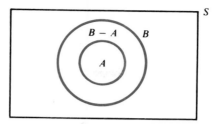

Figure 3

Suppose A is a subset of B. That is, $A \subset B$. From Figure 3, we see that $B = A \cup (B - A)$. Since A and $B - A$ are mutually exclusive, we find that $P(A) + P(B - A) = P(B)$. But $P(B - A) \geq 0$ by Axiom 3. Thus we have the following.

> **8.** If $A \subset B$, then $P(A) \leq P(B)$.

(8)

Now, consider the Venn diagram in Figure 4. We see that

$$A = (A - B) \cup (A \cap B)$$

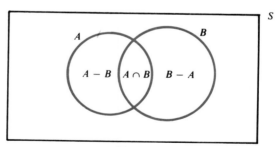

Figure 4

so that, since the sets on the right are disjoint,

$$P(A) = P(A - B) + P(A \cap B)$$

or

> **9.** $P(A - B) = P(A) - P(A \cap B)$.

(9)

We can use (9) to obtain an extremely useful formula. From Figure 4 we see that

$$A \cup B = (A - B) \cup (A \cap B) \cup (B - A)$$

so that, since the sets $A - B$, $A \cap B$, and $B - A$ are mutually disjoint,

$$P(A \cup B) = P(A - B) + P(A \cap B) + P(B - A)$$

$$= \underbrace{P(A) - P(A \cap B)}_{P(A - B)} + P(A \cap B) + \underbrace{P(B) - P(A \cap B)}_{P(B - A)}$$

using (9) twice

or, after canceling,

> **10.** $P(A \cup B) = P(A) + P(B) - P(A \cap B)$. (10)

Equation (10) can be obtained in a different way. Look at Figure 5. As we have indicated, we may think of probabilities as areas. Then, in computing the area of $A \cup B$, we add the area of A to the area of B. But in doing so, we have added in the area of $A \cap B$ twice (since $A \cap B$ is contained in both A and B). Thus we must subtract it once in order to obtain the correct area.

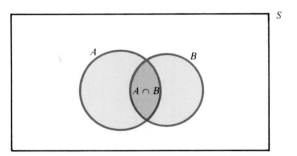

Figure 5

Warning 1. Students often make the error of computing $P(A \cup B) = P(A) + P(B)$. This is correct *only* if $P(A \cap B) = 0$. For example, consider the experiment of flipping a coin twice and counting the number of heads. Let $A = \{\text{at least 1 head}\}$ and $B = \{\text{exactly 1 head}\}$. Then $A \cup B = \{1 \text{ or } 2 \text{ heads}\}$ and $P(A \cup B) = P(1 \text{ head}) + P(2 \text{ heads}) = \frac{1}{2} + \frac{1}{4} = \frac{3}{4}$. But $P(A) = \frac{3}{4}$ and $P(B) = \frac{1}{2}$, so $P(A) + P(B) = \frac{5}{4}$. Clearly, in this example $P(A \cup B) \neq P(A) + P(B)$. This also suggests another warning.

Warning 2. If you ever obtain a probability that is greater than 1 or less than 0, then start again. Any such answer violates Axiom 1 or Axiom 2 and is clearly wrong.

Example 5 In a market survey of toothpaste users, 50% of those surveyed preferred a mint-flavored toothpaste, 25% preferred a toothpaste with a visible stripe, and 40% of those who preferred the stripe also preferred the mint flavor.

(a) What is the probability that an individual chosen at random from those surveyed prefers the mint flavor or the stripe or both?

(b) What is the probability that the chosen individual prefers the stripe but not the mint flavor?

Solution The information given is represented in the Venn diagram in Figure 6. We see that $P(\text{mint}) = 0.5$ and $P(\text{stripe}) = 0.25$. Also, 40% of the stripe people were mint people and 40% of 25% $= 0.4 \times 0.25 = 0.1 = 10\%$. Thus

$$P(\text{mint} \cap \text{stripe}) = 0.1.$$

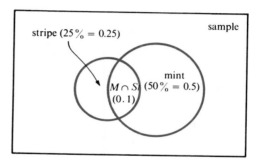

Figure 6

(a) The event {mint or stripe or both} is the event {mint $\cup$ stripe}. We have, from (10),

$$P(\text{mint} \cup \text{stripe}) = P(\text{mint}) + P(\text{stripe}) - P(\text{mint} \cap \text{stripe})$$
$$= 0.5 + 0.25 - 0.1 = 0.65$$

(b) The event {stripe but not mint} is the event {stripe $-$ mint} and, from (9),

$$P(\text{stripe} - \text{mint}) = P(\text{stripe}) - P(\text{stripe} \cap \text{mint})$$
$$= 0.25 - 0.1 = 0.15.$$

Remark. Probability theory often seems mysterious to the student who first encounters it because of the introduction of new terms and formulas. However, a great deal of elementary probability theory is common sense. We could solve Example 5, for example, without using any probability theory at all. It's only a question of drawing the right picture. In Figure 7 we insert percentages easily obtained from the statement of the problem. The answer to part (a) is now obtained by observing that mint or stripe occurs in 40% + 10% + 15% = 65% of all cases. The answer to (b) is even more direct. You will often find that a good

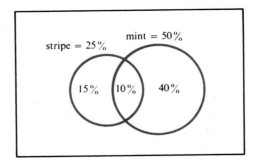

Figure 7

picture (which will almost always be a Venn diagram or a tree diagram) will be extremely useful in solving many problems in probability.

Example 6   A tank contains three fish: A, B, and C. Pellets of food are infrequently placed into the tank. Each time a pellet is dropped, the fish compete for it. Suppose that over a long period of time it is observed that either fish A or B is successful $\frac{1}{2}$ of the time and that either A or C is successful $\frac{3}{4}$ of the time.

(a)   What is the probability that fish A is successful?
(b)   Which fish is the best fed?

Solution   Let A be the event that fish A gets the food, and so on. Since only one of the fish can devour each particle of food, we find that $P(A \cap B) = P(A \cap C) = P(B \cap C) = 0$. We are given $P(A \cup B) = \frac{1}{2}$ and $P(A \cup C) = \frac{3}{4}$. Thus

$$\tfrac{1}{2} = P(A \cup B) = P(A) + P(B) - P(A \cap B) = P(A) + P(B)$$

and

$$\tfrac{3}{4} = P(A \cup C) = P(A) + P(C) - P(A \cap C) = P(A) + P(C).$$

Since one fish is always successful, we also have

$$1 = P(A) + P(B) + P(C).$$

Thus

$$P(B) = \tfrac{1}{2} - P(A), \qquad P(C) = \tfrac{3}{4} - P(A)$$

and

$$1 = P(A) + \overbrace{\tfrac{1}{2} - P(A)}^{P(B)} + \overbrace{\tfrac{3}{4} - P(A)}^{P(C)} = \tfrac{5}{4} - P(A),$$

so that $P(A) = \frac{1}{4}$. Similarly, $P(B) = \frac{1}{4}$ and $P(C) = \frac{1}{2}$. Fish C gets twice as much to eat as each of the other two.

We close this section with a discussion of *odds*.

Odds.

Let E be an event in a sample space S and suppose that

$$P(E) = \frac{a}{a + b}.$$

Then we say that **the odds in favor of E** are a to b and **the odds against E** are b to a.

Example 7 A horse wins $\frac{2}{9}$ of its races. What are the odds against the horse winning in a randomly chosen race?

Solution Let E be the event {the horse wins}. Then

$$P(E) = \frac{2}{9} = \frac{2}{2 + 7}.$$

Hence $a = 2, b = 7$, and the odds against the horse winning are 7 to 2. The odds in favor of the horse winning are 2 to 7.

Example 8 The odds against the San Francisco Giants winning the next National League pennant are 10 to 1. What is the probability that the Giants will win?

Solution We reason as follows: The odds mean that the Giants will not win the pennant 10 times as often as they win it. This means that out of 11 tries, the Giants will win once. Thus

$$P(\text{Giants win the pennant}) = \tfrac{1}{11}.$$

Alternatively, we have $b = 10$ and $a = 1$, so

$$P(\text{Giants win the pennant}) = \frac{a}{a + b} = \frac{1}{1 + 10} = \frac{1}{11}.$$

PROBLEMS 5.5 Problems 1–8 refer to the experiment of throwing two dice and recording the sum of the numbers that turn up.

1. Find the probability of each of the simple events in the sample space.

2. Find the probability that the number recorded is even.

3. Find the probability of a number greater than 7.

4. Find the probability of a number less than or equal to 5.

5. Find the probability that the number is between 4 and 9 inclusive.

6. Let $A = \{\text{number} > 7\}$. Find $P(A^c)$.

7. Let $A = \{\text{number} \leq 6\}$ and $B = \{4 \leq \text{number} \leq 9\}$. Find $P(A \cup B)$.

8. With A and B as in Problem 7, find each value.
   (a) $P(A - B)$
   (b) $P(B - A)$

9. Let A and B be events in a sample space with $P(A) = 0.52$, $P(B) = 0.28$, and $P(A \cap B) = 0.13$. Find each value.
   (a) $P(A \cup B)$
   (b) $P(A - B)$
   (c) $P(B - A)$
   (d) $P(A^c)$
   (e) $P(B^c)$

10. Let A and B be events in a sample space S with $A \cup B = S$, $P(A) = 0.72$ and $P(B) = 0.44$. Find $P(A \cap B)$.

11. Let A and B be events in a sample space with $P(A) = 0.55$, $P(B) = 0.28$, and $P(A \cup B) = 0.67$. Find each value.
    (a) $P(A \cap B)$
    (b) $P(A - B)$
    (c) $P(B - A)$
    (d) $P(A^c)$
    (e) $P(B^c)$

In Problems 12–18, determine whether the events listed are mutually exclusive. If not, explain why not.

12. Throw a die: {even number}, {odd number}.

13. Throw a die: {even number}, $\{<3\}$.

14. Throw a die: $\{\leq 4\}$, $\{\geq 4\}$.

15. Deal a poker hand: {flush}, {straight}.

16. A refrigerator thermostat is inspected: {unable to withstand sufficiently low temperatures}, {inaccurate in the range $-10°C$ to $0°C$}.

17. A tree in a forest is examined: {gypsy moth infestation}, {below-average annual growth}.

18. A coin is tossed: {head}, {tail}.

19. A production manager for a tire company has determined that the average retail demand (per retail outlet) of the company's steel-belted radials are as given in the table.

Demand	Probability
0	0.10
1	0.08
2	0.15
3	0.23
4	0.14
5	0.08
6	0.07
7	0.06
8 or more	0.09

Compute $P(\text{demand} \leq 5)$.

20. In Problem 19, find $P(2 < \text{demand} \leq 6)$.

21. In Problem 19, let $A = \{\text{demand} \leq 5\}$ and $B = \{2 < \text{demand} \leq 6\}$. Find each value.

(a) $P(A^c)$

(b) $P(B^c)$

(c) $P(A \cup B)$

(d) $P(B - A)$

22. A radio manufacturer determined that of 20,000 radios produced, 250 had a defect in the tuning adjustment and 150 had a defect in the volume adjustment. Moreover, 50 of the radios had both defects. A radio is chosen at random.

(a) What is the probability that the radio has at least one of the two defects?

(b) What is the probability that it has the tuning defect but not the volume defect?

(c) What is the probability it has the volume defect but not the tuning defect?

(d) What is the probability that it has neither defect?

*23. Three people are to be chosen randomly from among 12 people, 4 of whom have a cold.

(a) What is the sample space and what is the probability of each simple event?

(b) What is the probability that none of the 3 chosen people has a cold?

(c) What is the probability that at least 1 of the 3 chosen has a cold?

(d) What is the probability that at least 1 of the 3 chosen people does not have a cold?

*24. A family is known to have five children. Let A be the event that the three oldest children are boys and B be the event that the three youngest children are boys.

(a) Describe in words A^c, $A \cap B$, $A \cup B$, and $A - B$.

(b) Calculate the probabilities of the six events in this problem.

25. How many children must a couple plan to have in order that the probability of at least one boy is 90%?

*26. How many children must a couple plan to have in order that the probability of at least one girl and one boy will be greater than 70%?

27. In 1975 the five leading causes of death among Americans were heart diseases (716,215 deaths), cancer (365,693), stroke (194,038), accidents (103,030), and influenza and pneumonia (55,664). What is the probability that a person who died in 1975 from one of these causes actually died from a stroke?

28. In order for a bill to become a federal law, it has to be passed by both the House and Senate and signed by the president. A lobbyist estimates that a certain bill has a probability of 0.4 of being passed by the House, 0.65 of being passed by the Senate and 0.8 of being passed either by the House *or* the Senate. What is the probability that the bill will reach the president's desk for a signature?

****29.** Let A, B, and C be events in a finite sample space S. Show that

$$P(A \cup B \cup C) = P(A) + P(B) + P(C) - P(A \cap B) - P(A \cap C)$$
$$- P(B \cap C) + P(A \cap B \cap C).$$

[*Hint:* Look at a Venn diagram before doing any computations.]

In Problems 30–34 find the odds in favor of and against the given event.

30. An event E with $P(E) = \frac{1}{4}$.

31. An event E with $P(E) = \frac{3}{8}$.

32. An event E with $P(E) = \frac{9}{13}$.

33. A sum of 7 when two dice are thrown.

34. Three heads when a coin is tossed three times.

35. If the odds against E occurring are 12 to 5, find $P(E)$.

36. If the odds in favor of E occurring are 11 to 4, find $P(E)$.

37. Every time a horse runs, the odds against it winning are 3 to 2. Assuming that the odds reflect the horse's true performance, what percentage of the time does the horse win?

5.6 Conditional Probability

The probabilities that we assign to events depend on the information that is known about them. Between the extremes of no information and complete information, there are many levels of partial information which, if known, must be taken into account in the calculation of probabilities.

To illustrate the fundamental idea of this section, consider the following three-part problem: Suppose a family has two children and assume that boys and girls are equally likely at birth.

(a) What is the probability that both children are boys?

(b) If it is known that at least one of the children is a boy, what is the probability that both are boys?

(c) If the oldest child is known to be a boy, what is the probability that both are boys?

All the problems ask the same question but, as we shall see, have different answers.

In (a) the sample space is the equiprobable space $S = \{BB, BG, GB, GG\}$ and $P(\text{both boys}) = P(BB) = \frac{1}{4}$.

(b) Since we know that there is at least one boy, the sample space is reduced to $S_1 = \{BB, BG, GB\}$ and, because these events are equally likely, $P(BB) = \frac{1}{3}$.

(c) Given that the oldest child is a boy, the sample space is now $S_2 = \{BB, BG\}$, so $P(BB) = \frac{1}{2}$.

Note that in (b) and (c) the result of giving more information was to *reduce* the sample space. We now generalize these ideas.

Conditional Probability Let E be an event in the probability space S with $P(E) > 0$. Then the **conditional probability** of an event A in S given that E has occurred is written $P(A|E)$ and is given by

$$P(A|E) = \frac{P(A \cap E)}{P(E)}. \tag{1}$$

Before giving further examples of computations of conditional probability, let us try to see why the definition of conditional probability is reasonable. As we have done before, we think of probability as area. Look at Figure 1, in

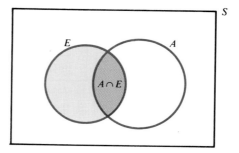

Figure 1

which $P(S) =$ area of $S = 1$, $P(E) =$ area of E, and $P(A) =$ area of A. When we ask for $P(A|E)$, we are asking for the probability that A occurs *given that E has already occurred*. Knowing that E has already occurred *reduces our sample space* to the set E. Then the probability that A also occurs can be thought of as the "relative area" of $A \cap E$ compared to the area of E. This relative area is $P(A \cap E)/P(E)$.

Example 1 Two fair dice are thrown.

(a) What is the probability that at least one of the numbers showing is a 2?

(b) If the sum is 6, what is the probability that one of the numbers is a 2?

Solution

(a) The sample space can be taken as the equiprobable space consisting of the 36 pairs (x, y) where $1 \le x \le 6$ and $1 \le y \le 6$. There are 11 ways to have at least one 2: $(2, 1), (2, 2), \ldots, (2, 6), (1, 2), (3, 2), \ldots, (6, 2)$. Thus

$$P(\text{at least 1 two}) = \tfrac{11}{36}.$$

(b)   We seek

$$P(\text{at least one 2}|\text{sum of 6}) = \frac{P(\text{at least one 2} \cap \text{sum of 6})}{P(\text{sum of 6})}.$$

The event {at least one 2} ∩ {sum of 6} = {(4, 2), (2, 4)} with probability $\frac{2}{36}$. A sum of 6 can occur in five ways, {(1, 5), (5, 1), (2, 4), (4, 2), (3, 3)}, so $P\{\text{sum of 6}\} = \frac{5}{36}$. Thus

$$P(\text{at least one 2}|\text{sum of 6}) = \frac{\frac{2}{36}}{\frac{5}{36}} = \frac{2}{5}.$$

There is a simpler way to solve this problem. Knowing that the sum is 6, the only possibilities are (1, 5), (5, 1), (2, 4), (4, 2), (3, 3). These five events are equally likely, so $P(\text{at least one 2}) = \frac{2}{5}$. The reduced sample space is illustrated in Figure 2.

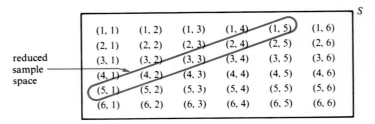

Figure 2

Remark.   *It is often the case that conditional probability problems can be solved more easily by first considering a reduced sample space.*

Example 2   In a survey of cable television viewers it was found that 35% liked an all-sports program, 15% liked an all-news program, and 10% liked both. A person from the sample is chosen at random.

(a)   If the person liked the all-news program, what is the probability that he or she also liked the all-sports program?

(b)   If the person liked the all-sports program, what is the probability that he or she also liked the all-news program?

(c)   What is the probability that the person liked at least one of the programs?

Solution   We will solve this problem two ways: by formula and by picture. You can decide which is easier.

(a) $P(AS|AN) = \dfrac{P(AS \cap AN)}{P(AN)} = \dfrac{0.10}{0.15} = \dfrac{10}{15} = \dfrac{2}{3}$

(b) $P(AN|AS) = \dfrac{P(AN \cap AS)}{P(AS)} = \dfrac{0.10}{0.35} = \dfrac{10}{35} = \dfrac{2}{7}$

(c) $P(AN \cup AS) = P(AN) + P(AS) - P(AN \cap AS)$

$$= 0.15 + 0.35 - 0.10 = 0.4$$

Those are the answers. Let's do it another way. In Figure 3 the situation is depicted in a Venn diagram. To compute $P(AS|AN)$, we first note that the sample space has been reduced to the all-news circle. Then we ask what percentage of the all-news circle is taken up by the part that's also in the all-sports circle. The answer is $0.10/0.15 = \frac{2}{3}$. Similarly, we find that $P(AN|AS) = 0.10/0.35 = \frac{2}{7}$. The answers are the same. The advantage of the second method is that you can see what is going on.

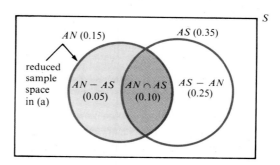

Figure 3

Example 3   A Senate committee contains six Republicans and five Democrats. Three members are chosen at random for a subcommittee.

(a)   What is the probability that all three are Republicans?

(b)   What is the probability that all three are Republicans if it is known that at least one member of the subcommittee is a Republican?

Solution

(a)   This is a typical problem involving combinations. From Sections 5.3 and 5.4, the answer is given by

$$P(3R) = \frac{\text{number of ways of choosing 3 Republicans from 6 Republicans}}{\text{number of ways of choosing 3 subcommittee members from 11}}$$
$$\text{committee members}$$

$$= \frac{\binom{6}{3}}{\binom{11}{3}} = \frac{20}{165} = \frac{4}{33} \approx 0.1212$$

(b)   We seek $P(3R|\text{at least }1R) = P(3R \cap \text{at least }1R)/P(\text{at least }1R)$. The event $\{3R \cap \text{at least }1R\} = \{3R\}$ and $P(3R) = \frac{20}{165}$. There are two ways to compute $P(\text{at least }1R)$. The hardest way is to add: $P(\text{at least }1R) = P(1R) +$

$P(2R) + P(3R)$. A much easier way is to note that the complementary event of {at least $1R$} is {no R} = {3 Democrats}. But

$$P(3D) = \frac{\binom{5}{3}}{\binom{11}{3}} = \frac{10}{165}$$

so that

$$P(\text{at least } 1R) = 1 - \tfrac{10}{165} = \tfrac{155}{165}.$$

Thus

$$P(3R \,|\, \text{at least } 1R) = \frac{\frac{20}{165}}{\frac{155}{165}} = \frac{20}{155} = \frac{4}{31} \approx 0.1290.$$

Note. Here the sample space is reduced to the 155 choices of a subcommittee that contain at least one Republican. However, in this problem that information simplifies computations only very slightly.

Example 4 An employment agency specializes in technical workers and categorizes prospective employees according to area of expertise and years of experience. The data for 100 selected candidates are given in the table.

	Field			
Experience	Electrical engineer	Computer scientist	Chemist	**Total**
Under 2 years	10	24	10	44
2–5 years	18	8	12	38
More than 5 years	6	10	2	18
Total	34	42	24	100

One candidate is chosen at random. Find each probability.

(a) That the candidate has more than 5 years' experience given that he is a computer scientist.

(b) That the candidate is an electrical engineer given that he has 2–5 years' experience.

(c) That the candidate is a chemist *or* has less than 2 years' experience.

Solution

(a) $P(>5 \,|\, CS) = P(>5 \cap CS)/P(CS)$. Ten people have more than 5 years' experience and are computer scientists. Thus $P(>5 \cap CS) = \frac{10}{100}$. Similarly, $P(CS) = \frac{42}{100}$. Thus

$$P(>5|CS) = \frac{\frac{10}{100}}{\frac{42}{100}} = \frac{10}{42} = \frac{5}{21}.$$

However, it is much simpler to observe that the sample space has been reduced to the 42 computer scientists and, of these, 10 have more than 5 years' experience. From that, directly, $P(>5|CS) = \frac{10}{42} = \frac{5}{21}$.

(b) Here, 38 people have 2–5 years' experience and, of these, 18 are electrical engineers. Thus $P(EE|2\text{--}5) = \frac{18}{38} = \frac{9}{19}$. Of course, we can solve this using the formula for conditional probability, but there is no reason to do so.

(c) There are two ways to solve this. The hard way is, as before, to use a formula:

$$P(C \cup <2) = P(C) + P(<2) - P(C \cap <2)$$

$$= \frac{24}{100} + \frac{44}{100} - \frac{10}{100} = \frac{58}{100} = 0.58.$$

The easy way is to count. From the table there are $10 + 12 + 2 + 10 + 24 = 58$ people who are chemists, or have less than 2 years' experience, or both. Thus, directly, $P(C \cup <2) = \frac{58}{100} = 0.58$.

In using the conditional probability formula (1), we need to know $P(A \cap E)$ to compute $P(A|E)$. However, we often know $P(A|E)$ and need to compute $P(A \cap E)$. Here the conditional probability formula is also useful. Since $P(A|B) = P(A \cap B)/P(B)$ and $P(B|A) = P(B \cap A)/P(A)$, we obtain the following.

Multiplication of Probabilities

$$\boxed{P(A \cap B) = P(A|B)P(B) = P(B|A)P(A)} \qquad (2)$$

Example 5 Ranch A has 1000 head of cattle and Ranch B has 2000 head of cattle. There is an outbreak of hoof-and-mouth disease on both ranches. Of the cattle, $\frac{1}{5}$ on Ranch A are infected and $\frac{1}{4}$ on Ranch B are infected. One cow is chosen at random. What is the probability that the chosen cow comes from Ranch A and has the disease?

Solution Let H be the event {has hoof and mouth disease}. Then we seek $P(A \cap H)$. From (2), $P(A \cap H) = P(H|A)P(A)$. There are 3000 head of cattle in total and 1000 of these come from Ranch A. Thus $P(A) = \frac{1000}{3000} = \frac{1}{3}$. Also, $\frac{1}{5}$ of the cattle on Ranch A have the disease. This means that $P(H|A) = \frac{1}{5}$. Thus

$$P(A \cap H) = \frac{1}{5} \cdot \frac{1}{3} = \frac{1}{15}.$$

Note. We also have, from (2), $P(A \cap H) = P(A|H)P(H)$. The problem here is that we are *not* given $P(A|H)$, so that this formula will not do us any good.

We can also solve the problem using the tree diagram given in Figure 4. The number $\frac{1}{5}$ in the diagram represents $P(H|A)$. Similarly, $\frac{3}{4}$ represents $P(\text{not } H|B)$. In the figure, to get $P(A \cap H)$, we multiply the numbers in the branch of the tree linking A and H: $P(A \cap H) = \frac{1}{3} \cdot \frac{1}{5} = \frac{1}{15}$.

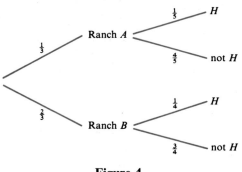

Figure 4

Remark. Don't lose sight of what we just did. *We used the multiplication rule in (2) to justify multiplying along the branches of a tree diagram!* This can be extended to more complicated tree diagrams. In almost all cases, these will enable us to solve some very complicated problems.

Example 6 In Example 5, suppose that 70% of the infected cattle on Ranch A and 60% on Ranch B are less than 1 year old. What is the probability that a randomly chosen cow comes from Ranch B, has the disease, and is at least 1 year old?

Solution We seek $P(B \cap H \cap \geq 1)$. There are formulas for things like this, but we won't worry about them. Rather, we will draw a tree diagram for this three-part experiment (see Figure 5). All the numbers in the diagram come directly from the statement of the problem. Then, multiplying through the $B, H, \geq 1$ branch of the tree, we obtain

$$P(B \cap H \cap \geq 1) = \tfrac{2}{3} \cdot \tfrac{1}{4} \cdot (0.4) = \tfrac{2}{3} \cdot \tfrac{1}{4} \cdot \tfrac{2}{5} = \tfrac{4}{60} = \tfrac{1}{15}.$$

Remark 1. As in so many problems in probability, we can also solve this problem by counting: 2000 cattle come from Ranch B, $\frac{1}{4}$, or 500, of them have the disease, and 40% or 200 of these are at least 1 year old. Thus of a total of 3000 cattle, $\frac{200}{3000} = \frac{1}{15}$ of them are from Ranch B, have the disease, and are at least 1 year old.

Remark 2. We could solve the problem by counting even if we did not know the actual number of cattle on Ranch B. All we need to know is that $P(B) = \frac{2}{3}$. Then we can pick some number of total cattle and assign $\frac{2}{3}$ of them to Ranch B.

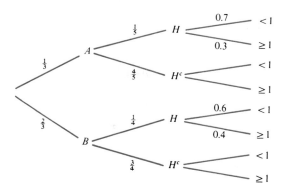

Figure 5

For example, if there are 24,000 head, we find that Ranch B has 16,000. Then $\frac{1}{4}$ of these, or 4000, have the disease; 40%, or 1600, also are at least 1 year old, and $P(B \cap H \cap \geq 1) = 1600/24,000 = \frac{1}{15}$.

Remark 3. Probability theory is often not as difficult as it seems at first!

We close this section by discussing a well-known problem in probability.

Birthday Problem *If n people are in a room, what is the probability that at least two of them have the same birthday?*

Let E be the event {at least two people have the same birthday}. We will compute $P(E^c)$, where E^c is the event {all n people have different birthdays}. We start with two people. Then the probability that they do not have the same birthdays is the same as the probability that the birthday of the second person is not the same as that of the first. There are 365 days in the year (we assume that no one in the group was born on February 29) and 364 ways to choose a different birthday for the second person. Thus

$$P(2 \text{ people have different birthdays}) = \tfrac{364}{365}.$$

Suppose now that there are three people. The situation is depicted in Figure 6. Multiplying through the bottom branches of the tree, we find that

$$P(3 \text{ people have different birthdays}) = \tfrac{364}{365} \cdot \tfrac{363}{365} \approx 0.9918$$

so P(at least 2 of 3 people have the same birthday) $\approx 1 - 0.9918 = 0.0082$. Similarly,

$$P(4 \text{ people have different birthdays}) = \tfrac{364}{365} \cdot \tfrac{363}{365} \cdot \tfrac{362}{365} \approx 0.9836$$

and

$$P(\text{at least 2 of 4 people have the same birthday}) \approx 0.0164.$$

In general,

$$P(n \text{ people have different birthdays}) = \frac{364 \cdot 363 \cdots (365 - n + 1)}{365^{n-1}} \qquad (3)$$

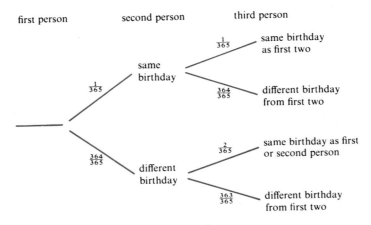

first person · second person · third person

same birthday $\frac{1}{365}$ → same birthday as first two
$\frac{1}{365}$ same birthday $\frac{364}{365}$ → different birthday from first two
$\frac{364}{365}$ different birthday $\frac{2}{365}$ → same birthday as first or second person
different birthday $\frac{363}{365}$ → different birthday from first two

Figure 6

It turns out that this number is less than $\frac{1}{2}$ if $n \geq 23$. In the table, some values obtained from (3) are given. Remember, we must take the number obtained in (3) and subtract it from 1 to obtain the probability that two or more people have common birthdays.

Number of People	Probability that two or more have the same birthday	Number of people	Probability that two or more have the same birthday
3	0.0082	23	0.5073
5	0.0271	24	0.5383
10	0.1169	25	0.5687
15	0.2529	30	0.7063
18	0.3469	40	0.8912
20	0.4114	50	0.9703
22	0.4757	70	0.9992

PROBLEMS 5.6   In Problems 1–4, assume that A, B, C, and D are events in a finite probability space.

1. $P(A) = 0.6$, $P(B) = 0.3$ and $P(A \cap B) = 0.2$. Compute $P(A|B)$ and $P(B|A)$. Illustrate with a Venn diagram.

2. $P(A) = \frac{1}{2}$, $P(B) = \frac{2}{3}$, and $P(A \cap B) = \frac{1}{3}$. Compute $P(A|B)$ and $P(B|A)$ and illustrate with a Venn diagram.

3. $P(A) = 0.27$, $P(B) = 0.44$, and $P(A \cap B) = 0$. Compute $P(A|B)$ and $P(B|A)$ and illustrate with a Venn diagram.

4. $P(A) = \frac{1}{2}$, $P(B) = \frac{1}{3}$, and $P(A \cap B) = \frac{1}{6}$. Compute $P(A|B)$ and $P(B|A)$ and illustrate with a Venn diagram.

5. Two fair dice are thrown. What is the probability that the sum is 8 if at least one of the dice shows a 3?

6. Two fair dice are thrown.

(a) What is the probability that at least one 5 turns up?

(b) What is the probability that at least one 5 turns up if the sum is greater than 7?

7. A family has three children. What is the probability it has two girls if it is known that at least one of the children is a girl?

8. A family has three children. What is the probability that it has at least one boy if it is known that the oldest child is a girl?

9. A family has five children. Define A to be the event that the three oldest children are all of the same sex and define B to be the event that the three youngest children are all of the same sex.

(a) Describe in words $A|B$ and $B|A$.

(b) Determine the probabilities of the four events in this problem.

*10. (a) What is the probability that a child from a family of four has an older brother? [*Hint*: Draw a tree diagram and treat the cases of the first, second, third, and fourth child in the family separately.]

(b) What is the probability that a child from a family of four has an older brother if it is known that the child has at least one brother?

11. To study the relation between heart disease and smoking, a large group S of people is examined. Define H to be the subset of people with heart disease and define C to be the subset of people who smoke regularly.

(a) Define in words the subsets $H \cap C$, $H - C$, $C - H$, and $H^c \cap C^c$.

(b) If it is known that $P(H) = 0.2$, $P(C) = 0.5$, and $P(C|H) = 0.6$, determine the probabilities of the four events corresponding to the subsets in (a).

12. Suppose the weather in a certain region obeys the following rules: Days are either sunny, snowy, or rainy. No two consecutive days are the same. Sun is followed by snow with 30% probability, snow is followed by rain with 50% probability and rain is followed by sun with 25% probability. If today is sunny, what sort of a prediction can be made for the day after tomorrow? [*Hint*: Use a tree diagram.]

13. A recent survey of gun-control preferences in an Eastern state resulted in the following data.

	Ban on handguns		
Sex	**For**	**Against**	**Undecided**
Male	50	30	20
Female	60	40	10

A person was selected at random from those surveyed. What is the probability that the person was male if the person opposed a ban on handguns?

14. In Problem 13, what is the probability that a randomly chosen male is in favor of a ban on handguns?

15. In Problem 13, what is the probability that the randomly chosen individual is either female, or undecided, or both?

16. In a certain city 15% of the population smoke an average of one pack of cigarettes a day or more, 3% have serious respiratory problems, and 1% smoke heavily and have serious respiratory problems. If a person chosen randomly from this population is observed to have a serious respiratory problem, what is the probability that he or she is a heavy smoker?

17. In Problem 15, what percentage of heavy smokers have serious respiratory problems?

18. What is the probability of being dealt a flush in poker if the deck is arranged so that the first three cards dealt to you are hearts?

*19. What is the probability that a poker hand contains a full house if it is known that it contains at least three nines?

20. A chimpanzee is placed at a toy typewriter with the letters A, B, C, D, and E. The chimpanzee types four keys at random.

  (a) What is the probability that the word $BEAD$ is typed?

  (b) What is the probability that the word $BEAD$ is typed if it is known that the C key doesn't work?

  (c) What is the probability that the word $BEAD$ is typed if the chimpanzee doesn't hit the same letter twice?

21. An accounting quiz contains five multiple-choice questions worth 20 points each, with four possible answers for each question. A passing grade is 60 points. If a student guesses on all five questions, what is the probability that he or she receives a score of 100% if it is known that the student passed?

22. A shoe manufacturer has two plants. The second plant manufactures three times as many shoes as the first. It is known that 0.2% of the shoes of the first plant and 0.3% of the shoes of the second plant are defective. If a shoe is chosen at random from one of the two plants, what is the probability that it is a defective shoe manufactured in the second plant?

23. In Problem 22, suppose that 40% of the shoes in the first plant and 30% of the shoes in the second plant have heels at least 1 inch thick. What is the probability that the randomly chosen shoe comes from the first plant, is not defective, and has a heel less than 1 inch thick?

24. Twenty-five percent of the automobiles driven in a western state are Japanese, 60% are American, and 15% are of other origin. Forty percent of the Japanese cars, 10% of the American cars, and 30% of the other cars have a standard shift. If a car is chosen at random for inspection, what is the probability that it is

  (a) a Japanese car with a standard transmission?

  (b) an American car with automatic transmission?

*25. In Problem 24, what is the probability that the randomly selected car has a standard transmission? [*Hint:* Add three probabilities.]

26. A market survey conducted in three North American cities determined preferences for NODIRT detergent. The responses are given in the table.

Response	New York	Atlanta	Montréal
Liked	58	42	38
Disliked	37	51	21
No opinion	16	6	12

A person was chosen at random.

(a) What is the probability that the person liked NODIRT?

(b) What is the probability that the person lived in Montréal?

(c) What is the probability that the person disliked NODIRT if he lived in Atlanta?

(d) What is the probability that the person lived in New York if he liked NODIRT?

 27. What is the probability that of ten people in a room, exactly two have the same birthday?

28. What is the probability that among ten people in the room, exactly three have the same birthday?

29. The Indian National Institute of Nutrition in Canada estimates that children with three or more older brothers and sisters constitute 34% of the child population but account for 61% of all cases of malnutrition. Suppose that 20% of all children suffer from malnutrition. Among the children with three or more older brothers and sisters, what proportion suffer from malnutrition?

30. A California fruit inspector inspected tomatoes for possible infestation of Mediterranean fruit fly. He rejected 20% of all shipments. It was determined by further examination that he rejected 90% of all infested shipments, but only 15% of the shipments were indeed infested. What is the probability that a given shipment was both infested and rejected?

31. Seeds are planted on a farm. The seeds will germinate 80% of the time if there is sufficient rain. There is sufficient rain 65% of the time. What is the probability of both sufficient rain and germination?

*32. One card is chosen at random from a standard deck of 52 cards. If the card is not a spade, it is returned to the deck, which is shuffled; another card is then chosen. This continues until a spade is chosen.

(a) What is the probability that more than 4 cards will be chosen?

(b) What is the probability that the third card will be the first spade if the first card is a diamond?

33. Nine black balls, numbered 1 to 9, are placed in a box. Nine red balls, also numbered 1 to 9, are placed in a second box. Two balls, one from each box, are drawn at random.

(a) Find the probability that the sum of the numbers showing is 10.

(b) Find the probability that the sum is 10 if it is known that at least one of the numbers drawn is a 6.

*34. A drunk comes home and tries to unlock the front door. The drunk has eight keys on a key chain and tries them randomly, never trying the same key twice, until the door opens. What is the probability that the drunk tries exactly four keys?

*35. In a population, 70% of male children and 75% of female children survive to maturity. A couple plans to have three children. Determine the probability that

(a) all three children survive to maturity.

(b) at least one male child survives to maturity.

5.7 Independent Events

We saw in the last section that the probability that an event occurs may be affected if it is known that another event occurred. For instance, we saw in Example 5.6.1 on page 261 that when two dice are thrown, the probability of obtaining at least one 2 changes from $\frac{11}{36}$ to $\frac{2}{5}$ when we assume that the event {sum is 6} occurs.

The conditional probability $P(A\,|\,B)$ gives the probability that A occurs when it is known that B occurs. In many important applications, however, the probability of the event A is *not* affected by the occurrence of the event B.

Two Independent Events   The events A and B in the finite probability space S are said to be **independent** if $P(A\,|\,B) = P(A)$. In other words, the probability that A occurs is not affected by the occurrence (or nonoccurrence) of B.

Example 1   Suppose a fair coin is tossed twice. Let A be the event {head on first toss} and B be the event {head on second toss}. It seems that A and B are independent since each toss of the coin is unaffected by the results of other tosses if the coin is fair. We can verify this directly using the equiprobable space $\{HH, HT, TH, TT\}$. Then $A \cap B = \{HH\}$, $P(A) = \frac{1}{2}$, $P(B) = \frac{1}{2}$, and $P(A \cap B) = \frac{1}{4}$. We observe that

$$P(A\,|\,B) = \frac{P(A \cap B)}{P(B)} = \frac{\frac{1}{4}}{\frac{1}{2}} = \frac{1}{2} = P(A)$$

so that A and B are, indeed, independent.

Before continuing, we note that if A and B are independent, then

$$P(A) = P(A\,|\,B) = \frac{P(A \cap B)}{P(B)}. \tag{1}$$

So, multiplying the equations in (1) by $P(B)$, we obtain

$$\boxed{P(A \cap B) = P(A)P(B) \qquad \text{if } A \text{ and } B \text{ are independent.}} \tag{2}$$

Equation (2) is often used as a test for independence. In some books, it is given as the *definition* of independence. Note that in Example 1, $P(A) = \frac{1}{2}$, $P(B) = \frac{1}{2}$, and $P(A)P(B) = \frac{1}{4} = P(A \cap B)$.

Example 2   In a study of lung disease, 10,000 people over the age of 60 are examined. It is found that 4000 of this group have been steady smokers. Among the smokers, 1800 have serious lung disorders. Among the nonsmokers 1500 have serious lung disorders. Are smoking and lung cancer independent events?

Solution   Let S denote a smoker and L a person with a serious lung disorder. Then $P(S) = 4000/10,000 = 0.4$, $P(L) = 3300/10,000 = 0.33$, and $P(S \cap L)$ $1800/10,000 = 0.18$. But $P(L)P(S) = (0.4)(0.33) = 0.132 \neq 0.18$, so the events are *not* independent. Alternatively, we may compute

$$P(S\,|\,L) = \frac{P(S \cap L)}{P(L)} = \frac{0.18}{0.33} = 0.55 \neq P(S) = 0.33.$$

When the subject of independence is introduced, some people have difficulty distinguishing between *independent* events and *mutually exclusive* events.

Mutually exclusive events are events that cannot occur simultaneously. For example, consider the experiment of flipping a fair coin once with the sample space $\{H, T\}$. Then $\{H\}$ and $\{T\}$ are mutually exclusive because you *cannot* get a head and a tail on the same flip of the coin. For the same reason, they are *not* independent since $P(H|T) = 0 \neq P(H)$ (once a tail turns up, it is impossible that a head also turns up).

On the other hand, as we saw in Example 1, if we flip the same coin twice, the events $A = \{H$ on first flip$\}$ and $B = \{H$ on second flip$\}$ are independent. However, they are not mutually exclusive since $A \cap B = \{HH\} \neq \phi$.

Warning. To determine whether two events $A \cap B$ are independent, it is necessary to show that $P(A|B) = P(A)$ or $P(A \cap B) = P(A)P(B)$. Any other method will be incorrect.

Example 3 In a survey of cable television viewers (see Example 5.6.2 on page 262), 35% liked an all-sports program, 15% liked an all-news program, and 10% liked both. A person is chosen at random from among those surveyed. Are the events {likes all sports} and {likes all news} independent?

Solution In Example 5.6.2 we found that $P(AS|AN) = \frac{2}{3}$. But $P(AS) = 0.35 \neq \frac{2}{3}$. Thus the events are not independent. Alternatively, we can compute

$$P(AS)P(AN) = 0.35 \times 0.15 = 0.0525 \neq P(AS \cap AN) = 0.1.$$

Since $P(AS|AN) > P(AS)$, we conclude that knowing that a person likes the all-news program makes him or her more likely to like the all-sports program also.

PROBLEMS 5.7   In Problems 1–5, A and B are events in a finite probability space. Based on the information given, determine whether they are independent.

1. $P(A) = \frac{1}{3}$, $P(B) = \frac{1}{3}$, $P(A \cap B) = \frac{1}{6}$

2. $P(A) = 0.4$, $P(B) = 0.5$, $P(A \cap B) = 0.2$

3. $P(A) = 0.6$, $P(B) = 0.5$, $P(A \cap B) = 0.3$

4. $P(A) = 0.7$, $P(B) = 0.7$, $P(A \cap B) = 0.35$

5. $P(A|B) = 0.4$, $P(A \cap B) = 0.2$, $P(B|A) = 0.5$

6. $P(A) = 0.4$, $P(B) = 0.3$, $P(A \cup B) = 0.7$

7. Two dice are thrown. Are the events {sum is 7} and {sum $\geq$ 2} independent events?

8. Two dice are thrown. Are the events {sum is even} and {sum $>$ 4} independent events?

9. A couple has three children. Let A be the event {first child is a girl} and B be the event {at least one child is a boy}. Are A and B independent?

10. In a certain midwestern city during the month of August, the probability of a thunderstorm is 0.25 and the probability of hail is 0.10 any given day. The probability of hail during a thunderstorm is 0.3. Are the events {hail} and {thunderstorm} independent?

11. It has been determined that 40% of the students entering a certain university will graduate and that 25% will get married while still at the university. Assuming that marriage and graduation are independent, compute each probability.

(a) An entering student will graduate married.

(b) An entering student will neither graduate nor get married while in attendance.

(c) An entering student will graduate unmarried.

12. In the market survey of Problem 5.6.26 on page 270, determine whether the events {New York} and {disliked NODIRT} are independent.

13. In Problem 5.6.24 on page 270, determine whether the events {car is Japanese} and {car has a standard transmission} are independent.

14. A group of men and women were polled to determine their political party preferences. The results are given in the table.

Sex	Republican	Democrat	Independent
Men	10	8	2
Women	40	25	15

Are the events {male} and {Republican preference} independent?

15. In a study of respiratory conditions, 25,000 males over the age of 50 are examined. It is found that 16,000 have not had regular physical exercise since the age of 25 and, among the 16,000, 6000 have respiratory conditions requiring treatment. Among the 9000 that have had regular exercise, 2000 have respiratory conditions requiring treatment. From these observations, are regular exercise E and respiratory condition requiring treatment R independent events?

16. In a study of physical exercise and nutrition, 40,000 males over the age of 40 are examined. It is found that 28,000 have not had regular physical exercise since the age of 20 and, among the 28,000, 12,000 have inadequate nutrition. In the study, a total of 16,000 men are found to have inadequate nutrition. Define E and N to be the events that a person chosen at random from the study population has had regular exercise and adequate nutrition, respectively. Are E and N independent events?

****17.** Let A and B be two events in a sample space S with $A \neq S$. Show that the following three things cannot all be true: (1) $P(A) = P(B)$; (2) $P(A \cup B) = 1$; (3) A and B are independent.

Three events A, B, and C in sample space S are said to be *independent* if the following conditions hold.

1. $P(A \cap B \cap C) = P(A)P(B)P(C)$

2. $P(A \cap B) = P(A)P(B)$

3. $P(A \cap C) = P(A)P(C)$

4. $P(B \cap C) = P(B)P(C)$

Use this definition in Problems 18–20.

18. A coin is flipped three times. Let A be the event {head on first toss}, B be the event {head on second toss}, and C be the event {tail on third toss}. Are A, B, and C independent?

19. Two committees contain three men and three women each. One member is chosen randomly from each for a joint subcommittee. Let A be the event {the person chosen from the first committee is a man}, B the event {the person chosen from the second committee is a woman}, and C the event {both people chosen are of the same sex}. Show that A and B, A and C, and B and C are independent, but that the three events A, B, and C are not independent.

20. A supervisor for a large construction firm is currently managing three projects. The probabilities that the projects will be completed according to schedule are 0.8, 0.7, and 0.9, respectively. If the completion times are independent, what is the probability that all three projects will be completed on time?

21. A box contains six chips numbered 2, 3, 6, 7, 8, and 9. One chip is drawn at random. Let A be the event {an even number is drawn} and B the event {a number greater than 6 is drawn}.

(a) Find $P(A)$, $P(B)$, and $P(A \cap B)$.

(b) Find $P(A|B)$ and $P(B|A)$.

(c) Are A and B mutually exclusive events?

(d) Are A and B independent events?

22. The Technicomp Company manufactures personal computers. It hires a market researcher to help determine the needs of its buyers based on years of education and whether the occupation is technical or nontechnical. The market researcher finds the information given in the table.

	Occupation	
Education	Technical (E)	Nontechnical (F)
High school (A)	0	12
2 years of college (B)	4	24
4 years of college (C)	34	15
Post-graduate work (D)	16	4

(a) Determine $P(A|E)$, $P(B|F)$, $P(C|E)$, $P(D|E)$, and $P(D|F)$.

(b) Among the events A, B, C, D, E, and F, find all pairs that are independent.

23. The symptoms S_1 and S_2 are often associated with a certain disease. Among victims of the disease, $P(S_1) = 0.5$, $P(S_2) = 0.3$, and $P(S_1|S_2) = 0.25$. Are the two symptoms independent events among victims of the disease?

*24. Suppose that A and B are independent. Show that A and B^c are independent.

5.8 Bayes' Probabilities

In this section we discuss a type of problem that arises quite often. When we introduced conditional probability, we computed the probability of an event A given that an event B was known to have occurred earlier. Now suppose we observe that the later event occurred and ask the probability that the earlier event occurred.

Example 1 A manufacturer receives 60% of its machine parts from supplier 1 and 40% of its machine parts from supplier 2. It has been observed that 2% of the parts from supplier 1 and 1% of the parts from supplier 2 are defective. A part is selected at random and is found to be defective. What is the probability that it came from supplier 1?

Solution We will solve this problem in two ways. First, we observe that we have a two-part experiment. In the first part, we choose a supplier: S_1 or S_2. In the second, we determine whether a given machine part from the supplier is good (G) or defective (D). We put all the information into the tree diagram of Figure 1.

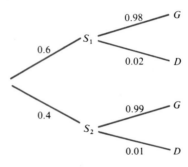

Figure 1

Before continuing, you should make sure you understand all the numbers in the diagram. For example, 0.02 is $P(D|S_1)$, which is given as 2% in the statement of the problem. Then $P(G|S_1) = 1 - P(D|S_1) = 0.98$. The question is to determine $P(S_1|D)$. From the conditional probability formula, we have

$$P(S_1|D) = \frac{P(S_1 \cap D)}{P(D)}.$$

The problem is that we do not yet know either $P(S_1 \cap D)$ or $P(D)$. But, from multiplying through the tree, we see that $P(S_1 \cap D) = (0.6)(0.2) = 0.012$. Alternatively, from equation (5.6.2) on page 265,

$$P(S_1 \cap D) = P(D|S_1)P(S_1) = (0.02)(0.6) = 0.012.$$

What about $P(D)$? We can get a defective part in two ways: either from supplier 1 or from supplier 2. A defective part from supplier 1 is in the set $S_1 \cap D$. A defective part from supplier 2 is in the set $S_2 \cap D$. Thus

$$D = (S_1 \cap D) \cup (S_2 \cap D)$$

and, since S_1 and S_2 are mutually exclusive (no part comes from both S_1 and S_2), we have

$$
\begin{aligned}
P(D) &= P(S_1 \cap D) + P(S_2 \cap D) \\
&= P(D|S_1)P(S_1) + P(D|S_2)P(S_2) \\
&= (0.02)(0.6) + (0.01)(0.4) \\
&= 0.012 + 0.004 = 0.016.
\end{aligned}
$$

These numbers could also be obtained directly by multiplying through the tree. Therefore

$$P(S_1|D) = \frac{0.012}{0.016} = \frac{3}{4} = 0.75.$$

Even in this complicated-looking problem, we can obtain the answer by drawing a picture. Suppose that 1000 machine parts are bought from the two suppliers (the actual number is irrelevant). Then 600 are bought from supplier 1 and 400 are bought from supplier 2. Since 2% of 600 is 12 and 1% of 400 is 4, we obtain the numbers given in Figure 2. From the figure, we see that there are 16 defective parts, 12 of which come from supplier 1. Thus $P(S_1|D) = \frac{12}{16} = \frac{3}{4}$.

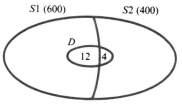

Figure 2

The "reverse" probability computed in the past example is called a **Bayes' probability**.[†] The nice thing about problems involving Bayes' probability is that they are almost all very similar—if you can do one, you can do them all.

We will now give a formula for computing Bayes' probabilities. We do so because many people like formulas. However, the two methods used to solve Example 1 show you what is going on and do not require any memorization.

Bayes' Formula   Let A and B be two events in a finite probability space S such that $A \cap B = \phi$ and $A \cup B = S$. Let E be any given event in S. Then

$$P(A|E) = \frac{P(E|A)P(A)}{P(E|A)P(A) + P(E|B)P(B)} \tag{1}$$

and

$$P(B|E) = \frac{P(E|B)P(B)}{P(E|A)P(A) + P(E|B)P(B)} \tag{2}$$

[†] Named after the English clergyman and mathematician Thomas Bayes (1702–1761). The result was published in 1761.

Why is this true? Look at the tree diagram in Figure 3. Then, reasoning exactly as in Example 1, we have

$$P(A|E) = \frac{P(A \cap E)}{P(E)} = \frac{P(E|A)P(A)}{P(E \cap A) + P(E \cap B)}$$

$$= \frac{P(E|A)P(A)}{P(E|A)P(A) + P(E|B)P(B)}.$$

A similar computation works for $P(B|E)$.

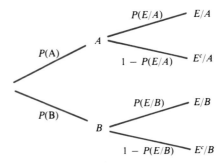

Figure 3

Note 1.   When $A \cap B = \phi$ and $A \cup B = S$, we say that A and B **partition** S.

Note 2.   A similar result holds when $A \cup B \cup C = S$, and $A \cap B = A \cap C = B \cap C = \phi$. Then Bayes' formula becomes

$$P(A|E) = \frac{P(E|A)P(A)}{P(E|A)P(A) + P(E|B)P(B) + P(E|C)P(C)}. \qquad (3)$$

Remark.   Formulas (1), (2) and (3) are there for those who want them. However, tree diagrams and pictures are much handier.

Example 2   A certain disease that occurs in 5% of a population is very difficult to diagnose. One crude test for the disease gives a positive result (indicating the presence of the disease) in 60% of cases when the patient has the disease and in 30% of the cases when the patient does not have the disease. Suppose that the test gives a positive result for a particular patient. What is the probability that the patient has the disease?

Solution   We will solve this problem three ways.

Method 1.   Let D indicate the presence of the disease, D^c indicate its absence,

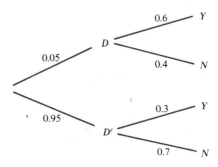

Figure 4

Y denote a positive result (yes), and N denote a negative result on the test. Then all the data are given in the tree diagram of Figure 4. We then compute

$$P(D \mid Y) = \frac{P(D \cap Y)}{P(Y)} = \frac{(0.05)(0.6)}{(0.05)(0.6) + (0.95)(0.3).}$$

$$= \frac{0.03}{0.03 + 0.285} = \frac{30}{315} = \frac{2}{21} \approx 0.095.$$

Note. As in Example 1, we made use of the fact that $P(Y) = P(Y \cap D) + P(Y \cap D^c)$ and then multiplied through the tree to obtain these probabilities.
Method 2. We draw a picture (Figure 5), assigning (arbitrarily) the number 1000 to the population. Then 50 have the disease and 950 do not. Sixty percent of 50 is 30 and 30% of 950 is 285. Then there will be $30 + 285 = 315$ positive results, of which 30 have the disease. Thus

$$P(D \mid Y) = \frac{30}{315} = \frac{2}{21}.$$

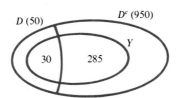

Figure 5

Method 3. We'll use formula (1) with Y instead of E, D instead of A, and D^c instead of B.

$$P(D \mid Y) = \frac{P(Y \mid D)P(D)}{P(Y \mid D)P(D) + P(Y \mid D^c)P(D^c)}$$

$$= \frac{(0.6)(0.05)}{(0.6)(0.05) + (0.3)(0.95)}$$

$$= \frac{2}{21}$$

as in Method 1.

We have derived a result that is important enough to be mentioned explicitly.

Multiplication Theorem   If A and B form a partition of S and if E is any other event, then

$$P(E) = P(E|A)P(A) + P(E|B)P(B). \qquad (4)$$

Example 3 In Example 2 we computed

$$P(Y) = P(Y|D)P(D) + P(Y|D^c)P(D^c)$$
$$= (0.6)(0.05) + (0.3)(0.95) = 0.315.$$

Example 4 In a large population, 60% of the people have systolic blood pressure less than 120 mm, 25% are between 120 mm and 150 mm, and 15% are above 150 mm. In these three groups 20%, 30%, and 50%, respectively, have heart conditions for which medical treatment is recommended.

(a) What part of the population have heart conditions for which medical treatment is recommended?

(b) A person chosen at random from the population is found to require treatment for a heart condition. What is the probability that this person has systolic blood pressure less than 120 mm?

Solution
Method 1.   We put the information into a tree diagram (Figure 6). Here H is the event {heart condition for which medical treatment is recommended}.

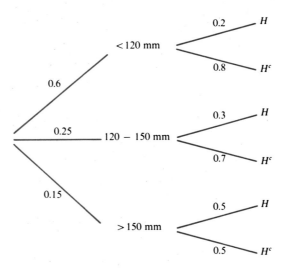

Figure 6

(a) All people with heart conditions are in one of the groups <120, $120–150$, or >150, so that

$$H = (H \cap <120) \cup (H \cap 120\text{–}150) \cup (H \cap >150)$$

and

$$P(H) = P(H \cap <120) + P(H \cap 120\text{–}150) + P(H \cap >150).$$

Multiplying through the tree, we obtain

$$P(H) = (0.6)(0.2) + (0.25)(0.3) + (0.15)(0.5)$$

$$= 0.12 + 0.075 + 0.075 = 0.27.$$

Thus 27% of the population have heart conditions requiring treatment.

(b) $$P(<120|H) = \frac{P(<120 \cap H)}{P(H)} = \frac{(0.6)(0.2)}{0.27} = \frac{12}{27} = \frac{4}{9} \approx 0.4444.$$

Method 2. Yes, we can do all this with a picture. Again, assume that there are 1000 people in the population. The percentages in the problem give us the numbers in Figure 7.

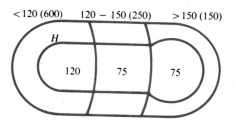

Figure 7

(a) There are 270 people with heart problems out of a population of 1000, so $P(H) = \frac{270}{1000} = 0.27$.

(b) Of those in H, 120 are also in the <120 group. Thus

$$P(<120|H) = \tfrac{120}{270} = \tfrac{4}{9}.$$

Method 3. Use formula (3) with H instead of E, <120 instead of A, $120–150$ instead of B, and >150 instead of C. This yields

$$P(<120|H) = \frac{P(H|<120)P(<120)}{P(H|<120)P(<120) + P(H|120\text{–}150)P(120\text{–}150)}$$
$$+ P(H|>150)P(>150)$$

$$= \frac{(0.2)(0.6)}{(0.2)(0.6) + (0.3)(0.25) + (0.5)(0.15)} = \frac{0.12}{0.27} = \frac{4}{9}.$$

We close this section by giving a general statement of Bayes' formula when the sample space is partitioned into n subsets.

Bayes' Formula—General Version   Let the events $A_1, A_2, \ldots, A_n$ form a **partition** of the sample space S. By this we mean

1. $A_i \cap A_j = \phi$ for $i \neq j$.

2. $A_1 \cup A_2 \cup \cdots \cup A_n = S$.

Let E be an event in S. Then

$$P(A_i|E) = \frac{P(E|A_i)P(A_i)}{P(E|A_1)P(A_1) + P(E|A_2)P(A_2) + \cdots + P(E|A_n)P(A_n)} \quad (5)$$
$$\text{for } i = 1, 2, \ldots, n.$$

PROBLEMS 5.8

1. A certain disease occurs in 3% of a large population and is often associated with a symptom S. The symptom is present in 90% of the people with the disease and in 20% of the people who do not have the disease. A person chosen at random from the population is observed to have the symptom. What is the probability that this person has the disease?

2. On any given day, a student is either well (probability 0.96) or ill. If the student is well, she will attend mathematics class with probability 0.98, and if she is ill, the probability of attendance is 0.20.

    (a)   What is the probability that the student is ill when attending the class?

    (b)   If the student is absent from class, what is the probability that she is well?

3. In North America, about 8% of males and 0.6% of females are red-green colorblind. A person is chosen at random from a group of 55 males and 45 females. (The group of 100 people has been chosen at random from the entire population.) If the person chosen is not red-green colorblind, what is the probability that this person is male?

4. An oil wildcatter has determined that he has a probability of 0.40 of striking oil on one of his properties. He orders a seismic survey that, in the past, has predicted oil in 75% of the cases where oil was actually present and predicted oil in 20% of the cases where there was no oil present.

    (a)   What is the probability that the test will indicate oil?

    (b)   What is the probability that there is oil if the test results are negative?

5. A sales organization gives an aptitude test for prospective employees. Past experience showed that 30% of applicants made good sales representatives. Of the good reps, 90% had scored well on the test; of the poorer reps, 40% had scored well. If an applicant scores well on the test, what is the probability that he or she will make a good sales representative?

6. A treatment for a disease produces a cure in 75% of cases. Among the patients cured, 20% develop other complications from the treatment. Among those not cured, 40% develop such complications.

    (a)   What proportion of the patients treated develop other complications?

    (b)   What proportion of the patients who develop complications from the treatment have been cured by the treatment?

7. A laboratory animal is either healthy (probability 0.9) or it is not. If it is healthy, it will be able to perform a certain task in 75% of all attempts. If it is not healthy, it

will be able to perform the task in 40% of all attempts. Suppose that the animal is observed and it fails to complete the task. What is the probability that it is healthy?

8. Between 1955 and 1966, the fraction of American men over 18 who smoked cigarettes declined from 57% to 51%, while the fraction of women smokers increased from 28% to 33%. Assuming that the population over 18 was 51% women and 49% men throughout this period, what were the probabilities in 1955 and 1966 that a smoker chosen at random was a man?

9. In a large population, equal numbers of people have black and brown hair. It is observed that 30% of the people with black hair have blue eyes, as do 50% of the people with brown hair. A person is chosen at random from those with black or brown hair and is observed to have blue eyes. What is the probability that this person has black hair?

10. Machines L, M, and N produce 25%, 30%, and 45% of the output of a factory. The proportions of output that are defective are 1%, 2%, and 3%, respectively. An item of output is chosen at random and is found to be defective. What are the probabilities that it came from L, M, and N?

11. A laboratory rat is trained to perform four tasks, each within a time limit of 5 minutes. When a task is completed within the time limit, the rat pushes a lever and receives a pellet of food. The probabilities of successful completion of the various tasks within the time limit are 0.8, 0.6, 0.4, and 0.2, respectively. Suppose that the rat begins a task chosen at random and is observed 5 minutes later with a pellet of food. What is the probability that the rat completed the first task? The second task?

12. Two cages contain experimental mice. One cage contains five brown and six white mice; the second cage contains two brown and five white mice. A cage is selected at random and a mouse chosen at random is removed from this cage. If the mouse is brown, what is the probability that it came from the second cage?

*13. To test the contagiousness of several strains of bacteria, a large number of guinea pigs are kept in pairs in separate cages. One guinea pig of each pair is then infected with one of strain I, strain II, or strain III (equal numbers of guinea pigs being used for each strain). The fractions of healthy guinea pigs that become infected are $\frac{1}{3}, \frac{1}{4}$, and $\frac{3}{4}$. A cage is chosen at random and it is observed that both animals have been infected. What are the probabilities that the infection was caused by strains I, II, and III?

14. The California Sunlovers (a football team) and the Wisconsin Icemen have had a long rivalry. The California team wins with probability 0.8 when the weather is warm and with probability 0.3 when it is cold. During the football season, it is warm 90% of the time in California and 20% of the time in Wisconsin. Suppose that the California team lost the last game with Wisconsin. (Assume that equal numbers of games are played in each state.)

(a) What is the probability that the game was played in California?

(b) What is the probability that the weather was cold?

*15. A vaccine produces immunity against rubella in 95% of cases. Suppose that, in a large population, 30% have been vaccinated. Suppose also that a vaccinated person without immunity has the same probability of contracting rubella as an unvaccinated person. What is the probability that a person who contracts rubella has been vaccinated?

*16. In one industry it is estimated that 3% of the labor force are alcoholics with an absentee rate three times that of other workers. If a worker chosen at random is absent from work, what is the probability that this worker is an alcoholic?

17. In a large nursing home, it is estimated that 50% of the men and 30% of the women have serious heart disorders. There are twice as many women as men in the institution. A patient chosen at random has a serious heart disorder. What is the probability that this patient is a man?

18. A large population is divided into two groups of equal size. One group receives a special diet high in unsaturated fats and the control group eats a normal diet high in saturated fats. After 10 years on these diets, the incidences of cardiovascular disease in the two groups are 31% and 48%, respectively. A person is chosen from the population at random and is found to have cardiovascular disease. What is the probability that this person belongs to the control group?

19. Rubella may result in major congenital malformations in children if it is contracted by the mother during the early stages of pregnancy. The chance of malformation is estimated to be 45%, 20%, and 5% if rubella is contracted during the first, second, and third months of pregnancy, respectively. Suppose that the probability of contracting rubella is the same in any month of pregnancy and that a child is born with major congenital malformations resulting from rubella. What is the probability that the mother contracted rubella during the first month of pregnancy?

20. It is estimated that male cigarette smokers over the age of 40 are ten times as likely to die of lung cancer as are male nonsmokers. Assuming that 60% of this population are smokers, what is the probability that a male who dies of lung cancer had been a smoker?

21. (a) The probability that a moose will survive the winter is estimated to be 80% if the moose is healthy and 30% if the moose is not healthy. If 20% of the moose population is not healthy, what part of the population will survive the winter?

　　 (b) If wolves kill 80% of the healthy moose and 70% of the sick moose that do not survive the winter, what part of the moose population is killed by wolves during the winter?

*22. A rare disease occurs in 0.1% of a population and is very difficult to diagnose. One crude test for the disease gives a positive result (indicating the presence of the disease) in 75% of cases when the patient has the disease and in 25% of cases when the patient does not have the disease. Suppose that the test gives a positive result for a person chosen at random. The test is then given again to this person and gives a negative result. Assuming that the tests are independent, what is the probability that this person has the disease?

23. It is estimated that 1 male child in 700 on the average is born with an extra Y-chromosome and that these children are 20 times as likely to show highly aggressive behavior. Accepting these figures, suppose that a boy is known to be highly aggressive. What is the probability that this child has an extra Y-chromosome?

5.9 Binomial Experiments and Bernoulli Trials

We have, so far, been concerned with the probabilities assigned to the outcomes of a trial of an experiment. In this section we are interested in the probabilities of the possible outcomes when the same experiment is performed several times in succession. The model we shall discuss, called the **binomial probability model**, was first described by the noted Swiss mathematician Jacob Bernoulli (1654–1705) in about 1700.

In the binomial probability model we have an experiment in which there are only two possible outcomes: success or failure. We then perform the experiment several times. This kind of procedure occurs very often, as the following example illustrates.

Example 1   The following are examples of the binomial probability model.

(a)   Flip a coin several times and count the number of heads.

(b)   Examine a number of trees for gypsy moth infestation and count the number of infected trees.

(c)   Poll a group of people on the issue of increased arms control and count the number of people in favor.

(d)   Test 1000 automobiles for a manufacturer's defect and count the number of cars with the defect.

(e)   Test a number of people for high blood pressure and count the number of people with high blood pressure.

Note that in each case there are two possibilities on each trial of the experiment: in (a), heads or tails; in (b), infected or not infected; in (c), in favor of increased arms control or not in favor; in (d), has the defect or doesn't have it; in (e), has high blood pressure or doesn't have it.

An experiment in which there are only two possible outcomes is called a **binomial experiment**. Repeated binomial experiments are called *Bernoulli trials*.

BERNOULLI TRIALS

Several repetitions of an experiment, called **repeated trials** of the experiment, are **Bernoulli trials** if the following hold:

1. There are only two possible outcomes in each trial of the experiment: success and failure.

2. The probabilities of success and failure do not change from trial to trial.

3. The trials are independent; that is, the probabilities of of success or failure on any one trial are not affected by the outcome of any previous trial of the experiment.

It turns out that the calculation of various probabilities is fairly easy when we have Bernoulli trials. Before giving general rules, however, we give an example.

Example 2   A fair coin is flipped ten times. What is the probability of obtaining exactly six heads?

Solution We first note that we do have Bernoulli trials since there are two possibilities (heads or tails) and the probability of each is always $\frac{1}{2}$. Now, what is the probability of, for instance, $HHHHHHTTTT$? The probability of the first head is $\frac{1}{2}$. In fact, the probability of every H or T in the string is $\frac{1}{2}$. Thus, since the trials are independent, we can multiply to obtain

$$P(HHHHHHTTTT) = \underbrace{\tfrac{1}{2} \cdot \tfrac{1}{2} \cdot \tfrac{1}{2} \cdot \tfrac{1}{2} \cdot \tfrac{1}{2} \cdot \tfrac{1}{2}}_{6 \text{ heads}} \underbrace{\tfrac{1}{2} \cdot \tfrac{1}{2} \cdot \tfrac{1}{2} \cdot \tfrac{1}{2}}_{4 \text{ tails}} = (\tfrac{1}{2})^{10} = \tfrac{1}{1024}.$$

But we are not done. There are lots of ways to get 6 heads. Two others are $HTHTHTHTHH$ and $TTHTHHHHTHH$. Each string that has 6 heads and 4 tails has probability $\frac{1}{1024}$ of occurring. How many such strings are there? To answer that, we go back to our trick of putting things in chairs. Every string of interest has 6 heads. Think of the H's and T's in the string as sitting on 10 chairs. We must put H's in 6 of these chairs. Each different choice of the 6 chairs gives us another way of obtaining 6 heads. But we know that the number of ways of choosing 6 chairs from among 10 chairs is $\binom{10}{6}$. Thus

$$P(6 \text{ heads}) = \binom{10}{6}\left(\frac{1}{2}\right)^{10} = \frac{210}{1024} \approx 0.2051.$$

We now use the technique of Example 2 to obtain a formula for solving all problems involving Bernoulli trials. Suppose we perform a binomial experiment n times and ask for the probability of k successes (S) where $P(S) = p$. Then $P(\text{failure}) = P(F) = 1 - p$. One string that has k successes is

$$\underbrace{SSSS \cdots S}_{k \text{ times}} \underbrace{FF \cdots F}_{n - k \text{ times}} \tag{1}$$

Since $P(S) = p$ and $P(F) = 1 - p$, and since the trials are independent, the probability of the event in (1) is

$$\underbrace{p \cdot p \cdots p}_{k \, p\text{'s}}\underbrace{(1 - p)(1 - p) \cdots (1 - p)}_{n - k \,\, (1 - p)\text{'s}} = p^k(1 - p)^{n-k}$$

How many ways can we get k S's out of n trials? As before, we pick k "chairs" out of n chairs and seat k S's on them. This can be done in $\binom{n}{k}$ ways. Thus $P(k \text{ successes}) = \binom{n}{k}p^k(1 - p)^{n-k}$. We summarize this result.

Probabilities in Bernoulli Trials If a binomial experiment is performed n times with a probability p of success in each trial, then the probability of obtaining exactly k successes, denoted by $B(n, k, p)$, is given by

$$\boxed{B(n, k, p) = \binom{n}{k}p^k(1 - p)^{n-k}.} \tag{2}$$

Example 3 An automobile manufacturer determines that 5% of the cars leaving one of his assembly plants have defects that must be repaired before shipment to retail outlets. If 12 cars are selected at random from those leaving the assembly plant, find the probability that

(a)   exactly 2 have defects.

(b)   none has a defect.

(c)   at least one has a defect.

Solution

(a)   Here we have Bernoulli trials with $n = 12$, $k = 2$, $p = 0.05$, and $1 - p = 0.95$ (the word *success* is used very loosely here). Then $B(12, 2, 0.05) = \binom{12}{2}(0.05)^2(0.95)^{10} \approx (66)(0.0025)(0.5987) \approx 0.0988$.

(b)   There are two ways to solve this. No defects mean that all are good and $P(\text{good}) = 0.95$. Since the trials are independent,

$$P(12 \text{ good}) = (0.95)^{12} \approx 0.54.$$

Or, we can use the formula with $n = 12$, $k = 0$, and $P = 0.05$:

$$B(12, 0, 0.05) = \binom{12}{0}(0.05)^0(0.95)^{12} = (0.95)^{12} \approx 0.54.$$

(c)   The event {at least one has a defect} is the complement of the event {no defects}. Thus, from part (b),

$$P(\text{at least one defect}) = 1 - P(\text{no defects}) \approx 1 - 0.54 = 0.46.$$

 Example 4 In a medical study it was found that the arterioles (blood vessels) of heavy smokers had a degenerative thickening in 90% of all cases. Six heavy smokers are chosen at random from the population.

(a)   What is the probability that exactly four have degenerative thickening of the arterioles?

(b)   What is the probability that at least four have it?

(c)   What is the probability that at least one has it?

(d)   If at least four have it, what is the probability that all six have it?

Solution   These are Bernoulli trials with $n = 6$, $p = 0.9$, and $1 - p = 0.1$.

(a) $B(6, 4, 0.9) = \binom{6}{4}(0.9)^4(0.1)^2 = 15(0.6561)(0.01) \approx 0.0984$

(b) $P(\text{at least } 4) = P(4) + P(5) + P(6)$

$$= B(6, 4, 0.9) + B(6, 5, 0.9) + B(6, 6, 0.9)$$

$$\approx 0.0984 + \binom{6}{5}(0.9)^5(0.1)^1 + \binom{6}{6}(0.9)^6(0.1)^0$$

$$\approx 0.0984 + 6(0.5905)(0.1) + 0.5314$$

$$= \underbrace{0.0984}_{P(4)} + \underbrace{0.3543}_{P(5)} + \underbrace{0.5314}_{P(6)} = 0.9841$$

(c) $P(\text{at least one}) = 1 - P(0) = 1 - 0.1^6 = 0.999999$

(d) This is a conditional probability problem.

$$P(6\,|\,\text{at least } 4) = \frac{P(6 \cap \text{at least } 4)}{P(\text{at least } 4)} = \frac{P(6)}{P(\text{at least } 4)}.$$

In (b) we found that $P(6) \approx 0.5314$ and $P(\text{at least } 4) \approx 0.9841$. Thus

$$P(6\,|\,\text{at least } 4) \approx \frac{0.5314}{0.9841} \approx 0.54.$$

PROBLEMS 5.9 In Problems 1–10, compute the binomial probability for the given values of n, k, and p. Use a calculator where necessary.

1. $B(6, 4, \frac{1}{2})$

2. $B(10, 3, \frac{1}{3})$

3. $B(10, 7, \frac{2}{3})$

4. $B(8, 4, 0.4)$

5. $B(12, 0, 0.6)$

6. $B(12, 12, 0.4)$

7. $B(8, 5, 0.12)$

8. $B(9, 3, 0.38)$

9. $B(100, 3, 0.01)$

10. $B(100, 99, 0.99)$

11. It is known that 80% of all students will pass the course for which this book is intended. Out of a group of eight students taking the course, find the probability that

   (a) exactly three will pass.

   (b) at least six will pass.

   (c) not all will pass.

12. A fair coin is flipped 15 times. What is the probability of each event?

   (a) exactly 12 heads.

   (b) no more than 3 tails.

   *(c) more heads than tails.

13. In a given jungle 15% of the mosquitos carry malaria. Seven mosquitos are captured. Find each probability.

   (a) 2 are carriers.

   (b) At least one is a carrier.

   (c) 2 are carriers if it is known that at least one is a carrier.

14. Colorblindness affects 1% of a large population. Suppose that n people are chosen at random. What is the probability that none of the n people is colorblind?

15. In Problem 14, how large must n be in order that this probability be less than 10%? [*Hint:* By trial and error find the smallest value of n such that $B(n, 0, 0.01) < 0.1$.]

16. A meteorologist applies for a research grant to travel to Spain to test the theory that the rain in Spain falls mainly on the plain. She plans to conduct her observations during a part of the year when the probability of rain on any given day on the plain is 20%. (Assume that this probability is independent of the previous weather.) How many days must she plan to spend in Spain to be 99% certain that she will observe rain?

17. In Problem 16 suppose that the meteorologist has been awarded research funds for 15 days in Spain and that, after 10 days, no rain has been observed. What is the probability that her trip will be a failure; that is, that no rain will be observed?

18. In a population of drosophila (fruit flies), 20% have a certain wing mutation. If six flies are chosen at random from the population, find the probability that

   (a)   two have the mutation.

   (b)   at least two have the mutation.

   (c)   fewer than five have the mutation.

19. A general who has won five battles in a row is usually considered to be a great military leader. Assuming that the outcome of battles is determined by chance with 50% probabilities of success and failure and that each general fights exactly five battles, what proportion of generals are great military leaders?

20. Caffeine and benzedrine are stimulants that seem to have some ability to counteract the depressing effects of alcohol. In a test of their relative effectiveness, 40 volunteers each consume 6 oz of alcohol. The volunteers are then divided into 20 pairs and one member of each pair receives benzedrine, while the second member receives caffeine. As measured by certain tests, benzedrine brings about a more rapid recovery in all 20 pairs. What is the probability of obtaining this result if there is no difference in the effects of caffeine and benzedrine? [*Hint:* If there is no difference, then assuming that one drug works faster in each pair, the probability that benzedrine works faster in one pair is $\frac{1}{2}$].

*21. A certain disease is present in 4% of a large population. One test for this disease yields a positive result in 80% of people with the disease and in 15% of people who do not have the disease.

   (a)   If the test is given to a person chosen at random from the population, what is the probability of a "correct" result—positive if the disease is present and negative otherwise?

   (b)   If a person with the disease is given the test five times, what is the probability of exactly three negative results?

   (c)   If the test is given five times with a negative result exactly three times, what is the probability that the person has the disease? (In parts (b) and (c), assume that test results are independent of previous test results.)

*22. Obesity affects about 20% of a large population. A survey indicates that 60% of those affected do not believe they have this problem. Twelve people are chosen at random from the population and examined for obesity. Determine the probability that exactly three people are obese and at least two of the three do not believe they are obese.

23. The failure rate for a cardiac arrest alarm in an intensive care bed is estimated to be 1 in 1000. One hospital estimates that the alarm will be used 500 times in a 1-month period.

(a) What is the probability of no failures in this period?

(b) What is the probability of at least two failures?

24. A sales firm has found that 25% of all applicants for positions make successful sales representatives. The firm needs to hire 10 permanent sales representatives, and 25 have applied and have been hired on a trial basis. What is the probability that the firm will reach its quota?

25. In traditional farming societies, it is extremely important for families to have a male child survive to adulthood.

(a) Suppose that 75% of male children survive to adulthood. How many children must the family plan to have in order that, with probability 95%, at least one male child survives to adulthood?

(b) Repeat part (a) under the assumption that 50% of male children survive to adulthood.

26. It is observed that 80% of all persons with a certain chronic disease exhibit a specific symptom A and that 40% of those with the symptom exhibit it in the acute form B. Four persons are chosen at random from a large group of people with the disease.

(a) What is the probability that none of the four persons has the symptom in the acute form B?

(b) What is the probability that exactly two persons do not have the symptom A?

27. Hypertension (above-normal blood pressure) affects about 15% of Canadians. A survey indicates that about 55% of those affected do not realize they have hypertension. Twenty people are chosen at random from the population and examined for hypertension. Determine the probability that exactly four people of the group have the disease and all four do not realize that they have hypertension.

Review Exercises for Chapter 5

In Exercises 1–6, a procedure or activity is described. In each exercise determine the number of possible outcomes.

1. A coin is flipped four times.

2. Three dice are thrown.

3. A drugstore purchases five brands of toothpaste, six brands of shampoo, and eight brands of deodorant.

4. Nine people run for three positions in the city council.

5. Nine people run for chairperson, vice-chairperson, and secretary of the city council.

6. A family has five children.

7. How many five-letter words can be made from the letters of the word *PRIMATE*?

8. Evaluate.

    (a) $P_{7,3}$ (b) $P_{8,4}$ (c) $P_{9,8}$ (d) $P_{9,1}$ (e) $P_{10,6}$

9. Evaluate.

    (a) $\binom{9}{4}$ (b) $\binom{7}{5}$ (c) $\binom{9}{8}$ (d) $\binom{9}{1}$ (e) $\binom{10}{6}$

10. A scrabble player has seven letters and wishes to play a five-letter word. If she chooses to test each possible five-letter permutation and if each test takes $1\frac{1}{2}$ seconds, how long will it take her to complete her play?

11. Twelve heavyweight contenders are ranked by the World Boxing Association (WBA). How many rankings are possible?

12. If in Exercise 11, the WBA ranks only the top 5 contenders among the 12, how many rankings are possible?

13. How many distinct permutations are there of the letters of the word *BANANAS*?

14. A textbook has four chapters on probability and six on statistics. If an instructor wishes to cover two probability chapters and three statistics chapters, how many choices does he have?

15. How many poker hands contain a flush?

16. What is the probability of being dealt a flush?

17. What is the probability of being dealt a flush if two of the five cards dealt are known to be clubs?

18. John and Mary are on a six-person committee that is selecting a chairperson and vice-chairperson at random. What is the probability that Mary will be chosen chairperson and John vice-chairperson?

19. In Exercise 18, what is the probability of the Mary-John outcome if it is known that Erik and Kerstin, two other members of the committee, have not been selected for either position?

20. In Exercise 18 the committee selects instead two members for a subcommittee. What is the probability that Mary and John will be selected?

21. Of 50 stocks on the Amercian Stock Exchange, 18 advanced and 13 declined. In how many ways could this happen?

22. A chimpanzee is placed in front of a toy typewriter with the 26 letters of the alphabet in front of him. If he punches 5 letters at random (without repetition) what is the probability that he will spell out the word *CHIMP*?

23. If a student guesses on a ten-question, true-false test, what is the probability of each event?

   (a) He gets 100%?

   (b) He gets exactly three right?

   (c) He gets at least one right?

   (d) He gets at most eight right?

24. What is the probability of getting exactly two 5's if four dice are thrown?

25. Four people, labeled P_1, P_2, P_3, and P_4, are to be tested for blood type (A, B, AB, O).

   (a) How many different distributions of blood type are possible among these four people?

   (b) In how many of these distributions is there exactly one person of each type?

26. A family has four children. Define A to be the event that at least two children are girls and B to be the event that at least two children are boys.

   (a) Define the events $A \cup B$, $A \cap B$, $A - B$, and $B - A$.

   (b) Determine the probabilities of the six events in this problem.

27. A family has four children.

   (a) What is the probability that all the children are of the same sex?

   (b) What is the probability that there is no girl in this family older than any boy?

   (c) If it is known that at least one of the children is a boy, what is the probability that at least three of the children are boys?

28. A study at the General Hospital indicated that 20% of all patients suffered adverse effects from drugs during their stay in the hospital. Five patients are chosen at random.

   (a) What is the probability that at least two of the five people suffer adverse effects from drugs?

   (b) If it is known that at least one of the five people suffers an adverse drug reaction, what is the probability that at least two suffer such reactions?

29. In a group of 11 people, 7 are right-handed and 4 are left-handed. Five people are to be chosen at random from this group.

   (a) Determine the sample space and the probability of a simple event in this experiment.

   (b) What is the probability that the 4 left-handed people are chosen?

30. A certain disease occurring in 5% of a population is very difficult to diagnose. One crude test for the disease gives a positive result (indicating the presence of the disease) in 60% of cases when the patient has the disease and in 30% of the cases when the patient does not have the disease. Suppose the test is applied twice to a particular individual and suppose it leads first to a positive and then to a negative result. Assuming the tests to be independent, what is the probability that the individual has the disease?

31. Group A consists of 40 men and 50 women, group B contains 40 men and 80 women, and group C contains 80 men and 10 women. One person is chosen at random from each of the three groups.

   (a) What is the probability that exactly one woman is chosen?

   (b) If exactly one woman is chosen, what is the probability that she comes from group C?

32. For the following experiments, determine the sample spaces and probabilities of simple events.

   (a) Choose five integers at random from 1 to 100; repetitions not allowed.

   (b) Choose six women and two men at random from ten women and five men.

33. Two dice are thrown. Find each probability.

   (a) The sum of the numbers showing is 8.

   (b) The sum is 8 given that at least one of the numbers is a 5.

   (c) At least one of the numbers is a 5 given that the sum is 8.

34. Let A and B be events in a sample space with $P(A) = 0.6$, $P(B) = 0.4$, and $P(A \cap B) = 0.3$. Find each value.

   (a) $P(A \cup B)$

   (b) $P(A - B)$

   (c) $P(B - A)$

   (d) $P(A|B)$

   (e) $P(B|A)$

(f) $P(A^c)$

(g) $P(B^c)$

35.   Let A and B be events in a sample space with $P(A) = 0.48$, $P(B) = 0.36$, and $P(A \cup B) = 0.82$. Find each value.

(a) $P(A \cap B)$

(b) $P(A - B)$

(c) $P(B \mid A)$

36.   The demand for Rolls Royce automobiles in Los Angeles in a given month is estimated by the dealer as follows.

Demand	Probability
0	0.15
1	0.10
2	0.20
3	0.25
4	0.10
5	0.05
6 or more	0.15

Compute $P(\text{demand} > 2)$.

37.   In Exercise 36, find $P(1 \le \text{demand} < 5)$.

38.   In Exercise 36, let $A = \{\text{demand} \le 3\}$ and $B = \{1 \le \text{demand} < 5\}$. Find

(a) $P(A^c)$

(b) $P(B^c)$

(c) $P(A \cup B)$

(d) $P(B - A)$

39.   In Exercise 22, what is the probability that the word $CHIMP$ is typed out if the letters A, B, U, V, W, X, Y, and Z are broken?

40.   If $P(A) = 0.6$, $P(B) = 0.5$, and $P(A \cap B) = 0.5$, are A and B independent?

41.   If $P(A) = 0.2$, $P(B) = 0.3$, and $P(A \cup B) = 0.5$, are A and B independent?

42.   A coin is flipped three times. Let $A = \{\text{tail on first toss}\}$ and $B = \{\text{head on third toss}\}$.

(a)   Are A and B independent?

(b)   Are A and B mutually exclusive?

43.   Give two events in the sample space of throwing one die that are mutually exclusive but not independent.

44.   A plant receives 60% of its parts from factory 1 and 40% from factory 2. The percentages of defects from factory 1 and factory 2 are, respectively, 3% and 2%.

(a)   What percentage of parts received from the plant are defective?

(b)   If a part is defective, what is the probability that it came from factory 2?

45. In Exercise 44, if ten parts come from factory 2, what is the probability that at least one of them will be defective?

46. It is estimated that 90% of the individuals who invest in commodities lose money. If eight people invest in commodities, find the probability that

(a) 6 of them lose money.

(b) at least 6 of them lose money.

47. In a large population 30% have black hair, 45% have brown hair, and 25% have blond hair. Seventy percent of the black-haired people have brown eyes, as do 60% of the brown-haired people and 35% of the blonds. All others have blue eyes. A person is chosen at random.

(a) What is the probability that he or she has blue eyes?

(b) If the person has blue eyes, what is the probability that he or she has brown hair?

(c) If the person has brown eyes, what is the probability that he or she has blond hair?

48. In a study of a large population of people over 40 years of age, it is found that 30% are heavy smokers and 10% are heavy drinkers. It is also found that half of the heavy drinkers are also heavy smokers.

(a) Are heavy smoking and heavy drinking independent events for this population?

(b) What proportion of heavy smokers are also heavy drinkers?

6 INTRODUCTION TO STATISTICS

6.1 Random Variables

In many experiments, numbers are assigned to the elementary events in the sample space. For example, we could toss a coin three times and record the number of heads. Or the experiment might be to choose 10 persons from 1000 people and then test them for a certain disease. For the second experiment, we could associate to the outcome the number of persons from the 10 chosen who have the disease.

Random Variable   A **random variable** X associated with a finite probability space S is a rule which assigns a real number to each simple event in S. The set of numbers the random variable takes on is called the **range** of the random variable.

Example 1 Toss a coin three times in a row. The sample space $S = \{HHH, HHT, HTH, HTT, THH, THT, TTH, TTT\}$. One random variable X associated with this experiment is the rule that assigns to each elementary event the number of heads. For example, X assigns the number 2 to HTH and the number 1 to TTH. The range of X is the set $\{0, 1, 2, 3\}$. The values taken by the random variable X are given in Table 1 (see page 296).

Example 2   A roulette wheel is numbered 1 to 36 with slots marked 0 and 00 added. The odd numbers are red, the even numbers are black, and the slots 0 and 00 are neither red nor black. A player bets $1 on red. If a red number turns up, he wins $1. If anything else turns up, he loses $1. The experiment here is spinning a ball around the wheel and observing the slot in which it lands. The sample space of this experiment consists of the numbers from 1 to 36, 0, and 00.

TABLE 1

Outcome	Value taken by X
HHH	3
HHT	2
HTH	2
HTT	1
THH	2
THT	1
TTH	1
TTT	0

An appropriate random variable associated with the experiment assigns the number $+1$ if a red number turns up and the number -1 if any other number turns up. The range of this random variable is the set $\{1, -1\}$. For example, X assigns the number 1 to the outcome 15 and the number -1 to the outcomes 27 and 00. The values taken by this random variable are given in Table 2.

TABLE 2

Outcome	Value taken by X	Outcome	Value taken by X	Outcome	Value taken by X
0	-1	12	-1	25	1
00	-1	13	1	26	-1
1	1	14	-1	27	1
2	-1	15	1	28	-1
3	1	16	-1	29	1
4	-1	17	1	30	-1
5	1	18	-1	31	1
6	-1	19	1	32	-1
7	1	20	-1	33	1
8	-1	21	1	34	-1
9	1	22	-1	35	1
10	-1	23	1	36	-1
11	1	24	-1		

Example 3   An electronics store sells portable radios and cassette tape players. Each week the number of units sold is recorded. No more than 100 tape players are sold in a given week. The number of tape players sold is a random variable with range $\{0, 1, 2, 3, \ldots, 100\}$.

Example 4   The children in a sixth grade classroom are measured by the school nurse and each child's height is recorded. No child is shorter than 3 ft, 6 in.

(42 in.) or taller than 5 ft, 4 in. (64 in.). If heights are taken to the nearest $\frac{1}{4}$ in., then the height of each child is a random variable with range

$$\{42, 42\tfrac{1}{4}, 42\tfrac{1}{2}, \ldots, 63\tfrac{1}{2}, 63\tfrac{3}{4}, 64\}.$$

Example 5 A binomial experiment (see Section 5.9) is performed n times. The number of successes is a random variable, called the **binomial random variable**, with range $\{0, 1, 2, \ldots, n\}$.

Example 6 A coin is tossed five times. The number of heads that turn up is a random variable X with range $\{0, 1, 2, 3, 4, 5\}$. For example, X assigns the number 3 to the outcome $THTHH$ and the number 1 to the outcome $THTTT$.

We will often be interested in determining the probability that a random variable takes on a certain value.

Probability Function A **probability function** p associated to a random variable X is defined by the rule

$$\boxed{p(x) = P(X = x)} \tag{1}$$

That is, $p(x)$ is the probability that the random variable X takes on the value x.

Note. There might be some confusion between the capital X and the lower case x. The X stands for the random variable (which is a rule) and the x denotes a real number. With this understanding, equation (1) is read

$p(x)$ *is the probability that the random variable X takes the value x.*

For example, $p(2) = P(X = 2)$ is the probability that X takes the value 2.

Example 7 In the coin tossing experiment of Example 1, we know that $P(0 \text{ heads}) = \frac{1}{8} = P(3 \text{ heads})$ and $P(1 \text{ head}) = \frac{3}{8} = P(2 \text{ heads})$. Thus the probability function is given by $p(0) = \frac{1}{8}$, $p(1) = \frac{3}{8}$, $p(2) = \frac{3}{8}$, and $p(3) = \frac{1}{8}$. But that is not all. We also have $p(6) = 0$, $p(-7) = 0$, and $p(1.623) = 0$, for example. This follows because we cannot flip 3 coins and get 6, -7, or 1.623 heads.

Remark. As Example 7 illustrates, a probability function is defined for all real numbers. However, the value taken by the probability function will often be zero.

Example 8 In the roulette wheel of Example 2, $p(\text{red}) = p(\text{odd number}) = \frac{18}{38} \approx 0.4737$ and $p(\text{not red}) = \frac{20}{38} \approx 0.5263$. Thus $p(1) = \frac{18}{38}$, $p(-1) = \frac{20}{38}$, and $p(x) = 0$ if x is not one of the numbers 1 or -1.

Example 9   For the binomial experiment of Example 5, $p(k) = \binom{n}{k} p^k (1 - p)^{n-k}$ for $k = 0, 1, 2, 3, \ldots, n$ with p the probability of success. (Look again at equation (5.9.2) on page 286 if this notation is unclear to you.)

Remark.   In Example 3 we described the problem of a retailer (in this case an electronics store) trying to determine demand. This demand is a random variable with range $\{0, 1, 2, \ldots, 100\}$. But here, as is common, the retailer does not usually know the probability function associated with that random variable, at least when the store is opened. Only after the store has been in operation for a long time can the retailer begin to know, in advance, approximately what demand will be. In Example 3, then, the probability function will be learned only after experimentation and observation. We therefore observe

> the determination of a probability function is often a problem in statistics.

We return to this problem in Section 6.3.

Example 10   In a certain Southern city in July, it rains on 20% of the days, it is cloudy (without rain) on 25% of the days, and it is sunny on 55% of the days. Consider the experiment of choosing a day in July at random and observing the weather in this city. Associate a random variable X to the outcomes of this experiment by defining $X = -3$ if it rains, $X = -1$ if it is cloudy, and $X = 4$ if it is sunny. The range of X is $\{-3, -1, 4\}$. The probability function $p(x) = 0$ unless $x = -3, -1$, or 4. From the information given, we have $p(-3) = 0.2$, $p(-1) = 0.25$, and $p(4) = 0.55$.

Remark.   In Example 10 we were able to determine the probability function because *a statistical survey had already been performed.* Some individuals kept weather records in the city over a period of many years.

In many problems we will be interested in knowing the probability that a random variable takes on values less than or equal to a certain number x. For example, if we choose 10 persons from 1000 people and test them for a certain disease, we may be interested in the probability that no more than 5 of the 10 persons have the disease.

Example 11   A cosmetics manufacturer has determined that 60% of the women who buy hair spray have brown hair. Four of his customers are chosen at random. What is the probability that no more than three of the four women chosen have brown hair?

Solution Define the random variable X to be the number of women chosen with brown hair. Then $P(X \leq 3) = p(0) + p(1) + P(2) + p(3) = 1 - p(4)$. But $p(4) = P(\text{all brown}) = (0.6)^4 = 0.1296$. Thus $P(X \leq 3) = 1 - 0.1296 = 0.8704$.

The last example leads to the following useful definition.

Distribution Function The **distribution function** F associated with a random variable X is the rule given by

$$F(x) = P(X \leq x) \tag{2}$$

where x can be any real number.

Remark. Equation (2) is read

" $F(x)$ is the probability that the random variable X " (which, remember, is a rule) " takes a value less than or equal to x " (which is a real number).

For example, $F(2)$ is the probability that X takes on a value less than or equal to 2.

Example 12 Consider the experiment of throwing two dice. The sample space S of this experiment consists of 36 outcomes: $(1, 1), (1, 2), \ldots, (6, 6)$. Suppose that we are interested in the sum of the outcomes on the two dice. This defines a random variable X with range space $X(S) = \{2, 3, 4, \ldots, 12\}$. By a counting argument, we obtain $p(x)$ and $F(x)$ for this random variable X. These functions are given in Table 3.

TABLE 3

X	2	3	4	5	6	7	8	9	10	11	12
$p(x)$	$\frac{1}{36}$	$\frac{2}{36}$	$\frac{3}{36}$	$\frac{4}{36}$	$\frac{5}{36}$	$\frac{6}{36}$	$\frac{5}{36}$	$\frac{4}{36}$	$\frac{3}{36}$	$\frac{2}{36}$	$\frac{1}{36}$
$F(x)$	$\frac{1}{36}$	$\frac{3}{36}$	$\frac{6}{36}$	$\frac{10}{36}$	$\frac{15}{36}$	$\frac{21}{36}$	$\frac{26}{36}$	$\frac{30}{36}$	$\frac{33}{36}$	$\frac{35}{36}$	$\frac{36}{36}$

The distribution function $F(x)$ is defined for all $x \in \mathbb{R}$. In the example, $F(\frac{5}{2}) = P(X \leq \frac{5}{2}) = \frac{1}{36}$, since the probability that X is less than or equal to $\frac{5}{2}$ is equal to the probability that $X = 2$. Similarly, $F(6.75) = P(X \leq 6.75) = P(X \leq 6) = \frac{15}{36}$ and $F(1000) = P(X \leq 1000) = 1$. In fact, it should be evident that as x increases, $F(x)$ increases in steps from 0 to 1. This distribution function is illustrated in Figure 1.

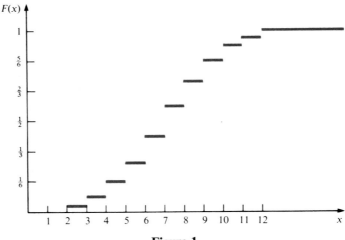

Figure 1

In Section 6.6 we deal with a special type of distribution function called the *normal distribution*. In that section we will see why, in many practical situations, all the interesting statistical information is given by using a distribution function.

PROBLEMS 6.1

1. A family plans to have three children. Let X denote the random variable that counts the number of boys.

(a) What is the range of X?

(b) What is the probability function for X?

(c) What is the distribution function for X?

(d) Find $F(4)$.

(e) Find $P(X \geq 7)$.

(f) Find $P(1.65 \leq X \leq 2.85)$.

(g) Sketch the distribution function.

2. Consider an experiment with three elementary outcomes, E_1, E_2, and E_3, each with probability $\frac{1}{3}$. Define a random variable X by $X(E_1) = -1$, $X(E_2) = 0$, and $X(E_3) = 1$.

(a) What is the range of the random variable X?

(b) What is the probability function for X?

(c) What is the distribution function for X?

(d) Find $P(X \geq 0)$.

(e) Find $F(3)$.

(f) Sketch the distribution function.

3. Consider the experiment of choosing an integer from 1 to 10 at random. Define the random variable X by $X = 0$ if the number chosen is 3, 5, 6, or 9; $X = 1$ if the number chosen is 2, 4, 8, or 10; and $X = 2$ if the number chosen is 1 or 7.

(a) What is the sample space of this experiment? What are the simple events?

(b) What is the range of X?

(c) Determine the probability function and the distribution function for X.

(d) Find $P(X \geq 0)$.

(e) Find $F(2)$.

(f) Sketch the distribution function.

4. A fair coin is tossed four times. Let X be the number of heads obtained.

(a) What is the range of this random variable?

(b) Determine the probability function and the distribution function of X.

(c) Find $F(7)$.

(d) Find $F(1.65)$.

(e) Sketch the distribution function.

5. A pair of dice are thrown five times in succession. Define the random variable X to be the number of times that a 7 is thrown.

(a) What is the range of X?

(b) What are the probability function and distribution function of X?

(c) Find $F(3.2)$.

(d) Find $F(6.99)$.

(e) Sketch the distribution function.

6. A department-store manager has estimated that on 25% of Saturdays, total sales are under $100,000, on 35% of Saturdays they are between $100,000 and $200,000, on 30% of Saturdays they are between $200,000 and $300,000, and on 10% of Saturdays they are over $300,000. Let X be the random variable that assigns the number 1 to sales under $100,000, 2 to sales between $100,000 and $200,000, 3 to sales between $200,000 and $300,000, and 4 to sales over $300,000.

(a) What is the range of X?

(b) What are the probability and distribution functions of X?

(c) Find $F(3.2)$.

(d) Find $F(-1.6)$.

(e) Sketch the distribution function.

7. In a large drosophila (fruit fly) population, 25% of the flies have a wing mutation. Let X be the number of flies that have the mutation in a random sample of five chosen from the population. Determine the range, the probability function, and the distribution function of the random variable X.

8. An experiment is completed successfully in 80% of all attempts. Let X denote the number of successful experiments in a series of eight repetitions. Determine the range, the probability function, and the distribution function of X.

9. The probability that a first-year student of English will know the meaning of the word *adumbrate* is estimated to be 10%. In an English class of 20 first-year students, define X to be the number of students who can define this word correctly.

(a) What is the range, the probability function, and the distribution function of X?

(b) What is the probability that $X \leq 4$?

(c) What is the probability that $X \geq 4$?

10. A rat is being trained to go through a maze. There are five possible paths, only one of which leads through the maze. Assume that the rat chooses a path at random until the correct path is chosen; also assume that an incorrect path will not be chosen a second time. Define a random variable X to be the number of incorrect paths chosen.

(a)   What are the range, probability function, and distribution function of X?

(b)   What is the probability that $X \geq 3$? That $X > 0$?

11. In a gambling game, A has the first move and then B follows. The rules of the game are such that either competitor can win on any move with probability p. They continue playing until one of them wins. Define X to be the number of moves taken by A until the game ends and define Y to be the number of moves taken by B in the game.

(a)   What are the ranges of X and Y?

(b)   Determine $P(X = 2)$, $P(Y = 1)$, and $P(Y > 2)$.

˙12. Two-thirds of the children in a large school are absent because of an epidemic of influenza. In one class of 25 students, the teacher calls the roll. Define X to be the number of students called until one responds.

(a)   What is the range of X?

(b)   What is the probability that $X = 10$; that is, the tenth child is the first one to be present?

(c)   Determine $P(X \leq 2)$ and $P(X \geq 2)$.

13. A low-iodine diet produces enlarged thyroid glands in 60% of the animals in a large population. Four enlarged glands are needed for an experiment.

(a)   What is the probability that four animals chosen at random will have enlarged thyroid glands?

˙(b)   Define X to be the number of animals selected at random until four of the selected animals have enlarged thyroids. What are the range, probability function, and distribution function of X?

(c)   Determine $P(X = 7)$ and $P(X > 5)$.

14. A storekeeper estimates that the probability of selling n loaves of bread on any given day is of the form $p(n) = kn$ for $n = 0, 1, 2, \ldots, 25$ and $p(n) = k(50 - n)$ for $n = 26, 27, \ldots, 50$.

(a)   For what value of the constant k is $p(n)$ the probability function of a random variable? Define the random variable.

(b)   What is the probability that on a given day the storekeeper will sell fewer than 26 loaves? Between 22 and 28 loaves? More than 45 loaves?

˙(c)   If the storekeeper wishes to meet the demand for bread on at least 95% of all days, how many loaves should be stocked?

15. In an experiment, mice are to be tested for left or right preference. The mice pass through a maze, in which ten choices must be made between turning left and turning right. Define X to be the random variable equal to the number of left turns made by a mouse.

(a)   What is the range of X?

(b)   If mice choose at random between left and right, what is the probability function $p(x)$ of the random variable X?

(c)   If the probability that a mouse chooses a left path is 0.6, what is the probability that exactly six left paths and four right paths are chosen?

16. Consider the experiment of choosing an odd integer from 7 to 17 at random. Define the random variable X by $X = n - 12$ if n is the chosen integer.

(a) What is the sample space of this experiment and the range of X?

(b) Determine the probability function and distribution function.

*17. Consider the experiment of choosing six trees at random from a group of three maple trees, four elm trees, and five oak trees. Define X to be the random variable equal to the number of maple trees chosen. Determine the range, probability function, and distribution function of X.

18. Consider the experiment of throwing two fair dice. Define a random variable X to be n if the number n turns up on both dice and to be -1 otherwise. Determine the range and probability function of X.

19. A biased coin with probability $\frac{2}{3}$ of heads is tossed three times in succession. Define a random variable X to be the number of heads minus the number of tails obtained in the three tosses. Determine the range of X and its probability function.

20. Consider the experiment of choosing at random an integer from -3 to $+3$. Define a random variable X to be equal to k^2 when the integer chosen is k.

(a) Determine the range of X.

(b) Determine the probability and distribution functions for X.

21. In a large drosophila (fruit fly) population, 40% of the flies belong to strain A, 30% to B, and 30% to C. Four flies are chosen at random from this population. Define X to be the random variable equal to the number of strain A flies chosen. Determine the range, probability function, and distribution function of X.

22. Consider the experiment of flipping a fair coin repeatedly until the first head appears. Let X be the random variable counting the number of flips.

(a) Find the range of X.

(b) Determine the probability function of X.

(c) Find $p(2)$, $p(4)$, and $p(7)$.

(d) Find $F(2)$, $F(4)$, and $F(7)$.

23. A salesman finds that he can sell an expensive machine in 30% of all attempts. He makes sales calls until he has sold one machine. Then he goes home. Define a random variable X to be the number of calls he makes until he sells a machine.

(a) What is the range of X?

(b) Find $p(1)$, $p(3)$, $p(6)$, and $p(n)$, where $n \geq 1$ is an integer.

(c) Find $F(1)$, $F(3)$, and $F(6)$.

*(d) Find $F(x)$ for every real number x.

*24. Problems 22 and 23 are problems involving a **geometric distribution**. In a geometric distribution, a binomial experiment is performed until the first success occurs, where p is the probability of success. A geometric random variable is the random variable that counts the number of times the experiment is performed.

(a) Show that, for a geometric distribution, the probability function is given by $p(n) = (1 - p)^{n-1}p$.

**(b) Show that the distribution function takes the values

$$F(n) = 1 - (1 - p)^n$$

for every positive integer n.

6.2 Expected Value of a Random Variable

Consider the experiment of flipping a fair coin four times and counting the number of heads. You might get no heads, one head, two heads, three heads, or four heads; however, it is reasonable to expect that if the experiment is repeated many times, you will get an average of two heads. That is, on the average, half the tosses (two) will be heads.

This intuitive idea leads to one of the central concepts in probability and statistics theory, that of *expected value*. The expected value of a random variable can be thought of as a weighted average of the possible values taken by that random variable.

Expected Value of a Random Variable　The **expected value** $E(X)$ of a random variable X is given by

$$E(X) = x_1 p(x_1) + x_2 p(x_2) + \cdots + x_n p(x_n) \qquad (1)$$

where $x_1, x_2, \ldots, x_n$ are the numbers in the range of X and $p(x_1), p(x_2), \ldots, p(x_n)$ are their corresponding probabilities.

Note.　The expected value is sometimes called the **expectation**, or **mean**, of X and is usually denoted by the symbol μ.[†]

Example 1　Consider the random variable X associated with the experiment of flipping a fair coin three times and counting the number of heads. Compute $E(X)$.

Solution　The values taken by X and their probabilities are given in Table 1. Then, from (1),

TABLE 1

Value taken by X: x	0	1	2	3
Probability: p(x)	$\frac{1}{8}$	$\frac{3}{8}$	$\frac{3}{8}$	$\frac{1}{8}$

$$\begin{aligned} \mu = E(X) &= 0p(0) + 1p(1) + 2p(2) + 3p(3) \\ &= 0 \cdot \tfrac{1}{8} + 1 \cdot \tfrac{3}{8} + 2 \cdot \tfrac{3}{8} + 3 \cdot \tfrac{1}{8} = \tfrac{12}{8} = \tfrac{3}{2}. \end{aligned}$$

Note that this makes sense intuitively since it is reasonable to expect that, on the average, half the tosses will be heads, and half of 3 is $\frac{3}{2}$.

[†] This is the lowercase Greek letter mu.

Remark. Example 1 illustrates an important fact: The expected value μ is an average that, in some problems, could not be taken on by the random variable. Certainly it is impossible to flip a coin three times and obtain exactly $1\frac{1}{2}$ heads. But if you flip a coin three times and repeat that experiment many times, you will obtain an *average* of $1\frac{1}{2}$ heads. Try it!

Example 2 A woman plays roulette and bets $1 on red (see Example 6.1.2 on page 295). What is her expected return?

Solution The random variable X takes the value 1 with probability $\frac{18}{38}$ and -1 with probability $\frac{20}{38}$. Thus

$$E(X) = (1)\tfrac{18}{38} + (-1)\tfrac{20}{38} = -\tfrac{2}{38} = -\tfrac{1}{19} \approx -0.0526 \approx -5\cancel{c}.$$

This means that, while the woman will win or lose $1 on each spin of the wheel, she will lose an average of 5¢ each time she plays.

Example 3 The Great Woods Insurance Company sells a policy that pays $50,000 in case of accidental death. According to its latest figures (1976), the rate of accidental death in the United States is 46.9 per 100,000 population. What annual premium must it charge to break even (on average)? What must it charge to make a profit?

Solution Let Z denote the premium that is to be charged. Consider the experiment of selling one policy to an individual in a given year. Let X be the random variable representing Great Woods' profit (or loss). We are being asked to find the value of Z such that $E(X) = 0$. There are two possibilities: The insured individual does or does not die accidentally. If the person does not die, the company gains the premium of $Z. If the person dies accidentally, the company loses (pays) $50,000, but still receives the $Z premium. The *net loss* is, therefore, $(50,000 − Z). The given information is summarized in Table 2. Then

TABLE 2 Gains or Losses by Great Woods Insurance

	the minus sign indicates a loss	
Value taken by X	$-(50,000 - Z)$	Z
Probability	$\dfrac{46.9}{100,000}$	$1 - \dfrac{46.9}{100,000} = \dfrac{99,953.1}{100,000}$

$$\mu = E(X) = -(50,000 - Z) \cdot \frac{46.9}{100,000} + Z\left(\frac{99,953.1}{100,000}\right)$$

$$= -(50,000)\left(\frac{46.9}{100,000}\right) + Z\left(\frac{46.9}{100,000}\right) + Z\left(\frac{99,953.1}{100,000}\right)$$

$$= -23.45 + Z$$

The expected value is zero when $Z = \$23.45$. This is the premium required to break even over the long run. Any larger premium will provide a profit. For example, if the company charges $30, then $-23.45 + 30 = \$6.55$. This means that the company will earn an average of $6.55 per year on each person insured.

Let us make sure we understand the meaning of the number $6.55 computed above. (This is the expected value of the "profit" random variable X when a premium of $30 is charged.) The company can never make a profit of $6.55 on any one individual. It either gains $30 if the individual does not die accidentally or it loses $49,970 ($50,000 − $30) if the individual does die accidentally. The number $6.55 is an average gain. Thus, for example, if the company insures 100,000 persons, it should earn $6.55 × 100,000 = \$655,000$. Of course, there is *no guarantee* that it will earn $655,000. If there are more deaths than average, the company will earn less, and vice versa. If, for example, 100 insured individuals die accidentally in one year, then the company will earn

$$49,900(30) - 100(49,970) = -\$3,500,000.$$

This is a loss of $3\frac{1}{2}$ million! This illustrates the importance of accurate statistical information (and also explains why insurance premiums sometimes seem excessively high).

Example 4 Compute the expected value of the random variable described in Example 6.1.10 (see page 298).

Solution Here X takes the values -3, -1, and 4 with probabilities 0.2, 0.25, and 0.55, respectively. Thus

$$E(X) = -3(0.2) + (-1)(0.25) + 4(0.55) = 1.35$$

Example 5 What is the expected outcome when two dice are thrown and their sum is recorded?

Solution The outcomes and their probabilities are given in Example 6.1.12 on page 299. We have

$$\mu = 2 \cdot \tfrac{1}{36} + 3 \cdot \tfrac{2}{36} + 4 \cdot \tfrac{3}{36} + 5 \cdot \tfrac{4}{36} + 6 \cdot \tfrac{5}{36} + 7 \cdot \tfrac{6}{36} + 8 \cdot \tfrac{5}{36}$$
$$+ 9 \cdot \tfrac{4}{36} + 10 \cdot \tfrac{3}{36} + 11 \cdot \tfrac{2}{36} + 12 \cdot \tfrac{1}{36}$$
$$= \tfrac{1}{36}(2 + 6 + 12 + 20 + 30 + 42 + 40 + 36 + 30 + 22 + 12) = \tfrac{252}{36} = 7.$$

Explain why this result is not surprising.

Example 6 Gambler's Ruin. We return again to the roulette wheel of Example 2. Many gamblers like to play according to a system. For reasons we will soon see, one popular system is called *doubling up*, or **gambler's ruin**. In this system our gambler bets $1 on red, for instance. If she wins, she's ahead $1 and she bets $1 on the next spin. If she loses, she bets $2 on red so that if she wins, she'll be ahead $1. If she loses, she'll be $3 behind, so she bets $4; if she loses again, she'll be $7 behind so she bets $8, and so on. Every time she wins, she wins $1. It seems that she can't lose because, assuming that the roulette wheel is fair, red must come up at least once in a while, can't it? (Casino owners love this kind of reasoning.) Where's the flaw? The problems are that (1) our gambler does not have an unlimited amount of money and (2) there is a maximum amount you can bet on any one spin of the wheel. Let us suppose that our gambler has approximately $1000 or that the legal betting limit is $1000 (any other numbers will lead to the same result). If our gambler has a string of bad luck, her losses will look like this:

$$\$1 + \$2 + \$4 + \$8 + \$16 + \$32 + \$64 + \$128 + \$256 + \$512 = \$1023.$$

That is, if she loses 10 times in a row, she'll either be broke or unable to make the next bet of $1024, or both. Since the wheel is assumed to be fair, the spins are independent and the probability of losing ten times in a row is $(\frac{20}{38})^{10} \approx 0.001631$. There are two possibilities:

she will lose $1023 with probability 0.001631

or

she will win $1 with probability $1 - 0.001631 = 0.998369$.

Thus

$$\mu = \text{expected gain (or expected value)}$$
$$= 1(0.998369) - 1023(0.001631) \approx -\$0.67.$$

This means that if our gambler plays this system over many spins of the roulette wheel, she will lose an average of 67¢ each time she plays. Here we define a "play" to mean betting until she either wins $1 or loses $1023. Many gamblers have lost lots of money playing this seemingly foolproof system.

For binomial random variables, the expected value can be easily computed. For example, if you flip a fair coin 100 times, you expect 50 heads. If 25% of a population have blond hair and 80 people are selected at random, then we would expect that 20 of them had blond hair.

Expected Value, or Mean, of a Binomial Random Variable

> If the random variable X represents the number of successes when a binomial experiment is performed n times with probability p of success, then
>
> $$\mu = E(X) = np. \tag{2}$$

We will not derive formula (2) although, as the examples in the last paragraph suggest, it does make a great deal of intuitive sense.

Example 7 A factory produces batteries, of which 0.2% have been found to be defective. If 300 batteries are selected at random, what is the expected number of defective ones?

Solution Selecting a battery and determining whether it is defective is a binomial experiment with $n = 300$ and $p = 0.002$. Thus

$$\mu = np = (300)(0.002) = 0.6.$$

Warning. The formula $\mu = np$ can be used only when dealing with a binomial random variable. Don't try to apply it in other situations.

The expected value, or mean, of a binomial random variable can be interpreted in a number of interesting ways. Some of these are given in Table 3.

TABLE 3   Interpretation of Binomial Mean

n trials of the experiment	Probability of	Mean = expected number of
An insured individual.	An accidental death.	Accidental deaths among those insured.
Manufacture a product.	A defect in a product.	Defective products.
Make a sales call.	A sale.	Sales.
Test a person for a disease.	Having the disease.	Individuals having the disease.
Observe the weather each day for a month.	A rainy day.	Rainy days in the month.

We now turn to one final example of expected value.

Example 8   Decision Making.   An investor has $10,000 to invest and has a choice of two investments. If he invests his money in second mortgages, he knows that he can earn $2000 with 60% probability and $1000 with 40% probability. On the other hand, if he invests in soybean futures, he can make $20,000 with 30% probability and lose $5000 with 70% probability. Which investment gives him the largest expected gain?

Solution   For the second mortgages,

$$E(X) = 2000(0.6) + 1000(0.4) = \$1600.$$

For the soybean futures,

$$E(X) = 20{,}000(0.3) + (-5000)(0.7) = \$2500.$$

Thus he has a larger expected return by investing in soybeans. (However, if he cannot afford a loss under any circumstances, he would then invest in the mortgages.)

We close this section by noting that when an experiment is performed, the expected value of the relevant random variable often is unknown and can only be estimated experimentally. This important estimate, called the *sample mean*, is discussed in detail in Section 6.4.

PROBLEMS 6.2 In Problems 1–18, determine the expected value of the given random variable defined in Problem Set 6.1 (see page 300).

1.	Problem 6.1.1.	**2.**	Problem 6.1.2.	**3.**	Problem 6.1.3.
4.	Problem 6.1.4.	**5.**	Problem 6.1.5.	**6.**	Problem 6.1.6.
7.	Problem 6.1.7.	**8.**	Problem 6.1.8.	**9.**	Problem 6.1.9.
10.	Problem 6.1.10.	*11.	Problem 6.1.12.	**12.**	Problem 6.1.15.
13.	Problem 6.1.16.	*14.	Problem 6.1.17.	**15.**	Problem 6.1.18.
16.	Problem 6.1.19.	**17.**	Problem 6.1.20.	**18.**	Problem 6.1.21.

*19. Determine $E(X)$ and $E(Y)$ in Problem 6.1.11 on page 302.

20. In Canada the accidental death rate in 1976 was 30.0 per 100,000 population. What premium should the Great Woods Insurance Company of Example 3 charge in order to break even on its average $50,000 policy?

21. In Problem 20, what is the expected profit if the company sells 25,000 policies with a premium of $25?

22. Answer the question of Problem 20 if the company markets its policy in Belgium, where the 1976 accidental death rate was 60.5 per 100,000.

23. The number of telephone calls received during the lunch hour ($12:00-1:00$) of a small business has the probability function given in the table.

Number of calls	Probability
0	0.05
1	0.15
2	0.20
3	0.35
4	0.10
5	0.10
6	0.05

What is the expected number of calls during the lunch hour?

*24. A house committee has ten Democrats and eight Republicans. A committee of five is chosen at random. What is the expected number of Democrats on the committee?

25. The Hicks Plumbing Company pays $1000 to bid on a large contract. If its bid is accepted, it estimates its profit at $25,000. If it is not accepted, it forfeits the $1000. Assuming that eight plumbing firms submit bids and that all are equally likely to be successful, compute the expected profit for Hicks Plumbing.

26. Kerstin Sena is interested in investing $10,000 in a risky venture where there is a 25 % chance of making $10,000 in 1 year, a 40 % chance of making $5000, and a 35 % chance of losing $8000. She also has the option of investing in a safe savings certificate paying 15 % per annum. Which investment would give her the largest expected return?

*27. At each of the Yeehaw State University home football games, local high school students sell programs. These programs can be purchased by the students for a cost of $1, and the selling price is $1.50. Any unsold programs are worthless after the game, so the students suffer a loss. The number of programs an individual student can sell depends upon the size of the crowd for that game. Since many fans purchase tickets at the gate, there is no way to know in advance how large the crowd will be for any given game. In studying past attendance records, a local program salesperson has determined that there is a sellout 50 % of the time, a 90 % capacity crowd 30 % of the time, and an 80 % capacity crowd 20 % of the time. Her sales records show that where there is a capacity crowd, she can sell 200 programs; when there is a 90 % capacity crowd, she can sell 150 programs; and when there is an 80 % capacity crowd, she can sell only 100 programs. If you were a friend of the salesperson, how many programs would you suggest she buy for each home game?

*28. Steve Shrier of the Bishop Produce Company has started subscribing to a special weather service, which costs him $10 per 3-day period. The company that markets this service has provided Steve with information on how well its service would have performed had it been available over the past 5 years. This information is in the form of the percentage of instances that the company would have predicted correctly each type of weather that occurred. Using the table below, advise Steve whether or not he should subscribe to the weather prediction service.

Weather Prediction Probabilities

Weather that would have been predicted	Weather that actually occurred		
	Bad (%)	Unsettled (%)	Good (%)
Bad	50	30	20
Unsettled	20	40	40
Good	30	30	40

*29. A dairy farmer sells milk by the gallon to a wholesaler. However, he occasionally sets up a stand on the road near his farm to sell gallons of milk to passersby on Saturdays. It costs him $1 to produce a gallon of milk and the wholesaler pays him $1.50 per gallon. However, he can sell the milk on the road for $2 a gallon. Past meteorological data suggest that it rains on 20 % of the days, is cloudy on 25 % of the days, and is sunny on 55 % of the days. From experience, the farmer knows that he can sell 150 gallons on sunny days, 100 on cloudy days, and 70 on rainy days. Any milk put up for sale that remains unsold must be thrown out. The farmer must make a decision 1 week in advance so he has no way of predicting the weather on the sale day. How many gallons should he set aside in order to maximize his expected profit?

30. As a gimmick to lure customers, a car dealer offers a free raffle ticket on an $8000 automobile to the first 100 customers who purchase the car from him at that price. However, his competitor across town sells the same car for $7900. What is the expected value of each free raffle ticket?

The expected value of a geometrically distributed random variable (see Problem 6.1.24 on page 303 is $1/p$, where p is the probability of success on one trial of the experiment. Use this fact in Problems 31–34.

31. A coin is tossed until the first head appears. What is the expected number of tosses?

32. What is the expected number of sales calls made by the salesperson in Problem 6.1.23 on page 303?

33. A certain type of skin graft is successful in 40 % of all cases. The skin graft is performed on a patient several times in succession until a graft takes. What is the expected number of grafts performed?

*34. An employee of a large corporation is nearing retirement and must chose one of the following four options for disbursement of her pension benefits.

(a) A $50,000 lump sum payment.

(b) $5000 a year for the rest of her life.

(c) $10,000 a year for each of the next 10 years that she is alive and nothing thereafter.

(d) $7500 a year for the next 10 years, whether or not she is alive, and nothing thereafter.

The woman estimates that the probability of dying in any one year is 0.05. Which pension plan will provide her and her heirs with the largest expected value?

35. A drunk comes home and tries to enter his house. He has eight keys on his key ring and tries them randomly until he finds the one that opens his door. Assuming that he does not try the same key twice, what is the expected number of keys he will try?

36. Answer the question of Problem 35 assuming that the man is so drunk that he chooses a key randomly whether or not he has tried that key before.

Problems 37–39 were taken from recent CPA exams in accounting.

37. (May 1975) The Stat Company wants more information on the demand for its products. The following data are relevant.

Units demanded	Probability of unit demand	Total cost of units demanded
0	0.10	$ 0
1	0.15	1.00
2	0.20	2.00
3	0.40	3.00
4	0.10	4.00
5	0.05	5.00

What is the total expected value or payoff with perfect information?

(a) $2.40

(b) $7.40

(c) $9.00

(d) $9.15

38. (May 1979) The Polly Company wishes to determine the amount of safety stock that it should maintain for product D that will result in the lowest cost. The following information is available.

Stockout cost	$80 per occurrence
Carrying cost of safety stock	$2 per unit
Number of purchase orders	5 per year

The available options open to Polly are as follows.

Units of safety stock	Probability of running out of safety stock
10	50%
20	40%
30	30%
40	20%
50	10%
55	5%

Find the number of units of safety stock that will result in the lowest cost.

(a)   20

(b)   40

(c)   50

(d)   55

39. (November 1975) Your client, a charity, is planning a carnival to raise money. The charity has permission from the local authorities to have games of chance. For one of these games, the player draws one card from a standard deck of 52 cards. If the player draws a jack of hearts, jack of diamonds, or jack of spades, he is paid $6.50. If he draws any card of clubs, he is paid $2.50. Assume X equals the price your client should charge per draw so that the long-run expected value of this game is zero. Which one of the following equations should be used to determine that price?

(a) $\frac{3}{52}(6.50 - X) + \frac{13}{52}(2.50 - X) = X$

(b) $\frac{3}{52}(6.50 - X) + \frac{13}{52}(2.50 - X) = \frac{36}{52}X$

(c) $\frac{3}{52}(6.50) + \frac{13}{52}(2.50) = \frac{36}{52}X$

(d) $6.50 - 2.50 = \frac{36}{52}X$

6.3 Graphing Data

Let us look again at some of the examples discussed earlier in this book: Example 5.5.2 (page 251), where a California aircraft manufacturer has determined probabilities for the number of orders for the coming year for its largest wide-body jet; Example 5.6.2 (page 262), where a survey of cable television viewers found that 35% liked an all-sports program, 15% liked an all-news program, and 10% liked both; Example 5.8.1 (page 276), in which a manufacturer receives 60% of its machine parts from supplier 1 and 40% from supplier 2 and it has been observed that 2% of the parts from supplier 1 and 1% of the parts from supplier 2 are defective; and Example 6.1.10 (page 298), where in a certain Southern city in July, it rains on 20% of the days, it is cloudy (without rain) on 25% of the days, and it is sunny on 55% of the days.

Each of these four problems involves the collection and analysis of data. The aircraft manufacturer can try to determine probabilities only after studying sales records and certain factors affecting sales over a long period of time. Conclusions can be drawn after observations covering a short time interval, but probabilities then are unlikely to be useful.

Similarly, we can make predictions about the weather in that Southern city only if we have confidence that the computed percentages of rainy, cloudy, and sunny days are based on data covering a large number of Julys. The large number is necessary so that unusually sunny or rainy periods do not significantly affect the percentages.

The collection and analysis of data is the basis of statistics. It is the essential meaning of statistics. Obviously, a complete or even extensive discussion of these topics would take us far beyond the introductory nature of this book. However, we can present some elementary ways to represent and analyze collected data. In this section we discuss ways to graph data that have been collected. In Sections 6.4 and 6.5, we indicate some of the kinds of information that can be obtained from these data.

A statistician uses the word **population** to refer to something that he or she wishes to study. Thus we may have a population of fish, a population of light bulbs, or a population of football games, for example. A **sample** is some randomly chosen subset of the population to be studied. A **statistic** is a quantitative measure of some characteristic of the sample (such as the average value taken, for example).

Example 1 There are many trout in a large lake and a biologist wishes to determine their average weight. To do so she captures 50 trout and weighs them. This set of trout is her sample and each weight entered is a piece of data. The average weight of the trout in the sample is a statistic.

Example 2 An electronics firm manufactures light bulbs and the owner of the firm wishes to determine how long the average bulb burns. He selects 80 bulbs from his inventory and keeps them lit until they burn out. He records the number of hours each stays lit. The selection of 50 bulbs is his sample and each recorded bulb life (measured in hours) is a piece of data. The average bulb life of the bulbs in the sample is a statistic.

These two examples are fairly typical and they raise some important questions.

Question 1. How do we make sense of the collected data?

Question 2. What is a meaningful measure of the average or central value of these data?

Question 3. How do we measure the variation from this central value?

Question 4. How big a sample is necessary in order to be able to draw meaningful conclusions? That is, what should the *size of the sample* be?

We give some answers to Question 1 in this section and address Questions 2 and 3 in Sections 6.4 and 6.5, respectively. The answer to Question 4 is far more difficult and is best left to a course in statistics.

Before continuing, we ask an even more basic question: Why sample a population at all? The answer has to do with overwhelming numbers. If there are 100,000 fish in a lake, it would be impossible to weigh them all (because of limitations of time and money). If there are 100 million potential voters, it is impossible to poll them all. If it takes one pollster 15 minutes to poll one voter, 32 voters could be polled in an 8-hour day. Thus it would take 10,000 pollsters 312.5 days (close to a year) to complete one survey. If each worker received $50 a day, the total cost would be $156.25 million. That's what we mean by the word *impossible*.

Since it is rarely practical to measure everything in a population, it is necessary to take a sample. We now describe a simple sample and illustrate some of the ways the data can be graphically depicted.

Suppose that the manufacturer of Example 2 has 100,000 light bulbs in stock and wishes to determine the average bulb life. He burns a sample of 80 bulbs and accumulates the data given in Table 1. There are many ways to represent these

TABLE 1 Bulb Life (in Hours) of 80 Light Bulbs in a Sample

303	359	267	391	342	305	279	417
315	335	314	286	337	363	354	331
383	363	355	262	374	352	402	336
240	247	446	300	316	297	326	301
377	355	348	349	286	347	257	342
353	273	304	409	313	311	322	289
328	370	345	314	298	401	317	353
371	432	252	307	412	332	313	212
351	371	328	340	358	312	339	311
281	334	306	380	371	285	421	336

data. One way is to depict each datum as a point in the xy-plane with the x-axis being the bulb life (in hours) and the y-axis being the **frequency** of each bulb life. The frequency of a piece of data is the number of times that datum occurs. If we do this, we get a graph, part of which is given in Figure 1. The most obvious

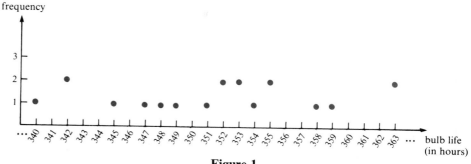

Figure 1

thing about this part of a graph is that it does not seem very interesting. We can see that, for example, one light bulb burned out after 340 hours, none after 341 hours, and two after 342 hours. But there does not appear to be any pattern from which we can draw conclusions.

Since there is very little difference between 340 hours and 341 hours, for example, let us instead depict our data by *grouping* the data. Since the data are numbers, the groups will be intervals of real numbers. In choosing these intervals, there are three rules that must always be observed:

RULES FOR CHOOSING INTERVALS

1. Every number or datum can be placed in some interval.

2. No number can be placed into more than one interval.

3. The intervals must be of equal length.

As long as we follow these rules, our groups can be constructed. There are obviously many choices. We could have the single group 200–500 or 300 groups 200–201, 201–202, 202–203, . . . , 499–500. However, these two groupings will not be very useful as the first is too coarse and the second too fine. Here we will use intervals each having length 20. We must be able to include the smallest number, 212, and the largest, 446. A natural choice, then, includes the following 12 intervals.

210–230

230–250

250–270

270–290

290–310

310–330

330–350

350–370

370–390

390–410

410–430

430–450

The trouble with this choice is that Rule 2 is violated. For example, the number 370 appears in two groups. To solve this problem, we use endpoints which cannot be data:

209.5–229.5

229.5–249.5

249.5–269.5

269.5–289.5

289.5–309.5

309.5–329.5

329.5–349.5

349.5–369.5

369.5–389.5

389.5–409.5

409.5–429.5

429.5–449.5

Now we count the number of data that fall into each group. This is the **frequency** of the group. We also compute the **relative frequency** of each group by dividing the frequency by 80.

$$\text{Relative frequency} = \frac{\text{frequency}}{\text{size of the sample}} \tag{1}$$

All this information is given in the **frequency table**, Table 2. The last column represents the **percentage of occurrences**, which is defined by

$$\text{percentage of occurrences} = \text{relative frequency} \times 100. \qquad (2)$$

Note. The relative frequencies can be thought of as approximations to probabilities. If, for example, we know that 10% of all light bulbs burn out between 369.5 hours and 389.5 hours, then the probability is 0.1 that one of the 100,000 light bulbs chosen at random will burn out after between 369.5 hours and 389.5 hours. Of course, we do not know that 10% of all 100,000 bulbs will burn out in that interval. We only know that 10% of our sample did so.

TABLE 2 Frequency Table for Grouped Light-Bulb Data

Group or interval	Count (by tally marks)	Frequency	Relative frequency (frequency)/80	Percent of occurrences
209.5–229.5	|	1	0.0125	1.25
229.5–249.5	||	2	0.025	2.5
249.5–269.5	||||	4	0.05	5
269.5–289.5	₸ ||	7	0.0875	8.75
289.5–309.5	₸ ||||	9	0.1125	11.25
309.5–329.5	₸ ₸ ||||	14	0.175	17.5
329.5–349.5	₸ ₸ ₸	15	0.1875	18.75
349.5–369.5	₸ ₸ |	11	0.1375	13.75
369.5–389.5	₸ |||	8	0.1	10
389.5–409.5	||||	4	0.05	5
409.5–429.5	|||	3	0.0375	3.75
429.5–449.5	||	2	0.025	2.5

Sum 1.00   Sum 100%

Example 3 If one light bulb is drawn at random from the sample, what is the probability that it will burn out

(a) between 289.5 and 369.5 hours?

(b) between 229.5 and 309.5 hours?

Solution

(a) From Table 2 we see that the relative frequency of the interval

289.5–369.5 is 0.1125 + 0.175 + 0.1875 + 0.1375 = 0.6125.

This is the answer. Another way to see this is to observe that 9 + 14 + 15 + 11 = 49 bulbs have a bulb life between 289.5 and 369.5 hours and $\frac{49}{80} = 0.6125$.

(b)   Here

$$0.025 + 0.05 + 0.0875 + 0.1125 = 0.275$$

is the relative frequency of the interval 229.5–309.5. Also, $2 + 4 + 7 + 9 = 22$ and $\frac{22}{80} = 0.275$.

There are two common ways to depict the information in Table 2. The first is to draw a **frequency histogram**. The frequency histogram consists of rectangles. The base of each rectangle equals the length of each interval chosen (in this case 20) and the height represents the frequency or relative frequency of the group. The frequency histogram for the data of our example is given in Figure 2.

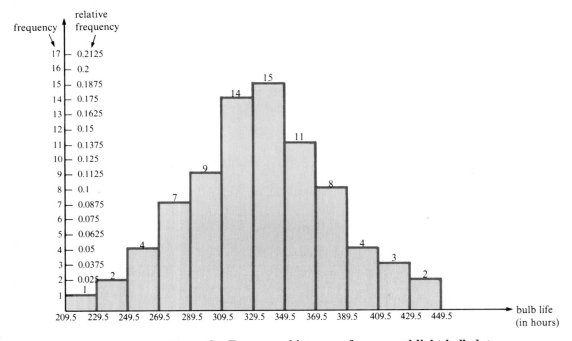

Figure 2   Frequency histogram for grouped light-bulb data.

Now we can see some sort of pattern. The histogram in Figure 2 gives us some idea of the *shape of the probability distribution* of light-bulb lives. The figure suggests that the average light-bulb life is somewhere around 330 hours. It also suggests that the further an interval is from the "central" values 310–350, the less likely it is that a randomly chosen light bulb has a life in that interval.

The shape of the distribution can also be suggested by a **frequency diagram**, or **frequency polygon**. Here the x-axis contains the midpoints of the intervals and the y-axis gives the frequency (or relative frequency) of that group. The resulting points are then connected by line segments. The frequency diagram for the light-bulb data is given in Figure 3.

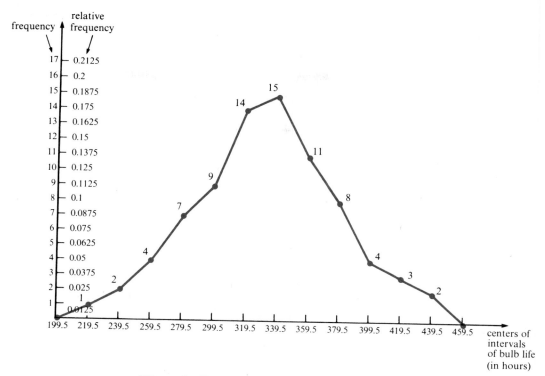

Figure 3   Frequency diagram for the grouped light-bulb data.

The frequency histogram in Figure 2 can be interpreted in a very interesting way. Let us relabel the x-axis so that the length of the base of each rectangle in Figure 2 is equal to 1. Then, if the height of each rectangle is taken to be the relative frequency, the area of a rectangle is equal to the relative frequency of the interval it represents (since area = base × height) and, as we have seen, relative frequency represents probability. Thus *the area of a given rectangle in a frequency histogram represents the probability that a randomly chosen member of the sample will lie in the group corresponding to that rectangle.*

Moreover, if we want to find the probability that a randomly chosen light bulb lies in the interval 289.5–369.5, say, (as in Example 3), then we simply add the areas of the four rectangles covering this set of intervals. Thus, suitably interpreted, we see that *area under a frequency histogram represents probability.* This fact will be very useful to us in Section 6.6.

The answer to Example 3(a) is illustrated by the shaded region in Figure 4.

We will often be interested in determining the percentage of a sample that lies above or below a certain value. We have seen that group frequencies take values very much like those of a probability function. Similarly, *cumulative frequencies* take values like those of a distribution function (see equation (6.1.2), page 299).

Cumulative Frequency   When a sample is taken, the **cumulative frequency** of a number is the part of the sample that takes values less than or equal to that number. If the sample

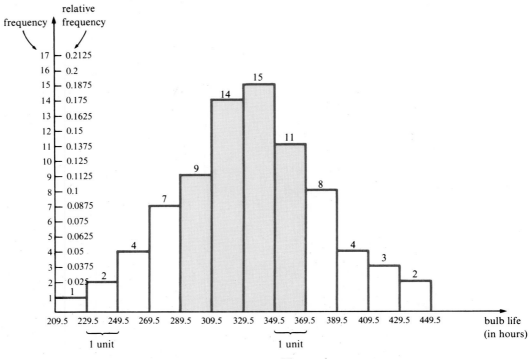

Figure 4

is divided into intervals, the cumulative frequency of an interval is the part of the sample that takes values in that interval and all preceeding intervals.

The **cumulative frequency table** for the light-bulb data is given in Table 3.

We see, for example, that 14 light bulbs had bulb lives less than 289.5 hours and 71 had bulb lives less than 389.5 hours. As in (1) and (2), we have

$$\text{relative cumulative frequency} = \frac{\text{cumulative frequency}}{\text{size of the sample}} \qquad (3)$$

and

$$\text{cumulative percentage of occurrences} = \text{relative cumulative frequency} \times 100. \qquad (4)$$

Example 4   A bulb is drawn at random from the sample. What is the probability that it takes a value

(a)   less than 329.5?

(b)   less than 369.5?

(c)   less than 600?

TABLE 3 Cumulative Frequency Table for Grouped Light-Bulb Data

Group or interval	Frequency	Cumulative frequency	Relative cumulative frequency (cumulative frequency)/80	Cumulative percentage of occurrences
209.5–229.5	1	1	0.0125	1.25
229.5–249.5	2	3	0.0375	3.75
249.5–269.5	4	7	0.0875	8.75
269.5–289.5	7	14	0.175	17.5
289.5–309.5	9	23	0.2875	28.75
309.5–329.5	14	37	0.4675	46.75
329.5–349.5	15	52	0.65	65
349.5–369.5	11	63	0.7875	78.75
369.5–389.5	8	71	0.8875	88.75
389.5–409.5	4	75	0.9375	93.75
409.5–429.5	3	78	0.975	97.5
429.5–449.5	2	80	1.000	100

Solution

(a) This question can be rephrased: What proportion of outcomes have values less than 329.5? The answer is the relative cumulative frequency of 329.5, which is—by Table 3—equal to 0.4675.

(b) The relative cumulative frequency of 369.5 is 0.7875.

(c) All values are less than 600, so that the relative cumulative frequency of 600 is 1.

We can also draw a **cumulative frequency histogram** and a **cumulative frequency diagram** (or **polygon**) for our light-bulb data. These are given in Figures 5 and 6.

Figure 6 is very revealing. It is the graph of a rule that takes values starting at zero and increasing to 1. Compare this with the graph of the distribution function given in Figure 6.1.1 on page 300.

At this point you should reread Section 6.1. In Section 6.1 we discussed *theoretical* probability and distribution functions. That is, we computed probabilities based on known information. In this section we discussed relative frequencies and cumulative relative frequencies that represent *empirical* or *experimental* probabilities. That is, we started with no information about probabilities and then estimated these probabilities as proportions of the total number of observations. The material in Section 6.1 really belongs to the realm of probability theory, while the material in this section is clearly statistical. However, the two sections contain very similar ideas, which is why it is often so difficult to separate probability and statistics.

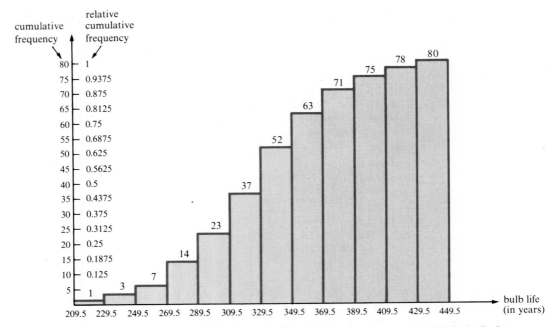

Figure 5   Cumulative frequency histogram for grouped light-bulb data.

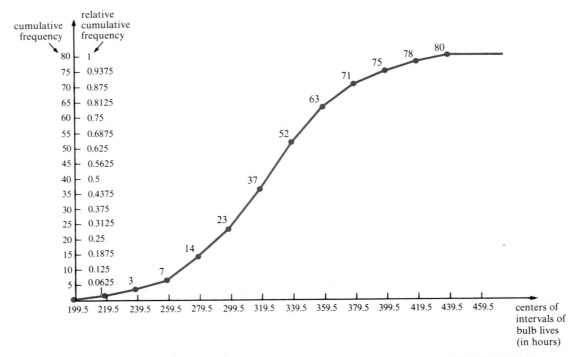

Figure 6   Cumulative frequency diagram for grouped light-bulb data.

In the next section we discuss the empirical (experimental) version of expected value.

PROBLEMS 6.3

1. Place the light-bulb data into 24 intervals, each of length 10, starting at 209.5 hours.
   (a) Construct new frequency tables and cumulative frequency tables.
   (b) Sketch a frequency histogram and frequency diagram for these new data groupings.
   (c) Construct a new cumulative frequency histogram and a new cumulative frequency diagram.

2. Answer the question of Problem 1 if the data are placed in six intervals of length 40 beginning at 209.5 hours.

3. The following outcomes were obtained when a die was tossed 50 times.

5	3	3	2	6	5	4	5	5	2
2	1	6	3	4	2	3	1	3	3
3	5	1	2	2	2	5	5	1	5
6	4	5	1	5	6	2	4	2	4
5	5	2	5	2	1	2	6	5	2

   (a) Construct a frequency table.
   (b) Construct a frequency histogram.
   (c) Construct a frequency diagram.
   (d) Looking at the frequency diagram, do you believe the die to be a fair one?
   (e) Compute the probability that one toss randomly selected from the sample took a value between 3 and 5 inclusive.
   (f) Compute the probability that the toss in (e) took a value less than or equal to 4.

4. Answer the questions of Problem 3 using the following data.

1	5	6	2	4	1	3	1	2	2
3	2	4	6	5	6	6	4	3	5
5	1	5	1	4	6	3	6	1	4
2	3	3	6	1	2	2	4	3	3
4	6	3	4	5	4	5	2	6	5

5. Sixty pupils in the fifth grade are measured and their heights, to the nearest centimeter, are recorded (1 in. = 2.5 cm). The results are given below.

137	141	138	104	99	124
102	122	121	132	132	136
117	126	119	136	123	131
123	111	92	124	131	103
128	104	126	145	118	118
120	143	148	131	125	127
108	110	121	152	122	134
122	130	120	103	112	121
138	117	133	124	143	131
115	146	134	141	130	112

(a) Construct a frequency and cumulative frequency table for these data using seven intervals of length 10, starting at 90.5 cm.

(b) Draw frequency and cumulative frequency histograms.

(c) Draw frequency and cumulative frequency diagrams.

(d) Compute the probability that a pupil randomly selected from the sample had a height in the range 110.5–140.5.

(e) Compute the probability that the pupil of part (d) was less than 130.5 cm tall.

6. Answer parts (a), (b), and (c) of Problem 5 if the data are grouped into 14 intervals of length 5, again starting at 90.5 cm.

7. Seventy students took a mathematics exam. Their scores were as follows.

71	24	78	100	47	38	47
68	56	36	72	65	47	68
51	52	77	80	61	69	64
83	76	61	17	56	96	73
76	70	72	36	82	26	51
52	68	50	52	91	58	75
63	81	61	66	28	43	49
71	93	58	38	43	85	70
89	47	8	88	64	53	80
94	68	42	45	57	60	58

(a) Construct a frequency and cumulative frequency table for these data using ten intervals of length 10, starting at 0.5.

(b) Draw frequency and cumulative frequency histograms.

(c) Draw frequency and cumulative frequency diagrams.

(d) Compute the probability that a randomly selected student got a score in the range 30.5–60.5.

(e) Compute the probability that a randomly selected student got a score less than 71.

8. Answer parts (a), (b), and (c) of Problem 7 if the data are grouped into five intervals.

9. Answer parts (a), (b), and (c) of Problem 7 if the data are grouped into 20 intervals.

10. A consumer agency measured the fuel consumption (in terms of miles per gallon) obtained by driving 100 medium-sized cars on an open highway. The measurements (to the nearest tenth of a mile) are given below.

22.6	19.2	21.5	14.1	20.2	22.7	20.9	22.9	22.4	16.8
20.8	22.9	19.9	19.3	20.5	17.6	21.8	19.2	20.7	22.5
22.6	21.5	21.2	15.5	20.3	21.9	16.3	16.3	21.9	17.4
21.3	23.0	15.2	18.8	21.5	21.7	20.5	19.1	23.8	19.3
16.6	19.3	14.7	16.8	20.6	18.7	20.5	15.4	19.3	21.9
17.3	18.9	21.7	16.8	21.9	21.1	18.3	21.6	21.8	15.1
21.9	17.2	16.2	14.3	15.1	21.0	19.1	18.3	20.4	18.3
20.2	19.7	14.5	20.5	18.6	22.7	19.0	17.0	19.6	22.7
20.1	18.0	22.1	21.3	16.5	21.9	21.6	23.9	20.8	22.3
18.5	19.3	21.4	20.4	18.6	21.6	19.8	19.2	22.8	18.7

(a)   Construct a frequency and cumulative frequency table for these data using five intervals of length 2, starting at 14.05 mi/gal.

(b)   Construct a frequency histogram for these data.

(c)   Construct a cumulative frequency histogram.

11.   Repeat the steps of Problem 10 using ten intervals of length 1.

12.   Repeat the steps of Problem 10 using 20 intervals of length 0.5.

13.   Compare the frequency histograms in Problems 10, 11, and 12. Which one gives the best indication of the shape of the distribution? Explain why this problem suggests that care must be taken in choosing the number of groups if meaningful conclusions are to be drawn.

6.4 Measures of Central Tendency: Sample Mean and Median

When we have a random variable for which the probability function is known, then its expected value can be computed as in Section 6.2. In Section 6.2 we said that the expected value μ represents the expected average value taken by the random variable. This number can be computed without doing any sampling. Thus we know that if we flip a fair coin 60 times, the expected number of heads is 30. We do not actually have to toss the coin to compute this. This is why we say that the expected value of a random variable is a *theoretical* concept.

Now we turn the process around. As in Section 6.3, we take a sample and record the results. We do not, when we start, have any idea what the average value taken by our sample will be.

Sample Mean

The **sample mean** or, simply, **mean**, of a sample is the arithmetic average of the values taken by the sample. That is, if the sample takes the n values $x_1, x_2, \ldots, x_n$, then the sample mean is equal to $(x_1 + x_2 + \cdots + x_n)/n$.

Notation.   The sample mean is usually denoted by $\bar{x}$; thus we have

$$\bar{x} = \frac{x_1 + x_2 + \cdots + x_n}{n} \qquad (1)$$

Example 1   Compute the sample mean of the numbers 5, 8, 9, 7, 2, 5, 3, 8.

Solution

$$\bar{x} = \frac{5 + 8 + 9 + 7 + 2 + 5 + 3 + 8}{8} = \frac{47}{8} = 5.875$$

Remark.   As with expected value, the sample mean, in some cases, will not be equal to any of the values taken by the sample.

Example 2   Compute the mean of the bulb lives of the light-bulb sample of Section 6.3.

Solution   There are 80 numbers given in Table 1 on page 314. Their sum is 26,614. Thus $\bar{x} = 26,614/80 = 332.675$.

Notation.   The Greek letter sigma, Σ, which stands for the word *sum*, is often used to denote a sum. Thus the sum $x_1 + x_2 + \cdots + x_n$ is denoted as Σx. Then we have

$$\bar{x} = \frac{\Sigma x}{n} \qquad (2)$$

where n is the size of the sample.

The last example suggests that finding the mean of a large sample can be quite tedious (it is no fun to find the sum of 80 numbers—even with the aid of a calculator). To simplify calculations, another statistic is often computed: the *sample mean of grouped data*.

Sample Mean of Grouped Data   Let m_i denote the **midpoint** of the ith interval that forms the group and let f_i denote the frequency of that group. Then, if there are k groups, the **sample mean of the grouped data**, also denoted by $\bar{x}$, is given by

$$\bar{x} = \frac{\Sigma mf}{n} = \frac{m_1 f_1 + m_2 f_2 + \cdots + m_k f_k}{n}. \qquad (3)$$

Remark.   When data are grouped, the sample mean is computed as if all the numbers in an interval take the same value (the midpoint of that interval).

 Example 3   Compute the sample mean of the grouped light-bulb data given in Table 6.3.2 on page 317.

Solution   We alter Table 6.3.2 slightly.
    Thus, from (3),

$$\bar{x} = \frac{26,680}{80} = 333.5$$

Note that the sample mean of the grouped data is easier to compute than the ordinary sample mean, but results in a bit of inaccuracy. Here the error is $332.675 - 333.5 = -0.875$, which is a *relative error* of $0.875/332.675 \approx 0.00263 \approx 0.26\%$. The relative errors introduced by grouping the data will rarely be very large.

TABLE 1   Frequency Table for Light-Bulb Data

Group or interval	Midpoint of interval (m_i)	Frequency (f_i)	Product $(m_i f_i)$
209.5–229.5	219.5	1	219.5
229.5–249.5	239.5	2	479
249.5–269.5	259.5	4	1038
269.5–289.5	279.5	7	1956.5
289.5–309.5	299.5	9	2695.5
309.5–329.5	319.5	14	4473
329.5–349.5	339.5	15	5092.5
349.5–369.5	359.5	11	3954.5
369.5–389.5	379.5	8	3036
389.5–409.5	399.5	4	1598
409.5–429.5	419.5	3	1258.5
429.8–449.5	439.5	2	879
			$\Sigma mf = 26{,}680$

In Figure 1 we reconstruct the histogram in Figure 6.3.2 on page 318. If we think of the shaded area as weight, then the geometrical object would balance if a triangular wedge were placed at the point $\bar{x} = 333.5$. That is, the sample mean of the grouped data can be thought of as a *balancing point*.

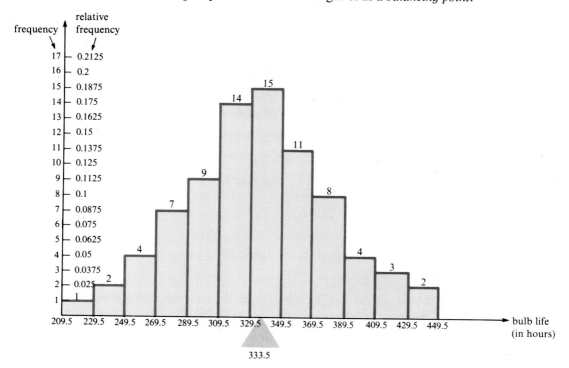

Figure 1

Example 4 A professional basketball team has ten players, nine of whom earn $30,000. The tenth is a superstar who earns $1 million. Compute the sample mean of these salaries.

Solution Here

$$\bar{x} = \frac{9 \cdot 30,000 + 1,000,000}{10} = \frac{1,270,000}{10} = \$127,000.$$

The answer in the last problem is correct, but quite ridiculous. The reporter who walked into the team's locker room and mentioned the average salary of $127,000 would be lucky to be only laughed out of the room. Clearly, another measure of average is needed here.

Median of a Sample Arrange the numbers in a sample in increasing order. Then, if there are an odd number of data, the number in the middle is called the **median** of the sample. If there are an even number of data, then the median is the average of the two numbers in the middle.

Example 5 Compute the median of the salaries in Example 4.

Solution We write the ten salaries in increasing order.

$30,000

$30,000

$30,000

$30,000

$30,000⎫
        ⎬ Middle values.
$30,000⎭

$30,000

$30,000

$30,000

$1,000,000

The average of the two middle values is 30,000 and this is the median. The effect of taking the median rather than the mean in this problem is to decrease the importance of the *extreme value* $1,000,000. Clearly, the number $30,000 is much more representative of the players' salaries than is the sample mean.

Example 6 Compute the median and sample mean of the numbers 57, 31, 46, 38, 68, 43, 52, 51, and 65.

Solution   We arrange the numbers in increasing order.

$$31$$
$$38$$
$$43$$
$$46$$
$$51 \quad \longleftarrow \text{Middle number.}$$
$$52$$
$$57$$
$$65$$
$$68$$

We see that the median is 51. Then, we find that

$$\bar{x} = \frac{31 + 38 + 43 + 46 + 51 + 52 + 57 + 65 + 68}{9} = \frac{451}{9} \approx 50.11.$$

Example 7   Compute the median of the light-bulb data given in Table 6.3.1 on page 314.

Solution   There are 80 light bulbs. The two middle values represent the 40th and 41st light bulbs (when arranged in order of increasing bulb life). From Table 6.3.2 on page 317, we see that there are 37 light bulbs with lives less than 329.5 hours and 28 with lives more than 369.5 hours. Thus the 40th and 41st bulbs have lives in the interval 329.5–349.5. Writing these in increasing order, we have

$$331 \leftarrow \text{38th bulb}$$
$$332 \leftarrow \text{39th bulb}$$
$$334 \leftarrow \text{40th bulb}$$
$$335 \leftarrow \text{41st bulb}$$
$$336 \leftarrow \text{42nd bulb}$$
$$336 \leftarrow \text{43rd bulb}$$

$\left.\begin{array}{c} 334 \leftarrow \text{40th bulb} \\ 335 \leftarrow \text{41st bulb} \end{array}\right\}$ Middle values.

The average of the two middle values is $(334 + 335)/2 = 334.5$, and this is the median. Note that, in this case, the median is close to the sample mean of 332.675.

When a sample is analyzed, the statistic used most often is the sample mean. But, as we have seen, the median is sometimes more representative. Whether to use the mean or the median as representative of the "average" is a matter of common sense. The mean will give distorted values when there are a few extreme values, as with the players' salaries of Example 4.

It has often been said that statistics can be used (and misused) to prove almost anything. As an example, the mean United States per capita income in 1979 was $8706.[†] This means that the average family of four had an income of $8706 × 4 = $34,824. However, the median income for a family of four in 1979 was only $18,492.[‡] This is quite a difference! Which one do you think is more realistic? Evidently, the higher per capita income is influenced by a relatively small number of very wealthy individuals (like the superstar of Example 4). The median ignores these large incomes and ignores very small incomes as well. In this case, the median is a better representative of national prosperity.

PROBLEMS 6.4

In Problems 1–11, compute the sample mean and median of the given data.

1. 3, 8, 6, 4, 9, 7, 11
2. 14, 23, 46, 59, 16, 21, 27, 31
3. 1.6, 1.8, 0.9, 0.4, 3.2, 1.5, 1.6
4. 1534, 1713, 1419, 1036, 1814, 1521, 1632
5. 1, 2, 4, 6, 3, 7, 6, 985, 4
6. 3, 51, 48, 95, 2, 101, 49, 46, 53, 98, 5
7. 1.62, 1.42, 1.57, 1.83, 1.96, 1.47, 1.65, 1.72
8. −5, −6, 2, 5, −13, 8, −7, 7, 0
9. 0, 100, 0, 100, 0, 100, 0, 100
10. 0, 100, 0, 100, 0, 100, 0, 100, 0
11. 100, 0, 100, 0, 100, 0, 100, 0, 100
12. Look at Problems 9, 10, and 11. What possible drawback to relying solely on the median is suggested by these problems?
13. Compute the sample mean and median of the ungrouped data in Problem 6.3.5 on page 323.
14. Compute the sample mean of the grouped data of Problem 6.3.5(a) on page 324.
15. Compute the sample mean of the grouped data of Problem 6.3.5 (page 323) with the grouping of Problem 6.3.6.
16. Compute the sample mean and median of the ungrouped data in Problem 6.3.7 on page 324.
17. Compute the sample mean of the grouped data of Problem 6.3.7(a).
18. Compute the sample mean and median of the ungrouped data of Problem 6.3.10 on page 324.
19. Compute the sample means of each of the groupings for the data of Problem 6.3.10 given in Problems 6.3.10(a), 6.3.11, and 6.3.12.

[†] Source: U.S. Dept. of Commerce, Bureau of Economic Analysis. (From the *1981 Hammond Almanac.*)
[‡] Source: U.S. Bureau of the Census. (From the *1981 Hammond Almanac.*)

In Problems 20–29, compute the sample mean and median of the data given in the table.

20. State Per Capita Income in 1978

State	Income
Alaska	$10,849
California	8916
Pennsylvania	7444
Mississippi	5582
Massachusetts	7926
Georgia	6779
Utah	6594
Montana	6915
Nevada	9377
Oregon	8076

21. Federal Grants to States in 1979

State	Grant (in millions of dollars)
Indiana	$1398.8
New York	8870.3
Maine	503.1
Hawaii	407.7
Oklahoma	945.7
Texas	3588.2
New Jersey	2715.4
Louisiana	1509.3
Idaho	315.4
New Mexico	535.4
Nebraska	473.4

22. Companies with Largest Number of Common Stockholders (Early 1980)

Company	Stockholders
American Telephone and Telegraph	2,978,000
General Motors	1,219,000
IBM	697,000
Exxon	686,000
General Electric	527,000
General Telephone and Electronics	462,000
Texaco	415,000
Gulf Oil	343,000
Southern Company	341,000
Ford Motor	337,000

23. Unemployment Rate in United States (1965–1976)

Year	Unemployment rate
1965	4.5
1966	3.8
1967	3.8
1968	3.6
1969	3.5
1970	4.9
1971	5.9
1972	5.6
1973	4.9
1974	5.6
1975	8.5
1976	7.7

24.
Average Union Hourly Scale (1977)

Type of worker	Average hourly salary 1977
(Building) journeymen	$10.44
(Building) helpers and laborers	8.03
(Printing) book and job	7.91
(Printing) newspapers	8.74
(Local trucking) drivers	8.09
(Local trucking) helpers	7.28
Local transit workers	7.12

25.
Adjusted Hospital Expense per Inpatient Day: Community Hospitals

Size of community hospitals	Average expense per inpatient day (1978)
6–24 beds	$152.34
24–49 beds	151.68
50–99 beds	149.68
100–199 beds	172.50
200–299 beds	186.94
300–399 beds	200.70
400–499 beds	208.78
500 or more beds	228.53

26.
Diameters of Planets

Planet	Diameter (in terms of Earth's diameter: 7926 miles)
Mercury	0.38
Venus	0.95
Earth	1.00
Mars	0.53
Jupiter	11.20
Saturn	9.36
Uranus	4.06
Neptune	3.88
Pluto	0.23

27.

College and University Graduates (1965–1978)

School year	Number of degrees, (bachelor's, master's, doctorate)
1965–1966	709,832
1966–1967	768,871
1967–1968	866,548
1968–1969	984,129
1969–1970	1,065,391
1970–1971	1,140,292
1971–1972	1,215,680
1972–1973	1,270,528
1973–1974	1,310,441
1974–1975	1,305,382
1975–1976	1,334,230
1976–1977	1,334,304
1977–1978	1,331,536

28. ## Winning Times in Olympics 100-Meter Free-Style Championships

Year	Time (minutes/seconds)
1896	1:22.2
1904	1:02.8
1908	1:05.6
1912	1:03.4
1920	1:01.4
1924	59.0
1928	58.6
1932	58.2
1936	57.5
1948	57.3
1952	57.4
1956	55.4
1960	55.2
1964	53.4
1968	52.2
1972	51.2
1976	50.0

29. ## National League Pennant Winners—Percentage of Games Won

Year	Club	Percent
1880	Chicago	0.798
1885	Chicago	0.777
1890	Brooklyn	0.667
1895	Baltimore	0.669
1900	Brooklyn	0.603
1905	New York	0.686
1910	Chicago	0.675
1915	Philadelphia	0.592
1920	Brooklyn	0.604
1925	Pittsburgh	0.621
1930	St. Louis	0.597
1935	Chicago	0.649
1940	Cincinnati	0.654
1945	Chicago	0.636
1950	Philadelphia	0.591
1955	Brooklyn	0.641
1960	Pittsburgh	0.617
1965	Los Angeles	0.599
1970	Cincinnati	0.630
1975	Cincinnati	0.667
1980	Philadelphia	0.562

The **range** of a sample is the total spread of the numbers; that is, it is the difference between the largest and smallest values. For example, the range of the numbers 4, 3, 8, 11, 9, 6, and 7 is $11 - 3 = 8$. In Problems 30–36, compute the range of the sample in the given problem.

30. Problem 1.

31. Problem 4.

32. Problem 7.

33. Problem 10.

34. Problem 6.3.5 on page 323.

35. Problem 6.3.7 on page 324.

36. Problem 6.3.10 on page 324.

6.5 Distance Measures of Dispersion: Standard Deviation and Sample Variance

Consider the following two experiments.

Experiment 1. Flip a fair coin 100 times and count the number of heads.

Experiment 2. Put the numbers 0, 1, 2, ..., 100 on 101 pieces of paper and put the pieces of paper into a bowl. Choose one piece of paper at random and observe the number that appears.

Both experiments have similar properties. The expected value of each is 50. If we repeat each experiment many times, then the sample mean and median of each sample will be close to or equal 50. However, there is a big difference between the two experiments. When we flip a fair coin 100 times, we may get 47 heads or 54 heads, but the probability of getting 8 heads or 95 heads is extremely small. On the other hand, in Experiment 2 the outcomes 8 and 95 are just as likely as the outcomes 47 and 54. Each has probability $\frac{1}{101}$.

In order to differentiate between these very different types of experiments, it is necessary to measure the deviation from the mean. In Experiment 1 it is likely to be smaller than in Experiment 2.

There are many ways to measure deviations from the average. Those universally used by statisticians are related to the notion of distance. To see the relationship between distance and this measure of deviation from the average, consider two points (x_1, y_1) and (x_2, y_2) in the xy-plane (see Figure 1). According to equation (2.1.1) on page 37, the distance d between them is given by

$$d = \sqrt{(x_1 - x_2)^2 + (y_1 - y_2)^2}$$

so that

$$d^2 = (x_1 - x_2)^2 + (y_1 - y_2)^2.$$

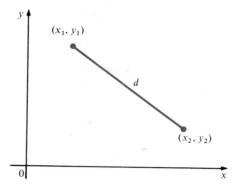

Figure 1

Variance of a Sample Suppose that a sample takes the n values $x_1, x_2, \ldots, x_n$ with sample mean $\bar{x}$; then the **sample variance**, denoted by s^2, is given by

$$s^2 = \frac{(x_1 - \bar{x})^2 + (x_2 - \bar{x})^2 + \cdots + (x_n - \bar{x})^2}{n - 1}.$$ (1)

Remark 1. Note that the denominator in (1) is $n - 1$, not n. This is so that $E(s^2)$ will equal the theoretical variance of the true (unknown) population being studied. We need not worry about this here.

Remark 2. The quantity $x_1 - \bar{x}$ is the difference between the datum x_i and the sample mean $\bar{x}$. Thus the sum in the numerator in (1) is the sum of the squares of the differences between the x_i's and $\bar{x}$. The difference between x_i and $\bar{x}$ is often called the **deviation** of x_i from the mean, $\bar{x}$. With this terminology, the sample variance of a sample can be thought of as the *average square deviation from the mean.*

Remark 3. Using the summation notation introduced in Section 6.4, we have

$$s^2 = \frac{\Sigma(x - \bar{x})^2}{n - 1}.$$ (2)

Before giving examples, we define the most commonly used measure of deviation from the mean.

Standard Deviation The **standard deviation** of a sample, denoted by s, is the square root of the sample variance. That is,

$$s = \sqrt{\frac{(x - \bar{x})^2 + (x_2 - \bar{x})^2 + \cdots + (x_n - \bar{x})^2}{n - 1}}$$ (3)

or, using the summation notation,

$$s = \sqrt{\frac{\Sigma(x - \bar{x})^2}{n - 1}}.$$

(4)

Example 1   Compute the sample variance and standard deviation of the data 5, 8, 11, 13, 6, 5, 3, and 9.

Solution   The sample mean of these data is

$$\bar{x} = \frac{5 + 8 + 11 + 13 + 6 + 5 + 3 + 9}{8} = \frac{60}{8} = 7.5.$$

Then

$$s^2 = \frac{(5 - 7.5)^2 + (8 - 7.5)^2 + (11 - 7.5)^2 + (13 - 7.5)^2 + (6 - 7.5)^2 \\ + (5 - 7.5)^2 + (3 - 7.5)^2 + (9 - 7.5)^2}{8 - 1}$$

$$= \frac{(-2.5)^2 + (0.5)^2 + (3.5)^2 + (5.5)^2 + (-1.5)^2 + (-2.5)^2 + (-4.5)^2 + (1.5)^2}{7}$$

$$= \frac{6.25 + 0.25 + 12.25 + 30.25 + 2.25 + 6.25 + 20.25 + 2.25}{7}$$

$$= \frac{80}{7} \approx 11.4286$$

and

$$s = \sqrt{s^2} \approx \sqrt{11.4286} \approx 3.3806.$$

Example 2   Experiment 1 (on page 334) is repeated ten times. The number of heads (out of 100 flips) is given below:

48	57	41	44	49
53	48	56	52	46

Compute the sample variance and standard deviation of this sample.

Solution   We must first compute the sample mean. We have

$$\bar{x} = \frac{48 + 57 + 41 + 44 + 49 + 53 + 48 + 56 + 52 + 46}{10} = \frac{494}{10} = 49.4.$$

Then

$$s^2 = \frac{\begin{matrix}(48 - 49.4)^2 + (57 - 49.4)^2 + (41 - 49.4)^2 + (44 - 49.4)^2 \\ + (49 - 49.4)^2 + (53 - 49.4)^2 + (48 - 49.4)^2 + (56 - 49.4)^2 \\ + (52 - 49.4)^2 + (46 - 49.4)^2\end{matrix}}{10 - 1}$$

$$= \frac{\begin{matrix}(-1.4)^2 + 7.6^2 + (-8.4)^2 + (-5.4)^2 + 0.4^2 + 3.6^2 + (-1.4)^2 \\ + 6.6^2 + 2.6^2 + (-3.4)^2\end{matrix}}{9}$$

$$= \frac{1.96 + 57.76 + 70.56 + 29.16 + 0.16 + 12.96 + 1.96 + 43.56 + 6.76 + 11.56}{9}$$

$$= \frac{236.4}{9} \approx 26.27.$$

Thus $s = \sqrt{s^2} \approx \sqrt{26.27} \approx 5.125.$

Example 3 A slip of paper is drawn at random from the bowl of Experiment 2 (on page 334) and the number on the paper is recorded. The paper is put back in the bowl and the experiment is repeated for a total of ten times. The following numbers were drawn.

31	98	9	52	63
85	17	58	39	70

Determine the sample variance and standard deviation of this sample.

Solution   The sample mean $\bar{x}$ is given by

$$\bar{x} = \frac{31 + 98 + 9 + 52 + 63 + 85 + 17 + 58 + 39 + 70}{10} = \frac{522}{10} = 52.2.$$

Then

$$s^2 = \frac{\begin{matrix}(31 - 52.2)^2 + (98 - 52.2)^2 + (9 - 52.2)^2 + (52 - 52.2)^2 \\ + (63 - 52.2)^2 + (85 - 52.2)^2 + (17 - 52.2)^2 + (58 - 52.2)^2 \\ + (39 - 52.2)^2 + (70 - 52.2)^2\end{matrix}}{10 - 1}$$

$$= \frac{7369.6}{9} \approx 818.84.$$

Thus $s \approx \sqrt{818.84} \approx 28.62.$

Examples 2 and 3 (when confirmed by additional experiments) reinforce our intuitive claim that the average deviation from the mean in Experiment 1 is much smaller than the average deviation from the mean in Experiment 2.

Remark. We will show later in this section that, if the two experiments are repeated a very large number of times, the average standard deviation of Experiment 1 will be 5 and that of Experiment 2 will be $\sqrt{850} \approx 29.15$.

We can also define the sample variance and standard deviation of grouped data. Reasoning as in Section 6.4, we assume that each number in an interval is equal to the midpoint of that interval.

Sample Variance and Standard Deviation of Grouped Data The **sample variance** of grouped data is given by

$$s^2 = \frac{\Sigma(m - \bar{x})^2 f}{n - 1} = \frac{(m_1 - \bar{x})^2 f_1 + (m_2 - \bar{x})^2 f_2 + \cdots + (m_k - \bar{x})^2 f_k}{n - 1} \tag{5}$$

and the **standard deviation** is given by

$$s = \sqrt{\frac{\Sigma(m - \bar{x})^2 f}{n - 1}} = \sqrt{\frac{(m_1 - \bar{x})^2 f_1 + (m_2 - \bar{x})^2 f_2 + \cdots + (m_k - \bar{x})^2 f_k}{n - 1}} \tag{6}$$

Remark. See the material above equation (6.4.3) on page 326 for an explanation of notation.

 Example 4 Compute the sample variance and standard deviation for the grouped light-bulb data in Table 6.3.2 on page 317.

Solution In Example 6.4.3 on page 326, we computed the sample mean $\bar{x} = 333.5$. Then

	Frequency of interval 209.5–229.5	Frequency of interval 229.5–249.5	Frequency of interval 249.5–269.5
	↓	↓	↓

$$s^2 = \frac{\begin{aligned}(219.5 - 333.5)^2(1) \quad &+ (239.5 - 333.5)^2(2) + (259.5 - 333.5)^2(4) \\ + (279.5 - 333.5)^2(7) \quad &+ (299.5 - 333.5)^2(9) + (319.5 - 333.5)^2(14) \\ + (339.5 - 333.5)^2(15) &+ (359.5 - 333.5)^2(11) + (379.5 - 333.5)^2(8) \\ + (399.5 - 333.5)^2(4) &+ (419.5 - 333.5)^2(3) + (439.5 - 333.5)^2(2)\end{aligned}}{80 - 1}$$

$$= \frac{173,120}{79} \approx 2191.4.$$

The standard deviation is given by

$$s \approx \sqrt{2191.4} \approx 46.8.$$

We could also compute the standard deviation for the ungrouped data given in Table 6.3.1 on page 314, but that would be extremely tedious (we would have to add 80 squares and divide by 79).

We have seen the difference between the expected value, or mean, of a random variable, denoted by μ, and the sample mean of a sample, denoted by $\bar{x}$. The first is theoretical, the second empirical. Similarly, we can define the variance and standard deviation of a random variable. These, like μ, are theoretical quantities.

Variance of a Random Variable The **variance of a random variable** X, denoted by σ^2, is the expected value of the square deviation from the mean. We have

$$\sigma^2 = E(X - \mu)^2 = \Sigma(x - \mu)^2 p(x)$$
$$= (x_1 - \mu)^2 p(x_1) + (x_2 - \mu)^2 p(x_2) + \cdots + (x_n - \mu)^2 p(x_n), \tag{7}$$

where X takes the values $x_1, x_2, \ldots, x_n$ and $p(x)$ is its probability function.

Standard Deviation of a Random Variable The **standard deviation of a random variable** X, denoted by σ, is the square root of its variance. That is,

$$\sigma = \sqrt{\Sigma(x - \mu)^2 p(x)}$$
$$= \sqrt{(x_1 - \mu)^2 p(x_1) + (x_2 - \mu)^2 p(x_2) + \cdots + (x_n - \mu)^2 p(x_n)}. \tag{8}$$

Example 5 Compute the variance and standard deviation of the weather problem in Example 6.1.10 on page 298.

Solution Here the random variable X takes the values -3, -1, and 4 with $p(-3) = 0.2$, $p(-1) = 0.25$, and $p(4) = 0.55$. In Example 6.2.4 on page 306, we found that $\mu = E(X) = 1.35$. Then

$$\sigma^2 = (-3 - 1.35)^2 p(-3) + (-1 - 1.35)^2 p(-1) + (4 - 1.35)^2 p(4)$$
$$= (-4.35)^2 (0.2) + (-2.35)^2 (0.25) + 2.65^2 (0.55)$$
$$= (18.9225)(0.2) + (5.5225)(0.25) + (7.0225)(0.55) = 9.0275$$

and

$$\sigma = \sqrt{9.0275} \approx 3.0046.$$

Example 6 A fair coin is flipped 3 times. Find the standard deviation of the random variable that computes the number of heads.

Solution There are four outcomes: 0, 1, 2, 3, with $p(0) = \frac{1}{8}$, $p(1) = \frac{3}{8}$, $p(2) = \frac{3}{8}$, and $p(3) = \frac{1}{8}$. Since $\mu = \frac{3}{2}$ (explain why!), we have

$$\sigma^2 = (0 - \tfrac{3}{2})^2 p(0) + (1 - \tfrac{3}{2})^2 p(1) + (2 - \tfrac{3}{2})^2 p(2) + (3 - \tfrac{3}{2})^2 p(3)$$
$$= \tfrac{9}{4} \cdot \tfrac{1}{8} + \tfrac{1}{4} \cdot \tfrac{3}{8} + \tfrac{1}{4} \cdot \tfrac{3}{8} + \tfrac{9}{4} \cdot \tfrac{1}{8} = \tfrac{24}{32} = \tfrac{3}{4}$$

and

$$\sigma = \sqrt{\frac{3}{4}} = \frac{\sqrt{3}}{2}.$$

The last example was an example of a binomial distribution with $n = 3$ and $p = \frac{1}{2}$. It is always very easy to compute the variance and standard deviation of a binomial random variable. The following fact is given without proof.

If X is a binomial random variable, then the variance and standard deviation of X are given by

$$\sigma^2 = np(1 - p) \tag{9}$$

and

$$\sigma = \sqrt{np(1 - p)}. \tag{10}$$

Note that in Example 6, $np(1 - p) = 3(\frac{1}{2})(1 - \frac{1}{2}) = \frac{3}{4}$.

Example 7   Experiment 1 on page 334 (coin tossing) is a binomial experiment with $n = 100$ and $p = \frac{1}{2}$. Thus $\sigma^2 = 100(\frac{1}{2})(1 - \frac{1}{2}) = 25$ and $\sigma = 5$. These are close to the sample variance and standard deviation obtained in Example 2.

Example 8   A factory produces batteries, of which 0.2% have been found to be defective. If 300 batteries are selected at random, let X denote the number of defective ones. Find the variance and standard deviation of X.

Solution   Since X is a binomial random variable with $n = 300$ and $p = 0.002$,

$$\sigma^2 = 300(0.002)(1 - 0.002) = (300)(0.002)(0.998) = 0.5988$$

and

$$\sigma = \sqrt{0.5988} \approx 0.7738.$$

Example 9   Compute the variance and standard deviation of the random variable in Experiment 2 on page 334.

Solution   Here X takes the values $0, 1, 2, \ldots, 100$ with $p(n) = \frac{1}{101}$ for $n = 0, 1, 2, \ldots, 100$ and $\mu = 50$. Therefore

$$\sigma^2 = (0 - 50)^2 \cdot \tfrac{1}{101} + (1 - 50)^2 \cdot \tfrac{1}{101} + (2 - 50)^2 \cdot \tfrac{1}{101} + \cdots +$$
$$(99 - 50)^2 \cdot \tfrac{1}{101} + (100 - 50)^2 \cdot \tfrac{1}{101}.$$

If we arrange these terms, we see that the square of every integer from 0 to 50 is included twice ($(0 - 50)^2 = (-50)^2 = 50^2$ and $(100 - 50)^2 = 50^2$; similarly, $(12 - 50)^2 = (-38)^2 = 38^2$ and $(88 - 50)^2 = 38^2$). Thus

$$\sigma^2 = \tfrac{2}{101}(1^2 + 2^2 + 3^2 + \cdots + 50^2)^\dagger = \tfrac{2}{101}(42{,}925) = 850$$

---

† There are two ways to add these. First, we can use brute force. Second, we can use an algebraic formula that says that the sum of the first n squares is given by $n(n + 1)(2n + 1)/6$. Here $n = 50$, so our sum is $50(51)(101)/6 = 42{,}925$.

and

$$\sigma = \sqrt{850} \approx 29.15.$$

Computations of sample and theoretical variances and standard deviations can be very tedious and are best done with a calculator. There are some statistical calculators that will compute the sample mean, variance, and standard deviation of a set of data. All you need to do is enter the data.

The standard deviation of a random variable is interesting in that it gives us a quantitative measure of the dispersion from the mean. But this is not its only use; a knowledge of the standard deviation is necessary for an astonishingly large number of statistical applications. We will see perhaps the most important of these in the next section.

PROBLEMS 6.5

In Problems 1–9, compute the sample variance and standard deviation of the given sample. (You computed their sample means in Problems 6.4.1–9 on page 330.)

1. 3, 8, 6, 4, 9, 7, 11
2. 14, 23, 46, 59, 16, 21, 27, 31
3. 1.6, 1.8, 0.9, 0.4, 3.2, 1.5, 1.6
4. 1534, 1713, 1419, 1036, 1814, 1521, 1632
5. 1, 2, 4, 6, 3, 7, 6, 985, 4
6. 3, 51, 48, 95, 2, 101, 49, 46, 53, 98, 5
7. 1.62, 1.42, 1.57, 1.83, 1.96, 1.47, 1.65, 1.72
8. $-5, -6, 2, 5, -13, 8, -7, 7, 0$
9. 0, 100, 0, 100, 0, 100, 0, 100
10. Take a fair die and throw it 15 times. Compute the sample mean, sample variance, and standard deviation of your sample.
11. Observe the daily high temperature in your city for each of the next ten days. Compute the sample mean, sample variance, and standard deviation of your sample.
12. Look at the scores on your last eight math tests. Compute the sample mean, sample variance, and standard deviation of your scores.

In Problems 13–27, compute the sample variance and standard deviation of the sample in the given problem from Section 6.4 on page 330.

13.	Problem 6.4.14.	14.	Problem 6.4.16.
15.	Problem 6.4.15.	16.	Problem 6.4.18.
17.	Problem 6.4.17.	18.	Problem 6.4.20.
19.	Problem 6.4.21.	20.	Problem 6.4.22.
21.	Problem 6.4.23.	22.	Problem 6.4.24.
23.	Problem 6.4.25.	24.	Problem 6.4.26.

25. Problem 6.4.27.   **26.** Problem 6.4.28.

27. Problem 6.4.29.

In Problems 28–42, compute the theoretical variance and standard deviation of the given random variable defined in Problem Sets 6.1 and 6.2 on pages 300 and 309.

28. Problem 6.1.1.   **29.** Problem 6.1.2.

30. Problem 6.1.3.   **31.** Problem 6.1.4.

32. Problem 6.1.5.   **33.** Problem 6.1.6.

34. Problem 6.1.7.   **35.** Problem 6.1.8.

36. Problem 6.1.9.   **37.** Problem 6.1.10.

38. Problem 6.1.16.   **39.** Problem 6.1.19.

40. Problem 6.1.20.   **41.** Problem 6.2.23.

42. Problem 6.2.25.

6.6 The Normal Distribution

In this section we discuss what is certainly the most important probability distribution used in statistics, the *normal distribution*. To introduce it, we begin with a simple example. Suppose a fair coin is flipped six times. Let X be the random variable representing the number of heads obtained. X takes the values 0, 1, 2, 3, 4, 5, 6. Its probability function is given in the table.

Number of heads	$p(k) = \binom{6}{k}\left(\frac{1}{2}\right)^6 = \frac{1}{64}\binom{6}{k}$
0	$\frac{1}{64}$
1	$\frac{6}{64}$
2	$\frac{15}{64}$
3	$\frac{20}{64}$
4	$\frac{15}{64}$
5	$\frac{6}{64}$
6	$\frac{1}{64}$

Suppose now that we repeat this experiment a large number of times. We can then expect that $\frac{1}{64}$ of the time we will get no heads, $\frac{6}{64}$ of the time we will get 1 head, $\frac{15}{64}$ of the time we will get 2 heads, and so on. Figure 1 shows a frequency histogram of these *hypothetical* data. We now draw a smooth curve through the corners of the rectangles in the figure, as in Figure 2.

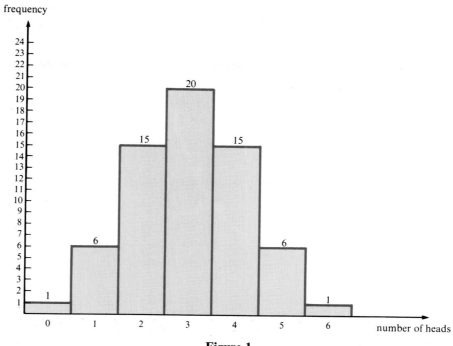

Figure 1

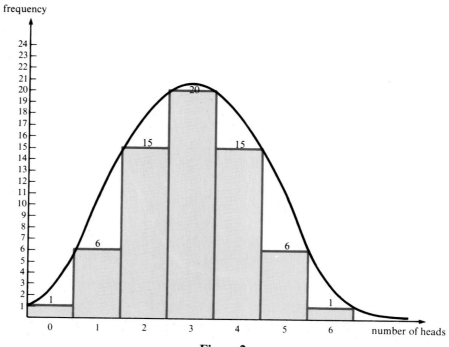

Figure 2

The bell-shaped curve in Figure 2, called a **normal curve**, is the curve of a **normal distribution**. In Figure 3 we have drawn a histogram and associated normal curve for a binomial distribution with $n = 12$ and $p = \frac{1}{3}$.

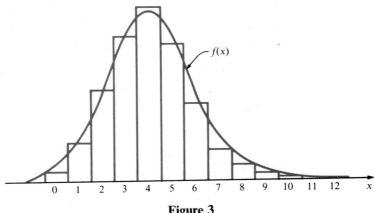

Figure 3

In Section 6.3 we made an observation relating the area under a histogram and probability (see page 319). In particular, suppose we choose an interval on the x-axis. Then the area of the part of a relative frequency histogram that lies over the interval is equal to the probability that a number chosen randomly from the sample will lie in the interval. Thus, since the smooth normal curve seems to approximate the binomial histogram fairly well, it seems that we can estimate binomial probabilities by computing the area under the normal curve and above certain intervals on the x-axis. We will say more about this later.

The normal distribution, represented by a bell-shaped curve, was originally developed as a result of studies of errors that occur in a variety of experiments. It is now widely known that many physical, economic, and biological phenomena have probability distributions approximated by a normal distribution. By this we mean that the probability that a measurement falls in the interval $[a, b]$,

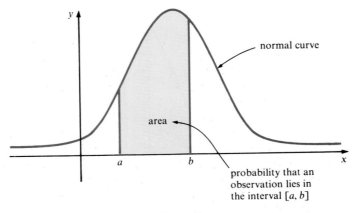

Figure 4

say, is approximately equal to the area under that part of the particular normal curve lying over the interval $[a, b]$. This is illustrated in Figure 4.

Phenomena that are normally distributed include the diameters of Granny Smith apples, the scores on standardized tests (such as the Scholastic Aptitude Test (SAT)), the heights of 10-year-old girls in the United States, the length of newborn trout, I.Q. scores, the weight of loaves of bread baked in a bakery, and so on.

The reason why so many things are normally distributed is given later in this section. Our immediate problem now it to determine how to compute areas under normal curves. One specific normal curve is given in Figure 5. The bell-shaped curve in the figure can be described algebraically by the equation

$$y = \frac{1}{\sigma\sqrt{2\pi}} e^{-(x-\mu)^2/2\sigma^2} \tag{1}$$

where μ is the mean of the entire population, σ is the standard deviation,

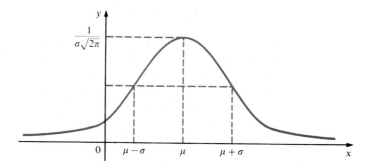

Figure 5 Typical normal curve.

$\pi \approx 3.1416$, and $e \approx 2.7183$ (we will not discuss the number e any further in this book). Here are some facts about normal curves:

1. The curve is symmetrical about the line $x = \mu$. The mean is at the center.

2. The curve is always bell-shaped and is completely determined by the mean and variance of the population.

3. The smaller the variance, the steeper the curve.

4. The area under the curve is always 1.

Fact 4 follows immediately from the fact that the area under the normal curve represents probability. Fact 3 is illustrated in Figure 6, which shows three normal curves, all with mean 50 but with different variances.

Unit Normal Curve   The **unit normal curve**, or **standard normal curve**, representing the **unit normal distribution** is the normal curve with mean 0 and standard deviation 1. A standard normal curve is shown in Figure 7.

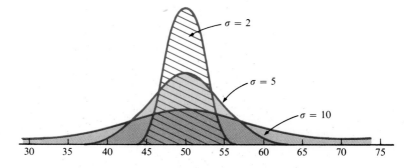

Figure 6   Three normal curves with μ = 50.

In Figure 7 the variable is called *z* rather than *x*. It is common that the unit normal random variable is denoted by *Z*. Areas under the unit normal curve are given in Table 5 (at the back of the book). We note two important facts before using these tables.

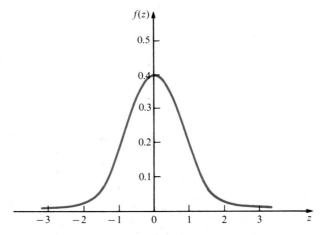

Figure 7   Unit normal curve (μ = 0, σ = 1).

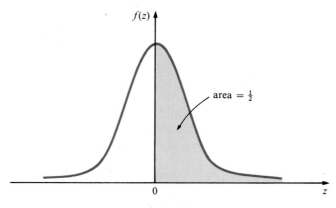

Figure 8

5. Since the area under the curve is 1 and since the curve is symmetric about the y-axis (the line $x = 0$), the area under the unit normal curve from 0 to ∞ is $\frac{1}{2}$. This is illustrated in Figure 8.

6. Because of symmetry, the area under the unit normal curve from $-z_0$ to 0 is the same as the area under the curve from 0 to z_0. This is illustrated in Figure 9.

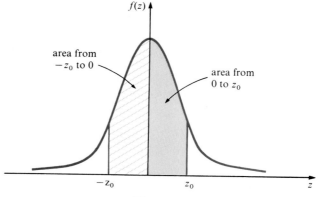

Figure 9

The following four examples show how Table 5 can be used to compute areas under the unit normal curve.

Example 1   Compute the area under the unit normal curve between 0 and 1.25.

Solution   The area requested is the shaded area in Figure 10. In Table 5 we move down the left-hand column until we find 1.2 and then move to the right until we are under the heading 0.05. We find the number 0.3944. This is the desired area.

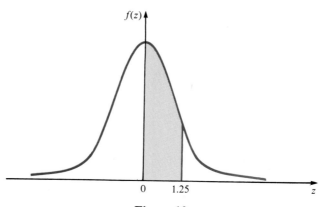

Figure 10

Remark. In Example 1 we answered the following question: What is the probability that a normally distributed random variable with mean 0 and variance 1 takes a value between 0 and 1.25? We found that $P(0 \leq Z \leq 1.25) = 0.3944$.

Example 2   Compute $P(Z \geq 2.13)$.

Solution   We now seek the shaded area of Figure 11. From Table 5 we know

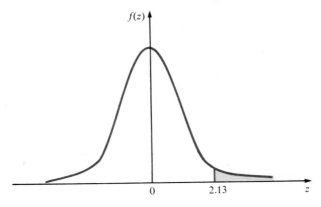

Figure 11

that the area from 0 to 2.13 is 0.4838. Now we want the area from 2.13 to ∞. Since, from Fact 5, the area from 0 to ∞ is 0.5, we see that

$$\text{area from 2.13 to } \infty = 0.5 - \text{area from 0 to 2.13}$$

or

$$P(Z \geq 2.13) = 0.5 - 0.4838 = 0.0162.$$

Example 3   Compute $P(Z \leq 1.2)$.

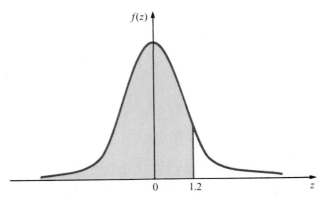

Figure 12

Solution   The area sought is illustrated in Figure 12. We have

area from $-\infty$ to 1.2 = area from $-\infty$ to 0 plus area from 0 to 1.20

From Fact 5
$$= 0.5 + 0.3849 = 0.8849.$$

Thus $P(Z \leq 1.2) = 0.8849$.

Example 4   Compute $P(-1.04 \leq Z \leq 0.67)$.

Solution   The area sought is sketched in Figure 13. We have

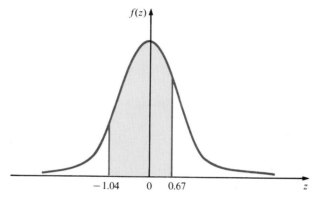

Figure 13

area from -1.04 to 0.67 = area from -1.04 to 0 plus area from 0 to 0.67

From Fact 6
$$= \text{area from 0 to 1.04 plus area from 0 to 0.67}$$

From Table 5
$$= 0.3508 + 0.2486 = 0.5994.$$

We now know how to compute probabilities for a unit normal distribution. What about other normal distributions? The answer is given below.

> If X is a normal random variable with mean μ and standard deviation σ, then
>
> $$Z = \frac{X - \mu}{\sigma} \qquad (2)$$
>
> is a unit normal random variable.

That is, every normal distribution can be converted to a unit normal distribution by subtracting the mean and dividing by the standard deviation.

Example 5   The random variable X is normally distributed with mean 5 and standard deviation 2. Find each probability.

(a) $P(5 \leq X \leq 6.5)$

(b) $P(X \leq 5.8)$

(c) $P(X \geq 3)$

(d) $P(2 \leq X \leq 7.4)$

Solution   Since $\mu = 5$ and $\sigma = 2$, we know that $Z = (X - 5)/2$ is a unit normal random variable. To transfer from X to Z, we take each number, subtract 5, and divide the result by 2. Then we can use Table 5.

(a) $P(5 \leq X \leq 6.5) = P\left(\dfrac{5 - 5}{2} \leq \dfrac{X - 5}{2} \leq \dfrac{6.5 - 5}{2}\right) = P(0 \leq Z \leq 0.75).$

This is the area given in Figure 14 and we see that, from Table 5, this area is

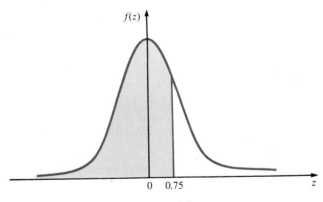

Figure 14

0.2734. Thus

$$P(5 \leq X \leq 6.5) = 0.2734.$$

(b) $$P(X \leq 5.8) = P\left(\dfrac{X - 5}{2} \leq \dfrac{5.8 - 5}{2}\right) = P(Z \leq 0.4)$$

From Figure 15 we see that

$$P(Z \leq 0.4) = 0.5 + P(0 \leq Z \leq 0.4) = 0.5 + 0.1554 = 0.6554$$

From Table 5

(c) $$P(X \geq 3) = P\left(\dfrac{X - 5}{2} \geq \dfrac{3 - 5}{2}\right) = P(Z \geq -1)$$

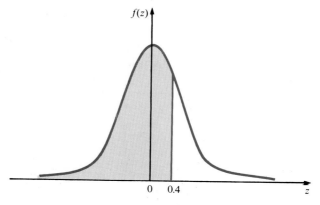

Figure 15

From Figure 16 and Fact 5,

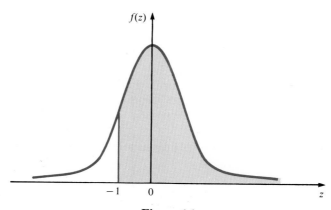

Figure 16

$$P(Z \geq -1) = P(-1 \leq Z \leq 0) + P(Z \geq 0)$$

From Table 5

$$= P(0 \leq Z \leq 1) + 0.5 = 0.3413 + 0.5 = 0.8413$$

(d)
$$P(2 \leq X \leq 7.4) = P\left(\frac{2 - 5}{2} \leq Z \leq \frac{7.4 - 5}{2}\right)$$

$$= P(-1.5 \leq Z \leq 1.2) = P(-1.5 \leq Z \leq 0)$$
$$+ P(0 \leq Z \leq 1.2)$$

From Fact 6

$$= P(0 \leq Z \leq 1.5) + P(0 \leq Z \leq 1.2)$$

From Table 5

$$= 0.4332 + 0.3849 = 0.8181$$

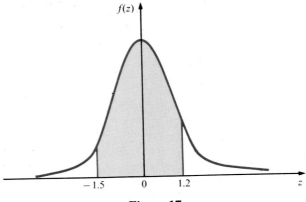

Figure 17

Example 6   The time T required for digestion of food by a certain protozoan is a normal random variable with mean 31 min and standard deviation 5 min.

(a)   What is the probability that a unit of food will be digested in less than 35 min?

(b)   What is the probability that a unit of food will be digested between 28 min and 32 min?

Solution   The random variable $Z = (T - 31)/5$ is a unit normal random variable.

(a) $P(T \leq 35) = P\left(\dfrac{T - 31}{5} \leq \dfrac{35 - 31}{5}\right) = P(Z \leq 0.8)$ From Table 5

$= P(Z \leq 0) + P(0 \leq Z \leq 0.8) = 0.5 + 0.2881 = 0.7881$

(b) $P(28 \leq T \leq 32) = P\left(\dfrac{28 - 31}{5} \leq \dfrac{T - 31}{5} \leq \dfrac{32 - 31}{5}\right)$ From Table 5

$= P(-0.6 \leq Z \leq 0.2) = 0.2257 + 0.0793 = 0.305$

Example 7   The life of a light bulb is normally distributed with mean 300 hours and standard deviation 40 hours.

(a)   What percentage of light bulbs in a sample will last more than 350 hours?

(b)   What percentage will last less than 220 hours?

Solution The random variable L of light-bulb life is normally distributed with $\mu = 300$ and $\sigma = 40$.

(a)
$$P(L \geq 350) = P\left(\frac{L - 300}{40} \geq \frac{350 - 300}{40}\right)$$
$$= P(Z \geq 1.25) = 0.5 - P(0 \leq Z \leq 1.25)$$

From Table 5

$$= 0.5 - 0.3944 = 0.1056$$

Thus 10.56% of the light bulbs will last more than 350 hours.

(b)
$$P(L \leq 220) = P\left(Z \leq \frac{220 - 300}{40}\right) = P(Z \leq -2)$$

By symmetry

$$= P(Z \geq 2) = 0.5 - P(0 \leq Z \leq 2).$$

From Table 5

$$= 0.5 - 0.4772 = 0.0228.$$

Therefore 2.28% of the bulbs will burn out in less than 220 hours.

Example 8 Each section of the Scholastic Aptitude Test (SAT) is scored from 200 to 800. The scores are normally distributed with $\mu = 500$ and $\sigma = 100$.

(a) What percentage of students will score better than 750?

(b) What percentage will score under 400?

(c) What percentage will score between 450 and 675?

Solution The SAT score S is normally distributed with $\mu = 500$ and $\sigma = 100$.

(a)
$$P(S \geq 750) = P\left(\frac{S - 500}{100} \geq \frac{750 - 500}{100}\right) = P(Z \geq 2.5)$$

From Table 5

$$= 0.5 - P(0 \leq Z \leq 2.5) = 0.5 - 0.4938 = 0.0062$$

Thus 0.62% (less than 1%) of the students will score over 750.

(b)
$$P(S \leq 400) = P\left(Z \leq \frac{400 - 500}{100}\right) = P(Z \leq -1)$$

From Table 5

$$= P(Z \geq 1) = 0.5 - 0.3413 = 0.1587$$

This means that 15.87% of the students will score under 400.

(c) $$P(450 \le S \le 675) = P\left(\frac{450 - 500}{100} \le Z \le \frac{675 - 500}{100}\right)$$

$$= P(-0.5 \le Z \le 1.75) = P(0 \le Z \le 0.5)$$

$$+ P(0 \le Z \le 1.75)$$

↙ From Table 5

$$= 0.1915 + 0.4599 = 0.6514$$

Thus 65.14% (about two-thirds) of the students will have scores in the range 450–675.

You might ask, "Okay, I can compute probabilities if I know that something is normally distributed, but how do I know that in practice?" There are several answers to this question. First, there are statistical tests that can be used to determine whether some set of data is normally distributed. These are covered in most statistics courses. Second, there is a fundamental result in statistics known as the **central limit theorem**, which states that the means of *any* collection of samples are approximately normally distributed if the sample size is large enough.

We can discuss a third answer here. In the introduction to this section, we showed the normal distribution as an approximation to the binomial distribution. This approximation becomes better and better as n becomes larger.

> The normal distribution with $\mu = np$ and $\sigma = \sqrt{np(1 - p)}$ is a good approximation to the binomial distribution if n is large.

TABLE 1   Comparison of Binomial Distribution with $n = 20$ and $p = 0.5$ and Normal Distribution with $\mu = 10$ and $\sigma = \sqrt{5}$. (Values given are probabilities of k successes in 20 trials of the experiment.)

k (Number of successes)	$\binom{20}{k}\left(\frac{1}{2}\right)^{20}$ (Binomial)	$\frac{1}{\sqrt{2\pi}\sqrt{5}} e^{-(k - 10^2)/10}$ (Normal)
10	0.175	0.179
9	0.160	0.161
8	0.120	0.119
7	0.074	0.073
6	0.037	0.036
5	0.015	0.015

Note that the values np and $\sqrt{np(1-p)}$ are the mean and standard deviation of a binomial random variable, as discussed in Sections 6.4 and 6.5. Table 1 and Figure 18 compare the binomial distributions with $n = 20$ and $p = 0.5$.

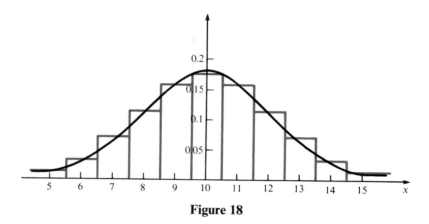

Figure 18

Example 9   It is known that 1.5% of the resistors produced by an electronics manufacturer are defective. Among 10,000 resistors, what is the probability that

(a)   more than 160 are defective?

(b)   fewer than 130 are defective?

Solution

(a)   We can solve this two ways. First, we can use the binomial distribution directly. We have

$$P(X \geq 160) = P(X = 160) + P(X = 161) + P(X = 162) + \cdots$$
$$+ P(X = 10{,}000)$$
$$= \binom{10{,}000}{160}(0.015)^{160}(0.985)^{9840}$$
$$+ \binom{10{,}000}{161}(0.015)^{161}(0.985)^{9839} + \cdots.$$

This computation would take several days (or weeks) even if you had a calculator, and it is only manageable with a programmable calculator or a computer. However, using the normal approximation to the binomial, it becomes much easier. Here we have a normal random variable with $\mu = np = (10{,}000)(0.15) = 150$ and

$$\sigma = \sqrt{np(1-p)}$$
$$= \sqrt{10{,}000(0.015)(0.985)} = \sqrt{147.75} \approx 12.555.$$

Thus

$$P(X \geq 160) = P\left(\frac{X - 150}{12.155} \geq \frac{160 - 150}{12.155}\right)$$

$$\approx P(Z \geq 0.82) = 0.5 - P(0 \leq Z \leq 0.82)$$

From Table 5

$$= 0.5 - 0.2939 = 0.2061.$$

(b) $$P(X \leq 130) = P\left(\frac{X - 150}{12.155} \leq \frac{130 - 150}{12.155}\right)$$

$$\approx P(Z \leq -1.65) = P(Z \geq 1.65)$$

From Table 5

$$= 0.5 - P(0 \leq Z \leq 1.65) = 0.5 - 0.4505$$

$$= 0.0495$$

Warning. Do not use the normal approximation to the binomial if the product np is small. A good rule of thumb is to use the normal approximation if $np \geq 10$.

Example 10 Consider a binomial random variable with $n = 5$ and $p = \frac{1}{2}$ (so that $np = 2.5$). Then

$$P(X \geq 4) \approx P(4) + P(5) = \binom{5}{4}(\tfrac{1}{2})^5 + (\tfrac{1}{2})^5 = \tfrac{6}{32} = \tfrac{3}{16} \approx 0.1875.$$

Using the normal approximation with $\mu = np = 2.5$ and $\sigma = \sqrt{np(1 - p)} = \sqrt{\frac{5}{4}} = \sqrt{5}/2 \approx 1.1$, we have

$$P(X \geq 4) = P\left(\frac{X - 2.5}{1.1} \geq \frac{4 - 2.5}{1.1}\right) = P(Z \geq 1.36)$$

From Table 5

$$= 0.5 - P(0 \leq Z \leq 1.36) = 0.5 - 0.4131$$

$$= 0.0869.$$

Note that our normal approximation is very poor indeed. The correct probability has been approximated by a number that is off by a relative error $(0.1875 - 0.0869)/0.1875 \approx 0.54 = 54\%$.

We now make some further observations about normal random variables. If X is normally distributed with mean μ and standard deviation σ, then

$$P(\mu - \sigma \leq X \leq \mu + \sigma) = P\left(-1 \leq \frac{X - \mu}{\sigma} \leq 1\right)$$

$$= P(-1 \leq Z \leq 1) = 0.3413 + 0.3413 = 0.6813.$$

That is,

there is a 68% probability that a normally distributed random variable will take values within one standard deviation of its mean.

This is illustrated in Figure 19. Similarly,

$$P(\mu - 2\sigma \le X \le \mu + 2\sigma) = P\left(-2 \le \frac{X - \mu}{\sigma} \le 2\right)$$
$$= P(-2 \le Z \le 2) = 0.4772 + 0.4772 = 0.9544$$

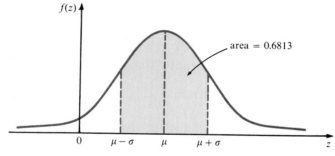

Figure 19

Thus

there is a 95% probability that a normally distributed random variable will take values within two standard deviations of its mean.

This is illustrated in Figure 20.

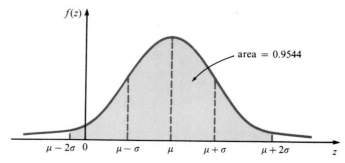

Figure 20

Example 11   In Example 7 we find that 95.44 % of all the light bulbs will burn out in the interval from $300 - 2 \cdot 40$ to $300 + 2 \cdot 40$; that is, between 200 hours and 380 hours.

We now turn to one final type of problem.

Example 12   In Example 7, after how many hours will 30 % of the light bulbs have burned out?

Solution   We are seeking a number k such that $P(X \leq k) = 0.3$—that is, 30 % of the bulbs have a life of less than or equal to k hours. Then

$$P(X \leq k) = P\left(\frac{X - 300}{400} \leq \frac{k - 300}{40}\right) = P\left(Z \leq \frac{k - 300}{40}\right) = 0.3.$$

Let us find a number z_0 such that

$$P(Z \leq z_0) = 0.3.$$

Look at Figure 21. Clearly, z_0 is negative. Now the unshaded area in the figure to

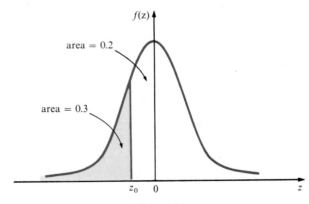

Figure 21

the left of the y-axis has area 0.2. From Table 5 we see that the z-value 0.52 gives an area 0.1985 (the next closest to 0.2 is 0.2019, which is farther away from 2 than 0.1985.) Then $P(Z \leq -0.52) \approx 0.3$. This implies that

$$\frac{k - 300}{40} \approx -0.52, \quad \text{or} \quad k - 300 \approx -20.8 \quad \text{and} \quad k \approx 279.2.$$

That is, 30 % of the bulbs will have burned out after 279.2 hours.

Example 13   The weight of adult males in a certain city is normally distributed with mean 160 lb and standard deviation 20 lb. What weight interval *centered at the mean* includes 80 % of the males?

Solution   We seek a number x such that

$$P(160 - x \le W \le 160 + x) = 0.8.$$

Then

$$P\left(\frac{(160 - x) - 160}{20} \le \frac{W - 160}{20} \le \frac{(160 + x) - 160)}{20}\right)$$

$$= P\left(-\frac{x}{20} \le Z \le \frac{x}{20}\right) = 0.8.$$

We seek a number z_0 so that the shaded area in Figure 22 is equal to 0.8. We

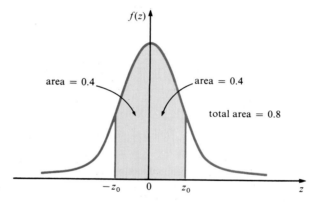

Figure 22

find, from Table 5, that

$$P(0 \le Z \le 1.28) = 0.3997 \approx 0.4.$$

Thus $x/20 = z_0 \approx 1.28$ and $x \approx (1.28)(20) = 25.6$ lb. This means that 80% of the adult males in the population have weights between $160 - 25.6 = 134.4$ lb and $160 + 25.6 = 185.6$ lb.

PROBLEMS 6.6   In Problems 1–15, find the area under the unit normal curve over the given interval.

1. $0 \le Z \le 1.2$

2. $0 \le Z \le 2.3$

3. $-1.63 \le Z \le 0$

4. $-0.73 \le Z \le 0$

5. $Z \ge 1.62$

6. $Z \ge 2.31$

7. $Z \ge -1.2$

8. $Z \le 0$

9. $Z \le -1.2$

10. $Z \le 2$

11. $Z \le 1.34$

12. $-0.22 \le Z \le 0.45$

13. $1.1 \le Z \le 1.65$

14. $-1.38 \le Z \le 1.02$

15. $-2.56 \le Z \le 0.56$

In Problems 16–30, find the requested probability for the normally distributed random variable with the given mean and standard deviation.

16. $P(60 \leq X \leq 70); \mu = 60, \sigma = 8$

17. $P(X \geq 80); \mu = 60, \sigma = 8$

18. $P(X \leq 50); \mu = 60, \sigma = 8$

19. $P(40 \leq X \leq 66); \mu = 60, \sigma = 8$

20. $P(X \geq 1.5); \mu = 1.4, \sigma = 0.2$

21. $P(X \leq 1.3); \mu = 1.4, \sigma = 0.2$

22. $P(1 \leq X \leq 2); \mu = 1.4, \sigma = 0.2$

23. $P(X \geq 6); \mu = 4, \sigma = 1$

24. $P(350 \leq X \leq 530); \mu = 430, \sigma = 80$

25. $P(X \geq -8); \mu = -9, \sigma = 2$

26. $P(-11 \leq X \leq -8.5); \mu = -9, \sigma = 2$

27. $P(0.0013 \leq X \leq 0.0017); \mu = 0.0016, \sigma = 0.0004$

28. $P(X \geq 0.0013); \mu = 0.0016, \sigma = 0.0004$

29. $P(X \leq 17,000,000); \mu = 16,500,000, \sigma = 400,000$

30. $P(16,000,000 \leq X \leq 18,250,000); \mu = 16,500,000, \sigma = 400,000$

31. Suppose that X is a normal random variable with mean 5 and variance 4.

(a) Define the corresponding unit normal random variable.

(b) Determine $P(3 \leq X)$, $P(X \leq 6)$, and $P(2 \leq X \leq 8)$.

(c) Determine numbers a and b such that $P(X \leq a) = 0.95$ and $P(X \leq b) = 0.99$.

32. The acidity of human blood measured on the pH scale is a normal random variable with mean 7.4 and standard deviation 0.2. What is the probability that the pH level is (a) greater than 7.43? (b) between 7.35 and 7.45?

33. The annual rainfall in a certain region is a normally distributed random variable with mean 30 in. and standard deviation 2 in. What is the probability of more than 31 in. of rain in a given year?

34. The adult weight of a certain mammal is a normally distributed random variable with mean 100 lb and standard deviation 8 lb. What are the probabilities that a randomly chosen animal from this species weighs

(a) less than 90 lb?

(b) between 95 and 105 lb?

(c) more than 110 lb?

35. The diastolic blood pressure in hypertensive women has been estimated to have an average value of 98 millimeters with a standard deviation of 15 millimeters. Assuming that the diastolic blood pressure is a normal random variable, estimate the probabilities that it is

(a) less than 89 millimeters.

(b) more than 104 millimeters.

(c) between 86 and 101 millimeters.

(d) if it is known that the diastolic blood pressure is greater than 104 millimeters, what is the probability that it is more than 110 millimeters?

36. The probability of success on a single trial of a binomial experiment is $p = 0.3$. By using a normal approximation, estimate the probability of obtaining more than 40 successes in 100 trials of the experiment.

37. A disease affects 0.1% of a large population. Define X to be the number of people who have the disease in a group of 100,000 people chosen at random from the population. By using a normal approximation, estimate the probability that (a) at least 80 people have the disease, and (b) the probability that no more than 130 people have the disease.

38. In a large population of drosophila (fruit flies), 25% have a wing mutation. A sample of 300 flies is chosen at random. Estimate the probability that at least 60 and no more than 90 flies in the sample have the wing mutation.

39. The frequency of tuberculosis in a large population is estimated to be 0.04%. What is the probability that, among 1,000,000 people chosen at random from the population, at least 360 and no more than 440 have tuberculosis?

40. The length of adult fish of a certain species has been estimated to have an average value of 65 centimeters with a standard deviation of 5 centimeters. Assuming a normal distribution, what is the probability that a given fish is longer than 70 centimeters? Shorter than 55 centimeters?

*41. (a) An observation is made of the normal random variable X with mean μ and variance σ^2. If it is known that the observed value is less than μ, what is the probability that it is less than $\mu - \sigma$?

(b) If a fish of Problem 40 is known to be shorter than the average, what is the probability that it is shorter than 60 centimeters?

*42. In two large school systems in neighboring cities, the intelligence quotients of equal numbers of third-grade schoolchildren were measured by standard tests. In system I, the average was 100 with standard deviation 10; in system II, the average was 105 with standard deviation 12. A child chosen at random was found to have an intelligence quotient over 120. Assuming normal distributions, what is the probability that this child came from the second school system? [*Hint*: Use Bayes' theorem.]

43. About 1 child in 700 is born with Down's syndrome. In one state there are 34,300 births recorded in 1 year. Estimate the probability that at least 56 cases of Down's syndrome occur among these recorded births.

44. The annual incidence of tuberculosis in the United States is 14.8 per 100,000. What is the probability that, among 10,000 people chosen at random, there are at least 3 cases of tuberculosis in a given year?

*45. A random variable X is normally distributed with mean μ and variance σ^2. Determine $P(X \geq \mu + 2\sigma \mid X \geq \mu)$ and $P(X \geq \mu + 2\sigma \mid X \geq \mu + \sigma)$.

*46. Suppose that a process produces widgets of length μ with standard deviation σ. A **quality control chart** consists of independent repeated observations of widget length. If about 95% of the observations fall within the 2σ tolerance limit about μ, the process is said to be "in control." However, a "run" of points on the quality control chart which fall outside the limits would lead us to suspect that μ, σ, or both had changed. What is the probability, for fixed μ and σ, that three measurements out of four successive measurements chosen at random give results outside of the 2σ tolerance limits? (Assume a normal distribution of widget length.)

47. The length of mature female fish of a certain species is a normal random variable with mean 18 centimeters and a standard deviation 2 centimeters.

(a) What proportion of such fish are at least 17 centimeters long?

(b) What proportion of fish at least 17 centimeters long are longer than 18 centimeters?

*48. Continuing with the situation described in Problem 47, what is the probability that the lengths of 4 out of 5 randomly captured fish are less than 14 centimeters or greater than 22 centimeters?

49. Monthly fuel bills for a large business are normally distributed with a mean of $2000 and a variance of $1600. In a randomly chosen month, what is the probability that

    (a) the bill is between $1800 and $2300?

    (b) the bill is less than $2200?

*50. How much money must the business of Problem 49 budget each month for fuel in order to be 98% certain that the fuel bill does not exceed the budget?

*51. In Problem 35, what interval of diastolic blood pressures centered at the mean includes 85% of all hypertensive women?

*52. If the fish in Problem 40 are grouped according to size, one group contains the smallest 15% of fish. What is the maximum length of a fish in that group?

*53. The maximum height of a river during the spring run-off is normally distributed with mean 48 in. and standard deviation 12 in. The average height of the river bank is 50 in. How high a levee must be built in order to be 99% certain that the river will not overflow the levee?

*54. The average time between buses in a certain city is 13 min with a standard deviation of 4 min. An individual always seems just to miss a bus. Assume a normal distribution.

    (a) What percentage of the time will this individual wait less than 10 minutes?

    (b) What percentage of the time will this individual wait more than 15 minutes?

    (c) The individual will wait less than how many minutes at most 20% of the time?

55. If X is a normal random variable with mean μ and variance σ^2, the 95% and 99% **probability intervals** for X are the ranges $\mu - a \leq X \leq \mu + a$ and $\mu - b \leq X \leq \mu + b$, where a and b are chosen so that $P(\mu - a \leq X \leq \mu + a) = 0.95$ and $P(\mu - b \leq X \leq \mu + b) = 0.99$. Show that $a \approx 1.96\sigma$ and $b \approx 2.58\sigma$.

56. Suppose that X is a normal random variable with mean 25 and variance 4. What are the 95% and 99% probability intervals for X?

57. What are the 95% and 99% probability intervals for the annual rainfall in Problem 33?

58. What are the 95% and 99% probability intervals for the fuel bills of Problem 49?

Review Exercises for Chapter 6

1. A family plans to have four children. Let X denote the random variable that counts the number of girls. Assume that boys and girls are equally likely.

    (a) What is the range of X?

    (b) What are the probability and distributions functions for X?

    (c) Find $F(5)$.

    (d) Find $F(2.7)$.

    (e) Sketch the distribution function.

    (f) Compute $E(X)$.

    (g) Compute the variance and standard deviation of X.

2. Consider an experiment with four elementary outcomes E_1, E_2, E_3, and E_4 with $P(E_1) = \frac{1}{6}$, $P(E_2) = \frac{1}{4}$, $P(E_3) = \frac{1}{2}$, and $P(E_4) = \frac{1}{12}$. Define a random variable X by $X(E_1) = -1$, $X(E_2) = 2$, $X(E_3) = -3$, and $X(E_4) = 7$.

(a) What is the range of X?

(b) What are the probability and distribution functions for X?

(c) Find $F(2)$.

(d) Find $F(-1.1)$.

(e) Find $F(7)$.

(f) Sketch the distribution function.

(g) Compute $E(X)$.

(h) Compute the variance and standard deviation of X.

3. A pair of dice are thrown four times in succession. Define a random variable X to be the number of times a sum of 10 is thrown.

(a) What is the range of X?

(b) What are the probability and distribution functions for X?

(c) Compute $E(X)$.

(d) Compute the variance and standard deviation of X.

4. A customer has to choose among three cars in a showroom. The probabilities that she will choose car A, B, or C are estimated to be $\frac{1}{3}$, $\frac{1}{2}$, and $\frac{1}{6}$. The salesperson will make $600 on a sale of car A, $400 on car B, and $1100 on car C.

(a) What is the salesperson's expected profit?

(b) What is the standard deviation of this profit?

5. The number of sewing machines sold in a given day by a sewing machine retailer has the probability function given below.

Number of machines	Probability
0	0.10
1	0.18
2	0.28
3	0.22
4	0.16
5	0.06

What is the expected number of sewing machines sold in a given day?

6. What is the standard deviation of the random variable of Exercise 5?

7. The following outcomes were obtained when two dice were thrown 45 times.

2	8	12	3	8
4	6	7	6	7
6	7	5	7	10
7	5	6	4	12
11	6	5	10	3
7	9	4	12	7
9	2	8	7	4
10	8	7	11	5
5	6	5	8	9

(a) Construct a frequency table.

(b) Construct a frequency histogram (including relative frequencies).

(c) Construct a frequency diagram.

(d) Find the sample mean.

(e) Find the sample variance and standard deviation.

8. Eighty male construction workers are weighed as part of a medical experiment. Their weights (in pounds) are given below.

160	136	144	152	161	174	144	158
172	173	173	173	192	156	183	188
188	164	163	165	140	175	164	173
136	189	192	170	176	150	199	168
174	212	200	173	180	184	173	191
159	162	226	156	166	197	186	182
175	171	178	176	186	172	193	203
166	178	182	201	163	178	185	147
183	199	146	189	212	204	155	178
163	150	185	214	204	179	208	162

(a) Construct a frequency and a cumulative frequency table for these data using ten intervals of length 10, starting at 129.5 lb.

(b) Draw frequency and cumulative frequency histograms (including relative frequencies).

(c) Determine the sample mean of the grouped data.

(d) Compute the sample mean of the ungrouped data.

In Exercises 9–14, compute the sample mean, median, sample variance, and standard deviation of the data.

9. 4, 7, 8, 9, 5, 3, 8

10. 1.6, 1.2, 2.1, 2.3, 1.8, 4.1, 0.6, 3.5

11. 23, 28, 34, 41, 26, 51, 16, 43

12. 1325, 4161, 2304, 3162, 4093, 2985, 3614

13. 0.132, 0.205, 0.306, 0.285, 0.416, 0.372, 0.316

14. −8, −6.2, 0, 1.6, 4.3, −11.3, 6.2, −3.5

In Exercises 15–19, find the area under the unit normal curve over the given interval.

15. $0 \leq Z \leq 1.63$

16. $Z \geq 3$

17. $Z \leq -0.6$

18. $-1.2 \leq Z \leq 0.78$

19. $-0.45 \leq Z \leq 1.03$

In Exercises 20–24, find the requested probability for the normally distributed random variable with the given mean and standard deviation.

20. $P(10 \leq X \leq 24); \mu = 17, \sigma = 4$

21. $P(X \geq 20); \mu = 17, \sigma = 4$

22. $P(X \leq 15); \mu = 17, \sigma = 4$

23. $P(1.6 \leq X \leq 2.3); \mu = 2, \sigma = 0.25$

24. $P(2500 \leq X \leq 2800); \mu = 2600, \sigma = 200$

25. The adult weight of a certain mammal is normally distributed with mean 40 lb and standard deviation 5 lb. What is the probability that a randomly chosen animal from this species weighs

(a) less than 38 lb?

(b) more than 46 lb?

(c) between 37 and 44 lb?

26. In Exercise 25, find a number α such that 95% of the animals of the species have weights between $(40 - \alpha)$ lb and $(40 + \alpha)$ lb.

27. A manufacturer makes spark plugs, of which $\frac{1}{2}$% are defective. Of 20,000 spark plugs, what is the probability that

(a) more than 125 are defective?

(b) fewer than 80 are defective?

(c) between 90 and 110 are defective?

28. One person in every 250 in a certain country has impaired hearing. Ten thousand individuals in the country are selected at random.

(a) What is the expected number with defective hearing?

(b) What is the standard deviation?

(c) What is the probability that between 30 and 60 have impaired hearing?

(d) What is the probability that more than 40 have impaired hearing?

(e) What is the probability that fewer than 35 have impaired hearing?

7 MARKOV CHAINS AND GAME THEORY

7.1 Markov Chains

In this section we discuss a mathematical technique that is used to model a large variety of processes in business and the social, biological, and physical sciences. The technique, that of *Markov chains*, was developed by the Russian mathematician A. A. Markov (1856–1922) in 1906. Initially, Markov chains were used to analyze processes in physics and meteorology. One early application was to forecast weather patterns. More recent applications include the analysis of commodity price movements, the maintenance of high performance machinery, the behavior of laboratory animals, consumer product selection, the length of lines at supermarkets and airports, inventory size and variety, and plant administration.

Before giving general definitions, we begin with an example.

Example 1 The Gourmet Catering Company has 40% of the catering business in a certain medium-size city. Its only competitor, Delicious Food Services (DFS) has the other 60%. To become more competitive, Gourmet engages an advertising firm to bolster its image. During an extensive advertising campaign, monthly sales figures are compiled. It is found that 90% of Gourmet's customers return to Gourmet the following month, while 20% of DFS's customers switch to Gourmet.

(a) What percentage of customers use each service after 1 month?

(b) What percentage use each service after 2 months?

(c) What is the long term share of the market for each service?

Solution   We will solve this problem in several steps. First, we introduce some terminology. In the language of Markov chains, the market for catering services in our city is a **system** in which there are two **states**: Gourmet and DFS. A catering customer is in state Gourmet if he or she uses Gourmet's services. Otherwise he or she is in state DFS. The probabilities of moving from one state to another are called **transition probabilities**. For our problem, we indicate the transition probabilities in Figure 1. The four probabilities in the right-hand branches of the tree are all conditional probabilities. Writing these out, we have

$$P(\text{Gourmet}|\text{Gourmet}) = 0.9$$

$$P(\text{DFS}|\text{Gourmet}) = 0.1$$

$$P(\text{Gourmet}|\text{DFS}) = 0.2$$

$$P(\text{DFS}|\text{DFS}) = 0.8$$

(a)   Using the multiplication theorem (equation (5.8.4) on page 280) or simply multiplying through the tree, we obtain (denoting a preference for Gourmet after one month by G_1 and a preference for DFS after one month by DFS_1),

$$P(G_1) = P(G_1|G)P(G) + P(G_1|\text{DFS})P(\text{DFS})$$

$$= (0.9)(0.4) + (0.2)(0.6) = 0.36 + 0.12 = 0.48$$

and

$$P(\text{DFS}_1) = P(\text{DFS}_1|G)P(G) + P(\text{DFS}_1|\text{DFS})P(\text{DFS})$$

$$= (0.1)(0.4) + (0.8)(0.6) = 0.04 + 0.48 = 0.52$$

Thus, after 1 month, 48 % of the catering customers choose Gourmet and 52 % choose DFS. Note that $0.48 + 0.52 = 1$.

We can obtain this answer in another way. We define the **initial probability vector** $\mathbf{p}_0$ by $\mathbf{p}_0 = (0.4 \quad 0.6)$. We define the **transition matrix** T by

$$T = \begin{array}{c} \\ G \\ \text{DFS} \end{array}\!\!\begin{array}{cc} G & \text{DFS} \\ \left(\begin{array}{cc} 0.9 & 0.1 \\ 0.2 & 0.8 \end{array}\right) \end{array}$$

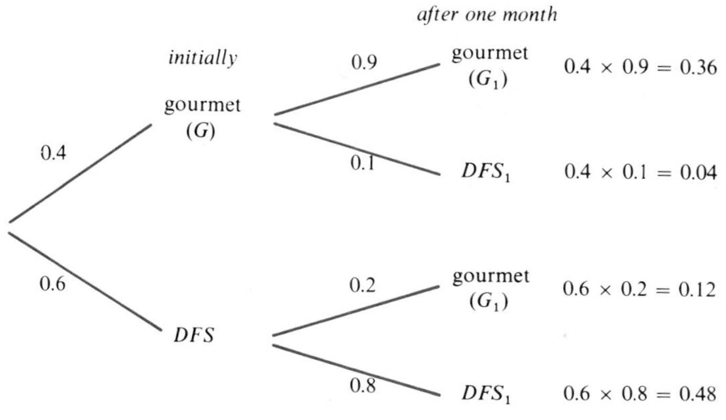

Figure 1

The transition matrix exhibits the probabilities of passing from one state to another during one trial of the experiment. Thus, for example, the 1,2 component of T is the probability of going from state 1 (Gourmet) to state 2 (DFS) in one month.

Now observe that

$$\mathbf{p}_0 T = (0.4 \quad 0.6)\begin{pmatrix} 0.9 & 0.1 \\ 0.2 & 0.8 \end{pmatrix} = (0.48 \quad 0.52) = \mathbf{p}_1,$$

the probability vector of proportions after one month. You should explain why taking the product $\mathbf{p}_0 T$ gives the same result as multiplying through the tree diagram.

(b) There are two ways to obtain $\mathbf{p}_2$, the vector of proportions after 2 months. First, we can draw a tree diagram as in Figure 2. Multiplying through the tree, we obtain

$$P(G_2) = (0.4)(0.9)(0.9) + (0.4)(0.1)(0.2) + (0.6)(0.2)(0.9)$$
$$+ (0.6)(0.8)(0.2) = 0.536$$

and

$$P(\text{DFS}_2) = (0.4)(0.9)(0.1) + (0.4)(0.1)(0.8) + (0.6)(0.2)(0.1)$$
$$+ (0.6)(0.8)(0.8) = 0.464.$$

Thus $\mathbf{p}_2$, the vector of proportions after 2 months, is given by

$$\mathbf{p}_2 = (0.536 \quad 0.464).$$

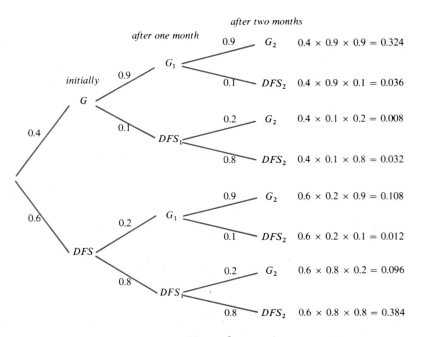

Figure 2

Note that $0.536 + 0.464 = 1$. Alternatively, by reasoning as in part (a), we find that

$$\mathbf{p}_2 = \mathbf{p}_1 T = (0.48 \quad 0.52)\begin{pmatrix} 0.9 & 0.1 \\ 0.2 & 0.8 \end{pmatrix} = (0.536 \quad 0.464).$$

(c) From parts (a) and (b), we conclude that

$$\mathbf{p}_1 = \mathbf{p}_0 T$$
$$\mathbf{p}_2 = \mathbf{p}_1 T = (\mathbf{p}_0 T)T = \mathbf{p}_0 T^2$$
$$\mathbf{p}_3 = \mathbf{p}_2 T = (\mathbf{p}_0 T^2)T = \mathbf{p}_0 T^3$$
$$\mathbf{p}_4 = \mathbf{p}_3 T = \mathbf{p}_0 T^4$$

and so on. For example,

$$\mathbf{p}_3 = \mathbf{p}_2 T = (0.536 \quad 0.464)\begin{pmatrix} 0.9 & 0.1 \\ 0.2 & 0.8 \end{pmatrix} = (0.5752 \quad 0.4248)$$

Continuing on a programmable calculator, we obtain the results in Table 1. It appears that as n(the number of months) increases, the proportions approach the **fixed probability vector**

$$\mathbf{t} = (0.6666\ldots \quad 0.3333\ldots) = (\tfrac{2}{3} \quad \tfrac{1}{3}).$$

This is called a **fixed vector** for the probability matrix T because

$$\mathbf{t}T = (\tfrac{2}{3} \quad \tfrac{1}{3})\begin{pmatrix} 0.9 & 0.1 \\ 0.2 & 0.8 \end{pmatrix} = (\tfrac{2}{3} \quad \tfrac{1}{3}) = \mathbf{t}.$$

That is, the vector $\mathbf{t}$ is unchanged when multiplied on the right by the matrix T. Thus we conclude that the long-run proportions of shares are $\tfrac{2}{3}$, or 67%, of the market for Gourmet and $\tfrac{1}{3}$, or 33%, of the market for DFS.

TABLE 1

Month n	Probability (or proportion) vector after n months
(Initial) 0	(0.4 0.6)
1	(0.48 0.52)
2	(0.536 0.464)
3	(0.5752 0.4248)
4	(0.60264 0.39736)
5	(0.621848 0.378152)
7	(0.64470552 0.35529448)
10	(0.6591339934 0.3408660066)
15	(0.6654006503 0.3345993497)
20	(0.6664538873 0.3335461127)
25	(0.6666309048 0.3333690952)
30	(0.6666606562 0.3333393438)
40	(0.6666664969 0.3333335031)

Before giving additional examples, we shall discuss some general properties of a Markov chain.

Probability Vector The n-component row vector $\mathbf{p} = (p_1 \quad p_2 \quad \cdots \quad p_n)$ is a **probability vector** if all its components are nonnegative and the sum of the components is 1:

$$p_1 + p_2 + \cdots + p_n = 1.$$

Example 2   The following are probability vectors.

(a) $(\frac{1}{5} \quad \frac{4}{5})$ (b) $(\frac{1}{3} \quad \frac{1}{3} \quad \frac{1}{3})$ (c) $(\frac{1}{8} \quad \frac{1}{2} \quad \frac{3}{8})$ (d) $(\frac{1}{6} \quad \frac{1}{3} \quad \frac{1}{12} \quad \frac{5}{12})$

(e) $(0 \quad \frac{1}{2} \quad 0 \quad \frac{1}{2})$ (f) $(0 \quad \frac{2}{5} \quad \frac{1}{3} \quad 0 \quad \frac{4}{15})$

Probability Matrix The $n \times n$ matrix $P = (p_{ij})$ is a **probability matrix** if each of its rows is a probability vector. This means that all of its components are nonnegative and the sum of the components in each row of P is 1.

Example 3   The following are probability matrices.

(a) $\begin{pmatrix} 1 & 0 \\ \frac{1}{2} & \frac{1}{2} \end{pmatrix}$ (b) $\begin{pmatrix} \frac{1}{2} & \frac{1}{2} \\ \frac{1}{4} & \frac{3}{4} \end{pmatrix}$ (c) $\begin{pmatrix} \frac{1}{4} & \frac{1}{4} & \frac{1}{2} \\ 0 & 1 & 0 \\ \frac{1}{3} & \frac{1}{3} & \frac{1}{3} \end{pmatrix}$ (d) $\begin{pmatrix} 1 & 0 & 0 \\ \frac{1}{4} & \frac{1}{3} & \frac{5}{12} \\ 0 & 0 & 1 \end{pmatrix}$

(e) $\begin{pmatrix} 0 & \frac{1}{2} & 0 & \frac{1}{2} \\ \frac{1}{6} & \frac{1}{3} & \frac{1}{12} & \frac{5}{12} \\ \frac{1}{4} & \frac{1}{4} & \frac{1}{4} & \frac{1}{4} \\ \frac{1}{7} & \frac{3}{7} & \frac{2}{7} & \frac{1}{7} \end{pmatrix}$

The following facts are not difficult to prove.

> **1.** The product of a probability vector (on the left) and a probability matrix (on the right) is a probability vector.
>
> **2.** The product of two probability matrices is a probability matrix.

Example 4   Let $\mathbf{p} = (\frac{1}{8} \quad \frac{1}{2} \quad \frac{3}{8})$ and $P = \begin{pmatrix} \frac{1}{4} & \frac{1}{4} & \frac{1}{2} \\ 0 & 1 & 0 \\ \frac{1}{3} & \frac{1}{3} & \frac{1}{3} \end{pmatrix}$. Verify that $\mathbf{p}P$ is a probability vector.

Solution

$$\mathbf{p}P = \begin{pmatrix} \frac{1}{8} & \frac{1}{2} & \frac{3}{8} \end{pmatrix} \begin{pmatrix} \frac{1}{4} & \frac{1}{4} & \frac{1}{2} \\ 0 & 1 & 0 \\ \frac{1}{3} & \frac{1}{3} & \frac{1}{3} \end{pmatrix} = \begin{pmatrix} \frac{5}{32} & \frac{21}{32} & \frac{3}{16} \end{pmatrix}$$

is a probability vector since $\frac{5}{32} + \frac{21}{32} + \frac{3}{16} = 1$. This illustrates Fact 1.

Example 5 Let $P = \begin{pmatrix} \frac{1}{4} & \frac{1}{4} & \frac{1}{2} \\ 0 & 1 & 0 \\ \frac{1}{3} & \frac{1}{3} & \frac{1}{3} \end{pmatrix}$ and $Q = \begin{pmatrix} 1 & 0 & 0 \\ \frac{1}{4} & \frac{1}{3} & \frac{5}{12} \\ 0 & 0 & 1 \end{pmatrix}$. Verify that PQ is a probability matrix.

Solution

$$PQ = \begin{pmatrix} \frac{1}{4} & \frac{1}{4} & \frac{1}{2} \\ 0 & 1 & 0 \\ \frac{1}{3} & \frac{1}{3} & \frac{1}{3} \end{pmatrix} \begin{pmatrix} 1 & 0 & 0 \\ \frac{1}{4} & \frac{1}{3} & \frac{5}{12} \\ 0 & 0 & 1 \end{pmatrix} = \begin{pmatrix} \frac{5}{16} & \frac{1}{12} & \frac{29}{48} \\ \frac{1}{4} & \frac{1}{3} & \frac{5}{12} \\ \frac{5}{12} & \frac{1}{9} & \frac{17}{36} \end{pmatrix}$$

The components of PQ are all nonnegative. Also, $\frac{5}{16} + \frac{1}{12} + \frac{29}{48} = \frac{1}{4} + \frac{1}{3} + \frac{5}{12} = \frac{5}{12} + \frac{1}{9} + \frac{17}{36} = 1$. Thus PQ is a probability matrix. This illustrates Fact 2.

To define a Markov chain, consider an experiment with a finite sample space $S = \{E_1, E_2, \ldots, E_n\}$. Consider a sequence (or **chain**) of trials of this experiment. The experiment is said to be in the **state** E_i on the mth trial if E_i is the outcome of the mth trial of the experiment.

Markov Chain A sequence of trials of an experiment is a **Markov chain** if

(a) the outcome of the mth trial depends only on the outcome of the $(m - 1)$st trial and not on the outcome of earlier trials, and

(b) the probability of going from state E_i to state E_j in two successive trials of the experiment does not change.

For example, if the probability of going from state E_2 to E_4 on the third trial is 0.6, then the probability of going from E_2 to E_4 on the fourth or tenth or fiftieth trial is also 0.6.

Note that in Example 1 we assumed that the probability that a customer chose Gourmet in one month depended only on the catering firm he or she had chosen in the previous month.

If the weather today depends only on the weather yesterday, then observing and predicting weather problems is a problem involving Markov chains. If the probability of choosing a particular make of car on your next car purchase depends only on what car you own now, then the analysis of sales' patterns is a problem involving Markov chains.

On the other hand, if the weather today is determined by the weather over the past several days, then we no longer have a Markov chain. Similarly, if in making your next car purchase you take into account the last three cars you have owned, then the analysis of car patterns does not involve a Markov chain. This follows because the probability of choosing one make of car on the mth trial of the experiment (your mth car) depends on your choices for the $(m-1)$st, $(m-2)$nd and $(m-3)$rd cars (your last three choices).

A Markov chain is characterized by the probabilities that the system goes from one state to any other state on successive trials.

Transition Matrix of a Markov Chain The **transition matrix** of a Markov chain is the $n \times n$ probability matrix $T = (p_{ij})$ whose ijth component p_{ij} is the probability that the system goes from state E_i to state E_j on successive trials of the experiment.

Example 6 In Example 1 the transition matrix is $T = \begin{pmatrix} 0.9 & 0.1 \\ 0.2 & 0.8 \end{pmatrix}$. This means, for example, that the probability of going from state 1 (Gourmet) to state 2 (DFS) is 0.1.

Let T be the transition matrix of a Markov chain. Since the product of two probability matrices is a probability matrix, it follows that the matrices $T^2 = TT$, $T^3 = T^2 T$, T^4, T^5, ... are all probability matrices.

Regular Matrix and Regular Markov Chain A probability matrix T is **regular** if all the components of at least one of its powers T^m are strictly positive (greater than zero). A Markov chain is **regular** if its transition matrix is regular.

Example 7 Which of the following matrices are transition matrices for a regular Markov chain?

(a) $\begin{pmatrix} \frac{1}{2} & \frac{1}{2} \\ \frac{1}{3} & \frac{2}{3} \end{pmatrix}$ (b) $\begin{pmatrix} \frac{1}{2} & \frac{1}{2} \\ 0 & 1 \end{pmatrix}$ (c) $\begin{pmatrix} 0 & \frac{1}{2} & \frac{1}{2} \\ \frac{1}{2} & 0 & \frac{1}{2} \\ \frac{1}{2} & \frac{1}{2} & 0 \end{pmatrix}$ (d) $\begin{pmatrix} \frac{3}{5} & \frac{1}{5} & \frac{1}{5} \\ \frac{1}{4} & \frac{1}{2} & \frac{1}{4} \\ \frac{1}{4} & \frac{1}{4} & \frac{1}{2} \end{pmatrix}$

Solution

(a) $T = \begin{pmatrix} \frac{1}{2} & \frac{1}{2} \\ \frac{1}{3} & \frac{2}{3} \end{pmatrix}$ is regular because all its components are positive.

(b) $T^2 = \begin{pmatrix} \frac{1}{2} & \frac{1}{2} \\ 0 & 1 \end{pmatrix} \begin{pmatrix} \frac{1}{2} & \frac{1}{2} \\ 0 & 1 \end{pmatrix} = \begin{pmatrix} \frac{1}{4} & \frac{3}{4} \\ 0 & 1 \end{pmatrix}$,

$T^3 = \begin{pmatrix} \frac{1}{8} & \frac{7}{8} \\ 0 & 1 \end{pmatrix}$, $T^4 = \begin{pmatrix} \frac{1}{16} & \frac{15}{16} \\ 0 & 1 \end{pmatrix}$, and so on. Since there is always a zero in the 2,1 position in T^m, we conclude that T is *not* regular.

(c) $T^2 = \begin{pmatrix} 0 & \frac{1}{2} & \frac{1}{2} \\ \frac{1}{2} & 0 & \frac{1}{2} \\ \frac{1}{2} & \frac{1}{2} & 0 \end{pmatrix} \begin{pmatrix} 0 & \frac{1}{2} & \frac{1}{2} \\ \frac{1}{2} & 0 & \frac{1}{2} \\ \frac{1}{2} & \frac{1}{2} & 0 \end{pmatrix} = \begin{pmatrix} \frac{1}{2} & \frac{1}{4} & \frac{1}{4} \\ \frac{1}{4} & \frac{1}{2} & \frac{1}{4} \\ \frac{1}{4} & \frac{1}{4} & \frac{1}{2} \end{pmatrix}$,

so T is regular.

(d) The components are all positive so T is regular.

Why do we study regular Markov chains? In Example 1 we saw that the regular Markov chain (regular because $T = \begin{pmatrix} 0.9 & 0.1 \\ 0.2 & 0.8 \end{pmatrix}$ has all positive components) had a fixed probability vector $\mathbf{t} = (\frac{2}{3} \quad \frac{1}{3})$ and that the probability vector $\mathbf{p}_n = \mathbf{p}_0 T^n$ got closer to $\mathbf{t}$ as n got large. This is not an accident.

> If T is a regular probability matrix, then there is a unique probability vector $\mathbf{t}$ such that $\mathbf{t}T = \mathbf{t}$. Moreover, for *any* probability vector $\mathbf{p}$, the probability vector $\mathbf{p}T^n$ gets closer and closer to $\mathbf{t}$ as n increases. The **fixed vector t** is called the **stationary distribution** of the regular Markov chain whose transition matrix is T. Moreover, as n gets large, each row of T^n gets close to the fixed vector $\mathbf{t}$.   (1)

Example 8   Find fixed vectors for each regular matrix.

(a) $\begin{pmatrix} \frac{1}{2} & \frac{1}{2} \\ \frac{1}{3} & \frac{2}{3} \end{pmatrix}$ (b) $\begin{pmatrix} \frac{3}{5} & \frac{1}{5} & \frac{1}{5} \\ \frac{1}{4} & \frac{1}{2} & \frac{1}{4} \\ \frac{1}{4} & \frac{1}{4} & \frac{1}{2} \end{pmatrix}$

Solution   We seek a probability vector $\mathbf{t}$ such that $\mathbf{t}T = \mathbf{t}$. If $\mathbf{t} = (x \quad y)$, we solve the equation

$$(x \quad y)\begin{pmatrix} \frac{1}{2} & \frac{1}{3} \\ \frac{1}{3} & \frac{2}{3} \end{pmatrix} = (x \quad y)$$

or

$$(\tfrac{1}{2}x + \tfrac{1}{3}y \quad \tfrac{1}{2}x + \tfrac{2}{3}y) = (x \quad y)$$

Equating components, we obtain

$$\tfrac{1}{2}x + \tfrac{1}{3}y = x$$

$$\tfrac{1}{2}x + \tfrac{2}{3}y = y$$

or

$$-\tfrac{1}{2}x + \tfrac{1}{3}y = 0$$

$$\tfrac{1}{2}x - \tfrac{1}{3}y = 0$$

Also, since **t** is a probability vector, we must have $x + y = 1$. This leads to the system

$$-\tfrac{1}{2}x + \tfrac{1}{3}y = 0$$

$$\tfrac{1}{2}x - \tfrac{1}{3}y = 0$$

$$x + \ y = 1$$

Row reducing, we obtain, successively,

Recall that $M_1(-2)$ means multiply the first row by -2; $A_{1,\,2}(-\tfrac{1}{2})$ means multiply the first row by $-\tfrac{1}{2}$ and add it to the second row; $P_{2,\,3}$ means interchange (permute) the second and third rows.

$$\left(\begin{array}{rr|r} -\tfrac{1}{2} & \tfrac{1}{3} & 0 \\ \tfrac{1}{2} & -\tfrac{1}{3} & 0 \\ 1 & 1 & 1 \end{array}\right) \xrightarrow{M_1(-2)} \left(\begin{array}{rr|r} 1 & -\tfrac{2}{3} & 0 \\ \tfrac{1}{2} & -\tfrac{1}{3} & 0 \\ 1 & 1 & 1 \end{array}\right) \xrightarrow[A_{1,\,3}(-1)]{A_{1,\,2}(-\tfrac{1}{2})} \left(\begin{array}{rr|r} 1 & -\tfrac{2}{3} & 0 \\ 0 & 0 & 0 \\ 0 & \tfrac{5}{3} & 1 \end{array}\right)$$

$$\xrightarrow{M_3(\tfrac{3}{5})} \left(\begin{array}{rr|r} 1 & -\tfrac{2}{3} & 0 \\ 0 & 0 & 0 \\ 0 & 1 & \tfrac{3}{5} \end{array}\right) \xrightarrow{A_{3,\,1}(\tfrac{2}{3})} \left(\begin{array}{rr|r} 1 & 0 & \tfrac{2}{5} \\ 0 & 0 & 0 \\ 0 & 1 & \tfrac{3}{5} \end{array}\right) \xrightarrow{P_{2,\,3}} \left(\begin{array}{rr|r} 1 & 0 & \tfrac{2}{5} \\ 0 & 1 & \tfrac{3}{5} \\ 0 & 0 & 0 \end{array}\right).$$

Thus $x = \tfrac{2}{5}$, $y = \tfrac{3}{5}$, and the unique probability vector $\mathbf{t} = (\tfrac{2}{5}\ \ \tfrac{3}{5})$.

Check. $\mathbf{t}T = (\tfrac{2}{5}\ \ \tfrac{3}{5})\begin{pmatrix} \tfrac{1}{2} & \tfrac{1}{2} \\ \tfrac{1}{3} & \tfrac{2}{3} \end{pmatrix} = (\tfrac{2}{5}\ \ \tfrac{3}{5}) = \mathbf{t}.$

(b)   We solve

$$\mathbf{t}(x \quad y \quad z) = \mathbf{t}T = (x \quad y \quad z)\begin{pmatrix} \tfrac{3}{5} & \tfrac{1}{5} & \tfrac{1}{5} \\ \tfrac{1}{4} & \tfrac{1}{2} & \tfrac{1}{4} \\ \tfrac{1}{4} & \tfrac{1}{4} & \tfrac{1}{2} \end{pmatrix},$$

or

$$(x \quad y \quad z) = (\tfrac{3}{5}x + \tfrac{1}{4}y + \tfrac{1}{4}z \quad \tfrac{1}{5}x + \tfrac{1}{2}y + \tfrac{1}{4}z \quad \tfrac{1}{5}x + \tfrac{1}{4}y + \tfrac{1}{2}z),$$

or

$$\tfrac{3}{5}x + \tfrac{1}{4}y + \tfrac{1}{4}z = x$$

$$\tfrac{1}{5}x + \tfrac{1}{2}y + \tfrac{1}{4}z = y$$

$$\tfrac{1}{5}x + \tfrac{1}{4}y + \tfrac{1}{2}z = z.$$

Then, together with the condition $x + y + z = 1$, we obtain the system

$$x + y + z = 1$$

$$-\tfrac{2}{5}x + \tfrac{1}{4}y + \tfrac{1}{4}z = 0$$

$$\tfrac{1}{5}x - \tfrac{1}{2}y + \tfrac{1}{4}z = 0$$

$$\tfrac{1}{5}x + \tfrac{1}{4}y - \tfrac{1}{2}z = 0.$$

We now row-reduce.

$$\begin{pmatrix} 1 & 1 & 1 & | & 1 \\ -\tfrac{2}{5} & \tfrac{1}{4} & \tfrac{1}{4} & | & 0 \\ \tfrac{1}{5} & -\tfrac{1}{2} & \tfrac{1}{4} & | & 0 \\ \tfrac{1}{5} & \tfrac{1}{4} & -\tfrac{1}{2} & | & 0 \end{pmatrix} \xrightarrow[\substack{A_{1,3}(-\tfrac{1}{5}) \\ A_{1,4}(-\tfrac{1}{5})}]{A_{1,2}(\tfrac{2}{5})} \begin{pmatrix} 1 & 1 & 1 & | & 1 \\ 0 & \tfrac{13}{20} & \tfrac{13}{20} & | & \tfrac{2}{5} \\ 0 & -\tfrac{7}{10} & \tfrac{1}{20} & | & -\tfrac{1}{5} \\ 0 & \tfrac{1}{20} & -\tfrac{7}{10} & | & -\tfrac{1}{5} \end{pmatrix}$$

$$\xrightarrow{M_2(\tfrac{20}{13})} \begin{pmatrix} 1 & 1 & 1 & | & 1 \\ 0 & 1 & 1 & | & \tfrac{8}{13} \\ 0 & -\tfrac{7}{10} & \tfrac{1}{20} & | & -\tfrac{1}{5} \\ 0 & \tfrac{1}{20} & -\tfrac{7}{10} & | & -\tfrac{1}{5} \end{pmatrix}$$

$$\xrightarrow[\substack{A_{2,3}(\tfrac{7}{10}) \\ A_{2,4}(-\tfrac{1}{20})}]{A_{2,1}(-1)} \begin{pmatrix} 1 & 0 & 0 & | & \tfrac{5}{13} \\ 0 & 1 & 1 & | & \tfrac{8}{13} \\ 0 & 0 & \tfrac{3}{4} & | & \tfrac{3}{13} \\ 0 & 0 & -\tfrac{3}{4} & | & -\tfrac{3}{13} \end{pmatrix}$$

$$\xrightarrow{M_3(\tfrac{4}{3})} \begin{pmatrix} 1 & 0 & 0 & | & \tfrac{5}{13} \\ 0 & 1 & 1 & | & \tfrac{8}{13} \\ 0 & 0 & 1 & | & \tfrac{4}{13} \\ 0 & 0 & -\tfrac{3}{4} & | & -\tfrac{3}{13} \end{pmatrix}$$

$$\xrightarrow[A_{3,4}(\tfrac{3}{4})]{A_{3,2}(-1)} \begin{pmatrix} 1 & 0 & 0 & | & \tfrac{5}{13} \\ 0 & 1 & 0 & | & \tfrac{4}{13} \\ 0 & 0 & 1 & | & \tfrac{4}{13} \\ 0 & 0 & 0 & | & 0 \end{pmatrix}$$

Thus $x = \tfrac{5}{13}$, $y = \tfrac{4}{13}$, $z = \tfrac{4}{13}$, and $\mathbf{t} = (\tfrac{5}{13} \quad \tfrac{4}{13} \quad \tfrac{4}{13})$.

Check. $\mathbf{t}T = (\tfrac{5}{13} \quad \tfrac{4}{13} \quad \tfrac{4}{13}) \begin{pmatrix} \tfrac{3}{5} & \tfrac{1}{5} & \tfrac{1}{5} \\ \tfrac{1}{4} & \tfrac{1}{2} & \tfrac{1}{4} \\ \tfrac{1}{4} & \tfrac{1}{4} & \tfrac{1}{2} \end{pmatrix} = (\tfrac{5}{13} \quad \tfrac{4}{13} \quad \tfrac{4}{13}).$

Remark. For the matrix in part (a) of this example, we use a calculator to compute some powers of T^m.

$$T = \begin{pmatrix} \frac{1}{2} & \frac{1}{2} \\ \frac{1}{3} & \frac{2}{3} \end{pmatrix} = \begin{pmatrix} 0.5 & 0.5 \\ 0.3333\ldots & 0.6666\ldots \end{pmatrix}$$

$$T^2 = \begin{pmatrix} \frac{5}{12} & \frac{7}{12} \\ \frac{7}{18} & \frac{11}{18} \end{pmatrix} = \begin{pmatrix} 0.4166\ldots & 0.5833\ldots \\ 0.3888\ldots & 0.6111\ldots \end{pmatrix}$$

$$T^4 = T^2T^2 \approx \begin{pmatrix} 0.400462963 & 0.599537037 \\ 0.399691358 & 0.600308642 \end{pmatrix}$$

$$T^8 = T^4T^4 \approx \begin{pmatrix} 0.4000003572 & 0.5999996428 \\ 0.3999997618 & 0.6000002382 \end{pmatrix}$$

$$T^{16} = T^8T^8 \approx \begin{pmatrix} 0.4 & 0.6 \\ 0.4 & 0.6 \end{pmatrix} = \begin{pmatrix} \frac{2}{5} & \frac{3}{5} \\ \frac{2}{5} & \frac{3}{5} \end{pmatrix}$$

$$T^m \approx \begin{pmatrix} 0.4 & 0.6 \\ 0.4 & 0.6 \end{pmatrix} \qquad \text{for } m > 16 \qquad \text{check it!}$$

The last two matrices are correct to ten decimal places.

Now let $\mathbf{p} = (a \quad b)$ be any probability vector. Then for $m > 16$,

$$\mathbf{p}T^m = (a \quad b)\begin{pmatrix} 0.4 & 0.6 \\ 0.4 & 0.6 \end{pmatrix} = (0.4a + 0.4b \quad 0.6a + 0.6b) = (0.4(a + b) \quad 0.6(a + b)).$$

Since $a + b = 1$, we see that, to ten decimal places, $\mathbf{p}T^m = (0.4 \quad 0.6) = (\frac{2}{5} \quad \frac{3}{5})$ for *any* initial probability vector $\mathbf{p}$. This illustrates the big result in (1).

Example 9 A mouse is placed in a box divided into three compartments, as shown in Figure 3. In the absence of other information, it is reasonable to assume that the mouse will choose a door at random to move from one compartment to

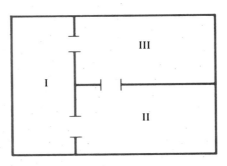

Figure 3

another. If this assumption is valid, in the long run the mouse can be expected to be in each of the three compartments equally often. Wecker described a series of experiments with prairie deer mice that can be analyzed in these terms.[†] By allowing the mice to travel among ten compartments, half in an open field and half in a wooded area, Wecker was able to study the strength of the preference of prairie deer mice for the field habitat over the wooded habitat.

We can analyze the mouse situation in terms of a Markov chain and show that, in the long run, our intuition that the mouse will spend equal amounts of time in each of the three compartments is valid. The three obvious states in this example are compartments I, II, and III. Since all doors are equally likely to be chosen, the probability is $\frac{1}{2}$ that the mouse will move to one of the two compartments it is not in. This gives us the transition matrix

$$T = \begin{pmatrix} 0 & \frac{1}{2} & \frac{1}{2} \\ \frac{1}{2} & 0 & \frac{1}{2} \\ \frac{1}{2} & \frac{1}{2} & 0 \end{pmatrix}$$

Since

$$T^2 = \begin{pmatrix} \frac{1}{2} & \frac{1}{4} & \frac{1}{4} \\ \frac{1}{4} & \frac{1}{2} & \frac{1}{4} \\ \frac{1}{4} & \frac{1}{4} & \frac{1}{2} \end{pmatrix},$$

T is regular and has a unique fixed probability vector $\mathbf{t}$. If $\mathbf{t} = (x \quad y \quad z)$, then $x + y + z = 1$ and

$$(x \quad y \quad z)\begin{pmatrix} 0 & \frac{1}{2} & \frac{1}{2} \\ \frac{1}{2} & 0 & \frac{1}{2} \\ \frac{1}{2} & \frac{1}{2} & 0 \end{pmatrix} = (x \quad y \quad z)$$

or

$$\frac{1}{2}y + \frac{1}{2}z = x$$
$$\frac{1}{2}x \quad\quad + \frac{1}{2}z = y$$
$$\frac{1}{2}x + \frac{1}{2}y \quad\quad = z$$

or

$$x + y + z = 1$$
$$-x + \frac{1}{2}y + \frac{1}{2}z = 0$$
$$\frac{1}{2}x - y + \frac{1}{2}z = 0$$
$$\frac{1}{2}x + \frac{1}{2}y - z = 0$$

[†] S. C. Wecker, "Habitat Selection," *Scientific American*, 211 (Oct. 1964): 109–116.

Then, row-reducing, we obtain

$$
\begin{pmatrix}
1 & 1 & 1 & 1 \\
-1 & \frac{1}{2} & \frac{1}{2} & 0 \\
\frac{1}{2} & -1 & \frac{1}{2} & 0 \\
\frac{1}{2} & \frac{1}{2} & -1 & 0
\end{pmatrix}
\xrightarrow[\substack{A_{1,3}(-\frac{1}{2}) \\ A_{1,4}(-\frac{1}{2})}]{A_{1,2}(1)}
\begin{pmatrix}
1 & 1 & 1 & 1 \\
0 & \frac{3}{2} & \frac{3}{2} & 1 \\
0 & -\frac{3}{2} & 0 & -\frac{1}{2} \\
0 & 0 & -\frac{3}{2} & -\frac{1}{2}
\end{pmatrix}
$$

$$
\xrightarrow{M_2(\frac{2}{3})}
\begin{pmatrix}
1 & 1 & 1 & 1 \\
0 & 1 & 1 & \frac{2}{3} \\
0 & -\frac{3}{2} & 0 & -\frac{1}{2} \\
0 & 0 & -\frac{3}{2} & -\frac{1}{2}
\end{pmatrix}
$$

$$
\xrightarrow[\substack{A_{2,3}(\frac{3}{2})}]{A_{2,1}(-1)}
\begin{pmatrix}
1 & 0 & 0 & \frac{1}{3} \\
0 & 1 & 1 & \frac{2}{3} \\
0 & 0 & \frac{3}{2} & \frac{1}{2} \\
0 & 0 & -\frac{3}{2} & -\frac{1}{2}
\end{pmatrix}
$$

$$
\xrightarrow{M_3(\frac{2}{3})}
\begin{pmatrix}
1 & 0 & 0 & \frac{1}{3} \\
0 & 1 & 1 & \frac{2}{3} \\
0 & 0 & 1 & \frac{1}{3} \\
0 & 0 & -\frac{3}{2} & -\frac{1}{2}
\end{pmatrix}
$$

$$
\xrightarrow[\substack{A_{3,4}(\frac{3}{2})}]{A_{3,2}(-1)}
\begin{pmatrix}
1 & 0 & 0 & \frac{1}{3} \\
0 & 1 & 0 & \frac{1}{3} \\
0 & 0 & 1 & \frac{1}{3} \\
0 & 0 & 0 & 0
\end{pmatrix}.
$$

Thus $\mathbf{t} = (\frac{1}{3} \quad \frac{1}{3} \quad \frac{1}{3})$, which means that, as expected, the mouse will spend equal amounts of time in each of the three compartments.

Example 10   The weather in Montreal is good, indifferent, or bad on any given day. If the weather is good today, it will be good tomorrow with probability 0.60, indifferent with probability 0.20, and bad with probability 0.20. If the weather is indifferent today, tomorrow it will be good, indifferent, or bad with probabilities 0.25, 0.50, and 0.25, respectively. Finally, if the weather is bad today, the probabilities are 0.25, 0.25, and 0.50 of good, indifferent, or bad weather tomorrow. This can be described as a Markov chain of trials of an experiment with three outcomes E_1, E_2, and E_3, corresponding to good, indifferent, and bad weather on any given day. The transition matrix for this Markov chain is

$$
P = \begin{matrix} & \overset{\text{To}}{\underset{\text{From}}{\diagdown}} \\ G \\ I \\ B \end{matrix}
\begin{matrix} G & I & B \end{matrix}
\begin{pmatrix}
0.60 & 0.20 & 0.20 \\
0.25 & 0.50 & 0.25 \\
0.25 & 0.25 & 0.50
\end{pmatrix}
=
\begin{pmatrix}
\frac{3}{5} & \frac{1}{5} & \frac{1}{5} \\
\frac{1}{4} & \frac{1}{2} & \frac{1}{4} \\
\frac{1}{4} & \frac{1}{4} & \frac{1}{2}
\end{pmatrix}.
$$

This is regular. In Example 8 we computed the fixed vector $(\frac{5}{13} \ \frac{4}{13} \ \frac{4}{13})$. This means that, in the long run, the weather in Montreal will be good $\frac{5}{13} \approx 38\%$ of the time, indifferent $\frac{4}{13} \approx 31\%$ of the time, and bad 31% of the time.

Note. We were able to solve this problem only by assuming that today's weather is affected only by yesterday's weather—not by what happened on previous days.

PROBLEMS 7.1

In Problems 1–10, determine whether the given vector is a probability vector.

1. $(\frac{1}{2} \ \frac{1}{2})$

2. $(\frac{1}{3} \ \frac{1}{3})$

3. $(\frac{1}{5} \ \frac{1}{5} \ \frac{1}{5} \ \frac{1}{5})$

4. $(\frac{1}{4} \ \frac{1}{4} \ \frac{1}{4} \ \frac{1}{4})$

5. $(\frac{1}{2} \ -\frac{1}{2} \ 1)$

6. $(\frac{1}{3} \ \frac{1}{4} \ \frac{5}{12})$

7. $(1 \ 0 \ 0 \ 0)$

8. $(0 \ 1 \ 0 \ 1)$

9. $(0.235 \ 0.361 \ 0.162 \ 0.242)$

10. $(\frac{1}{10} \ \frac{2}{10} \ \frac{3}{10} \ \frac{4}{10} \ \frac{5}{10})$

In Problems 11–20, determine whether the given matrix is a probability matrix.

11. $\begin{pmatrix} \frac{1}{2} & \frac{1}{2} & 0 \\ \frac{1}{2} & \frac{1}{2} & 0 \\ \frac{1}{2} & \frac{1}{2} & 0 \end{pmatrix}$

12. $\begin{pmatrix} \frac{1}{2} & \frac{1}{2} & \frac{1}{2} \\ \frac{1}{2} & \frac{1}{2} & \frac{1}{2} \\ \frac{1}{2} & \frac{1}{2} & \frac{1}{2} \end{pmatrix}$

13. $\begin{pmatrix} 1 & 0 & 0 \\ 0 & 1 & 0 \\ 0 & 0 & 1 \end{pmatrix}$

14. $\begin{pmatrix} \frac{1}{2} & 1 & -\frac{1}{2} \\ \frac{2}{3} & \frac{2}{3} & -\frac{1}{3} \\ -\frac{1}{5} & \frac{3}{5} & \frac{3}{5} \end{pmatrix}$

15. $\begin{pmatrix} 0.99 & 0.02 & -0.01 \\ 0 & 1 & 0 \\ 0.98 & 0.1 & 0.1 \end{pmatrix}$

16. $\begin{pmatrix} \frac{1}{2} & \frac{1}{4} & \frac{1}{8} & \frac{1}{8} \\ 0 & \frac{1}{2} & \frac{1}{2} & 0 \\ \frac{1}{2} & \frac{1}{4} & \frac{1}{8} & \frac{1}{8} \end{pmatrix}$

17. $\begin{pmatrix} \frac{1}{3} & \frac{1}{3} & \frac{1}{3} \\ \frac{1}{2} & 0 & \frac{1}{2} \\ 1 & 0 & 0 \end{pmatrix}$

18. $\begin{pmatrix} 1 & 0 & 0 \\ 1 & 0 & 0 \\ 0 & 0 & 0 \end{pmatrix}$

19. $\begin{pmatrix} \frac{1}{2} & \frac{1}{2} & 0 \\ \frac{1}{4} & \frac{1}{4} & \frac{1}{2} \\ \frac{1}{8} & \frac{1}{8} & \frac{3}{4} \end{pmatrix}$

20. $\begin{pmatrix} \frac{3}{5} & \frac{1}{5} & \frac{1}{5} & 0 \\ \frac{1}{6} & \frac{1}{3} & \frac{1}{4} & \frac{1}{4} \\ \frac{1}{7} & \frac{2}{7} & \frac{3}{7} & \frac{1}{7} \\ \frac{2}{11} & \frac{5}{11} & \frac{1}{11} & \frac{3}{11} \end{pmatrix}$

In Problems 21–30, a transition matrix T and an initial probability vector $\mathbf{p}_0$ are given. Compute $\mathbf{p}_1$, $\mathbf{p}_2$, and $\mathbf{p}_3$.

21. $T = \begin{pmatrix} \frac{1}{2} & \frac{1}{2} \\ \frac{3}{4} & \frac{1}{4} \end{pmatrix}$; $\mathbf{p}_0 = (\frac{1}{2} \ \frac{1}{2})$

22. $T = \begin{pmatrix} \frac{1}{2} & \frac{1}{2} \\ \frac{3}{4} & \frac{1}{4} \end{pmatrix}$; $\mathbf{p}_0 = (1 \ 0)$

23. $T = \begin{pmatrix} \frac{1}{8} & \frac{7}{8} \\ \frac{2}{3} & \frac{1}{3} \end{pmatrix}$; $\mathbf{p}_0 = (0 \ 1)$

24. $T = \begin{pmatrix} \frac{1}{8} & \frac{7}{8} \\ \frac{2}{3} & \frac{1}{3} \end{pmatrix}; \mathbf{p}_0 = (\frac{2}{5} \quad \frac{3}{5})$

25. $T = \begin{pmatrix} 0 & 1 & 0 \\ \frac{1}{2} & 0 & \frac{1}{2} \\ \frac{1}{3} & \frac{1}{3} & \frac{1}{3} \end{pmatrix}; \mathbf{p}_0 = (\frac{1}{2} \quad \frac{3}{8} \quad \frac{1}{8})$

26. $T = \begin{pmatrix} \frac{1}{3} & \frac{1}{2} & \frac{1}{6} \\ 1 & 0 & 0 \\ \frac{1}{4} & \frac{1}{4} & \frac{1}{2} \end{pmatrix}; \mathbf{p}_0 = (\frac{1}{5} \quad \frac{2}{5} \quad \frac{2}{5})$

27. $T = \begin{pmatrix} \frac{1}{8} & \frac{3}{4} & \frac{1}{8} \\ \frac{1}{2} & \frac{1}{2} & 0 \\ \frac{1}{5} & \frac{1}{5} & \frac{3}{5} \end{pmatrix}; \mathbf{p}_0 = (\frac{2}{7} \quad \frac{1}{7} \quad \frac{4}{7})$

28. $T = \begin{pmatrix} 0 & 0 & 1 \\ 0 & 1 & 0 \\ 1 & 0 & 0 \end{pmatrix}; \mathbf{p}_0 = (1 \quad 0 \quad 0)$

29. $T = \begin{pmatrix} 0 & 0 & 1 \\ 0 & 1 & 0 \\ 1 & 0 & 0 \end{pmatrix}; \mathbf{p}_0 = (\frac{1}{3} \quad \frac{1}{3} \quad \frac{1}{3})$

30. $T = \begin{pmatrix} 0.16 & 0.49 & 0.35 \\ 0.23 & 0.58 & 0.19 \\ 0.37 & 0.21 & 0.42 \end{pmatrix}; \mathbf{p}_0 = (0.31 \quad 0.44 \quad 0.25)$

In Problems 31–40, determine whether the given probability matrix is regular. If it is regular, find its unique fixed probability vector.

31. $\begin{pmatrix} \frac{1}{4} & \frac{3}{4} \\ \frac{1}{2} & \frac{1}{2} \end{pmatrix}$

32. $\begin{pmatrix} \frac{3}{5} & \frac{2}{5} \\ \frac{2}{5} & \frac{3}{5} \end{pmatrix}$

33. $\begin{pmatrix} 1 & 0 \\ \frac{3}{4} & \frac{1}{4} \end{pmatrix}$

34. $\begin{pmatrix} \frac{1}{5} & \frac{4}{5} \\ \frac{2}{3} & \frac{1}{3} \end{pmatrix}$

*35. $\begin{pmatrix} a & 1-a \\ b & 1-b \end{pmatrix}$, where $0 < a < 1$ and $0 < b < 1$

36. $\begin{pmatrix} \frac{1}{2} & 0 & \frac{1}{2} \\ \frac{1}{3} & \frac{1}{3} & \frac{1}{3} \\ \frac{1}{4} & \frac{1}{2} & \frac{1}{4} \end{pmatrix}$

37. $\begin{pmatrix} 1 & 0 & 0 \\ 0 & \frac{1}{2} & \frac{1}{2} \\ \frac{3}{5} & \frac{1}{5} & \frac{1}{5} \end{pmatrix}$

38. $\begin{pmatrix} 0.2 & 0.3 & 0.5 \\ 0.4 & 0.4 & 0.2 \\ 0.3 & 0.6 & 0.1 \end{pmatrix}$

39. $\begin{pmatrix} 0.12 & 0.37 & 0.51 \\ 0.43 & 0.19 & 0.38 \\ 0.26 & 0.59 & 0.15 \end{pmatrix}$

40. $\begin{pmatrix} 0.283 & 0 & 0.162 & 0.555 \\ 0 & 0.217 & 0.498 & 0.285 \\ 0.361 & 0.203 & 0.092 & 0.344 \\ 0.085 & 0.416 & 0.122 & 0.377 \end{pmatrix}$

41. (a) Calculate a fixed probability vector for the matrix

$$T = \begin{pmatrix} \frac{2}{3} & \frac{1}{6} & \frac{1}{6} \\ \frac{1}{4} & \frac{1}{2} & \frac{1}{4} \\ 0 & 0 & 1 \end{pmatrix}$$

(b) Is this vector unique?

42. (a) Calculate a fixed probability vector for the matrix

$$T = \begin{pmatrix} \frac{1}{2} & \frac{1}{4} & \frac{1}{4} \\ 0 & 1 & 0 \\ 0 & 0 & 1 \end{pmatrix}$$

(b) Is this vector unique?

43. On any given day, a person is either healthy or ill. If the person is healthy today, the probability that she will be healthy tomorrow is estimated to be 98%. If the person is ill today, the probability that she will be healthy tomorrow is 30%. Describe the sequence of states of health as a Markov chain. What is the transition matrix?

44. If the person of Problem 43 is ill today, what are the probabilities that she will recover tomorrow, 2 days from now, and 3 days from now?

45. On what percentage of days will the person of Problem 43 be healthy?

46. Consider the experiment of placing a mouse in the box drawn in Figure 4.

(a) Assuming that the mouse is equally likely to choose any door to leave a compartment, describe this experiment as a Markov chain and determine the transition matrix.

(b) Find the percentage of time that, in the long run, the mouse will spend in each compartment.

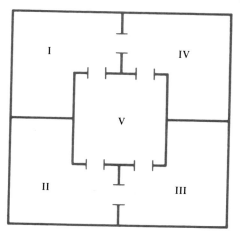

Figure 4

47. An examination consists of 100 true-or-false questions. For an average student, the examination is such that if a question is answered correctly, the probability that the next question is answered correctly is $\frac{3}{4}$. Similarly, if a question is answered incorrectly, the probability that the next question will be answered correctly is $\frac{1}{4}$. Estimate the average score on this examination, assuming that the first question is answered correctly.

48. A laboratory animal has a choice of three foods available in standard units. After lengthy observation, it is found that if the animal chooses one food on one trial, it will choose the same food on the next trial with probability 50% and it will choose the other foods on the next trial with equal probabilities 25%.

   (a) Describe this process as a Markov chain and determine the transition matrix.

   (b) Show that, in the long run, equal quantities of the three foods are consumed.

49. There are five tests in a certain course. The possible grades on each test are A, B, C, D, and E. It is estimated that the probability is 60% that a student will obtain the same grade as on the previous test and 10% for each of the other possibilities. Describe this process as a Markov chain. What is the transition matrix?

50. (a) If a student receives an A on the first test of Problem 49, what is the probability that he will receive a grade of C on the third test?

   (b) If a student receives a B on the second test, what is the probability that all his grades on the remaining three tests are B?

51. The UDrive Car Rental Company operates in California and charges a surcharge for cars rented at one location and dropped off at another. To determine the proper charge, the state was divided into three areas: north, central and south. The following probabilities of a car being picked up in one area and dropped off in the same or a different area were determined.

To From	North	Central	South
North	0.6	0.3	0.1
Central	0.2	0.5	0.3
South	0.1	0.2	0.7

Describe this process as a Markov chain and determine the transition matrix.

52. In Problem 51, determine the percentages of cars that, in the long run, will be in each of the three areas.

53. Jeffrey Newman is a sales representative for a large textbook company. He sells textbooks in an area that is subdivided into four regions. He visits college campuses in one region each week and never visits the same region in two consecutive weeks. If he sells in region I this week, he has a 70% chance of going to Region II and a 30% chance of going to region IV next week. If he sells in region III this week, he will sell in region II next week. If he sells in regions II or IV this week, he will sell in one of the other regions with equal probability next week. Describe Mr. Newman's travelling schedule as a Markov chain and find the transition matrix.

54. In Problem 53, what percentage of the time will Mr. Newman spend in each region in the long run?

55. In Problem 54, Mr. Newman's company spends $700, $650, $580, and $820, respectively, to pay his weekly expenses in the four regions. What will be the company's average weekly expense, in the long run?

56. A company has to choose between two copying machines for lease. The machines make copies which are indistinguishable. Each machine is either working or not working. According to past service records it has been determined that if machine

I is working one day, the probability is 0.95 that it will be working the next day. If it is not working, the probability is 0.75 that it will be working the next day. If machine II is working today, the probability is 0.9 that it will be working tomorrow. If it is not working today, the probability is 0.8 that it will be working tomorrow. Which machine should the company lease?

 57. In state senate elections in a certain state, it has been determined that a person will switch votes according to the following probabilities.

From \ To	Democrat	Republican	Independent
Democrat	0.7	0.2	0.1
Republican	0.35	0.6	0.05
Independent	0.4	0.3	0.3

Over the long run, what percentages of voters will vote for a Democrat, a Republican, and an Independent?

7.2 Absorbing Markov Chains

In the mouse experiment of Example 7.1.9 (see Figure 7.1.3 on page 376), we introduce the modification that the mouse is removed from the box when it reaches compartment III. This new experimental procedure can be described as a Markov chain if we agree to say that the system remains in the third state on all trials after it reaches the third state. States with this property are very common in applications of Markov chains.

Absorbing State   A state E_i of a Markov chain is **absorbing** if, once the system reaches the state E_i on some trial, the system remains in the state E_i on all future trials.

Absorbing Markov Chain   A Markov chain is **absorbing** if it has one or more absorbing states and if it is possible to reach an absorbing state from every nonabsorbing state.

Note.   It may be that it is necessary to pass through several nonabsorbing states before reaching an absorbing state.

If the state E_i is absorbing, the probability of transition from E_i to E_i is 1. In other words, the state E_i is absorbing if and only if $p_{ii} = 1$. The number of absorbing states of an absorbing Markov chain is equal to the number of ones on the diagonal of its transition matrix. The nonabsorbing states of an absorbing Markov chain are called the **transient states**. The probability that the system is in a transient state decreases as the number of trials increases.

Example 1 If the mouse (Example 7.1.9 on page 376) remains in the third compartment once it reaches there, the transition matrix is

$$T = \begin{pmatrix} 0 & \frac{1}{2} & \frac{1}{2} \\ \frac{1}{2} & 0 & \frac{1}{2} \\ 0 & 0 & 1 \end{pmatrix}$$

Here E_1 and E_2 are transient states and E_3 is an absorbing state. We note that it is possible to reach E_3 from either E_1 or E_2, and therefore the Markov chain is absorbing.

Example 2 The matrix

$$T = \begin{pmatrix} \frac{1}{2} & 0 & \frac{1}{2} \\ 0 & 1 & 0 \\ \frac{1}{3} & \frac{1}{3} & \frac{1}{3} \end{pmatrix}$$

is the transition matrix of an absorbing Markov chain with three states. The second state is absorbing. It can be reached from E_3 on one trial and from E_1 after two trials (first go from E_1 to E_3 with probability $\frac{1}{2}$ and then from E_3 to E_2 with probability $\frac{1}{3}$).

Example 3 Consider a game in which two players each begin with two marbles. On each play of the game, the first player has probability p of winning one marble and probability $q = 1 - p$ of losing one marble. The game ends when either player has lost all his marbles. Describe this game as a Markov chain.

Solution This game has five states, E_0, E_1, E_2, E_3, and E_4, corresponding to the first player having 0, 1, 2, 3, or 4 marbles. The states E_0 and E_4 are absorbing and the states E_1, E_2, and E_3 are nonabsorbing, or transient. The transition matrix is

$$T = \begin{pmatrix} p_{00} & p_{01} & p_{02} & p_{03} & p_{04} \\ p_{10} & p_{11} & p_{12} & p_{13} & p_{14} \\ p_{20} & p_{21} & p_{22} & p_{23} & p_{24} \\ p_{30} & p_{31} & p_{32} & p_{33} & p_{34} \\ p_{40} & p_{41} & p_{42} & p_{43} & p_{44} \end{pmatrix} = \begin{pmatrix} 1 & 0 & 0 & 0 & 0 \\ q & 0 & p & 0 & 0 \\ 0 & q & 0 & p & 0 \\ 0 & 0 & q & 0 & p \\ 0 & 0 & 0 & 0 & 1 \end{pmatrix}$$

The system begins in state E_2; that is, the initial probability distribution is $\mathbf{p_0} = (0 \quad 0 \quad 1 \quad 0 \quad 0)$. On succeeding plays, the successive probability distributions are $\mathbf{p_1} = \mathbf{p_0} T = (0 \quad q \quad 0 \quad p \quad 0)$, $\mathbf{p_2} = (q^2 \quad 0 \quad 2pq \quad 0 \quad p^2)$, $\mathbf{p_3} = (q^2 \quad 2pq^2 \quad 0 \quad 2p^2q \quad p^2)$, $\mathbf{p_4} = (q^2 + 2pq^3 \quad 0 \quad 4p^2q^2 \quad 0 \quad p^2 + 2p^3q), \ldots$. After many trials, the probability that the system is in one of the transient states becomes very small. This means that after a large number of plays, it is almost certain that one of the players has lost all his marbles. For examples, if $p = \frac{2}{3}$, then $q = \frac{1}{3}$ and

$$\mathbf{p_4} = ((\tfrac{1}{3})^2 + 2(\tfrac{2}{3})(\tfrac{1}{3})^3 \quad 0 \quad 4(\tfrac{2}{3})^2(\tfrac{1}{3})^2 \quad 0 \quad (\tfrac{2}{3})^2 + 2(\tfrac{2}{3})^3(\tfrac{1}{3}))$$
$$= (\tfrac{7}{27} \quad 0 \quad \tfrac{8}{81} \quad 0 \quad \tfrac{52}{81}).$$

These are the probabilities of being in a transient state after 4 plays

---

The result of the last example suggests that when dealing with an absorbing Markov chain, the system will eventually be in an absorbing state. We shall not prove this fact here, but we will make use of it in the remainder of the section. Two questions naturally arise.

1. How long, on the average, will the system be in some transient state before passing into an absorbing state?

2. If there is more than one absorbing state, what are the long-run probabilities of ending up in each absorbing state?

There is a technique used for answering these questions. The proof that this technique works is difficult and we shall not discuss it here. Rather, we illustrate the technique on a specific matrix and then use the technique in other examples.

Consider the matrix

$$T = \begin{pmatrix} \frac{1}{2} & \frac{1}{8} & \frac{1}{8} & \frac{1}{4} \\ 0 & 1 & 0 & 0 \\ \frac{1}{3} & \frac{1}{6} & \frac{3}{8} & \frac{1}{8} \\ 0 & 0 & 0 & 1 \end{pmatrix}$$

This is the transition matrix of an absorbing Markov chain with states 2 and 4 absorbing.

Step 1. From the transition matrix T, delete the rows corresponding to absorbing states; call the new matrix T'.

$$T' = \begin{pmatrix} \frac{1}{2} & \frac{1}{8} & \frac{1}{8} & \frac{1}{4} \\ \frac{1}{3} & \frac{1}{6} & \frac{3}{8} & \frac{1}{8} \end{pmatrix}$$

Step 2. Divide the matrix T' into absorbing and nonabsorbing states. Call the matrix corresponding to absorbing states S and the one corresponding to nonabsorbing states R.

In our example, the second and fourth columns of T' represent the absorbing states:

Nonabsorbing states   Absorbing states

$$\begin{pmatrix} \frac{1}{2} & \frac{1}{8} & \frac{1}{8} & \frac{1}{4} \\ \frac{1}{3} & \frac{1}{6} & \frac{3}{8} & \frac{1}{8} \end{pmatrix}.$$

Thus

$$S = \begin{pmatrix} \frac{1}{8} & \frac{1}{4} \\ \frac{1}{6} & \frac{1}{8} \end{pmatrix} \quad \text{and} \quad R = \begin{pmatrix} \frac{1}{2} & \frac{1}{8} \\ \frac{1}{3} & \frac{3}{8} \end{pmatrix}.$$

Step 3.   Compute the matrix $Q = (I - R)^{-1}$. Here

$$I - R = \begin{pmatrix} 1 & 0 \\ 0 & 1 \end{pmatrix} - \begin{pmatrix} \frac{1}{2} & \frac{1}{8} \\ \frac{1}{3} & \frac{3}{8} \end{pmatrix} = \begin{pmatrix} \frac{1}{2} & -\frac{1}{8} \\ -\frac{1}{3} & \frac{5}{8} \end{pmatrix}$$

and

$$Q = (I - R)^{-1} = \frac{48}{13}\begin{pmatrix} \frac{5}{8} & \frac{1}{8} \\ \frac{1}{3} & \frac{1}{2} \end{pmatrix} = \begin{pmatrix} \frac{30}{13} & \frac{6}{13} \\ \frac{16}{13} & \frac{24}{13} \end{pmatrix}. \quad \text{See page 122.}$$

Note.   We call Q the **fundamental matrix** for the absorbing Markov chain. We can now answer our questions.

Step 4.   The component q_{ij} of Q is interpreted as follows: If we start in transient state i, q_{ij} is the expected number of times we will be in transient state j before we reach some absorbing state. In our example, the number $q_{21} = \frac{16}{13}$ means that if we start in the second transient state (which is state 3 in our problem), then we will return to transient state 1 an average of $\frac{16}{13} \approx 1.23$ times before passing to an absorbing state (state 2 or state 4). Also, since $\frac{16}{13} + \frac{24}{13} = \frac{40}{13} \approx 3.08$, we will be in some transient state an average of 3.08 times if we start in transient state 2.

Step 5.   Compute the probability matrix $A = QS$. Here

$$A = QS = \frac{1}{13}\begin{pmatrix} 30 & 6 \\ 16 & 24 \end{pmatrix}\begin{pmatrix} \frac{1}{8} & \frac{1}{4} \\ \frac{1}{6} & \frac{1}{8} \end{pmatrix}$$

$$= \frac{1}{13}\begin{pmatrix} \frac{19}{4} & \frac{33}{4} \\ 6 & 7 \end{pmatrix} = \begin{pmatrix} \frac{19}{52} & \frac{33}{52} \\ \frac{6}{13} & \frac{7}{13} \end{pmatrix} \approx \begin{pmatrix} 0.3654 & 0.6346 \\ 0.4615 & 0.5385 \end{pmatrix}.$$

Note that A is a probability matrix.

To interpret the components in A, we add labels.

	Absorbing state 1	Absorbing state 2
Transient state 1	0.3654	0.6346
Transient state 2	0.4615	0.5385

This means that if we start in transient state 1, we will end up in absorbing state 1 for $0.3654 = 36.54\%$ of the time and in absorbing state 2 for $0.6346 = 63.46\%$ of the time. A similar interpretation is given to the second row of A.

We now apply our technique to a practical situation.

 Example 4   The Centerville Vo-Tech is a 2-year, state-supported school specializing in vocational training. Every 2 years the State Board of Regents presents a budget for Centerville to the State Legislature. The budget is based both on the number of students attending and the number who are expected to graduate. To compute these numbers, students are grouped into four categories: first year, second year, graduate, and dropout. The last category includes those who transfer to another school.

After collecting data over several years, the school is able to predict the proportions of students who will move from one category to another in a given year. These data are given in Table 1. If there are 2000 first-year students and 1500 second-year students at Centerville this year, how many of these students will eventually graduate?

TABLE 1

To From	First year	Second year	Graduate	Dropout
First year	0.15	0.65	0	0.2
Second year	0	0.1	0.75	0.15
Graduate	0	0	1	0
Dropout	0	0	0	1

Solution We can describe this process as an absorbing Markov chain with four states, two of which (graduate and dropout) are absorbing. The transition matrix for this Markov chain is

$$T = \begin{array}{c} \\ \text{1st} \\ \text{2nd} \\ G \\ D \end{array}\begin{pmatrix} \overset{\text{To 1st}\quad \text{2nd}\quad G\quad D}{0.15} & 0.65 & 0 & 0.2 \\ 0 & 0.1 & 0.75 & 0.15 \\ 0 & 0 & 1 & 0 \\ 0 & 0 & 0 & 1 \end{pmatrix}.$$

Step 1.
$$T' = \begin{pmatrix} 0.15 & 0.65 & 0 & 0.2 \\ 0 & 0.1 & 0.75 & 0.15 \end{pmatrix}$$

Step 2.
$$R = \begin{pmatrix} 0.15 & 0.65 \\ 0 & 0.1 \end{pmatrix}, \quad S = \begin{pmatrix} 0 & 0.2 \\ 0.75 & 0.15 \end{pmatrix}$$

Step 3.
$$I - R = \begin{pmatrix} 1 & 0 \\ 0 & 1 \end{pmatrix} - \begin{pmatrix} 0.15 & 0.65 \\ 0 & 0.1 \end{pmatrix} = \begin{pmatrix} 0.85 & -0.65 \\ 0 & 0.9 \end{pmatrix}$$

and

$$Q = (I - R)^{-1} \approx \begin{pmatrix} 1.17647 & 0.84967 \\ 0 & 1.11111 \end{pmatrix}$$

Step 4. The numbers in Q mean that the average student now in his or her first year will spend approximately 1.18 years in the first year and 0.85 years in the second year before reaching an absorbing state (graduation or dropout). The average second-year student will spend 1.11 years in his or her second year before graduating or dropping out.

Step 5.
$$A = QS \approx \begin{pmatrix} 1.17647 & 0.84967 \\ 0 & 1.11111 \end{pmatrix} \begin{pmatrix} 0 & 0.2 \\ 0.75 & 0.15 \end{pmatrix}$$

$$\approx \begin{pmatrix} 0.63726 & 0.36274 \\ 0.83333 & 0.16667 \end{pmatrix}$$

This means that 63.7% of first-year students will graduate (into absorbing state 1) and 36.3% will drop out, while 83.3% and 16.7% of second-year students will graduate and drop out, respectively. Now, to answer our questions:
Of 2000 first-year students,

$$2000 \times 0.63726 = 1275 \text{ will graduate.}$$

Of 1500 second-year students,

$$1500 \times 0.83333 = 1250 \text{ will graduate.}$$

Note. In Step 5 we computed the number who will graduate eventually, while in Step 4 we determined how long "eventually" really is for the average student.

Example 5 In Example 1 on page 384, E_1 and E_2 are transient states, while E_3 is absorbing. Thus, since $T = \begin{pmatrix} 0 & \frac{1}{2} & \frac{1}{2} \\ \frac{1}{2} & 0 & \frac{1}{2} \\ 0 & 0 & 1 \end{pmatrix}$, we have

$$T' = \begin{pmatrix} 0 & \frac{1}{2} & \frac{1}{2} \\ \frac{1}{2} & 0 & \frac{1}{2} \end{pmatrix}, \quad R = \begin{pmatrix} 0 & \frac{1}{2} \\ \frac{1}{2} & 0 \end{pmatrix} \quad \text{and} \quad S = \begin{pmatrix} \frac{1}{2} \\ \frac{1}{2} \end{pmatrix}.$$

Since there is only one absorbing state, it is evident that we must reach that state from all nonabsorbing states, so A should be the matrix $\begin{pmatrix} 1 \\ 1 \end{pmatrix}$. We can use this fact to check our calculations.

Now $\quad I - R = \begin{pmatrix} 1 & -\frac{1}{2} \\ -\frac{1}{2} & 1 \end{pmatrix} \quad \text{and} \quad Q = (I - R)^{-1} = \begin{pmatrix} \frac{4}{3} & \frac{2}{3} \\ \frac{2}{3} & \frac{4}{3} \end{pmatrix}.$

The meaning of $q_{12} = \frac{2}{3}$, for example, is that if the system begins in state E_1, the expected number of times the system will be in state E_2 is $\frac{2}{3}$. Similarly, $q_{11} + q_{12} = \frac{4}{3} + \frac{2}{3} = 2$. In other words, if the mouse is in compartment I on the first move, the expected number of moves (including the first move) before reaching compartment III is 2. Finally,

$$A = QS = \begin{pmatrix} \frac{4}{3} & \frac{2}{3} \\ \frac{2}{3} & \frac{4}{3} \end{pmatrix} \begin{pmatrix} \frac{1}{2} \\ \frac{1}{2} \end{pmatrix} = \begin{pmatrix} 1 \\ 1 \end{pmatrix},$$

as expected.

Example 6   In Example 3 the transition matrix for our game is

$$T = \begin{pmatrix} 1 & 0 & 0 & 0 & 0 \\ q & 0 & p & 0 & 0 \\ 0 & q & 0 & p & 0 \\ 0 & 0 & q & 0 & p \\ 0 & 0 & 0 & 0 & 1 \end{pmatrix}.$$

So

$$T' = \begin{pmatrix} q & 0 & p & 0 & 0 \\ 0 & q & 0 & p & 0 \\ 0 & 0 & q & 0 & p \end{pmatrix}.$$

Since the first and fifth states are absorbing, we have

$$R = \begin{pmatrix} 0 & p & 0 \\ q & 0 & p \\ 0 & q & 0 \end{pmatrix} \quad \text{and} \quad S = \begin{pmatrix} q & 0 \\ 0 & 0 \\ 0 & p \end{pmatrix}.$$

For simplicity, consider the case $p = q = \frac{1}{2}$. Then

$$R = \begin{pmatrix} 0 & \frac{1}{2} & 0 \\ \frac{1}{2} & 0 & \frac{1}{2} \\ 0 & \frac{1}{2} & 0 \end{pmatrix}, \qquad S = \begin{pmatrix} \frac{1}{2} & 0 \\ 0 & 0 \\ 0 & \frac{1}{2} \end{pmatrix},$$

$$I - R = \begin{pmatrix} 1 & -\frac{1}{2} & 0 \\ -\frac{1}{2} & 1 & -\frac{1}{2} \\ 0 & -\frac{1}{2} & 1 \end{pmatrix} \quad \text{and} \quad (I - R)^{-1} = \begin{pmatrix} \frac{3}{2} & 1 & \frac{1}{2} \\ 1 & 2 & 1 \\ \frac{1}{2} & 1 & \frac{3}{2} \end{pmatrix} = Q.$$

This tells us, for example, that if the first player begins with two marbles, then he can expect the game to last $q_{21} + q_{22} + q_{23} = 1 + 2 + 1 = 4$ moves before he either wins everything or loses everything. Finally,

$$A = QS = \begin{pmatrix} \frac{3}{2} & 1 & \frac{1}{2} \\ 1 & 2 & 1 \\ \frac{1}{2} & 1 & \frac{3}{2} \end{pmatrix} \begin{pmatrix} \frac{1}{2} & 0 \\ 0 & 0 \\ 0 & \frac{1}{2} \end{pmatrix} = \begin{pmatrix} \frac{3}{4} & \frac{1}{4} \\ \frac{1}{2} & \frac{1}{2} \\ \frac{1}{4} & \frac{3}{4} \end{pmatrix}.$$

Thus if player 1 starts with 1 marble (transient state 1), he will end up with no marbles (absorbing state 1) $\frac{3}{4}$ of the time and with all 4 marbles (absorbing state 2) $\frac{1}{4}$ of the time.

Examples 3 and 6 illustrate a game whose outcome is not predictable. The result of each play or move of the game is determined by chance. In Sections 7.4 and 7.5 we study other types of games in which the skill of the players can affect the outcome.

PROBLEMS 7.2 In Problems 1–10, a matrix is given. Determine whether the matrix is the transition matrix of an absorbing Markov chain. If so, find the number of absorbing states.

1. $\begin{pmatrix} \frac{1}{2} & \frac{1}{2} \\ 0 & 1 \end{pmatrix}$ **2.** $\begin{pmatrix} \frac{1}{2} & \frac{1}{2} \\ 1 & 0 \end{pmatrix}$ **3.** $\begin{pmatrix} 1 & 0 \\ 0 & 1 \end{pmatrix}$

4. $\begin{pmatrix} 1 & 0 & 0 \\ 0 & 0 & 1 \\ 0 & 1 & 0 \end{pmatrix}$ **5.** $\begin{pmatrix} 0 & 1 & 0 \\ 1 & 0 & 0 \\ 0 & 0 & 1 \end{pmatrix}$ **6.** $\begin{pmatrix} \frac{1}{3} & \frac{1}{3} & \frac{1}{3} \\ 0 & 0 & 1 \\ \frac{1}{3} & \frac{1}{3} & \frac{1}{3} \end{pmatrix}$

7. $\begin{pmatrix} \frac{1}{2} & \frac{1}{6} & \frac{1}{3} \\ 0 & 1 & 0 \\ 0 & 0 & 1 \end{pmatrix}$ **8.** $\begin{pmatrix} \frac{1}{2} & \frac{1}{2} & 0 & 0 \\ 1 & 0 & 0 & 0 \\ 0 & 0 & 1 & 0 \\ 0 & 0 & \frac{1}{2} & \frac{1}{2} \end{pmatrix}$ **9.** $\begin{pmatrix} \frac{1}{2} & \frac{1}{3} & \frac{1}{6} & 0 \\ 0 & 1 & 0 & 0 \\ 0 & 0 & 1 & 0 \\ \frac{1}{2} & \frac{3}{8} & 0 & \frac{1}{8} \end{pmatrix}$

10. $\begin{pmatrix} \frac{1}{5} & \frac{4}{5} & 0 & 0 \\ 0 & \frac{1}{3} & \frac{1}{3} & \frac{1}{3} \\ 0 & 0 & 1 & 0 \\ 0 & 0 & 0 & 1 \end{pmatrix}$

In Problems 11–20, the transition matrix of an absorbing Markov chain is given. Determine the matrices T', R, S, Q, and A.

11. $\begin{pmatrix} \frac{1}{3} & \frac{2}{3} \\ 0 & 1 \end{pmatrix}$ **12.** $\begin{pmatrix} 1 & 0 \\ \frac{3}{4} & \frac{1}{4} \end{pmatrix}$

13. $\begin{pmatrix} \frac{1}{3} & \frac{1}{3} & \frac{1}{3} \\ \frac{1}{2} & 0 & \frac{1}{2} \\ 0 & 0 & 1 \end{pmatrix}$ **14.** $\begin{pmatrix} 0.2 & 0.7 & 0.1 \\ 0.6 & 0.1 & 0.3 \\ 0 & 0 & 1 \end{pmatrix}$

15. $\begin{pmatrix} 0.4 & 0.4 & 0.2 \\ 0 & 1 & 0 \\ 0 & 0 & 1 \end{pmatrix}$ **16.** $\begin{pmatrix} \frac{1}{2} & \frac{1}{3} & \frac{1}{6} & 0 \\ \frac{1}{4} & \frac{1}{4} & \frac{1}{4} & \frac{1}{4} \\ 0 & 0 & 1 & 0 \\ 0 & 0 & 0 & 1 \end{pmatrix}$

17. $\begin{pmatrix} 1 & 0 & 0 & 0 \\ \frac{1}{3} & \frac{1}{3} & \frac{1}{6} & \frac{1}{6} \\ \frac{1}{2} & 0 & 0 & \frac{1}{2} \\ 0 & 0 & 0 & 1 \end{pmatrix}$ **18.** $\begin{pmatrix} 0.21 & 0.46 & 0.13 & 0.20 \\ 0 & 1 & 0 & 0 \\ 0.31 & 0.25 & 0.21 & 0.23 \\ 0 & 0 & 0 & 1 \end{pmatrix}$

19. $\begin{pmatrix} \frac{1}{8} & \frac{1}{4} & \frac{1}{8} & \frac{1}{8} & \frac{3}{8} \\ \frac{1}{7} & \frac{2}{7} & \frac{1}{7} & \frac{2}{7} & \frac{1}{7} \\ \frac{1}{4} & \frac{1}{2} & 0 & \frac{1}{8} & \frac{1}{8} \\ 0 & 0 & 0 & 1 & 0 \\ 0 & 0 & 0 & 0 & 1 \end{pmatrix}$ **20.** $\begin{pmatrix} 0.17 & 0.23 & 0.15 & 0.32 & 0.13 \\ 0 & 1 & 0 & 0 & 0 \\ 0.15 & 0.21 & 0 & 0.38 & 0.26 \\ 0 & 0 & 0 & 1 & 0 \\ 0 & 0 & 0 & 0 & 1 \end{pmatrix}$

21. A laboratory animal must complete a certain task to receive a unit of food. The probability of successful completion of the task on any trial is $\frac{4}{5}$. Suppose that the animal repeats the task until it receives a total of four units of food. Describe this process as an absorbing Markov chain with five states. What is the transition matrix?

22. What is the expected number of times that the task of Problem 21 will be repeated until four successful completions of the task?

*23. Two gamblers, G_1 and G_2, are playing a game; the probability that G_1 wins on each move is $\frac{3}{7}$. Suppose that G_1 begins with \$7 and G_2 begins with \$1. On each play, \$1 is bet by both players and the game continues until one player has lost all his money.

(a) Describe this game as an absorbing Markov chain with nine states.

(b) What are the probability distributions of the states after one play and after two plays?

(c) What is the probability that G_1 will win?

24. In the game of Problem 23, suppose that G_2 bets all his money on every play of the game.

(a) Describe this game as a Markov chain with five states.

(b) What is the expected number of plays of this game?

(c) What is the probability that G_1 will win?

25. A laboratory animal must choose one of four panels to receive food. Panels I and II, if chosen, result in a very small amount of food. Panels III and IV both yield much larger amounts of food. If either panel III or panel IV is chosen on a trial, it is observed that the same panel is chosen on all future trials. If either panel I or panel II is chosen on a trial, then any of the four panels may be chosen on the next trial with equal probability. Describe this process as an absorbing Markov chain with four states. If panel I is chosen on the first trial, what is the expected number of trials before either panel III or panel IV is chosen?

26. The Zephyr Electronics Co. manufactures portable cassette tape players. Before releasing a tape deck to sales, the deck is inspected. The inspection categories are not functioning (NF), fair, good, and excellent. The NF decks are thrown out, while the excellent decks are immediately released to sales. The fair and good decks are sent back for adjustment and then tested again. The proportions of fair and good decks that change categories are given in the table.

To From	NF	Fair	Good	Excellent
Fair	0.05	0.25	0.35	0.35
Good	0	0.15	0.2	0.65

(a) Describe this testing process as an absorbing Markov chain and compute the transition matrix.

(b) How many times, on average, will a deck that tested fair be tested again?

(c) How many times, on average, will a deck that tested good be tested again?

(d) What is the probability that a fair deck will eventually be thrown out?

(e) What is the probability that a fair deck will eventually be released to sales?

(f) Of 30,000 decks that originally tested good, how many will eventually be released to sales?

27. The Mastervise Corporation issues credit cards to 20,000 individuals in a certain state. Its auditor finds that 2356 of these did not pay their bills last month. It is company policy that if bills have not been paid for 3 months in a row, the card privileges are revoked and the bill is given to a collection agency. Past Mastervise experience has shown the collection experience in the table.

From \ To	1 month delinquent	2 months delinquent	3 months delinquent	Paid up	Revoked card
1 month delinquent	0	0.35	0	0.65	0
2 months delinquent	0	0	0.4	0.6	0
3 months delinquent	0	0	0	0.3	0.7

   (a) Describe this process as an absorbing Markov chain and find its transition matrix.

   (b) How many of the 2356 delinquent payers will have their cards revoked?

28. We refer to Problem 7.1.46 on page 381. Assume that once a mouse reaches either compartment II or compartment IV, it stays there.

   (a) If a mouse starts in compartment I, how many compartments will it visit, on the average?

   (b) What percentage of the mice that start in compartment III will end up in compartment IV?

29. In Example 4 it is discovered that 30% of first-year students repeat the first year, 45% reach the second year, and 25% drop out. If all other data remain the same, how many of the present 2000 first-year students will graduate?

*30. (a) Consider a game played between team A and team B with a total of n players on the two teams. On each play of the game, one team gains a player from the other team and the game continues until a team has lost all its members. If there are k players on team A and $n - k$ players on team B, the probability that team A gains one player on the next play is $(k/n)^2$. Describe this game as an absorbing Markov chain. Are the rules "fair"? (This is an elementary model of species competition.)

   (b) For the above game, suppose that $k = 3$ and $n = 6$. What is the probability distribution of the teams after one play? After two plays?

7.3 Two Applications of Markov Chains: Queuing Theory and a Model in Psychology (Optional)

In this section we describe two interesting applications of the theory of Markov chains. The first application, to *queuing theory*, requires only the material in Section 7.1. The second application requires the material in both Sections 7.1 and 7.2.

QUEUING THEORY

A queue[†] is a line, as in "let's queue up to buy tickets."[††] **Queuing theory** is the branch of mathematics that describes waiting lines. There are many applications of this theory. For example, a supermarket manager wants to ensure that (1) he does not lose business by having too many people growing impatient waiting in long check-out lines, and (2) he does not lose money by paying check-out employees who are standing idle with no customers to serve. Another example is furnished by the airline traveler. If you have ever bought your ticket at the airport, you had to (1) stand in a queue to buy your ticket, (2) stand in a queue to pass through security, (3) wait as the plane on which you were flying joined a queue of planes waiting to take off, (4) do the same upon landing, (5) wait for your luggage, and (6) wait in a queue for a taxi.

In these and similar situations, it is of great interest to answer this question: What is the percentage of time that there will be n people in a queue, where $n = 0, 1, 2, 3, \ldots$? For obvious reasons, the manager of a supermarket or the supervisor in any mode of travel needs an answer to this question in order to maintain reasonable levels of both customer satisfaction and profit. A general analysis of queuing models is beyond the scope of this text. However, we will discuss a simplified model to show how the theory of Markov chains can be used to obtain an answer to the question posed above.

We describe our model using as an example a single queue at an airline ticket counter. We make the following assumptions.

1. In a time period of 1 minute, there is a probability of $\frac{1}{3}$ that one person will join the queue and a probability of $\frac{2}{3}$ that no one will join the queue. Thus we are assuming that no more than one person will join the queue in any 1-minute interval.

2. If a person is being served in one time period, then the probability is $\frac{3}{8}$ that the person will have received his or her ticket and be out of the queue in the next time period.

3. All probabilities are independent of what has happened in earlier time periods.

4. A person cannot be served in the same time period in which he or she joined the queue.

5. At most one person can be served in a single 1-minute time period.

6. Because of an anticongestion policy, the queue is closed if there are four people in it. That is, there is a maximum of four people in the queue at any one time.

We can view this very simplified model as a Markov chain with five states, where the state of the system is defined as the number of people in the queue,

---

[†] Pronounced Q.
[††] You are more likely to hear this phrase in England than in the United States.

including the person being served. The five states are represented by the numbers 0, 1, 2, 3, and 4. The transition matrix for this Markov chain is given by

$$
T = \begin{array}{c} \\ 0 \\ 1 \\ 2 \\ 3 \\ 4 \end{array}
\begin{array}{ccccc}
0 & 1 & 2 & 3 & 4 \\
\left(\begin{array}{ccccc}
\frac{2}{3} & \frac{1}{3} & 0 & 0 & 0 \\
\frac{1}{4} & \frac{13}{24} & \frac{5}{24} & 0 & 0 \\
0 & \frac{1}{4} & \frac{13}{24} & \frac{5}{24} & 0 \\
0 & 0 & \frac{1}{4} & \frac{13}{24} & \frac{5}{24} \\
0 & 0 & 0 & \frac{3}{8} & \frac{5}{8}
\end{array} \right)
\end{array}
$$

We explain this as follows.

1. In the first row we have $p_{00} = \frac{2}{3}$ and $p_{01} = \frac{1}{3}$. We can go from state 0 to state 0 only if no one joins the queue. By Assumption 1, this event has probability $\frac{2}{3}$; Assumption 1 also explains why $p_{01} = \frac{1}{3}$.

2. In the second row we have $p_{10} = \frac{1}{4}$, $p_{11} = \frac{13}{24}$, and $p_{12} = \frac{5}{24}$. To go from state 1 to state 0, two independent events must occur: First, the person in the line is served, and second, no new person joins the queue. The probability of the first event is $\frac{3}{8}$ (by Assumption 2) and the probability of the second event is $\frac{2}{3}$ (by Assumption 1). Thus

$$p_{10} = \frac{3}{8} \cdot \frac{2}{3} = \frac{1}{4}.$$

Next, to go from state 1 to state 1, one of two mutually exclusive things can happen:

(a) No one is served and no one joins the queue; or

(b) One person is served and one person joins the queue. The probability of (a) is $\frac{5}{8} \cdot \frac{2}{3}$ and the probability of (b) is $\frac{3}{8} \cdot \frac{1}{3}$. Thus

$$p_{11} = \frac{5}{8} \cdot \frac{2}{3} + \frac{3}{8} \cdot \frac{1}{3} = \frac{10}{24} + \frac{3}{24} = \frac{13}{24}.$$

Finally, to go from state 1 to state 2, we must have no one served (with probability $\frac{5}{8}$) and one person join the queue (with probability $\frac{1}{3}$). Thus

$$p_{12} = \frac{5}{8} \cdot \frac{1}{3} = \frac{5}{24}.$$

Note that

$$p_{10} + p_{11} + p_{12} = \frac{1}{4} + \frac{13}{24} + \frac{5}{24} = 1.$$

3. Rows 3 and 4 are obtained just like row 2.

4. In the fifth row, we have $p_{43} = \frac{3}{8}$ and $p_{44} = \frac{5}{8}$. To go from state 4 to state 3, we must have one person served (note that no one can join the now-closed queue according to Assumption 6). Similarly, to go from state 4 to state 4 we must have no one served (with probability $\frac{5}{8}$).

Recall that a Markov chain is regular if some power T^m of its transition matrix T has strictly positive components. Using a calculator, we compute the first four powers of T to four decimal places.

$$T = \begin{pmatrix} 0.6667 & 0.3333 & 0.0000 & 0.0000 & 0.0000 \\ 0.2500 & 0.5417 & 0.2083 & 0.0000 & 0.0000 \\ 0.0000 & 0.2500 & 0.5417 & 0.2083 & 0.0000 \\ 0.0000 & 0.0000 & 0.2500 & 0.5417 & 0.2083 \\ 0.0000 & 0.0000 & 0.0000 & 0.3750 & 0.6750 \end{pmatrix}$$

$$T^2 = \begin{pmatrix} 0.5278 & 0.4028 & 0.0694 & 0.0000 & 0.0000 \\ 0.3021 & 0.4288 & 0.2257 & 0.0434 & 0.0000 \\ 0.0625 & 0.2709 & 0.3976 & 0.2257 & 0.0434 \\ 0.0000 & 0.0625 & 0.2709 & 0.4236 & 0.2430 \\ 0.0000 & 0.0000 & 0.0938 & 0.4375 & 0.4687 \end{pmatrix}$$

$$T^3 = \begin{pmatrix} 0.4526 & 0.4115 & 0.1215 & 0.0145 & 0.0000 \\ 0.3086 & 0.3894 & 0.2224 & 0.0705 & 0.0090 \\ 0.1094 & 0.2670 & 0.3282 & 0.2214 & 0.0741 \\ 0.0156 & 0.1016 & 0.2657 & 0.3770 & 0.2401 \\ 0.0000 & 0.0235 & 0.1602 & 0.4323 & 0.3841 \end{pmatrix}$$

$$T^4 = \begin{pmatrix} 0.4046 & 0.4041 & 0.1552 & 0.0332 & 0.0030 \\ 0.3031 & 0.3694 & 0.2192 & 0.0879 & 0.0203 \\ 0.1397 & 0.2631 & 0.2888 & 0.2161 & 0.0924 \\ 0.0358 & 0.1267 & 0.2593 & 0.3496 & 0.2286 \\ 0.0059 & 0.0528 & 0.1998 & 0.4116 & 0.3301 \end{pmatrix}$$

Since T^4 has strictly positive components, we see that our Markov chain is regular.

We now recall the result (see (1) on page 373) which states that since T is a regular matrix, there is a unique probability vector $\mathbf{t}$ whose components are all positive such that $\mathbf{t}T = \mathbf{t}$. Moreover, the matrices T^m approach (as m gets large) a matrix P, each of whose rows is equal to the vector $\mathbf{t}$. In our case the components of $\mathbf{t}$ are the long-range probabilities of the length of the queue. To find $\mathbf{t}$, we start with

$$\mathbf{t}T = \mathbf{t}$$

or

$$(p_0 \ \ p_1 \ \ p_2 \ \ p_3 \ \ p_4)\begin{pmatrix} \frac{2}{3} & \frac{1}{3} & 0 & 0 & 0 \\ \frac{1}{4} & \frac{13}{24} & \frac{5}{24} & 0 & 0 \\ 0 & \frac{1}{4} & \frac{13}{24} & \frac{5}{24} & 0 \\ 0 & 0 & \frac{1}{4} & \frac{13}{24} & \frac{5}{24} \\ 0 & 0 & 0 & \frac{3}{8} & \frac{5}{8} \end{pmatrix} = (p_0 \ \ p_1 \ \ p_2 \ \ p_3 \ \ p_4)$$

or

$$\tfrac{2}{3}p_0 + \tfrac{1}{4}\,p_1 \qquad\qquad\qquad\;\; = p_0$$
$$\tfrac{1}{3}p_0 + \tfrac{13}{24}p_1 + \tfrac{1}{4}\,p_2 \qquad\qquad = p_1$$
$$\tfrac{5}{24}p_1 + \tfrac{13}{24}p_2 + \tfrac{1}{4}\,p_3 \qquad\; = p_2$$
$$\tfrac{5}{24}p_2 + \tfrac{13}{24}p_3 + \tfrac{3}{8}p_4 = p_3$$
$$\tfrac{5}{24}p_3 + \tfrac{5}{8}p_4 = p_4.$$

Since **t** is a probability vector, we also have

$$p_0 + p_1 + p_2 + p_3 + p_4 = 1.$$

Combining terms, we obtain the system

$$-\tfrac{1}{3}p_0 + \tfrac{1}{4}\,p_1 \qquad\qquad\qquad\;\; = 0$$
$$\tfrac{1}{3}p_0 - \tfrac{11}{24}p_1 + \tfrac{1}{4}\,p_2 \qquad\qquad = 0$$
$$\tfrac{5}{24}p_1 - \tfrac{11}{24}p_2 + \tfrac{1}{4}\,p_3 \qquad\; = 0$$
$$\tfrac{5}{24}p_2 - \tfrac{11}{24}p_3 + \tfrac{3}{8}p_4 = 0$$
$$\tfrac{5}{24}p_3 - \tfrac{3}{8}p_4 = 0$$
$$p_0 + \;\; p_1 + \;\; p_2 + \;\; p_3 + \;\; p_4 = 1.$$

We now write the system in augmented matrix form and solve by row reduction. To simplify matters, we start by multiplying the first five equations by 24.

$$\begin{pmatrix} -8 & 6 & 0 & 0 & 0 & 0 \\ 8 & -11 & 6 & 0 & 0 & 0 \\ 0 & 5 & -11 & 6 & 0 & 0 \\ 0 & 0 & 5 & -11 & 9 & 0 \\ 0 & 0 & 0 & 5 & -9 & 0 \\ 1 & 1 & 1 & 1 & 1 & 1 \end{pmatrix}$$

$$\xrightarrow{M_1(-\tfrac{1}{8})} \begin{pmatrix} 1 & -0.75 & 0 & 0 & 0 & 0 \\ 8 & -11 & 6 & 0 & 0 & 0 \\ 0 & 5 & -11 & 6 & 0 & 0 \\ 0 & 0 & 5 & -11 & 9 & 0 \\ 0 & 0 & 0 & 5 & -9 & 0 \\ 1 & 1 & 1 & 1 & 1 & 1 \end{pmatrix}$$

$$\xrightarrow[A_{1,6}(-1)]{A_{1,2}(-8)} \begin{pmatrix} 1 & -0.75 & 0 & 0 & 0 & 0 \\ 0 & -5 & 6 & 0 & 0 & 0 \\ 0 & 5 & -11 & 6 & 0 & 0 \\ 0 & 0 & 5 & -11 & 9 & 0 \\ 0 & 0 & 0 & 5 & -9 & 0 \\ 0 & 1.75 & 1 & 1 & 1 & 1 \end{pmatrix}$$

$$\xrightarrow{M_2(-\frac{1}{5})} \begin{pmatrix} 1 & -0.75 & 0 & 0 & 0 & | & 0 \\ 0 & 1 & -1.2 & 0 & 0 & | & 0 \\ 0 & 5 & -11 & 6 & 0 & | & 0 \\ 0 & 0 & 5 & -11 & 9 & | & 0 \\ 0 & 0 & 0 & 5 & -9 & | & 0 \\ 0 & 1.75 & 1 & 1 & 1 & | & 1 \end{pmatrix}$$

$$\begin{array}{c} A_{2,1}(0.75) \\ A_{2,3}(-5) \\ A_{2,6}(-1.75) \\ \xrightarrow{\hspace{2cm}} \end{array} \begin{pmatrix} 1 & 0 & -0.9 & 0 & 0 & | & 0 \\ 0 & 1 & -1.2 & 0 & 0 & | & 0 \\ 0 & 0 & -5 & 6 & 0 & | & 0 \\ 0 & 0 & 5 & -11 & 9 & | & 0 \\ 0 & 0 & 0 & 5 & -9 & | & 0 \\ 0 & 0 & 3.1 & 1 & 1 & | & 1 \end{pmatrix}$$

$$\xrightarrow{M_3(-\frac{1}{5})} \begin{pmatrix} 1 & 0 & -0.9 & 0 & 0 & | & 0 \\ 0 & 1 & -1.2 & 0 & 0 & | & 0 \\ 0 & 0 & 1 & -1.2 & 0 & | & 0 \\ 0 & 0 & 5 & -11 & 9 & | & 0 \\ 0 & 0 & 0 & 5 & -9 & | & 0 \\ 0 & 0 & 3.1 & 1 & 1 & | & 1 \end{pmatrix}$$

$$\begin{array}{c} A_{3,1}(0.9) \\ A_{3,2}(1.2) \\ A_{3,4}(-5) \\ A_{3,6}(-3.1) \\ \xrightarrow{\hspace{2cm}} \end{array} \begin{pmatrix} 1 & 0 & 0 & -1.08 & 0 & | & 0 \\ 0 & 1 & 0 & -1.44 & 0 & | & 0 \\ 0 & 0 & 1 & -1.2 & 0 & | & 0 \\ 0 & 0 & 0 & -5 & 9 & | & 0 \\ 0 & 0 & 0 & 5 & -9 & | & 0 \\ 0 & 0 & 0 & 4.72 & 1 & | & 1 \end{pmatrix}$$

Since the fifth row is the negative of the fourth, we delete it and continue by dividing the fourth row by -5.

$$\xrightarrow{M_4(-\frac{1}{5})} \begin{pmatrix} 1 & 0 & 0 & -1.08 & 0 & | & 0 \\ 0 & 1 & 0 & -1.44 & 0 & | & 0 \\ 0 & 0 & 1 & -1.2 & 0 & | & 0 \\ 0 & 0 & 0 & 1 & -1.8 & | & 0 \\ 0 & 0 & 0 & 4.72 & 1 & | & 1 \end{pmatrix}$$

$$\begin{array}{c} A_{4,1}(1.08) \\ A_{4,2}(1.44) \\ A_{4,3}(1.2) \\ A_{4,5}(-4.72) \\ \xrightarrow{\hspace{2cm}} \end{array} \begin{pmatrix} 1 & 0 & 0 & 0 & -1.944 & | & 0 \\ 0 & 1 & 0 & 0 & -2.592 & | & 0 \\ 0 & 0 & 1 & 0 & -2.16 & | & 0 \\ 0 & 0 & 0 & 1 & -1.8 & | & 0 \\ 0 & 0 & 0 & 0 & 9.496 & | & 1 \end{pmatrix}$$

To this point, all numbers are exact. We will carry the remaining computations to five decimal places of accuracy.

$$\xrightarrow{M_5\left(\frac{1}{9.496}\right)} \begin{pmatrix} 1 & 0 & 0 & 0 & -1.944 & 0 \\ 0 & 1 & 0 & 0 & -2.592 & 0 \\ 0 & 0 & 1 & 0 & -2.16 & 0 \\ 0 & 0 & 0 & 1 & -1.8 & 0 \\ 0 & 0 & 0 & 0 & 1 & 0.10531 \end{pmatrix}$$

$$\xrightarrow[\substack{A_{5,1}(1.944) \\ A_{5,2}(2.592) \\ A_{5,3}(2.16) \\ A_{5,4}(1.8)}]{} \begin{pmatrix} 1 & 0 & 0 & 0 & 0 & 0.20472 \\ 0 & 1 & 0 & 0 & 0 & 0.27296 \\ 0 & 0 & 1 & 0 & 0 & 0.22746 \\ 0 & 0 & 0 & 1 & 0 & 0.18955 \\ 0 & 0 & 0 & 0 & 1 & 0.10531 \end{pmatrix}$$

Thus, the fixed probability vector is given by

$$(p_0 \quad p_1 \quad p_2 \quad p_3 \quad p_4) = (0.20472 \quad 0.27296 \quad 0.22746 \quad 0.18955 \quad 0.10531).$$

This means, for example, that the ticket agent is idle about 20% of the time (since $p_0 \approx 0.20$). Moreover, the line is closed about 10% of the time, since $p_4 \approx 0.10$, and this means that there are four people in the queue about 10% of the time. Finally, we have one person in the queue about 27% of the time, two people about 23% of the time, and three people about 19% of the time.

Again, we stress that we have dealt with a very simplified model. Nevertheless, this example shows us the power of Markov chains combined with matrix theory to analyze interesting situations.

A MODEL IN PSYCHOLOGY: PAIRED-ASSOCIATE LEARNING

As our second application we present a simple model for analyzing a learning process. We proceed with caution because none but the most elementary and simplified activities of the mind can be treated with any sort of workable model. The model we describe was first analyzed mathematically by G. H. Bower[†] and is called **paired-associate learning**.

In Bower's model there is an experimenter and one or more subjects. The experimenter designs two sets: a stimulus set and a response set. For each item s in the stimulus set S, there is a correct response r in the response set R. Thus, in effect, the experimenter creates a set of ordered pairs of the form (s, r), where $s \in S$ and $r \in R$. In Bower's original experiment, S and R each contained ten items. For example, S could contain ten letters, and R, ten numbers randomly assigned to the ten letters. In a different context, S could contain ten Spanish words, and R, their English equivalents. Or, as a final example, S could contain

---

[†] G. H. Bower, "Applications of a Model to Paired Associate Learning," *Psychometrica*, 26, (1961): 255–280.

ten nonsense syllables (dah, dum, uph, and so on) and R, a set of additional syllables, one associated with each of the syllables in S.[†]

Once S and R are created, the experiment can begin. On each trial of the experiment, the experimenter shows (or tells) an element $s \in S$ to the subject, and the subject responds with an element $r \in R$. Of course, the subject tries to respond with the correct answer, the r that is associated with the given s. If the subject gives the correct response, then a 0 is recorded. If the subject gives an incorrect response, then a 1 is recorded, and the experimenter tells or shows the subject the correct response.

It is assumed that, initially, the subject does not know the correct answers and must guess. If the subject guesses incorrectly, he will be given the correct answer and will, presumably, begin to learn. The experiment continues until the subject has been able to give the correct responses to the given stimuli two times in succession. At that point it is assumed that the subject has learned the answers completely.

In order to analyze this model, we must make five more-or-less reasonable assumptions. We refer to stimulus-response pairs in the form (s, r).

Assumptions.

1. Each stimulus-response pair is in one of two states on any trial of the experiment: *conditioned* (C) or *guessing* (G). It is in state C if the subject has learned the correct response (r) to the given question or stimulus (s). It is in state G if the subject does not know the correct response and must guess.

2. Initially, all pairs are in state G. That is, we assume that when the experiment begins the subject does not know any of the correct responses.

3. If an item is in state C on the nth trial of the experiment, then it remains in state C on all future trials. That is, once the subject has learned the proper response to a given stimulus, he does not forget it.

4. If an item is in state G on the nth trial of the experiment, then the probability is $c > 0$ that it will be in state C on the $(n + 1)$st trial of the experiment. That is, c is the probability that the subject will learn the correct response before the next trial of the experiment.

Note. Assumption 4 is the least plausible of the assumptions made so far. Here we are assuming that (a) all items are equally easy (or difficult) to learn, and (b) the subject is not more likely to learn the correct response after many guesses than after just one guess. It is hoped that if the subjects are human beings of normal intelligence, the probabilities of learning will increase after each guess and after each time the subject is told the correct answer.

[†]In Bower's original experiment the stimuli consisted of ten pairs of letters, and the responses were the numbers 1 or 2. That is, the subjects were given a pair of letters and asked to reply "1" or "2." After each trial, the subjects were told whether or not their answers were correct. If the subjects were able to answer correctly for each of the ten stimuli twice in a row, they were assumed to have learned the correct responses and the experiment was concluded.

5.   If the item is in state G and there are N possible responses, the probability is $1/N$ that the subject will guess the correct one. That is, in the guessing state all possible responses are equally likely to be chosen.

With these assumptions, we can regard the process of learning the correct response for each stimulus as a Markov chain with states C and G.[†] The transition matrix for this Markov chain is

$$T = \begin{matrix} & \begin{matrix} C & \;\; G \end{matrix} \\ \begin{matrix} C \\ G \end{matrix} & \begin{pmatrix} 1 & 0 \\ c & 1-c \end{pmatrix}, \end{matrix} \qquad (1)$$

since—by Assumption 3—once we are in C, we stay in C. Also—by Assumption 4—if we are in G, then the probability is c that we will be in C on the next trial of the experiment.

Using the transition matrix given by (1), it is easy to calculate the probability that an item is in states C or G after any trial of the experiment. First, by Assumption 2, the initial probability distribution is

$$\mathbf{p}_0 = (0 \quad 1). \qquad (2)$$

Now

$$T^2 = \begin{pmatrix} 1 & 0 \\ c & 1-c \end{pmatrix}\begin{pmatrix} 1 & 0 \\ c & 1-c \end{pmatrix} = \begin{pmatrix} 1 & 0 \\ c[1+(1-c)] & (1-c)^2 \end{pmatrix},$$

$$T^3 = \begin{pmatrix} 1 & 0 \\ c[1+(1-c)+(1-c)^2] & (1-c)^3 \end{pmatrix}, \dots$$

Continuing in this way (see Problem 18), we can show that

$$T^n = \begin{pmatrix} 1 & 0 \\ c[1+(1-c)+(1-c)^2+\cdots+(1-c)^{n-1}] & (1-c)^n \end{pmatrix}. \qquad (3)$$

The formula for the sum of a geometric progression is (see Problem 19)

$$1+x+x^2+\cdots+x^{n-1} = \frac{1-x^n}{1-x}, \qquad \text{for } x \neq 1. \qquad (4)$$

Thus

$$1+(1-c)+(1-c)^2+\cdots+(1-c)^{n-1} = \frac{1-(1-c)^n}{1-(1-c)} = \frac{1-(1-c)^n}{c} \qquad (5)$$

and, inserting (5) into (3), we have

$$T^n = \begin{pmatrix} 1 & 0 \\ 1-(1-c)^n & (1-c)^n \end{pmatrix}. \qquad (6)$$

[†] Note that there is one Markov chain for each stimulus-response item (s, r). Thus, if there are ten (s, r) pairs, as in Bower's experiments, then there are ten Markov chains. However, under our assumptions, all ten Markov chains will have identical properties.

The probability that an item is in state C or G after the nth trial is

$$\mathbf{p}_n = \mathbf{p}_0\, T^n = (0 \quad 1)\begin{pmatrix} 1 & 0 \\ 1-(1-c)^n & (1-c)^n \end{pmatrix} = (1-(1-c)^n \quad (1-c)^n). \quad (7)$$

Formula (7) is very revealing. Since $0 < c \le 1$ (by Assumption 4), we have $0 \le 1 - c < 1$. Thus $(1-c)^n$ gets closer and closer to zero as n gets larger. This tells us that so long as c is not zero, the correct response to the stimulus will eventually be learned. The larger the value of c, the faster it will be learned.

Example 1 Let $c = 0.25$. Then $1 - c = 0.75$. The table gives the probabilities that the item will be in state G or C after each trial.

Trial	$P(C) = 1 - (0.75)^n$	$P(G) = (0.75)^n$
0	1	0
1	0.25	0.75
2	0.4375	0.5625
3	0.5781	0.4219
4	0.6836	0.3164
5	0.7627	0.2373
6	0.8220	0.1780
7	0.8665	0.1335
8	0.8999	0.1001
9	0.9249	0.0751
10	0.9437	0.0563
11	0.9578	0.0422
12	0.9683	0.0317
13	0.9762	0.0238
14	0.9822	0.0178
15	0.9866	0.0134
16	0.9900	0.0100
17	0.9925	0.0075
18	0.9944	0.0056
19	0.9958	0.0042
20	0.9968	0.0032

Note that after the sixteenth trial, it is 99% certain that the subject has learned the correct response.

It is interesting to analyze this Markov chain in terms of the theory of absorbing Markov chains (see Section 7.2). This Markov chain is absorbing because the state C is absorbing. We now use the notation of Section 7.2. We have

$$\begin{matrix} & G & C \end{matrix}$$
$$T' = (1-c \quad c),$$

R is the 1×1 matrix $(1 - c)$ and $I - R = 1 - (1 - c) = c$, so that

$$Q = (I - R)^{-1} = 1/c.$$

That is, the subject will be guessing—or in the transient state G—an average of $1/c$ times before learning the correct response.

Example 2 In Example 1, $c = 0.25$, so that the subject will be guessing an average of $1/0.25 = 4$ times before learning the correct response.

We now turn to a more complicated (and more interesting) situation. In the model already discussed we did not distinguish between guessing correctly and guessing incorrectly for an item in state G. If we wish to distinguish between these possibilities, we need to analyze the model by means of a Markov chain with three states.

$$E_1: \quad \text{guessing correctly}$$

$$E_2: \quad \text{guessing incorrectly}$$

$$E_3: \quad \text{conditioned}$$

The transition matrix for this Markov chain is

$$
T = \begin{array}{c} \\ E_1 \\ E_2 \\ E_3 \end{array}
\begin{array}{c} E_1 \qquad\qquad E_2 \qquad\qquad E_3 \end{array}
\left(
\begin{array}{ccc}
\dfrac{(1-c)}{N} & (1-c)\left(1 - \dfrac{1}{N}\right) & c \\[2ex]
\dfrac{(1-c)}{N} & (1-c)\left(1 - \dfrac{1}{N}\right) & c \\[2ex]
0 & 0 & 1
\end{array}
\right). \tag{8}
$$

We explain this as follows: p_{11} is the probability that the item will start in E_1 and stay in E_1. To do this, the item must start and remain in a guessing state (with probability $1 - c$) and the response must be guessed correctly (with probability $1/N$, by Assumption 5). Thus,

$$p_{11} = (1-c)\left(\frac{1}{N}\right) = \frac{1-c}{N}.$$

Similarly, $p_{21} = (1-c)/N$. Next, p_{12} is the probability that the item will start and stay in a guessing state (probability $1 - c$) and the subject will guess incorrectly (with probability $1 - 1/N$). Thus $p_{12} = (1-c)(1 - 1/N)$. The same reasoning shows that $p_{22} = (1-c)(1 - 1/N)$. Finally, the other probabilities follow from the simpler model discussed earlier. From the assumptions, it is apparent that the initial probability distribution is

$$\mathbf{p}_0 = \left(\frac{1}{N} \quad 1 - \frac{1}{N} \quad 0\right).$$

In our new Markov chain the transient states are E_1 and E_2 while state E_3 is absorbing. The matrix R is given by

$$
R = \left(
\begin{array}{cc}
\dfrac{1-c}{N} & (1-c)\left(1 - \dfrac{1}{N}\right) \\[2ex]
\dfrac{1-c}{N} & (1-c)\left(1 - \dfrac{1}{N}\right)
\end{array}
\right) \tag{9}
$$

Example 3   We consider the experiment of Example 1 with $N = 5$. Then, as $c = 0.25$, we have (from (8) and (9) and the fact that $(1 - c)/N = (1 - 0.25)/5 = 0.75/5 = 0.15$),

$$T = \begin{pmatrix} 0.15 & 0.6 & 0.25 \\ 0.15 & 0.6 & 0.25 \\ 0 & 0 & 1 \end{pmatrix}, \qquad R = \begin{pmatrix} 0.15 & 0.6 \\ 0.15 & 0.6 \end{pmatrix}, \qquad I - R = \begin{pmatrix} 0.85 & -0.6 \\ -0.15 & 0.4 \end{pmatrix}$$

and

$$Q = (I - R)^{-1} = \frac{1}{0.25} \begin{pmatrix} 0.4 & 0.6 \\ 0.15 & 0.85 \end{pmatrix} = \begin{pmatrix} 1.6 & 2.4 \\ 0.6 & 3.4 \end{pmatrix}.$$

Initially, the subject is in state E_1 (guesses correctly) with probability $\frac{1}{5} = 0.2$ and in state E_2 with probability 0.8. Thus the expected number of times he will be in state E_1 is

$$P(E_1) \times \text{(expected number of times in } E_1 \text{ when starting in } E_1\text{)}$$
$$+ P(E_2) \times \text{(expected number of times in } E_1 \text{ when starting in } E_2\text{)}$$
$$= (0.2)(1.6) + (0.8)(0.6) = 0.8.$$

Similarly, the expected number of times in E_2 is

$$(0.2)(2.4) + (0.8)(3.4) = 3.2.$$

That is, the subject will guess correctly an average of 0.8 times and incorrectly an average of 3.2 times before he has learned the correct response.

PROBLEMS 7.3

1. In the queuing model described in this section, assume that the probability is $\frac{1}{3}$ that one person will join the queue and $\frac{2}{3}$ that no one will. Assume that the probability is $\frac{1}{2}$ that a person being served will be out of the queue in one time period. Keep Assumptions 3–6.

   (a) Find the matrix associated with the Markov chain inherent in the model.

   (b) Show that the Markov chain is regular.

   (c) Find the long-range probabilities that n people will be in line for $n = 0, 1, 2, 3,$ and 4.

2. In a supermarket, the number of people joining the express check-out line (people with fewer than ten items) is random but averages seven people every 10 minutes between noon and 2 P.M. A person being served will have left the line 1 minute later with probability $\frac{3}{4}$. If five people are in the line, the line is closed. Estimate the percentage of time that

   (a) the check-out person will be idle.

   (b) the line will be closed.

   (c) exactly three people will be in the line.

3. Answer the questions in Problem 2 if the line is closed when six people are in the line and all other facts remain the same.

4. A student is taking a reading course in French. She learns vocabulary by putting a French word on one side of a card and its English equivalent on the other side. There are ten cards. We make the following assumptions.

    (i) Initially, the student knows the ten English words but has no idea what French word goes with which English one. She is guessing.

    (ii) The student is shown a French word, gives an English equivalent and, if correct, is shown the other side of the card. The student either knows the word or is guessing.

    (iii) If the student is guessing, she chooses one of the ten English answers at random.

    (iv) Once the student learns a word, she remembers it.

    (v) If she guesses (correctly or incorrectly) and is then shown the correct word, the probability is 0.3 that she will know the English equivalent of the French word she has just seen the next time she sees it.

    With the above assumptions, find the transition matrix of the Markov chain with states K (for knowing a given word) and G (for guessing).

5. In Problem 4, what is the probability that the student will have learned the correct translation of a given French word before guessing four times?

6. In Problem 4, how many times must the experiment be repeated until the probability is at least 98% that the student will have learned the word?

7. In Problem 4, what is the expected number of times the student will guess until she knows the word?

8. In Problem 4, find the transition matrix of the associated Markov chain with three states: guessing correctly, guessing incorrectly, and knowing.

9. In Problem 8, what is the expected number of times the student will guess incorrectly until she learns a given word?

10. In Problem 8, what is the expected number of times the student will guess correctly until she knows the word?

11. In an experiment, a monkey is placed in front of four doors. Each door has a blue, red, green, or brown marker. Behind the green door is a banana. The monkey is allowed to open one of the doors. If it chooses the green door, it is rewarded by getting the banana. If it chooses any other door, it is shown, but not given, the banana. The experiment is repeated by moving around the markers and again putting the banana behind the green door. The object is to determine how long it takes the monkey to figure out that it is not the door, but the color, that determines the location of the reward. Previous experiments with other monkeys have indicated that the monkey has about a 40% chance of learning what is going on after first guessing. Use the assumptions of Problem 4 and find the transition matrix of this experiment.

12. In Problem 11, what is the probability that the monkey will have figured out the rules before its third attempt at the banana?

13. In Problem 11, how many tries should the monkey get in order to be at least 95% certain that it has learned how to play?

14. In Problem 11, how many guesses, on the average, will the monkey make until it knows the rules?

15. In Problem 11, find the transition matrix of the associated Markov chain with three states: guessing correctly, guessing incorrectly, and knowing the rules.

16. In Problem 15, what is the expected number of times the monkey will guess incorrectly until it learns how to play?

17. In Problem 15, what is the expected number of times the monkey will guess correctly?

*18. Show that formula (3) is true.

19. Let $S = 1 + x + x^2 + \cdots + x^n$.

   (a) Compute xS.

   (b) Compute $(1 - x)S$

   (c) Show that $S = (1 - x^{n+1})/(1 - x)$ if $x \neq 1$.

20. In the model in this chapter, we discussed two-state and three-state Markov chains. Let $E(G)$, $E(G_C)$, and $E(G_I)$ denote the expected number of guesses in the two-state Markov chain, the expected number of correct guesses in the three-state Markov chain, and the expected number of incorrect guesses in the three-state Markov chain, respectively. Show that

$$E(G_C) + E(G_I) = E(G).$$

**21. If T is given by (8), show that

$$T^n = \begin{pmatrix} \frac{1}{N}(1-c)^n & \left(1 - \frac{1}{N}\right)(1-c)^n & 1 - (1-c)^n \\ \frac{1}{N}(1-c)^n & \left(1 - \frac{1}{N}\right)(1-c)^n & 1 - (1-c)^n \\ 0 & 0 & 1 \end{pmatrix}.$$

7.4 Two-Person Games: Pure Strategies

The modern theory of games was developed in the 1940s by John von Neumann and Oskar Morgenstern[†] to provide a general mathematical framework for economics. The principal ideas of this theory were abstracted from ordinary games such as chess, bridge, solitaire, dominoes, and checkers. The general theory was developed without direct reference to any particular game. The theory of games can be applied to the analysis of any competitive behavior including ordinary games, economics, warfare, and biological competition. In the study of biological competition, the theory of games provides a useful conceptual framework for understanding behavior.[††]

Many very familiar games have opponents or competitors who make a sequence of moves according to the rules of the game. In some games, the successive moves are made with complete information about the opponent's opportunities (chess). In other games, moves are made with incomplete information (bridge). A player may decide his moves purely by chance (by tossing a coin, for example) or the player may choose his moves deliberately from all possible moves. The game may terminate after a finite number of moves with a winner and a loser. There is generally a return to the winner of the game, which may be a cash payment or merely the satisfaction of winning. (The reward to a species playing the game of ecology is to be allowed to continue to play).

[†]John von Neumann and Oskar Morgenstern, *The Theory of Games and Economic Behavior*, (Princeton, N.J.: Princeton University Press, 1944).
[††]R. C. Lowontin, "Evolution and the Theory of Games," *Journal of Theoretical Biology*, 1 (1961):382–403.
L. B. Slobodkin, "The Strategy of Evolution," *American Scientist*, 52 (1964):342–357.

A game is characterized by its rules. In some cases, the game may be so complicated that discovering its rules may be a considerable achievement. Consider the problem of determining the rules of chess by watching it being played. After four or five games, the principal rules would be apparent, but it would be necessary to observe many more games to determine all the rules. Analogously, the complex interactions of a human social system or of an ecosystem can be thought of as the unfolding of a game with a very large number of players; its rules are not completely understood.

Once the rules of the game are known, the problem is to determine how the players should choose their moves and what the consequences of these moves are. In other words, the players must determine their strategies by analyzing the rules of the game. The final outcome of a game will usually be critically dependent on the choices of moves of all players. For complex games, it may be impossible to analyze all possibilities, and, in this case, the players must rely on experience, intuition, or simple trial and error to determine their moves.

In this and the next section, we study a simple two-person game in considerable detail. The concepts introduced and the results derived for this game form a model for the analysis of more general games. We begin with three examples of two-person games.

Example 1 Matching Pennies. Two players, R and C, each place a penny in front of themselves and, in doing so, determine whether a head (H) or a tail (T) is facing up. Neither player knows, initially, whether the other player has selected a head or a tail. Then the pennies are exposed. If the pennies match (that is, both are heads or both are tails), then player R pays player C $1. If the pennies do not match, then player C pays player R $1.

We can describe this game in terms of a **payoff matrix**.

$$\begin{array}{cc} & \begin{array}{c} \textit{Player C} \\ \hline \begin{array}{cc} H & \quad T \end{array} \end{array} \\ \textit{Player R} \begin{array}{c} H \\ T \end{array} & \begin{pmatrix} -1 & 1 \\ 1 & -1 \end{pmatrix} \end{array}$$

Thus, for example, if player R selects H and player C selects T, then player R gains 1 unit (in this case $1). If players R and C both select H, then player R loses 1 unit.

Example 2 A Business Game. Big Save and Giant Foods are the only two supermarkets in Central City. The retail food market is shared equally by these two companies. Because of increased costs, Big Save wishes to increase its prices. However, if it does so, it fears a loss of business to Giant Foods. On the other hand, if it lowers its prices and Giant Foods raises its prices, then the resulting gain in customers will more than offset the reduced profit per item. Both companies have three choices: raise prices, make no change, and lower prices.

Big Save can control its own prices but has no control over what Giant Foods does. In order to help decide what to do, it hires an independent market research analyst who obtains the data in Table 1. The numbers in the table

TABLE 1   Marketing Choices for Competing Supermarkets

Big Save's choices \ Giant Food's choices	(*I*) Increase prices	(*S*) Keep prices the same	(*D*) Decrease prices
(*I*) Increase prices	2	−2	−7
(*S*) Keep prices the same	6	0	−3
(*D*) Decrease prices	10	5	3

represent percentage increases or decreases. For example, if Big Save keeps prices the same and Giant Foods lowers its prices, then Big Save will *lose* 3% of the total number of customers to Giant Foods. If Big Save lowers its prices and Giant Foods raises its prices, then Big Save will gain 10% of the market.

We can represent these data in the following payoff matrix.

$$\begin{array}{c} \quad\quad\quad\quad\quad Giant\ Foods \\ \quad\quad\quad\quad I \quad\ S \quad\ D \\ Big\ Save\ \begin{array}{c} I \\ S \\ D \end{array} \begin{pmatrix} 2 & -2 & -7 \\ 6 & 0 & -3 \\ 10 & 5 & 3 \end{pmatrix} \end{array}$$

What should each supermarket do? We answer this question later in this section.

Example 3   A War Game.[†]   During World War II a critical battle, called the Battle of the Bismarck Sea, was fought for control of New Guinea. The allied leader, General Kenney, had intelligence reports which indicated that the Japanese would move a troop and supply convoy from the port of Rabaul, at the eastern tip of the island of New Britain, to Lae, which lies just west of New Britain on New Guinea. The Japanese leader had two choices: take a route that passed north of New Britain or one that passed south. On the northern route, poor visibility was almost certain, while on the southern route the weather was likely to be clear. With either choice the trip would take 3 days.

[†]This example is adapted from an example in *Games and Decisions* by R. Duncan Luce and Howard Raiffa, (New York: Wiley, 1958) pp. 64–65.

General Kenney had the choice of concentrating the bulk of his reconnaissance aircraft on one route or the other. Once sighted, the convoy could be bombed until its arrival at Lae. In days of bombing time. Kenney's staff estimated the outcomes for the various choices given in Table 2.

TABLE 2   Choices for Japanese and Allies (Number of Bombing Days)

Allied choices \\ Japanese choices	Northern route	Southern route
Northern route	2	2
Southern route	1	3

Again, these options can be represented by a payoff matrix.

$$\begin{array}{cc} & \begin{array}{cc} \textit{Japanese} \\ \textit{choices} \end{array} \\ \textit{Allied choices} & \begin{array}{c} N \\ S \end{array} \begin{array}{cc} N & S \\ \begin{pmatrix} 2 & 2 \\ 1 & 3 \end{pmatrix} \end{array} \end{array}$$

Which routes should have been chosen? If the Japanese convoy went north, then it would expose itself to either 1 or 2 days of bombing. If it went south, it would face 2 or 3 days of bombing. Going north certainly looks better. From General Kenney's point of view, by concentrating his forces in the north, he could guarantee at least 2 days of bombing; in the south, he could guarantee only 1 day.

It turns out that both commanders chose the northern route. And, as we will soon see, these choices are consistent with those predicted by game theory.

These examples lead to the following definition.

Matrix Game

Let $A = (a_{ij})$ be an $m \times n$ matrix. Consider a game determined by A played between two competitors R and C (rows and columns) according to the following rules.

1.   On each move of the game, R chooses one of the m rows of A and C chooses one of the n columns of A. These choices are made simultaneously, and neither competitor knows in advance the choice (or move) of the other competitor.

2. If R chooses the *i*th row and C chooses the *j*th column, then C pays R an amount a_{ij}. If a_{ij} is negative, this is interpreted to mean that C receives an amount $-a_{ij}$ from R.

This game is the $m \times n$ **matrix game** determined by the $m \times n$ matrix $A = (a_{ij})$.

The matrix game may terminate after one move or it may continue for any number of moves. The matrix $A = (a_{ij})$ of the game is called the **game matrix**, or the **payoff matrix**.

The following examples illustrate how matrix games may be analyzed.

Example 4 Describe the matrix games with each payoff matrix.

(a) $\quad A = \begin{pmatrix} 1 & 2 \\ -2 & 3 \end{pmatrix}$ (b) $\quad B = \begin{pmatrix} 1 & 0 & 1 & 0 \\ 0 & -1 & 2 & 0 \\ -1 & 0 & 3 & 1 \end{pmatrix}$

Solution

(a) In this 2×2 matrix game, R and C each have two choices. If R chooses the first row, R gains 1 unit if C chooses the first column and 2 units if C chooses the second column. If R chooses the second row, he loses 2 units if C chooses the first column and gains 3 units if C chooses the second column. If C is playing rationally, C will choose the first column. In this case, R should choose the first row. With these choices, R guarantees that he will win at least 1 unit and C guarantees that he will lose no more than 1 unit.

(b) In this 3×4 matrix game, R has three choices and C has four choices. By analyzing all possible choices, it is clear that C will do best by choosing the second column. By this choice, C guarantees that no loss will be suffered; R will do best by choosing the first row. If these choices are made, there is no payment between the players.

The general $m \times n$ matrix game is an example of a **two-person, zero-sum game**, since there are two competitors and the sum of their winnings is zero. The gains of one competitor are the losses of the other.

The two players of the $m \times n$ matrix game $A = (a_{ij})$ must analyze their possible moves and decide which rows or columns to play on successive moves. A **pure strategy** for R (or C) is the decision to play the same row (or column) on every move of the game. Player R (or C) is said to be using a **mixed strategy** if more than one row (or column) is chosen on different moves of the game. If both players employ pure strategies, the outcome of each move is exactly the same and the game is completely predictable. For example, if R always chooses the *i*th row and C always chooses the *j*th column, then on every play of the game R receives a_{ij} units from C. When mixed strategies are used by one or both players, the game is more complicated. For example, if R decides to play a mixed strategy, he will randomize the choice of rows in order to increase his return.

When will a pure strategy be used? When will a mixed strategy be used? To answer these questions, we must be more precise about what we mean by a strategy.

Strategy

A **strategy** for R in the $m \times n$ matrix game $A = (a_{ij})$ is an m-component probability vector $\mathbf{p} = (p_1 \quad p_2 \quad \cdots \quad p_m)$, where p_i is the probability that R plays the ith row for $i = 1, 2, \ldots, m$. A strategy for C is an n-component probability vector

$$\mathbf{q} = \begin{pmatrix} q_1 \\ q_2 \\ \vdots \\ q_n \end{pmatrix}$$

where q_j is the probability that C plays the jth column for $j = 1, 2, \ldots n$.

The players R and C must choose their strategies $\mathbf{p}$ and $\mathbf{q}$. In other words, they must choose the probabilities p_i and q_j that will determine how often they play the various rows and columns. For example, if R and C play the first row and first column of A on every move, they are playing the pure strategies $\mathbf{p} = (1 \quad 0 \quad \cdots \quad 0)$ and

$$\mathbf{q} = \begin{pmatrix} 1 \\ 0 \\ \vdots \\ 0 \end{pmatrix}$$

If R and C play all rows and columns with equal probabilities, they are playing the mixed strategies $\mathbf{p} = (1/m \quad 1/m \quad \cdots \quad 1/m)$ and

$$\mathbf{q} = \begin{pmatrix} \dfrac{1}{n} \\ \dfrac{1}{n} \\ \vdots \\ \dfrac{1}{n} \end{pmatrix}.$$

Every m-component probability vector is a possible strategy for R, and every n-component probability vector is a possible strategy for C.

To see when a pure strategy might be used, consider the following example.

Example 5   Consider the matrix game whose payoff matrix is

$$A = \begin{pmatrix} 6 & 2 & 3 \\ 6 & 5 & 4 \\ 7 & -1 & 2 \\ 2 & 6 & 1 \end{pmatrix}$$

What strategies should R and C adopt?

Solution   R will play to make the smallest amount he will win as large as possible (read that last sentence again). If R plays row 1, R will win at least 2 units, no matter which column C plays. If R plays row 2, R will win at least 4 units. Similarly, if R plays row 3 or row 4, R will gain at least -1 or 1 unit, respectively. Thus, his maximum, minimum return is 4 units.

But what should C play? C wants to minimize his maximum loss. If C plays column 1, he can lose as many as 7 units; in column 2 C can lose at most 6 units and in column 3 C can lose at most 4 units. We write these numbers below:

$$
\begin{array}{ccc}
 & \text{C} & \text{Row minimum} \\
\text{R}\begin{pmatrix} 6 & 2 & 3 \\ 6 & 5 & 4 \\ 7 & -1 & 2 \\ 2 & 6 & 1 \end{pmatrix} & & \begin{array}{c} 2 \\ 4 \\ -1 \\ 1 \end{array}
\end{array}
$$

Column maximum $\longrightarrow$ 7   6   4

The number 4 in the 2,3 position is a minimum in its row and a maximum in its column. Such a number is called a **saddle point** for the payoff matrix A. In Section 7.5 we show that when the number a_{ij} is a saddle point, the optimum strategies for R and C are for R to play the ith row and C to play the jth column. Thus, in the example, R should adopt the pure strategy of playing the second row ($\mathbf{p} = (0 \quad 1 \quad 0 \quad 0)$) and C should adopt the pure strategy of playing the third column

$$
\mathbf{q} = \begin{pmatrix} 0 \\ 0 \\ 1 \end{pmatrix}.
$$

Before leaving this example, we make an observation that can simplify our computations. Every number in the first row of A is less than or equal to the corresponding component in the second row of A. That is, $6 \leq 6$, $2 < 5$, and $3 < 4$. So R will *never* choose the first row because R can always do better by choosing the second row. The first row is called a **recessive row**. Similarly, every number in the first column of A is greater than the corresponding number in the third column of A; that is, $6 > 3$, $6 > 4$, $7 > 2$, and $2 > 1$. Hence C will never choose the first column, for then C would certainly lose more than if he had chosen the third column (remember, R's gain is C's loss). The first column is called a **recessive column**.

We can eliminate the recessive rows and columns from further consideration. If we do this, we obtain the new payoff matrix A'.

$$
A' = \begin{pmatrix} 5 & 4 \\ -1 & 2 \\ 6 & 1 \end{pmatrix}
$$

As before, 4 is a minimum in its row and a maximum in its column and is a saddle point for A'. The second row is recessive, so we can reduce the matrix further to obtain $A'' = \begin{pmatrix} 5 & 4 \\ 6 & 1 \end{pmatrix}$; then $A''' = \begin{pmatrix} 4 \\ 1 \end{pmatrix}$ since the first column of A'' is recessive. Continuing, we see that the second row of A''' is recessive, so $A^{iv} = (4)$. Obviously, this is as far as we can go.

We now outline a general strategy for playing a matrix game in those cases where there is a saddle point.

DETERMINING PURE STRATEGIES FOR A MATRIX GAME

Step 1.   Eliminate all recessive rows and columns.

Step 2.   Find the smallest number in each row. This is called the **row minimum**.

Step 3.   Find the largest number in each column. This is called the **column maximum**.

Step 4.   Look for a **saddle point**; this is a number that is both a row minimum and a column maximum.[†] If a_{ij} is a saddle point, then R should play row i and C should play column j. In this case the matrix game is said to be **strictly determined**.

Step 5.   If there is no saddle point, then either R or C (or both) should play a mixed strategy. The game is not strictly determined.

Remark.   In Problem 23 you are asked to show that if a_{ij} and a_{kl} are saddle points for A, then $a_{ij} = a_{kl}$. In this case the same result will hold in Step 4 if either saddle point is used.

Example 6   Determine whether the game whose payoff matrix is given is strictly determined. If it is, determine optimal strategies for R and C.

(a) $\begin{pmatrix} 3 & 0 & 5 \\ 2 & 1 & 3 \\ 2 & -1 & -2 \end{pmatrix}$ (b) $\begin{pmatrix} 1 & 6 & 2 \\ 4 & 2 & 3 \\ 5 & 1 & 6 \end{pmatrix}$ (c) $\begin{pmatrix} 1 & 0 & 5 \\ 1 & -5 & 0 \end{pmatrix}$

[†]A saddle point is sometimes called the **minimax** solution to the matrix game.

Solution

(a) Since each number in row 3 is less than or equal to the corresponding number in row 2, row 3 is recessive. Similarly, each number in column 1 is greater than the corresponding number in column 2, so column 1 is recessive. Eliminating row 3 and column 1, we obtain

$$\begin{pmatrix} 0 & 5 \\ \textcircled{1} & 3 \end{pmatrix}. \qquad \begin{array}{c} \text{Row minimum} \\ \downarrow \\ 0 \\ 1 \end{array}$$

Column maximum $\longrightarrow$ 1   5

We see that 1 is the minimum in its row and maximum in its column, so 1 is a saddle point and the game is strictly determined. Because 1 is the 2,2 component of the payoff matrix, the optimal strategies are $\mathbf{p} = (0 \quad 1 \quad 0)$ and

$$\mathbf{q} = \begin{pmatrix} 0 \\ 1 \\ 0 \end{pmatrix}.$$

That is, **R** plays row 2 and **C** plays column 2. [†]

(b) There are no recessive rows or columns. Rewriting the payoff matrix, we have

$$\begin{pmatrix} 1 & 6 & 2 \\ 4 & 2 & 3 \\ 5 & 1 & 6 \end{pmatrix}. \qquad \begin{array}{c} \text{Row minimum} \\ \downarrow \\ 1 \\ 2 \\ 1 \end{array}$$

Column maximum $\longrightarrow$ 5   6   6

Here there is no saddle point because no number is both the minimum in its row and the maximum in its column. The game is *not* strictly determined and a mixed strategy is called for.

(c) Row 2 is recessive, as is column 1. The game matrix reduces to (0   5) and 0 is a saddle point. Note that there are two zeros in the payoff matrix but only the zero in the 1,2 position is a saddle point. The game is strictly determined and the optimal strategies are $\mathbf{p} = (1 \quad 0)$ and

$$\mathbf{q} = \begin{pmatrix} 0 \\ 1 \\ 0 \end{pmatrix}.$$

That is, **R** plays row 1 and **C** plays column 2.

[†] We note further that in $\begin{pmatrix} 0 & 5 \\ 1 & 3 \end{pmatrix}$ the second column is recessive so the matrix reduces to $\begin{pmatrix} 0 \\ 1 \end{pmatrix}$. But now the first row is recessive and we end up with the 1 × 1 matrix (1). This is our saddle point.

Example 7   In the matching pennies game of Example 1, the payoff matrix is

$$
\begin{array}{cc}
 & \begin{array}{cc} H & T \end{array} \quad \text{Row minimum} \\
\begin{array}{c} H \\ T \end{array} &
\begin{pmatrix} -1 & 1 \\ 1 & -1 \end{pmatrix}.
\begin{array}{c} -1 \\ -1 \end{array}
\end{array}
$$

Column maximum $\longrightarrow$ 1 1

There are no saddle points, so a mixed strategy is called for. We will find the optimal strategy in the next section.

Example 8   In the business game of Example 2, the payoff matrix is

$$
\begin{array}{cc}
 & \begin{array}{ccc} I & S & D \end{array} \quad \text{Row minimum} \\
\begin{array}{c} I \\ S \\ D \end{array} &
\begin{pmatrix} 2 & -2 & -7 \\ 6 & 0 & -3 \\ 10 & 5 & 3 \end{pmatrix}.
\begin{array}{c} -7 \\ -3 \\ 3 \end{array}
\end{array}
$$

Column maximum $\longrightarrow$ 10 5 3

Here 3 is a saddle point, so **Big Save** should play row 3 and **Giant Foods** should play column 3. That is, if the data are correct, they should both decrease prices. Note that rows 1 and 2 and columns 1 and 2 are recessive, so that the game matrix reduces to the 1×1 payoff matrix (3).

Example 9   In the war game of Example 3, the payoff matrix is

$$
\begin{array}{cc}
 & \begin{array}{cc} \text{North} & \text{South} \end{array} \\
\begin{array}{c} \text{North} \\ \text{South} \end{array} &
\begin{pmatrix} 2 & 2 \\ 1 & 3 \end{pmatrix}.
\end{array}
$$

The 2 in the 1,1 position is a saddle point, so both commanders should choose the northern route. This is indeed what happened.

In the next section we use a theorem of von Neumann to show why pure strategies should always be used when there is a saddle point. In that section we also show how to find optimal strategies when a matrix game is not strictly determined.

PROBLEMS 7.4   In Problems 1–10, the payoff matrix for a game is given. Determine whether the game is strictly determined and, if it is, find the optimal strategies for R and C.

1.
$$
\begin{pmatrix} 1 & 1 & 1 \\ 2 & 2 & 3 \\ -2 & 4 & 5 \end{pmatrix}
$$

2.
$$
\begin{pmatrix} 0 & 0 & 0 \\ 1 & 1 & 2 \\ -3 & 3 & 4 \end{pmatrix}
$$

3.
$$
\begin{pmatrix} 4 & 6 & -2 \\ 3 & 5 & 7 \\ 2 & -3 & 1 \end{pmatrix}
$$

4.
$$
\begin{pmatrix} 5 & 6 & 4 & 7 \\ 3 & 2 & 3 & 1 \\ 4 & 6 & 2 & 5 \end{pmatrix}
$$

5. $\begin{pmatrix} 1 & 6 \\ 2 & 4 \\ 3 & 5 \end{pmatrix}$

6. $\begin{pmatrix} 2 & -3 & -2 & 0 \\ -4 & -6 & -1 & 2 \\ 0 & -3 & 2 & 5 \end{pmatrix}$

7. $\begin{pmatrix} 2 & -3 & -2 & 0 \\ -4 & -6 & -1 & 2 \\ 0 & -2 & 2 & -5 \end{pmatrix}$

8. $\begin{pmatrix} 1 & 6 & 2 & 4 \\ -1 & 3 & 7 & 5 \\ 2 & 1 & 7 & 8 \\ 3 & 0 & -5 & 4 \end{pmatrix}$

9. $\begin{pmatrix} 4 & 2 & 1 & 3 \\ 6 & 4 & 3 & 5 \\ 2 & 1 & 3 & 6 \\ 1 & 4 & 2 & 0 \end{pmatrix}$

10. $\begin{pmatrix} 1 & 0 & 0 & 1 \\ 0 & 1 & 0 & 1 \\ 1 & 1 & 0 & 0 \\ 1 & 1 & 0 & 1 \end{pmatrix}$

In Problems 11–20, formulate each situation as a two-person matrix game and write the payoff matrix for that game. If the game is strictly determined, find the optimal strategy for each player.

11. Two people each put out, simultaneously, one or two fingers. If the total number of fingers is even, R pays C that number of dollars. If it is odd, C pays R that number of dollars.

12. Repeat Problem 11, except that each player puts out four or five fingers.

13. Repeat Problem 11, except that each player puts out one, two, or three fingers.

14. Repeat Problem 11, except that each player puts out one, two, three, four, or five fingers.

15. Two gas stations compete for business in a small town. Station A has determined that if it increases its prices, it will lose 1% of the market if B increases its prices, 3% of the market if B makes no change, and 11% of the market if B lowers its prices. If A makes no change, then it gains 4% if B increases and loses 5% if B decreases. There is no change in market share if both stations make no change in prices. Finally, if A decreases its prices, it gains 9% if B raises its prices, gains 3% if B holds the line, and loses 1% if B also decreases its prices.

16. A new shopping mall is being constructed in Centerville. Two western-clothing stores, Vince's and Readywear, share the market in Centerville and surrounding areas. If one of the stores moves into the new mall, it will win 80% of the market. If both or neither go in, they will continue to divide the market equally.

17. A Congressional district is divided into two regions. A Democrat and a Republican are running for office. The regions have 60,000 voters and 40,000 voters, respectively. There are 2 days remaining in the campaign and each candidate can spend 0, 1, or 2 days in each region. Political analysts have estimated that if the candidates spend the same number of days in a region, they will split the votes in that region. However, if one candidate spends 1 or 2 more days in a region than his opponent, he will get 53% or 57% of the votes in that region, respectively.

18. In experiments, ravens and parakeets have learned to recognize numbers up to 7. The following experiment is proposed: The diet of a raven R and a parakeet C is to be determined by a matrix game. Each bird is shown three cards, labeled with 2, 4, and 7 dots. If each bird chooses the same card, then R receives from the diet of C a number of worms equal to twice the number of dots on the card. If the cards chosen are different, then C receives from the diet of R a number of worms equal to the difference of the number of dots on the cards. Assume that the moves are made independently.

19. Peter wants to call his girlfriend, Roberta. He is planning to call at night when the price of a 3-minute, station-to-station call is $2 and a 3-minute, person-to-person call is $4.50. If no one is home, he knows he can reach her tomorrow at her place of work and he will pay the day rate of $3 for a 3-minute, station-to-station call. If he calls station-to-station and Roberta is home, he saves money. On the other hand, if Roberta's roommate answers and Roberta is not home, he loses money.

20. A farmer grows tomatoes. The longer the tomatoes stay on the vine, the redder they get and the higher the price he can charge per bushel. On the other hand, if frost sets in, some of the tomatoes will be ruined and his average price per bushel will decrease. If he picks his tomatoes on August 25, he can be sure of no frost and will get an average of $8 per bushel. If he waits until September 5, he will get $11 per bushel if there is no frost and $5 per bushel if there is.

21. The Maxigrip Tire Company has a dispute with its union. Each group (management and labor) has a choice among four bargaining positions: tough, hold firm; a reasonable approach based on logic; leave it to the lawyers to seek a legal solution to the dispute; be conciliatory. A labor-relations expert has determined that the company will pay higher weekly salaries according to the position both it and its union adopt. The various possibilities are given in the table.

Weekly Increase to Tire Company per Worker (in Dollars)

Company position \ Union position	Tough	Logical	Legal	Conciliatory
Tough	20	24	15	17
Logical	35	30	26	28
Legal	15	23	20	30
Conciliatory	40	35	23	28

What should the company and the union do?

22. The University of Montana (UM) plays Montana State University (MSU) every year in football. The UM quarterback on one down has a choice of five plays: halfback run, fullback run, short pass, long pass, and draw play. The MSU defense has four choices: normal defense, prevent defense against a long pass, prevent defense against a short

pass, blitz. One coach estimated the approximate number of yards to be gained with all combinations of offensive and defensive options. These estimates are given in the table.

Expected Yards Gained Defense

UM \ MSU	Normal	Prevent (short)	Prevent (long)	Blitz
HB run	2	4	8	6
FB run	5	5	7	9
Short pass	4	0	6	−2
Long pass	8	3	0	−4
Draw play	−1	2	4	10

What should each team do?

23. Show that if a payoff matrix has two saddle points a_{ij} and a_{kl}, then $a_{ij} = a_{kl}$. [*Hint*: Write the matrix A and explain what is implied by the fact that a_{ij} is the minimum in the ith row and maximum in the jth column. Do the same for a_{kl}.]

7.5 Two-Person Games: Mixed Strategies

In Section 7.4 we showed how to determine optimal strategies when a game is strictly determined. In order to show that a strategy is optimal, however, it is necessary to answer the question "optimal with respect to what?" In this section we show how to compute the expected return to players in a game. Then a strategy is optimal for R if it optimizes R's return. We begin with an example.

Example 1 Consider the game whose payoff matrix is $A = \begin{pmatrix} 3 & 2 \\ -2 & 4 \end{pmatrix}$. What is the expected return to R if R adopts the strategy $\mathbf{p} = (\frac{1}{3} \quad \frac{2}{3})$ and C adopts the strategy $\mathbf{q} = \begin{pmatrix} \frac{2}{5} \\ \frac{3}{5} \end{pmatrix}$?

Solution Here R has four possible returns: 3, 2, −2, and 4. It will receive 3 units, for example, if R chooses the first row (with probability $\frac{1}{3}$) and C chooses the first column (with probability $\frac{2}{5}$). Since it is assumed that neither R nor C knows what the other is doing, the events {1st row} and {1st column} are independent. Therefore

$$P(\text{1st row} \cap \text{1st column}) = P(\text{1st row}) \cdot P(\text{1st column}) = \tfrac{1}{3} \cdot \tfrac{2}{5} = \tfrac{2}{15}$$

and we have

$$P(3) = P(\text{return of 3}) = \tfrac{2}{15}$$

Similarly

$$P(2) = P(\text{1st row} \cap \text{2nd column}) = \tfrac{1}{3} \cdot \tfrac{3}{5} = \tfrac{1}{5},$$

$$P(-2) = P(\text{2nd row} \cap \text{1st column}) = \tfrac{2}{3} \cdot \tfrac{2}{5} = \tfrac{4}{15},$$

and

$$P(4) = P(\text{2nd row} \cap \text{2nd column}) = \tfrac{2}{3} \cdot \tfrac{3}{5} = \tfrac{2}{5}.$$

Now, if $E(\mathbf{p}, \mathbf{q})$ denotes the expected value of the random variable which takes values equal to the possible returns to R, we have[†]

$$E(\mathbf{p}, \mathbf{q}) = 3P(3) + 2P(2) + (-2)P(-2) + 4P(4)$$
$$= 3 \cdot \tfrac{2}{15} + 2 \cdot \tfrac{1}{5} - 2 \cdot \tfrac{4}{15} + 4 \cdot \tfrac{2}{5} = \tfrac{28}{15} \approx 1.87$$

Now, we observe that

$$\mathbf{p}A\mathbf{q} = (\tfrac{1}{3} \quad \tfrac{2}{3}) \begin{pmatrix} 3 & 2 \\ -2 & 4 \end{pmatrix} \begin{pmatrix} \tfrac{2}{5} \\ \tfrac{3}{5} \end{pmatrix}$$

$$= (\tfrac{1}{3} \quad \tfrac{2}{3}) \begin{pmatrix} \tfrac{12}{5} \\ \tfrac{8}{5} \end{pmatrix} = \tfrac{12}{15} + \tfrac{16}{15} = \tfrac{28}{15} \approx 1.87$$

Example 1 can be generalized.

Expected Return   Suppose that R uses strategy $\mathbf{p}$ and that C uses strategy $\mathbf{q}$ for the game whose payoff matrix is the $m \times n$ matrix A. Then the **expected return** to R, denoted by $E(\mathbf{p}, \mathbf{q})$, is given by

$$\boxed{E(\mathbf{p}, \mathbf{q}) = \text{expected return} = \mathbf{p}A\mathbf{q}.} \tag{1}$$

Note.   Since A is an $m \times n$ matrix and $\mathbf{q}$ is an n-vector (an $n \times 1$ matrix), the product $A\mathbf{q}$ is an $m \times 1$ matrix and $\mathbf{p}A\mathbf{q}$ is then a $(1 \times m) \times (m \times 1) = $ a 1×1 matrix or, simply, a real number.

[†] We assume in this computation that you are familiar with the material in Section 6.2.

Example 2   What is the expected return to R in the 3 × 4 matrix game

$$A = \begin{pmatrix} 2 & -1 & 3 & 0 \\ 3 & -2 & -1 & 2 \\ 1 & 4 & 3 & -3 \end{pmatrix}$$

(a)   if R adopts the strategy $(\frac{1}{3} \quad \frac{1}{2} \quad \frac{1}{6})$ and C adopts the strategy $\begin{pmatrix} \frac{1}{4} \\ \frac{1}{8} \\ \frac{1}{8} \\ \frac{1}{2} \end{pmatrix}$?

(b)   if R and C choose the 2nd row and 3rd column, respectively?

Solution

(a) $E(\mathbf{p}, \mathbf{q}) = (\frac{1}{3} \quad \frac{1}{2} \quad \frac{1}{6}) \begin{pmatrix} 2 & -1 & 3 & 0 \\ 3 & -2 & -1 & 2 \\ 1 & 4 & 3 & -3 \end{pmatrix} \begin{pmatrix} \frac{1}{4} \\ \frac{1}{8} \\ \frac{1}{8} \\ \frac{1}{2} \end{pmatrix}$

$$= (\frac{1}{3} \quad \frac{1}{2} \quad \frac{1}{6}) \begin{pmatrix} \frac{3}{4} \\ \frac{11}{8} \\ -\frac{3}{8} \end{pmatrix} = \frac{7}{8} = 0.875$$

(b)   Here $\mathbf{p} = (0 \quad 1 \quad 0)$ and $\mathbf{q} = \begin{pmatrix} 0 \\ 0 \\ 1 \\ 0 \end{pmatrix}$. Then

$$E(\mathbf{p}, \mathbf{q}) = (0 \quad 1 \quad 0) \begin{pmatrix} 2 & -1 & 3 & 0 \\ 3 & -2 & -1 & 2 \\ 1 & 4 & 3 & -3 \end{pmatrix} \begin{pmatrix} 0 \\ 0 \\ 1 \\ 0 \end{pmatrix} = (0 \quad 1 \quad 0) \begin{pmatrix} 3 \\ -1 \\ 3 \end{pmatrix} = -1.$$

This is, of course, the 2,3 component of *A*.

The basic question remains: How do R and C choose strategies? This question is partially answered by a fundamental result discovered by von Neumann.

VON NEUMANN'S THEOREM

For any *m* × *n* matrix game *A*, there exist strategies $\mathbf{p}_0$ and $\mathbf{q}_0$ and a number *v* such that $E(\mathbf{p}_0, \mathbf{q}) \geq v$ for any strategy $\mathbf{q}$ and $E(\mathbf{p}, \mathbf{q}_0) \leq v$ for any strategy $\mathbf{p}$. The strategies $\mathbf{p}_0$ and $\mathbf{q}_0$ are called **optimal strategies** for R and C, respectively, and the number *v* is called the **value** of the game.

Why are $\mathbf{p}_0$ and $\mathbf{q}_0$ optimal? Because if R chooses $\mathbf{p}_0$, then he knows he will win at least v units. If he chooses any other strategy, then C can choose $\mathbf{q}_0$ and ensure that R will win *at most* v units. Thus, assuming that C plays wisely, R can do best by choosing the strategy $\mathbf{p}_0$. Similar reasoning shows that the optimal strategy for C is $\mathbf{q}_0$.

Using von Neumann's theorem, it is possible to show that

> if a_{ij} is a saddle point, then the value of the game is a_{ij} and optimal strategies are the pure strategies of playing row i and column j with probability 1.

This justifies what we did in the last section.

Von Neumann's theorem is limited in that it tells us what optimal strategies and the value of a game are, but it does not tell us how to compute them. It turns out that it is fairly easy to compute these when the game matrix is a 2×2 matrix. Other cases are more difficult and will not be considered here.

Let

$$A = \begin{pmatrix} a_{11} & a_{12} \\ a_{21} & a_{22} \end{pmatrix}. \tag{2}$$

Optimal Strategies for a 2 × 2 Matrix Game   If the matrix in (2) is not strictly determined, then the **optimal strategies** for R and C are

$$\mathbf{p}_0 = \left(\frac{a_{22} - a_{21}}{a_{11} + a_{22} - a_{12} - a_{21}} \quad \frac{a_{11} - a_{12}}{a_{11} + a_{22} - a_{12} - a_{21}} \right) \tag{3}$$

and

$$\mathbf{q}_0 = \begin{pmatrix} \dfrac{a_{22} - a_{12}}{a_{11} + a_{22} - a_{12} - a_{21}} \\[2ex] \dfrac{a_{11} - a_{21}}{a_{11} + a_{22} - a_{12} - a_{21}} \end{pmatrix}. \tag{4}$$

The value of the game is

$$v = \frac{a_{11}a_{22} - a_{12}a_{21}}{a_{11} + a_{22} - a_{12} - a_{21}}. \tag{5}$$

In Problem 44 you are asked to show why equations (3), (4), and (5) are valid.

Example 3   Determine optimal strategies and the value of the matching pennies game of Example 7.4.1 on page 406.

Solution The game matrix is

$$
\begin{array}{c c}
 & \begin{array}{c c} H & T \end{array} \\
\begin{array}{c} H \\ T \end{array} & \begin{pmatrix} -1 & 1 \\ 1 & -1 \end{pmatrix}.
\end{array}
$$

Then $a_{11} = -1, a_{12} = 1, a_{21} = 1, a_{22} = -1$ and, from (3), (4), and (5),

$$
\mathbf{p}_0 = \left(\frac{-1 - 1}{-1 - 1 - 1 - 1} \quad \frac{-1 - 1}{-1 - 1 - 1 - 1} \right) = \left(\frac{1}{2} \quad \frac{1}{2} \right),
$$

$$
\mathbf{q}_0 = \begin{pmatrix} \dfrac{-1 - 1}{-1 - 1 - 1 - 1} \\[2ex] \dfrac{-1 - 1}{-1 - 1 - 1 - 1} \end{pmatrix} = \begin{pmatrix} \dfrac{1}{2} \\[2ex] \dfrac{1}{2} \end{pmatrix}
$$

and

$$
v = \frac{(-1)(-1) - (1)(1)}{-1 - 1 - 1 - 1} = 0.
$$

Thus, in this very simple game, R and C do best by choosing a head or a tail with equal probability. The value of this game is zero.

Fair Game

A game is **fair** if its value is zero.

The matching pennies game is an example of a fair game.

Example 4 In the war game of Example 7.4.3 on page 407, let us modify our assumptions to obtain the given table.

TABLE 1 Choice for Japanese and Allies (Number of Bombing Days)

Allied choices \ Japanese choices	Northern route	Southern route
Northern route	2	1
Southern route	1	3

What are the optimal strategies now?

Solution   The game matrix is $\begin{pmatrix} 2 & 1 \\ 1 & 3 \end{pmatrix}$ and there is no saddle point, so the game is not strictly determined. We have $a_{11} = 2$, $a_{12} = 1 = a_{21}$, and $a_{22} = 3$. Thus

$$\mathbf{p}_0 = \left(\frac{3-1}{2+3-1-1} \quad \frac{2-1}{2+3-1-1} \right) = \left(\frac{2}{3} \quad \frac{1}{3} \right),$$

$$\mathbf{q}_0 = \begin{pmatrix} \dfrac{3-1}{2+3-1-1} \\ \dfrac{2-1}{2+3-1-1} \end{pmatrix} = \begin{pmatrix} \dfrac{2}{3} \\ \dfrac{1}{3} \end{pmatrix}$$

and

$$v = \left(\frac{2 \cdot 3 - 1 \cdot 1}{2 + 3 - 1 - 1} \right) = \frac{5}{3}.$$

This means that if this game were repeated many times, both commanders should choose the northern route $\frac{2}{3}$ of the time and the southern route $\frac{1}{3}$ of the time. There will be an average of $\frac{5}{3}$ days of bombing. Of course, this procedure will not be done more than once. Therefore the correct interpretation of the vector $\mathbf{p}_0 = (\frac{2}{3} \quad \frac{1}{3})$ is that General Kenney should go north with $\frac{2}{3}$ probability. He could do this, for example, but putting two N's and one S in a box and selecting one letter at random. His expected number of bombing days would then be $\frac{5}{3}$ although, of course, he could only have 1, 2, or 3 bombing days.

Example 5   Assessing Risk in Medical Procedures.   As is well known, many medical procedures involve substantial risk to the patient and should only be undertaken when the patient is exposed to greater risk if no treatment is given. How do we decide in a given situation which risk is greater? This problem is complicated further when it is not completely certain that the patient does have the disease suspected. For example, surgery is often undertaken to remove tumors, even when there is only a relatively small probability that the tumor will be found to be malignant. How large must this probability be before surgery can be recommended?

To analyze this question, suppose that the probability that a patient has a particular disease is q_1. (This probability has been determined by performing various tests.) The treatment for this disease is a serious operation. If the patient has the disease but does not have the operation, he can expect to live 5 years, but if the patient has the operation, he can expect to live 20 years. If the patient does not have the disease, he can expect to live 25 years with the operation and 30 years without the operation. The decision to have the operation or not clearly depends on q_1, the probability that the patient has the disease. If $q_1 = 0$, the patient does not have the disease and should not have the operation. If $q_1 = 1$, the patient has the disease and should have the operation. What is the smallest value of q_1 for which the operation is advisable?

Solution   This problem can be analyzed as a matrix game. Define

$$A = \begin{pmatrix} 20 & 25 \\ 5 & 30 \end{pmatrix}$$

to be the game matrix. The patient "plays" the rows; row I corresponds to having the operation and row II corresponds to not having the operation. The opponent, nature, plays the columns; column I corresponds to the disease and column II corresponds to no disease. The strategy of nature is

$$\mathbf{q} = \begin{pmatrix} q_1 \\ 1 - q_1 \end{pmatrix},$$

where q_1 is the probability that the patient has the disease. The patient must play a pure strategy, but for the moment let us suppose that the patient's strategy is $\mathbf{p} = (p_1 \quad p_2)$. The expected return (in years of life) is

$$E(\mathbf{p}, \mathbf{q}) = (p_1 \quad 1 - p_1) \begin{pmatrix} 20 & 25 \\ 5 & 30 \end{pmatrix} \begin{pmatrix} q_1 \\ 1 - q_1 \end{pmatrix}$$

$$= (p_1 \quad 1 - p_1) \begin{pmatrix} 20q_1 + 25(1 - q_1) \\ 5q_1 + 30(1 - q_1) \end{pmatrix}$$

$$= 20p_1 q_1 - 5p_1 - 25q_1 + 30.$$

If the patient has the operation, then $\mathbf{p} = (1 \quad 0)$ and $E(\mathbf{p}, \mathbf{q}) = 25 - 5q_1$. If he does not have the operation, then $\mathbf{p} = (0 \quad 1)$ and $E(\mathbf{p}, \mathbf{q}) = 30 - 25q_1$. The patient should have the operation if $25 - 5q_1 > 30 - 25q_1$—that is, if $20q_1 > 5$, or $q_1 > 0.25$. This means that the patient should have the operation if the probability that he has the disease is greater than 25%. If the information available indicates that the probability of the disease is, for instance, 15%, the operation should not be performed. More information must be obtained before the operation can be recommended.

Example 6   Consider the game whose payoff matrix is

$$A = \begin{pmatrix} 4 & 5 & 7 \\ 1 & 2 & 4 \\ 5 & 8 & 3 \end{pmatrix}.$$

This is not a 2×2 game and it has no saddle point, so it is not strictly determined. However, row 2 is recessive. R can always do better by choosing row 1 and so will never choose row 2. Similarly, column 2 is recessive and will never be chosen by C. Eliminating the recessive row and column, we obtain

$$A' = \begin{pmatrix} 4 & 7 \\ 5 & 3 \end{pmatrix}$$

The optimal strategies and value of this game are, from (3), (4), and (5),

$$\mathbf{p}_0' = (\tfrac{2}{5} \quad \tfrac{3}{5}), \qquad \mathbf{q}_0' = \begin{pmatrix} \tfrac{4}{5} \\ \tfrac{1}{5} \end{pmatrix} \quad \text{and} \quad v = \tfrac{23}{5}.$$

Thus

$$\mathbf{p}_0 = (\tfrac{2}{5} \quad 0 \quad \tfrac{3}{5}) \quad \text{and} \quad \mathbf{q}_0 = \begin{pmatrix} \tfrac{4}{5} \\ 0 \\ \tfrac{1}{5} \end{pmatrix}$$

are the optimal strategies for A. The value of the game is still $\tfrac{23}{5}$.

We close this section with a wonderful example of the use of game theory applied to philosophy.[†]

Example 7   Free Will.   Do we have a role in directing our own lives? Or are we destined to play out a preordained plan, moving along a conveyor belt of fate? One of Martin Gardner's columns that generated a considerable response from readers was his discussion of free will and determinism, illustrated by what is known as **Newcomb's paradox**.

The paradox involves the ability of a "Superior Being" to predict behavior accurately. This Being may be a psychic, a time traveler, God, or any other prescient entity.

Before you are two boxes. Box A contains $1000. Box B contains either $1 million or nothing. You may take what is in both boxes, or you may take only what is in Box B. Sometime before you make your decision, the Superior Being makes a prediction about your choice. If the Being believes that you will take only Box B, then he will put $1 million in it. If he predicts you will take both boxes, he will leave B empty. The Being is not necessarily infallible, but he is an extremely accurate predictor.

What is your choice? If you take both boxes, the Being will have predicted this and left Box B empty. If you take only Box B, you will get $1 million. The Being has never in your experience been wrong. Taking Box B seems to be the correct choice.

	Being predicts B	Being predicts A and B
Take B	$1,000,000	$0
Take A & B	$1,001,000	$1,000

---

† By William Foster Allman; excerpted by permission of *Science '82* magazine, July–August, 1981, pages 36–37. Copyright American Association for the Advancement of Science.

But consider another argument. The Being made his prediction a week ago or perhaps even the day you were born. Either he put $1 million in Box B or he didn't, and the money will not disappear. It therefore makes more sense to take both, for if Box B has money in it, you get $1,001,000; if not, at least you have $1000. If you take only Box B, there is even a remote chance you would get nothing.

Gardner suggests one way to consider the paradox is to use what is called in game theory a payoff matrix. If you consider the result of both choices in light of the Being's predictions, it appears that taking both boxes is the best choice, for that option gives the highest maximum payoff ($1,001,000) as well as the highest minimum payoff ($1000). Note that $1000 is a saddle point.

But another approach of game theory, multiplying the various outcomes of each choice by the probability that they will happen results in the opposite conclusion. Even if you assume the predictor is correct a mere 90% of the time, the expected return of taking both boxes is 90% of $1000 plus 10% of $1,001,000, equalling $101,000. In other words, if you played the game ten times, your average take would be $101,000. Taking only Box B yields 90% of $1,000,000 plus 10% of zero, equalling $900,000. It is best to take only Box B even if the Being were correct only a little more than half the time.

The paradox is grounded in the conflicting concepts of free will and determinism. For those who feel that their choices are wholly committed to a present future or are wholly dependent on what the 19th century pragmatist William James called the "push of the past," taking Box B is the only choice, if that is what it can be called, because an omniscient Being or time traveler will be aware of these forces.

For those who believe that independent choices can be made—that there exists, no matter how infinitesimally, a freedom of action beyond that ordained by the past or set in the future—taking both boxes is the proper course of action. Suppose you have observed 1000 such trials prior to your decision, and every person who took only Box B received $1 million, and those who took both boxes found B empty. Even so, a friend able to see the contents of both boxes before you make your choice would, regardless of what was in them, advise you to take both, for whether Box B contained $1 million or not, you would still come out ahead by $1000.

Issac Asimov, in a response to Gardner, wrote: "I would, without hesitation, take both boxes. I am myself a determinist, but it is perfectly clear to me that any human being worthy of being considered a human being (including most certainly myself) would prefer free will, if such a thing could exist.... You will then, at least, have expressed your willingness to gamble on [God's] nonomniscience and on your own free will.... If you take only the second box, however, you get your damned million and not only are you a slave but also you have demonstrated your willingness to be a slave for that million, and you are not someone I recognize as human."

It appears that Martin Gardner takes more delight in posing paradoxes like this than solving them. "I'm an indeterminist, which is to say I'm not a determinist," he says. "I don't think the paradox has properly been explained away yet, but I'm not going to lose any sleep over it."

PROBLEMS 7.5 In Problems 1–10, the payoff matrix of a game is given. Find the expected return to R with the given pair of strategies **p** and **q**.

1. $A = \begin{pmatrix} 5 & 3 \\ 3 & 6 \end{pmatrix}$; $\mathbf{p} = (\frac{1}{3} \quad \frac{2}{3})$, $\mathbf{q} = \begin{pmatrix} \frac{3}{4} \\ \frac{1}{4} \end{pmatrix}$

2. $A = \begin{pmatrix} 5 & 3 \\ 4 & 6 \end{pmatrix}$; $\mathbf{p} = (1 \quad 0)$, $\mathbf{q} = \begin{pmatrix} \frac{1}{2} \\ \frac{1}{2} \end{pmatrix}$

3. $A = \begin{pmatrix} 1 & 3 & -1 \\ 2 & 0 & 4 \end{pmatrix}$; $\mathbf{p} = (\frac{1}{5} \quad \frac{4}{5})$, $\mathbf{q} = \begin{pmatrix} \frac{1}{4} \\ \frac{1}{2} \\ \frac{1}{4} \end{pmatrix}$

4. $A = \begin{pmatrix} 1 & 3 & -1 \\ 2 & 0 & 4 \end{pmatrix}$; $\mathbf{p} = (0 \quad 1)$, $\mathbf{q} = \begin{pmatrix} \frac{1}{3} \\ \frac{1}{3} \\ \frac{1}{3} \end{pmatrix}$

5. $A = \begin{pmatrix} 1 & 3 & -1 \\ 2 & 0 & 4 \end{pmatrix}$; $\mathbf{p} = (\frac{2}{3} \quad \frac{1}{3})$, $\mathbf{q} = \begin{pmatrix} 0 \\ 1 \\ 0 \end{pmatrix}$

6. $A = \begin{pmatrix} 2 & -1 & 4 \\ 5 & 7 & 3 \\ 6 & 2 & -2 \end{pmatrix}$; $\mathbf{p} = (\frac{1}{3} \quad \frac{1}{3} \quad \frac{1}{3})$, $\mathbf{q} = \begin{pmatrix} \frac{1}{3} \\ \frac{1}{3} \\ \frac{1}{3} \end{pmatrix}$

7. $A = \begin{pmatrix} 2 & -1 & 4 \\ 5 & 7 & 3 \\ 6 & 2 & -2 \end{pmatrix}$; $\mathbf{p} = (\frac{1}{2} \quad \frac{1}{3} \quad \frac{1}{6})$, $\mathbf{q} = \begin{pmatrix} 0 \\ 0 \\ 1 \end{pmatrix}$

8. $A = \begin{pmatrix} 2 & -1 & 4 \\ 5 & 7 & 3 \\ 6 & 2 & -2 \end{pmatrix}$; $\mathbf{p} = (0 \quad \frac{1}{2} \quad \frac{1}{2})$, $\mathbf{q} = \begin{pmatrix} \frac{1}{5} \\ \frac{2}{5} \\ \frac{2}{5} \end{pmatrix}$

9. $A = \begin{pmatrix} -3 & 4 & 2 & 3 \\ -2 & 6 & 1 & 5 \\ -4 & 0 & 6 & 7 \end{pmatrix}$; $\mathbf{p} = (\frac{1}{4} \quad 0 \quad \frac{3}{4})$, $\mathbf{q} = \begin{pmatrix} \frac{1}{10} \\ \frac{1}{5} \\ \frac{3}{10} \\ \frac{2}{5} \end{pmatrix}$

10. $A = \begin{pmatrix} -3 & 4 & 2 & 3 \\ -2 & 6 & 1 & 5 \\ -4 & 0 & 6 & 7 \end{pmatrix}$; $\mathbf{p} = (\frac{1}{5} \quad \frac{3}{5} \quad \frac{1}{5})$, $\mathbf{q} = \begin{pmatrix} \frac{1}{4} \\ \frac{1}{4} \\ \frac{1}{2} \\ 0 \end{pmatrix}$

In Problems 11–23, determine optimal strategies and the value of the given matrix game. Which games are fair?

11. $\begin{pmatrix} 2 & -1 \\ -1 & 2 \end{pmatrix}$

12. $\begin{pmatrix} 1 & 3 \\ 2 & -1 \end{pmatrix}$

13. $\begin{pmatrix} 1 & 0 \\ -2 & 2 \end{pmatrix}$

14. $\begin{pmatrix} \frac{1}{2} & -\frac{1}{2} \\ -\frac{1}{2} & \frac{1}{2} \end{pmatrix}$

15. $\begin{pmatrix} 0 & 1 \\ -1 & 0 \end{pmatrix}$

16. $\begin{pmatrix} 1 & -1 \\ -1 & 1 \end{pmatrix}$

17. $\begin{pmatrix} -1 & -1 \\ 0 & -2 \end{pmatrix}$

18. $\begin{pmatrix} \frac{1}{2} & \frac{3}{2} \\ \frac{1}{2} & 1 \end{pmatrix}$

19. $\begin{pmatrix} 3 & 5 \\ 2 & 6 \end{pmatrix}$

20. $\begin{pmatrix} 5 & -3 \\ -5 & 3 \end{pmatrix}$ **21.** $\begin{pmatrix} 7 & 4 & 2 \\ 5 & 3 & 6 \end{pmatrix}$ **22.** $\begin{pmatrix} 5 & 3 & 2 \\ 3 & 2 & 6 \\ 4 & 0 & 1 \end{pmatrix}$

23. $\begin{pmatrix} 5 & 8 & 4 \\ 3 & 7 & 1 \\ 4 & 6 & 9 \\ 5 & 7 & 2 \end{pmatrix}$

24. Suppose that the game matrix in Example 5 is

$$A = \begin{pmatrix} 10 & 39 \\ 5 & 40 \end{pmatrix}$$

What is the minimum value of the probability that the patient has the disease for which the operation can be recommended?

25. The model of medical decision-making in Example 5 can be studied more generally. Define the game matrix

$$A = \begin{pmatrix} a_{11} & a_{12} \\ a_{21} & a_{22} \end{pmatrix},$$

where a_{11}, a_{12}, a_{21}, and a_{22} are the expected lengths of life if the patient has the operation and the disease, the operation and no disease, the disease and no operation, or no disease and no operation, respectively. Assuming that the patient wishes to maximize his expected length of life, show that the operation can be recommended if the probability that the patient has the disease is greater than

$$\frac{a_{22} - a_{12}}{a_{11} + a_{22} - a_{12} - a_{21}}$$

[*Hint*: Compare the patient's expected life with the operation to his expected length of life without the operation.]

26. Show that the game of Problem 7.4.11 on page 415 is unfair. What is its value? What strategy should R adopt?

27. Answer the questions of Problem 26 for the game of Problem 7.4.12 on page 415.

28. In Problem 7.4.15 on page 415, what should the owner of gas station A do if she knows that the owner of gas station B will flip a coin to determine whether to raise or lower prices?

29. Find the optimal strategy for the farmer in Problem 7.4.20 on page 416 if the probability of frost between August 25 and September 5 is 0.5.

30. Answer the question in Problem 29 if the probability of frost is 0.2.

31. Answer the question in Problem 29 if the probability of frost is 0.9.

32. In a certain farming region, the average weather during the growing season is either cool or hot. Two crops are to be planted on a farm of 1500 acres. If the growing season is cool, the expected profits are $20 per acre from crop I and $10 per acre from crop II. If the growing season is hot, the expected profits are $10 per acre from crop I and $30 per acre from crop II. Describe the competition between the farmer and the weather as a matrix game. If no information is available concerning the probabilities of hot or cool weather, what is the optimal strategy of the farmer?

33. Suppose that the weather in Problem 32 is equally likely to be hot or cool. How many acres of each crop should the farmer plant?

34. In an experiment, a monkey must choose one of three panels to receive a banana. On each trial of the experiment, the experimenter places either two bananas behind panel I or one banana behind both panel II and panel III. (These are the two "moves" of the experimenter.) Describe this experiment as a 3×2 matrix game. What are the optimal strategies of the monkey and the experimenter?

35. If all components of a game matrix are positive, show that the value of the game is positive.

*36. Given two $m \times n$ matrix games $A = (a_{ij})$ and $B = (b_{ij})$ such that $b_{ij} = a_{ij} + k$ for all i and j, show that the value of the game of B is equal to the value of the game of A plus the constant k. Show that the optimal strategies for R and C are the same for B as for A. (The games of A and B are said to be **equivalent matrix games**.)

37. What are the optimal strategies and values of the equivalent matrix games

$$A = \begin{pmatrix} 3 & 2 \\ 2 & -3 \end{pmatrix} \quad \text{and} \quad B = \begin{pmatrix} 5 & 4 \\ 4 & -1 \end{pmatrix}?$$

38. What are the optimal strategies and values of the equivalent matrix games

$$A = \begin{pmatrix} 1 & 1 & 1 \\ 2 & 2 & 3 \\ -2 & 4 & 5 \end{pmatrix} \quad \text{and} \quad B = \begin{pmatrix} 0 & 0 & 0 \\ 1 & 1 & 2 \\ -3 & 3 & 4 \end{pmatrix}?$$

39. As the variable t increases from 0 to 1, how do the optimal strategies change in the following games?

(a) $\begin{pmatrix} t & 0 \\ 0 & 1-t \end{pmatrix}$ (b) $\begin{pmatrix} 1 & 2t \\ 0 & t \end{pmatrix}$ (c) $\begin{pmatrix} t & t^2 \\ \frac{1}{2} & \frac{1}{4} \end{pmatrix}$

40. Determine the values of the 2×2 matrix games of Problem 39 as functions of t for $0 \le t \le 1$. For which values of t are these games fair?

41. (a) Show that the 2×2 matrix game

$$\begin{pmatrix} t & 1-t \\ 1-t & t \end{pmatrix}$$

is not strictly determined for any value of t.

(b) Show that the value of this game is a constant independent of t. Find that constant.

*42. (a) Suppose that every component of $\mathbf{p}_0 A$ is greater than or equal to v. Show that $E(\mathbf{p}_0, \mathbf{q}) = \mathbf{p}_0 A\mathbf{q} \ge v$. [*Hint*: Use the fact that the components of $\mathbf{q}$ have a sum of 1.]

(b) If $E(\mathbf{p}_0, \mathbf{q}) \ge v$ for every strategy $\mathbf{q}$, show that every component of $\mathbf{p}_0 A$ is greater than or equal to v. [*Hint*: Show that the kth component of $\mathbf{p}_0 A$ is equal to $E(\mathbf{p}_0, \mathbf{q})$, where $\mathbf{q}$ is the column vector with a 1 in the kth position and a 0 everywhere else.]

*43. By following a similar procedure as in Problem 42, show that every component of $A\mathbf{q}_0$ is less than or equal to v if and only if $E(\mathbf{p}, \mathbf{q}_0) \le v$ for every strategy $\mathbf{p}$.

*44. Assume $\mathbf{p}_0$, $\mathbf{q}_0$, and v are given by (3), (4) and (5), respectively.

(a) Show that $\mathbf{p}_0 A \ge v$ and $A\mathbf{q}_0 \le v$.

(b) Use the results of Problems 42 and 43 to conclude that (3) and (4) do indeed provide optimal strategies and that v, given by (5), is the value of the 2×2 matrix game.

Review Exercises for Chapter 7

In Exercises 1–4, determine whether the given vector is a probability vector.

1. $(\frac{1}{2} \quad \frac{1}{2} \quad \frac{1}{2})$ **2.** $(\frac{1}{3} \quad \frac{1}{3} \quad \frac{1}{3})$ **3.** $(\frac{1}{5} \quad \frac{2}{5} \quad \frac{2}{5})$ **4.** $(-\frac{1}{5} \quad \frac{3}{5} \quad \frac{3}{5})$

In Exercises 5–9, determine whether the given matrix is a probability matrix.

5. $\begin{pmatrix} \frac{1}{5} & \frac{2}{5} & \frac{2}{5} \\ \frac{1}{4} & \frac{1}{2} & \frac{1}{4} \end{pmatrix}$
  6. $\begin{pmatrix} \frac{1}{3} & \frac{1}{3} & \frac{1}{3} \\ \frac{1}{4} & \frac{1}{4} & \frac{1}{4} \\ \frac{1}{2} & \frac{1}{2} & \frac{1}{2} \end{pmatrix}$
  7. $\begin{pmatrix} \frac{1}{7} & \frac{2}{7} & \frac{3}{7} & \frac{1}{7} \\ \frac{1}{3} & \frac{1}{9} & \frac{2}{9} & \frac{1}{3} \\ \frac{1}{8} & \frac{3}{4} & 0 & \frac{1}{8} \end{pmatrix}$

8. $\begin{pmatrix} 0 & 1 & 0 & 0 \\ 0 & 1 & 0 & 0 \\ 0 & 0 & 0 & 1 \end{pmatrix}$
  9. $\begin{pmatrix} 0.2 & 0.3 & 0.5 \\ 0.1 & 0.6 & 0.4 \end{pmatrix}$

In Exercises 10–15, the transition matrix T of a Markov chain and an initial probability vector $\mathbf{p}_0$ are given. (a) Compute $\mathbf{p}_1$, $\mathbf{p}_2$ and $\mathbf{p}_3$. (b) Determine whether the matrix is regular. (c) If it is regular, find its fixed vector.

10. $T = \begin{pmatrix} \frac{1}{3} & \frac{2}{3} \\ \frac{3}{4} & \frac{1}{4} \end{pmatrix}; \mathbf{p}_0 = (\frac{1}{5} \quad \frac{4}{5})$

11. $T = \begin{pmatrix} \frac{1}{8} & \frac{7}{8} \\ 0 & 1 \end{pmatrix}; \mathbf{p}_0 = (\frac{1}{2} \quad \frac{1}{2})$

12. $T = \begin{pmatrix} \frac{1}{2} & \frac{1}{4} & \frac{1}{4} \\ \frac{1}{3} & 0 & \frac{2}{3} \\ \frac{3}{5} & \frac{2}{5} & 0 \end{pmatrix}; \mathbf{p}_0 = (\frac{1}{2} \quad 0 \quad \frac{1}{2})$

13. $T = \begin{pmatrix} \frac{1}{2} & 0 & \frac{1}{2} \\ 0 & \frac{1}{2} & \frac{1}{2} \\ \frac{1}{2} & \frac{1}{2} & 0 \end{pmatrix}; \mathbf{p}_0 = (\frac{1}{3} \quad \frac{1}{2} \quad \frac{1}{6})$

14. $T = \begin{pmatrix} 0.2 & 0.4 & 0.4 \\ 0.3 & 0.3 & 0.4 \\ 0.5 & 0.2 & 0.3 \end{pmatrix}; \mathbf{p}_0 = (0.3 \quad 0.2 \quad 0.5)$

 15. $T = \begin{pmatrix} 0.38 & 0.43 & 0.19 \\ 0.16 & 0.57 & 0.27 \\ 0.29 & 0.36 & 0.35 \end{pmatrix}; \mathbf{p}_0 = (0.17 \quad 0.49 \quad 0.34)$

In Exercises 16–22, the transition matrix for an absorbing Markov chain is given. (a) Find the number of absorbing states. (b) Determine the matrices T', R, S, Q, and A.

16. $\begin{pmatrix} \frac{3}{4} & \frac{1}{4} \\ 0 & 1 \end{pmatrix}$
  17. $\begin{pmatrix} 1 & 0 \\ \frac{1}{2} & \frac{1}{2} \end{pmatrix}$

18. $\begin{pmatrix} \frac{1}{4} & \frac{1}{4} & \frac{1}{2} \\ \frac{1}{3} & \frac{1}{3} & \frac{1}{3} \\ 0 & 0 & 1 \end{pmatrix}$
  19. $\begin{pmatrix} 1 & 0 & 0 \\ \frac{1}{5} & \frac{2}{5} & \frac{2}{5} \\ 0 & 0 & 1 \end{pmatrix}$

20. $\begin{pmatrix} 0.6 & 0.3 & 0.1 \\ 0 & 1 & 0 \\ 0.4 & 0.1 & 0.5 \end{pmatrix}$

21. $\begin{pmatrix} \frac{1}{3} & \frac{1}{3} & \frac{1}{6} & \frac{1}{6} \\ 0 & 1 & 0 & 0 \\ \frac{1}{2} & 0 & \frac{1}{4} & \frac{1}{4} \\ 0 & 0 & 0 & 1 \end{pmatrix}$

22. $\begin{pmatrix} 0.13 & 0.34 & 0.25 & 0.28 \\ 0.23 & 0.41 & 0.09 & 0.27 \\ 0 & 0 & 1 & 0 \\ 0.24 & 0.36 & 0.28 & 0.12 \end{pmatrix}$

23. A student is studying Russian. He learns vocabulary by putting a Russian word on one side of a card and its English equivalent on the other side. There are 12 cards. We make the following assumptions.

   (i) Initially, the student knows the 12 English words but has no idea which Russian word goes with a given English word. He is guessing.

   (ii) The student is shown a Russian word, gives an English equivalent and, if incorrect, is shown the other side of the card. The student either knows the word or is guessing.

   (iii) If guessing, he chooses one of the 12 English answers at random.

   (iv) Once he learns a word, he remembers it.

   (v) If he guesses (correctly or incorrectly) and is then shown the correct word, the probability is 0.25 that he will know the English equivalent of the Russian word he has just seen the next time he sees it.

   With the above assumptions, find the transition matrix of the Markov chain with states K (for knowing a given word) or G (for guessing).

24. In Exercise 23, what is the probability that the student will have learned the correct translation of a given Russian word before guessing four times?

25. In Exercise 23, how many times must the experiment be repeated until the probability is at least 99% that the student will have learned the word?

26. In Exercise 23, what is the expected number of times the student will guess until he knows the word?

27. In Exercise 23, find the transition matrix of the associated Markov chain with three states: guessing correctly, guessing incorrectly, and knowing.

28. In Exercise 27, what is the expected number of times the student will guess incorrectly until he learns a given word?

29. In Exercise 27, what is the expected number of times the student will guess correctly until he knows the word?

30. At a bank, the number of people who join a teller's line is random, but averages 4 people every 10 minutes between noon and 1 P.M. A person being served will have left the line 1 minute later with probability $\frac{2}{3}$. If more than four people are in the line, the line is closed. Estimate the percentage of time that (a) the teller's line will be empty; and (b) the line will be closed.

In Exercises 31–36, the payoff matrix for a game is given. Determine whether the game is strictly determined and, if it is, find the optimal strategies for R and C.

31. $\begin{pmatrix} 1 & 3 \\ 2 & 2 \end{pmatrix}$

32. $\begin{pmatrix} 1 & 3 \\ 4 & 2 \end{pmatrix}$

33. $\begin{pmatrix} 1 & 3 \\ 2 & 4 \end{pmatrix}$

34. $\begin{pmatrix} 5 & 7 & -3 \\ -2 & 4 & 6 \\ 8 & 5 & 7 \end{pmatrix}$
  35. $\begin{pmatrix} 5 & 1 & -3 \\ 4 & 2 & 0 \\ 7 & 6 & 1 \end{pmatrix}$
  36. $\begin{pmatrix} 1 & 6 & 2 & 4 \\ 3 & 6 & 9 & 5 \\ 0 & -1 & 2 & 4 \end{pmatrix}$

In Exercises 37–40, the payoff matrix of a game is given. Find the expected return to R with the given pair of strategies **p** and **q**.

37. $A = \begin{pmatrix} 6 & 2 \\ 4 & 1 \end{pmatrix}; \mathbf{p} = (\frac{1}{2} \quad \frac{1}{2}), \mathbf{q} = \begin{pmatrix} \frac{3}{4} \\ \frac{1}{4} \end{pmatrix}$

38. $A = \begin{pmatrix} 1 & 6 & 2 \\ 3 & -1 & 5 \end{pmatrix}; \mathbf{p} = (\frac{1}{3} \quad \frac{2}{3}); \mathbf{q} = \begin{pmatrix} \frac{1}{4} \\ \frac{1}{2} \\ \frac{1}{4} \end{pmatrix}$

39. $A = \begin{pmatrix} 1 & 6 & 2 \\ 3 & 0 & -2 \\ 4 & -1 & -6 \end{pmatrix}; \mathbf{p} = (\frac{1}{5} \quad \frac{2}{5} \quad \frac{2}{5}); \mathbf{q} = \begin{pmatrix} \frac{1}{2} \\ 0 \\ \frac{1}{2} \end{pmatrix}$

40. $A = \begin{pmatrix} 5 & -1 & 2 \\ 3 & 0 & 4 \\ 6 & 2 & 5 \end{pmatrix}; \mathbf{p} = (1 \quad 0 \quad 0); \mathbf{q} = \begin{pmatrix} \frac{1}{7} \\ \frac{2}{7} \\ \frac{4}{7} \end{pmatrix}$

In Exercises 41–45, the payoff matrix of a game is given. Find the optimal strategies for R and C and the value of the game.

41. $\begin{pmatrix} 3 & 6 \\ 2 & 4 \end{pmatrix}$
  42. $\begin{pmatrix} 3 & 6 \\ 4 & 2 \end{pmatrix}$
  43. $\begin{pmatrix} -1 & 3 \\ 5 & 2 \end{pmatrix}$

44. $\begin{pmatrix} 8 & 4 & 2 \\ -1 & 3 & 2 \\ 5 & -2 & -6 \end{pmatrix}$
  45. $\begin{pmatrix} 3 & -1 & 5 \\ -1 & -6 & -3 \\ -4 & 0 & 2 \end{pmatrix}$

8 BUSINESS MATHEMATICS

In this chapter we shall discuss some topics common to almost all business dealings. We begin by discussing simple and compound interest and we then expand on that material. Our goal is to be able to answer such typical questions as the following.

1. If $1000 is invested at 8% interest compounded semiannually, what will it be worth in 5 years?

2. How long does it take an investment to double at an annual interest rate of 8%?

3. If I invest $50 per month at 7% interest compounded monthly, how much will I have after 10 years?

4. If I have a note which will pay $10,000 in 5 years, what is the note worth today assuming an interest rate of 6% compounded quarterly?

5. If I buy a $70,000 house, put $10,000 down, and obtain a 30-year mortgage for the balance at a 9% annual interest rate, what will be my monthly payments?

Many of the numbers in this chapter can be obtained from tables that appear at the back of the book. However, most of the answers to questions posed in this part can be obtained fairly quickly with a scientific, hand-held calculator. Therefore, in most cases we will work the examples with a calculator.

8.1 Simple and Compound Interest

If money is invested for a certain period of time, then interest may be paid in two ways. The amount invested is called the **principal**. In the first method, the interest for a given period is paid to the investor and the principal remains the same. This is the method of *simple interest* payments.

The simple interest method is fairly common. The term *living off one's investments* often applies to people who support themselves with the periodic interest payments from their investments. That is, the people are deriving their incomes from simple interest. As another example, suppose a university graduate endows a scholarship for his alma mater. This is often done by donating a large sum of money to the university which, in term, invests the money in a "safe" way and uses the simple interest for a scholarship. The principal, which in this case is the original donation, is never touched. As a third example, simple interest is paid on U.S. Government Series H bonds and certain types of municipal bonds.

Suppose P dollars (the *principal*) are invested in an enterprise (which may be a bank, bonds, or a common stock) with an annual interest rate of i. **Simple interest** is the amount earned on the P dollars over a period of time. If the P dollars are invested for t years, then the simple interest I is given by $I = Pit$.

Simple Interest

$$\boxed{I = Pit} \tag{1}$$

where

$$I = \text{total interest paid}$$
$$P = \text{principal}$$
$$i = \text{annual interest rate}$$
$$t = \text{time in years.}$$

Example 1 If $1000 is invested for 5 years with an interest rate of 6%, then $i = 0.06$ and the simple interest earned is

$$I = (\$1000)(0.06)(5) = \$300.$$

Often, formula (1) is written as

$$\boxed{I = Prt} \tag{2}$$

where P stands for principal and r stands for the *rate* of interest. That is,

$$\text{simple interest paid} = \text{principal} \times \text{rate} \times \text{time.}$$

Example 2 David Hilbert, a graduate of Notre Dame, leaves $250,000 to his school. If the university invests the money at an annual interest rate of 8%, how many $2500 scholarships can it give each year in Hilbert's name?

Solution After one year, the interest received is

$$I = (250,000)(0.08)(1) = \$20,000.$$

Thus a total of $20,000 is available for scholarships each year and the total number of $2500 scholarships is $20,000/$2500 = 8. Note that each year the original $250,000 is still available for investment. The principal is left unchanged, although the total number of scholarships will vary as the interest rate varies.

The second method of paying interest is the *compound interest* method. Here the interest for each time period is added to the principal before interest is computed for the next time period. This method applies whenever the periodic interest payments are not withdrawn. Examples of compound interest are investments in saving accounts and U.S. Government Series E bonds.

Compound interest is interest paid on the interest previously earned as well as on the original investment. Suppose that interest is paid annually. Then if P dollars are invested, the interest after one year is iP dollars and the original investment is now worth

$$\underset{\substack{\uparrow \\ \text{Original} \\ \text{principal}}}{P} + \underset{\substack{\uparrow \\ \text{Interest on} \\ \text{principal}}}{iP} = \underset{\substack{\uparrow \\ \text{Value of investment} \\ \text{after 1 year}}}{P(1 + i)} \text{ dollars}$$

What is the investment worth after 2 years? The annual interest on $P(1 + i)$ dollars is $i[P(1 + i)]$ so that, after 2 years, the investment is worth

$$\underset{\substack{\uparrow \\ \text{Value after} \\ \text{1 year}}}{P(1 + i)} + \underset{\substack{\uparrow \\ \text{Interest on value} \\ \text{after 1 year}}}{iP(1 + i)} = P(1 + i)(1 + i) = \underset{\substack{\uparrow \\ \text{Value after} \\ \text{2 years}}}{P(1 + i)^2}$$

Let $A(t)$ denote the value (*amount*) of the investment after t years with an interest rate of i. Then we have the following formula.

Compound Interest Formula: Annual Compounding

$$\boxed{A(t) = P(1 + i)^t} \tag{3}$$

where

$$P = \text{original principal}$$

$$i = \text{annual interest rate}$$

$$t = \text{number of years investment is held}$$

$$A(t) = \text{amount (in dollars) after } t \text{ years.}$$

 Example 3   If the interest in Example 1 is compounded annually, then after 5 years the investment is worth

$$A(5) = 1000(1 + 0.06)^5 \approx 1000(1.33823) \approx \$1338.23.$$

The actual interest paid is $338.23.

Example 3 was done using a calculator with a $\boxed{y^x}$ button. The only other reasonable way to solve problems of this type is to use a table. This is, in fact, the way this type of problem was almost always solved before people had calculators. In Table 1 at the back of the book we provide values of $(1 + i)^t$ for i ranging from $\frac{1}{2}\%$ to 18% and t ranging from 1 to 50 years.

Example 4   What is the value after 20 years of a $5000 investment earning 10% interest compounded annually?

Solution   Using $t = 20$ and $i = 10\%$ in Table 1, we find that $(1 + 0.10)^{20} \approx 6.7275$. Thus, from (3),

$$A(20) = 5000(1.1)^{20} \approx (5000)(6.7275) = \$33{,}637.50.$$

Example 5   If $2500 is invested for 6 years at 9% compounded annually, how much interest is earned?

Solution   Using Table 1 or a calculator, we find that

$$A(6) = 2500(1 + 0.09)^6 \approx 2500(1.6771) = \$4192.75.$$

The interest earned is the difference between the value after 6 years and the original investment. Thus interest is $4192.75 - 2500 = \$1692.75$.

In practice, interest is compounded more frequently than annually. If it is paid m times a year, then in each interest period the rate of interest is i/m and in t years there are tm pay periods. Then, similar to formula (3), we have the following.

Compound Interest Formula: Compounding *m* Times a Year

$$\boxed{A(t) = P\left(1 + \frac{i}{m}\right)^{mt}} \tag{4}$$

where

$$P = \text{original principal}$$
$$i = \text{annual interest rate}$$
$$t = \text{number of years investment is held}$$
$$m = \text{number of times interest is compounded each year}$$
$$A(t) = \text{amount (in dollars) after } t \text{ years.}$$

Example 6   If the interest in Example 1 is compounded quarterly (four times a year) then after 5 years the investment is worth

$$A(5) = 1000\left(1 + \frac{0.06}{4}\right)^{(4)(5)} = 1000(1.015)^{20} = 1000(1.34686) = \$1346.86.$$

The interest paid is now \$346.86.

 Example 7   If the interest in Example 5 is compounded monthly, how much interest is earned?

Solution   Here $P = 2500$, $i = 0.09$ and $m = 12$ (times a year) so that

$$A(6) = 2500\left(1 + \frac{0.09}{12}\right)^{(12)(6)} = 2500(1.0075)^{72} \approx 2500(1.71255) = \$4281.38.$$

The interest earned is $\$4281.38 - 2500 = \$1781.38$.

As the preceding examples indicate, the more frequently interest is compounded, the more the investment increases in value. In Table 1, we show the value after 10 years and the interest earned on a \$1000 investment at 8% annual interest for different numbers of payment periods each year.

Table 1 is revealing. It suggests that while there is a considerable difference when we change from annual to semiannual compounding (a difference in this

TABLE 1   Value of a \$1000 Investment Compounded *m* Times a Year for 10 Years at an Annual Rate of 8%

m = number of times interest is compounded each year	Value of \$1000 after 10 years at 8% interest (\$)	Total interest earned (\$)
1 (annually)	2158.92	1158.92
2 (semiannually)	2191.12	1191.12
3	2202.34	1202.34
4 (quarterly)	2208.04	1208.04
8	2216.72	1216.72
12 (monthly)	2219.64	1219.64
24	2222.58	1222.58
52 (weekly)	2224.17	1224.17
100	2224.83	1224.83
365 (daily)	2225.35	1225.35
1000	2225.47	1225.47
8760 (hourly)	2225.53	1225.53
525,600 (each minute)	2225.54	1225.54

example of $32.20), the difference becomes negligible as we increase the number of interest periods. For example, the difference between monthly compounding and hourly compounding is only $5.89. The numbers in Table 1 suggest that, after a point, little is gained by increasing the number of annual pay periods.

Many bank advertisements contain statements like "our 8% savings plans carries an effective interest rate or yield of $8\frac{1}{3}\%$." The **effective interest rate**, or **yield**, is the rate of simple interest received over a 1-year period. For example, $100 would be worth $108 if that sum is invested for 1 year at 8% interest compounded annually. But if it is compounded quarterly, for instance, then it is worth

$$100(1.02)^4 = \$108.24$$

after 1 year. Thus the interest paid is $8.24 and the effective interest rate is $8.24\% \approx 8\frac{1}{4}\%$.

We see that for most problems there are two rates of interest: the *quoted* rate and the *effective* rate. The first of these is often called the **nominal** rate of interest. Thus, as we have seen, a nominal rate of 8% provides an effective rate of 8.24% when interest is compounded quarterly.

Example 8   If money is invested at a nominal rate of 15% compounded monthly, what is the effective rate of interest?

Solution   Starting with P dollars, there will be $P(1 + 0.15/12)^{12} \approx 1.161P$ dollars after one year. Thus P dollars will have grown by approximately 16.1% after 1 year. This is the effective rate of interest.

Example 9   The First National Bank advertises that its savings accounts pay $6\frac{1}{4}\%$ compounded quarterly. The Western State Bank pays $6\frac{1}{8}\%$ compounded daily. Where should you put your money?

Solution   The problem is to determine which bank gives the higher effective interest rate. For the First National Bank, $1 will be worth

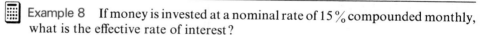

$$\left(1 + \frac{0.0625}{4}\right)^4 \approx \$1.064$$

after 1 year, so that the effective interest rate is 6.4%. For the Western State Bank, $1 will be worth

$$\left(1 + \frac{0.06125}{365}\right)^{365} \approx 1.0632$$

after 1 year with an effective interest rate of 6.32%. Clearly you are better off at First National.

Table 2 gives the effective interest rates if a sum is invested at a nominal rate of 8% compounded m times a year.

TABLE 2

m = number of times interest is paid per year	Effective interest rate (based on 8%) (%)
1	8.000
2	8.160
4	8.243
8	8.286
12	8.300
24	8.314
52	8.322
365	8.32776
1000	8.32836
10,000	8.32867
1,000,000	8.32871

We close this section by solving a type of problem posed at the beginning of chapter.

 Example 10   If interest is compounded quarterly at 6%, how long will it take an investment to double?

Solution   Suppose that P dollars are invested initially. Then we need to determine a value of t such that $A(t) = 2P$. That is,

$$2P = P\left(1 + \frac{0.06}{4}\right)^{4t}$$

$$2P = (1.015)^{4t}$$

or

$$(1.015)^{4t} = 2.$$

There are at least two ways to complete this problem.

Method 1   Trial and Error   We find a value of t such that $(1.015)^{4t} = 2$ by testing different values on a calculator. One set of results is given in Table 3. We see that, to two decimal places, $t = 11.64$. That is, it takes 11.64 years for money to double if it is invested at 6% compounded quarterly.

TABLE 3

t	4t	(1.015)^{4t}	
1	4	1.0614	
5	20	1.3469	
10	40	1.8140 ⎱	← so 10 < t < 15
15	60	2.4432 ⎰	
12	48	2.0435 ⎱	
11	44	1.9253 ⎰	← so 11 < t < 12
11.5	46	1.9835	
11.6	46.4	1.9954 ⎱	
11.7	46.8	2.0073 ⎰	← so 11.6 < t < 11.7
11.65	46.6	2.0013	
11.64	46.56	2.0001 ⎱	← so 11.635 < t < 11.64
11.635	46.54	1.9995 ⎰	

Method 2   Method of Logarithms   There is a much faster way to obtain the answer if you have studied logarithms. If you have not, skip this.

We take logarithms to the base 10.

$$(1.015)^{4t} = 2$$

$\log(1.015)^{4t} = \log 2$ We took the logarithms of both sides

$4t \log(1.015) = \log 2$ Since log $x^y = y$ log x

$$t = \frac{\log 2}{4 \log 1.015}$$ we divided by 4 log 1.015.

$$t \approx \frac{0.30103}{4(0.006466)}$$ We pressed the ⬚ log ⬚ button on our calculator.

$$t \approx \frac{0.30103}{0.025864} \approx 11.639$$

Rounded, this gives us the same answer: 11.64 years.

▦ **Example 11**   At an annual 12% rate of inflation, how long will it take for average prices to double? Assume daily compounding.

Solution   We rephrase the question: After how many years will an investment double if 12% interest compounded 365 times a year is paid? Then we must find a number t such that

$$2P = A(t) = P\left(1 + \frac{0.12}{365}\right)^{365t}$$

or

$$(1.000328767)^{365t} = 2$$

Taking common logarithms (logs to the base 10), we obtain

$$365t \log(1.00028767) = \log 2$$

and

$$t = \frac{\log 2}{365 \log 1.000328767} \approx \frac{0.3010299957}{365(0.0001427582)}$$

$$\approx \frac{0.3010299957}{0.052106743} \approx 5.777$$

That is, it takes approximately 5.8 years for prices to double at 12% inflation. We could also obtain this answer by trial and error.

In Table 4 we show how long it takes money to double at various interest rates assuming daily compounding.

TABLE 4

Interest rate (%)	Doubling time (years)	Interest rate (%)	Doubling time (years)
1	69.3	8	8.7
2	34.7	10	6.9
3	23.1	12	5.8
4	17.3	15	4.6
5	13.9	18	3.9
6	11.6	25	2.8
7	9.9	50	1.4

PROBLEMS 8.1   In Problems 1–6, P dollars are invested at $i\%$ interest for t years. Compute the simple interest paid.

1. $P = \$500, i = 8\%, t = 5$

2. $P = \$2000, i = 4\frac{1}{2}\%, t = 8$

3. $P = \$1500, i = 6\%, t = 2\frac{1}{2}$

4. $P = \$10,000, i = 8\%, t = 6$

5. $P = \$25,000, i = 9.3\%, t = 7.4$

6. $P = \$15,000, i = 12\%, t = 10$

In Problems 7–20, compute the value of an investment after t years and the total interest paid if P dollars are invested at a nominal interest rate of $i\%$, compounded m times a year.

7. $P = \$5000, i = 5\%, t = 5, m = 1$

8. $P = \$5000, i = 5\%, t = 5, m = 4$

9. $P = \$5000, i = 5\%, t = 5, m = 12$

10. $P = \$5000, i = 5\%, t = 5, m = 365$

11. $P = \$8000, i = 11\%, t = 4, m = 1$

12. $P = \$8000, i = 11\%, t = 4, m = 4$

13. $P = \$8000, i = 11\%, t = 4, m = 12$

14. $P = \$8000, i = 11\%, t = 4, m = 100$

15. $P = \$8000, i = 11\%, t = 4, m = 525,600$

16. $P = \$10,000, i = 7\frac{1}{2}, t = 10, m = 1$

17. $P = \$10,000, i = 7\frac{1}{2}\%, t = 10, m = 4$

18. $P = \$10,000, i = 7\frac{1}{2}\%, t = 10, m = 12$

19. $P = \$10,000, i = 7\frac{1}{2}\%, t = 10, m = 365$

20. $P = \$10,000, i = 7\frac{1}{2}\%, t = 10, m = 8760$

21. What is the simple interest paid on $5000 invested at 9% for 4 years?

22. A benefactor endows a chair of economics at a major midwestern university. He contributes $500,000. It the university can invest the money at $8\frac{1}{2}\%$ simple interest, what annual salary can it pay the holder of the chair?

23. Calculate the percentage increase in return on investment if P dollars are invested for 10 years at 6% compounded annually and quarterly.

24. As a gimmick to lure depositors, some banks offer 5% interest compounded daily in comparison to their competitors, who offer $5\frac{1}{8}\%$ compounded annually. Which bank would you choose?

25. Suppose a competitor in Problem 24 now compounds $5\frac{1}{8}\%$ semiannually. Which bank would you choose?

26. If $10,000 is invested in bonds yielding 9% compounded quarterly, what will the bonds be worth in 8 years?

*27. A certain government bond sells for $750 and can be redeemed for $1000 in 8 years. Assuming quarterly compounding, what is the nominal rate of interest paid?

28. How long would it take an investment to increase by half if it is invested at 4% compounded monthly?

*29. What must be the nominal interest rate in order that an investment triple in 15 years if interest is compounded semiannually?

30. If money is invested at 10% compounded daily, what is the effective interest rate?

31. Find the effective interest rate of 12% compounded

    (a) semiannually.

    (b) quarterly.

    (c) monthly.

    (d) daily.

32. Answer the questions of Problem 31 if the rate of interest is 4%.

33. How long will it take an investment to triple (assuming quarterly compounding) for each interest rate?

    (a) 2%

    (b) 5%

    (c) 8%

    (d) 10%

    (e) 15%

34. Answer the questions of Problem 33 if interest is compounded monthly.

35. Answer the questions of Problem 33 if interest is compounded annually.

*36. In 1920 the consumer price index (CPI) for perishable goods was 213.4 with 1913 assigned the base rate of 100. Assuming that inflation was constant on a monthly basis (that is, assume monthly compounding), find the annual rate of inflation of the price of perishable goods between 1913 and 1920.

*37. In 1920 the CPI for construction materials was 262.0. Again, with the CPI = 100 in 1913, find the annual rate of inflation of the price of construction materials between 1913 and 1920.

*38. The average annual earnings of full-time employees in the United States was $4743 in 1960 and $7564 in 1970. Assuming a monthly increase in earnings at a constant rate during the period 1960–1970, what was the annual increase in wages?

*39. The CPI was 88.7 in 1960 and 116.3 in 1970 (1967 = 100). Assuming a monthly increase in prices at a constant rate between 1960 and 1970, find the annual rate of inflation.

*40. The *real* increase in earnings is defined as the percentage increase in wages minus the rate of inflation. Using the data in Problems 38 and 39, determine the real annual increase in the average U.S. workers' earnings between 1960 and 1970.

*41. Assume that the figures in Problem 39 hold but that the CPI in 1970 is unknown. What would it be if it were known that workers between 1960 and 1970 enjoyed no gain or loss in real income?

8.2 Present Value

An extremely important concept in economics is that of the *present value* of money. As the money one has today will earn interest, it follows that in order to collect $1000 in 2 years, for instance, one needs *less* than $1000 today. We can use the compound interest formulas ((8.1.3) on page 434 and (8.1.4) on page 435) to calculate the amount needed today, called *present value*.

If we invest P dollars today, then after t years the investment is worth $A(t) = P(1 + i/m)^{mt}$ dollars if interest is compounded m times a year. Thus present value may be calculated by solving for P.

$$A(t) = P\left(1 + \frac{i}{m}\right)^{mt} \qquad \text{This is formula (8.1.4) on page 435.}$$

$$P = \frac{A(t)}{(1 + i/m)^{mt}} = A(t)\left(1 + \frac{i}{m}\right)^{-mt} \qquad \text{We divided by } \left(1 + \frac{i}{m}\right)^{mt}$$

Present Value

The **present value** P of an investment worth $A(t)$ in t years is given by

$$P = A(t)\left(1 + \frac{i}{m}\right)^{-mt} \tag{1}$$

where

$$i = \text{annual interest rate}$$

$$m = \text{number of times interest is compounded each year}$$

$$t = \text{number of years}$$

$$A(t) = \text{amount (in dollars) to be collected or paid in } t \text{ years.}$$

 Example 1 What is the present value of $1000 after 5 years at 6% compounded semiannually?

Solution We have $t = 5$, $A(t) = \$1000$, $i = 0.06$, and $m = 2$ so that

$$P = 1000\left(1 + \frac{0.06}{2}\right)^{-2(5)} = 1000(1.03)^{-10} = \$744.09.$$

Put another way, we would have to invest $744.09 now at 6% interest to have $1000 in 5 years.

 Example 2 A manufacturer receives a note promising payment of $25,000 in 3 years. She needs capital now so she sells the note to a bank, which reimburses her based on an annual interest rate of 7% compounded monthly. How much money does the manufacturer receive? [*Note*: The act of purchasing something for its present value is sometimes called **discounting**, and the interest rate i is called the **discount rate**.]

Solution She will receive the present value of the $25,000 based on an interest rate of 7%. Here $t = 3$, $A(t) = \$25,000$, $i = 0.07$, and $m = 12$, (monthly compounding = 12 times a year) so

$$P = A(3)\left(1 + \frac{0.07}{12}\right)^{-12(3)} \approx 25,000(1.005833)^{-36} = \$20,277.22$$

 Example 3 If the manufacturer in Example 2 receives $18,000 from the bank, how much interest is she effectively paying for the loan?

Solution Here the present value is given to us as $18,000. The only unknown is i. Thus we have

$$18,000 = 25,000\left(1 + \frac{i}{12}\right)^{-36}$$

or

$$\left(1 + \frac{i}{12}\right)^{-36} = \frac{18,000}{25,000} = 0.72$$

As in Example 8.1.10 on page 438, we can solve this by trial and error or by using logarithms. We will solve it here by trial and error. The results are given in Table 1.

TABLE 1

i	$\left(1 + \dfrac{i}{12}\right)^{-36}$	
0.02	0.9418	
0.04	0.8871	
0.06	0.8356	
0.1	0.7417 $\Big\}$	$0.1 < i < 0.12$
0.12	0.6989	
0.11	0.72001	
0.111	0.71787	

We conclude that the manufacturer is paying a discount rate of 11%.

There are standard tables used for calculating present value. In Table 2 at the back of the book we provide tables of $(1 + i)^{-t}$ for values of i between $\frac{1}{2}$% and 18% and values of t ranging from 1 to 50.

Example 4 Mr. and Mrs. Grabowski know that in 9 years they will need $20,000 to send their daughter to college. They can now invest in a savings certificate which pays a nominal rate of 8% interest compounded quarterly. They decide to invest whatever is required at once. How much should they invest?

Solution The amount they invest now is supposed to be worth $20,000 in 9 years. Thus the amount invested is the present value of $20,000, compounded quarterly at a nominal rate of 8%. From (1), we have

$$P = 20,000\left(1 + \frac{0.08}{4}\right)^{-4(9)} = (20,000)(1.02)^{-36}.$$

From Table 2 (at the back of the book), we have $(1.02)^{-36} = 0.490223$, so that

$$P = (20,000)(0.490223) = \$9804.46.$$

Thus $9804.46 invested now at 8% interest compounded quarterly will be worth $20,000 in 9 years.

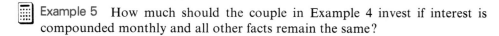 **Example 5** How much should the couple in Example 4 invest if interest is compounded monthly and all other facts remain the same?

Solution From (1) we obtain

$$P = 20,000\left(1 + \frac{0.08}{12}\right)^{-12(9)} \approx 20,000(1.0066667)^{-108}$$

$$\approx 20,000(0.487915) = \$9758.30$$

As the last two examples illustrate, the present value of a fixed sum of money at a fixed time in the future and at a fixed nominal interest rate decreases as the frequency of compounding increases. To illustrate this fact more clearly, Table 2 gives the present value P of $10,000 in 10 years at a nominal rate of 10% compounded m times a year.

TABLE 2

m	$P = 10,000\left(1 + \dfrac{0.10}{m}\right)^{-10m}$ ($)
1	3855.43
2	3768.89
4	3724.31
6	3709.24
12	3694.07
52	3682.33
365	3679.30

Example 6 Mrs. Albini bought a $700 television set and paid $300 down. She signed a note agreeing to pay the additional $400 in two years. The next day the salesperson called her and said that if she paid an additional $350 immediately, she would owe nothing more. Mrs. Albini had $1000 in a special savings account paying 9%, compounded quarterly. Should she have accepted the offer?

Solution The question is whether $400 in two years has a present value of more or less than $350. If more, she should accept; if less, she should decline because, if she accepts, she would be paying $350 for a note worth less than $350. At 9% compounded quarterly, $400 in 2 years has a present value of

$$400\left(1 + \frac{0.09}{4}\right)^{-8} = \$334.78.$$

Thus she should turn down the offer. Looking at this problem another way, $350 now will be worth $350(1 + 0.09/4)^8 = \$418.19$ in 2 years. What might look like a good deal would result in a loss of $18.19.

PROBLEMS 8.2 In Problems 1–15, compute the present value of $A(t)$ dollars, t years in the future, at a nominal rate of i% compounded m times a year.

1. $A(t) = \$5000$, $t = 6$, $i = 7\%$, $m = 1$

2. $A(t) = \$5000$, $t = 6$, $i = 7\%$, $m = 4$

3. $A(t) = \$5000$, $t = 6$, $i = 7\%$, $m = 12$

4. $A(t) = \$8000, t = 10, i = 11\frac{3}{4}\%, m = 1$

5. $A(t) = \$8000, t = 10, i = 11\frac{3}{4}\%, m = 2$

6. $A(t) = \$8000, t = 10, i = 11\frac{3}{4}\%, m = 4$

7. $A(t) = \$8000, t = 10, i = 11\frac{3}{4}\%, m = 12$

8. $A(t) = \$8000, t = 10, i = 11\frac{3}{4}\%, m = 52$

9. $A(t) = \$2500, t = 4, i = 5\frac{1}{2}\%, m = 1$

10. $A(t) = \$2500, t = 4, i = 5\frac{1}{2}\%, m = 4$

11. $A(t) = \$2500, t = 4, i = 5\frac{1}{2}\%, m = 12$

12. $A(t) = \$1,300,000, t = 25, i = 6\%, m = 1$

13. $A(t) = \$1,300,000, t = 25, i = 6\%, m = 4$

14. $A(t) = \$1,300,000, t = 25, i = 6\%, m = 12$

15. $A(t) = \$1,300,000, t = 25, i = 6\%, m = 365$

16. A sales firm receives a cash payment of $8000 for a $15,000 note due in 5 years. Assuming semiannual compounding, what interest is it paying?

17. The population of a certain city has grown at a steady rate of $\frac{3}{4}\%$ a year for the last 50 years. In 1980 the population was 725,000. What was the population in 1930? Assume monthly compounding.

18. An oilman buys a large estate near Dallas. The total cost is $2,000,000. He pays $1,000,000 immediately and signs a note agreeing to pay the additional $1,000,000 in 5 years. One year later the purchaser agrees to return the note in exchange for $700,000 cash. The oilman's investments earn him an average return of 8% compounded monthly. Should he accept the offer?

19. Mr. Collier holds a $10,000 note due in 3 years and needs cash now. Mr. Mersault has cash which he can invest at 7% compounded annually. Mr. Collier offers the note to Mr. Mersault. What is a fair purchase price?

20. The note in Problem 19 was sold for $7500. Assuming quarterly compounding, what was the discount rate?

21. In Problem 20, compute the discount rate assuming monthly compounding.

*22. A decision must be made for a capital equipment purchase. Using an interest rate of 7%, what is the maximum reasonable purchase price for a machine whose earnings over the next 5 years are projected to be $1000, $1500, $1800, $1300, and $700? Assume a value of zero after 5 years and quarterly compounding.

23. A novelist writes a novel loaded with adventure and sex, which centers around the story of a 5-year old beagle named Bennette who can turn lead into gold. A publisher is eager to publish this book as it is expected to be a best seller. The author demands a cash advance based on first year royalties, which are anticipated to be $250,000. Because of the time needed to revise the novel and publish the book, the first royalty payments are not due for 3 years. The publisher must pay 11% compounded quarterly on any money she pays out now in anticipation of future earnings. If the author's demand is to be met, what is a reasonable cash advance?

8.3 Annuities

In Section 8.1 we saw how to compute the future value of an investment when a fixed sum of money was deposited in an account that pays interest compounded periodically. Often, however, people (and corporations) do not deposit large sums of money and then sit back and watch them grow. Rather, money is invested in smaller amounts at periodic intervals (for example, monthly deposits in a bank, annual life insurance premiums, installment loan payments, and so on).

Annuity

An **annuity** is a fixed amount of money that is paid or received at regular intervals. The time between successive payments of an annuity is its **payment interval** and the time from the beginning of the first interval to the end of the last interval is called its **term**. The value of the annuity after the last payment has been made or received is called the **future value** of the annuity. If payment is made and interest is computed at the *end* of each time period, then the annuity is called an **ordinary annuity**.

 Example 1   Suppose $1000 is invested in a savings plan at the end of each year and that 7% interest is paid, compounded annually. How much will be in the account after 4 years?

Solution   To find the value of the annuity after 4 years, we compute the value of each of the four payments after 4 years and then find the sum of these payments. First, $1000 deposited at the end of the first year will be earning interest for 3 years. Thus the $1000 will be worth

$$1000(1 + 0.08)^3 = \$1259.71.$$

Similarly, $1000 deposited at the end of the second year will be earning interest for 2 years. Thus it will be worth

$$1000(1.08)^2 = \$1166.40.$$

Continuing, we see that, after the fourth year, the $1000 invested at the end of the third year will be worth

$$1000(1.08) = \$1080.$$

Finally, the $1000 invested at the end of the fourth year will not earn any interest and so will be worth $1000. Summing, we have

future value of the annuity $= \$1259.71 + \$1166.40 + \$1080 + \$1000 = \$4506.11.$

As example 1 suggests, computing the future value of annuity by calculating the future value of each payment separately can be a tedious affair. Imagine trying to compute the future value of an annuity consisting of 360 payments (as in a 30-year mortgage). Fortunately, there is a much easier way to do it.

Suppose that B dollars are deposited or received at the end of each time period. An interest rate of i is paid at the same time. Let $A(n)$ denote the amount in the account after n time periods. Then after one period we deposit B dollars; after two periods the B dollars have now become $B(1 + i)$ and we deposit another B dollars to obtain $A(2) = B + B(1 + i)$. After 3 periods we have $A(3) = B + B(1 + i) + B(1 + i)^2$ and, after n periods,

$$A(n) = B + B(1 + i) + B(1 + i)^2 + \cdots + B(1 + i)^{n-1}$$
$$= B[1 + (1 + i) + (1 + i)^2 + \cdots + (1 + i)^{n-1}].$$

The expression in brackets can be written more succinctly. We will show you how to do this at the end of this section. We will show that

$$1 + (1 + i) + (1 + i)^2 + \cdots + (1 + i)^{n-1} = \frac{(1 + i)^n - 1}{i}.$$

Thus we have the following.

Future Value of an Annuity: Annual Compounding The **future value $A(n)$ of an annuity with interest compounded annually** is given by

$$A(n) = \frac{B[(1 + i)^n - 1]}{i} \tag{1}$$

where

B = amount deposited at the end of each year

i = annual interest rate

n = number of years

$A(n)$ = future value (in dollars) of the annuity after n years.

 Example 1 (continued) We can solve Example 1 much more quickly if we use equation (1). We have

$$A(4) = \frac{1000[(1.08)^4 - 1]}{0.08} = \$4506.11.$$

If interest is compounded m times a year, a formula similar to (1) holds.

Future Value of an Annuity: Compounding m Times a Year The **future value $A(n)$ of an annuity with interest compounded m times a year** is given by

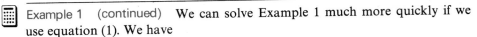

$$A(n) = B \left\{ \frac{[1 + (i/m)]^{mn} - 1}{[1 + (i/m)]^m - 1} \right\} \tag{2}$$

where B, i, and n are as before.

Example 2 If a man deposits $500 every 6 months and this is compounded quarterly at 6%, how much does he have after 10 years?

Solution Here $B = 500$ and $i = 0.03$, since the interval of deposit is $\frac{1}{2}$ year. Then $m = 2$ (2 payments every $\frac{1}{2}$ year), $n = 20$ (there are 20 semiannual deposits) and

$$A(20) = 500 \left\{ \frac{[1 + (0.03/2)]^{2(20)} - 1}{(1 + (0.03/2))^2 - 1} \right\}$$

$$= 500 \left(\frac{1.015^{40} - 1}{1.015^2 - 1} \right) \approx 500 \left(\frac{1.814018 - 1}{1.030225 - 1} \right) = 13,465.97.$$

Example 3 Betsy puts $2 a week in a bank's Christmas Club. The bank advertises that it pays $5\frac{1}{2}\%$ interest compounded daily. How much will Betsy have in her account after 1 year (52 weeks)?

Solution The payment interval in this problem is 1 week $= \frac{1}{52}$ year. Thus $i = 0.055/52$ and, since interest is compounded seven times in one period, $m = 7$. Therefore we have

$$A(52) = 2 \left[\frac{\left(1 + \dfrac{0.055/52}{7}\right)^{(7)52} - 1}{\left(1 + \dfrac{0.055/52}{7}\right)^{7} - 1} \right]$$

$$\approx 2 \left[\frac{(1.000151099)^{364} - 1}{(1.000151099)^{7} - 1} \right] \approx 2 \left[\frac{0.056536263}{0.001058173} \right] = \$106.86.$$

Note that the total interest earned is \$2.86, since \$104 is invested.

There are standard tables used for computing the future value of an annuity. First, we introduce some notation. We denote the future value of an ordinary annuity by $s_{\overline{n}|i}$. We have

$$s_{\overline{n}|i} = \begin{cases} \text{future value of ordinary annuity consisting of } n \text{ payments of \$1} \\ \text{with an interest rate } i \text{ paid at the end of each period.} \end{cases}$$

Recall that an ordinary annuity is one in which payments and interest are computed at the same time. That is, interest is compounded once at the end of each time period. Thus, setting $B = 1$ in equation (1), we obtain

$$s_{\overline{n}|i} = \frac{(1 + i)^n - 1}{i}. \tag{3}$$

Also, if B dollars are invested each time period in an ordinary annuity, we have, from (1),

$$\text{future value of annuity} = Bs_{\overline{n}|i}. \tag{4}$$

Values of $s_{\overline{n}|i}$ are given for various values of n and i in Table 3 at the end of the book.

Example 4 A corporation invests \$25,000 at the end of each year in a second mortgage account paying 10% interest compounded annually. Using Table 3, determine how much money the corporation will have in the account after 8 years.

Solution   This is an ordinary annuity with $n = 8$ and $i = 0.10$. Thus, using the table, we obtain

$$\text{future value} = 25{,}000 s_{\overline{8}|0.10} = 25{,}000(11.4359) = \$285{,}897.50.$$

Suppose money is received, instead of being paid out, at periodic intervals. (For example, you may own an apartment building where monthly rents are received.) What is the present value of such payments? We suppose that an interest rate of i is paid over a fixed time period and assume that B_1 dollars are expected after the first time period, B_2 dollars after the second, and so on. Then the present values are $P_1 = B_1/(1 + i)$, $P_2 = B_2/(1 + i)^2$, and so on, so that the total present value over n time periods is

$$P = \frac{B_1}{1 + i} + \frac{B_2}{(1 + i)^2} + \cdots + \frac{B_n}{(1 + i)^n} \tag{5}$$

If all payments are equal, then $B_1 = B_2 = \cdots = B_n = B$ and

$$P = B\left[\frac{1}{(1 + i)} + \frac{1}{(1 + i)^2} + \cdots + \frac{1}{(1 + i)^n}\right]$$

$$= \frac{B}{1 + i}\left[1 + \frac{1}{(1 + i)} + \frac{1}{(1 + i)^2} + \cdots + \frac{1}{(1 + i)^{n-1}}\right]$$

It can be shown that

$$\frac{1}{1 + i}\left[1 + \frac{1}{1 + i} + \frac{1}{(1 + i)^2} + \cdots + \frac{1}{(1 + i)^{n-1}}\right]$$

$$= \frac{1}{1 + i}\left[\frac{1 - \left(\frac{1}{1 + i}\right)^n}{1 - \frac{1}{1 + i}}\right] = \frac{1 - \left(\frac{1}{1 + i}\right)^n}{i}$$

Thus we have the following.

Present Value of an Annuity   The **present value P of an ordinary annuity** is given by

$$\boxed{P = \frac{B}{i}\left[1 - \left(\frac{1}{1 + i}\right)^n\right]} \tag{6}$$

where

$B = $ the amount paid at the end of each period

$i = $ the interest rate paid each period

$n = $ the number of periods.

Example 5   What is the present value of an annuity that would pay $3000 a year for 20 years, assuming an interest rate of 6% compounded annually?

Solution $A(20) = \dfrac{3000}{0.06}\left(1 - \dfrac{1}{1.06^{20}}\right) = \$34,409.76$

There is also a special notation for the present value of an ordinary annuity. It is denoted by $a_{\overline{n}|i}$ and we have

$$a_{\overline{n}|i} = \begin{cases} \text{present value of an ordinary annuity consisting} \\ \text{of payments of \$1 with an interest rate } i \text{ paid} \\ \text{at the end of each of } n \text{ periods.} \end{cases}$$

From (6), we obtain (setting $B = 1$)

$$a_{\overline{n}|i} = \frac{1 - (1 + i)^{-n}}{i}. \tag{7}$$

If B dollars are paid at the end of each time period, then we have, from (6),

$$\boxed{\text{present value of an ordinary annuity} = Ba_{\overline{n}|i}.} \tag{8}$$

Values of $a_{\overline{n}|i}$ are given for various values of n and i in Table 4 at the end of the book.

Example 6   Using Table 4, determine the present value of an ordinary annuity consisting of payments of $500 every 3 months for 5 years into an account paying 8% interest compounded quarterly.

Solution   Here $B = 500$, $n = 20$ (5 years at 4 payments a year) and $i = 0.02$ (the quarterly interest). Thus, from (8) and Table 4,

$$\text{present value} = 500a_{\overline{20}|0.02} = 500(16.3514) = \$8175.70.$$

We now describe a type of annuity that is similar, but not identical to an ordinary annuity. In an ordinary annuity payments are made at the end of a time period. An **annuity due** is an annuity in which payments are made at the **beginning** of the time period. Examples of annuities due are deposits in a savings account, rent payments, and payments of an insurance premium.

The term of an annuity due is from the first payment to the end of one period after the last payment. This means that an annuity due draws interest for one more period than a corresponding ordinary annuity.

Example 7   What will be the final amount in a savings account if $200 is deposited every 6 months for 5 years and the account pays interest at the rate of 6%, compounded semiannually?

Solution   This is an annuity due. There are 10 payments of $200 each. Suppose, for a moment, that there was one additional payment at the end of the tenth period. Then we would have an ordinary annuity with $n = 11$, $B = 200$, and $i = 0.03$ (the semiannual interest). From (3), (4), and Table 3 or a calculator, the future value of this annuity is

$$200\left[\frac{(1 + 0.03)^{11} - 1}{0.03}\right] = 200(12.8078) = \$2561.56.$$

However, the last $200 (which earns no interest) was never really deposited. So we must subtract it from our total to obtain

future value of the annuity due = $2561.56 − $200 = 2361.56.

The technique used in the last example can be used to find the future value of any annuity due.

> To find the future value of an annuity due with n payments of B dollars at an interest rate of i paid each period:
>
> **1.** Compute $Bs_{\overline{n+1}|i}$.
>
> **2.** Subtract B.

Putting these together, we obtain

$$\text{future value of an annuity due} = B(s_{\overline{n+1}|i} - 1). \qquad (9)$$

 Example 8   The Ajax Mortgage Corporation owns an apartment building that provides $10,000 a month in rents. It needs money now and wants to assign the rents for the next 4 years (48 months) to a finance company in exchange for cash. How much money should the finance company give the corporation if its corporate interest rates are 12% compounded monthly?

Solution   Since rent is payable at the beginning of each month, the finance company is buying a 4-year annuity due with an annual interest rate of 12%. A fair price is the present value of that annuity.

The first $10,000, paid immediately, has a present value of $10,000. The other 47 rental payments comprise an ordinary annuity with $B = 10,000$, $n = 47$, and $i = 0.01$. Thus

$$\text{present value of the annuity due} = \$10,000 + 10,000a_{\overline{47}|0.01}$$

$$= 10,000\left[1 + \frac{1 - (1.01)^{-47}}{0.01}\right]$$

$$\approx 10,000[1 + 37.3537] = 10,000(38.3537)$$

$$= \$383,537.$$

Note that over 4 years the finance company will receive $48(10,000) = \$480,000$ so that Ajax will be paying $\$480,000 - \$383,537 = \$96,463$ in interest for the privilege of receiving its rents in advance.

Remark. If $383,537 were invested for 4 years at 12% compounded monthly, then it would be worth $383,537(1.01)^{48} = \$618,348$. Thus, at first sight, it seems that the finance company is making a bad deal (it only receives $480,000 in rents). But this is not the case because the company is, remember, receiving an annuity due and can receive interest on the earlier rents collected. From (9), the future value of this annuity is

$$10,000(s_{\overline{49}|0.01} - 1) = 10,000\left(\frac{(1.01)^{49} - 1}{0.01} - 1\right)$$

$$\approx 10,000(61.8348) = \$618,348.$$

We close this section by obtaining a formula for the sum

$$S = 1 + (1 + i) + (1 + i)^2 + \cdots + (1 + i)^{n-1}. \tag{10}$$

Multiplying both sides of (10) by $1 + i$ gives

$$(1 + i)S = (1 + i)[1 + (1 + i) + (1 + i)^2 + \cdots + (1 + i)^{n-1}]$$

$$= (1 + i) + (1 + i)^2 + \cdots + (1 + i)^{n-1} + (1 + i)^n. \tag{11}$$

Subtracting (11) from (10), we have

$$S - (1 + i)S = 1 - (1 + i)^n \qquad \text{All other terms cancel.}$$

So

$$S - S - iS = 1 - (1 + i)^n$$

or

$$-iS = 1 - (1 + i)^n \tag{12}$$

and

$$S = \frac{1 - (1 + i)^n}{-i} = \frac{(1 + i)^n - 1}{i}. \qquad \text{We multiplied numerator and denominator by } -1.$$

A similar "trick" can be used to show that

$$1 + \frac{1}{1+i} + \frac{1}{(1+i)^2} + \cdots + \frac{1}{(1+i)^{n-1}} = \frac{1 - \left(\dfrac{1}{1+i}\right)^n}{1 - \dfrac{1}{1+i}}.$$

PROBLEMS 8.3

In Problems 1–10, find the future value of an ordinary annuity with payments of B dollars, over n periods, and at an interest rate of i per period.

1. $B = \$500, n = 10, i = 0.02$
2. $B = \$500, n = 10, i = 0.03$
3. $B = \$500, n = 10, i = 0.10$
4. $B = \$500, n = 8, i = 0.03$
5. $B = 2500, n = 25, i = 0.06$
6. $B = \$625, n = 30, i = 0.025$
7. $B = \$8000, n = 18, i = 0.017$
8. $B = \$3, n = 104, i = 0.06/52$
9. $B = \$10, n = 520, i = 0.08/52$
10. $B = \$3785, n = 27, i = 0.0375$

In Problems 11–20, find the present value of an ordinary annuity with payments of B dollars, over n periods, and at an interest rate of i per period.

11. $B = 500, n = 10, i = 0.02$
12. $B = \$500, n = 10, i = 0.03$
13. $B = \$500, n = 10, i = 0.10$
14. $B = \$1500, n = 8, i = 0.03$
15. $B = 2500, n = 25, i = 0.06$
16. $B = \$625, n = 30, i = 0.025$
17. $B = \$8000, n = 18, i = 0.017$
18. $B = \$3, n = 104, i = 0.06/52$
19. $B = \$10, n = 520, i = 0.08/52$
20. $B = \$3785, n = 27, i = 0.0375$

21. Find the future value of an annuity of $1 paid at the end of each year for 10 years if it earns interest at 5%, compounded annually.

22. Find the future value of an ordinary annuity of $1 paid every 6 months for 3 years if the interest rate is 6%, compounded semiannually.

23. Find the future value of an annuity of $1300 paid at the end of each quarter for 5 years if interest is earned at a rate of 6%, compounded quarterly.

24. Find the future value of an annuity of $50 paid every 6 months for 10 years if it earns 4%, compounded semiannually.

25. Find the future value of an ordinary annuity of $80 paid quarterly for 3 years if the interest rate is 8%, compounded quarterly.

26. Find the future value of an ordinary annuity of $300 paid quarterly for 5 years if the interest rate is 12%, compounded quarterly.

27. Find the future value of an ordinary annuity of $40 paid at the end of each 6-month period for 10 years if the interest rate is 6%, compounded semiannually.

28. Find the future value of an ordinary annuity of $75 paid every 6 months for 7 years if the interest rate is 9%, compounded semiannually.

29. Mr. Rider plans to invest $300 in a savings account at the end of each year for 12 years. If the account pays $5\frac{1}{2}\%$, compounded annually, how much will he have at the end of 12 years?

30. Mrs. Eliot saves $500 each 6 months and invests it at 7%, compounded semiannually. How much will she have at the end of 8 years?

31. Find the present value of an annuity of $100 paid at the end of each year for 17 years if the interest rate is $7\frac{1}{2}\%$, compounded annually.

32. Find the present value of an annuity of $800 paid at the end of each year for 15 years if the interest rate is $8\frac{3}{4}\%$, compounded annually.

33. Find the present value of annuity of $100 paid at the end of each 6-month period for 8 years if the interest rate is 6%, compounded semiannually.

34. Find the present value of an annuity which pays $300 at the end of each 6-month period for 6 years if the interest rate is 6%, compounded semiannually.

35. A woman buying a home paid $30,000 down and agreed to make 18 quarterly payments of $500 beginning in 3 months. If money is worth 9%, compounded quarterly, how much should the house cost if she paid for it in cash?

36. Richard Vitale inherited a sum of money. He wants to purchase an annuity which will give him $1000 at the end of each 6-month period for 9 years. If money is worth 10%, compounded semiannually, how much will he have to pay for the annuity?

37. Is it more profitable to buy an automobile for $3900 cash or to pay $800 down and $400 each quarter for 2 years, if money is worth 8%, compounded quarterly?

38. Find the future value of an annuity due of $1 payable at the beginning of each year for 5 years if the interest rate is 7%, compounded annually.

39. Find the future value of an annuity due of $1 payable at the beginning of each 6-month period for 7 years if the interest rate is 6%, compounded semiannually.

40. Find the future value of an annuity due of $1000 each year for 10 years at 3%, compounded annually.

41. Find the future value of an annuity due of $1500 each year for 6 years if the interest rate is 5%, compounded annually.

42. Find the future value of an annuity due of $200 paid at the beginning of each 6-month period for 8 years if the interest rate is 6%, compounded semiannually.

43. For 3 years, $400 is placed in a savings account at the beginning of each 6-month period. If the account pays interest at 5%, compounded semiannually, how much will be in the account at the end of the 3 years?

44. Mary Ellen deposits $500 in an account at the beginning of each year for 19 years. If the account pays interest at the rate of 6%, compounded annually, how much will she have in her account at the end of 19 years?

45. A house is rented for $360 per quarter, with each quarter's rent payable in advance. If money is worth 8%, compounded quarterly, what is the cash value of the rent for 5 years?

46. What is the present value of the annuity of Problem 42?

47. What is the present value of the annuity of Problem 43?

Problems 48 and 49 have been taken from recent CPA exams.

48. (November 1979)   On May 1, 1979, a company sold some machinery to another company. The two companies entered into an installment sales contract at a predetermined interest rate. The contract required five equal annual payments with the first payment due on May 1, 1979. What present value concept is appropriate for this situation?

(a)   Present value of an annuity due of $1 for five periods.

(b)   Present value of an ordinary annuity of $1 for five periods.

(c)   Future amount of an annuity of $1 for five periods.

(d)   Future amount of $1 for five periods.

49. (May 1980) Virginia Company invested in a four-year project. Virginia's expected rate of return is 10%. Additional information on the project is as follows.

Year	Cash inflow from operations, net of income taxes	Present value of $1 at 10%
1	$4,000	0.909
2	4,400	0.826
3	4,800	0.751
4	5,200	0.683

Assuming a positive net present value of $1000, what was the amount of the original investment?

(a) $2552

(b) $4552

(c) $13,427

(d) $17,400

8.4 Sinking Funds and Amortization

In this section we look at two interesting applications of the future and present value of an annuity. We introduce the first application by means of an example.

 Example 1 Mr. Carter has a 3-year-old daughter. He estimates that when his daughter enters college in 15 years, he will need $26,000 to put her through. He decides to deposit a certain amount of money into a bank account paying 6% compounded quarterly at the end of every 3 months. What must his quarterly payment be in order that he have $26,000 in 15 years?

Solution Mr. Carter will be paying into an ordinary annuity whose future value in 15 years is supposed to be $26,000. The future value of an ordinary annuity is, by equations (8.3.1) and (8.3.3) on pages 448 and 449, given by

$$A(n) = \text{future value} = \frac{B[(1 + i)^n - 1]}{i} = Bs_{\overline{n}|i}. \tag{1}$$

In this problem $A(n) = \$26,000$, $n = 60$ (4 payments a year for 15 years) and $i = 0.06/4 = 0.015$. The only unknown is the quarterly payment B. Thus, from (1),

$$B = \frac{26,000}{s_{\overline{60}|0.015}} = 26,000\left[\frac{0.015}{(1.015)^{60} - 1}\right] \approx 26,000\left(\frac{0.015}{1.44322}\right)$$

$$\approx 26,000(0.010393) \approx \$270.22.$$

Thus Mr. Carter must make quarterly payments of $270.22 to achieve his goal.
We can check this answer. An annuity of $270.22 every 3 months with $i = 0.015$ and $n = 60$ has a future value of

$$270.22\left[\frac{(1.015)^{60} - 1}{0.015}\right] \approx 270.22(96.2147) \approx \$25,999.14$$

(we lost 86¢ when we rounded).

We now define a term that is often used in business.

Sinking Fund

A **sinking fund** is an account into which periodic deposits are made so that a fixed sum of money may be paid on the date of maturity.

As in Example 1, we assume that the deposits are made regularly at the end of each period and that interest is paid at the time of each deposit. Then the sinking fund is an ordinary annuity with a fixed future value. The problem is then to determine the payment necessary to arrive at that value.

Let S denote the future value of the annuity. Then, from (1),

$$S = Bs_{\overline{n}|i} = B\left[\frac{(1 + i)^n - 1}{i}\right]$$

and we have

the regular payment B required in a sinking fund to have a future value S after n time periods, where interest i is paid at the end of each period, is given by

$$B = \frac{S}{s_{\overline{n}|i}} = \frac{iS}{(1 + i)^n - 1}. \qquad (2)$$

Example 2   How much will have to be invested at the end of each year at 8%, compounded annually, to pay off a debt of $75,000 after 10 years?

Solution   Here $S = \$75,000$, $i = 0.08$, and $n = 10$, so that, from (2) and Table 3 (at the back of the book) or a calculator,

$$B = \frac{75,000}{14.4866} \approx \$5177.20.$$

 Example 3   In most eastern European countries people cannot buy things on credit. In order to buy a car (or even a house), for example, it is necessary to pay cash. Members of the Gomulka family in Warsaw want to buy a car that, they estimate, will cost 250,000 zlotys in 8 years. They open an account in a state bank that pays 4% interest compounded monthly. What monthly deposit must the

Gomulkas make into the account in order to have enough money to buy their car in 8 years?

Solution The Gomulkas are establishing a sinking fund with $S = 250{,}000$, $i = 0.04/12$ and $n = 12 \cdot 8 = 96$. Thus their monthly payment is

$$B = \frac{(250{,}000)(0.04/12)}{(1 + 0.04/12)^{96} - 1} \approx \frac{833.3333}{0.376395} \approx 2214 \text{ zlotys.}$$

We now turn to the second application. In long-term debts with relatively large amounts of money involved, a borrower will often retire the debt by making regular payments until the debt and the interest on it are repaid. Examples include paying off a car, a home, or a loan used for business expansion.

 Example 4 Barbara Moorman buys a new car which sells for $8000. She agrees to pay for the car over 4 years by making 48 payments, one at the end of each month, at an interest rate of 12% compounded monthly. What will her monthly payments be?

Solution Ms. Moorman is paying into a regular annuity. The $8000 she receives immediately is the present value of that annuity. From equations (8.3.7) and (8.3.8) on page 451,

$$\text{present value of the annuity} = P = B\left[\frac{1 - (1 + i)^{-n}}{i}\right] = Ba_{\overline{n}|i}, \qquad (3)$$

where B is the periodic payment. In this problem $P = \$8000$, $i = 0.01$ (the monthly interest), and $n = 48$. Therefore, from (3),

$$8000 = B\left[\frac{1 - (1.01)^{-48}}{0.01}\right]$$

or

$$B = \frac{(8000)(0.01)}{1 - (1.01)^{-48}} \approx \frac{80}{0.37974} \approx \$210.67.$$

Thus Ms. Moorman will pay $210.67 a month for 48 months to pay for her car.

Amortization

The process of paying off a debt by systematically making partial payments until the debt and the interest are repaid is called **amortization**.

Remark 1. When a debt is amortized, the payments may be regular or irregular, equal or unequal. However, in this text, we will discuss only amortized loans with regular, equal payments so that we can apply our knowledge of ordinary annuities.

Remark 2. If it makes things easier, think of an amortized loan as an installment loan.

From (3), we obtain the following fact.

> The regular payment B required to pay off an amortized loan of P dollars in n time periods where interest is i per period is given by
>
> $$B = \frac{P}{a_{\overline{n}|i}} = \frac{iP}{1 - (1 + i)^{-n}}.$$ (4)

Example 5 Mr. Goldberg buys a country house for $60,000. He pays $5000 down and agrees to amortize the remaining debt with quarterly payments over the next 5 years. If interest is 8%, compounded quarterly, how large will his quarterly payments be?

Solution Here $P = \$60,000 - \$5000 = \$55,000, n = 20,$ and $i = 0.08/4 = 0.02$. Thus, from (4) and Table 4 (at the back of the book),

$$B = \frac{55,000}{a_{\overline{20}|0.02}} \approx \frac{55,000}{16.3514} \approx \$3363.63.$$

Example 6 Joe Ivy graduated from college and was immediately given a brand new VISA card. Flushed with a sense of economic power, Joe charged $3500 until his bank caught up with him and canceled the card. Joe agreed to pay the money back in regular monthly installments over a 3-year period. VISA charged 18% annual interest, compounded monthly. What was Joe's monthly payment?

Solution Here $P = \$3500, i = 0.18/12 = 0.015,$ and $n = 36$. Thus, from (4),

$$B = \frac{(0.015)(3500)}{1 - (1.015)^{-36}} \approx \frac{52.5}{0.41491} \approx \$126.53.$$

If you are paying off an amortized loan, you will often want to know the **outstanding principal** remaining on the loan. This is the present value of the annuity based on the number of payments still be to made. It is the amount to be paid if the loan is to be retired at once.

Example 7 What is the outstanding principal on Mr. Goldberg's loan (in Example 5) after his 12th payment?

Solution Since there are a total of 20 payments, there are 8 payments remaining. We must compute the present value of an annuity with $B = \$3363.63, n = 8,$ and $i = 0.02$. From (3) and Table 4 (at the back of the book), we have

$$P = 3363.63a_{\overline{8}|0.02} \approx 3363.63(7.32548) \approx \$24,640.20.$$

Thus it would cost Mr. Goldberg $24,640.20 if he wanted to pay off the loan after 3 years. This compares with 8 additional payments of $3363.63, which equals $26,909.04.

 Example 8   After making 8 monthly payments, Joe Ivy (Example 6) invents a car engine that runs on pickles. He receives $10 million for the patent and immediately pays off VISA. How much does he pay?

Solution   We must compute the outstanding principal on the annuity with $36 - 8 = 28$ payments remaining. Since $B = \$126.53$, $i = 0.015$, and $n = 28$, we obtain, from (3),

$$P = 126.53 \left[\frac{1 - (1.015)^{-28}}{0.015} \right] \approx 126.53(22.7267) \approx \$2875.61.$$

This compares with 28 payments of $126.53, which add up to $3542.84.

PROBLEMS 8.4

In Problems 1–10, a sinking fund is established to pay S dollars in n periods at an interest rate i paid at the end of each period. If payments are made at the end of each period, find the size of each payment.

1. $S = \$10,000, n = 10, i = 6\%$
2. $S = \$10,000, n = 10, i = 8\%$
3. $S = \$10,000, n = 10, i = 12\%$
4. $S = \$10,000, n = 10, i = 18\%$
5. $S = \$25,000, n = 60, i = 2\%$
6. $S = \$5000, n = 6, i = 4\%$
7. $S = \$20,000, n = 120, i = 1\%$
8. $S = \$8000, n = 60, i = 0.8\%$
9. $S = \$75,000, n = 30, i = 2.75\%$
10. $S = 50,000, n = 24, i = 4.5\%$

In Problems 11–20, determine the periodic payment required to pay off an amortized loan with an initial balance of S dollars, in n periods, with a periodic interest rate i.

11. $S = \$10,000, n = 10, i = 6\%$
12. $S = \$10,000, n = 10, i = 8\%$
13. $S = \$10,000, n = 10, i = 12\%$
14. $S = \$10,000, n = 10, i = 18\%$
15. $S = \$25,000, n = 60, i = 2\%$
16. $S = \$5000, n = 6, i = 4\%$
17. $S = \$20,000, n = 120, i = 1\%$
18. $S = \$8000, n = 60, i = 0.8\%$
19. $S = \$75,000, n = 30, i = 2.75\%$
20. $S = \$40,000, n = 24, i = 4.5\%$

21. A company establishes a sinking fund to discharge a debt of $75,000 due in 8 years by making equal semiannual deposits, the first due in 6 months. If its investment pays 6%, compounded semiannually, what is the size of the deposits?

22. A sinking fund is established to discharge a debt of $80,000 in 10 years. If deposits are made at the end of each 6-month period and interest is paid at the rate of 6%, compounded semiannually, what is the size of each deposit?

23. A debt of $10,000 is to be amortized by equal payments at the end of each year for 5 years. The interest charged is 8%, compounded annually.

   (a) Find the periodic payment.

   (b) Find the outstanding principal immediately after the third payment.

24. A debt of $15,000 is to be amortized by equal payments at the end of each year for 10 years. The interest charged is 6%, compounded annually.
    (a) Find the periodic payment.
    (b) Find the outstanding principal immediately after the eighth payment.

25. A debt of $8000 is to be amortized with eight equal semiannual payments. The interest rate is 9%, compounded semiannually.
    (a) What is the size of each payment?
    (b) What is the outstanding principal immediately after the fifth payment?

26. A loan of $10,000 is to be amortized with ten equal quarterly payments. The interest rate is 8%, compounded quarterly.
    (a) What is the periodic payment?
    (b) What is the outstanding principal after the sixth payment?

27. A man buys a house for $25,000. He pays $10,000 down and amortizes the rest of the debt with semiannual payments over the next 10 years. If the interest rate on the debt is 6%, compounded semiannually, what will be the size of the payments?

28. The Wilde Welding Company purchased $10,000 worth of equipment by making a $2000 down payment and promising to pay the remainder of the cost in semiannual payments over the next 4 years. If the interest rate on the debt is 13%, compounded semiannually, what will be the size of the payments?

29. John Polo buys a yacht for $48,000. He pays $10,000 down and agrees to amortize the remainder by making semiannual payments for 6 years at 10% interest, compounded semiannually. After $2\frac{1}{2}$ years, he decides to pay off the outstanding principal.
    (a) What are his semiannual payments?
    (b) How much does he pay in $2\frac{1}{2}$ years?

30. The Smiths put $12,000 down on a new house and make monthly payments of $475, based on a 30-year mortgage of 10%, compounded monthly. How much did they pay for the house?

31. How much interest will the Smiths in Problem 30 pay over a 30-year period?

8.5 Bonds (Optional)

A **bond** is a promissory note, which promises the investor a fixed sum of money at a future time and periodic interest payments.

A study of the rules under which people buy and sell bonds is, in some sense, a study of terminology. Once we have learned the right words, we see that all the computations involving bonds are applications of material covered in Sections 8.1, 8.2, and 8.3.

When the bond is issued, it carries with it a *maturity date*, a *coupon rate*, and a *face value* (or denomination).

1. The **face value** is the amount paid by the initial bond holder. It is, in essence, the amount of money loaned by the bondholder to the business that issued the bond. The face values of most bonds are multiples of $1000.

2. The **maturity date** is the date at which face value of the bond (the amount of money initially borrowed) is repaid to the bondholder.

3. The **coupon rate** (or **nominal rate**) is the annual rate of interest paid periodically to the bondholder. It is expressed as a percentage of the bond's face value.

Example 1   Mrs. Dvorak buys a bond in the Stainless Steel Company with a face value of $5000, a coupon rate of 7% paid semiannually, and a maturity date 10 years from the date of purchase. How much interest will she collect in 10 years?

Solution   The interest paid is simple interest. The semiannual interest is $3\frac{1}{2}\% = 0.035$. Each 6 months Mrs. Dvorak receives $(5000)(0.035) = \$175$ and, since she receives 20 interest payments, the total interest paid is $(175)(20) = \$3500$. In addition, at the maturity date she gets her $5000 back.

Unfortunately, things are not as simple as the last example suggests. Usually many years elapse between the issue date and the maturity date of a bond.[†] During that time interest rates change, and so do the prices people are willing to pay for the bond. The coupon rate of the bond is fixed during the lifetime of the bond. If current interest rates are less than the coupon rate, then an investor will be willing to pay more for the bond. If interest rates are higher, then the investor will pay less.

The **bond purchase**, or **book value**, price is expressed as a percentage of the face value of a bond. Thus a $1000 bond that sells for $780 will have a listed price of 78 (the percent sign is left off). Such a bond is said to be sold at a **discount**. If the same bond sells for $1150, then the bond price is 115. The bond is now being sold at a **premium**. If the bond purchase price is equal to the face value, the bond is said to be sold **at par**.

The **current yield**, or **effective interest rate**, of a bond is the annual interest payment divided by the bond price.

Example 2   What is the current yield of a $1000 bond sold at 78 with a coupon rate of 6%?

Solution   A $1000 bond with a coupon rate of 6% will pay $60 per year. At a price of 78 the bond will sell for $780 (78% of 1000). Thus

$$\text{yield} = \tfrac{60}{780} = 0.0769 \approx 7.7\%.$$

In Table 1 we reproduce an excerpt from the New York Stock Exchange Bond Trading report that summarizes the week ending September 5 in a recent year. We explain the entries in the table for the first bond listed.

(a), (b)   Give the high and low bond price for a 12-month period. The high of $100\frac{1}{2}$ means that a $1000 bond sold for as much as $1005. The low of 78 means that the bond sold for as little as $780.

(c)   Gives the name of the company issuing the bond. AMF stands for the AMF Corporation (American Machine and Foundry).

[†] Traditionally, this period has been in the neighborhood of 25 to 30 years. However, with recent unpredictable increases and decreases in interest rates, many companies are issuing at least some of their bonds for periods of 10 years or less. For an interesting discussion of this topic, see the article by Rosalyn Retkwa on page 2 of Section 3 in the *New York Times* of September 7, 1980.

TABLE 1   Corporation

12-month High	Low	Bonds	Cur. Yld	Sls in $1,000	High	Low	Last	Net Chg.
a	b	c d	e	f			g	
$100\frac{1}{2}$	78	AMF 10s85	11.0	10	91	91	91	$-\frac{1}{8}$
88	59	AMInt $9\frac{3}{8}$95	13.4	14	71	$70\frac{1}{8}$	$70\frac{1}{8}$	$+\frac{1}{2}$
86	61	APL $10\frac{3}{4}$97	14.5	40	74	$72\frac{1}{2}$	74	$+2\frac{1}{8}$
66	$41\frac{7}{8}$	ARA $4\frac{5}{8}$s96	cv	2	$57\frac{1}{4}$	$57\frac{1}{4}$	$57\frac{1}{4}$	$+1$
70	$59\frac{1}{8}$	ATO $4\frac{3}{8}$s87	cv	4	67	66	67	.....
90	$74\frac{3}{4}$	ATO $9\frac{3}{8}$86	11.0	12	85	$84\frac{7}{8}$	85	.....
$96\frac{1}{2}$	80	AbbtL 9.2s99	10.9	9	$84\frac{3}{8}$	$84\frac{3}{8}$	$84\frac{3}{8}$	$+1\frac{7}{8}$
88	$63\frac{1}{2}$	AetnLf $8\frac{1}{8}$07	11.8	17	69	$65\frac{1}{2}$	69	$+2$
$102\frac{1}{4}$	$97\frac{1}{2}$	AlaBn 9.65s99	9.7	32	$99\frac{3}{4}$	$98\frac{7}{8}$	$99\frac{3}{4}$	$+2\frac{1}{4}$
99	80	AlaBnc $9\frac{1}{4}$84	10.2	34	$93\frac{1}{2}$	$91\frac{1}{2}$	$93\frac{1}{2}$	$+2$
75	69	AlaP $3\frac{1}{2}$s84	4.2	5	$74\frac{3}{8}$	$74\frac{1}{4}$	$74\frac{1}{4}$	$+1\frac{1}{4}$
85	60	AlaP 9s2000	13.5	2	$66\frac{1}{2}$	$66\frac{1}{2}$	$66\frac{1}{2}$	.....
$81\frac{5}{8}$	$56\frac{1}{8}$	AlaP $8\frac{1}{2}$s01	13.7	5	62	62	62	$+\frac{7}{8}$
$76\frac{7}{8}$	$52\frac{1}{8}$	AlaP $7\frac{7}{8}$s02	12.7	12	$62\frac{1}{4}$	$61\frac{1}{4}$	$62\frac{1}{4}$	$+1\frac{1}{2}$
$76\frac{1}{4}$	52	AlaP $7\frac{3}{4}$s02	13.3	11	$58\frac{1}{4}$	$58\frac{1}{8}$	$58\frac{1}{4}$	.....
$82\frac{1}{2}$	59	AlaP $8\frac{7}{8}$s03	13.6	40	$67\frac{1}{2}$	$64\frac{1}{2}$	$65\frac{1}{4}$	$+\frac{3}{4}$
80	56	AlaP $8\frac{1}{4}$s03	12.9	10	64	$60\frac{1}{8}$	64	$+2$
$92\frac{1}{2}$	$63\frac{1}{4}$	AlaP $9\frac{3}{4}$s04	13.2	90	74	$71\frac{1}{2}$	74	$+2$
100	72	AlaP $10\frac{7}{8}$05	13.7	15	$79\frac{1}{4}$	$79\frac{1}{4}$	$79\frac{1}{4}$	$-1\frac{3}{4}$
$99\frac{1}{4}$	$70\frac{1}{8}$	AlaP $10\frac{1}{2}$05	13.7	23	$79\frac{1}{8}$	$76\frac{3}{8}$	$76\frac{3}{8}$	$-2\frac{5}{8}$
$84\frac{5}{8}$	58	AlaP $8\frac{7}{8}$06	13.4	25	$66\frac{3}{8}$	$66\frac{3}{8}$	$66\frac{3}{8}$	$-\frac{1}{8}$
83	57	AlaP $8\frac{3}{4}$07	13.7	41	$64\frac{1}{8}$	$63\frac{1}{4}$	64	$+\frac{3}{4}$
90	$68\frac{1}{4}$	AlaP $8\frac{5}{8}$87	11.0	5	79	79	79	$+1\frac{1}{8}$
$87\frac{3}{4}$	61	AlaP $9\frac{1}{4}$07	13.4	10	69	69	69	.....
$90\frac{1}{2}$	$59\frac{5}{8}$	AlaP $9\frac{1}{4}$08	12.8	15	74	72	74	$+4$
90	62	AlaP $9\frac{5}{8}$08	13.3	22	$72\frac{1}{2}$	$68\frac{3}{4}$	$72\frac{1}{2}$	$+3\frac{3}{4}$
$93\frac{1}{8}$	90	AlaP $12\frac{5}{8}$10	14.0	71	91	$90\frac{1}{2}$	$90\frac{1}{2}$	$-\frac{1}{4}$
103	82	Alskin $12\frac{3}{4}$99	14.2	6	$89\frac{7}{8}$	87	$89\frac{7}{8}$	.....

(d)   Gives coupon rate and maturity date. 10s85 means that the bond has a 10% coupon rate with a maturity date of 1985.

(e)   Gives the current yield: 11.0 means 11%.

(f)   Gives sales; 10 $1000 bonds were sold during the week.

(g)   Gives weekly closing bond price. The weekly closing bond price was 91. This means that the bond sold for $910 on September 5 of that year.

Note.   At a coupon rate of 10%, the $1000 bond will pay $100 per year. Since the purchase price is $910, we have

$$\text{yield} = \tfrac{100}{910} = 0.1099 \approx 0.11 = 11\%.$$

This explains the 11% yield in column (e).

The price of a bond is, as we have seen, affected by current interest rates. We can compute the market price by using the procedure outlined in the next example.

 Example 3 A bond has a face value of $1000 and a coupon rate of 7%; it will mature in 12 years. The bond pays interest semiannually. Current interest rates are 10%. What bond price will provide a yield of 10%?

Solution The purchaser of the bond is, in effect, buying two things: a note due to pay $1000 in 12 years and an annuity consisting of 24 semiannual payments. Thus, in order to yield 10%, the bond price will be the sum of the present value of the $1000 due and the present value of the annuity, based on 10% compounded semiannually.

The present value of the $1000 due is given by (see equation (8.2.1) on page 443)

$$P(\$1000) = 1000\left(1 + \frac{0.10}{2}\right)^{-2(12)} \approx \$310.$$

Every 6 months the annuity pays $3\frac{1}{2}\%$ of $1000 = $35. There are 24 such payments so, as 10% annually equals 5% semiannually, the present value of the annuity is (see equation (8.3.6) on page 451)

$$P\text{ (annuity)} = \frac{35}{0.05}\left[1 - \left(\frac{1}{1.05}\right)^{24}\right] \approx \$483.$$

Thus, in order to yield 10%,

$$\text{bond price} = P(\$1000) + P(\text{annuity}) = \$310 + \$483 = \$793.$$

This bond is bought at a discount.

 Example 4 In early 1980, the prime interest rate[†] reached 20%. Later in that year, the rates fell to 12%. Suppose that in March, a corporation issued $1000 bonds with a coupon rate of 18% and a maturity date of March, 1995. What would be a fair price for the bond in September, 1980, assuming a yield of 12% if interest is paid quarterly?

Solution In September, 1980, there were $14\frac{1}{2}$ years remaining until maturity. The present value of $1000 in $14\frac{1}{2}$ years, compounded quarterly at 12%, is

$$P(\$1000) = 1000\left(1 + \frac{0.12}{4}\right)^{-(14.5)(4)} = 1000(1.03)^{-58} \approx \$180.$$

---

† This is the rate the banks charge their best and most reliable borrowers.

At a coupon rate of 18 %, the bond pays \$45 each quarter. There are $14\frac{1}{2} \times 4 = 58$ payments due in this annuity. The present value of the annuity is

$$P(\text{annuity}) = \frac{45}{0.03}\left[1 - \left(\frac{1}{1.03}\right)^{58}\right] \approx 1230.$$

Thus, in order to yield 12 %, the bond value must be \$180 + \$1230 = \$1410. The bond would sell at the premium listed price of 141.

PROBLEMS 8.5

In Problems 1–4, determine how much interest is paid in n years for a bond with face value F and coupon rate i.

  1. $F = \$3000, n = 10, i = 6\%$

  2. $F = \$1000, n = 25, i = 8\%$

  3. $F = \$10,000, n = 30, i = 7\frac{1}{2}\%$

  4. $F = \$2500, n = 16, i = 11\frac{1}{2}\%$

In Problems 5–10, determine the yield of a \$1000 bond with a coupon rate of i and a price D (given as a percentage of the face value).

  5. $D = 80, i = 6\%$

  6. $D = 110, i = 7\frac{1}{2}\%$

  7. $D = 95, i = 10\%$

  8. $D = 135, i = 9\frac{1}{2}\%$

  9. $D = 182, i = 4\frac{3}{4}\%$

  10. $D = 57, i = 8\frac{1}{4}\%$

 In Problems 11–20, determine the purchase price of a \$1000 bond due to mature in n years if the coupon rate is i and the yield is Y. Assume that interest is paid and compounded semiannually.

  11. $i = 6\%, Y = 7\%, n = 10$

  12. $i = 6\%, Y = 7\%, n = 20$

  13. $i = 6\%, Y = 7\%, n = 6$

  14. $i = 8\%, Y = 6\%, n = 10$

  15. $i = 8\%, Y = 6\%, n = 20$

  16. $i = 8\%, y = 6\%, n = 6$

  17. $i = 6\frac{3}{4}\%, Y = 7\frac{1}{2}\%, n = 15$

  18. $i = 8.6\%, Y = 7.2\%, n = 30$

  19. $i = 12\%, Y = 11.4\%, n = 25$

  20. $i = 11.4\%, Y = 12\%, n = 25$

 21. Mr. Conigliaro buys a 15-year, \$5000 bond with a coupon rate of 9 % that pays interest quarterly. Each time he receives an interest payment, he puts the money into a bank account paying 6 % compounded quarterly. How much money will Mr. Conigliaro have in the account after 15 years?

 22. What is the purchase price of a \$10,000 bond due to mature in 1 year that has a coupon rate of 9 % and pays interest monthly if an investor expects a yield of 12 %, compounded monthly?

 ·23. A **serial bond** is a bond that is redeemed in installments rather than all at one time. Suppose an \$8000 serial bond has a coupon rate of 6 % with interest paid semiannually. It is redeemed by payments of \$3000 in 8 years and \$5000 in 15 years. What is the purchase price if it is to yield 8 %, compounded semiannually?

 ·24. Answer the question in Problem 23 if the bond is to yield 5 %.

 ·25. Find the purchase price of a \$25,000 serial bond yielding $8\frac{1}{2}\%$ if the coupon rate is $7\frac{3}{4}\%$ and the bond is redeemed by payments of \$6000 in 6 years, \$9000 in 10 years, and \$10,000 in 16 years.

 ·26. Answer the question in Problem 25 if the yield is only $6\frac{1}{4}\%$.

▦ Review Exercises for Chapter 8

1. What is the simple interest paid on $10,000 invested at 6% for 7 years?

2. What is the simple interest paid on $8000 invested at $7\frac{1}{2}$% for 12 years?

In Problems 3–10, compute the value of an investment after t years and the total interest paid if P dollars is invested at a nominal rate of i%, compounded m times a year.

3. $P = \$6000$, $i = 4\frac{1}{2}$%, $t = 6$, $m = 1$

4. $P = \$750$, $i = 8$%, $t = 10$, $m = 4$.

5. $P = \$10,000$, $i = 6\frac{1}{2}$%, $t = 4$, $m = 2$

6. $P = \$3000$, $i = 18$%, $t = 5$, $m = 12$

7. $P = \$8000$, $i = 12$%, $t = 6$, $m = 12$

8. $P = \$25,000$, $i = 7\frac{3}{4}$%, $t = 15$, $m = 4$

9. $P = \$7500$, $i = 6$%, $t = 10$, $m = 52$

10. $P = \$3500$, $i = 8\frac{1}{2}$%, $t = 7\frac{1}{2}$, $m = 365$

11. If money is invested at 8% compounded quarterly, what is the effective interest rate?

12. What is the effective interest rate of 8% compounded monthly?

13. How long will it take an investment to double (assuming quarterly compounding) if the interest rate is

  (a) 3%?

  (b) $6\frac{1}{2}$%?

  (c) 8%?

  (d) 12%?

14. Answer the questions of Exercise 13 if money is compounded monthly.

15. Answer the questions of Exercise 13 if money is compounded daily.

In Exercises 16–21, compute the present value of A dollars, t years in the future, at a nominal rate of i%, compounded m times a year.

16. $A = \$10,000$, $t = 8$, $i = 5$%, $m = 1$

17. $A = \$10,000$, $t = 8$, $i = 5$%, $m = 4$

18. $A = \$10,000$, $t = 8$, $i = 5$%, $m = 12$

19. $A = \$6500$, $t = 15$, $i = 7\frac{3}{4}$%, $m = 1$

20. $A = \$6500$, $t = 15$, $i = 7\frac{3}{4}$%, $m = 2$

21. $A = \$6500$, $t = 15$, $i = 7\frac{3}{4}$%, $m = 4$

22. Mr. Berger holds a $15,000 note due in 4 years, which he sells to Mrs. Nordstrom. The purchase price is based on a rate of 10% compounded quarterly. What is the purchase price?

23. The note in Exercise 22 was sold for $10,200. Assuming quarterly compounding, what was the discount rate?

24. In Exercise 23, compute the discount rate assuming monthly compounding.

In Exercises 25–30, find the future value of an ordinary annuity with payments of B dollars, over n periods and with interest rate i per period.

25. $B = \$1000, n = 8, i = 0.03$ **26.** $B = \$1000, n = 8, i = 0.07$

27. $B = \$35, n = 36, i = 0.01$ **28.** $B = \$35, n = 36, i = 0.015$

29. $B = \$475, n = 360, i = \dfrac{0.085}{12}$ **30.** $B = \$475, n = 360, i = 0.01$

In Exercises 31–36, find the present value of an ordinary annuity with payments of B dollars, over n periods and with interest rate i per period.

31. $B = \$1000, n = 8, i = 0.03$ **32.** $B = \$1000, n = 8, i = 0.07$

33. $B = \$35, n = 36, i = 0.01$ **34.** $B = \$35, n = 36, i = 0.015$

35. $B = \$475, n = 360, i = \dfrac{0.085}{12}$ **36.** $B = \$475, n = 360, i = 0.01$

37. Find the future value of an ordinary annuity of $100 paid at the end of each quarter for 8 years if interest is earned at a rate of 6%, compounded quarterly.

38. Find the future value of an ordinary annuity of $325 paid every 6 months for 12 years if the interest rate is $11\frac{1}{2}\%$, compounded semiannually.

39. Find the future value of an annuity due of $100 payable at the beginning of each year for 8 years if the interest rate is 9%, compounded annually.

40. Find the future value of an annuity due of $250 payable at the beginning of each month for 6 years if the interest rate is 6%, compounded monthly.

In Exercises 41–44, a sinking fund is established to pay S dollars in n periods with an interest rate of $i\%$ paid at the end of each period. If payments are made at the end of each period, find the size of the payment.

41. $S = \$15,000, n = 12, i = 3\%$ **42.** $S = \$25,000, n = 60, i = 1\%$

43. $S = \$4200, n = 36, i = 1\frac{1}{2}\%$ **44.** $S = \$6000, n = 10, i = 5\%$

In Exercises 45–48, determine the periodic payment required to pay off an amortized loan with an initial balance of S dollars, in n periods, with a periodic interest rate of $i\%$.

45. $S = \$15,000, n = 12, i = 3\%$ **46.** $S = \$25,000, n = 60, i = 1\%$

47. $S = \$4200, n = 36, i = 1\frac{1}{2}\%$ **48.** $S = \$6000, n = 10, i = 5\%$

49. A sinking fund is established to discharge a debt of $35,000 in 8 years. If deposits are made at the end of each 6-month period and interest is paid at the rate of $8\frac{1}{2}\%$, compounded semiannually, what is the size of each deposit?

50. A loan of $6500 is to be amortized with 20 equal quarterly payments. The interest rate is 11.4%, compounded quarterly.

(a) What is the periodic payment?

(b) What is the outstanding principal after the 12th payment?

51. A loan of $13,500 is to be amortized with equal monthly payments over a 10-year period. The interest rate is $7\frac{3}{4}\%$, compounded monthly.

    (a)   What is the periodic payment?

    (b)   What is the outstanding principal after 2 years?

52. How much interest is paid in 4 years for a $2000 bond with a coupon rate of 7%?

53. How much interest is paid in 15 years for a $5000 bond with a coupon rate of $8\frac{1}{2}\%$?

54. What is the yield of a $1000 bond purchased for $850 if the coupon rate is 6%?

55. What is the yield of a $5000 bond purchased for $6250 if the coupon rate is 10%?

56. What is the purchase price of a $1000 bond due to mature in 15 years if the coupon rate is 8%, paid semiannually, and the yield is 9%?

57. What is the purchase price of a $2500 bond due to mature in 12 years if the coupon rate is 10%, paid semiannually, and the yield is $8\frac{3}{4}\%$?

58. What is the purchase price of a $10,000 serial bond with a coupon rate of 8% paid quarterly and a yield of 9% compounded quarterly, if $4000 is to be redeemed in 4 years and $6000 is to be redeemed in 10 years?

9 FUNCTIONS AND GRAPHS

9.1 Functions and Graphs

Let us return to the equation of a straight line (see Section 2.3). For example, consider the line whose equation is $y = 3x + 5$. This line can be thought of as the set of all ordered pairs (or points) (x, y) such that $y = 3x + 5$. The important fact here is that, for every real number x, there is a *unique* real number y such that $y = 3x + 5$ and the ordered pair (x, y) is on the line. We generalize this idea in the following way.

Function

Let X and Y be sets of real numbers. A **function**, f, is a rule that assigns to each number x in X a single number $f(x)$ in Y. X is called the **domain** of f. The set of images of elements of X is called the **range** of f, or range f.

Remark. Simply put, a function is a rule that assigns to every x in the domain of f a unique number y in the range of f. We will usually write this as

$$y = f(x), \tag{1}$$

which is read "y equals f of x."[†]

When the domain of a function is not given, we usually take the domain to be the set of values for which equation (1) makes sense. For example, let f be the function given by $f(x) = 1/x$. Since the expression $1/x$ is not defined for $x = 0$, the number 0 is not in the domain of f. However, $1/x$ is defined for any $x \neq 0$, so that the domain of f is the set of all real numbers except zero. This

---

[†] This notation was first used by the great Swiss mathematician Leonhard Euler (1707–1783) in the *Commentarii Academia Petropolitanae* (Petersburg Commentaries), published in 1734–1735.

can be written as $\mathbb{R} - \{0\}$. The range of f is also $\mathbb{R} - \{0\}$ because $1/x$ can take on any real number except 0. To see this, let r be a number with $r \neq 0$. Then if

$$x = \frac{1}{r},$$

$$\frac{1}{x} = \frac{1}{(1/r)} = r.$$

Example 1   Let $f(x) = \dfrac{1}{x-3}$.

(a)   Find the domain of f.

(b)   Evaluate $f(1)$, $f(-1)$, and $f(5)$.

(c)   Find the range of f.

Solution

(a) f is defined as long as we are not dividing by 0. The denominator is zero when $x - 3 = 0$ or $x = 3$. Thus domain of $f = \mathbb{R} - \{3\}$.

(b)
$$f(1) = \frac{1}{1-3} = \frac{1}{-2} = -\frac{1}{2}$$

$$f(-1) = \frac{1}{-1-3} = \frac{1}{-4} = -\frac{1}{4}$$

$$f(5) = \frac{1}{5-3} = \frac{1}{2}$$

(c)   Suppose that $r \neq 0$. Let us try to find an x such that $\dfrac{1}{x-3} = r$. Then

$$1 = r(x - 3) \qquad \text{We multiplied by } x - 3$$

$$1 = rx - 3r$$

$$rx = 1 + 3r$$

$$x = \frac{1 + 3r}{r}. \qquad \text{Valid because } r \neq 0$$

For example, if $r = 20$,

$$x = \frac{1 + 3r}{r} = \frac{1 + 60}{20} = \frac{61}{20}$$

and

$$\frac{1}{x-3} = \frac{1}{\frac{61}{20} - 3} = \frac{1}{\frac{61}{20} - \frac{60}{20}} = \frac{1}{\frac{1}{20}} = 20.$$

Thus if $r \neq 0$, there is an x such that $\dfrac{1}{x-3} = r$. This means that the range of f is $\mathbb{R} - \{0\}$.

Example 2 Let $f(x) = \sqrt{3x + 4}$.

(a) Find the domain of f.

(b) Evaluate $f(0)$, $f(-1)$, $f(-2)$, and $f(10)$.

(c) Find the range of f.

Solution

(a) We cannot take the square root of a negative number, so f is defined if

$$3x + 4 \geq 0$$

or

$$3x \geq -4$$

or

$$x \geq -\tfrac{4}{3}.$$

Thus

$$\text{domain of } f = [-\tfrac{4}{3}, \infty).$$

(b) $f(0) = \sqrt{3 \cdot 0 + 4} = \sqrt{4} = 2$

$f(-1) = \sqrt{3(-1) + 4} = \sqrt{-3 + 4} = \sqrt{1} = 1$

$f(-2)$ is not defined since -2 is not in the domain of f $[3(-2) + 4 = -6 + 4 = -2 < 0]$

$f(10) = \sqrt{3 \cdot 10 + 4} = \sqrt{34}$

(c) $\sqrt{x}$ denotes the positive square root, so $f(x) = \sqrt{3x + 4} \geq 0$ for every x and

$$\text{range of } f = [0, \infty),$$

which is denoted $\mathbb{R}^+$.

Example 3 Let $f(x) = x^2 - 4x + 1$. Find

(a) domain of f,

(b) $f(2)$ and $f(-5)$,

(c) range f.

Solution

(a) f is defined for every real number, so domain of $f = \mathbb{R}$.

(b)   Since $f(x) = x^2 - 4x + 1$, substituting 2 for x gives us

$$f(2) = 2^2 - 4 \cdot 2 + 1 = 4 - 8 + 1 = -3$$
$$f(-5) = (-5)^2 - 4(-5) + 1 = 25 + 20 + 1 = 46.$$

(c)   This part is more difficult. To answer the question, we complete the square:[†]

$$x^2 - 4x + 1 = (x^2 - 4x + 4) - 4 + 1 = (x - 2)^2 - 3.$$

Since $(x - 2)^2 \geq 0$ (a square is always nonnegative),

$$x^2 - 4x + 1 = (x - 2)^2 - 3 \geq -3.$$

Thus

$$\text{range} f = [-3, \infty),$$

since $(x - 2)^2$ can equal any nonnegative real number.

At this point there are essentially three reasons for restricting the domain of a function:

(a)   you cannot divide by zero (see Example 1);

(b)   you cannot take an even root (square root, fourth root, etc.) of a negative number (see Example 2);

(c)   the domain is restricted by the nature of the business or physical problem under consideration (see Example 9 on page 476).

You will see a fourth kind of restriction when we discuss logarithmic functions in Section 9.5.

Graph of a Function   The **graph** of the function f is the set of ordered pairs $\{(x, f(x)): x \in \text{domain}$ of $f\}$. You saw in Section 2.2 how to obtain the graph of a straight line. Obtaining the graphs of other types of functions is often more difficult.

Example 4   Let $y = f(x) = x^2$. This rule constitutes a function whose domain is $\mathbb{R}$ because each real number has a unique square. The range of f equals $\{(x: x \geq 0)\} = \mathbb{R}^+$, the set of nonnegative real numbers, since the square of any real number is nonnegative. The graph of this function is obtained by plotting all points of the form $(x, y) = (x, x^2)$.[‡] First we note that as $f(x) = x^2, f(x) = f(-x)$ since $(-x)^2 = x^2$. Thus, it is necessary only to calculate $f(x)$ for $x \geq 0$. For every $x > 0$, there is a value of $x < 0$ that gives the same value of y. In this situation we say that the function is **symmetric** about the y-axis. Some values for $f(x)$ are shown in Table 1. The graph drawn in Figure 1 is the graph of a **parabola**.

---

† The technique of completing the square is discussed in Appendix A1.2.

‡ Of course, we can't plot *all* points (there are an infinite number of them). Rather, we plot some sample points and assume they can be connected to obtain the sketch of the graph.

TABLE 1

x	0	$\frac{1}{2}$	1	$\frac{3}{2}$	2	$\frac{5}{2}$	3	4	5
$f(x) = x^2$	0	$\frac{1}{4}$	1	$\frac{9}{4}$	4	$\frac{25}{4}$	9	16	25

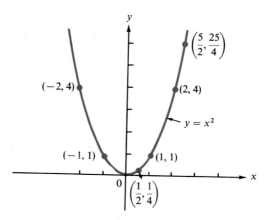

Figure 1

Example 5   Graph the function $f(x) = \sqrt{x}$.

Solution   First observe that f is defined only for $x \geq 0$. Table 2 gives values of $\sqrt{x}$. We plot these points

TABLE 2

x	0	0.5	1	2	3	4	5	10	15	20	25
$\sqrt{x}$	0	0.707	1	1.414	1.732	2	2.236	3.162	3.873	4.472	5

and then join them to obtain the graph in Figure 2.

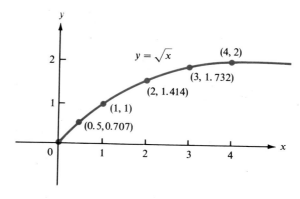

Figure 2

Example 6   Graph the function $f(x) = x^3$.

Solution f is defined for all real numbers (domain of $f = \mathbb{R}$), and the cube of a negative number is negative, so we plot both positive and negative values for x. Some representative values are given in Table 3. We plot these points and join them to obtain the graphs in Figure 3.

TABLE 3

x	0	$\frac{1}{2}$	$-\frac{1}{2}$	1	-1	2	-2	3	-3
x^3	0	$\frac{1}{8}$	$-\frac{1}{8}$	1	-1	8	-8	27	-27

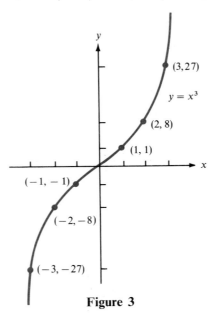

Figure 3

In all the examples considered so far, we gave you a function and asked questions about it. However, not everything that at first looks like a function actually is a function. We illustrate this in the next example. Remember that to have a function, there must be a single value of $f(x)$ for each x in the domain of f.

Example 7   Consider the equation $y^2 = x$. The rule $f(x) = y$ where $y^2 = x$ does *not* determine y as a function of x, since for every $x > 0$, there are *two* values of y such that $y^2 = x$; namely, $y = \sqrt{x}$ and $y = -\sqrt{x}$. For example, if $x = 4$, then $y = 2$ and $y = -2$; both satisfy $y^2 = 4$. However, if we specify one of these values, say $g(x) = \sqrt{x}$, then we have a function. Here, domain of $g = \mathbb{R}^+$ and range of $g = \mathbb{R}^+$. We could obtain a second function, h, by choosing the negative square root. That is, the rule defined by $h(x) = -\sqrt{x}$ is a function with domain $\mathbb{R}^+$ and range $\mathbb{R}^-$ (the nonpositive real numbers).

We can look at these things in a different way. The graph of the relation $y^2 = x$ is given in Figure 4. The figure shows that for every positive x, there are two y's such that (x, y) is on the graph. This evidently violates the rule that, for y to be a function of x, there must be a unique y for every x.

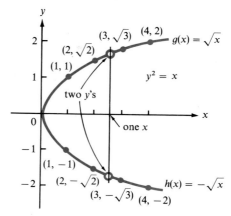

Figure 4

Example 8   For the following equations, determine whether y can be written as a function of x.

(a) $3x - 8y = 10$

(b) $x^2 y = 3$

(c) $y^2 - 2x^2 = 4$

(d) $x^3 + y^3 = 2$

Solution

(a)   We can rewrite this equation as

$$8y = 3x - 10 \quad \text{or} \quad y = \tfrac{1}{8}(3x - 10).$$

For every real number x, there is exactly one value for $\tfrac{1}{8}(3x - 10)$. For example, if $x = 0$, $y = -\tfrac{10}{8} = -\tfrac{5}{4}$; if $x = -5$, $y = \tfrac{1}{8}(-15 - 10) = -\tfrac{25}{8}$. So the answer is yes.

(b)   Here we may write $y = 3/x^2$ and, for every nonzero x, $3/x^2$ is a positive real number. So again the answer is yes.

(c)   Solving, we see that $y^2 = 4 - 2x^2$ or $y = \pm\sqrt{4 + 2x^2}$. Now, for any real number x, there are two values of y (one positive, one negative) such that $y^2 - 2x^2 = 4$. Thus y cannot be written as a function of x.

(d)   Here $y^3 = 2 - x^3$, or $y = \sqrt[3]{2 - x^3}$. The cube root of a positive number is positive and the cube root of a negative number is negative, so we can, indeed, write y as a function of x.

Suppose that $y = f(x) = mx + b$ for some numbers m and b. Then f is called a **linear function** or a **straight-line function**. We have already seen how linear functions can occur in applications. For instance, in Example 2.3.10 on page 54 we discussed the total cost and total revenue functions, which in that example were linear. Since we now know how to find the equation of a straight line, we can find total cost and total revenue functions, assuming that they are linear, if we know two "points" on the line.

Example 9   The manager of a shoelace company has found that it costs $2770 to produce 1000 pairs of shoelaces and $3310 to produce 3000 pairs. Find the total cost function, assuming that it is linear.

Solution   If C denotes cost and q denotes the number of pairs of shoelaces manufactured, then we seek the linear function $C(q)$ given that $C(1000) = 2770$ and $C(3000) = 3310$. We have assumed that C is linear, and we know that two points on this line (in the qC-plane) are $(1000, 2770)$ and $(3000, 3310)$. Using the techniques of Section 2.3 to find the slope of this line, we obtain

$$m = \frac{C - 2770}{q - 1000} = \frac{3310 - 2770}{3000 - 1000} = \frac{540}{2000} = \frac{27}{100} = 0.27.$$

Thus,

$$C - 2770 = 0.27(q - 1000) = 0.27q - 270$$

and

$$C = 0.27q + 2500.$$

Using the definitions in Example 1.3.5 on page 17, we see that the shoelace manufacturer has fixed costs of $2500 and variable costs of 27¢ per pair. Note that domain of $C = \{q: q \geq 0,\ q \text{ is an integer}\}$. This is a natural restriction caused by the fact that it is impossible to produce a negative or fractional number of pairs of shoelaces.

This total cost function is graphed in Figure 5.

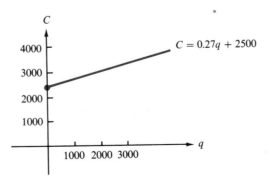

Figure 5

In the next example we show a different way to define a function.

Example 10   Let

$$f(x) = \begin{cases} 1, & x \geq 0 \\ 2, & x < 0 \end{cases}$$

It is perfectly legitimate to define a function in "pieces," as we have done, so long as for each x in the domain of f there is a unique y in the range. A graph of this function is given in Figure 6. We have domain of $f = \mathbb{R}$ and range of $f = \{1, 2\}$.

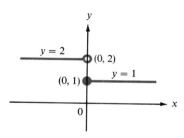

Figure 6

PROBLEMS 9.1   In Problems 1–10 evaluate the given function at the given values.

1. $f(x) = 1/(1 + x)$; $f(0)$, $f(1)$, $f(-2)$, and $f(-5)$.

2. $f(x) = 1 + \sqrt{x}$; $f(0)$, $f(1)$, $f(16)$, and $f(25)$.

3. $f(x) = 3x^2 + 1$; $f(0)$, $f(-3)$, $f(2)$, and $f(10)$.

4. $f(x) = 1/2x^3$; $f(1)$, $f(\frac{1}{2})$, and $f(-3)$.

5. $f(x) = x^4$; $f(0)$, $f(2)$, $f(-2)$, and $f(\sqrt{5})$.

6. $g(t) = t/(t - 2)$; $g(0)$, $g(1)$, $g(-1)$, and $g(3)$.

7. $g(t) = \sqrt{t + 1}$; $g(0)$, $g(-1)$, $g(3)$, and $g(7)$.

8. $h(z) = \sqrt[3]{z}$; $h(0)$, $h(8)$, $h(-\frac{1}{27})$, and $h(1000)$.

9. $h(z) = 1 + z + z^2$; $h(0)$, $h(2)$, $h(\frac{1}{3})$, and $h(-\frac{1}{2})$.

10. $h(z) = z^3 + 2z^2 - 3z + 5$; $h(0)$, $h(1)$, $h(-1)$, and $h(2)$.

In Problems 11–22 an equation involving x and y is given. Determine whether or not y can be written as a function of x.

11. $2x + 3y = 6$

12. $\dfrac{x}{y} = 2$

13. $x^2 - 3y = 4$

14. $x - 3y^2 = 4$

15. $x^2 + y^2 = 4$

16. $x^2 - y^2 = 1$

17. $\sqrt{x+y} = 1$

18. $y^2 + xy + 1 = 0$ [*Hint*: Use the quadratic formula.]

19. $y^3 - x = 0$ **20.** $y^4 - x = 0$

21. $y = |x|$ **22.** $y^2 = \dfrac{x}{x+1}$

23. Explain why the equation $y^n - x = 0$ allows us to write y as a function of x if n is an odd integer but does not if n is an even integer. [*Hint*: First solve Problems 19 and 20.]

In Problems 24–37 find the domain and range of the given function.

24. $y = f(x) = 2x - 3$ **25.** $s = g(t) = 4t - 5$

26. $y = f(x) = 3x^2 - 1$ **27.** $v = h(u) = \dfrac{1}{u^2}$

28. $y = f(x) = x^3$ **29.** $y = f(x) = \dfrac{1}{x+1}$

30. $s = g(t) = t^2 + 4t + 4$ **31.** $y = f(x) = \sqrt{x^3 - 1}$

32. $v = h(u) = |u - 2|$ **33.** $y = f(x) = \dfrac{1}{|x|}$

34. $y = f(x) = \dfrac{1}{|x+2|}$ **35.** $y = \begin{cases} x, & x \ge 0 \\ -x, & x < 0 \end{cases}$

36. $y = \begin{cases} x, & x \ge 1 \\ 1, & x < 1 \end{cases}$ **37.** $y = \begin{cases} x^3, & x > 0 \\ x^2, & x \le 0 \end{cases}$

In Problems 38–43 sketch the graph of the given function by plotting some points and then connecting them. Use a calculator where marked.

38. $f(x) = (x-1)^2$ **39.** $f(x) = \sqrt{3x - 4}$ **40.** $f(x) = -2x^2$

41. $f(x) = 1 + 2x^2$ **42.** $f(x) = \sqrt[3]{x}$

43. $f(x) = x^2 - 4x + 7$

44. Let $f(x) = \dfrac{1}{x-1}$. Find $f(t^2)$ and $f(3t + 2)$.

45. Let $f(x) = x^2$. Find $f(x + \Delta x)$ and $[f(x + \Delta x) - f(x)]/\Delta x$, where x denotes an arbitrary real number.[†] Simplify your answer.

*46. Let $f(x) = \sqrt{x}$. Show, assuming that $\Delta x \ne 0$, that

$$\frac{f(x + \Delta x) - f(x)}{\Delta x} = \frac{1}{\sqrt{x + \Delta x} + \sqrt{x}}.$$

[*Hint*: Multiply and divide by $\sqrt{x + \Delta x} + \sqrt{x}$.]

[†] Here Δx, read "delta x," denotes a small change in x. It does not stand for the number Δ times the number x.

47. Let $f(x) = |x|/x$. Show that $f(x) = \begin{cases} 1, & x > 0 \\ -1, & x < 0 \end{cases}$. Find the domain and range of f.

48. Describe a computational rule for expressing Fahrenheit temperature as a function of Centigrade temperature. [*Hint*: Pure water at sea level boils at 100°C and 212°F, and it freezes at 0°C and 32°F; the graph of this function is a straight line.]

49. Consider the set of all rectangles whose perimeters are 50 cm. Once the width, W, of any one rectangle is measured, it is possible to compute the area of the rectangle. Verify this by producing an explicit expression for area A as a function of width W. Find the domain and the range of your function.

*__50.__ A spotlight shines on a screen; between them is an obstruction that casts a shadow (see Figure 7). Suppose the screen is vertical, 20 m wide by 15 m high, and 50 m from the spotlight. Also suppose the obstruction is a square, 1 m on a side, and is parallel to the screen. Express the area of the shadow as a function of the distance from the light to the obstruction.

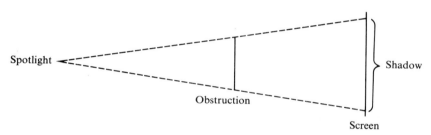

Spotlight

Obstruction

Screen

Shadow

Figure 7

*__51.__ A baseball diamond is a square, 90 feet long on each side. Casey runs a constant 30 ft/sec whether he hits a ground ball or a home run. Today, in his first at-bat, he hit a home run. Write an expression for the function that measures his line-of-sight distance from second base as a function of the time, t in sec, after he left home plate.

52. Let $f(x)$ be the fifth decimal place of the decimal expansion of x. For example, $f(\frac{1}{64}) = f(0.015624) = 2, f(98.786543210) = 4, f(-78.90123456) = 3$, etc. Find the domain and range of f.

53. Alec, on vacation in Canada, found that he got a 12% premium on his U.S. money. When he returned, he discovered there was a 12% discount on converting his Canadian money back into U.S. currency. Describe each conversion function. Show that, after converting both ways, Alec lost money.

54. The shoelace manufacturer of Example 9 modernized his plant and then found that it cost him $1160 to produce 2000 pairs of shoelaces and $1700 to produce 5000 pairs. Find his new total cost function assuming that it is linear. Determine his fixed and variable costs.

55. A woman in Iowa buys corn from farmers and sells it from a roadside stand. She pays 50¢ a dozen for the corn and sells it for 80¢ a dozen. Her fixed costs for maintenance of the stand and wages for additional help average $40 per day. Assuming that she is able to sell all the corn she buys, determine her profit function.

56. Suppose that the woman in Problem 55 does not always sell all the corn she buys, but any unsold corn can be sold to a canner for 28¢ a dozen. Determine the profit function.

57. The Dow Jones closing averages for industrial stocks are given in Figure 8 for the 3-month period from April 15 to July 15, 1983. Let April 15 be day 1 and July 15 be day 92. Let $A(t)$ be the closing average on day t. Find (a) $A(1)$, (b) $A(8)$, (c) $A(30)$, (d) $A(60)$, and (e) $A(88)$.

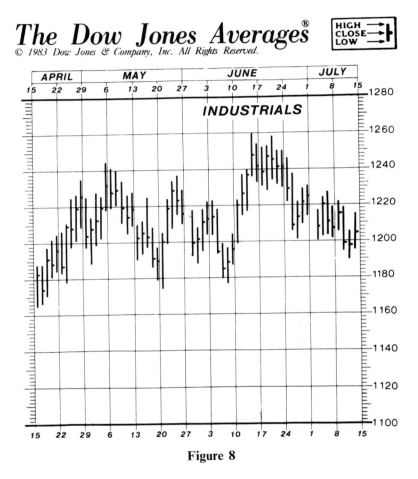

Figure 8

9.2 Operations with Functions

We begin this section by showing how functions can be added, subtracted, multiplied, and divided. Let f and g be two functions. Then

(a) the sum $f + g$ is defined by

$$(f + g)(x) = f(x) + g(x)$$

(b)   the difference $f - g$ is defined by

$$\boxed{(f - g)(x) = f(x) - g(x)}$$ (2)

(c)   the product $f \cdot g$ is defined by

$$\boxed{(f \cdot g)(x) = f(x)g(x)}$$ (3)

(d)   the quotient f/g is defined by

$$\boxed{\left(\frac{f}{g}\right)(x) = \frac{f(x)}{g(x)}}$$ (4)

Furthermore, $f + g$, $f - g$, and $f \cdot g$ are defined for each x for which both f and g are defined. Finally, f/g is defined whenever both f and g are defined and $g(x) \neq 0$ (so that we do not divide by zero).

Example 1   Let $f(x) = \sqrt{x + 1}$ and $g(x) = \sqrt{4 - x^2}$. Since domain of $f = [-1, \infty)$ and domain of $g = [-2, 2]$, we have domain of $(f + g) = $ domain of $(f - g) = $ domain of $(f \cdot g) = [-1, \infty) \cap [-2, 2] = [-1, 2]$ and domain   of $(f/g) = [-1, 2] - \{x : \sqrt{4 - x^2} = 0\} = [-1, 2] - \{-2, 2\} = [-1, 2)$. The functions are

(a) $(f + g)(x) = f(x) + g(x) = \sqrt{x + 1} + \sqrt{4 - x^2}$;

(b) $(f - g)(x) = f(x) - g(x) = \sqrt{x + 1} - \sqrt{4 - x^2}$;

(c) $(f \cdot g)(x) = f(x) \cdot g(x) = \sqrt{x + 1} \cdot \sqrt{4 - x^2} = \sqrt{(x + 1)(4 - x^2)}$;

(d) $\left(\dfrac{f}{g}\right)(x) = \dfrac{f(x)}{g(x)} = \dfrac{\sqrt{x + 1}}{\sqrt{4 - x^2}} = \sqrt{\dfrac{x + 1}{4 - x^2}}$.

Example 2   The Universal Card Company (UCC) produces greeting cards. One of its most popular birthday cards can be produced for 16¢ apiece (for labor and materials) plus fixed costs of $1500 per month (for plant, utilities, managerial salaries, and so on). The production manager of UCC has determined the following monthly **demand function** for this particular card:

$$x = 20{,}000 \ (1.50 - p),$$

where p is the retail price of the card and x is the number of cards that can be sold at that price. For example, at a price of $p = \$1.00$, $x = 20{,}000 \times (1.50 - 1.00) = 20{,}000(0.50) = 10{,}000$ cards. If $p = 70¢ = \$0.70$, $x = 20{,}000 \times (1.50 - 0.70) = 20{,}000(0.80) = 16{,}000$ cards.

(a)   Find the total cost function C for this particular card.

(b)   Find the total revenue function R.

(c) Find the total profit function P.

(d) Determine the price UCC should set to maximize its monthly profit. What is that maximum profit?

Solution

(a) Total cost = fixed costs + variable costs

$$= 1500 + (\text{cost per unit}) \cdot (\text{number of units produced})$$
$$= 1500 + (0.16)x = 1500 + (0.16)20{,}000\ (1.50 - p)$$
$$= 1500 + 3200\ (1.50 - p)$$
$$= 1500 + 4800 - 3200p = 6300 - 3200p.$$

That is,

$$C(p) = 6300 - 3200p$$

is the total cost when a retail price of p is charged. We can also write

$$C(x) = 1500 + 0.16x.$$

Both are total cost functions. Usually, however, it is more convenient to write everything in terms of the price, p.

(b) Total revenue = (price received per item) $\cdot$ (number of items sold)

$$= px = p(20{,}000)(1.50 - p) = p(30{,}000 - 20{,}000p)$$
$$= -20{,}000p^2 + 30{,}000p.$$

So

$$R(p) = -20{,}000p^2 + 30{,}000p$$

is the total amount received when a price p is charged.

Note We can also write R in terms of x. Since

$$x = 20{,}000\ (1.50 - p),$$

we have

$$\frac{x}{20{,}000} = 1.5 - p \quad \text{or} \quad p = 1.5 - \frac{x}{20{,}000}$$

and

$$R(x) = px = \left(1.5 - \frac{x}{20{,}000}\right)x = 1.5x - \frac{x^2}{20{,}000}.$$

(c) Total profit = total revenue − total cost, so

$$P(p) = R(p) - C(p) = -20{,}000p^2 + 30{,}000p - (6300 - 3200p)$$
$$= -20{,}000p^2 + 30{,}000p - 6300 + 3200p$$
$$= -20{,}000p^2 + 33{,}200p - 6300 = -20{,}000(p^2 - 1.66p + 0.315).$$

In Table 1 we give the total profit for various retail prices. Numbers in parentheses represent losses.

TABLE 1

Retail price p	Profit $P = -20{,}000(p^2 - 1.66p + 0.315)$
$0.15	($1770)
0.20	(460)
0.30	1860
0.40	3780
0.50	5300
0.60	6420
0.70	7140
0.80	7460
0.90	7380
1.00	6900
1.10	6020
1.20	4740
1.30	3060
1.40	980
1.45	(210)
1.50	(1500)

The last entry confirms that, at a price $p = \$1.50$, no cards will be sold, so the only costs will be the fixed costs of $1500. With no revenue, this represents a loss of $1500.

(d) From Table 1 we suspect that the most profitable price will be between 70¢ and 90¢. To confirm this, we complete the square.

$$\underbrace{0.83 = \tfrac{1}{2}(1.66) \text{ and } (p - 0.83)^2 = p^2 - 1.66 + (0.83)^2}$$
$$\downarrow$$

$$P(p) = -20{,}000(p^2 - 1.66p + 0.315) - 20{,}000[(p - 0.83)^2 - (0.83)^2 + 0.315]$$
$$= -20{,}000[(p - 0.83)^2 - 0.6889 + 0.315]$$
$$= -20{,}000[(p - 0.83)^2 - 0.3739]$$

Since $(p - 0.83)^2 \geq 0$, we see that profit is maximized when $p = \$0.83$ $= 83¢$. The maximum profit is, then,

$$P(0.83) = -20{,}000(-0.3739) = \$7478.$$

Composite Function You will often need to deal with functions of functions. If f and g are functions, then their **composite function**, $f \circ g$, is defined by

$$(f \circ g)(x) = f(g(x)) \qquad (5)$$

and domain of $(f \circ g) = \{x : x \in \text{domain of } (g) \text{ and } g(x) \in \text{domain of } (f)\}$. That is, $(f \circ g)(x)$ is defined for every x such that $g(x)$ and $f(g(x))$ are defined.

Example 3 Let $f(x) = \sqrt{x}$ and $g(x) = x^2 + 1$. Then

$$(f \circ g)(x) = f(g(x)) = f(x^2 + 1) = \sqrt{x^2 + 1}$$

and

$$(g \circ f)(x) = g(f(x)) = g(\sqrt{x}) = (\sqrt{x})^2 + 1 = x + 1.$$

Now domain of $f = \mathbb{R}^+$, domain of $g = \mathbb{R}$, and we have

$$\text{domain of } f \circ g = \{x : g(x) = x^2 + 1 \in \text{domain of } f\}.$$

But since $x^2 + 1 > 0$, $x^2 + 1 \in$ domain of f for every real x so that domain of $f \circ g = \mathbb{R}$. On the other hand, domain of $g \circ f = \mathbb{R}^+$ since f is defined only for $x \geq 0$.

Warning. It is *not* true, in general, that $(f \circ g)(x) = (g \circ f)(x)$. This is illustrated in Examples 2 and 3.

Example 4 Let $f(x) = 3x - 4$ and $g(x) = x^3$. Then

$$(f \circ g)(x) = f(g(x)) = f(x^3) = 3x^3 - 4$$

and

$$(g \circ f)(x) = g(f(x)) = g(3x^3 - 4) = (3x - 4)^3.$$

Here domain of $f \circ g =$ domain of $g \circ f = \mathbb{R}$. Note that the functions $f \circ g$ and $g \circ f$ are quite different.

Example 5 Let $f(x) = 2x + 1$. Find a function $g(x)$ such that $(f \circ g)(x) = x^3$.

Solution We must have $(f \circ g)(x) = f(g(x)) = 2g(x) + 1 = x^3$. Then $2g(x) = x^3 - 1$ and $g(x) = \dfrac{x^3 - 1}{2}$.

PROBLEMS 9.2 In Problems 1–8 two functions, f and g, are given. Determine the functions $f + g$, $f - g$, $f \cdot g$, and f/g and find their respective domains.

1. $f(x) = 2x - 5$, $g(x) = -4x$

2. $f(x) = x^2$, $g(x) = x + 1$

3. $f(x) = \sqrt{x + 2}$, $g(x) = \sqrt{2 - x}$

4. $f(x) = x^3 + x$, $g(x) = \dfrac{1}{\sqrt{x + 1}}$

5. $f(x) = 1 + x^5$, $g(x) = 1 - |x|$

6. $f(x) = \sqrt{1 + x}$, $g(x) = \dfrac{1}{x^5}$

7. $f(x) = \sqrt[5]{x + 2}$, $g(x) = \sqrt[4]{x - 3}$

8. $f(x) = \dfrac{x}{x + 1}$, $g(x) = \dfrac{x - 1}{x}$

In Problems 9–16 find $f \circ g$ and $g \circ f$ and determine the domain of each.

9. $f(x) = x + 1$, $g(x) = 2x$

10. $f(x) = x^2$, $g(x) = 2x + 3$

11. $f(x) = 3x + 5$, $g(x) = 5x + 2$

12. $f(x) = \sqrt{x + 1}$, $g(x) = x^4$

13. $f(x) = \dfrac{x}{x + 2}$, $g(x) = \dfrac{x - 1}{x}$

14. $f(x) = |x|$, $g(x) = -x$

15. $f(x) = \sqrt{1 - x}$, $g(x) = \sqrt{x - 1}$

16. $f(x) = \begin{cases} x, & x \geq 0 \\ 2x, & x < 0 \end{cases}$, $g(x) = \begin{cases} -3x, & x \geq 0 \\ 5x, & x < 0 \end{cases}$

17. Let $f(x) = 2x + 4$ and $g(x) = \frac{1}{2}x - 2$. Show that $(f \circ g)(x) = (g \circ f)(x) = x$. (When this occurs, we say that f and g are **inverse functions**.)

18. If $f(x) = -3x + 2$, find a function g such that $(f \circ g)(x) = (g \circ f)(x) = x$.

19. If $f(x) = x^2$, find two functions g such that $(f \circ g)(x) = x^2 - 10x + 25$.

20. Let $h(x) = 1/\sqrt{x^2 + 1}$. Determine two functions f and g such that $f \circ g = h$.

21. Let $k(x) = (1 + \sqrt{x})^{5/7}$. Find the domain of k. Determine three functions f, g, and h such that $f \circ g \circ h = k$.

22. Let $h(x) = x^2 + x$, and let $f_1(x) = x^2 - x$, $g_1(x) = x + 1$, $f_2(x) = x^2 + 3x + 2$, and $g_2(x) = x - 1$. Show that $f_1 \circ g_1 = f_2 \circ g_2 = h$. This illustrates the fact that there is often more than one way to write a given function as the composition of two other functions.

23. Let f and g be the following linear functions:

$$f(x) = ax + b;$$

$$g(x) = cx + d.$$

Find conditions on a and b in order that $f \circ g = g \circ f$.

*24. Each of the following functions satisfies an equation of the form $(f \circ f)(x) = x$ or $(f \circ f \circ f)(x) = x$ or $(f \circ f \circ f \circ f)(x) = x$, and so on. For each function, discover what type of equation is appropriate.

(a) $A(x) = \sqrt[3]{1 - x^3}$

(b) $B(x) = \sqrt[7]{23 - x^7}$

(c) $C(x) = 1 - \dfrac{1}{x}$, domain $= \mathbb{R} - \{0, 1\}$

(d) $D(x) = 1/(1 - x)$, domain $= \mathbb{R} - \{0, 1\}$

(e) $E(x) = (x + 1)/(x - 1)$, domain $= \mathbb{R} - \{1\}$

(f) $F(x) = (x - 1)/(x + 1)$, domain $= \mathbb{R} - \{-1, 0, 1\}$

(g) $G(x) = \dfrac{4x - 1}{4x + 2}$, domain $= \mathbb{R} - \{-\tfrac{1}{2}, 0, \tfrac{1}{4}, \tfrac{1}{2}, 1\}$

25. A manufacturer of designer shirts determines that the demand function for her shirts is $x = 400(50 - p)$, where p is the wholesale price she charges per shirt and x is the number of shirts she can sell at that price. Note that, as is common, the higher the price, the fewer shirts she can sell. Assume that the manufacturer's fixed cost is $8000 and her material and labor costs amount to $8 per shirt.

(a) Determine the total cost function, C, as a function of p.

(b) Determine the total revenue function, R, as a function of p.

(c) Determine the profit function, P, as a function of p.

(d) By completing the square, determine the price that yields the greatest profit. What is this maximum profit (or minimum loss)?

9.3 Shifting the Graphs of Functions

Although more advanced methods (see Sections 11.1 and 11.2) are needed to obtain the graphs of most functions (without plotting a large number of points), there are some techniques that make it a relatively simple matter to sketch certain functions based on known graphs. As an illustration of what we have in mind, consider these six functions:

(a) $f(x) = x^2$,

(b) $g(x) = x^2 + 1$,

(c) $h(x) = x^2 - 1$,

(d) $f(x) = (x - 1)^2$,

(e) $k(x) = (x + 1)^2$,

(f) $l(x) = -x^2$.

They are all graphed in Figure 1. In Figure 1a, we have used the graph of $y = x^2$ obtained in Figure 9.1.1 on page 473. To graph $y = x^2 + 1$ in Figure 1b, we simply add 1 unit to every y value obtained in Figure 1a; that is, we shift the graph of $y = x^2$ up 1 unit. Analogously, for Figure 1c we simply shift the graph of $y = x^2$ down 1 unit to obtain the graph of $y = x^2 - 1$. The analysis of the graph in Figure 1d is a little trickier. Since, for example, $y = 0$ when $x = 0$ for the function $y = x^2$, then $y = 0$ when $x = 1$ for the function $y = (x - 1)^2$. Similarly, $y = 4$ when $x = -2$ if $y = x^2$, and $y = 4$ when $x = -1$ if

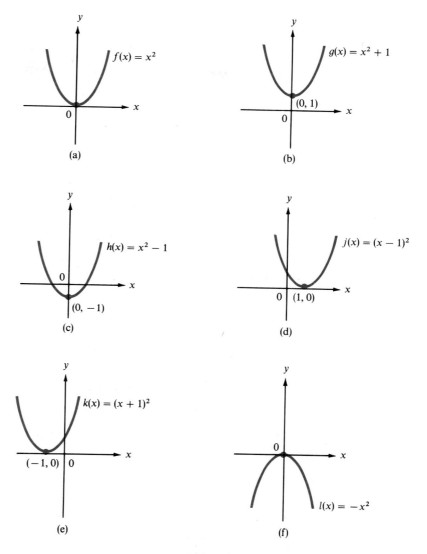

Figure 1

$y = (x - 1)^2$. By continuing in this manner, you can see that y values in the graph of $y = x^2$ are the same as y values in the graph of $y = (x - 1)^2$, except that they are achieved 1 unit to the right on the x-axis. Some representative values are given in Table 1. Thus, we find that the graph of $y = (x - 1)^2$ is the graph of $y = x^2$ *shifted 1 unit to the right*. Similarly, in Figure 1e we find that the graph of $y = (x + 1)^2$ is the graph of $y = x^2$ *shifted 1 unit to the left*. Some values of g are given in Table 2. Finally, in Figure 1f, to obtain the graph of $y = -x^2$, note that each y value is replaced by its negative so that the graph of $y = -x^2$ is the graph of $y = x^2$ *reflected through the x-axis* (that is, turned upside down).

TABLE 1

x	x^2	$(x-1)^2$
-5	25	36
-4	16	25
-3	9	16
-2	4	9
-1	1	4
0	0	1
1	1	0
2	4	1
3	9	4
4	16	9

TABLE 2

x	x^2	$(x+1)^2$
-5	25	16
-4	16	9
-3	9	4
-2	4	1
-1	1	0
0	0	1
1	1	4
2	4	9
3	9	16
4	16	25

In general, we have the following rules: Let $y = f(x)$. Then, to obtain the graph of

1. $y = f(x) + c$, shift the graph of $y = f(x)$ *up c* units if $c > 0$ and *down $|c|$* units if $c < 0$;

2. $y = f(x - c)$, shift the graph of $y = f(x)$ *to the right c* units if $c > 0$ and *to the left $|c|$* units if $c < 0$;

3. $y = -f(x)$, *reflect* the graph of $y = f(x)$ *through the x-axis*;

4. $y = f(-x)$, *reflect* the graph of $y = f(x)$ *through the y-axis*.

Remark. Don't confuse $f(-x)$ and $-f(x)$. They are usually *not* the same. For example, if $f(x) = x^2$, then $f(-2) = (-2)^2 = 4$, but $-f(2) = -2^2 = -4$. If $f(x) = 2x + 3$, then $f(-5) = 2(-5) + 3 = -10 + 3 = -7$, but $-f(5) = -(2 \cdot 5 + 3) = -13$.

Example 1 The graph of $y = \sqrt{x}$ is given in Figure 2a. Then, using the rules above, the graphs of $\sqrt{x} + 3$, $\sqrt{x} - 2$, $\sqrt{x - 2}$, $\sqrt{x + 3} = \sqrt{x - (-3)}$, $-\sqrt{x}$, and $\sqrt{-x}$ are given in the other parts of Figure 2.

Example 2 The graph of a certain function, $f(x)$, is given in Figure 3a. Sketch the graph of $-f(3 - x)$.

Solution We do this in three steps:

(a) reflect through the y-axis to obtain the graph of $f(-x)$ (Figure 3b);

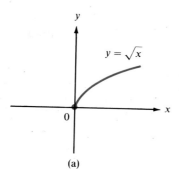

(a)

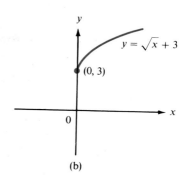

(b)

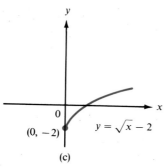

(c)

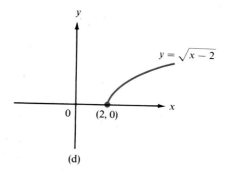

(d)

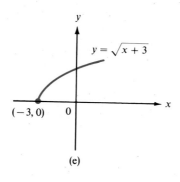

(e)

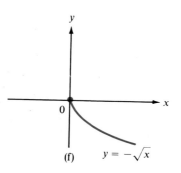

(f)

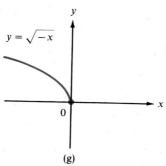

(g)

Figure 2

(b)   shift to the right 3 units to obtain the graph of $f(-(x-3)) = f(3-x)$ (Figure 3c);

(c)   reflect through the x-axis to obtain the graph of $y = -f(3-x)$ (Figure 3d).

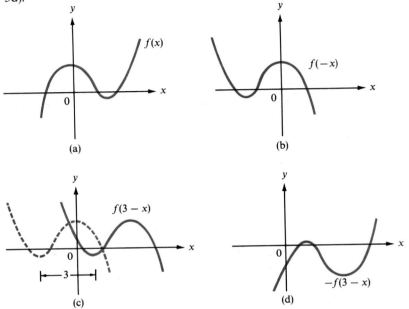

Figure 3

Example 3   Graph the parabola $y = x^2 - 10x + 22$.

Solution   Completing the square, we see that

$$x^2 - 10x + 22 = x^2 - 10x + 25 - 3 = (x-5)^2 - 3.$$

Thus, the graph of $x^2 - 10x + 22$ is obtained by shifting the graph of $y = x^2$ to the right 5 units and then down 3 units, as in Figure 4.

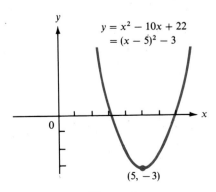

Figure 4

PROBLEMS 9.3   **1.** The graph of $f(x) = x^3$ is given in Figure 5. Sketch the graph of

(a) $(x - 2)^3$

(b) $-x^3$

(c) $(4 - x)^3 + 5$

2. The graph of $f(x) = 1/x$ is given in Figure 6. Sketch the graph of

(a) $\dfrac{1}{x + 3}$

(b) $2 - \dfrac{1}{x}$

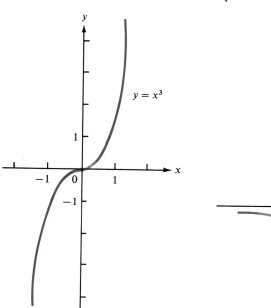

Figure 5

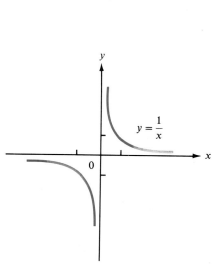

Figure 6

3. After completing the squares, sketch the graphs of these parabolas:

(a) $y = x^2 - 4x + 7$.

(b) $y = x^2 + 8x + 2$.

(c) $y = x^2 + 3x + 4$.

(d) $y = -x^2 + 2x - 3$.   [*Hint*: Write $-x^2 + 2x - 3 = -(x^2 - 2x + 3)$.]

(e) $y = -x^2 - 5x + 8$.

In each of Problems 4–12 the graph of a function is sketched. Obtain the graph of (a) $f(x-2)$, (b) $f(x+3)$, (c) $-f(x)$, (d) $f(-x)$, and (e) $f(2-x)+3$.

4.

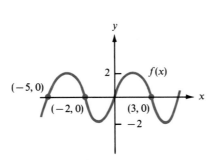

5.

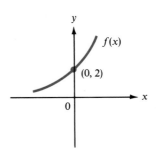

6.

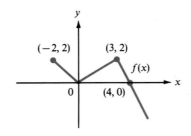

7.

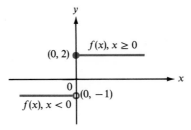

8.

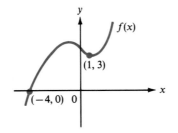

9.

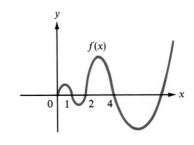

10.

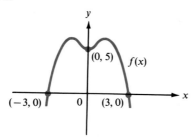

11.

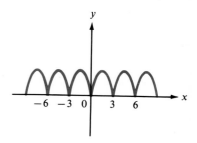

12.

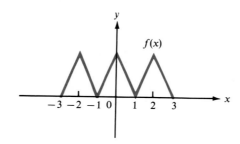

9.4 Exponential Functions

In this section we introduce one of the most important functions in mathematics. We begin by reviewing some algebraic facts.

Let a be a positive real number with $a \neq 1$. We define the number a^x in the following cases:

(a) $x = n$, *a positive integer.* Then

$$a^x = a^n = \underbrace{a \cdot a \cdot a \cdots a}_{n \text{ times}}$$

(b) $x = 0$. Then

$$a^x = a^0 = 1$$

(c) $x = -n$ *where n is a positive integer.* Then

$$a^x = a^{-n} = \frac{1}{a^n}$$

(d) $x = 1/n$ *where n is a positive integer.* Then

$$a^x = a^{1/n} = \text{the } n\text{th root of } a$$

(e) $x = a$ *positive rational number* m/n (*m and n are positive integers*). Then

$$a^x = a^{m/n} = (a^{1/n})^m$$

(f) $x = -m/n$, *a negative rational number.* Then

$$a^x = a^{-m/n} = \frac{1}{a^{m/n}}$$

Thus, $a^x (a > 0)$ is defined if x is a rational number.

If x is not a rational number, then we cannot compute a^x exactly. However, we can approximate a^x by first approximating x as a decimal and then computing a to the power of this decimal. With the aid of a calculator, this is quite easily done.

 Example 1 Use the procedure outlined above to approximate $4^{\sqrt{2}}$.

Solution We find that $\sqrt{2} = 1.414213562\cdots$. Thus, $\sqrt{2}$ can be approximated, successively, by 1, 1.4, 1.41, 1.414,..., and (since each of these numbers is a rational number) we can compute 4^1, $4^{1.4}$, and so on. Some results are given in Table 1.

TABLE 1

r	1	1.4	1.41	1.414	1.4142	1.414213562
4^r	4	6.964404506	7.06162397	7.100890698	7.102859756	7.102993298

Remark. The procedure described above really provides us with the definition of a^x when $a > 0$ and x is irrational. We simply define a^x as the "end result" or "limit" of a^r as r approximates x to more and more decimal places. We shall say a great deal more about limits in Section 10.2.

We can now define an exponential function.

Exponential Function Let a be a positive real number. Then the function $f(x) = a^x$ is called an **exponential function with base a**.

 Example 2 Sketch the graph of the function $y = 2^x$.

Solution We provide some values of 2^x in Table 2. We plot these values and then draw a curve joining the points to obtain the sketch in Figure 1.

TABLE 2

x	-10	-5	-2	-1	0	$\frac{1}{2}$	1	$\frac{3}{2}$	2	3	5	10
2^x	0.0001	0.03	0.25	0.5	1	1.4142	2	2.8284	4	8	32	1024

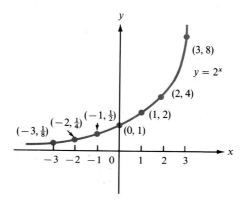

Figure 1

Example 3   Sketch the graph of the function $y = (\frac{1}{2})^x$.

Solution   We see that $(\frac{1}{2})^x = 1/2^x = 2^{-x}$. Thus, if $f(x) = 2^x$, $2^{-x} = f(-x)$, and from Rule 4 in Section 9.3 (see page 488), we obtain the graph of $(\frac{1}{2})^x$ by reflecting the graph of 2^x through the y-axis. This is done in Figure 2.

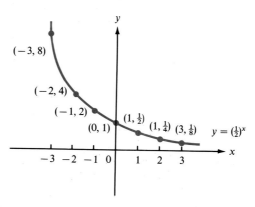

Figure 2

Example 4   Sketch the graph of $y = 10^x$.

Solution   We give some values of 10^x in Table 3 and draw the graph in Figure 3.

TABLE 3

x	-3	-2	-1	0	0.25	0.5	0.75	1	1.5	2	3
10^x	0.001	0.01	0.1	1	1.778	3.162	5.623	10	31.62	100	1000

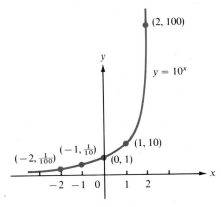

Figure 3

Example 5   Sketch the graph of $y = (\frac{1}{10})^x$.

Solution   As in Example 2, we can obtain the graph by reflecting the graph of 10^x through the y-axis. We do this is Figure 4.

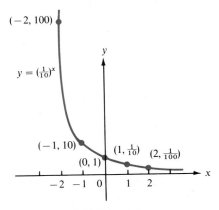

Figure 4

If you look at Figures 1 and 3, you may observe that the graphs of 2^x and 10^x are very similar. The only difference is that 10^x increases faster than 2^x as x increases and that 10^x decreases faster than 2^x as x becomes more negative. The functions $(\frac{1}{2})^x$ and $(\frac{1}{10})^x$ behave very similarly. These facts are not surprising after we observe that all exponential functions share a number of interesting properties. We cite some of these properties below. All of them follow from the material in Section 1.4 (on page 26).

> ## PROPERTIES OF EXPONENTIAL FUNCTIONS
>
> Let $a > 0$ and let x and y be real numbers. Then
>
> **1.** $a^x > 0$,
>
> **2.** $a^{-x} = \dfrac{1}{a^x}$,
>
> **3.** $a^{x+y} = a^x a^y$,
>
> **4.** $a^{x-y} = \dfrac{a^x}{a^y}$,
>
> **5.** $a^0 = 1$,
>
> **6.** $a^1 = a$,
>
> **7.** $(a^x)^y = a^{xy}$,
>
> **8.** if $a > 1$, a^x is an increasing function,
>
> **9.** if $0 < a < 1$, a^x is a decreasing function.

We need to say more about properties 8 and 9. A function, f, is *increasing* if $f(x)$ gets larger as x gets larger; that is, $f(x_2) > f(x_1)$ if $x_2 > x_1$. Similarly, f is *decreasing* if $f(x_2) < f(x_1)$ when $x_2 > x_1$. It is possible to prove that a^x is an increasing function if $a > 1$, but that requires techniques not available to us at this time. We shall indicate why this is true in Section 11.1. For now, the result should seem plausible, especially given the numbers in Tables 1 and 2.

In calculus, one particular exponential function is very important. This is the exponential function whose base is e.[†] The letter e is used to denote a certain irrational number (in much the same way the Greek letter π is used to denote the irrational number that is the ratio of the circumference to the diameter of a circle). An approximation of e, to 10 significant figures, is

$$e = 2.718281828. \tag{1}$$

Values of e^x are tabulated in Table 6 at the back of this book. If you have a scientific calculator, you can obtain values of e^x by pressing a button usually labeled $\boxed{e^x}$ or $\boxed{\exp x}$.[‡] In Table 4 below we give some sample approximate values of e^x. We provide a sketch of its graph in Figure 5a. Since $2 < e < 3$,

---

[†] The number e was discovered by the great Swiss mathematician and physicist Leonhard Euler (1707–1783).

[‡] As we shall see in the next section, e^x is the inverse of a function denoted $\ln x$. If your scientific calculator does not have an e^x key, then e^x can be obtained by pressing $\boxed{\text{INV}}$ followed by $\boxed{\ln x}$. For example, e^2 is obtained by the key sequence $\boxed{2}$ $\boxed{\text{INV}}$ $\boxed{\ln x}$ = 7.389056099.

the graph of e^x lies between the graphs of 2^x and 3^x. This is indicated in Figure 5b.

TABLE 4

x	-5	-3	-2	-1	-0.5	-0.25	0	0.25	0.5
e^x	0.0067	0.0498	0.1353	0.3679	0.6065	0.7788	1	1.284	1.6487

x	1	1.5	2	3	5	10
e^x	2.7183	4.4817	7.3891	20.086	148.41	22,026.5

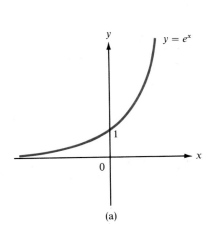

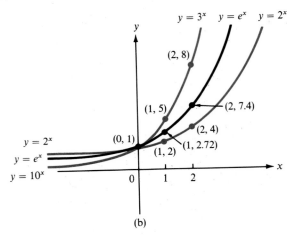

Figure 5

The number e arises in a wide variety of ways. One reason why this number is so important is given in Section 10.11. In Section 11.6 we give a large number of examples of the use of the exponential function. One application of e is given below.

In Section 8.1 we discussed compound interest and derived formulas for interest compounded m times a year [see formula (8.1.4) on page 435]. But many banks advertise that interest is paid *continuously* rather than at fixed intervals. In Section 11.6 we will derive the following formula:

Interest with Continuous Compounding

$$A(t) = Pe^{it} \qquad (2)$$

where
$$P = \text{original principal,}$$
$$i = \text{annual interest rate,}$$
$$t = \text{number of years investment is held,}$$
$$A(t) = \text{amount of money after } t \text{ years.}$$

Example 6   Suppose that $1000 is invested in a bank account paying 6%, compounded continuously. How much is in the account after 5 years? How much interest is paid during this period?

Solution $A(5) = 1000e^{(0.06)(5)} = 1000e^{0.3} = \$1349.86$. The interest paid is $349.86. Compare this to the result in Example 6 on page 436.

 Example 7 $5000 is invested in a bond yielding $8\frac{1}{2}\%$ annually. What will the bond be worth in 10 years if interest is compounded continuously?

Solution $A(t) = Pe^{it} = 5000e^{(0.085)(10)} = 5000e^{0.85} = \$11,698.23$.

PROBLEMS 9.4 In Problems 1–14 draw a sketch of the given exponential function.

1. $y = 3^x$ **2.** $y = (\frac{1}{3})^x$ **3.** $y = (\frac{1}{5})^x$

4. $y = 5^x$ **5.** $f(x) = (7.2)^x$ **6.** $f(x) = (0.623)^x$

7. $f(x) = 3 \cdot 2^x$ **8.** $f(x) = 4 \cdot 10^x$ **9.** $y = -2 \cdot 10^x$

10. $y = 10 \cdot 2^x$ **11.** $y = 2^{x+1}$ **12.** $y = 3^{x-2}$

13. $y = 3 \cdot 10^{x+1}$ **14.** $y = 4 \cdot 2^{1-x}$

In Problems 15–20 use a calculator to estimate the given number to as many decimal places of accuracy carried on the machine.

15. $e^{2.5}$ **16.** $10^{2.5}$ **17.** $e^{-0.6}$

18. $(\frac{1}{2})^{1.7}$ **19.** $3^{\sqrt{2}}$ **20.** $2^{\sqrt{3}}$

21. A sum of $5000 is invested at a return of 7% per year compounded continuously. What is the investment worth after 8 years?

22. As a gimmick to lure depositors, a bank offers 5% interest compounded continuously. Its competitor offers $5\frac{1}{8}\%$ compounded annually. Which bank would you choose?

23. Suppose the competitor in Problem 22 now compounds $5\frac{1}{8}\%$ semiannually. Which bank would you choose?

24. $10,000 is invested in bonds yielding 9% compounded continuously. What will the bonds be worth in 8 years?

25. An investor buys a bond that pays 12% annual interest compounded continuously. If she invests $10,000 now, what will her investment be worth in (a) 1 year? (b) 4 years? (c) 10 years?

*26. After how many years will the bond in Problem 25 be worth $20,000? [*Hint*: We will see an easy way to solve this problem in the next section. For now, use trial and error and give an answer to the nearest tenth of a year.]

27. The exponential e^x can be estimated for x in $[-\frac{1}{2}, \frac{1}{2}]$ by the formula

$$e^x \approx \left(\left\{ \left[\left(\frac{x}{5} + 1 \right) \frac{x}{4} + 1 \right] \frac{x}{3} + 1 \right\} \frac{x}{2} + 1 \right) x + 1.$$

(a)   Calculate an approximate value for $e^{0.13}$.

(b)   Calculate an approximate value for $e^{-0.37}$.

(c)   Calculate an approximate value for $e^{4.13}$.

(d)   Calculate an approximate value for $e^{-2.63}$.   [*Hint:* Use part (b).]

(e)   Calculate approximate values for $e^{4.82}$ and $e^{-1.44}$.

9.5 Logarithmic Functions

Let a^x be an exponential function. In Section 9.4 we said that such a function is increasing if $a > 1$ and decreasing if $0 < a < 1$. These facts enable us to define a new function, as we shall soon see.

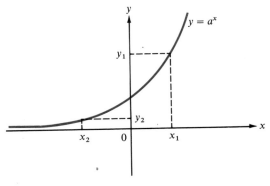

Figure 1

In Figure 1 we draw the graph of a typical exponential function for $a > 1$. The important fact about this function is that, for every real number $y > 0$, there is exactly one real number x such that $y = a^x$. In Figure 1 we have illustrated this with two typical values, y_1 and y_2, and corresponding x values, $x_1(>0)$ and $x_2(<0)$. In general, given a positive number y, the equation $a^x = y$ has a unique solution. We illustrate this in the following example.

Example 1   Solve for x: (a) $2^x = 8$, (b) $3^x = \frac{1}{9}$, and (c) $(\frac{1}{2})^x = 8$.

Solution   (a) We see that $2^x = 8$ if and only if $x = 3$. Similarly, in (b), $3^x = \frac{1}{9}$ if and only if $x = -2$, and in (c), $(\frac{1}{2})^x = 8$ if and only if $x = -3$ $[(\frac{1}{2})^{-3} = \frac{1}{(1/2)^3} = \frac{1}{(1/8)} = 8]$.

We now reverse the roles of x and y and define an important new function.†

Logarithm to the Base a   If $x = a^y$, then the **logarithm to the base a** of x is y. This is written

$$y = \log_a x. \tag{1}$$

---

† We reverse the roles of x and y so we can write the logarithmic function with x as the independent variable and y as the dependent variable.

The relationship between the exponential and logarithmic functions is illustrated in Figure 2. The graph of $\log_a x$ can immediately be obtained by turning the graph of $y = a^x$ on its side and flipping it over so as to interchange the positions of the x- and y-axes (see Figure 2). Figure 2b shows a typical graph of the logarithmic function for $a > 1$. The two functions $\log_a x$ and a^x are called *inverse* functions.

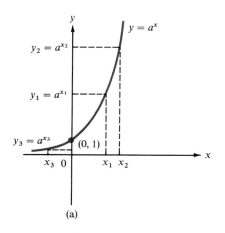

(a)

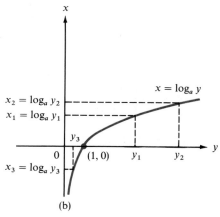

(b)

Figure 2

We stress the following facts about logarithmic functions.

$$y = \log_a x \text{ is equivalent to } x = a^y$$

and

$$y = a^x \text{ is equivalent to } x = \log_a y.$$

Think of $y = \log_a x$ as an answer to the question: To what power must a be raised to obtain the number x? This immediately implies that

$$a^{\log_a x} = x \text{ for every positive real number } x$$

and

$$\log_a a^x = x \text{ for every real number } x.$$

Remember that $\log_a x$ is only defined for $x > 0$ since the equation $a^y = x$ has no solution when $x \leq 0$. For example, $\log_2 (-1)$ is not defined since there is no real number y such that $2^y = -1$.[†] We have

$$\text{domain of } \log_a x = \{x : x > 0\}$$

---

† On page 472 we discussed three reasons for restricting the domain of a function. The logarithmic function provides a fourth reason.

Example 2 In Example 1 we found that (a) $\log_2 8 = 3$, (b) $\log_3 \frac{1}{9} = -2$, and (c) $\log_{1/2} 8 = -3$.

Example 3 The graphs of $y = \log_2 x$ and $y = \log_{1/2} x$ are given in Figure 3.

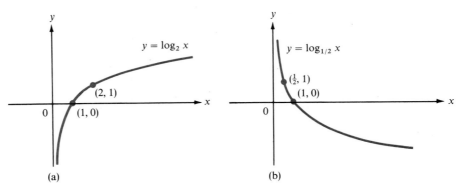

Figure 3

Example 4 Change to an exponential form: (a) $\log_9 3 = \frac{1}{2}$, and (b) $\log_4 64 = 3$.

Solution

(a) $\log_9 3 = \frac{1}{2}$ is equivalent to $9^{1/2} = 3$.

(b) $\log_4 64 = 3$ is equivalent to $4^3 = 64$.

Example 5 Change to a logarithmic form: (a) $5^4 = 625$, and (b) $(\frac{1}{2})^{-4} = 16$.

Solution

(a) $5^4 = 625$ is equivalent to $\log_5 625 = 4$.

(b) $(\frac{1}{2})^{-4} = 16$ is equivalent to $\log_{1/2} 16 = -4$.

Although any positive number can be used as a base for a logarithm, two bases are used almost exclusively. The first of these is base 10. Logarithms to the base 10 are called **common logarithms** and are denoted, simply, $\log x$. That is,

$$\log x = \log_{10} x \qquad (2)$$

Common logarithms were discovered by the Scottish mathematician John Napier, Baron of Merchiston (1550–1617). In one of Napier's earliest notes, the following table appears:

I	II	III	IIII	V	VI	VII	$\cdots$
1	2	4	8	16	32	64	128

You should note that the number represented by each roman numeral above is the logarithm to the base 2 of the number beneath it. Alternatively, 2 to the power of the roman numeral above is equal to the number below. Soon Napier discovered the importance of logarithms to the base 10 for computations, since our numbering system is founded on the base 10. Because he clearly foresaw the practical usefulness of logarithms in trigonometry and astronomy, he abandoned other mathematical pursuits and set himself the difficult task of producing a table of common logarithms—a task that took him 25 years to complete. When the tables were completed, they created great excitement on the European continent and were immediately used by two of the great astronomers of the day: the Dane Tycho Brahe and the German Johannes Kepler.

Until recently, common logarithms were frequently used for arithmetic computations. However, logarithms are now rarely used in this way because arithmetic computations can be carried out more accurately and much faster using a hand-held calculator. They are still used in certain scientific formulas (see, for example, Problems 51 and 52). Most scientific calculators have $\boxed{\log}$ buttons for computing common logarithms.

The second and more important base for logarithms is the base e. This is the base used in calculus. The number e was discussed in Section 9.4 on page 497. Logarithms to the base e are called **natural logarithms** and are denoted by ln x. That is,

$$\boxed{\ln x = \log_e x} \tag{3}$$

A table of natural logarithms appears at the back of the book (Table 7). All scientific calculators have $\boxed{\ln}$ buttons.

For the remainder of this book, the only logarithm we will need will be the natural logarithm. The function $y = \ln x$ has several properties. We list some of them here:

PROPERTIES OF ln x

Let x and y be positive real numbers. Then

1. $\ln xy = \ln x + \ln y$

2. $\ln \dfrac{x}{y} = \ln x - \ln y$

3. $\ln 1 = 0$

4. $\ln \dfrac{1}{x} = -\ln x$

5. $\ln e = 1$

6. $\ln x^y = y \ln x$

We indicate below why property (1) holds. The derivations of the other properties are left as problems (see Problems 61–65).

Let $u = \ln x$ and $v = \ln y$. Since $u = \ln x = \log_e x$, we have $x = e^u$. Similarly, $y = e^v$. Then

$$xy = e^u e^v = e^{u+v} \overset{\underset{\ln x \quad \ln y}{\downarrow \quad \downarrow}}{\text{ so that }} u + v = \ln xy.$$

Example 6   Given that $\ln 2 \approx 0.6931$ and $\ln 5 \approx 1.609$, estimate (a) $\ln 10$, (b) $\ln \frac{2}{5}$, (c) $\ln 8$, and (d) $\ln \frac{16}{25}$.

Solution

(a) $\ln 10 = \ln(2 \cdot 5) \overset{\underset{\text{From (1)}}{\downarrow}}{=} \ln 2 + \ln 5 \approx 0.6931 + 1.609 = 2.3021$

(b) $\ln \frac{2}{5} \overset{\underset{\text{From (2)}}{\downarrow}}{=} \ln 2 - \ln 5 \approx 0.6931 - 1.609 = -0.9159$

(c) $\ln 8 = \ln 2^3 \overset{\underset{\text{From (6)}}{\downarrow}}{=} 3 \ln 2 \approx 3(0.6931) = 2.0793$

(d) $\ln \frac{16}{25} = \ln 16 - \ln 25 = \ln 2^4 - \ln 5^2 = 4 \ln 2 - 2 \ln 5 \approx 4(0.6931)$
$$- 2(1.609)$$
$$= -0.4456$$

Example 7   Write these as a single logarithm:

(a) $\ln(x - 1) - 2 \ln(x + 5)$

(b) $\ln x + 2 \ln(x + 1) + 3 \ln(x + 2)$

Solution

(a) $\ln(x - 1) - 2 \ln(x + 5) \overset{\underset{\text{From (6)}}{\downarrow}}{=} \ln(x - 1) - \ln(x + 5)^2 \overset{\underset{\text{From (2)}}{\downarrow}}{=} \ln \dfrac{x - 1}{(x + 5)^2}$

(b) $\ln x + 2 \ln(x + 1) + 3 \ln(x + 2) \overset{\underset{\text{From (6)}}{\downarrow}}{=} \ln x + \ln(x + 1)^2 + \ln(x + 2)^3$
$$\overset{\underset{\text{From (1)}}{\downarrow}}{=} \ln[x(x + 1)^2(x + 2)^3].$$

We can rewrite the four statements on page 501 in terms of the natural logarithmic function:

> **(a′)** $y = \ln x$ means that $x = e^y$;
>
> **(b′)** $y = e^x$ means that $x = \ln y$;
>
> **(c′)** $e^{\ln x} = x$ for every positive real number x;
>
> **(d′)** $\ln e^x = x$ for every real number x.

Example 8   Solve the following equations for x:

(a) $e^{2(x-5)} = 30$ and

(b) $3 \ln x + \ln 5 = 4$.

Solution

(a)   We take the natural logarithms of both sides:

$$\ln e^{2(x-5)} = \ln 30$$

$$2(x - 5) = \ln 30 \qquad \text{From (d′)}$$

$$2x - 10 = \ln 30$$

From a calculator
↓

$$x = \frac{\ln 30 + 10}{2} = \tfrac{1}{2}\ln 30 + 5 \approx \tfrac{1}{2}(3.4) + 5 = 6.7.$$

Property (6)   Property (1)
↓   ↓

(b) $3 \ln x + \ln 5 = \ln x^3 + \ln 5 = \ln 5x^3$

so that

$$\ln 5x^3 = 4,$$

and, from (a′),

$$5x^3 = e^4$$

$$x^3 = \tfrac{1}{5}e^4$$

From a calculator
↓

$$x = (\tfrac{1}{5}e^4)^{1/3} = \frac{e^{4/3}}{5^{1/3}} = 2.2186.$$

To illustrate the great usefulness of the natural logarithm function, we begin with a problem starred in the last section (see Problem 9.4.26 on page 499).

Example 9   A bond pays 12% annual interest compounded continuously. If $10,000 is initially invested, when will the bond be worth $20,000?

Solution If $A(t)$ denotes the value of the bond after t years, then by equation (9.4.2) on page 498,

Original   Rate of
principal   interest

$$A(t) = Pe^{it} = 10{,}000e^{0.12t}.$$

Our problem is to determine a number t^* such that $A(t^*) = \$20{,}000$. Then

$$A(t^*) = 10{,}000e^{0.12t^*} = 20{,}000. \tag{4}$$

This leads to

$$e^{0.12t^*} = \frac{20{,}000}{10{,}000} = 2 \qquad \text{We divided both sides of (4) by 10,000}$$

$$\ln e^{0.12t^*} = \ln 2 \qquad \text{We took the natural logarithm of both sides}$$

$$0.12t^* = \ln 2 \qquad \text{From property (d')}$$

From Table 7 or a calculator

$$t^* = \frac{\ln 2}{0.12} \approx \frac{0.6931}{0.12} \approx 5.78 \text{ years.}$$

Thus we have shown that money invested at 12% annual interest, compounded continuously, doubles in approximately $5\frac{3}{4}$ years.

PROBLEMS 9.5 In Problems 1–5 change to an exponential form. For example, $\log_9 3 = \frac{1}{2}$ can be converted to $9^{1/2} = 3$.

1. $\log_{16} 4 = \frac{1}{2}$ **2.** $\log_2 32 = 5$ **3.** $\log_{1/2} 8 = -3$

4. $\log_3 \frac{1}{3} = -1$ **5.** $\log_{12} 1 = 0$

Change Problems 6–10 to a logarithmic form.

6. $3^4 = 81$ **7.** $(\frac{1}{2})^3 = \frac{1}{8}$ **8.** $(\frac{1}{3})^{-2} = 9$

9. $4^{-2} = \frac{1}{16}$ **10.** $17^1 = 17$

In Problems 11–50 solve for the unknown variable. (Do not use tables or a calculator.)

11. $y = \log_2 4$ **12.** $y = \log_4 16$ **13.** $y = \log_{1/3} 27$

14. $y = \frac{1}{3} \log_7 (\frac{1}{7})$ **15.** $y = \pi \log_\pi \left(\frac{1}{\pi^4} \right)$ **16.** $y = \log_{81} 3$

17. $y = \ln e^5$ **18.** $y = \log 0.01$ **19.** $y = \log_{1/4} 2$

20. $y = \log_a a \cdot \log_b b^2 \cdot \log_c c^3$ **21.** $y = \log_6 36 \log_{25} \frac{1}{5}$

22. $64 = x \log_{1/4} 64$ **23.** $2^{x^2} = 64$ **24.** $y = 1.3^{\log_{1.3} 48}$

25. $y = e^{\ln \sqrt{2}}$ **26.** $y = e^{\ln 14.6}$ **27.** $y = e^{\ln e^\pi}$

28. $\log_2 x^4 = 4$ **29.** $\log_x 64 = 3$ **30.** $\log_x 125 = -3$

31. $\log_x 32 = -5$ **32.** $y = \ln \dfrac{1}{e^{3.7}}$ **33.** $2e^x = 8$

34. $5e^{x-1} = 20$ **35.** $e^x e^{x+1} = 2$ **36.** $2 \ln x = 4$

37. $3 \ln 2x = 1$ **38.** $\ln 5x = 0$ **39.** $2 \ln x + 3 = 0$

40. $\ln(x - 2) = 2$ **41.** $e^{2x} e^{\ln 1/2} = 4$ **42.** $1 + \ln e^{2x} = 5$

43. $e^{x^2 + 2x - 8} = 1, x > 0$ **44.** $e^{x^2 + 2x - 8} = 1, x < 0$

45. $\ln x - \ln(x - 1) = 2$ **46.** $\ln 8x = \ln(x^2 - 20)$

47. $\ln(x + 3) = \ln(2x - 5)$ **48.** $\ln(\ln(x + 1)) = 0$

49. $\frac{1}{4} \ln x = 3$ **50.** $e^{e^x} = e$

51. A general psychophysical relationship was established in 1834 by the physiologist Ernst Weber[†] and given a more precise phrasing later by Gustav Fechner.[‡] By the Weber–Fechner law, $S = c \log(R + d)$, where S is the intensity of a sensation, R is the strength of the stimulus producing it, and c and d are constants. The Greek astronomer Ptolemy catalogued stars according to their visual brightness in six categories or **magnitudes**. A star of first magnitude was about 2.5 times as bright as a star of the second magnitude, which in turn was about 2.5 times as bright as a star of third category, and so on. Let b_n and b_m denote the apparent brightness of two stars having magnitudes n and m, respectively. Modern astronomers have established the Weber–Fechner law relating the relative brightness to the difference in magnitudes as

$$(m - n) = 2.5 \log\left(\frac{b_n}{b_m}\right).$$

(a) Using this formula, calculate the ratio of brightness for two stars of the second and fifth magnitudes, respectively.

(b) If star A is five times as bright to the naked eye as star B, what is the difference in their magnitudes?

(c) How much brighter is Sirius (magnitude 1.4) than a star of magnitude 21.5?

(d) The Nova Aquilae in a 2–3 day period in June 1918, increased in brightness about 45,000 times. How many magnitudes did it rise?

*(e) The bright star Castor appears to the naked eye as a single star but can be seen with the aid of a telescope to be really two stars whose magnitudes have been calculated to be 1.97 and 2.95. What is the magnitude of the two combined? [*Hint:* Brightness, but not magnitudes, can be added.]

52. The subjective impression of loudness can be described by a Weber–Fechner law. Let I denote the intensity of a sound. The least intense sound that can be heard is $I_0 = 10^{-12}$ watt/m^2 at a frequency of 1000 cycles/second (this is called the **threshold of audibility**). If L denotes the loudness of a sound measured in decibels, then $L = 10 \log(I/I_0)$.[§]

[†] Ernst Weber (1796–1878) was a German physiologist.
[‡] Gustav Fechner (1801–1887) was a German physicist.
[§] 1 decibel (dB) $= \frac{1}{10}$ Bel, named after Alexander Graham Bell (1847–1922), inventor of the telephone.

(a) If one sound has twice the intensity of another, what is the ratio of the perceived loudness of the two sounds?

(b) If one sound appears to be twice as loud as another, what is the ratio of their intensities?

(c) Ordinary conversation sounds six times as loud as a low whisper. What is the actual ratio of intensity of their sounds?

 53. Natural logarithms can be calculated on a hand calculator even if the calculator does not have an $\boxed{\ln}$ key. If $\frac{1}{2} \le x \le \frac{3}{2}$ and if $A \equiv (x - 1)/(x + 1)$, then a good approximation to $\ln x$ is given by

$$\ln x \approx \left[\left(\frac{3A^2}{5} + 1 \right) \cdot \frac{A^2}{3} + 1 \right] 2A, \qquad \frac{1}{2} \le x \le \frac{3}{2}.$$

(a) Use this formula to calculate $\ln 0.8$ and $\ln 1.2$.

(b) Using facts about logarithms, use the formula to calculate (approximately) $\ln 2 = \ln \left(\frac{3}{2} \cdot \frac{4}{3} \right)$.

(c) Using (b), calculate $\ln 3$ and $\ln 8$.

54. The quantity $n! = n(n - 1)(n - 2) \cdots 3 \cdot 2 \cdot 1$ grows very rapidly as n increases. According to **Stirling's formula**, when n is large,

$$n! \approx \sqrt{2\pi n} \left(\frac{n}{e} \right)^n.$$

Use Stirling's formula to estimate $100!$ and $200!$.

55. A sum of $10,000 is invested at a steady rate of return with interest compounded continuously. If the investment is worth $15,000 in 2 years, what is the annual interest rate?

56. A certain government bond sells for $750 and can be redeemed for $1000 in 8 years. Assuming continuous compounding, what is the rate of interest paid?

57. How long would it take an investment to increase by half if it is invested at 4% compounded continuously?

58. What must be the interest rate in order that an investment triple in 15 years if interest is continuously compounded?

59. If money is invested at 10% compounded continuously, what is the effective interest rate?

*60. What is the most a banker should pay for a $10,000 note due in 5 years if he can invest a like amount of money at 9% compounded annually?

61. Derive property (2) on page 503. [*Hint*: With u and v as before, show that $\dfrac{x}{y} = e^{u-v}$.]

62. Explain why $\ln 1 = 0$.

63. Use the results of Problems 61 and 62 to show that $\ln \dfrac{1}{x} = \ln 1 - \ln x = -\ln x$.

64. Explain why $\ln e = 1$.

65. Show that $\ln x^y = y \ln x$. [*Hint*: Let $u = \ln x$. Show that $x^y = e^{uy}$.]

Review Exercises for Chapter 9

In Exercises 1–9 determine whether the given equation defines a function, and if so, find its domain and range.

1. $4x - 2y = 5$

2. $\dfrac{x^2 - y}{2} = 4$

3. $\dfrac{y}{x} = 1$

4. $(x - 1)^2 + (y - 3)^2 = 4$

5. $y = \sqrt{x + 2}$

6. $3 = \dfrac{1 + x^2 + x^4}{2y}$

7. $y = \dfrac{x}{x^2 + 1}$

8. $y = \dfrac{x}{x^2 - 1}$

9. $y = \sqrt{x^2 - 6}.$

10. For $y = f(x) = \sqrt{x^2 - 4}$, calculate $f(2), f(-3), f(\sqrt{5})$, and $f(0)$.

11. If $y = f(x) = 1/x$, show that for $\Delta x \neq 0$,

$$\frac{f(x + \Delta x) - f(x)}{\Delta x} = -\frac{1}{x(x + \Delta x)}.$$

12. Let $f(x) = \sqrt{x + 1}$ and $g(x) = x^3$. Find $f + g, f - g, f \cdot g, g/f, f \circ g$, and $g \circ f$ and determine their respective domains.

13. Do the same for $f(x) = 1/x$ and $g(x) = x^2 - 4x + 3$.

14. For $f(x) = 4x - 6$, find a function $g(x)$ such that $(f \circ g)(x) = (g \circ f)(x) = x$.

15. The graph of the function $y = f(x)$ is given in Figure 1. Sketch the graph of $f(x - 3), f(x) - 5, f(-x), -f(x)$, and $4 - f(1 - x)$.

16. Do the same for the function graphed in Figure 2.

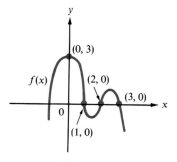

Figure 1

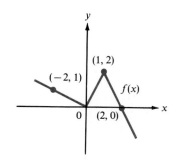

Figure 2

In Exercises 17–20 draw a sketch of the given exponential function.

17. 4^x

18. $\left(\tfrac{1}{4}\right)^x$

19. $3 \cdot 5^x$

20. $-2 \cdot 2^{-x}$

In Exercises 21–24 use a calculator to estimate the given number to as many decimal places of accuracy as your calculator carries.

 21. $e^{1.7}$　　 **22.** $10^{3.45}$　　**23.** $(\tfrac{1}{3})^{2.3}$　　**24.** $4^{\sqrt{5}}$

In Exercises 25–33 solve for the given variable.

25. $y = \log_3 9$

26. $y = \log_{1/3} 9$

27. $e^{2x} = 4$

28. $\ln(2x - 3) = 4$

29. $\ln(x + 1) - \ln(x - 3) = 1$

30. $3 \ln(x + 5) + 2 = 0$

31. $e^{\ln(2x + 5)} = 9$

32. $\ln e^{7x - 2} = 19$

33. $\ln \ln(10x) = 0$

34. A sum of $10,000 is invested at an interest rate of 6% compounded quarterly. What is the investment worth in 8 years?

35. What is the investment in Problem 34 worth if interest is compounded continuously?

10 THE DERIVATIVE

10.1 Introduction

From before the time of the great Greek scientist Archimedes (287–212 B.C.), mathematicians were concerned with the problem of finding the unique tangent line (if one exists) to a given curve at a given point on the curve. Some typical tangent lines at several different points are drawn in Figure 1.

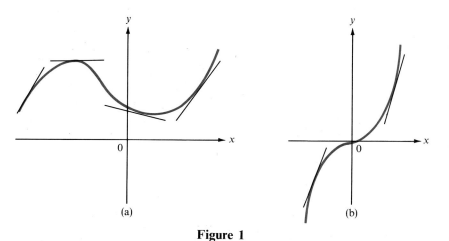

Figure 1

There are many reasons why it is useful to find the tangent to a curve. If, for example, the curve is the straight line $y = mx + b$, then the tangent line is the line itself. The slope of the line is, as we saw in Section 2.3 (see page 46), a measure of the relative rate of change of the x- and y-coordinates of points on the line as we move along the line. Thus, for example, if $y = 3x + 2$, then a 1-unit increase in x results in a 3-unit increase in y. If x represents time and y represents distance, then the slope 3 is the change of distance per unit of time, or velocity (velocity = distance/time).

As another example, consider the cost function given in Example 9.1.9 on page 476:

$$C = 0.27q + 2500.$$

Here, q represents the quantity manufactured and C represents the total cost. The slope 0.27 represents unit variable cost. That is, if we increase production by 1 unit, we increase the total cost by 0.27 units ($= \$0.27$).

How do we find the tangent line if the curve is not a straight line? This is the problem solved by Sir Isaac Newton (1642–1727) and Gottfried Leibniz (1646–1716), the two co-inventors of calculus. Newton and Leibniz showed how to find the slope of the tangent line to a point on a curve. This slope is called the **derivative** of the function at that point. In Section 10.5 we will show that the derivative at a point represents the rate of change of the function at that point.

If the function gives distance as a function of time, then the derivative represents velocity. If the function gives the total cost as a function of the quantity produced, then the derivative represents what we will term **marginal cost**.

In general, we shall see that the derivative represents rate of change in a variety of settings. Since many important concepts in business, economics, and the social, biological, and physical sciences involve quantities that are changing, it becomes evident that the idea of the derivative is one of the most important concepts in applied mathematics.

We will see in this chapter how to compute the derivatives of a wide variety of functions. In Chapter 11 we shall use derivatives to solve a number of interesting problems. Some examples of the types of problems that can be solved fairly easily with the aid of the derivative are listed below.

Problem 1. A manufacturer buys large quantities of a certain machine replacement part. He finds that his cost per unit decreases as the number of cases bought increases. He determines that a reasonable model for this is given by the formula

$$C(q) = 100 + 5q - 0.01q^2, \tag{1}$$

where q is the number of cases bought (up to 250 cases) and $C(q)$, measured in dollars, is the *total* cost of purchasing q cases. The 100 in equation (1) is a *fixed* cost, which does not depend on the number of cases bought. What are the *marginal* costs for various levels of purchase? (Example 10.5.3.)

Problem 2. A rope is attached to a pulley mounted on a 15 ft. tower. A worker can pull in rope at the rate of 2 ft./sec. How fast is the cart approaching the tower when it is 8 ft. from the tower (see Figure 2)? (Example 11.5.1.)

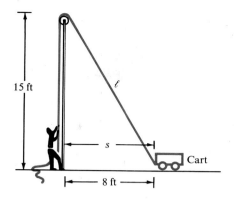

Figure 2

Problem 3. Sketch the curve $y = x^3 + 3x^2 - 9x - 10$. (Example 11.1.6.)

Problem 4. The Kertz Leasing Company leases fleets of new cars to large corporations. It charges $2000 per car per year. However, for contracts with a fleet size of more than 10 cars, the rental fee per car is discounted 1% for each car in the contract up to a maximum fleet size of 75 cars. How many cars leased to a single corporation in one year will produce maximum revenue and profit? (Example 11.3.8.)

10.2 Limits

The notion of a *limit* is central to the study of calculus. Before giving a formal definition, we illustrate a variety of limits.

Example 1   We begin by looking at the function

$$y = f(x) = x^2 + 3. \tag{1}$$

This function is graphed in Figure 1. What happens to $f(x)$ as x gets close to

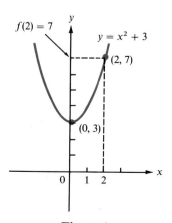

Figure 1

the value $x = 2$? To get an idea, look at Table 1, keeping in mind that x can get close to 2 from the right of 2 and from the left of 2 along the x-axis. It appears from the table that as x gets close to $x = 2$, $f(x) = x^2 + 3$ gets close to 7. This is not surprising since if we now calculate $f(x)$ at $x = 2$, we obtain $f(2) = 2^2 + 3 = 4 + 3 = 7$. In mathematical symbols we write

$$\lim_{x \to 2} (x^2 + 3) = 7.$$

TABLE 1

x	$f(x) = x^2 + 3$	x	$f(x) = x^2 + 3$
3	12	1	4
2.5	9.25	1.5	5.25
2.3	8.29	1.7	5.89
2.1	7.41	1.9	6.61
2.05	7.2025	1.95	6.8025
2.01	7.0401	1.99	6.9601
2.001	7.004001	1.999	6.996001
2.0001	7.00040001	1.9999	6.99960001

This is read "the limit as x approaches 2 (or tends to 2) of $x^2 + 3$ is equal to 7."

Note. In order to calculate this limit we did *not* have to evaluate $x^2 + 3$ at $x = 2$.

Example 2 What happens to the function $y = f(x) = \sqrt{2x - 6}$ as x gets close to $x = 5$?

Solution Since when $x = 5$, $\sqrt{2x - 6} = \sqrt{2 \cdot 5 - 6} = \sqrt{10 - 6} = \sqrt{4} = 2$, we might guess that as x gets close to 5, $\sqrt{2x - 6}$ gets close to 2. That this is indeed true is suggested by the computations in Table 2.

TABLE 2

x	$2x$	$2x - 6$	$\sqrt{2x - 6}$	x	$2x$	$2x - 6$	$\sqrt{2x - 6}$
6.0	12.0	6.0	2.449489743	4.0	8.0	2.0	1.414213562
5.5	11.0	5.0	2.236067977	4.5	9.0	3.0	1.732050808
5.1	10.2	4.2	2.049390153	4.9	9.8	3.8	1.949358869
5.01	10.02	4.02	2.004993766	4.99	9.98	3.98	1.994993734
5.001	10.002	4.002	2.000499938	4.999	9.998	3.998	1.999499937
5.0001	10.0002	4.0002	2.000049999	4.9999	9.9998	3.9998	1.999949999

Example 3   Consider the function

$$f(x) = \frac{x(x + 1)}{x}.$$

Since we cannot divide by zero, this function is defined for every real number except for $x = 0$. Since $x/x = 1$, we see that $f(x) = x + 1$ for all $x \neq 0$. This function is graphed in Figure 2. What happens to $f(x)$ as x approaches 0?

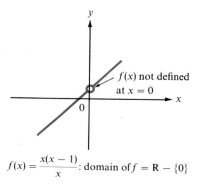

$f(x) = \dfrac{x(x - 1)}{x}$: domain of $f = R - \{0\}$

Figure 2

Again we illustrate this with a table of values (Table 3). As long as $x \neq 0$, we may use the fact that $f(x) = x + 1$. It is clear that as x gets close to 0, $f(x)$ gets close to 1. In mathematical notation we write

$$\lim_{x \to 0} \frac{x(x + 1)}{x} = 1.$$

TABLE 3

x	$f(x) = \dfrac{x(x + 1)}{x} = x + 1$	x	$f(x) = \dfrac{x(x + 1)}{x} = x + 1$
1	2	−1	0
0.5	1.5	−0.5	0.5
0.1	1.1	−0.1	0.9
0.05	1.05	−0.05	0.95
0.01	1.01	−0.01	0.99
0.001	1.001	−0.001	0.999

It is important to note that, for $f(x) = x(x + 1)/x$, it is still not permissible to set $x = 0$ because this would imply division by zero. However, we now know what happens to this function as x approaches zero. We can see why it is important that we are not required to evaluate $f(x)$ at $x = 0$ when we calculate the limit as x approaches zero.

Before giving further examples, we shall give a more formal definition of a limit. The definition given below is meant to appeal to your intuition. It is *not* a precise mathematical definition. In this section we hope that you will begin to get comfortable with the notion of limits and will acquire some facility in calculating them. More precise definitions of limits and proofs of standard limit theorems can be found in most engineering calculus textbooks.

Limit

Let L be a real number and suppose that $f(x)$ is defined on an open interval containing x_0, but not necessarily at x_0 itself. We say that the **limit** as x approaches x_0 of $f(x)$ is L, written

$$\lim_{x \to x_0} f(x) = L, \tag{2}$$

if, whenever x gets close to x_0 from either side with $x \neq x_0$, $f(x)$ gets close to L.

We insist that f be defined on an open interval (see page 11) containing the number x_0 except possibly at x_0 itself. This ensures that f is defined on both sides of x_0 (see Figure 3). It is necessary that $f(x)$ get close to L when x gets

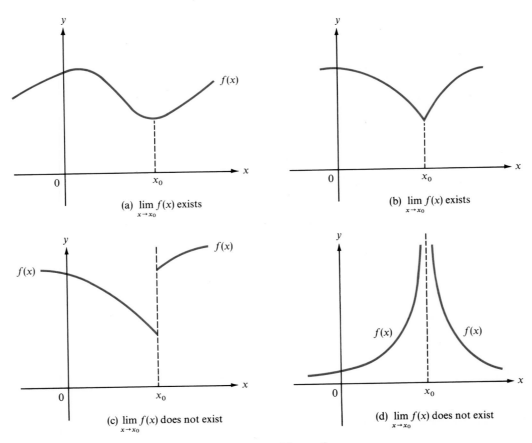

Figure 3

close to x_0 from either side. In Table 1, for example, $x^2 + 3$ gets close to 7 when x gets close to 2 from the right (the table on the left) and the left (the table on the right). Similarly, in Table 3, $f(x) = x(x + 1)/x$ gets close to 1 whether x approaches 0 from the left or the right.

The limit exists at x_0 in Figures 3a and 3b because $f(x)$ approaches the same value as we approach x_0 from the left or from the right. The limit does not exist at x_0 in Figure 3c because we get different values as we approach x_0 from different sides. In Figure 3d, the limit at x_0 does not exist because $f(x)$ becomes infinitely large as x approaches x_0.

Remark. It must be emphasized that, although we do not actually need to know what $f(x_0)$ is (in fact, $f(x_0)$ need not even exist), it is nevertheless often very helpful to know $f(x_0)$ in the actual computation of $\lim_{x \to x_0} f(x)$. It frequently happens that $\lim_{x \to x_0} f(x)$ indeed equals $f(x_0)$. However, we again emphasize that this is *not always* the case. In Example 3, we showed that $\lim_{x \to 0} f(x) = 1$ even though $f(0)$ did not exist.

Example 4   Calculate $\lim_{x \to 0} |x|$.

Solution   From Section 1.5 (page 29), we have

$$|x| = \begin{cases} x, & x \geq 0 \\ -x, & x < 0 \end{cases}.$$

If $x > 0$, then $|x| = x$, which tends to zero as $x \to 0$ from the right of 0. If $x < 0$, then $|x| = -x$, which again tends to zero as $x \to 0$ from the left of 0. Then, since we get the same answer when we approach zero from the left and from the right, we have

$$\lim_{x \to 0} |x| = 0.$$

This is pictured in Figure 4.

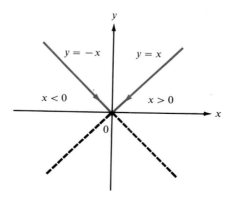

Figure 4

Example 5   Calculate $\lim\limits_{x \to 0} \dfrac{|x|}{x}$.

Solution   If $x > 0$, then $|x| = x$ so that $|x|/x = x/x = 1$. On the other hand, if $x < 0$, then $|x| = -x$ so that $|x|/x = -x/x = -1$. Note that $|x|/x$ is not defined at $x = 0$. The graph of $|x|/x$ is sketched in Figure 5. In sum, we have

$$\frac{|x|}{x} = \begin{cases} 1, & x > 0 \\ -1, & x < 0 \end{cases}.$$

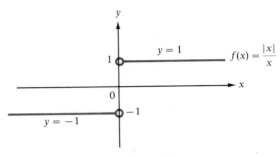

Figure 5

From Figure 5 we conclude that $f(x) = |x|/x$ has *no* limit as $x \to 0$. For $x > 0$, $f(x)$ remains at the constant value 1 and so approaches 1 as $x \to 0$; when $x < 0$, $f(x)$ remains at -1 and so approaches -1 as $x \to 0$.

Since the value of the limit has to be the same no matter from which direction we approach the value 0, we are left to conclude that there is no limit at 0. Of course, for any other value of x there is a limit. For example, $\lim_{x \to 2} |x|/x = 1$ since, near $x = 2$, $|x| = x$ and $|x|/x = 1$. Similarly, $\lim_{x \to -2} |x|/x = -1$.

The calculation of limits may be very tedious. Fortunately, there are a number of theorems which make limit computations much simpler. In the next section we shall show that for a wide variety of functions, called **continuous functions**, limits can be computed by evaluation; that is, $\lim_{x \to x_0} f(x) = f(x_0)$.

We now state several facts about limits. The proofs of these facts can be found in most engineering calculus books.

We saw in Example 1 that

$$\lim_{x \to 2} (x^2 + 3) = 2^2 + 3 = 7.$$

That is, the limit of $f(x) = x^2 + 3$ as x tends to 2 is equal to $f(x)$ evaluated at $x = 2$ (that is, $f(2)$). We can always compute a limit by *evaluation* when f is a polynomial.

THEOREM 1: LIMIT OF A POLYNOMIAL FUNCTION

Let $p(x) = c_0 + c_1 x + c_2 x^2 + c_3 x^3 + \cdots + c_n x^n$ be a *polynomial*, where $c_0, c_1, c_2, c_3, \ldots, c_n$ are real numbers and n is a fixed positive integer. Then

$$\lim_{x \to x_0} p(x) = p(x_0) = c_0 + c_1 x_0 + c_2 x_0^2 + c_3 x_0^3 + \cdots + c_n x_0^n. \tag{3}$$

Example 6 Calculate $\lim_{x \to 3} (x^3 - 2x + 6)$.

Solution $x^3 - 2x + 6$ is a polynomial. Hence

$$\lim_{x \to 3} (x^3 - 2x + 6) = 3^3 - 2 \cdot 3 + 6 = 27 - 6 + 6 = 27.$$

THEOREM 2: MULTIPLICATION OF A FUNCTION BY A CONSTANT

Let c be any real number and suppose that $\lim_{x \to x_0} f(x)$ exists. Then $\lim_{x \to x_0} cf(x)$ exists, and

$$\lim_{x \to x_0} cf(x) = c \lim_{x \to x_0} f(x). \tag{4}$$

Theorem 2 states that the limit of a constant times a function is equal to the product of that constant and the limit of the function.

Example 7 Calculate $\lim_{x \to 3} 5(x^3 - 2x + 6)$.

Solution We can find this limit two ways. We can multiply to find that $5(x^3 - 2x + 6) = 5x^3 - 10x + 30$ and then use Theorem 1. However, in Example 1 we calculated

$$\lim_{x \to 3} (x^3 - 2x + 6) = 27.$$

Therefore, using Theorem 2 we have

We use Theorem 2 here
$$\downarrow$$
$$\lim_{x \to 3} 5(x^3 - 2x + 6) = 5 \lim_{x \to 3} (x^3 - 2x + 6) = 5(27) = 135.$$

THEOREM 3: LIMIT OF THE SUM OF TWO FUNCTIONS

If $\lim_{x \to x_0} f(x)$ and $\lim_{x \to x_0} g(x)$ both exist, then $\lim_{x \to x_0} [f(x) + g(x)]$ exists, and

$$\lim_{x \to x_0} [f(x) + g(x)] = \lim_{x \to x_0} f(x) + \lim_{x \to x_0} g(x). \tag{5}$$

Theorem 3 states that the limit of the sum of two functions is equal to the sum of their limits.

Example 8   Calculate $\lim_{x\to 0}\left[\left(\dfrac{x(x+1)}{x}\right) + 4x^3 + 3\right]$.

$$\underset{\text{From Example 3}}{\downarrow} \qquad\qquad \underset{\text{From Theorem 1}}{\downarrow}$$

Solution $\lim_{x\to 0}(x(x+1)/x) = 1$ and $\lim_{x\to 0}[4x^3 + 3] = 4\cdot 0^3 + 3 = 3$. Hence

$$\lim_{x\to 0}\left[\frac{x(x+1)}{x} + 4x^3 + 3\right] = \lim_{x\to 0}\frac{x(x+1)}{x} + \lim_{x\to 0}(4x^3 + 3) = 1 + 3 = 4.$$

THEOREM 4:   LIMIT OF THE PRODUCT OF TWO FUNCTIONS

If $\lim_{x\to x_0} f(x)$ and $\lim_{x\to x_0} g(x)$ both exist, then $\lim_{x\to x_0}[f(x)\cdot g(x)]$ exists, and

$$\lim_{x\to x_0}[f(x)\cdot g(x)] = \left[\lim_{x\to x_0} f(x)\right]\cdot\left[\lim_{x\to x_0} g(x)\right]. \qquad (6)$$

Theorem 4 says that the limit of the product of two functions is the product of their limits.

Example 9   Calculate

$$\lim_{x\to 0}\frac{x(x+1)}{x}\cdot(4x^3 + 3).$$

Solution

$$\lim_{x\to 0}\frac{x(x+1)}{x}\cdot(4x^3 + 3) = \left[\lim_{x\to 0}\frac{x(x+1)}{x}\right]\cdot\left[\lim_{x\to 0}(4x^3 + 3)\right] = 1\cdot 3 = 4.$$

Example 10   Calculate $\lim_{x\to 1}(x^2 + 3x + 5)^2$.

$$\underset{\text{From Theorem 1}}{\downarrow}$$

Solution $\lim_{x\to 1}(x^2 + 3x + 5) = 1^2 + 3\cdot 1 + 5 = 9$. Hence

$$\lim_{x\to 1}(x^2 + 3x + 5)^2 = \lim_{x\to 1}(x^2 + 3x + 5)(x^2 + 3x + 5)$$

$$= \left[\lim_{x\to 1}(x^2 + 3x + 5)\right]\cdot\left[\lim_{x\to 1}(x^2 + 3x + 5)\right]$$

$$= 9\cdot 9 = 81$$

The result of Example 10 can be generalized to give us the useful rule below, which follows immediately from Theorem 4.

Power Rule for Limits   Let n be a positive integer. If $\lim_{x \to x_0} f(x)$ exists, then $\lim_{x \to x_0} [f(x)]^n$ exists, and

$$\lim_{x \to x_0} [f(x)]^n = \left[\lim_{x \to x_0} f(x) \right]^n$$

Example 11   Compute $\lim_{x \to 2} (x^2 + 1)^4$.

Solution $\lim_{x \to 2} (x^2 + 1) = 2^2 + 1 = 5.$ From Theorem 1

Thus,

$$\lim_{x \to 2} (x^2 + 1)^4 = \left[\lim_{x \to 2} (x^2 + 1) \right]^4 = 5^4 = 625.$$

THEOREM 5:   LIMIT OF THE QUOTIENT OF TWO FUNCTIONS

If $\lim_{x \to x_0} f(x)$ and $\lim_{x \to x_0} g(x)$ both exist and $\lim_{x \to x_0} g(x) \neq 0$, then $\lim_{x \to x_0} f(x)/g(x)$ exists, and

$$\lim_{x \to x_0} \frac{f(x)}{g(x)} = \frac{\lim_{x \to x_0} f(x)}{\lim_{x \to x_0} g(x)}. \tag{7}$$

Theorem 5 says that the limit of the quotient of two functions is the quotient of their limits, provided that the limit in the denominator function is not zero.

Example 12   Calculate $\lim_{x \to 3} (x + 1)/(x^2 - 2)$.

Solution

$$\lim_{x \to 3} (x + 1) = 3 + 1 = 4 \qquad \text{and} \qquad \lim_{x \to 3} (x^2 - 2) = 3^2 - 2 = 7.$$

Therefore,

$$\lim_{x \to 3} \frac{x + 1}{x^2 - 2} = \frac{\lim_{x \to 3} (x + 1)}{\lim_{x \to 3} (x^2 - 2)} = \frac{4}{7}.$$

Rational Function   A **rational function**, $r(x)$, is a function that can be written as the quotient of two polynomials; that is,

$$r(x) = \frac{p(x)}{q(x)} \tag{8}$$

where $p(x)$ and $q(x)$ are both polynomials.

For example, the function

$$r(x) = \frac{x+1}{x^2 - 2}$$

given in Example 12 is a rational function.

THEOREM 6: LIMIT OF A RATIONAL FUNCTION

Let $r(x) = p(x)/q(x)$ be a rational function with $q(x_0) \neq 0$. Then

$$\boxed{\lim_{x \to x_0} r(x) = \lim_{x \to x_0} \frac{p(x)}{q(x)} = \frac{p(x_0)}{q(x_0)} = r(x_0).}$$ (9)

Example 13 Calculate $\lim_{x \to 4} (x^3 - x^2 - 3)/(x^2 - 3x + 5)$.

Solution Here $q(x) = x^2 - 3x + 5$ and $q(4) = 16 - 12 + 5 = 9 \neq 0$. Therefore,

From Theorem 1
↓

$$\lim_{x \to 4} \frac{x^3 - x^2 - 3}{x^2 - 3x + 5} = \frac{4^3 - 4^2 - 3}{4^2 - 3 \cdot 4 + 5} = \frac{64 - 16 - 3}{16 - 12 + 5} = \frac{45}{9} = 5.$$

The next example illustrates the fact that Theorem 6 cannot be applied directly if $\lim_{x \to x_0} q(x) = 0$.

Example 14 Compute $\lim_{x \to 2} \dfrac{x^2 + 3x - 10}{x^2 - 4}$.

Solution

From Theorem 1
↓

$$\lim_{x \to 2} (x^2 - 4) = 2^2 - 4 = 0,$$

so we cannot apply Theorem 6. Also,

From Theorem 1
↓

$$\lim_{x \to 2} (x^2 + 3x - 10) = 2^2 + 3 \cdot 2 - 10 = 0$$

as well. However, since $x \neq 2$ in the computation of the limit as $x \to 2$, we can divide by $x - 2$ to obtain

$$\frac{x^2 + 3x - 10}{x^2 - 4} = \frac{(x - 2)(x + 5)}{(x - 2)(x + 2)} = \frac{x + 5}{x + 2}.$$

Thus,

$$\lim_{x \to 2} \frac{x^2 + 3x - 10}{x^2 - 4} = \lim_{x \to 2} \frac{x + 5}{x + 2} = \frac{2 + 5}{2 + 2} = \frac{7}{4}.$$

There are two other kinds of limits that frequently arise.

Infinite Limit

If $f(x)$ grows without bound in the positive direction as x gets close to the number x_0 from either side, then we say that **$f(x)$ tends to infinity as x approaches x_0**, and we write

$$\lim_{x \to x_0} f(x) = \infty.$$

Example 15 Compute $\lim\limits_{x \to 0} \dfrac{1}{x^2}$.

Solution From Table 4 we see that $\dfrac{1}{x^2}$ grows without bound as x approaches zero from either side. Thus,

$$\lim_{x \to 0} \frac{1}{x^2} = \infty.$$

TABLE 4

x	x^2	$\dfrac{1}{x^2}$	x	x^2	$\dfrac{1}{x^2}$
1	1	1	-1	1	1
0.5	0.25	4	-0.5	0.25	4
0.1	0.01	100	-0.1	0.01	100
0.01	0.0001	10,000	-0.01	0.0001	10,000
0.001	0.000001	1,000,000	-0.001	0.000001	1,000,000
0.0001	0.00000001	100,000,000	-0.0001	0.00000001	100,000,000

The graph of the function $f(x) = \dfrac{1}{x^2}$ is sketched in Figure 6. Notice that

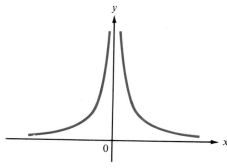

Figure 6

$\lim\limits_{x \to 0} \dfrac{1}{x^2} = \infty$ is illustrated graphically as the graph of f gets higher and higher as x approaches 0.

Limit at Infinity

The **limit as x approaches infinity of $f(x)$ is L**, written

$$\lim_{x \to \infty} f(x) = L,$$

if $f(x)$ is defined for all large values of x and if $f(x)$ gets close to L as x increases without bound.

Example 16   Compute $\lim\limits_{x \to \infty} \dfrac{1}{x^2}$.

Solution   As x gets large, x^2 gets large and $1/x^2$ gets small. In Table 5 we give values of x and $1/x^2$. From the table and from Figure 6, it is evident that

$$\lim_{x \to \infty} \frac{1}{x^2} = 0.$$

TABLE 5

x	x^2	$\dfrac{1}{x^2}$
1	1	1
5	25	0.04
10	100	0.01
100	10,000	0.0001
1,000	1,000,000	0.000001
10,000	100,000,000	0.00000001

 Example 17   Compute

$$\lim_{x \to \infty} \frac{2x + 3}{5x + 4}.$$

Solution

METHOD 1   We construct a table of values (Table 6). It seems from the table that

$$\lim_{x \to \infty} \frac{2x + 3}{5x + 4} = 0.4 = \frac{2}{5}.$$

TABLE 6

x	$2x + 3$	$5x + 4$	$\dfrac{2x + 3}{5x + 4}$
1	5	9	0.55555556
5	13	29	0.44827586
10	23	59	0.38983051
100	203	509	0.39882122
1,000	2,003	5,009	0.39988022
10,000	20,003	50,009	0.39998800
1,000,000	2,000,003	5,000,009	0.39999988

METHOD 2

$$\lim_{x \to \infty} \frac{2x + 3}{5x + 4} = \lim_{x \to \infty} \frac{\dfrac{2x + 3}{x}}{\dfrac{5x + 4}{x}}$$

We divided numerator and denominator by x.

$$= \lim_{x \to \infty} \frac{2 + \dfrac{3}{x}}{5 + \dfrac{4}{x}}.$$

But we can show, as in Example 16, that

$$\lim_{x \to \infty} \frac{3}{x} = 0 \quad \text{and} \quad \lim_{x \to \infty} \frac{4}{x} = 0.$$

Thus, the terms $\dfrac{3}{x}$ and $\dfrac{4}{x}$ become very small as x becomes large, so that

$$\lim_{x \to \infty} \frac{2x + 3}{5x + 4} = \lim_{x \to \infty} \frac{2 + \dfrac{3}{x}}{5 + \dfrac{4}{x}} = \frac{2}{5} = 0.4.$$

Example 18   Compute

$$\lim_{x \to \infty} \frac{3x^3 + 5x^2 - 9}{4x^3 - 3x + 16}.$$

Solution

$$\lim_{x \to \infty} \frac{3x^3 + 5x^2 - 9}{4x^3 - 3x + 16} = \lim_{x \to \infty} \frac{3 + \dfrac{5}{x} - \dfrac{9}{x^3}}{4 - \dfrac{3}{x^2} + \dfrac{16}{x^3}}.$$ We divided numerator and denominator by x^3.

Again, we can show that

$$\lim_{x \to \infty} \frac{5}{x} = \lim_{x \to \infty} \frac{-9}{x^3} = \lim_{x \to \infty} \frac{-3}{x^2} = \lim_{x \to \infty} \frac{16}{x^3} = 0.$$

Thus,

$$\lim_{x \to \infty} \frac{3x^3 + 5x^2 - 9}{4x^3 - 3x + 16} = \lim_{x \to \infty} \frac{3 + \dfrac{5}{x} - \dfrac{9}{x^3}}{4 - \dfrac{3}{x^2} + \dfrac{16}{x^3}} = \frac{3}{4}.$$

Example 19   The Easy Clean Corporation (ECC) is embarking on an extensive advertising campaign to market its new detergent. In the past, advertising has been very successful in increasing public awareness and sales of ECC's products. A senior advertising executive has estimated that, for the new product, profit (P) is related to advertising expenditures (x) according to the formula

$$P(x) = \frac{16x + 10}{x + 3}, \tag{10}$$

where x and P are measured in units of $100,000.

(a)   Show that, in equation (10), profit increases as advertising costs increase. This would confirm a fact about other products and make the model more believable.

(b)   Find an upper limit to the profit, if any exists.

Solution

(a)   We must show that if $x_2 > x_1$, then $P(x_2) > P(x_1)$. But

$$P(x_2) - P(x_1) = \frac{16x_2 + 10}{x_2 + 3} - \frac{16x_1 + 10}{x_1 + 3}$$

$$\frac{a}{b} + \frac{c}{d} = \frac{ad + bc}{bd}$$

$$= \frac{(16x_2 + 10)(x_1 + 3) - (16x_1 + 10)(x_2 + 3)}{(x_2 + 3)(x_1 + 3)}$$

$$= \frac{16x_2x_1 + 48x_2 + 10x_1 + 30 - 16x_1x_2 - 48x_1 - 10x_2 - 30}{(x_2 + 3)(x_1 + 3)}$$

$$= \frac{38x_2 - 38x_1}{(x_2 + 3)(x_1 + 3)} = \frac{38(x_2 - x_1)}{(x_2 + 3)(x_1 + 3)} > 0$$

since $x_2 > x_1$. Thus $P(x_2) > P(x_1)$.

(b)
$$\lim_{x \to \infty} P(x) = \lim_{x \to \infty} \frac{16x + 10}{x + 3} \overset{\substack{\text{Divide numerator and} \\ \text{denominator by } x. \\ \downarrow}}{=} \lim_{x \to \infty} \frac{16 + \dfrac{10}{x}}{1 + \dfrac{3}{x}} = 16.$$

Thus, the upper limit on profits is $16 \cdot \$100{,}000 = \$1.6$ million. This means that, after a while, large increases in advertising spending increase profits only by a very small amount. This is one example of the **law of diminishing returns**.
   The profit function (10) is sketched in Figure 7. In Section 11.1 (see Example 11.1.3) we will show much more easily that $P(x)$ increases as x increases.

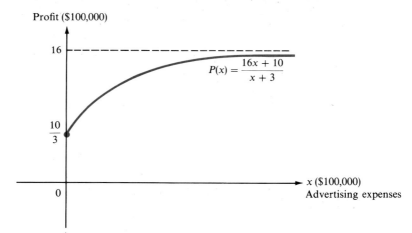

Figure 7

PROBLEMS 10.2
1. (a) Draw the graph of the function $f(x) = x + 7$.

   (b) Calculate $f(x)$ for $x = 3, 1, 2.5, 1.5, 2.1, 1.9, 2.01,$ and 1.99.

   (c) Calculate $\lim_{x \to 2} (x + 7)$.

2. (a) Draw the graph of the function $f(x) = x^2 - 4$ (see Section 9.3).

   (b) Calculate $f(x)$ for $x = 2, 0, 1.5, 0.5, 1.1, 0.9, 1.01,$ and 0.99.

   (c) Calculate $\lim_{x \to 1} (x^2 - 4)$.

3. (a) Draw the graph of the function $f(x) = x^2 - 3x + 4$.

   (b) Calculate $f(x)$ for $x = -0.5, -1.5, -0.9, -1.1, -0.99,$ and -1.01.

   (c) Calculate $\lim_{x \to -1} (x^2 - 3x + 4)$.

4. Let $f(x) = \dfrac{(x - 1)(x - 2)}{x - 1}$.

   (a) Explain why $f(x)$ is not defined for $x = 1$. (b) Calculate $\lim_{x \to 1} \dfrac{(x - 1)(x - 2)}{x - 1}$.

5. Let $f(x) = \dfrac{x^3 - 8}{x - 2}$.

    (a)   Explain why $f(x)$ is not defined at $x = 2$.

    (b)   Calculate $\displaystyle\lim_{x \to 2}\left(\dfrac{x^3 - 8}{x - 2}\right)$.

In Problems 6–24 calculate each limit if it exists, and explain why there is no limit if it does not exist.

6. $\displaystyle\lim_{x \to 5}(x^2 - 6)$

7. $\displaystyle\lim_{x \to 0}(x^3 + 17x + 45)$

8. $\displaystyle\lim_{x \to 0}(-x^5 + 17x^3 + 2x)$

9. $\displaystyle\lim_{x \to 0}\dfrac{1}{x^5 + 6x + 2}$

10. $\displaystyle\lim_{x \to 2}(x^4 - 9)$

11. $\displaystyle\lim_{x \to -1}\dfrac{(x + 1)^2}{x + 1}$

12. $\displaystyle\lim_{x \to 0}\dfrac{x^3}{x^2}$

13. $\displaystyle\lim_{x \to 4}(25 - x^2)^3$

14. $\displaystyle\lim_{x \to 4}(x^2 - 25)^3$

15. $\displaystyle\lim_{x \to 3}\dfrac{x^2 - 4x + 3}{x - 3}$

16. $\displaystyle\lim_{x \to 2}(x^3 - 8)^{27}$

17. $\displaystyle\lim_{x \to -2}\dfrac{x^2 + 6x + 8}{x + 2}$

18. $\displaystyle\lim_{x \to 1}\dfrac{x^4 - x}{x^3 - 1}$

19. $\displaystyle\lim_{x \to 1}\dfrac{\sqrt{x} - 1}{x - 1}$ [*Hint:* $a^2 - b^2 = (a + b)(a - b)$.]

20. $\displaystyle\lim_{x \to 2}\dfrac{1 - \sqrt{x/2}}{1 - (x/2)}$

*21. $\displaystyle\lim_{x \to 0}\dfrac{\sqrt{x + 1} - 1}{x}$

22. $\displaystyle\lim_{x \to 0}\dfrac{\dfrac{1}{x + 5} - \dfrac{1}{5}}{x}$

23. $\displaystyle\lim_{x \to 0}\dfrac{(x - 2)^3 + 8}{x}$

*24. $\displaystyle\lim_{x \to 0}\dfrac{\sqrt[3]{x + 27} - 3}{x}$ [*Hint:* $a^3 - b^3 = (a - b)(a^2 + ab + b^2)$.]

25. Let $f(x) = \dfrac{x^3 - 6x + 2}{x^2 + x + 9}$.

    (a)   Calculate $f(x)$ for $x = 3, 1, 2.5, 1.5, 2.1, 1.9, 2.01, 1.99, 2.001,$ and 1.999.

    (b)   Estimate $\displaystyle\lim_{x \to 2}\dfrac{x^3 - 6x + 2}{x^2 + x + 9}$.

    (c)   Calculate $f(2)$, and compare it with your estimate.

26. Let $f(x) = \dfrac{\sqrt{x^3 + 13}}{x + 8}$.

    (a)   Calculate $f(x)$ for $x = -1, -3, -1.5, -2.5, -1.9, -2.1, -1.99, -2.01, -1.999,$ and -2.001.

(b) Estimate $\lim\limits_{x \to -2} \dfrac{\sqrt{x^3 + 13}}{x + 8}$.

(c) Calculate $f(-2)$, and compare it with your estimate.

*27. (a) Graph the curve $y = x^2 + 3$.

(b) Draw (on your graph) the straight line joining the points $(1, 4)$ and $(2, 7)$.

(c) Draw the straight line joining the points $(1, 4)$ and $(1.5, 5.25)$.

(d) For any real number $\Delta x \neq 0$, what is represented by the quotient

$$\frac{[(1 + \Delta x)^2 + 3] - 4}{\Delta x}?$$

(e) Calculate $\lim\limits_{\Delta x \to 0} \dfrac{[(1 + \Delta x)^2 + 3] - 4}{\Delta x}$.

(f) What is the slope of the line tangent to the curve $y = x^2 + 3$ at the point $(1, 4)$?

*28. (a) Graph the curve $y = 5 - x^2$.

(b) Draw (on your graph) the straight line joining the points $(-3, -4)$ and $(-4, -11)$.

(c) Draw the straight line joining the points $(-3, -4)$ and $(-3.5, -7.25)$.

(d) For any real number $\Delta x \neq 0$, what is represented by the quotient

$$\frac{[5 - (-3 - \Delta x)^2] + 4}{-\Delta x}?$$

(e) Calculate $\lim\limits_{\Delta x \to 0} \dfrac{[5 - (-3 - \Delta x)^2] + 4}{-\Delta x}$.

(f) What is the slope of the line tangent to the curve $y = 5 - x^2$ at the point $(-3, -4)$?

29. (a) Graph the function $f(x) = |x - 3|$.

(b) Calculate $\lim\limits_{x \to 3} |x - 3|$.

30. (a) Graph the function $f(x) = \dfrac{|x + 3|}{x + 3}$.

(b) Explain why $\lim\limits_{x \to -3} \dfrac{|x + 3|}{x + 3}$ does not exist.

(c) Calculate $\lim\limits_{x \to 5} \dfrac{|x + 3|}{x + 3}$ and $\lim\limits_{x \to -5} \dfrac{|x + 3|}{x + 3}$.

31. Merlin strode into calculus class without fanfare and handed the participants the function f where $f(x) = 7x - 3$ and domain of $f = [0, 5)$. Merlin said he would close his eyes and cover his ears while, in turn, each person in the class chose a number s from the domain $[0, 5)$ and then redefined the value, $f(s)$, of the function there. When the class had done this, Merlin was tapped on the shoulder. He opened his eyes and ears and said, "Your modification of f is a new function.

Let's call it g. I don't know what you've done to f, so I can't draw a correct graph of the function g, but I do know that $\lim_{x\to 2} g(x) = 11$." Was Merlin right? Explain.

In Problems 32–49 use the limit theorems to help calculate the given limits.

32. $\lim_{x\to 3} (x^2 - 2x - 1)$

33. $\lim_{x\to -2} (-x^3 - x^2 - x - 1)$

34. $\lim_{x\to 1} (x^{50} - 1)$

35. $\lim_{x\to -1} (x^{49} + 1)$

36. $\lim_{x\to 5} 3\sqrt{x - 1}$

37. $\lim_{x\to 3} 5\sqrt{x^2 + 7}$

38. $\lim_{x\to -2} -4\sqrt{x + 3}$

39. $\lim_{x\to 1} 8(x^{100} + 2)$

40. $\lim_{x\to 5} (\sqrt{x - 1} + \sqrt{x^2 - 9})$

41. $\lim_{x\to -2} (1 + x + x^2 + x^3 + \sqrt{x^2 - 3})$

42. $\lim_{x\to -1} (x^9 + 2)^{33}$

43. $\lim_{x\to 4} (x^2 - x - 10)^7$

44. $\lim_{x\to 0} \dfrac{\sqrt{x + 1}}{\sqrt{x^2 - 3x + 4}}$

45. $\lim_{x\to -2} \dfrac{\sqrt{x^2 - 3}}{1 + x + x^2 + x^3}$

46. $\lim_{x\to 0} \dfrac{3}{x^5 + 3x^2 + 3}$

47. $\lim_{x\to -4} \dfrac{x^3 - x^2 - x + 1}{x^2 + 3}$

48. $\lim_{x\to 0} \dfrac{2x^2 + 5x + 1}{3x^5 - 9x + 2}$

49. $\lim_{x\to 0} \dfrac{x^{81} - x^{41} + 3}{23x^4 - 8x^7 + 5}$

In Problems 50–56 find the indicated limit, if it exists.

50. $\lim_{x\to 2} f(x)$ where $f(x) = \begin{cases} x - 2, & x > 2 \\ 0, & x \le 2 \end{cases}$

51. $\lim_{x\to 0} f(x)$ where $f(x) = \begin{cases} x + 3, & x \ge 0 \\ x - 3, & x < 0 \end{cases}$

52. $\lim_{x\to 0} f(x)$ where $f(x) = \begin{cases} x + 1, & x > 0 \\ x^3 + 1, & x < 0 \end{cases}$

53. $\lim_{x\to 1} f(x)$ where $f(x) = \begin{cases} x^4, & x < 1 \\ x^5, & x \ge 1 \end{cases}$

54. $\lim_{x\to 3} f(x)$ where $f(x) = \begin{cases} x^2 + 2, & x > 3 \\ 5x - 4, & x < 3 \end{cases}$

55. $\lim_{x\to 0} f(x)$ where $f(x) = \begin{cases} 0, & x < 0 \\ x^2, & 0 \le x \le 2 \\ 4, & x > 2 \end{cases}$

56. $\lim_{x\to 2} f(x)$ where $f(x)$ is as in Problem 55.

In Problems 57–60 find $\lim_{x\to 3} f(x)$ (if it exists) from the given graph.

57.

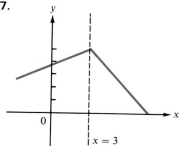

58.

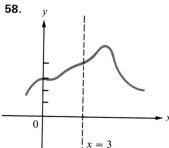

59.

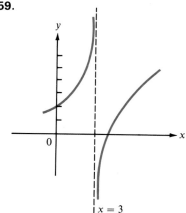

60.

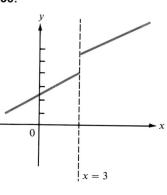

In Problems 61–81 calculate each limit (if it exists).

61. $\lim\limits_{x \to 0} \dfrac{1}{x^4}$

62. $\lim\limits_{x \to 0} \dfrac{1}{x^6}$

63. $\lim\limits_{x \to 0} \dfrac{1}{x^6 + x^{10}}$

64. $\lim\limits_{x \to 5} \dfrac{1}{(x - 5)^2}$

*65. $\lim\limits_{x \to 0} \dfrac{1}{x}$

66. $\lim\limits_{x \to -3} \dfrac{x - 4}{(x + 3)^2}$

67. $\lim\limits_{x \to 1} \dfrac{1}{(x - 1)^6}$

68. $\lim\limits_{x \to 0} \dfrac{1}{x^4 + x^8 + x^{12}}$

69. $\lim\limits_{x \to 0} \dfrac{x + x^2}{x^3 + x^4}$

70. $\lim\limits_{x \to \infty} \dfrac{1}{x + x^3}$

71. $\lim\limits_{x \to -\infty} \dfrac{x}{1 + x}$

72. $\lim\limits_{x \to \infty} \dfrac{1}{1 - \sqrt{x}}$

73. $\lim\limits_{x \to \infty} \dfrac{2x}{3x^3 + 4}$

74. $\lim\limits_{x \to -\infty} \dfrac{2x + 3}{3x + 2}$

75. $\lim\limits_{x \to \infty} \dfrac{5x - x^2}{3x + x^2}$

76. $\lim\limits_{x \to \infty} \dfrac{1 + \sqrt{x}}{1 - \sqrt{x}}$

77. $\lim\limits_{x \to \infty} \dfrac{2x^2 + 3x + 5}{3x^2 - x + 2}$

78. $\lim\limits_{x \to \infty} \dfrac{4x^4 + 1}{1 + 5x^4}$

79. $\lim\limits_{x \to \infty} \dfrac{x^5 - 3x + 4}{7x^6 + 8x^4 + 2}$

80. $\lim\limits_{x \to \infty} \dfrac{x^8 - 2x^5 + 3}{5x^4 + 3x + 1}$

81. $\lim\limits_{x \to \infty} \dfrac{x^8 - 1}{x^9 + 1}$

82. We have seen that $\lim_{x \to 0} 1/x^2 = \infty$. How small in absolute value must x be in order that $1/x^2 > 1,000,000$? $10,000,000$? $100,000,000$?

83. In Example 19, suppose that the profit function is estimated to be

$$P(x) = \frac{15x + 30}{3x + 17}.$$

(a) Show that if advertising expenses increase, profit increases.

(b) Find an upper limit on the profit.

10.3 Continuity

The concept of **continuity** is one of the central notions in mathematics. Intuitively, a function is continuous at a point if it is defined at that point and if its graph moves unbroken through that point. Figure 1 shows the graphs of six functions, three of which are continuous at x_0 and three of which are not.

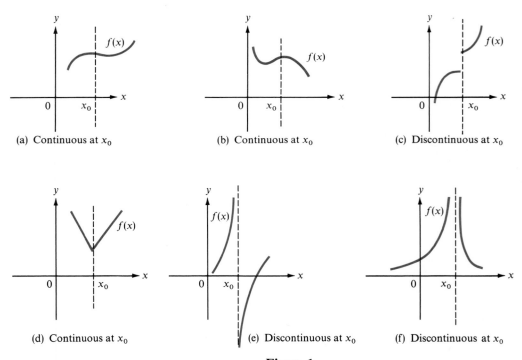

(a) Continuous at x_0 (b) Continuous at x_0 (c) Discontinuous at x_0

(d) Continuous at x_0 (e) Discontinuous at x_0 (f) Discontinuous at x_0

Figure 1

There are several equivalent definitions of continuity. The one we give here depends explicitly on limits.

Continuity at a Point   Let $f(x)$ be defined for every x in an open interval containing the number x_0. Then

> f is **continuous** at x_0 if all of the following three conditions hold:
>
> **1.** $f(x_0)$ exists (that is, x_0 is in the domain of f);
>
> **2.** $\lim_{x \to x_0} f(x)$ exists;
>
> **3.** $\lim_{x \to x_0} f(x) = f(x_0)$.

(1)

Remark. Condition (3) tells us that if a function f is continuous at x_0, then we can calculate $\lim_{x \to x_0} f(x)$ by evaluation. This is only one of the reasons continuous functions are so important. In the next few chapters we will see that a large majority of the functions we encounter in applications are indeed continuous.

Example 1 Let $f(x) = x^2$. Then, for any real number x_0,

$$\lim_{x \to x_0} f(x) = \lim_{x \to x_0} x^2 = x_0^2 = f(x_0),$$

so that f is continuous at every real number (see Figure 2).

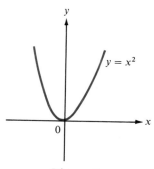

Figure 2

Example 2 Let $p(x) = c_0 + c_1 x + c_2 x^2 + c_3 x^3 + \cdots + c_n x^n$ be a polynomial. By Limit Theorem 1 on page 519,

$$\lim_{x \to x_0} p(x) = p(x_0)$$

(2)

for every real number x_0. Therefore,

> every polynomial is continuous at every real number.

(Note that this also shows that any constant function is continuous.)

Example 3 Let $r(x) = p(x)/q(x)$ be a rational function ($p(x)$ and $q(x)$ are polynomials). Then from Limit Theorem 5 on page 521, we have, if $q(x_0) \neq 0$, $\lim_{x \to x_0} r(x) = p(x_0)/q(x_0) = r(x_0)$, so that

> any rational function is continuous at all points x_0 at which the denominator, $q(x_0)$, is nonzero.

Example 4   Let

$$f(x) = \frac{x^5 + 3x^3 - 4x^2 + 5x - 2}{x^2 - 5x + 6}.$$

Here f is a rational function and is therefore continuous at any x for which the denominator, $x^2 - 5x + 6$, is not zero. Since $x^2 - 5x + 6 = (x - 3) \times (x - 2) = 0$ only when $x = 2$ or 3, f is continuous at all real numbers except at these two.

Discontinuous Function   A function that is not continuous at a point x_0 is called **discontinuous** at x_0.

As the examples above suggest, most commonly encountered functions are continuous. In this book, all the functions you meet will be continuous at every point except in one of three cases.

Discontinuity Case 1.   *We are dividing by zero.* This is exemplified by Example 4. The function is not continuous at $x = 2$ or $x = 3$.

Example 5   Let $f(x) = 1/x$. Then f is defined and continuous except at $x = 0$. Note that as x approaches zero, $f(x)$ "blows up." That is, f gets very large in the positive direction as x approaches zero from the positive side and very large in the negative direction as x approaches zero from the negative side. The function is sketched in Figure 3. The line $x = 0$ (the y-axis) is called a **vertical asymptote** for the graph of $f(x) = 1/x$.

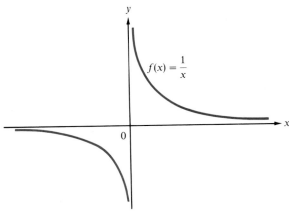

Figure 3

Example 6   Let $f(x) = (x^2 - 1)/(x - 1)$. Then f is not continuous at $x = 1$ because $f(1)$ does not exist, since $x - 1 = 0$ when $x = 1$. That is, domain of $f = \mathbb{R} - \{1\}$. However, since $x^2 - 1 = (x - 1)(x + 1)$, we see that

$$\lim_{x \to 1} \frac{x^2 - 1}{x - 1} = \lim_{x \to 1} (x + 1) = 2,$$

so that f does not "blow up" though its denominator is approaching zero.

Discontinuity Case 2.   *The function is not defined over a range of values.*

Example 7 $f(x) = \sqrt{x}$ is not continuous for $x < 0$ because the square-root function is not defined for negative values.

Discontinuity Case 3.   *The function is discontinuous if it is defined in pieces and has a "jump."*[†]

Example 8   A newspaper vendor finds that the wholesale price of the Centerville Times is 16¢ per copy if she purchases fewer than 100 copies each day. However, if the vendor purchases more than 100 copies, the price per copy drops to 14¢ per copy. Find the total cost function.

Solution   If q denotes the daily number of copies bought and C denotes the cost, then

$$C(q) = \begin{cases} 0.16q, & \text{if } q < 100 \\ 0.14q, & \text{if } q \geq 100 \end{cases}$$

This function is sketched in Figure 4. It is discontinuous at $q = 100$. Note that

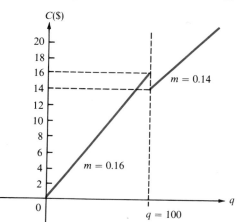

Figure 4

[†] A function defined in pieces could be continuous everywhere. For example, let $f(x) = |x| = \begin{cases} x, & x \geq 0 \\ -x, & x < 0 \end{cases}$. This function is defined in pieces but is continuous at every value of x since $\lim_{x \to 0} f(x) = 0 = f(0)$ (see Example 10.2.4 on page 517).

it "jumps" from the value 16 to the value 14 as q passes through the value 100. Note also that $C(q)$ is not continuous for $q < 0$ because it is not defined for $q < 0$.

Continuity over an Interval A function, f, is continuous over (or in) the open interval (a, b) if f is continuous at every point in that interval (a may be $-\infty$ and/or b may be ∞).

Example 9   From Example 2, we see that any polynomial is continuous in $(-\infty, \infty)$.

Example 10   Let $f(x) = \sqrt{x}$. Then f is continuous in the interval $(0, \infty)$.

Example 11   Let $f(x) = |x|$. Then

$$f(x) = \begin{cases} x, & x \geq 0 \\ -x, & x < 0 \end{cases}$$

This function is sketched in Figure 5. Since $\lim_{x \to 0} f(x) = 0 = f(0)$, we see that f is continuous in $(-\infty, \infty)$.

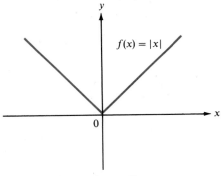

Figure 5

PROBLEMS 10.3   In Problems 1–14 find all points (if any) where the given function is discontinuous, and list the largest open interval or intervals over which it is continuous.

1. $f(x) = x^2 - 3$

2. $f(x) = \sqrt{x - 1}$

3. $f(x) = \dfrac{1}{4 - x}$

4. $f(x) = x^5 - x^3 + 2$

5. $f(x) = \dfrac{x}{x + 1}$

6. $f(x) = 2 - \dfrac{1}{x}$

7. $f(x) = x^{17} - 3x^{15} + 2$

8. $f(x) = x^{1/3}$

9. $f(x) = x^{1/4}$

10. $f(x) = \dfrac{1}{x + 2}$

11. $f(x) = \dfrac{-17x}{x^2 - 1}$

12. $f(x) = \dfrac{1}{(x - 10)^{15}}$

13. $f(x) = \dfrac{2x}{x^3 - 8}$

14. $f(x) = \dfrac{|x + 2|}{x + 2}$

15. For what values of α does the function $f(x) = (x^2 - 4)/(x - \alpha)$ *not* blow up at $x = \alpha$?

16. For what values of α does the function

$$f(x) = \frac{x^3 - 6x^2 + 11x - 6}{x - \alpha}$$

not blow up at $x = \alpha$?

17. Show that the function

$$f(x) = \begin{cases} \dfrac{x^3 - 1}{x - 1}, & x \neq 1 \\ 3, & x = 1 \end{cases}$$

is continuous on $(-\infty, \infty)$.

18. For what value of α is the function

$$f(x) = \begin{cases} \dfrac{x^4 - 1}{x - 1}, & x \neq 1 \\ \alpha, & x = 1 \end{cases}$$

continuous at $x = 1$?

*19. Let

$$f(x) = \begin{cases} x, & x \neq \text{an integer} \\ x^2, & x = \text{an integer} \end{cases}$$

Graph the function for $-3 \leq x \leq 3$. For what integer values of x is f continuous?

10.4 The Derivative as the Slope of a Curve

In Example 1.3.5 on page 16, we discussed the case of a shoelace manufacturer whose fixed cost was $2000 and who paid 34¢ in raw materials for each pair of shoelaces made. Then, as we found earlier, the manufacturer's **cost function** is given by

$$
\begin{array}{cc}
\text{Fixed} & \text{Variable} \\
\text{cost} & \text{cost} \\
\downarrow & \downarrow
\end{array}
$$

$$C(q) = 2000 + 0.34q. \tag{1}$$

This cost function is sketched in Figure 1. Equation (1) is the equation of a straight line with slope 0.34.

We can turn this problem around. Suppose that we are given a linear cost function and are asked to determine the cost per unit (variable cost) of the item produced. We can see from equation (1) that *the slope of a linear cost function is the unit (variable) cost.*

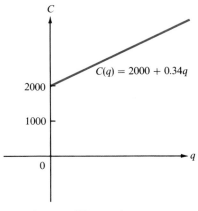

Figure 1

Example 1   The cost function for producing q compact cars is given by

$$C(q) = 125{,}000 + 3250q.$$

Find the variable cost.

Solution   The slope of the line $C = 125{,}000 + 3250q$ (which is already given in point slope form) is 3250. Thus, variable cost $= \$3250/\text{car}$.

If all the functions we ever encountered were linear, there would be no need to study calculus. However, many kinds of functions arise that are nonlinear. You will see many of these in this book. To answer most questions about such functions, it is necessary to develop new techniques. Let us first look at an example.

Example 2   A boy is standing on a bridge over a highway. The boy drops a rock from a point exactly 100 feet above the roadway. How fast is the rock traveling when it hits?

Solution   Let $s(t)$ denote the height of the rock above the road t seconds after the rock is released. Then it can be shown that, ignoring air resistance,

$$s(t) = 100 - 16t^2. \tag{2}$$

The rock hits the road when $s(t) = 0$. So, from equation (2), we have

$$0 = 100 - 16t^2$$

$$16t^2 = 100$$

$$t^2 = \tfrac{100}{16}$$

$$t = \sqrt{\tfrac{100}{16}} = \tfrac{10}{4} = \tfrac{5}{2} = 2\tfrac{1}{2} \text{ seconds.}$$

That is, the rock hits the road $2\tfrac{1}{2}$ seconds after it is released.

We might now reason as follows: The *average velocity* of a moving object is given by

$$\text{average velocity} = \frac{\text{distance traveled}}{\text{elapsed time}}.$$

So in our case,

$$\text{average velocity of the rock} = \frac{-100 \text{ ft}}{2.5 \text{ sec}} = -40 \text{ ft/second.}^{\dagger}$$

But this result doesn't answer our question. The expression -40 ft/sec represents the average velocity of the rock during the $2\frac{1}{2}$ seconds of its flight. When the rock is released by the boy, it isn't moving at all. As it falls, it gains speed until the moment of impact. Certainly the velocity on impact, after exactly $2\frac{1}{2}$ seconds, is greater than the average velocity. But how do we calculate this *instantaneous velocity*, as it is called, after exactly $2\frac{1}{2}$ seconds?

Let us begin by sketching, as in Figure 2, the graph of the function $s(t) = 100 - 16t^2$. We do so by plotting the points given in Table 1 and connecting them.

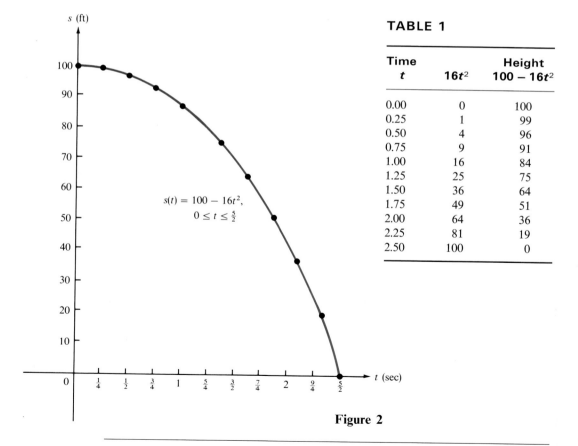

TABLE 1

Time t	$16t^2$	Height $100 - 16t^2$
0.00	0	100
0.25	1	99
0.50	4	96
0.75	9	91
1.00	16	84
1.25	25	75
1.50	36	64
1.75	49	51
2.00	64	36
2.25	81	19
2.50	100	0

Figure 2

---

† The minus sign indicates that the height is decreasing.

Note that $s(t)$ is defined only for $0 \leq t \leq \frac{5}{2}$, since neither negative time nor negative distance makes any sense in this problem.

Now let us attempt to compute the instantaneous velocity for any value of t in the interval $[0, 2.5]$. We enlarge the graph in Figure 2 to examine what happens near a point on the curve. The coordinates of any such point are $(t, 100 - 16t^2)$ (see Figure 3).

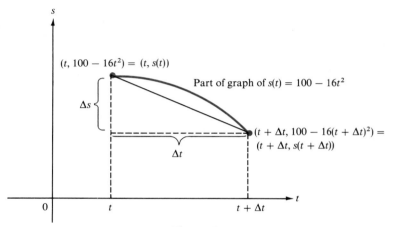

Figure 3

The Greek letter Δ is traditionally used to denote changes (differences). Thus Δt represents a period of time. Let us compute the average velocity of the rock in the time period t to $t + \Delta t$. We have, as before,

$$\text{average velocity} = \frac{\text{distance traveled}}{\text{elapsed time}}. \tag{3}$$

But in the time interval $[t, t + \Delta t]$ the rock has fallen from a height of $100 - 16t^2$ feet to a height of $100 - 16(t + \Delta t)^2$. Thus, from equation (3), we have

$$\text{average velocity} = \frac{[100 - 16(t + \Delta t)^2] - (100 - 16t^2)}{\Delta t}$$

$$= \frac{[100 - 16(t^2 + 2t\Delta t + \Delta t^2)] - 100 + 16t^2}{\Delta t}$$

$$= \frac{100 - 16t^2 - 32t\Delta t - 16\Delta t^2 - 100 + 16t^2}{\Delta t}$$

$$= \frac{-32t\Delta t - 16\Delta t^2}{\Delta t} = -32t - 16\, \Delta t.$$

We divided numerator and denominator by Δt.

But we can see from Figure 3 that, if Δs denotes the change in the height of the rock, then average velocity between t and $t + \Delta t = -32t - 16\,\Delta t$

$$= \frac{\Delta s}{\Delta t}$$

= slope of line joining points $(t, s(t))$ and $(t + \Delta t,\, s(t + \Delta t))$.

This line is called a **secant line**, and we see that

$$\text{slope of secant line} = -32t - 16\Delta t. \tag{4}$$

If Δt is very small, then over the time period t to $t + \Delta t$ the velocity changes, but *it does not change very much.* Thus,

$$-32t - 16\Delta t = \text{average velocity between } t \text{ and } t + \Delta t$$

$$\approx \text{instantaneous velocity at time } t. \tag{5}$$

But as Δt gets smaller and smaller, the approximation in equation (5) gets better and better. We have

instantaneous velocity = limiting value of average velocity
as Δt approaches 0 $\tag{6}$

or

$$\text{instantaneous velocity} = \lim_{\Delta t \to 0} (-32t - 16\Delta t) = -32t.$$

Thus

$$\text{instantaneous velocity at time } t = -32t, \tag{7}$$

and at impact (when $t = 2.5$),

$$\text{instantaneous velocity} = -32(2.5) = -80 \text{ ft/sec } (\approx 54.5 \text{ mi/hr}).$$

The minus sign indicates that the height is decreasing (the rock is going *down*), and it is decreasing considerably faster than the average velocity of 40 ft/sec.

Now observe that as Δt becomes smaller, the secant lines approach the line tangent to the curve at the point $(t, 100 - 16t^2)$ (see Figure 4). Therefore, *the slopes of secant lines approach the slope of the tangent line as Δt becomes smaller*, and we have, from equation (4),

$$\text{slope of tangent line at point } (t,\, 100 - 16t^2) = -32t. \tag{8}$$

Or, combining (7) and (8), we obtain

instantaneous velocity of falling rock = slope of line tangent to curve
$s = 100 - 16t^2$ at point $(t,\, 100 - 16t^2)$.

Thus, finding the tangent line to a curve has something to do with computing velocity. We will see many other applications of this idea in the chapters that follow.

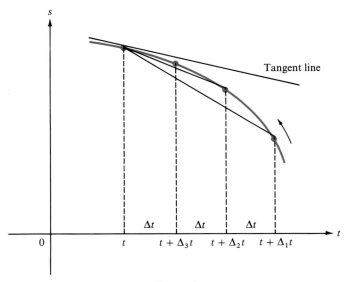

Figure 4

The slope of the unique tangent line (if one exists) to a point on the graph of a function is called the **derivative** of the function at that point. The result of Example 2 can be restated as follows:

> The derivative of the function $s(t) = 100t - 16t^2$ at the point $(t, 100t - 16t^2)$ is equal to $-32t$.

We now describe a procedure for computing the derivative of a function. We do this by defining a tangent line and showing how to compute its slope.

The Greeks knew how to find the line tangent to a circle at a given point, using the fact that, for a circle, the tangent line is perpendicular to the radius at the given point (see Figure 5). The Greeks also discovered how to construct tangent lines to other particular curves, and Archimedes himself devoted a major part of his book (*On Spirals*) to the tangent problem for a special curve called the *spiral of Archimedes*.

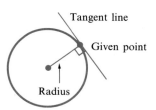

Figure 5

However, as more and more curves were studied, it became increasingly difficult to treat the large number of special cases, and a general method was sought for solving all such problems. Unfortunately, these early attempts met with failure. It wasn't until the independent discoveries of Isaac Newton (1642–1727) and Gottfried Leibniz (1646–1716) that the problem was resolved.

Part of the problem is that we do not yet know precisely what a tangent to a curve is if the curve is not a circle. The tangent line to a circle at a point on the circle hits the circle at that point and does not cross it. However, for other kinds of curves, there are other possibilities. Three such possibilities are exhibited in Figure 6.

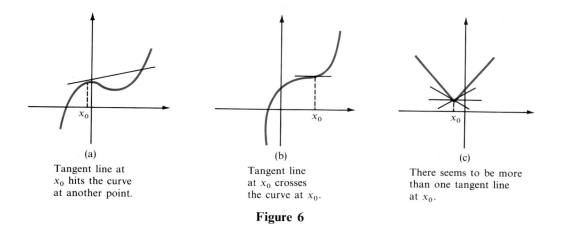

(a)
Tangent line at
x_0 hits the curve
at another point.

(b)
Tangent line
at x_0 crosses
the curve at x_0.

(c)
There seems to be more
than one tangent line
at x_0.

Figure 6

To solve our problem, we will use our intuitive idea of what a tangent line should be. This will finally lead to a definition of both a tangent line and a derivative.

The method we give below is essentially the method of Newton and Leibniz, which resolved the tangent problem posed so long ago by the Greek mathematicians.[†]

Let us consider the function $y = f(x)$, a part of whose graph is given in Figure 7. To calculate the derivative function, we must calculate the slope of the line tangent to the curve at each point of the curve at which there is a unique tangent line. Let $(x_0, f(x_0))$ be such a point. From now on we will assume that f is defined "near" x. If Δx is a small number (positive or negative) then $x_0 + \Delta x$ will be close to x_0. In moving from x_0 to $x_0 + \Delta x$, the values of f will move from $f(x_0)$ to $f(x_0 + \Delta x)$. Now look at the straight line in

---

[†] Isaac Newton, *Mathematical Principles of Natural Philosophy* (*Principia*), published in 1687, and Gottfried Leibniz, *A New Method for Maxima and Minima, and Also for Tangents, Which Is Not Obstructed by Irrational Quantities*, published in 1684.

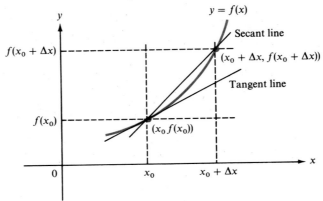

Figure 7

Figure 7, called a *secant line*, joining the points $(x_0, f(x_0))$ and $(x_0 + \Delta x, f(x_0 + \Delta x))$. What is its slope? If we define $\Delta y = f(x_0 + \Delta x) - f(x_0)$ and if we use m_s to denote the slope of such a secant line, we have, from Section 2.3,

$$m_s = \frac{\text{change in } y}{\text{change in } x} = \frac{f(x_0 + \Delta x) - f(x_0)}{(x_0 + \Delta x) - x_0} = \frac{f(x_0 + \Delta x) - f(x_0)}{\Delta x} = \frac{\Delta y}{\Delta x}.$$

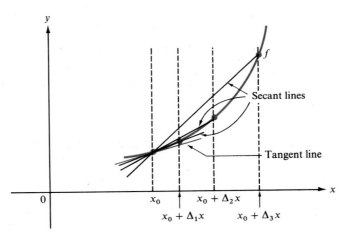

Figure 8

What does this have to do with the slope of the tangent line? The answer is suggested in Figure 8. From this illustration, we see that as Δx gets smaller, the secant line gets closer and closer to the tangent line. Put another way, as Δx approaches zero, the slope of the secant line approaches the slope of the tangent line. But the slope of the tangent line at the point $(x_0, f(x_0))$ is the derivative of f at x_0, denoted $f'(x_0)$. We therefore have

$$f'(x_0) = \lim_{\Delta x \to 0} m_s = \lim_{\Delta x \to 0} \frac{\Delta y}{\Delta x} = \lim_{\Delta x \to 0} \frac{f(x_0 + \Delta x) - f(x_0)}{\Delta x}. \tag{9}$$

We now make two important definitions.

Derivative at a Point If the limit (9) exists, the **derivative** of the function f at the point x_0 is given by

$$\text{derivative of } f \text{ at } x_0 = f'(x_0) = \lim_{\Delta x \to 0} \frac{f(x_0 + \Delta x) - f(x_0)}{\Delta x}. \tag{10}$$

If $f'(x_0)$ exists, then f is said to be **differentiable** at x_0.

In our definition of the derivative, we restricted our attention to a particular point, x_0. However, we can equally well attempt to take such a limit at any value of x and thereby obtain a new function.

Derivative Function The **derivative** f' of the function f is the function defined by

$$f'(x) = \lim_{\Delta x \to 0} \frac{f(x + \Delta x) - f(x)}{\Delta x}.$$

The domain of f' is the set of real numbers for which the limit in equation (10) exists.

Note. Since $f(x)$ is defined only if x is in the domain of f, we see that domain of $f' \subset$ domain of f. In other words, $f'(x)$ is not defined if $f(x)$ is not defined.

Remark. The definition of the derivative is given in terms of a limit and says nothing about tangent lines. However, we can use this definition to define a tangent line by saying that, if $f'(x_0)$ exists, then the **tangent line** to the curve $y = f(x)$ at the point $(x_0, f(x_0))$ is the *unique* line passing through $(x_0, f(x_0))$ with slope $f'(x_0)$.

Example 3 Let $y = f(x) = 3x + 5$. Calculate $f'(x)$.

Solution To solve this problem and the ones that follow, we simply use formula (10). For $f(x) = 3x + 5$, $f(x + \Delta x) = 3(x + \Delta x) + 5$. Then

$$f'(x) = \lim_{\Delta x \to 0} \frac{f(x + \Delta x) - f(x)}{\Delta x} = \lim_{\Delta x \to 0} \frac{[3(x + \Delta x) + 5] - [3x + 5]}{\Delta x}$$

$\Delta x \neq 0$, so we can divide by it.

$$= \lim_{\Delta x \to 0} \frac{3x + 3\Delta x + 5 - 3x - 5}{\Delta x} = \lim_{\Delta x \to 0} \frac{3\Delta x}{\Delta x} = \lim_{\Delta x \to 0} 3 = 3.$$

This answer is not surprising. It simply says that the slope of the line $y = 3x + 5$ is equal to the constant function 3.

Before giving further examples, if $y = f(x)$ we introduce the additional symbols dy/dx or df/dx to denote the derivative:

$$f'(x) = \frac{df}{dx} = \frac{dy}{dx} = \lim_{\Delta x \to 0} \frac{\Delta y}{\Delta x} = \lim_{\Delta x \to 0} \frac{f(x + \Delta x) - f(x)}{\Delta x}.$$

The symbol dy/dx is read "the derivative of y with respect to x." We emphasize that dy/dx is *not* a fraction. At this point the symbols dy and dx have no meaning of their own.

There are other notations for the derivative. We will often use the symbol y' or $y'(x)$ in place of f' or $f'(x)$. Thus, if $y = f(x)$, we may denote the derivative in four different ways:[†]

$$f' = y' = \frac{df}{dx} = \frac{dy}{dx}.$$

Example 4   Calculate the derivative of the function $y = x^2$. What is the equation of the line tangent to the graph of $y = x^2$ at the point $(3, 9)$?

Solution   For $y = f(x) = x^2$, $f(x + \Delta x) = (x + \Delta x)^2$. Then

$$\frac{dy}{dx} = \lim_{\Delta x \to 0} \frac{f(x + \Delta x) - f(x)}{\Delta x} = \lim_{\Delta x \to 0} \frac{(x + \Delta x)^2 - x^2}{\Delta x}$$

$$= \lim_{\Delta x \to 0} \frac{x^2 + 2x\Delta x + \Delta x^2 - x^2}{\Delta x} = \lim_{\Delta x \to 0} \frac{2x\Delta x + \Delta x^2}{\Delta x}$$

$$= \lim_{\Delta x \to 0} \frac{\Delta x(2x + \Delta x)}{\Delta x} = \lim_{\Delta x \to 0} (2x + \Delta x) = 2x.$$

At every point of the form $(x, f(x)) = (x, x^2)$, the slope of the line tangent to the curve is $2x$. For $x = 3$, $2x = 6$. Therefore, the slope of the tangent line at the point $(3, 9)$ is 6; that is, $f'(3) = 6$. We can now find the equation of the tangent line since it passes through the point $(3, 9)$ and has the slope 6. We have, if (x, y) is a point on the line,

$$\frac{\Delta y}{\Delta x} = \frac{y - 9}{x - 3} = 6 \quad \text{or} \quad y = 6x - 9.$$

It is interesting to see how the slopes of the secant lines approach the slope of the tangent line in this problem. For $x_0 = 3$, we have, from equation (10),

$$f'(3) = \lim_{\Delta x \to 0} \frac{(3 + \Delta x)^2 - 9}{\Delta x}.$$

The slope of a secant line is $[(3 + \Delta x)^2 - 9]/\Delta x$. Table 1 illustrates how quick-

---

[†] Newton (in England) and Leibniz (in Germany) independently discovered in the 1670s the equation for the slope of the tangent line given in this section. Newton used the symbol $\dot{y}$ (read "y dot") and Leibniz used the symbol dy/dx to indicate the derivative. There was a raging controversy, never resolved, over who made this momentous discovery first. The controversy was so intense that, to some extent, British mathematicians were alienated from mathematicians in the rest of Europe until well into the eighteenth century.

ly the slopes of secant lines approach the value 6, which is the slope of the tangent line, as $\Delta x \to 0$.

TABLE 1

Δx	$3 + \Delta x$	$(3 + \Delta x)^2$	$(3 + \Delta x)^2 - 9$	$\dfrac{(3 + \Delta x)^2 - 9}{\Delta x} = $ slope of secant line
0.5	3.5	12.25	3.25	6.5
0.1	3.1	9.61	0.61	6.1
0.01	3.01	9.0601	0.0601	6.01
0.0001	3.0001	9.00060001	0.00060001	6.0001
−0.5	2.5	6.25	−2.75	5.5
−0.1	2.9	8.41	−0.59	5.9
−0.01	2.99	8.9401	−0.0599	5.99
−0.0001	2.9999	8.99940001	−0.00059999	5.9999

Example 5 (a) Find the derivative of $y = \sqrt{x}$. (b) Calculate the slope of the tangent line at the point (9, 3).

Solution

(a) $f(x) = \sqrt{x}$ and $f(x + \Delta x) = \sqrt{x + \Delta x}$ so, from (10),

$$f'(x) = \lim_{\Delta x \to 0} \frac{f(x + \Delta x) - f(x)}{\Delta x} = \lim_{\Delta x \to 0} \frac{\sqrt{x + \Delta x} - \sqrt{x}}{\Delta x}$$

$$= \lim_{\Delta x \to 0} \frac{(\sqrt{x + \Delta x} - \sqrt{x})(\sqrt{x + \Delta x} + \sqrt{x})}{\Delta x(\sqrt{x + \Delta x} + \sqrt{x})} \qquad \text{We multiplied and divided by } \sqrt{x + \Delta x} + \sqrt{x}.$$

$$= \lim_{\Delta x \to 0} \frac{(\sqrt{x + \Delta x})^2 - (\sqrt{x})^2}{\Delta x(\sqrt{x + \Delta x} + \sqrt{x})} \qquad (a - b)(a + b) = a^2 - b^2$$

$$= \lim_{\Delta x \to 0} \frac{(x + \Delta x) - x}{\Delta x(\sqrt{x + \Delta x} + \sqrt{x})} = \lim_{\Delta x \to 0} \frac{\Delta x}{\Delta x(\sqrt{x + \Delta x} + \sqrt{x})}$$

$$= \lim_{\Delta x \to 0} \frac{1}{\sqrt{x + \Delta x} + \sqrt{x}} = \frac{1}{\sqrt{x} + \sqrt{x}} = \frac{1}{2\sqrt{x}}. \qquad (11)$$

(b)   If $x = 9$, then $f'(9) = 1/2\sqrt{9} = 1/(2 \cdot 3) = \frac{1}{6}$.

Note that, although the function $f(x) = \sqrt{x}$ is defined for $x \geq 0$, its derivative $1/(2\sqrt{x})$ is defined only for $x > 0$. For $x = 0$, the limit taken in the last step of equation (11) does not exist. Here we have

$$\lim_{x \to 0} f'(x) = \lim_{x \to 0} \frac{1}{2\sqrt{x}} = \infty.$$

In Section 2.3 we said that a vertical line had no slope. Sometimes vertical lines are said to have an *infinite* slope. Since $f'(x)$ is the slope of the tangent line and since, in our example, $\lim_{x \to 0} f'(x) = \infty$, we say that the graph of $y = \sqrt{x}$ has a **vertical tangent** at $x = 0$. The curve is sketched in Figure 9.

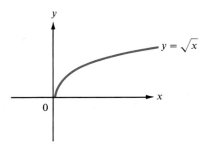

Figure 9

Example 6   Consider the function $y = |x|$. Since

$$|x| = \begin{cases} x, & x \geq 0 \\ -x, & x < 0 \end{cases},$$

we obtain the graph in Figure 10. To see if the graph of f has a tangent line at the point $(0, 0)$, we calculate

$$f'(0) = \lim_{\Delta x \to 0} \frac{f(0 + \Delta x) - f(0)}{\Delta x} = \lim_{\Delta x \to 0} \frac{|0 + \Delta x| - |0|}{\Delta x} = \lim_{\Delta x \to 0} \frac{|\Delta x|}{\Delta x}.$$

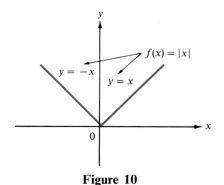

Figure 10

But as we saw in Example 10.2.5 on page 518, this limit does not exist, so that $|x|$ does not have a tangent line at $(0, 0)$. On the other hand, if $x \neq 0$ then the derivative does exist (see Problem 18). Note, too, that by the same reasoning, the curve in Figure 6(c) does not have a tangent line at x_0 because if it did, that tangent line would be unique.

As the examples in this section illustrate, the computation of a derivative can be very tedious if one is required to compute a limit each time. Fortunately, there are a variety of techniques that greatly simplify the computations of derivatives. We will discuss these techniques beginning in Section 10.6. At this point we know that

$$\frac{d}{dx}(x^2) = 2x$$

and

$$\frac{d}{dx}(\sqrt{x}) = \frac{d}{dx}x^{1/2} = \tfrac{1}{2}x^{-1/2} = \frac{1}{2\sqrt{x}}.$$

Here $\dfrac{d}{dx}(x^2)$ stands for the derivative of x^2 and $\dfrac{d}{dx}(\sqrt{x})$ denotes the derivative of $\sqrt{x}$.

We close this section by stating an important fact:

> Differentiable functions are continuous.

You are asked in Problem 20 to explain why this is true. Note, however, that continuous functions are not necessarily differentiable. As we saw in Example 6, $f(x) = |x|$ is continuous at 0 but is not differentiable there.

PROBLEMS 10.4

1. The cost function for producing q refrigerators is given by
$$C(q) = 14{,}500 + 280q.$$
Find the variable cost.

2. A tire manufacturer buys raw materials at the following prices:

   $16 per tire for the first 500 tires;

   $14 per tire for each additional tire.

   Find and sketch the manufacturer's total cost function.

3. What is the derivative of the total cost function in Problem 2?

4. (a) Find the derivative of the following total cost function:
$$C(q) = \begin{cases} 10{,}000 + 3q, & q \le 5{,}000 \\ 12{,}500 + 2.5q, & q > 5{,}000 \end{cases}$$

   (b) Sketch this function.

5. Consider the function $f(x) = 3x^2$.

   (a) For $x = 2$, calculate $f(x + \Delta x) = f(2 + \Delta x)$ for $\Delta x = 0.5$, $\Delta x = 0.1$, $\Delta x = 0.01$, $\Delta x = 0.001$, $\Delta x = -0.01$, and $\Delta x = -0.001$.

   (b) Calculate $[f(2 + \Delta x) - f(2)]/\Delta x$ for the values of Δx in part (a) and "guess" the value of $f'(2)$.

(c)   From the definition, calculate $f'(x)$, use this to compute $f'(2)$, and compare this with the answer you obtained in part (b).

(d)   What is the equation of the line tangent to the curve at the point $(2, 12)$?

6.   Carry out the steps in Problem 5 for the function $f(x) = 1/x$ at the point $(1, 1)$.

7.   Carry out the steps in Problem 5 for the function $f(x) = 5\sqrt{x}$ at the point $(1, 5)$.

In Problems 8–17 find the derivative of the given function and the equation of the tangent line to the curve at the given point.

8. $f(x) = 15x - 2$; $(1, 13)$ **9.** $f(x) = -4x + 6$; $(3, -6)$

10. $f(x) = 10x^2$; $(1, 10)$ **11.** $y = x^3$; $(2, 8)$

12. $y = x + x^2$; $(2, 6)$ **13.** $y = x^2 + 1$; $(1, 2)$

14. $y = x^2 + 5x + 3$; $(0, 3)$ **15.** $f(x) = x^2 - x + 2$; $(1, 2)$

16. $f(x) = x^3 + x^2$; $(2, 10)$ **17.** $f(x) = \dfrac{1}{x}$; $(\frac{1}{3}, 3)$

18.   Let $y = |x|$. Calculate dy/dx for $x \neq 0$.

19.   Show that if $y = mx + b$, then $dy/dx = m$.

20.   Suppose that $f'(x_0)$ exists (that is, f is differentiable at x_0).

(a)   Explain why, if Δx is small, $\dfrac{f(x_0 + \Delta x) - f(x_0)}{\Delta x} \approx f'(x_0)$ (where $\approx$ means "approximately equal to").

(b)   Use (a) to show that if Δx is small, $f(x_0 + \Delta x) - f(x_0) \approx f'(x_0)\Delta x$.

(c)   Use (b) to show that $\lim_{\Delta x \to 0} [f(x_0 + \Delta x) - f(x_0)] = 0$.

(d)   Explain why f is continuous at x_0.

10.5 The Derivative as a Rate of Change

In Section 10.4 we saw the relationship between the derivative and the velocity of a falling rock. In this section we see how the derivative represents the rate of change in a variety of interesting situations.

To begin the discussion, we again consider the line given by the equation

$$y = mx + b, \tag{1}$$

where m is the slope and b is the y-intercept (see Figure 1). Let us examine the slope more closely. It is defined as the change in y divided by the change in x:

$$m = \frac{\Delta y}{\Delta x}. \tag{2}$$

Implicit in this definition is the understanding that no matter which two points are chosen on the line, we obtain the same value for this ratio. Thus the slope of a straight line could instead be referred to as *the rate of change of y with respect to x*. It tells us how many units y changes for every one unit that x

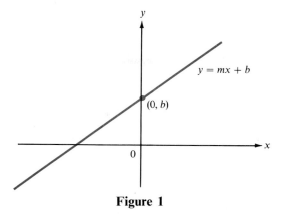

Figure 1

changes. In fact, instead of following Euclid in defining a straight line as the shortest distance between two points, we could define a straight line as a curve whose rate of change is constant.

Example 1   Let $y = 3x - 5$. Then in moving from the point $(1, -2)$ to the point $(2, 1)$ along the line, we see that as x has changed (increased) 1 unit, y has increased 3 units, corresponding to the slope $m = 3$. See Figure 2.

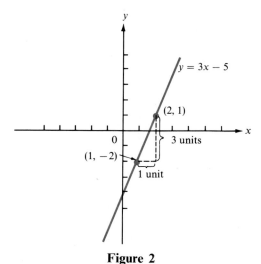

Figure 2

We would like to be able to calculate rates of change for functions that are not straight-line functions. We did that in Section 10.4. More generally, suppose that an object is dropped from rest at a given height. The distance, s, the object has dropped after t seconds (ignoring air resistance) is given by the formula

$$s = \tfrac{1}{2}gt^2, \tag{3}$$

where $g \approx 9.8$ m/sec$^2 \approx 32$ ft/sec^2 is the acceleration due to gravity. We now ask: What is the velocity of the object after 2 seconds? To answer this question, we first note that velocity is a rate of change. Whether measured in meters per second, feet per second, or miles per hour, velocity is the ratio of change in distance (meters, feet, miles) to the change in time (seconds, hours). The discussion in Section 10.4 suggests the following definition.

Instantaneous Velocity   Let $s(t)$ denote the distance traveled by a moving object in t seconds. Then the **instantaneous velocity** after t seconds, denoted ds/dt or $s'(t)$, is given by

$$s'(t) = \frac{ds}{dt} = \lim_{\Delta t \to 0} \frac{\Delta s}{\Delta t} = \lim_{\Delta t \to 0} \frac{s(t + \Delta t) - s(t)}{\Delta t}. \tag{4}$$

Note.   The quantity $[s(t + \Delta t) - s(t)]/\Delta t$ is the **average velocity** of the object between times t and $t + \Delta t$.

As you may have noticed, the velocity ds/dt given by equation (4) is the derivative of s with respect to t. Although our previous definitions of derivative involved the variables x and y, this concept does not change when we insert t in place of x and s in place of y. Thus, we can think of a derivative as a velocity or, more generally, as a rate of change. After Newton discovered the derivative, he used the word *fluxion* instead of velocity in his discussion of a moving object. (In technical terminology, a moving particle is a particle "in flux.")

Example 2   If an object has dropped a distance of $\frac{1}{2}gt^2 = 4.9t^2$ meters after t seconds, what is its velocity, ignoring air resistance, after exactly 2 seconds?

Solution $s(t) = 4.9t^2$ and $s(t + \Delta t) = 4.9(t + \Delta t)^2$. Thus

$$\frac{ds}{dt} = \lim_{\Delta t \to 0} \frac{4.9(t + \Delta t)^2 - 4.9t^2}{\Delta t} = 4.9 \lim_{\Delta t \to 0} \frac{(t + \Delta t)^2 - t^2}{\Delta t}$$

$$= 4.9 \lim_{\Delta t \to 0} \frac{t^2 + 2t\Delta t + \Delta t^2 - t^2}{\Delta t} = 4.9 \lim_{\Delta t \to 0} \frac{\Delta t(2t + \Delta t)}{\Delta t}$$

$$= 4.9 \lim_{\Delta t \to 0} (2t + \Delta t) = (4.9)(2t) = 9.8t.$$

Thus, after 2 seconds, the velocity is $(9.8)(2) = 19.6$ m/second.

The next example illustrates how the notion of rate of change arises in economics.

Example 3   A manufacturer buys large quantities of a certain machine replacement part. He finds that his cost depends on the number of cases bought at the same time, and the cost per unit decreases as the number of cases bought increases. He determines that a reasonable model for this is given by the formula

$$C(q) = 100 + 5q - 0.01q^2, \tag{5}$$

where q is the number of cases bought (up to 250 cases) and $C(q)$, measured in dollars, is the *total* cost of purchasing q cases. The 100 in (5) is a *fixed* cost that does not depend on the number of cases bought. What are the incremental and marginal costs for various levels of purchases?

Solution. **Incremental cost** is the cost per additional unit at a given level of purchase. **Marginal cost** is the *rate of change* of the cost with respect to the number of units purchased. These two concepts are different, as the average velocity of a falling object over a fixed period of time is different from the instantaneous velocity of the object at a fixed moment in time. For example, if the manufacturer buys 25 cases, the incremental cost is the cost for one *more* case, that is, the 26th case. This cost is not a constant. We can see this by calculating that one case costs $100 + 5 \cdot 1 - 0.01 = \$104.99$ and two cases cost $100 + 5 \cdot 2 - (0.01)4 = \$109.96$. Thus the incremental cost of buying the second case is $C(2) - C(1) = \$4.97$. On the other hand, for 100 cases it costs

$$100 + 5 \cdot 100 - (0.01)(100)^2 = 600 - 100 = \$500,$$

and to buy 101 cases it costs

$$100 + 5 \cdot 101 - (0.01)(101)^2 = 100 + 505 - (0.01)(10,201)$$
$$= 605 - 102.01 = \$502.99.$$

The incremental cost is now $C(101) - C(100) = \$2.99$. The hundred-and-first case is cheaper than the second.

Next, we compute the marginal cost (the rate of change of cost with respect to the quantity bought):

$$\text{marginal cost} = \frac{dC}{dq} = \lim_{\Delta q \to 0} \frac{C(q + \Delta q) - C(q)}{\Delta q}$$

$$= \lim_{\Delta q \to 0} \frac{[100 + 5(q + \Delta q) - 0.01(q + \Delta q)^2] - [100 + 5q - 0.01q^2]}{\Delta q}$$

$$= \lim_{\Delta q \to 0} \frac{5\Delta q - 0.01(2q\Delta q + \Delta q^2)}{\Delta q} = 5 - 0.02q.$$

Thus at $q = 10$ cases, the marginal cost is $5 - 0.2 = \$4.80$, whereas at $q = 100$ cases the marginal cost is $5 - 2 = \$3$. This confirms the manufacturer's statement that "the more he buys, the cheaper it gets."

Before leaving this example, we observe once again that the marginal cost at $q = 100$ ($= \$3$) is not equal to the incremental cost of buying the next unit when $q = 100$ ($= \$2.99$). Do you see why? The marginal cost (also called **instantaneous cost**) is like instantaneous velocity. The incremental cost is the average cost, averaged over 1 unit (between 100 cases and 101 cases). This is analogous to average velocity. Since the rate of change of cost is falling (5 − 0.02q decreases as q increases), we can expect that the average rate of change between q = 100 and q = 101 will be less than the instantaneous rate of change at q = 100.

Remark.   Example 3 discusses marginal cost. We can also define **marginal revenue** as the rate of change of the revenue function and **marginal profit** as the rate of change of the profit function.

PROBLEMS 10.5   **1.**   If a ball is thrown up into the air with an initial velocity of 75 ft/sec, its height (measured in feet) after t seconds is given by

$$h = 75t - 16t^2.$$

(a)   Tabulate the heights of the ball from t between 0 and 2 seconds in increments of 0.1 second and construct a table similar to Table 1.

(b)   What is the average velocity between 1.4 and 1.5 seconds?

(c)   What is the average velocity between 1.5 and 1.6 seconds?

(d)   Find an estimate for the instantaneous velocity, $v(1.5)$, after exactly 1.5 seconds.

(e)   Using $\Delta t = 0.01$, find a better estimate for $v(1.5)$.

(f)   Find $v(1.5)$ exactly and compare this with your estimates.

2.   The distance that an accelerating race car travels is given by $s = 0.6t^3$ where t is measured in minutes and s is measured in kilometers.

(a)   Tabulate the distance traveled by the car in increments of 0.1 minute from $t = 0$ to $t = 2.5$ minutes and construct a table similar to Table 1.

(b)   What is the average velocity of the car between 1.9 and 2.0 minutes?

(c)   What is the average velocity of the car between 2.0 and 2.1 minutes?

(d)   Estimate the instantaneous velocity, $v(2)$, at $t = 2.0$ minutes.

(e)   Using $t = 0.001$, find a better estimate for $v(2)$.

(f)   Calculate $v(2)$ exactly and compare this with your estimates.

In Problems 3–8 distance is given as a function of time. Find the instantaneous velocity at the indicated time.

3. $s = 1 + t + t^2, t = 4$ **4.** $s = t^3 - t^2 + 3, \quad t = 5$

5. $s = 1 + \sqrt{2t}, \quad t = 8$ **6.** $s = (1 + t)^2, \quad t = 2.5$

7. $s = 100t - 5t^2, \quad t = 6$ **8.** $s = t^4 - t^3 + t^2 - t + 5, \quad t = 3$

9.   Fuel in a rocket burns for $3\frac{1}{2}$ minutes. In the first t seconds the rocket reaches a height of $70t^2$ feet above the earth (for any t from 0 to 210 seconds). What is the velocity of the rocket (in ft/sec) after 3 seconds? after 10 seconds?

10. The manufacturer in Example 3 finds that his cost function for another machine part is given by

$$C(q) = 100q + 55.$$

What can you say about his marginal cost?

11. Assume that the cost function of Example 3 is given by

$$C(q) = 200 + 6q - 0.01q^2 + 0.01q^3.$$

(a) Find the marginal cost.

(b) Is the manufacturer better off buying in large quantities?

12. The price a manufacturer charges for his product depends on the quantity sold (see Example 9.2.2 on page 481) and is given by the demand function

$$p(q) = 20 - 0.0002q^2,$$

where p is the price charged per unit (in dollars) and q is the number of units sold. What is the marginal (instantaneous) change in price at a sales level of 10 units?

13. The revenue a manufacturer receives is the product of his unit price and the quantity sold. Thus, in Problem 12,

$$R = \text{revenue} = qp(q) = 20q - 0.0002q^3.$$

Compute the manufacturer's marginal revenue as a function of q.

14. In Problem 13,

(a) What is the difference in revenue between selling 10 and 11 items (the **incremental revenue**)?

(b) What is the marginal revenue at a sales level of 10 units?

(c) Explain the difference between the answers in (a) and (b).

10.6 Some Differentiation Formulas

In this chapter we have introduced the concepts of the limit and the derivative but found that the calculation of derivatives could be extremely difficult. An even more annoying problem was the seeming necessity to come up with a special trick (like multiplying and dividing by some quantity—see Example 5 on page 547) each time we took the limit in the process of computing a derivative.

In the remainder of this chapter we continue the discussion of properties of the derivatives of functions. We will be principally concerned with simplifying the process of differentiation so that it will no longer be necessary to deal with complicated limits.

In many of the sections of this chapter we will be deriving formulas for calculating derivatives. By the time you have completed the chapter, you will find that differentiation is not nearly so complicated as it now seems. You should be assured that the work involved in memorizing the appropriate formulas will pay dividends in the chapters to come.

We begin by giving a formula for calculating the derivative of $y = f(x) = x^n$, where n is a positive integer. First, let us look for a pattern. We have already calculated the following derivatives:

(a) $\dfrac{d}{dx}(x) = 1$ (since $y = x = 1 \cdot x + 0$ is the equation of a straight line with slope 1)

(b) $\dfrac{d}{dx}(x^2) = 2x$ (Example 10.4.4)

It is also true that

(c) $\dfrac{d}{dx}(x^3) = 3x^2$ (see Problem 35)

and

(d) $\dfrac{d}{dx}(x^4) = 4x^3$ (see Problem 36)

Do you see a pattern? The answer is given below.

Derivative of x^n   If n is a positive integer, then

$$\frac{d}{dx}(x^n) = nx^{n-1}. \tag{1}$$

Example 1   Let $f(x) = x^{17}$. Then $f'(x) = 17x^{16}$.

It turns out that formula (1) is valid when n is any real number.

Power Rule   If r is a real number, then

$$\frac{dy}{dx} = \frac{d}{dx}(x^r) = rx^{r-1}. \tag{2}$$

We shall indicate why formula (2) is valid in the problem sets of Sections 10.7 and 10.10 (see Problems 10.7.40, 10.10.35, and 10.10.36).

Example 2   Let $y = x^{-7}$. Calculate dy/dx.

Solution   Using formula (2), we obtain

$$\frac{dy}{dx} = -7x^{-8}.$$

Example 3   Let $y = 1/x$. Calculate dy/dx.

Solution $1/x = x^{-1}$, so that

$$\frac{d}{dx}\left(\frac{1}{x}\right) = \frac{d}{dx}(x^{-1}) = -1 \cdot x^{-2} = \frac{-1}{x^2}.$$

Example 4   From the power rule, we see that

$$\frac{d}{dx}(\sqrt{x}) = \frac{d}{dx}(x^{1/2}) = \frac{1}{2}x^{-1/2} = \frac{1}{2x^{1/2}} = \frac{1}{2\sqrt{x}},$$

as we computed in Example 10.4.5.

Example 5   Let $y = x^{1/5}$. Calculate dy/dx.

Solution

$$\frac{dy}{dx} = \frac{1}{5}x^{(1/5)-1} = \frac{1}{5}x^{-4/5}$$

Example 6   Let $y = x^{2/3}$. Calculate dy/dx.

Solution

$$\frac{dy}{dx} = \frac{2}{3}x^{(2/3)-1} = \frac{2}{3}x^{-1/3} = \frac{2}{3x^{1/3}}$$

We can compute derivatives of more complicated functions by deriving some more general formulas. We first give a result that states the obvious fact that a constant function does not change.

Derivative of a Constant Function

Let $f(x) = c$, a constant. Then $f'(x) = 0$.   (3)

This result is evident from looking at the graph of a constant function—a horizontal line with a slope of zero (see Figure 1).

It can be shown, although we shall not do so here, that the converse of formula (3) is true. That is,

if $f'(x) = 0$ for all x, then f is a constant function; that is, $f(x) = c$ where c is a number.   (4)

A number of results follow from the limit theorems of Section 10.2 and the definition of the derivative.

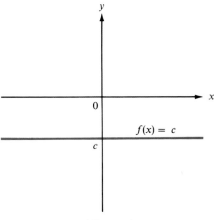

Figure 1

Derivative of a Constant Multiple of a Function   Let c be a constant. If f is differentiable, then cf is also differentiable, and

$$\frac{d}{dx}\,cf = c\,\frac{df}{dx}. \tag{5}$$

That is, the derivative of a constant times a function is the constant times the derivative of the function

We will prove this and the next result at the end of the section.

Example 7   Compute $\dfrac{d}{dx}\left(\dfrac{5}{3}x^7\right)$.

Solution $\dfrac{d}{dx}\left(\dfrac{5}{3}x^7\right) = \dfrac{5}{3}\dfrac{d}{dx}x^7 = \dfrac{5}{3}(7x^6) = \dfrac{35}{3}x^6$

Derivative of the Sum of Two Functions   Let f and g be differentiable. Then $f + g$ is also differentiable, and

$$(f + g)' = \frac{d}{dx}(f + g) = \frac{df}{dx} + \frac{dg}{dx} = f' + g'. \tag{6}$$

That is, the derivative of the sum of two differentiable functions is the sum of their derivatives

---

Example 8   Let $f(x) = 4x^3 + 3\sqrt{x}$. Find $f'(x)$.

Solution

$$\frac{d}{dx}(4x^3 + 3\sqrt{x}) = \frac{d}{dx}(4x^3) + \frac{d}{dx}(3\sqrt{x})$$

$$= 4\frac{d}{dx}(x^3) + 3\frac{d}{dx}(\sqrt{x}) = 4 \cdot 3x^2 + \frac{3}{2\sqrt{x}} = 12x^2 + \frac{3}{2\sqrt{x}}$$

The formula (6) can be extended to more than two functions.

Derivative of the Sum of *n* Functions Let $f_1, f_2, \ldots, f_n$ be n differentiable functions. Then $f_1 + f_2 + \cdots + f_n$ is differentiable, and

$$\frac{d}{dx}(f_1 + f_2 + \cdots + f_n) = \frac{df_1}{dx} + \frac{df_2}{dx} + \cdots + \frac{df_n}{dx}. \tag{7}$$

Example 9 Let $f(x) = 1 + x + x^2 + x^3$. Calculate df/dx.

Solution

$$\frac{df}{dx} = \frac{d}{dx}(1) + \frac{d}{dx}(x) + \frac{d}{dx}(x^2) + \frac{d}{dx}(x^3) = 0 + 1 + 2x + 3x^2$$

$$= 1 + 2x + 3x^2$$

Example 10 The cost function for a certain product is given by

$$C(q) = 500 + 3.5q - 0.02q^2.$$

Find the marginal cost function.

Solution

$$\text{Marginal cost} = \frac{dC}{dq} = \frac{d}{dq}(500 + 3.5q - 0.02q^2)$$

From (7)
↓

$$= \frac{d}{dq}(500) + \frac{d}{dq}(3.5q) + \frac{d}{dq}(-0.02q^2)$$

Derivative of a constant
↓

$$= 0 + 3.5 - 0.02\frac{d}{dq}q^2$$

From (5)

$$= 3.5 - 0.02(2q)$$

$$= 3.5 - 0.04q.$$

Proofs of Two Formulas

Equation (5):

Factor out the c.

$$\frac{d}{dx}(cf(x)) = \lim_{\Delta x \to 0} \frac{cf(x + \Delta x) - cf(x)}{\Delta x} \overset{\downarrow}{=} \lim_{\Delta x \to 0} c\left(\frac{f(x + \Delta x) - f(x)}{\Delta x}\right)$$

From Limit Theorem 2 on page 519

$$\overset{\downarrow}{=} c \lim_{\Delta x \to 0} \frac{f(x + \Delta x) - f(x)}{\Delta x} = c\frac{df}{dx}$$

Equation (6):

$$\frac{d}{dx}((f + g)(x)) = \lim_{\Delta x \to 0} \frac{(f + g)(x + \Delta x) - (f + g)(x)}{\Delta x}$$

$$= \lim_{\Delta x \to 0} \frac{f(x + \Delta x) + g(x + \Delta x) - f(x) - g(x)}{\Delta x}$$

Grouping the terms

$$\overset{\downarrow}{=} \lim_{\Delta x \to 0} \left[\left(\frac{f(x + \Delta x) - f(x)}{\Delta x}\right) + \left(\frac{g(x + \Delta x) - g(x)}{\Delta x}\right)\right]$$

From Limit Theorem 3 on page 519

$$\overset{\downarrow}{=} \lim_{\Delta x \to 0} \frac{f(x + \Delta x) - f(x)}{\Delta x} + \lim_{\Delta x \to 0} \frac{g(x + \Delta x) - g(x)}{\Delta x}$$

$$= \frac{df(x)}{dx} + \frac{dg(x)}{dx}$$

PROBLEMS 10.6 In Problems 1–25 calculate the derivative of the given function.

1. $f(x) = x^5$ **2.** $f(x) = 27$ **3.** $f(x) = 2x^4$

4. $f(x) = x^{1/5}$ **5.** $f(x) = \dfrac{1}{x^{2/3}}$ **6.** $g(t) = t^{5/4}$

7. $g(t) = t^{3/4}$ **8.** $h(z) = z^{203}$ **9.** $h(z) = \dfrac{1}{z^4}$

10. $h(z) = z^{-7/5}$ **11.** $v(r) = 12r^{12}$ **12.** $v(r) = \dfrac{3}{\sqrt{r}}$

13. $f(x) = 3x^2 + 19x + 2$ **14.** $g(t) = t^{10} - t^3$

15. $g(t) = t^5 + \sqrt{t}$ **16.** $g(t) = 1 - t + t^4 - t^7$

17. $h(z) = z^{100} + 100z^{10} + 10$ **18.** $h(z) = 27z^6 + 3z^5 + 4z$

19. $v(r) = 3r^8 - 8r^6 - 7r^4 + 2r^2 + 3$ **20.** $v(r) = -3r^{12} + 12r^3$

21. $f(x) = x^{3/4} - x^{7/8}$ **22.** $g(t) = 3t^{2/3} - \dfrac{4}{t} + \dfrac{5}{t^2}$

23. $f(t) = t^{-1} - 3t^{-4/3}$ **24.** $f(t) = \dfrac{5}{t^{3/5}} - 2$ **25.** $v(r) = \dfrac{1}{r} + \dfrac{2}{r^2} + \dfrac{3}{r^3}$

In Problems 26–34 find the equation of the line that is tangent to the given curve at the given point.

26. $y = x^4$, $(1, 1)$ **27.** $y = 3x^5 - 3x^3 + 1$, $(-1, 1)$

28. $y = 2x^7 - x^6 - x^3$, $(1, 0)$ **29.** $y = 5x^6 - x^4 + 2x^3$, $(1, 6)$

30. $y = 1 + x + x^2 + x^3 + x^4 + x^5$, $(0, 1)$

31. $y = 1 - x + 2x^2 - 3x^3 + 4x^4$, $(1, 3)$

32. $y = x^6 - 6\sqrt{x}$, $(1, -5)$ **33.** $y = \dfrac{1}{\sqrt{x}}$, $(1, 1)$

34. $y = -\dfrac{2}{x^3}$, $(2, -\tfrac{1}{4})$

35. Using the definition of the derivative, show that $\dfrac{d}{dx} x^3 = 3x^2$. [*Hint:* $(a + b)^3 = a^3 + 3a^2b + 3ab^2 + b^3$.]

36. Using the definition of the derivative, show that $\dfrac{d}{dx} x^4 = 4x^3$. [*Hint:* $(a + b)^4 = a^4 + 4a^3b + 6a^2b^2 + 4ab^3 + b^4$.]

In Problems 37–41 a total cost function, $C(q)$, or a total revenue function, $R(q)$, is given. Find the marginal cost or marginal revenue function.

37. $C(q) = 500 + 2q - 0.01q^2$ **38.** $C(q) = 12{,}500 + 250q - q^2$

39. $C(q) = 3 + 0.01q - 0.0002q^3$ **40.** $R(q) = 40q - 0.003q^3$

41. $R(q) = 2{,}000q - 0.05q^{5/2}$

In Problems 42–45 a distance function is given. Find the instantaneous velocity at the given value of t (in seconds) where distance, s, is measured in feet or meters.

42. $s(t) = 16t^2 + 45t + 80$; $t = 2$ (feet)

43. $s(t) = -4.9t^2 + 85t + 250$; $t = 5$ (meters)

44. $s(t) = \dfrac{1}{t} + \dfrac{4}{t^2} + 50$; $t = 2$ (meters)

45. $s(t) = -t^3 + 3t^2 + 8t + 50$; $t = 1$ (feet)

46. The total cost of producing q items of a certain product is given by $C(q) = 50 + 3q - 0.0015q^2 + 0.00002q^3$.

(a) Find the marginal cost function.

(b) Determine the marginal cost when $q = 50$.

(c) Determine the marginal cost when $q = 200$.

47. The total revenue received by the Happy Pizza Company when it sells q pizzas is given by

$$R(q) = 5q^{1/2} + 0.01q^2 - 10.$$

(a) Find the marginal revenue when 64 pizzas are sold.

(b) Find the marginal revenue when 100 pizzas are sold.

(c) If 100 pizzas are sold, find the average revenue received per pizza.

*48. A growing grapefruit with a diameter of $2k$ inches has a skin that is $k/12$ inches thick (the skin is included in the diameter of the grapefruit). What is the rate of growth of the volume of the skin (per unit growth in the radius) when the radius of the grapefruit is 3 inches?

49. Where, if anywhere, is the graph of $y = \sqrt{x}$ parallel to the line $\frac{1}{8}x - 8y = 1$?

*50. Let $f(x) = x^2$ and $g(x) = \frac{1}{3}x^3$. Each vertical line, $x = $ constant, meets the graph of f and the graph of g.

(a) On what vertical lines do the graphs of f and g have parallel tangents?

(b) On what horizontal lines do they have parallel tangents?

*51. Let $f(x) = (x - 1)^2 + 3 = x^2 - 2x + 4$ and $g(x) = -f(-x) = -x^2 - 2x - 4$. Find each line that is tangent to both of the graphs $y = f(x)$ and $y = g(x)$. [*Note*: A rough sketch indicates that there is at least one such line.]

52. Suppose that when an airplane takes off (starting from rest), the distance (in feet) it travels during the first few seconds is given by the formula

$$s = 1 + 4t + 6t^2.$$

How fast (in ft/sec) is the plane traveling after 10 seconds? after 20 seconds?

53. A petri dish contains two colonies of bacteria. The population of the first colony is given by $P_1(t) = 1000 + 50t - 20\sqrt{t}$, and the population of the second is given by $P_2(t) = 2000 + 30t^2 - 80t$, where t is measured in hours.

(a) Find a function that represents the *total* population of the two species.

(b) What is the instantaneous growth rate of the total population?

(c) How fast is the total population growing after 4 hours? after 16 hours?

*54. For the model of Problem 53, how fast is the first population growing when the second population is growing at a rate of 160 bacteria per hour?

*55. Show that the rate of change of the area of a circle with respect to its radius is equal to its circumference.

10.7 The Product and Quotient Rules

In this section we develop some additional rules to simplify the calculation of derivatives. To see why additional rules are needed, consider the problem of calculating the derivatives of

$$f(x) = \sqrt{x}(x^4 + 3) \qquad \text{or} \qquad g(x) = \frac{x^4 + 3}{\sqrt{x}}.$$

To carry out the calculations from the definition would be very tedious. However, you will shortly see that these calculations can be made rather simple.

Let f and g be two differentiable functions of x. What is the derivative of the product fg? It is easy to be led astray here. Limit Theorem 4 on page 520 states that

$$\lim_{x \to x_0} f(x)g(x) = \lim_{x \to x_0} f(x) \lim_{x \to x_0} g(x).$$

However, the derivative of the product is *not* equal to the product of the derivatives. That is,

$$\frac{d}{dx} fg \neq \frac{df}{dx} \cdot \frac{dg}{dx}.$$

Originally Leibniz, the co-discoverer of the derivative, thought that they were equal. But an easy example shows that this is false. Let $f(x) = x$ and $g(x) = x^2$. Then

$$(fg)(x) = f(x)g(x) = x^3 \quad \text{and} \quad \frac{d}{dx} fg = 3x^2.$$

But

$$\frac{df}{dx} = 1 \quad \text{and} \quad \frac{dg}{dx} = 2x,$$

so that

$$\frac{df}{dx} \cdot \frac{dg}{dx} = 1 \cdot 2x = 2x \neq 3x^2 = \frac{d}{dx} fg.$$

The correct formula, discovered after many false steps by both Leibniz and Newton, is given below.

Product Rule   Let f and g be differentiable at x. Then fg is differentiable at x and

$$\boxed{(fg)'(x) = \frac{d}{dx}(fg(x)) = f(x)\frac{dg(x)}{dx} + g(x)\frac{df(x)}{dx} = fg'(x) + gf'(x).} \quad (1)$$

Verbally, the product rule says that the derivative of the product of two functions is equal to the first times the derivative of the second plus the second times the derivative of the first.

At the end of this section we will prove that the product rule holds.

Example 1   Let $h(x) = \sqrt{x}(x^4 + 3)$. Calculate dh/dx.

Solution $h(x) = f(x)g(x)$, where $f(x) = \sqrt{x}$ and $g(x) = x^4 + 3$. Then

$$\frac{df}{dx} = \frac{1}{2\sqrt{x}} \quad \text{and} \quad \frac{dg}{dx} = \frac{d}{dx}x^4 + \frac{d}{dx}3 = 4x^3 + 0 = 4x^3,$$

so that

$$\frac{dh}{dx} = f\frac{dg}{dx} + g\frac{df}{dx} = \sqrt{x}(4x^3) + (x^4 + 3)\frac{1}{2\sqrt{x}}.$$

This is the correct answer, but we will use some algebra to simplify the result:

$$\frac{dh}{dx} = \frac{2\sqrt{x}\sqrt{x}(4x^3) + x^4 + 3}{2\sqrt{x}} = \frac{2x(4x^3) + x^4 + 3}{2\sqrt{x}} = \frac{9x^4 + 3}{2\sqrt{x}}.$$

Example 2   Let $h(t) = (t^2 + 2)(t^3 - 5)$. Compute $h'(t)$.

Solution   We could first multiply through and use the rules given in the last section. However, it is simpler to compute the derivative directly from the product rule. If $f(t) = t^2 + 2$ and $g(t) = t^3 - 5$, then $h(t) = f(t)g(t)$ and

$$h'(t) = f(t)g'(t) + f'(t)g(t)$$
$$= (t^2 + 2)(3t^2) + 2t(t^3 - 5).$$

This answer is correct as it stands. We can, if desired, multiply through and combine terms to obtain

$$h'(t) = 3t^4 + 6t^2 + 2t^4 - 10t = 5t^4 + 6t^2 - 10t.$$

Having now discussed the product of two functions, we turn to the quotient of functions.

Quotient Rule   Let f and g be differentiable at x. Then, if $g(x) \neq 0$, we have

$$\frac{d}{dx}\frac{f}{g} = \frac{g(x)(df/dx) - f(x)(dg/dx)}{g^2(x)} = \frac{gf' - fg'}{g^2}. \qquad (2)$$

The quotient rule states that the derivative of the quotient is equal to the denominator times the derivative of the numerator minus the numerator times the derivative of the denominator all over the denominator squared.
We will prove the quotient rule at the end of the section.

Example 3   Let $h(x) = x/(x - 1)$. Compute $h'(x)$.

Solution   With $f(x) = x$ and $g(x) = x - 1$, $h(x) = f(x)/g(x)$ and
$$h'(x) = \frac{g(x)f'(x) - f(x)g'(x)}{[g(x)]^2} = \frac{(x - 1)(1) - x(1)}{(x - 1)^2} = \frac{-1}{(x - 1)^2}.$$

Example 4   Let $h(x) = (x^4 + 3)/\sqrt{x}$. Calculate dh/dx.

Solution $h(x) = f(x)/g(x)$, where $f(x) = x^4 + 3$ and $g(x) = \sqrt{x}$. Thus

$$\frac{dh}{dx} = \frac{g(df/dx) - f(dg/dx)}{g^2} = \frac{\sqrt{x}(4x^3) - (x^4 + 3)(1/(2\sqrt{x}))}{(\sqrt{x})^2}$$

Now multiply numerator and denominator by $2\sqrt{x}$.

$$= \frac{2\sqrt{x}\sqrt{x}(4x^3) - x^4 - 3}{2\sqrt{x}} \cdot \frac{1}{x} = \frac{8x^4 - x^4 - 3}{2x^{3/2}} = \frac{7x^4 - 3}{2x^{3/2}}.$$

Example 5   Let $h(x) = (x^3 + x + 1)/(x^2 - 5)$. Calculate dh/dx.

Solution

$$\frac{dh}{dx} = \frac{(x^2 - 5)\dfrac{d}{dx}(x^3 + x + 1) - (x^3 + x + 1)\dfrac{d}{dx}(x^2 - 5)}{(x^2 - 5)^2}$$

$$= \frac{(x^2 - 5)(3x^2 + 1) - (x^3 + x + 1)(2x)}{(x^2 - 5)^2} = \frac{x^4 - 16x^2 - 2x - 5}{x^4 - 10x^2 + 25}$$

Example 6   Let $h(x) = x^{2/3}/(1 + x)$. Compute $h'(x)$.

Solution

$$\frac{dh}{dx} = \frac{(1 + x)\dfrac{d}{dx}(x^{2/3}) - x^{2/3}\dfrac{d}{dx}(1 + x)}{(1 + x)^2} = \frac{(1 + x)\frac{2}{3}x^{-2/3} - x^{2/3}}{(1 + x)^2}$$

This answer is correct. If we wish to simplify the answer, we obtain

Multiply top and bottom by $3x^{2/3}$.

$$\frac{dh}{dx} = \frac{\dfrac{2(1 + x)}{3x^{2/3}} - x^{2/3}}{(1 + x)^2} = \frac{\left(\dfrac{2(1 + x)}{3x^{2/3}} - x^{2/3}\right)3x^{2/3}}{(1 + x)^2 \cdot 3x^{2/3}} = \frac{2 + 2x - 3x^{4/3}}{3x^{2/3}(1 + x)^2}.$$

We now prove the product and quotient rules.

Proof of Product Rule

$$\frac{d}{dx}[fg(x)] = \lim_{\Delta x \to 0} \frac{f(x + \Delta x)g(x + \Delta x) - f(x)g(x)}{\Delta x}$$

This doesn't look very much like the derivative of anything. To continue, we will use the trick of adding and subtracting the term $f(x)g(x + \Delta x)$ in the numerator. As you shall see, this makes everything come out nicely. We have

$$\frac{d}{dx}[fg(x)]$$

Here are the additional terms.

$$= \lim_{\Delta x \to 0} \frac{f(x + \Delta x)g(x + \Delta x) - \overbrace{f(x)g(x + \Delta x) + f(x)g(x + \Delta x)} - f(x)g(x)}{\Delta x}$$

Using Limit Theorem 3 on page 519

$$= \lim_{\Delta x \to 0} \frac{g(x + \Delta x)(f(x + \Delta x) - f(x))}{\Delta x} + \lim_{\Delta x \to 0} \frac{f(x)(g(x + \Delta x) - g(x))}{\Delta x}$$

Using Limit Theorem 4 on page 520
$$\downarrow$$
$$= \lim_{\Delta x \to 0} g(x + \Delta x) \lim_{\Delta x \to 0} \frac{f(x + \Delta x) - f(x)}{\Delta x} + \lim_{\Delta x \to 0} f(x) \lim_{\Delta x \to 0} \frac{g(x + \Delta x) - g(x)}{\Delta x}$$

Using the definition of the derivative on page 545
$$\downarrow$$
$$= g(x) \frac{df(x)}{dx} + f(x) \frac{dg(x)}{dx}.$$

Here we have used the fact that $\lim_{\Delta x \to 0} g(x + \Delta x) = g(x)$. This follows from the fact that g is differentiable, so that g is continuous (see page 549).

Proof of the Quotient Rule

Multiply numerator and denominator by $g(x + \Delta x)g(x)$.

$$\frac{d}{dx} \frac{f}{g} = \lim_{\Delta x \to 0} \frac{\dfrac{f(x + \Delta x)}{g(x + \Delta x)} - \dfrac{f(x)}{g(x)}}{\Delta x} = \lim_{\Delta x \to 0} \frac{f(x + \Delta x)g(x) - f(x)g(x + \Delta x)}{\Delta x g(x + \Delta x)g(x)}$$

These terms sum to zero.
$$= \lim_{\Delta x \to 0} \frac{\overbrace{f(x + \Delta x)g(x) - f(x)g(x) + f(x)g(x)} - f(x)g(x + \Delta x)}{\Delta x g(x + \Delta x)g(x)}$$

Factor out $f(x)$ and $g(x)$.
$$\downarrow$$
$$= \lim_{\Delta x \to 0} \frac{g(x)\left(\dfrac{f(x + \Delta x) - f(x)}{\Delta x} \right) - f(x)\left(\dfrac{g(x + \Delta x) - g(x)}{\Delta x} \right)}{g(x + \Delta x)g(x)}$$

Use Limit Theorems 2, 3, and 5 on pages 519 and 521.
$$\downarrow$$
$$= \frac{\left\{ g(x) \lim_{\Delta x \to 0} \left(\dfrac{f(x + \Delta x) - f(x)}{\Delta x} \right) - f(x) \lim_{\Delta x \to 0} \left(\dfrac{g(x + \Delta x) - g(x)}{\Delta x} \right) \right\}}{g(x) \lim_{\Delta x \to 0} g(x + \Delta x)}$$

By the definition of the derivative
$$\downarrow$$
$$= \frac{g(x)f'(x) - f(x)g'(x)}{[g(x)]^2}$$

Again, we used the fact that g is continuous (being differentiable), so that $\lim_{\Delta x \to 0} g(x + \Delta x) = g(x)$.

PROBLEMS 10.7   In Problems 1–26 find the derivative of the given function.

1. $f(x) = 2x(x^2 + 1)$

2. $g(t) = t^3(1 + \sqrt{t})$

3. $f(x) = \dfrac{2x - 1}{5x - 3}$

4. $f(x) = \dfrac{x^2 + 2}{x^2 - 1}$

5. $s(t) = \dfrac{t^3}{1 + \sqrt{t}}$

6. $f(z) = \dfrac{1 + \sqrt{z}}{z^3}$

7. $f(x) = (1 + x + x^5)(2 - x + x^6)$

8. $f(x) = \dfrac{1 + x + x^5}{2 - x + x^6}$

9. $f(x) = \dfrac{2 - x + x^6}{1 + x + x^5}$

10. $g(t) = (1 + \sqrt{t})(1 - \sqrt{t})$

11. $g(t) = \dfrac{1 + \sqrt{t}}{1 - \sqrt{t}}$

12. $g(t) = \dfrac{1 - \sqrt{t}}{1 + \sqrt{t}}$

13. $p(v) = (v^3 - \sqrt{v})(v^2 + 2\sqrt{v})$

14. $p(v) = \dfrac{v^3 - \sqrt{v}}{v^2 + 2\sqrt{v}}$

15. $p(v) = \dfrac{v^2 + 2\sqrt{v}}{v^3 - \sqrt{v}}$

16. $f(x) = \dfrac{1 + x}{1 - x^{3/2}}$

17. $g(t) = (1 - x^{-3/4})(1 + x^3)$

18. $f(x) = (1 - \sqrt{x} - \sqrt[3]{x})(1 + x^{4/3})$

19. $m(r) = \dfrac{1}{\sqrt{r}}$

20. $f(x) = \dfrac{1}{x^5 + 3x}$

21. $g(t) = \dfrac{1}{\sqrt{t}}\left(\dfrac{1}{t^4 + 2}\right)$

22. $g(t) = \dfrac{\sqrt{t} - \dfrac{2}{\sqrt{t}}}{3t^3 + 4}$

23. $f(x) = \dfrac{1}{x^6}$

24. $g(t) = t^{-100}$

25. $f(x) = \dfrac{5}{7x^3}$

26. $p(v) = \dfrac{1 + v^{3/2}}{1 - \sqrt{v}}$

In Problems 27–30 find the equation of the tangent line to the curve passing through the given point.

27. $f(x) = 4x(x^5 + 1)$, $(1, 8)$

28. $g(t) = \dfrac{t^2}{1 + \sqrt{t}}$, $\left(4, \dfrac{16}{3}\right)$

29. $h(u) = \dfrac{1 + \sqrt{u}}{u^2}$, $(1, 2)$

30. $p(v) = (1 + \sqrt{v})(1 - \sqrt{v})$, $(1, 0)$

31. Let $v(x) = f(x)g(x)h(x)$.

(a) Using the product rule, show that
$$\frac{dv}{dx} = f\frac{d(gh)}{dx} + gh\frac{df}{dx}.$$

(b) Use part (a) to show that
$$\frac{dv}{dx} = fg\frac{dh}{dx} + fh\frac{dg}{dx} + gh\frac{df}{dx} = f'gh + fg'h + fgh'.$$

In Problems 32–34 use the result of Problem 31 to calculate the derivative of the given function.

32. $v(x) = x(1 + \sqrt{x})(1 - \sqrt{x})$

33. $v(x) = (x^2 + 1)(x^3 + 2)(x^4 + 3)$

34. $v(x) = x^{-2}(2 - 3\sqrt{x})(1 + x^3)$

35. Find the derivative of $fg/(f + g)$, where f and g are differentiable functions.

36. A demand function for a certain product is given by

$$p(q) = \frac{20}{1 + 0.2q}, \; 5 < q < 200.$$

(a) Find the rate of change of price with respect to change in demand.

(b) Find $p'(10)$ and $p'(100)$.

37. A demand function for a different product is given by

$$p(q) = \frac{30 + q}{10 + q + 0.02q^2}, \; 5 < q < 200.$$

Answer the questions of Problem 36.

38. According to **Poiseuille's law**, the resistance, R, of a blood vessel of length l and radius r is given by $R = \alpha l/r^4$, where α is a constant of proportionality determined by the viscosity of blood.[†] Assuming that the length of the vessel is kept constant while the radius increases, how fast is the resistance decreasing (as a function of the radius) when $r = 0.2$ mm?

39. A typical beginner's mistake is to compute the derivative of a product as if it equaled the product of the derivatives. If you tackle this problem, you'll discover that this is rarely the case. From the following functions, pick all pairs that satisfy the equation $(F \cdot G)' = F' \cdot G'$: $a(x) = 13$, $b(x) = x$, $c(x) = 1/x$, $d(x) = x + 1$, and $e(x) = 1/(x - 1)$.

40. Assume that we know the formula $\dfrac{d}{dx} x^n = nx^{n-1}$ for n a positive integer. Use the quotient rule to show that $\dfrac{d}{dx} x^{-n} = \dfrac{d}{dx} \left(\dfrac{1}{x^n} \right) = -nx^{-n-1}$.

10.8 The Chain Rule

In this section we derive a result that greatly increases the number of functions whose derivatives can easily be calculated. The idea behind the result is illustrated below.

Suppose that $y = f(u)$ is a function of u and that $u = g(x)$ is a function of x.[‡] Then $du/dx = g'(x)$ is the rate of change of u with respect to x while $dy/du = f'(u)$ is the rate of change of y with respect to u. We now may ask: What is the rate of change of y with respect to x? That is, what is dy/dx?

As an illustration of this idea, suppose that a particle is moving in the xy-plane in such a way that its x-coordinate is given by $x = 3t$, where t stands for time. For example, if t is measured in seconds and x is measured in feet, then in the x-direction, the particle is moving with a velocity of 3 ft/sec. That is, $dx/dt = 3$. In addition, suppose that we know that for every 1 unit change in

---

[†] Jean Louis Poiseuille (1799–1869) was a French physiologist.

[‡] So that $y(x) = (f \circ g)(x)$ where $f \circ g$ denotes the composition of f and g. See Section 9.2 (page 484).

the x-direction, the particle moves 4 units in the y-direction; that is, $dy/dx = 4$ (see Figure 1). Now we ask, what is the velocity of the particle, in feet per

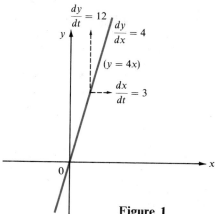

Figure 1

second, in the y-direction; that is, what is dy/dt? It is clear that for every 1 unit change in t, x will change 3 feet and so y will change $4 \cdot 3 = 12$ feet. That is, $dy/dt = 12$. We may write this as

$$\frac{dy}{dt} = \frac{dy}{dx}\frac{dx}{dt} = 4 \cdot 3 = 12 \text{ ft/sec.} \tag{1}$$

This result states simply that if x is changing 3 times as fast as t and if y is changing 4 times as fast as x, then y is changing $4 \cdot 3 = 12$ times as fast as t.

We now state a result that generalizes this example. The result, known as the **chain rule**, greatly facilitates the calculation of a wide variety of derivatives.

Chain Rule

Let g and f be differentiable functions. At every point x_0 such that $g(x_0)$ is defined, suppose that $f(g(x_0))$ is also defined. Then with $u = g(x)$, the composite function $y = (f \circ g)(x) = f[g(x)] = f(u)$ is a differentiable function of x, and

$$\boxed{\frac{dy}{dx} = \frac{d}{dx}(f \circ g)(x) = f'(g(x))g'(x) = \frac{dy}{du}\frac{du}{dx}.} \tag{2}$$

Example 1 Let $u = 3x - 6$, and let $y = 7u + 10$. Then $du/dx = 3$ and $dy/du = 7$, so that

$$\frac{dy}{dx} = \frac{dy}{du}\frac{du}{dx} = 7 \cdot 3 = 21.$$

To calculate dy/dx directly, we compute that $y = 7u + 10 = 7(3x - 6) + 10 = 21x - 32$. Therefore,

$$\frac{dy}{dx} = 21 = \frac{dy}{du}\frac{du}{dx}.$$

Example 2 In Example 1 we were able to calculate dy/dx directly. It is usually very difficult to do this, and this is why the chain rule is so useful. For example, let us calculate dy/dx where $y = \sqrt{x + x^2}$. If we define $u = x + x^2$, then $y = \sqrt{x + x^2} = \sqrt{u}$. Hence, by the chain rule, formula (2), since $dy/du = 1/2\sqrt{u}$ and $du/dx = 1 + 2x$, we have

$$\frac{dy}{dx} = \frac{dy}{du}\frac{du}{dx} = \frac{1}{2\sqrt{u}}(1 + 2x) = \frac{1}{2\sqrt{x + x^2}}(1 + 2x) = \frac{1 + 2x}{2\sqrt{x + x^2}}.$$

Note that the only other way to calculate this derivative would be to use the original definition, which in this example is very difficult.

Remark. The trick in using the chain rule to calculate the derivative dy/dx is to find a function $u(x)$ with the property that $y(x) = f(u(x))$, where both df/du and du/dx can be calculated without too much difficulty.

Example 3 Let $y = f(x) = 1/(x^3 + 1)^5$. Find dy/dx.

Solution We define $u = g(x) = x^3 + 1$. (There really is no other possibility.) Then $1/(x^3 + 1)^5 = 1/u^5$ and we know how to differentiate $1/u^5$ with respect to u. We have

$$\frac{dy}{du} = \frac{d}{du}\left(\frac{1}{u^5}\right) = \frac{-5}{u^6} \qquad \text{and} \qquad g'(x) = \frac{d}{dx}(x^3 + 1) = 3x^2.$$

We now use the chain rule to obtain

$$\frac{dy}{dx} = \frac{dy}{du}\frac{du}{dx} = \left(-\frac{5}{u^6}\right)(3x^2) = \frac{-15x^2}{(x^3 + 1)^6}.$$

An equivalent way to solve this problem is to write $y = f(g(x)) = 1/[g(x)]^5 = [g(x)]^{-5}$, where $g(x) = x^3 + 1$. Then $f'(g(x)) = -5/[g(x)]^6$ and

$$\frac{dy}{dx} = f'(g(x))g'(x) = -\frac{5}{[g(x)]^6}g'(x) = \frac{-15x^2}{(x^3 + 1)^6}.$$

Doing the problem this way avoids the need to introduce the variable u.

Example 4 Let $y = f(x) = (x^3 + 3)^{1/4}$. Find dy/dx.

Solution We define $u = x^3 + 3$. Then $y = u^{1/4}$, so that by the chain rule,

$$\frac{dy}{dx} = \frac{dy}{du}\frac{du}{dx} = \frac{1}{4}u^{-3/4}(3x^2) = \frac{1}{4}(x^3 + 3)^{-3/4}(3x^2).$$

We can generalize Example 4.

Power Rule   Let $g(x)$ be a differentiable function of x. Then, for any real number r,

$$\frac{d}{dx}[g(x)]^r = r[g(x)]^{r-1}g'(x).$$ (3)

Remark. The power rule (3) is a special case of the chain rule (2).

Example 5 Find the derivative of $(1 + \sqrt{x})^{100/3}$.

Solution If $g(x) = 1 + \sqrt{x}$, then $g'(x) = 1/2\sqrt{x}$ and

$$\frac{d}{dx}[g(x)]^{100/3} = \frac{100}{3}[g(x)]^{97/3}g'(x)$$

$$= \frac{100}{3}(1 + \sqrt{x})^{97/3}\cdot\frac{1}{2\sqrt{x}} = \frac{50}{3\sqrt{x}}(1 + \sqrt{x})^{97/3}.$$

At this point you can see that it is possible to differentiate a wide variety of functions. To aid you, we give in Table 1 a summary of the differentiation rules we have so far discussed. In the notation of the table, c stands for an arbitrary constant and $u(x)$ and $v(x)$ denote differentiable functions.

TABLE 1

Function $y = f(x)$	Its derivative $\frac{dy}{dx}$
I. c	$\frac{dc}{dx} = 0$
II. $cu(x)$	$\frac{d}{dx}cu(x) = c\frac{du}{dx}(x)$
III. $u(x) + v(x)$	$\frac{d}{dx}(u + v) = \frac{du}{dx} + \frac{dv}{dx}$
IV. x^r, r a real number	$\frac{d}{dx}x^r = rx^{r-1}$
V. $u(x)\cdot v(x)$	$\frac{d}{dx}uv(x) = u(x)\frac{dv(x)}{dx} + v(x)\frac{du(x)}{dx}$
VI. $\frac{u(x)}{v(x)}$, $v(x) \neq 0$	$\frac{d}{dx}\frac{u(x)}{v(x)} = \frac{v(x)\frac{du(x)}{dx} - u(x)\frac{dv(x)}{dx}}{(v(x))^2}$
VII. $u^r(x)$	$\frac{d}{dx}(u^r(x)) = ru^{r-1}(x)\frac{du}{dx}$
VIII. $f(g(x))$	$\frac{d}{dx}f(g(x)) = f'(g(x))g'(x)$

PROBLEMS 10.8 In Problems 1–25, use the chain rule to find the derivative of the given function.

1. $f(x) = (x + 1)^3$

2. $f(x) = (x^2 - 1)^2$

3. $f(x) = (\sqrt{x} + 2)^4$

4. $f(x) = (x^2 - x^3)^4$

5. $y = (1 + x^6)^6$

6. $y = (1 - x^2 + x^5)^3$

7. $y = (x^2 - 4x + 1)^5$

8. $y = \dfrac{1}{(\sqrt{x} - 3)^4}$

9. $s(t) = \left(\dfrac{t + 1}{t - 1}\right)^{3/5}$

10. $s(t) = (\sqrt{t} - t)^{7/2}$

11. $g(u) = (u^5 + u^4 + u^3 + u^2 + u + 1)^2$

12. $g(u) = \dfrac{5}{u^3 + u + 1}$

13. $h(y) = (y^2 + 3)^{-4}$

14. $h(y) = (y^3 - \sqrt{y} + 1)^{-17}$

15. $f(x) = (x^2 + 2)^5(x^4 + 3)^3$. [*Hint:* First use the product rule, then the chain rule.]

16. $f(x) = (x^4 + 1)^{1/2}(x^3 + 3)^4$

17. $s(t) = \dfrac{\sqrt{t^2 + 1}}{(t + 2)^4}$

18. $s(t) = \left(\dfrac{t^4 + 1}{t^4 - 1}\right)^{1/2}$

19. $g(u) = \dfrac{(u^2 + 1)^3(u^2 - 1)^2}{\sqrt{u - 2}}$

20. $g(x) = \dfrac{(x^2 + 1)^2(x^3 + 2)^3}{(x^4 + 3)^{1/2}}$

21. $f(x) = (1 + x^{4/3})^{2/3}$

22. $s(t) = (1 - t^3)^{17/2}$

23. $f(x) = \sqrt{x + \sqrt{1 + \sqrt{x}}}$. [*Hint:* Use the chain rule twice.]

24. $f(x) = \sqrt{x^2 + \sqrt{1 + x^2}}$

25. $h(y) = (y^{-2} + y^{-3} + y^{-7})^{-5}$

26. A missile travels along the path $y = 6(x - 3)^3 + 3x$. When $x = 1$ the missile flies off this path tangentially (that is, along the tangent line). Where is the missile when $x = 4$? [*Hint:* Find the equation of the tangent line at $x = 1$.]

27. The total cost function for a certain product is given by

$$C(q) = (30 + 1.5q)^{1.1}.$$

(a) Find the marginal cost function.

(b) What is the marginal cost at a level of production of $q = 100$ units?

*28. Let f and g be differentiable functions with $f[g(x)] = x$. Show that

$$f'[g(x)] = \dfrac{1}{g'(x)}.$$

(This formula is called the **differentiation rule for inverse functions**.)

29. Verify the result of Problem 28 in the following cases.

(a) $g(x) = 5x,\ f(x) = \frac{1}{5}x$

(b) $g(x) = 17x - 8,\ f(x) = \frac{1}{17}x + \frac{8}{17}$

(c) $g(x) = \sqrt{x},\ f(x) = x^2$

(d) $g(x) = x^2$ on $(-\infty, 0],\ f(x) = -\sqrt{x}$

30. In astronomy, the **luminosity** of a star is the star's total energy output. Loosely speaking, a star's luminosity is a measure of how bright the star would appear at the surface of the star. The **mass–luminosity relation** gives the approximate luminosity of a star as a function of its mass. It has been found experimentally that, approximately,

$$\frac{L}{L_0} = \left(\frac{M}{M_0}\right)^r,$$

where L and M are the luminosity and mass of the star and L_0 and M_0 denote the luminosity and mass of our sun.[†] The exponent r depends on the mass of the star, as shown in Table 2.

TABLE 2

Mass range, M/M_0	r
1.0–1.4	4.75
1.4–1.7	4.28
1.7–2.5	4.15
2.5–5	3.95
5–10	3.38
10–20	2.80
20–50	2.30
50–100	1.90

(a) In this model, how is the luminosity changing as a function of mass when the mass is 2 solar masses (that is, $2M_0$)?

(b) How is it changing at a mass of 8 solar masses?

(c) At a mass of 30 solar masses?

*(d) Writing L as a function of M, for what values of M does dL/dM not exist? How would you suggest altering the model so as to avoid discontinuities in this derivative?

*31. The equation of the circle centered at $(0,0)$ with radius r is given by $x^2 + y^2 = r^2$.

(a) Write y as a function of x for $y > 0$.

(b) Find the equation of the line tangent to the circle at a point (x_0, y_0) in the first quadrant.

[†] These data are based on stellar models computed by D. Ezer and A. Cameron in their paper "Early and main sequence evolution of stars in the range 0.5 to 100 solar masses," *Canadian Journal of Physics*, **45**, 3429–3460 (1967).

(c) Show that the line joining $(0, 0)$ and (x_0, y_0) (the *radial line*) is perpendicular to the tangent line at (x_0, y_0).

32. The total revenue received from selling q units of a certain product is given by

$$R(q) = \sqrt{100q - q^2}, \; 0 \le q \le 90.$$

(a) What is the marginal revenue when $q = 20$?

(b) What is the marginal revenue when $q = 70$?

(c) At what level of sales does total revenue begin to decrease?

*(d) What is the maximum revenue that can be received by selling this product? [*Hint*: Use the result of part (c).]

33. Answer the questions of Problem 32 using the revenue function $R(q) = 100{,}000/(q^2 - 80q) + 1000$, $10 \le q \le 75$.

**34. Suppose that f is differentiable on $(0, \infty)$ and that $f(A \cdot B) = f(A) + f(B)$ for any numbers $A, B > 0$. Show that $f'(x) = f'(1)/x$ for all $x > 0$.

10.9 Higher-Order Derivatives

Let $s(t)$ represent the distance an object has moved after t units of time have elapsed. Then, as we have seen, the derivative ds/dt evaluated at a time t may be interpreted as the instantaneous velocity of the object at that time. The velocity is the rate of change of position with respect to time. By definition, acceleration is the rate of change of velocity with respect to time. Thus acceleration can be thought of as the derivative of the derivative or, more simply, as the *second derivative* of the function representing position. If, for example, $s = \frac{1}{2}gt^2$ represents the position of a falling object, then

$$\frac{ds}{dt} = gt \text{ and } \frac{d}{dt}\left(\frac{ds}{dt}\right) = g,$$

which is the acceleration due to gravity.

We now generalize these ideas. Let $y = f(x)$ be a differentiable function. Then the derivative $y' = dy/dx = f'$ is also a function of x. This new function of x, f', may or may not be a differentiable function. If it is, we call the derivative of f' the **second derivative** of f (that is, the derivative of the derivative) and denote it by

$$f''. \tag{1}$$

There are other commonly used notations as well. They are

$$y'', (f')', \frac{d}{dx}\left(\frac{dy}{dx}\right), \text{ and } \frac{d^2y}{dx^2}. \tag{2}$$

The notations

$$f'', \frac{d^2y}{dx^2} \text{ and } y'' \tag{3}$$

will be used interchangeably in this book to denote the second derivative.

Similarly, if f'' exists, it might or might not be differentiable. If it is, then the derivative of f'' is called the **third derivative** of f and is denoted

$$f'''. \tag{4}$$

Alternate notations are

$$f''', \frac{d^3y}{dx^3}, \text{ and } y'''. \tag{5}$$

We can continue this definition indefinitely as long as each successive derivative is differentiable. After the third derivative, we avoid a clumsy succession of primes by using numerals to denote higher derivatives:

$$f^{(4)}, f^{(5)}, f^{(6)}, \dots. \tag{6}$$

Alternate notations are, for the successive derivatives,

$$f^{(4)}, \frac{d^4y}{dx^4}, \text{ and } y^{(4)} \tag{7}$$

$$f^{(5)}, \frac{d^5y}{dx}, \text{ and } y^{(5)} \tag{8}$$

$$\vdots \qquad \vdots$$

$$f^{(n)}, \frac{d^ny}{dx^n}, \text{ and } y^{(n)}. \tag{9}$$

We emphasize that *each higher-order derivative of $y = f(x)$ is a new function of x* (if it exists).

Example 1   Let $y = f(x) = x^3$. Then

$$\frac{dy}{dx} = f'(x) = 3x^2.$$

The second derivative is simply the derivative of the first derivative, so that

$$\frac{d^2y}{dx^2} = f''(x) = \frac{d}{dx} 3x^2 = 6x.$$

Similarly, the third derivative is the derivative of the second derivative, and we have

$$\frac{d^3y}{dx^3} = f'''(x) = \frac{d}{dx} 6x = 6.$$

Finally,

$$\frac{d^4y}{dx^4} = f^{(4)}(x) = \frac{d}{dx} 6 = 0.$$

Note that for $k \geq 4$, $f^{(k)}(x) = 0$ since the derivative of the zero function is zero.

Example 2   Let $y = f(x) = 1/x$. Then

$$\frac{dy}{dx} = f'(x) = -x^{-2} = -\frac{1}{x^2}, \frac{d^2y}{dx^2} = f''(x) = 2x^{-3} = \frac{2}{x^3},$$

$$\frac{d^3y}{dx^3} = f'''(x) = -6x^{-4} = -\frac{6}{x^4}, \text{ and so on.}$$

Notation. $f''(a)$ denotes the second derivative evaluated at $x = a$. For example, in Example 2,

$$f''(4) = \frac{2}{4^3} = \frac{2}{64} = \frac{1}{32}.$$

Analogously,

$$f'''(4) = -\frac{6}{4^4} = -\frac{6}{256} = -\frac{3}{128}.$$

Example 3   An object moves so that its position (in meters) is given by

$$s(t) = -2t^3 + 10t^2 + 8t + 200.$$

Find its acceleration after 2 seconds have elapsed.

Solution   Since acceleration is the second derivative of position, we need to calculate $s''(2)$. But

$$\frac{ds}{dt} = -6t^2 + 20t + 8 \quad \text{and} \quad \frac{d^2s}{dt^2} = -12t + 20.$$

Thus

$$s''(2) = -12(2) + 20 = -24 + 20 = -4 \text{ m/sec}^2.$$

This means that after 2 seconds the velocity of the object is decreasing (that is, the object is slowing down). Note, however, that after 1 second $s''(1) = -12 + 20 = 8$ m/sec^2, so the velocity of the object is increasing.

Although we have now defined derivatives of all orders, almost all of our major applications will involve first and/or second derivatives. Second derivatives are important for several reasons. In Section 11.2, for example, we shall show that the sign of the second derivative of a function tells us something about the shape of the graph of that function.

PROBLEMS 10.9   In Problems 1–15 find d^2y/dx^2 and d^3y/dx^3.

1. $y = 3$ **2.** $y = 17x + 1$ **3.** $y = 4x^2$

4. $y = 9x^3$ **5.** $y = \sqrt{x}$ **6.** $y = \dfrac{1}{\sqrt{x}}$

7. $y = (x + 1)^{2/3}$ **8.** $y = (x^2 + 1)^{1/2}$ **9.** $y = \sqrt{1 - x^2}$

10. $y = \dfrac{1 + x}{1 - x}$ **11.** $y = x^r$ (r a real number)   **12.** $y = \dfrac{1}{x^2}$

13. $y = ax^2 + bx + c$ **14.** $y = ax^3 + bx^2 + cx + d$ **15.** $y = \dfrac{1}{(x + 1)^5}$

*16. Show that, for any integer n,

$$\frac{d^n}{dx^n} x^n = n(n - 1)(n - 2) \cdots 3 \cdot 2 \cdot 1 = n!$$

17. Let $p(x) = a_n x^n + a_{n-1} x^{n-1} + \cdots + a_1 x + a_0$. Using Problem 16, show that

$$\frac{d^n p}{dx^n} = n! a_n \quad \text{and} \quad \frac{d^{n+1} p}{dx^{n+1}} = 0.$$

18. A rocket is shot upward in the earth's gravitational field so that its velocity at any time t is given by $v = 50t$. What is its acceleration?

19. A particle moves along a line so that its position along the line at time t is given by

$$s = 2t^3 - 4t^2 + 2t + 3.$$

The initial position is the position at $t = 0$.

(a) What is its initial position?

(b) What is its initial velocity?

(c) What is its initial acceleration?

(d) Show that the particle is initially decelerating.

(e) For what value of t does the particle stop decelerating and begin accelerating?

10.10 Implicit Differentiation

In most of the problems we have encountered, the variable y was given *explicitly* as a function of the variable x. For example, for each of the functions $y = 3x + 6$, $y = x^2$, $y = \sqrt{x + 3}$, $y = 1 + 2x + 4x^3$, and $y = (1 + 8x)^{3/2}$, the variable y appears alone on the left-hand side. Thus we may say "you give me an x and I'll tell you the value of $y = f(x)$." One exception to this is that the variables x and y are given *implicitly* in the equation of the circle of radius r centered at the origin:

$$x^2 + y^2 = r^2. \tag{1}$$

Here x and y are not given separately. In general, we say that x and y are given **implicitly** if neither one is expressed as an explicit function of the other.

Note. This is *not* to say that one variable *cannot* be solved explicitly in terms of the other.

Example 1 The following are examples in which the variables x and y are given implicitly:

(a) $x^3 + y^3 = 6xy^4$ (b) $(2x^{3/2} + y^{5/3})^{17} - 6y^5 = 2$

(c) $2xy(x + y)^{4/3} = 6x^{17/9}$ (d) $\dfrac{x + y}{\sqrt{x^2 - y^2}} = 16y^5$

(e) $xy = 1$ (here it is easy to solve for one variable in terms of the other).

For the example of the circle, it is possible to solve equation (1) explicitly for y in order to calculate dy/dx. However, it is very difficult or impossible to do the same thing for the functions (a)–(d) given in Example 1 (try it!). Nevertheless, the derivative dy/dx *may* exist. Can we calculate it?

To illustrate the answer to this question, let us again return to equation (1):

$$x^2 + y^2 = r^2.$$

For $y > 0$, we have

$$y = \sqrt{r^2 - x^2} = (r^2 - x^2)^{1/2},$$

so that, using the chain rule,

$$\frac{dy}{dx} = \frac{1}{2}(r^2 - x^2)^{-1/2}(-2x) = \frac{-x}{\sqrt{r^2 - x^2}} = \frac{-x}{y}.$$

We now calculate this another way. Assuming that y can be written as a function of x, we can write $y^2 = (g(x))^2$ for some function $g(x)$, which we assume to be unknown. Then, by the chain rule,

$$\frac{d}{dx}(y^2) = \frac{d}{dx}(g(x))^2 = 2g(x) \cdot g'(x) = 2y\frac{dy}{dx}. \tag{2}$$

Now taking the derivatives of both sides of (1) with respect to x and using (2), we obtain

$$2x + 2y\frac{dy}{dx} = \frac{d}{dx}r^2.$$

But $\dfrac{d}{dx}(r^2) = 0$ since r is a constant, so

$$2x + 2y\frac{dy}{dx} = 0. \tag{3}$$

We can now solve for dy/dx in equation (3):

$$\frac{dy}{dx} = -\frac{x}{y}. \tag{4}$$

If we do not know y as a function of x, then this is as far as we can go. However, since in this case we may choose $y = \sqrt{r^2 - x^2}$, we may write equation (4) as

$$\frac{dy}{dx} = \frac{-x}{\sqrt{r^2 - x^2}}.$$

Note. We should keep in mind that what makes this technique work is that we are *assuming* that y is a differentiable function of x. Thus we may calculate

$$\frac{d}{dx}(y^2) = 2y\frac{dy}{dx},$$

as in the last example.

The method we have used in the above calculation is called **implicit differentiation** and is the only way to calculate derivatives when it is impossible to solve for one variable in terms of the other. We begin by assuming that y is a differentiable function of x, without actually having a formula for the function. We proceed to find dy/dx by differentiating and then solving for it algebraically. We illustrate this method with a number of examples.

Example 2   Suppose that

$$x^2 + x^3 = y + y^4. \tag{5}$$

Find dy/dx.

Solution   By the chain rule,

$$\frac{d}{dx}y^4 = 4y^3\frac{d}{dx}y = 4y^3\frac{dy}{dx}.$$

Thus we may differentiate both sides of equation (5) with respect to x to obtain

$$\frac{d}{dx}x^2 + \frac{d}{dx}x^3 = \frac{d}{dx}y + \frac{d}{dx}y^4$$

or

$$2x + 3x^2 = \frac{dy}{dx} + 4y^3\frac{dy}{dx} = (1 + 4y^3)\frac{dy}{dx}.$$

Then

$$\frac{dy}{dx} = \frac{2x + 3x^2}{1 + 4y^3}.$$

At the point $(-1, 0)$, for example,[†]

$$\frac{dy}{dx} = \frac{2(-1) + 3\cdot 1}{1 + 0} = 1,$$

and the equation of the tangent line at that point is

$$y = x + 1.$$

[†]You should verify that $(-1, 0)$ is a point on the curve.

In this example we computed dy/dx. However, in some cases it might be useful to compute dx/dy (assuming that x can be written as a function of y). To do so, we again use the chain rule to find that

$$\frac{d}{dy} x^2 = 2x \frac{dx}{dy}$$

and

$$\frac{d}{dy} x^3 = 3x^2 \frac{dx}{dy}.$$

Then, differentiating both sides of (5) with respect to y yields

$$\frac{d}{dy} x^2 + \frac{d}{dy} x^3 = \frac{d}{dy} y + \frac{d}{dy} y^4$$

or

$$2x \frac{dx}{dy} + 3x^2 \frac{dx}{dy} = 1 + 4y^3 \quad \text{and} \quad \frac{dx}{dy} = \frac{1 + 4y^3}{2x + 3x^2}.$$

Note that $dx/dy = 1/(dy/dx)$. Although we will not prove this here, it is true that under certain hypotheses $dy/dx = 1/(dx/dy)$.

Example 3   Find the equation of the tangent line to the curve

$$x^4 + y^4 = 17 \tag{6}$$

at the point $(2, 1)$.

Solution   Since

$$\frac{d}{dx} y^4 = 4y^3 \frac{dy}{dx},$$

we may differentiate both sides of (6) with respect to x to obtain

$$\frac{d}{dx} x^4 + \frac{d}{dx} y^4 = \frac{d}{dx} 17 \quad \text{or} \quad 4x^3 + 4y^3 \frac{dy}{dx} = 0$$

(the derivative of a constant is zero). Then

$$4y^3 \frac{dy}{dx} = -4x^3 \quad \text{and} \quad \frac{dy}{dx} = -\frac{x^3}{y^3}.$$

At the point $(2, 1)$, $dy/dx = -8$, so that the equation of the tangent line is

$$\frac{y - 1}{x - 2} = -8 \quad \text{or} \quad 8x + y = 17.$$

Example 4   Let $(2x + 3y)/(x^2 + y) = 4$. Compute dy/dx.

Solution We use the quotient rule:

$$\frac{(x^2 + y) \dfrac{d}{dx}(2x + 3y) - (2x + 3y)\dfrac{d}{dx}(x^2 + y)}{(x^2 + y)^2} = \frac{d}{dx}(4) = 0.$$

Thus, multiplying both sides by $(x^2 + y)^2$, we obtain

$$(x^2 + y)\frac{d}{dx}(2x + 3y) - (2x + 3y)\frac{d}{dx}(x^2 + y) = 0$$

or

$$(x^2 + y)\left(2 + 3\frac{dy}{dx}\right) - (2x + 3y)\overbrace{\left(2xy + x^2\frac{dy}{dx}\right)}^{\text{Product rule}} = 0$$

or

$$2(x^2 + y) + 3(x^2 + y)\frac{dy}{dx} - (2x + 3y)(2xy) - x^2(2x + 3y)\frac{dy}{dx} = 0.$$

We solve for dy/dx:

$$[3(x^2 + y) - x^2(2x + 3y)]\frac{dy}{dx} = (2x + 3y)(2xy) - 2(x^2 + y)$$

or

$$(3x^2 + 3y - 2x^3 - 3x^2y)\frac{dy}{dx} = 4x^2y + 6xy^2 - 2x^2 - 2y$$

and, finally,

$$\frac{dy}{dx} = \frac{4x^2y + 6xy^2 - 2x^2 - 2y}{3x^2 + 3y - 2x^3 - 3x^2y}.$$

Let us return briefly to the circle $x^2 + y^2 = r^2$. For $y > 0$ we calculated $dy/dx = -x/y$, which is zero when $x = 0$ (and $y = r$). Thus for $x = 0$, the tangent line has slope zero and is horizontal (see Figure 1). Now let us consider x as a function of y (for $x > 0$) and differentiate implicitly with respect to y to obtain

$$2x\frac{dx}{dy} + 2y\frac{dy}{dy} = \frac{d}{dy}r^2 = 0 \qquad \text{and} \qquad \frac{dx}{dy} = -\frac{y}{x},$$

which is zero at the point $(r, 0)$. If $dx/dy = 0$ at a point, then the tangent line to the curve at that point is parallel to the y-axis. That is, the tangent line is vertical. This follows from the same reasoning that shows the tangent line is parallel to the x-axis at any point at which $dy/dx = 0$. In our example, we see that the tangent line to the curve $x^2 + y^2 = r^2$ at point $(0, r)$ is the line $x = r$. This line is vertical, as depicted in Figure 1.

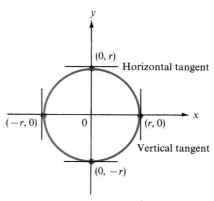

Figure 1

We generalize this example with the following rule:

---

1. If $dy/dx = 0$ at the point (x_0, y_0), then the graph of $y = f(x)$ has a **horizontal tangent** at that point, given by the straight line $y = y_0$.

2. If $dx/dy = 0$ at the point (x_0, y_0), then the graph of $y = f(x)$ has a **vertical tangent** at that point, given by the straight line $x = x_0$.

Note.   In both of these cases $dy/dx \neq 1/(dx/dy)$, since we cannot divide by zero.

PROBLEMS 10.10   In Problems 1–25 find dy/dx by implicit differentiation.

1. $x^3 + y^3 = 3$ **2.** $x^3 + y^3 = xy$

3. $\sqrt{x} + \sqrt{y} = 2$ **4.** $xy + x^2y^2 = x^5$

5. $\dfrac{1}{x} + \dfrac{1}{y} = 1$ **6.** $\dfrac{1}{x^2} - \dfrac{1}{y^2} = x + y$

7. $(x + y)^{1/2} = (x^2 + y)^{1/3}$ **8.** $xy + x^2y^2 + x^3y^3 = 2$

9. $\dfrac{1}{\sqrt{x^2 + y^2}} = 4$ **10.** $\sqrt{x^3 + xy + y^3} = 6$

11. $(3xy + 1)^5 = x^2$ **12.** $\dfrac{x + y}{x - y} = 2$

13. $(x + y)(x - y) = 7$ **14.** $\sqrt{xy^2 + yx^2} = 0$

15. $\dfrac{x^2 + y^2}{x^2 - y^2} = 4$

16. $x^{3/4} + y^{3/4} = 2$

17. $\dfrac{2xy + 1}{3xy - 1} = 2$

18. $x^{-1/2} + y^{-1/2} = 4$

19. $xy + x^2 y^2 = 2$

20. $\dfrac{x}{y} + \dfrac{x^2}{y^2} = 3$

21. $x^{-7/8} + y^{-7/8} = \frac{7}{8}$

22. $x^2 - \sqrt{xy} + y^2 = 6$

23. $(4x^2 y^2)^{1/5} = 1$

24. $(\sqrt{x} + \sqrt{y})(\sqrt[3]{x} - \sqrt[3]{y}) = -3$

25. $x^2 y^3 + x^3 y^2 = xy$

26. Find the equation of the line tangent to the curve $x^5 + y^5 = 2$ at the point $(1, 1)$.

27. Find the equation of the line tangent to the curve $(x + y)/(x - y) = 5$ at the point $(3, 2)$.

28. Find the equation of the line tangent to the curve $(x/y) - (4y/x) = 3$ at the point $(4, 1)$.

In Problems 29–34 find the points where the given curve has a vertical tangent. Also find the points where it has a horizontal tangent.

29. $\sqrt{x} + \sqrt{y} = 1$

30. $\dfrac{1}{x} + \dfrac{1}{y} = 1$

31. $xy = 1$

32. $(x - 3)^2 + (y - 4)^2 = 9$

33. $\dfrac{x^2}{a^2} - \dfrac{y^2}{b^2} = 1$

34. $\dfrac{x^2}{a^2} + \dfrac{y^2}{b^2} = 1$

35. Let $y = x^{1/n}$. Write $x = y^n$ and assume that y is a differentiable function of x. Use implicit differentiation to show that

$$\frac{dy}{dx} = \frac{1}{n} x^{1/n - 1}$$

36. Use the chain rule and the result of Problem 35 to show that $(d/dx)x^{m/n} = (m/n)x^{(m/n) - 1}$. [*Hint:* Write $x^{m/n} = (x^{1/n})^m$.]

10.11 Derivatives of Exponential and Logarithmic Functions

In Section 9.4 we introduced the number $e \approx 2.71828$. This number seems to be picked out of the air. Actually, the number e is defined as a limit:

$$e = \lim_{u \to 0} (1 + u)^{1/u}. \tag{1}$$

It may seem that $\lim_{u \to 0} (1 + u)^{1/u} = 1$ since $1 + u \to 1$ as $u \to 0$ and $1^r = 1$ for every real number r. However, as the numbers in Table 1 suggest, $(1 + u)^{1/u}$ does approach a number between 2 and 3 as $u \to 0$.

TABLE 1

u	$\dfrac{1}{u}$	$1 + u$	$(1 + u)^{1/u}$
1	1	2	2
0.5	2	1.5	2.25
0.2	5	1.2	2.48832
0.1	10	1.1	2.5937426
0.01	100	1.01	2.704813829
0.001	1,000	1.001	2.716923932
0.0001	10,000	1.0001	2.718145926
0.00001	100,000	1.00001	2.718268237
0.000001	1,000,000	1.000001	2.718268237
0.000000001	1,000,000,000	1.000000001	2.718281828

In Sections 9.4 and 9.5 we defined the functions a^x, e^x, $\log_a x$, and $\ln x = \log_e x$. The functions e^x and $\ln x$ are the ones most often encountered in applications, so we will show how to differentiate these. The following important formula is derived at the end of this section.

Derivative of ln x

$$\frac{d}{dx} \ln x = \frac{1}{x} \qquad (2)$$

We can use formula (2) to differentiate a wide variety of functions.

Example 1   Compute $\ln (1 + x^2)$.

Solution   Let $u = 1 + x^2$ and $y = \ln u$. Then, by the chain rule and formula (1),

$$\frac{d}{dx} \ln u = \frac{d}{dx} y = \frac{dy}{du}\frac{du}{dx} = \frac{1}{u}\frac{du}{dx} = \frac{1}{1 + x^2}(2x) = \frac{2x}{1 + x^2}$$

In general, we have the following:

Differentiation of a Logarithmic Function   Let u be a differentiable function of x. Then

$$\frac{d}{dx} \ln u = \frac{1}{u}\frac{du}{dx}. \qquad (3)$$

Example 2   Differentiate $y = \ln (x^3 + 3x + 1)$.

Solution   Applying our rule for the derivative of $\ln u$ with $u = x^3 + 3x + 1$,

we have

$$\frac{dy}{dx} = \frac{1}{u}\frac{du}{dx} = \frac{1}{x^3 + 3x + 1}\frac{d}{dx}(x^3 + 3x + 1)$$

$$= \frac{1}{x^3 + 3x + 1}(3x^2 + 3) = \frac{3x^2 + 3}{x^3 + 3x + 1}.$$

Note. $\ln(x^3 + 3x + 1)$ is defined only when $x^3 + 3x + 1 > 0$.

The next formula gives us a remarkable fact about the function e^x.

Derivative of e^x

$$\boxed{\frac{d}{dx}e^x = e^x} \tag{4}$$

That is, e^x is its own derivative!

Differentiation of an Exponential Function   Let u be a differentiable function of x. Then, from the chain rule, if $y = e^u$,

From (3)
$$\frac{dy}{dx} = \frac{dy}{du}\frac{du}{dx} \overset{\downarrow}{=} e^u\frac{du}{dx},$$

or

$$\boxed{\frac{d}{dx}e^u = e^u\frac{du}{dx}.} \tag{5}$$

Example 3   Find the derivative of e^{x^2}.

Solution   If $u = x^2$, then $\dfrac{du}{dx} = 2x$ and, from (4),

$$\frac{d}{dx}e^{x^2} = e^u\frac{du}{dx} = e^{x^2}\cdot 2x = 2xe^{x^2}.$$

Example 4   Find the derivative of $e^{\alpha x}$ where α is a constant.

Solution   Let $u = \alpha x$. Then $\dfrac{du}{dx} = \alpha$ and, from (5),

$$\frac{d}{dx}e^u = e^u\frac{du}{dx} = e^{\alpha x}\cdot\alpha = \alpha e^{\alpha x}.$$

Example 5   Find the second derivative of $\ln(1 + e^x)$.

Solution

$$\frac{d}{dx} \ln(1 + e^x) = \frac{1}{1 + e^x} \frac{d}{dx} (1 + e^x) = \frac{e^x}{1 + e^x}$$

Then

$$\frac{d^2}{dx^2} (1 + e^x) = \frac{d}{dx} \frac{e^x}{1 + e^x} = \frac{(1 + e^x)\frac{d}{dx} e^x - e^x \frac{d}{dx} (1 + e^x)}{(1 + e^x)^2}$$

$$= \frac{(1 + e^x)e^x - e^x(e^x)}{(1 + e^x)^2} = \frac{e^x}{(1 + e^x)^2}.$$

We will see many applications of exponential and logarithmic functions when we study a simple *differential equation* in Section 11.6.

Remark.   In this book we will not need the derivatives of $\log_a x$ and a^x when $a \neq e$. However, for the sake of completeness, we give these derivatives here.

$$\frac{d}{dx} \log_a x = \frac{1}{x} \log_a x \qquad (6)$$

$$\frac{d}{dx} a^x = a^x \ln a \qquad (7)$$

Logarithmic Differentiation   Sometimes we can simplify the differentiation of functions involving products, quotients, and exponents by using a process called **logarithmic differentiation**.

Example 6   Differentiate $y = \sqrt[4]{(x^3 + 1)/x^{7/9}}$.

Solution   We first take the natural logarithm of both sides:

$$\ln a^b = b \ln a \qquad \ln \frac{a}{b} = \ln a - \ln b$$

$$\ln y = \ln \left(\frac{x^3 + 1}{x^{7/9}} \right)^{1/4} = \frac{1}{4} \ln \left(\frac{x^3 + 1}{x^{7/9}} \right) = \frac{1}{4}[\ln(x^3 + 1) - \ln x^{7/9}]$$

$$= \frac{1}{4} \left[\ln(x^3 + 1) - \frac{7}{9} \ln x \right] = \frac{1}{4} \ln(x^3 + 1) - \frac{7}{36} \ln x. \qquad (8)$$

Now,

Chain rule

$$\frac{d}{dx} \ln y = \frac{1}{y} \frac{dy}{dx} \qquad \text{and} \qquad \frac{d}{dx} \ln(x^3 + 1) = \frac{1}{x^3 + 1} \frac{d}{dx} (x^3 + 1) = \frac{3x^2}{x^3 + 1}.$$

So we can differentiate both sides of equation (8) with respect to x to obtain

$$\frac{1}{y}\frac{dy}{dx} = \frac{3x^2}{4(x^3 + 1)} - \frac{7}{36x}$$

or

$$\frac{dy}{dx} = y\left(\frac{3x^2}{4(x^3 + 1)} - \frac{7}{36x}\right) = \left(\sqrt[4]{\frac{x^3 + 1}{x^{7/9}}}\right)\left(\frac{3x^2}{4(x^3 + 1)} - \frac{7}{36x}\right).$$

Here it was not necessary to use the quotient or power rule at all.

Example 7 Differentiate $y = x^x$.

Solution Taking natural logarithms, we have $\ln y = \ln x^x = x \ln x$. Then, using the chain rule on the left and the product rule on the right, we obtain

$$\frac{1}{y}\frac{dy}{dx} = x\frac{d}{dx}(\ln x) + \left[\frac{d}{dx}(x)\right]\ln x = x\cdot\frac{1}{x} + 1\cdot\ln x = 1 + \ln x,$$

so that

$$y = x^x$$
$$\downarrow$$
$$\frac{dy}{dx} = y(1 + \ln x) = x^x(1 + \ln x).$$

Note that, in this example, logarithmic differentiation provides the *only* way of obtaining the answer.

Derivation of Derivative of ln x: $\dfrac{d}{dx}\ln x = \dfrac{1}{x}$

Let $y = \ln x = \log_e x$. Then

$$\text{Since } \ln a - \ln b = \ln\frac{a}{b}$$
$$\downarrow$$
$$\frac{dy}{dx} = \lim_{\Delta x \to 0}\frac{\ln(x + \Delta x) - \ln x}{\Delta x} = \lim_{\Delta x \to 0}\frac{1}{\Delta x}\ln\left(\frac{x + \Delta x}{x}\right)$$

Multiply and divide by x.

$$\frac{x + \Delta x}{x} = 1 + \frac{\Delta x}{x}$$
$$\downarrow \qquad\qquad\qquad\qquad \downarrow$$
$$= \lim_{\Delta x \to 0}\frac{x}{x\Delta x}\ln\left(\frac{x + \Delta x}{x}\right) = \lim_{\Delta x \to 0}\frac{x}{x\Delta x}\ln\left(1 + \frac{\Delta x}{x}\right)$$

See Limit Theorem 2 on page 519.   Since $a \ln b = \ln b^a$

$$\downarrow \qquad\qquad\qquad\qquad\qquad \downarrow$$
$$= \frac{1}{x}\lim_{\Delta x \to 0}\frac{x}{\Delta x}\ln\left(1 + \frac{\Delta x}{x}\right) = \frac{1}{x}\lim_{\Delta x \to 0}\ln\left(1 + \frac{\Delta x}{x}\right)^{x/\Delta x}$$

We give a new name to the variable $x/\Delta x$. If $u = \Delta x/x$, then $x/\Delta x = 1/u$ and, for each fixed x, $u \to 0$ as $\Delta x \to 0$. Thus,

$$\frac{1}{x} \lim_{\Delta x \to 0} \ln\left(1 + \frac{\Delta x}{x}\right)^{x/\Delta x} = \frac{1}{x} \lim_{u \to 0} \ln(1 + u)^{1/u}.$$

Recall that

$$\lim_{u \to 0}(1 + u)^{1/u} = e,$$

from formula (1).

It follows from the continuity of $\ln x$, although we shall not prove it, that

$$\lim_{u \to 0} \ln(1 + u)^{1/u} = \ln\left[\lim_{u \to 0}(1 + u)^{1/u}\right].$$

Thus

Since $\ln e = 1$

$$\frac{dy}{dx} = \frac{1}{x} \lim_{u \to 0} \ln(1 + u)^{1/u} = \frac{1}{x} \ln\left[\lim_{u \to 0}(1 + u)^{1/u}\right] = \frac{1}{x} \ln e = \frac{1}{x}.$$

Derivation of Derivative of e^x: $\dfrac{d}{dx} e^x = e^x$

Let $y = e^x$. Then

$$\ln y = \ln e^x \qquad \text{We took the natural logarithm of both sides.}$$

$$\ln y = x \ln e \qquad \text{In } a^b = b \ln a$$

$$\ln y = x \qquad \text{In } e = 1$$

Now

$$\frac{d}{dx} \ln y = \frac{d}{dx} x = 1. \qquad \text{We differentiated both sides implicitly.} \qquad (9)$$

But

$$\frac{d}{dx} \ln y = \frac{1}{y} \frac{dy}{dx}.$$

Thus, from equation (9),

$$\frac{1}{y} \frac{dy}{dx} = 1$$

or

$$\frac{dy}{dx} = y = e^x.$$

PROBLEMS 10.11   In Problems 1–26 find the derivative of the given function.

1. $\ln(1 + x)$ 2. e^{3x} 3. e^{-x}

4. $\ln(1 - x^2)$ 5. $\ln(-x), x < 0$ 6. $\ln(1 + 5x)$

7. e^{1+5x} 8. $\ln(1 + x^5)$ 9. $e^{1/x}$

10. $e^{\sqrt{x}}$ 11. $\ln \ln x$

12. $e^{\ln x}$ [Simplify your answer.]   13. $\ln e^x$

14. $e^{\ln(x^5 + 6)}$ [*Hint:* This is easier than it looks.]

15. $\ln \dfrac{1 - x}{1 + x}$ 16. $\ln x(1 + x)$ 17. $(1 + \ln x)^4$

18. $e^x \ln(x + 1)$ 19. $\dfrac{x}{\ln x}$ 20. $\dfrac{\sqrt{x}}{\ln x}$

21. xe^{-x} 22. $\dfrac{e^x}{x}$ 23. $\dfrac{1}{e^{1-x}}$

24. $x \ln x$ 25. $x^2 e^x$ 26. $x \ln x - x$

27. The revenue a manufacturer receives when q units of a given product are sold is given by

$$R = 0.50q(e^{-0.001q}).$$

What is the marginal revenue when $q = 100$ units?

28. What is the velocity after 10 minutes of a particle whose equation of motion is

$$s(t) = 30 + 3t + 0.01t^2 + \ln t + e^{-2t^2}?$$

(Time t is measured in minutes, and s is measured in meters.)

29. Find the acceleration after 10 minutes of the particle in Problem 28.

In Problems 30–34 find the second derivative of the given function.

30. e^{x^2} 31. $\ln(1 + x)$ 32. e^{2+x} 33. $e^{1/x}$ 34. $\ln(1 + \sqrt{x})$

35. Show that

$$\frac{d^2}{dx^2} e^{kx} = k^2 e^{kx}$$

for every real number k.

36. Suppose that $e^x + e^{2y} = 4$. Use implicit differentiation to compute dy/dx and dx/dy.

In Problems 37–40 use logarithmic differentiation to compute $\dfrac{dy}{dx}$.

37. $y = \left(\dfrac{xe^x}{x^5 + 1}\right)^{4/3}$

38. $y = \sqrt[3]{\dfrac{x^2(x^4 - 3)}{(x + 1)}}$

39. $y = x^{2x}$

40. $y = x^{-x}$

Review Exercises for Chapter 10

1. Tabulate values of $f(x) = x^2 - 3x + 6$ for $x = 3$, 1, 2.5, 1.5, 2.1, 1.9, 2.01, and 1.99. What does your table tell you about $\lim_{x \to 2}(x^2 - 3x + 6)$?

2. Tabulate values of $f(x) = x^2 + 10x + 8$ for $x = -4$, -2, -3.5, -2.5, -3.1, -2.9, -3.01, and -2.99. What does your table tell you about $\lim_{x \to -3}(x^2 + 10x + 8)$?

In Exercises 3–24 compute the limit.

3. $\lim_{x \to 1} (x^3 - 3x + 2)$

4. $\lim_{x \to 5}(-x^3 + 17)$

5. $\lim_{x \to 3} \dfrac{x^4 - 2x + 1}{x^3 + 3x - 5}$

6. $\lim_{x \to -1} \dfrac{x^3 + x^2 + x + 1}{x^4 + x^3 + x^2 + x + 1}$

7. $\lim_{x \to 0} |x + 2|$

8. $\lim_{x \to 1} |x - 3|$

9. $\lim_{x \to -3} |x + 4|$

10. $\lim_{x \to 1} \dfrac{|x|}{x}$

11. $\lim_{x \to 3} \dfrac{(x - 3)(x - 4)}{x - 3}$

12. $\lim_{x \to 5} \dfrac{x^2 - 6x + 5}{x - 5}$

13. $\lim_{x \to 1} 23\sqrt{x - 17}$

14. $\lim_{x \to -1} (1 - x + x^2 - x^3 + x^4)$

15. $\lim_{x \to 4} \dfrac{x^2 + 9}{x^2 - 9}$

16. $\lim_{x \to -1} 5x^{250}$

17. $\lim_{x \to -1} 6x^{251}$

18. $\lim_{x \to 3} (x^2 + x - 8)^5$

19. $\lim_{x \to 0} \dfrac{x^8 - 7x^5 + x^3 - x^2 + 3}{x^{23} - 2x + 9}$

20. $\lim_{x \to 0} \dfrac{ax^2 + bx + c}{dx^2 + ex + f}$ (a, b, c, d, e, f are all nonzero real numbers.)

21. $\lim_{x \to \infty} \dfrac{x^2 + 2}{3 - x^2}$

22. $\lim_{x \to \infty} \dfrac{x^2}{1 + x^5}$

23. $\lim_{x \to \infty} \dfrac{3x^3 + 2x^2 - 5}{7x^3 - 5x^2 + 2x + 1}$

24. $\lim_{x \to \infty} \dfrac{x^3 + 3}{10x^2 + 5}$

25. Let
$$f(x) = \begin{cases} x^3, & x < 1 \\ x^4, & x > 1 \end{cases}$$
(a) Does $\lim_{x \to 1} f(x)$ exist?

(b) Is f continuous at 1?

26. Show that if
$$f(x) = \begin{cases} 2x + 3, & x \le -2, \\ x^2 - 5, & x > -2 \end{cases}$$
then $\lim_{x \to -2} f(x)$ exists. Calculate that limit.

In Exercises 27–35 find the open intervals in which each of the following functions is continuous.

27. $f(x) = 2\sqrt{x}$

28. $f(x) = 3\sqrt[3]{x}$

29. $f(x) = \dfrac{1}{x - 6}$

30. $f(x) = \dfrac{1}{x^2 - 6}$ **31.** $f(x) = \dfrac{x}{x^2 - 4}$ **32.** $f(x) = |x + 2|$

33. $f(x) = \dfrac{|x + 3|}{x + 3}$ **34.** $f(x) = \dfrac{x^2 - 9}{x + 3}$ **35.** $f(x) = \dfrac{x + 3}{x^2 - 9}$

36. If the distance a particle travels is given by $s(t) = t^3 + t^2 + 6$ kilometers after t hours, how fast is it traveling (in km/hr) after 2 hours?

37. A slaughterhouse purchases cattle from a ranch at a cost of $C(q) = 200 + 8q - 0.02q^2$, where q is the number of head of cattle bought at one time up to a maximum of 150. What is the house's marginal cost as a function of q? Does it pay to buy in large quantities?

In Exercises 38–65 calculate dy/dx.

38. $y = 3x + 4$ **39.** $y = 3x^2 - 6x + 2$

40. $y = x^3 - \sqrt{x}$ **41.** $y = x^{5/7}$

42. $y = x^5 - \sqrt[3]{x}$ **43.** $y = x^{350}$

44. $y = x^{2.3}$ **45.** $y = (4x)^{2/3}$

46. $y = (1 + x)^5$ **47.** $y = (1 + \sqrt{x})(1 - x^2)$

48. $y = \dfrac{1 + x}{1 - x}$ **49.** $y = e^{3x}$

50. $y = e^{1-x}$ **51.** $y = \ln(1 + 4x)$

52. $y = \ln(2 + x^3)$ **53.** $y = \dfrac{1 + e^x}{1 - e^x}$

54. $y = \dfrac{\sqrt{x + 3}}{\sqrt{x - 3}}$ **55.** $y = (1 + x + x^5)^{3/4}$

56. $y = \dfrac{1}{1 + \ln x}$ **57.** $y = \dfrac{1 + x + x^2}{1 + x + x^3}$

58. $y = (1 - x^4)(3 + x + x^3)$ **59.** $y = (1 + x)^4(1 - x^2)^{5/7}$

60. $x^4 + y^4 = 1$ **61.** $y^{2/3} - 3xy = 4$

62. $e^{4x} - e^{2y} = 3$ **63.** $\ln(1 + xy) = 4$

64. $y = \left[\dfrac{x(x^7 - 2)}{x^4 + 1}\right]^{7/15}$ **65.** $y = x^{x^2}$

In Exercises 66–69 find the equation of the tangent line to the given curve at the given point.

66. $y = x^3 - x + 4; (1, 4)$ **67.** $y = \dfrac{\sqrt{x^2 + 9}}{x^2 - 6}; \left(4, \dfrac{1}{2}\right)$

68. $y = (x^2 - 4)^2(\sqrt{x} + 3)^{1/2}; (1, 18)$ **69.** $xy^2 - yx^2 = 0; (1, 1)$

In Exercises 70–78 calculate the second and third derivatives of the given function.

70. $y = x^7 - 7x^6 + x^3 + 3$

71. $y = \sqrt{1 + x}$

72. $y = \dfrac{1}{1 + x}$

73. $y = \dfrac{x + 1}{x - 1}$

74. $y = \dfrac{x^2 - 3}{x^2 + 5}$

75. $y = \sqrt{1 + \sqrt{x}}$

76. $y = e^{x^3}$

77. $y = \ln(1 + x^2)$

78. $y = e^{20x}$

11 APPLICATIONS OF THE DERIVATIVE

11.1 Local Maxima and Minima and the First Derivative Test

Two of the most important applications of the derivative are finding the maximum and minimum values of a given function and obtaining the graphs of functions. These two applications are very closely related, as you shall see in this section and in Sections 11.2 and 11.3.

In this chapter, most of the functions that we deal with will be **smooth**. That is, the functions will be continuous and have continuous first and second derivatives except at points where we are dividing by zero. This is not much of a restriction, because virtually all the functions we have seen are smooth in intervals over which they are defined. The single exception to this was $f(x) = |x|$, which is continuous at 0 but not differentiable at 0 (see Example 10.4.6 on page 548).

Recall that a function is said to be **increasing** in (a, b) if, whenever x_1 and x_2 are in (a, b) and $x_2 > x_1$, then $f(x_2) > f(x_1)$. If $f(x_2) < f(x_1)$ for all x_1, x_2 in (a, b), with $x_2 > x_1$, then f is said to be **decreasing** in (a, b).

Consider the function whose graph is depicted in Figure 1. In the interval

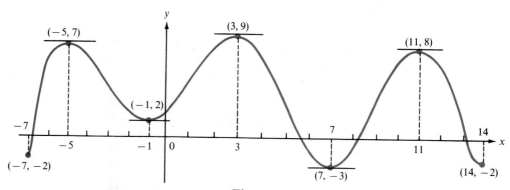

Figure 1

$(-7, -5)$ the function increases. In $(-5, -1)$ it decreases. From -1 to 3 the function again increases, and so on. Moreover, when $x = -5, -1, 3, 7,$ and 11, the tangent line is horizontal, so that the derivative is zero at those points.

In intervals where the function increases, the slope of the tangent to the curve (the derivative) is positive. Where the function decreases, the derivative is negative. The converse is also true.

INCREASING AND DECREASING FUNCTION

Increasing Function: If $f'(x) > 0$ for x in (a, b), then f **increases** in (a, b).

Decreasing Function: If $f'(x) < 0$ for x in (a, b), then f **decreases** in (a, b).

Example 1   In Figure 1, the function is increasing on the intervals $(-7, -5)$, $(-1, 3)$, and $(7, 11)$.

Example 2   In Figure 1 the function is decreasing on the intervals $(-5, -1)$, $(3, 7)$, and $(11, 14)$.

Example 3   In Example 10.2.19 on page 526 we discussed the profit function:

$$P(x) = \frac{16x + 10}{x + 3}.$$

Show that P is an increasing function.

Solution

Quotient rule
$$P'(x) = \overset{\downarrow}{\frac{(x + 3)(16) - (16x + 10)(1)}{(x + 3)^2}} = \frac{16x + 48 - 16x - 10}{(x + 3)^2}$$

$$= \frac{38}{(x + 3)^2} > 0 \quad \text{for all } x \geq 0.$$

Thus, P is an increasing function on $[0, \infty)$.

We now give an important definition:

CRITICAL POINT

Suppose that f is defined at x_0. Then x_0 is a **critical point** of f if $f'(x_0) = 0$ or $f'(x_0)$ does not exist.

Example 4   In Figure 1 the critical points are $-5, -1, 3, 7$, and 11 because f' evaluated at these points is zero. This follows because the tangent lines at these points are horizontal and horizontal lines have a slope of zero.

Knowing when a function increases or decreases is helpful for graphing the function.

Example 5   For what values of x is the function $f(x) = x^2 - 2x + 4$ increasing and decreasing? Find the critical point(s). Use this information to sketch the curve.

Solution

$$f'(x) = 2x - 2 = 2(x - 1)$$

We see that

$$2(x - 1) > 0 \quad \text{if } x > 1;$$
$$2(x - 1) = 0 \quad \text{if } x = 1;$$
$$2(x - 1) < 0 \quad \text{if } x < 1.$$

TABLE 1

	$f'(x)$	$f(x)$ is
$-\infty < x < 1$	$-$	Decreasing
$x = 1$	0	(critical point).
$1 < x < \infty$	$+$	Increasing.

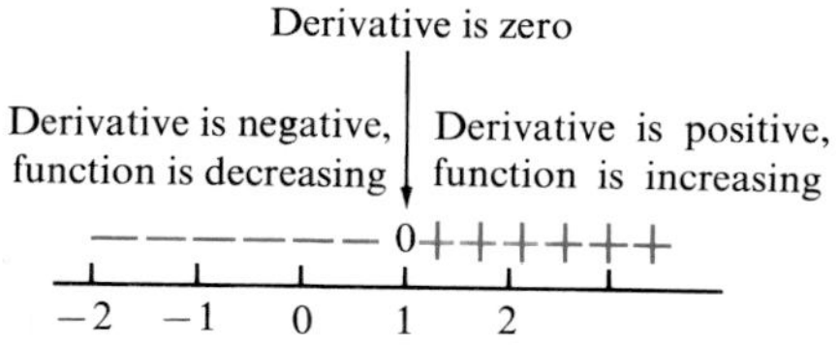

Derivative is zero

Derivative is negative, function is decreasing | Derivative is positive, function is increasing

$------\ 0{+}{+}{+}{+}{+}{+}$

$-2 \quad -1 \quad 0 \quad 1 \quad 2$

Figure 2

This information is summarized in Table 1 and in Figure 2. Since $f'(x) = 0$ only when $x = 1$, we see that 1 is the only critical point. Note that when $x = 1$, $f(x) = 1^2 - 2 \cdot 1 + 4 = 3$. From Table 1 you can see that

$$f \text{ decreases for } x < 1$$

and

$$f \text{ increases for } x > 1.$$

This implies that f takes its minimum value when $x = 1$ at the point $(1, 3)$. Also, you can see that when $x = 0$, $f(x) = 0^2 - 2 \cdot 0 + 4 = 4$. The point $(0, 4)$ is called the **y-intercept** of f. We draw the graph in Figure 3.

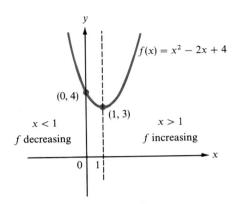

Figure 3

Example 6 Let $y = x^3 + 3x^2 - 9x - 10$. For what values of x is this function increasing and decreasing? Find all critical points and sketch the curve.

Solution We have

$$\frac{dy}{dx} = 3x^2 + 6x - 9 = 3(x^2 + 2x - 3) = 3(x + 3)(x - 1).$$

We see that

> when $x < -3$, $x + 3 < 0$ and $x - 1 < 0$;
> when $-3 < x < 1$, $x + 3 > 0$ and $x - 1 < 0$;
> when $x > 1$, $x + 3 > 0$ and $x - 1 > 0$.

Remember,

> a negative quantity times a negative quantity is a positive quantity;
> a negative quantity times a positive quantity is a negative quantity;
> a positive quantity times a positive quantity is a positive quantity.

This leads to the information in Table 2, which is also depicted in Figure 4. Also,

when $x = -3$, $y = (-3)^3 + 3(-3)^2 - 9(-3) - 10 = -27 + 27 + 27 - 10 = 17$;
when $x = 0$, $y = -10$ and $(0, -10)$ is the y-intercept;
when $x = 1$, $y = 1^3 + 3 \cdot 1^2 - 9 \cdot 1 - 10 = -15$.

TABLE 2

	$x + 3$	$x - 1$	$\dfrac{dy}{dx} = 3(x + 3)(x - 1)$	f is
$-\infty < x < -3$	$-$	$-$	+ (negative times negative)	Increasing
$x = -3$	0	-4	0	(critical point).
$-3 < x < 1$	$+$	$-$	$-$ (positive times negative)	Decreasing
$x = 1$	4	0	0	(critical point).
$1 < x < \infty$	$+$	$+$	+ (positive times positive)	Increasing.

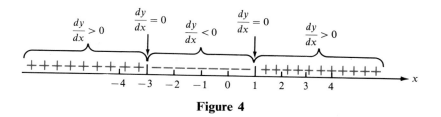

Figure 4

Putting this all together, we obtain the graph in Figure 5.

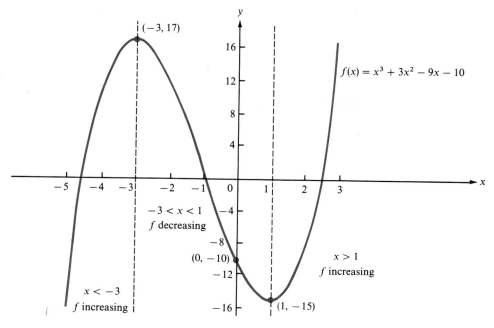

Figure 5

Notice from the curve that the point $(-3, 17)$ is a maximum point in the sense that "near" $x = -3$, y takes its largest value at $x = -3$. However, there is no *global* (or *absolute*) maximum value for the function, because as x increases beyond the value 1, y increases without bound. For example, if $x = 10$, $y = 10^3 + 3 \cdot 10^2 - 9 \cdot 10 - 10 = 1200$, which is much bigger than 17. In this setting, the point $(-3, 17)$ is called a *local maximum* (or *relative maximum*) in the sense that the function achieves its maximum value there for points *near* $(-3, 17)$. Similarly, we call the point $(1, -15)$ a *local minimum* (or *relative minimum*).

Maxima and Minima   The function f has

(a)   a **local maximum** at x_0 if f changes from increasing to decreasing at x_0;

(b)   a **local minimum** at x_0 if f changes from decreasing to increasing at x_0;

(c)   a **global maximum** at x_0 if $f(x_0) \geq f(x)$ for every x in the domain of f;

(d)   a **global minimum** at x_0 if $f(x_0) \leq f(x)$ for every x in the domain of f.

Note.   (a) In Example 6, f has a local but not a global maximum at $x = -3$. It also has a local but not global minimum at $x = 1$. This is evident from Figure 5.

(b) In Example 5, f has a local *and* global minimum at $x = 1$. There is no local or global maximum.

As we saw in Examples 5 and 6, whenever we had a local maximum or minimum, we also had a critical point. This is not a coincidence.

> If f has a local maximum or minimum at x_0, then x_0 is a critical point. (1)

Remark.   The converse of this result is not true, as the next example shows.

Example 7   Let $y = f(x) = x^3$. Then $f'(x) = 3x^2$, which is always positive except at the critical point $x = 0$. If $x < 0$, then $f(x) < 0$, and if $x > 0$, then $f(x) > 0$, so that the function increases from negative values to zero to positive values at $x = 0$ (see Figure 6) and has neither a local maximum nor a local minimum there. Thus, as this example shows, a critical point may be neither a local maximum nor a local minimum.

Example 8   Let $y = f(x) = |x|$. The graph is sketched in Figure 7. We know from Example 10.4.6 on page 548 that

$$f'(x) = \begin{cases} 1, & \text{if } x > 0 \\ -1, & \text{if } x < 0 \end{cases}$$

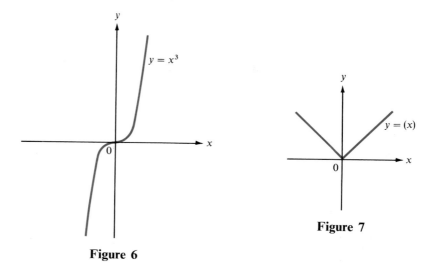

Figure 6

Figure 7

But $f'(0)$ does not exist, so 0 is a critical point. Clearly, the function has a local (and global) minimum at 0.

How do we determine if a critical point is a local maximum, a local minimum, or neither? Examples 5, 6, and 7 taken together illustrate the fact that at a critical point a function may have a local maximum, a local minimum, or neither. There are two ways to determine when a critical point is a local maximum or minimum. The first of these, called the **first derivative test**, is given here. The second is given in the next section.

First Derivative Test   Let x_0 be a critical point of a function f. Then

---

1. if $f'(x) > 0$ immediately to the left of x_0 and $f'(x) < 0$ immediately to the right of x_0, f has a local maximum at x_0.

2. if $f'(x) < 0$ immediately to the left of x_0 and $f'(x) > 0$ immediately to the right of x_0, f has a local minimum at x_0.

3. if $f'(x)$ has the same sign to both the immediate left and right of x_0, f has neither a local maximum nor a local minimum at x_0.

The first derivative test is illustrated in Figure 8, in which we have drawn three typical tangent lines in each of four cases. We see that f has a local

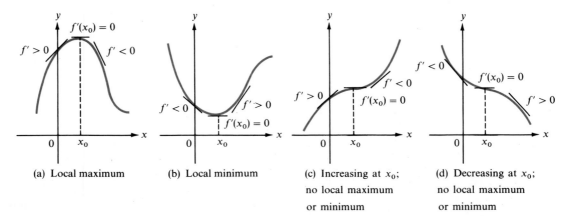

(a) Local maximum   (b) Local minimum   (c) Increasing at x_0;   (d) Decreasing at x_0;
 no local maximum   no local maximum
 or minimum   or minimum

Figure 8

maximum at the critical point, x_0, if the curve lies below the tangent lines near that point. Similarly, f has a local minimum if f lies above the tangent lines near that point. Finally, if neither of these "pictures" is valid near x_0, then x_0 has neither a local maximum nor a local minimum.

Example 9   Use the first derivative test to find the local maxima and/or local minima of $f(x) = x^2 - 4x + 5$.

Solution We must first find the critical points of f. Since $f'(x) = 2x - 4 = 2(x - 2)$, we see that $f'(x) = 0$ when $x = 2$. But $x - 2 < 0$ if $x < 2$ and $x - 2 > 0$ if $x > 2$ (see Figure 9). By the first derivative test f has a local

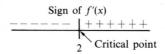

Figure 9

minimum at $x = 2$. Actually, because

Completing the square
↓
$$x^2 - 4x + 5 = (x - 2)^2 - 4 + 5 = (x - 2)^2 + 1 \geq 1,$$

we see that f has a global minimum of 1 at $x = 2$. The function is sketched in Figure 10.

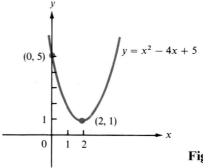

Figure 10

Example 6 (continued) Here $f'(x) = 3x^2 - 6x - 9 = 3(x + 3)(x - 1)$. We can use Table 2 on page 597 to describe the nature of the critical points $x = -3$ and $x = 1$. The sign of $f'(x)$ is given in Figure 11. Thus f has a local maximum at $x = -3$ and a local minimum at $x = 1$.

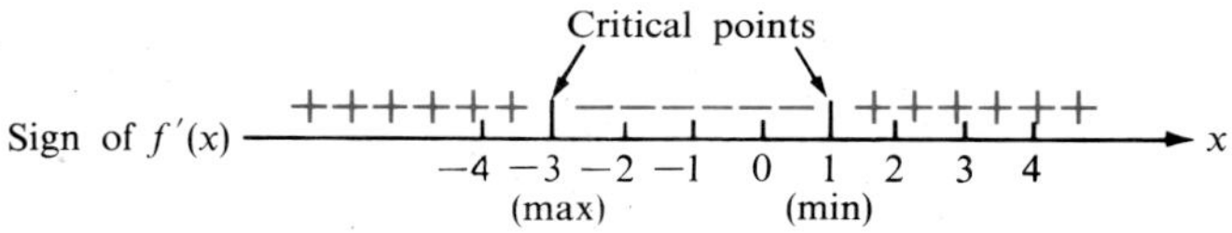

Critical points

Sign of $f'(x)$

Figure 11

Example 7 (continued) Here $f'(x) = 3x^2$, which is positive except at $x = 0$. Thus f has neither a local maximum nor a local minimum at 0.

We can use the facts of this section to illustrate properties of the functions e^x and $\ln x$.

Example 10 Sketch $y = e^x$.

Solution   See page 585.

$$\frac{dy}{dx} = \frac{d}{dx} e^x \overset{\downarrow}{=} e^x > 0 \quad \text{for all } x$$

Thus e^x is an increasing function. When $x = 0$, $y = e^0 = 1$, so $(0, 1)$ is the y-intercept. The function is sketched in Figure 12.

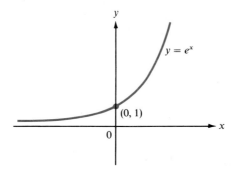

Figure 12

Example 11   Sketch $y = \ln x$.

Solution

See page 584.

$$\frac{dy}{dx} = \frac{d}{dx} \ln x = \overset{\downarrow}{\frac{1}{x}} > 0,$$

since $\ln x$ is defined only for $x > 0$. Thus $\ln x$ is also an increasing function of x. It has no y-intercept since $\ln 0$ is not defined. When $y = 0$, $\ln x = 0$ or $x = e^0 = 1$, so the x-intercept is $(1, 0)$. This function is sketched in Figure 13.

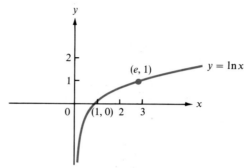

Figure 13

PROBLEMS 11.1   In Problems 1–21 (a) find the intervals over which the given function is increasing or decreasing; (b) find all critical points; (c) find the y-intercept (and the x-intercepts if convenient); (d) locate all local maxima and minima using the first derivative test; (e) sketch the curve.

1. $y = x^2 + x - 30$ **2.** $y = 2 - x^2$

3. $y = x^2 - 5x + 3$ **4.** $y = 3x^2 + 6x + 1$

5. $y = 1 - x - x^2$ **6.** $y = 6 - 8x^2$

7. $y = x - x^2$ **8.** $y = x^3 + 1$

9. $y = x^3 - 3x$ **10.** $y = x^3 - 12x + 10$

11. $y = x^3 - 3x^2 - 45x + 25$ **12.** $y = 4x^3 - 3x + 2$

13. $y = x^4$ **14.** $y = x^4 + 2$

15. $y = 1 - x^4$ **16.** $y = x^4 - 8x^2$

17. $y = x^4 - 4x^3 + 4x^2 + 1$ **18.** $y = e^{x-1}$

19. $y = 1 - e^x$ **20.** $y = \ln(2x + 3)$

21. $y = e^{x^2}$

22. Show that the function $f(x) = (7x + 3)/(5x + 8)$ is an increasing function on any interval over which it is defined.

23. Show that the function $f(x) = (2x + 5)/(3x + 7)$ is a decreasing function on any interval over which it is defined.

24. Sketch the total cost function (given in Example 10.5.3 on page 553)
$$C(q) = 100 + 5q - 0.01q^2.$$

25. Sketch the total cost function $C(q) = 500 + 45q + 2.1q^2 - 0.01q^3$.

In Problems 26–29 sketch a function with the given properties.

26. $f(3) = 2$, $f'(3) = 0$, $f'(x) > 0$ for $x < 3$ and $f'(x) < 0$ for $x > 3$

27. $f(-1) = -2$, $f'(-1) = 0$, $f'(x) < 0$ for $x < -1$ and $f'(x) > 0$ for $x > -1$

28. $f(0) = 3$, $f(4) = -1$, $f'(0) = 0$, $f'(4) = 0$, $f'(x) > 0$ for $x < 0$, $f'(x) < 0$ for $0 < x < 4$, $f'(x) > 0$ for $x > 4$

29. $f(-2) = 1$, $f(2) = -5$, $f(6) = 8$, $f'(-2) = f'(2) = f'(6) = 0$, $f'(x) > 0$ for $x < -2$, $f'(x) < 0$ for $-2 < x < 2$, $f'(x) > 0$ for $2 < x < 6$, $f'(x) < 0$ for $x > 6$

11.2 Concavity and the Second Derivative Test

Consider the graph of $f(x) = x^2 - 2x + 4$ given in Figure 1a (see Figure 11.1.3 on page 596). The derivative is $f'(x) = 2x - 2 = 2(x - 1)$. For $x < 1$, $f'(x) < 0$; at $x = 1$, $f'(x) = 0$; and for $x > 1$, $f'(x) > 0$. We see that the derivative function f' is increasing. When this occurs, the function is said to be **concave up**. Functions that are concave up have the shape of the curve in Figure 1a or 1b.

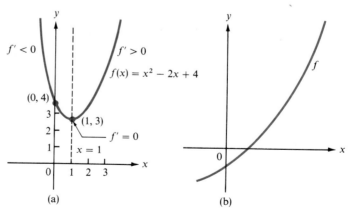

Figure 1

Now consider the function $y = -x^2$, whose graph is given in Figure 2a. For $x < 0$, $f'(x) > 0$, for $x = 0$, $f'(x) = 0$, and for $x > 0$, $f'(x) < 0$. Thus the derivative is decreasing and, in this case, the function is called **concave down**. Functions that are concave down have the shape of the curve in Figure 2a or 2b.

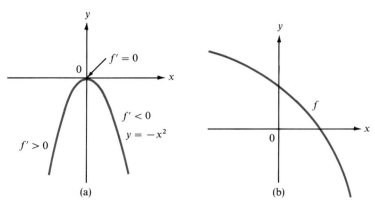

Figure 2

The more precise definition of this concept follows.

CONCAVITY

---

1. f is **concave up** in (a, b) if over that interval f' is an increasing function.
2. f is **concave down** in (a, b) if over that interval f' is a decreasing function.

Another way to think of concavity is suggested by Figure 3. Here we have drawn in some tangent lines. In Figure 3a, f is concave up and all points on the curve lie *above* the tangent lines. In 3b, f is concave down and all points

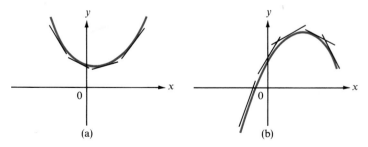

Figure 3

on the curve lie *below* the tangent lines. This information is very useful in curve plotting. If f is concave down in (a, b), then the type of behavior exhibited in Figure 4 is impossible. At the point c the curve lies above the tangent line but this is impossible since f is concave down. In general, if a curve is concave up or down in a certain interval, then it cannot "wiggle around" in that interval (that is, it cannot behave like the curve in Figure 4).

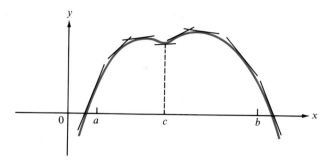

Figure 4

Many functions will alternate between concave up and concave down.

Point of Inflection The point $(x_0, f(x_0))$ is a **point of inflection** of f if f changes its direction of concavity at x_0.

Remark. As we shall see, f may have a point of inflection at $(x_0, f(x_0))$ in one of the following two cases:

(a) $f''(x_0) = 0$;

(b) $f''(x_0)$ does not exist.

Example 1 Consider the function graphed in Figure 5. The function is concave down in the intervals $(-6, -2)$, $(2, 8)$, and $(12, 18)$ and concave up in the intervals $(-2, 2)$ and $(8, 12)$. There are points of inflection at -2, 2, 8, and 12.

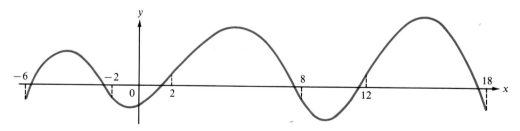

Figure 5

How can we tell when a function is concave up or concave down? The derivative of f' is f''. If $f'' > 0$, then f' has a positive derivative, and so we can conclude that f' is increasing. Therefore, f is concave up. If $f'' < 0$, then f' is decreasing and f is concave down. Thus if f'' exists, *we can determine the direction of concavity by simply looking at the sign of the second derivative.* We have therefore shown that the following is true when $f''(x)$ exists on the interval (a, b):

1. if $f''(x) > 0$ in (a, b), f is concave up in (a, b);
2. if $f''(x) < 0$ in (a, b), f is concave down in (a, b).

Example 2 Let $f(x) = x^3 - 2x$. Then $f'(x) = 3x^2 - 2$ and $f''(x) = 6x$. We see that $f''(x) < 0$ if $x < 0$ and $f''(x) > 0$ if $x > 0$. Thus, f is concave down if $x < 0$ and concave up if $x > 0$, so that $(0, 0)$ is a point of inflection. Note that $f''(0) = 0$. This function is sketched in Figure 6.

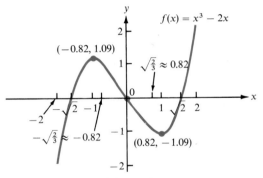

Figure 6

Example 3 Let $f(x) = x^{3/5}$. Then $f'(x) = (3/5)x^{-2/5}$ and $f''(x) = -(6/25)x^{-7/5}$. Here $f''(x) < 0$ for $x > 0$ and $f''(x) > 0$ for $x < 0$. Thus $(0, 0)$ is a point of inflection. Note that both $f'(x)$ and $f''(x)$ become very large as $x \to 0$; that is, neither $f'(0)$ nor $f''(0)$ exists.

Now suppose that f has a point of inflection at $(x_0, f(x_0))$. That means that at x_0, f' goes from increasing to decreasing (Figure 7a) or from decreasing to increasing (Figure 7b). In the first case, f' has a local maximum at x_0; in the

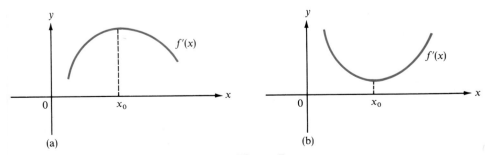

(a)   (b)

Figure 7

second case, f' has a local minimum at x_0. In either case, by the result shown in equation (1) on page 598, $(f')' = f'' = 0$ at x_0. Thus,

> at a point of inflection $(x_0, f(x_0))$, if $f''(x_0)$ exists, then $f''(x_0) = 0$.

Note. The converse of this result is not true in general. That is, if $f''(x_0) = 0$, then x_0 is not necessarily a point of inflection. For example, if $f(x) = x^4$, then $f''(x) = 12x^2 > 0$ for $x \neq 0$. Thus, f does not change concavity at 0, so 0 is *not* a point of inflection even though $f''(0) = 0$.

Finally, suppose that $f'(x_0) = 0$. If $f''(x_0) > 0$, then the curve is concave up at x_0. A glance at Figure 8a suggests that f has a local minimum at x_0. Similarly, if $f'(x_0) = 0$ and $f''(x_0) < 0$, then f has a local maximum at x_0. Thus, we have the following result.

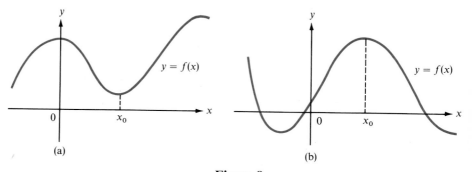

Figure 8

SECOND DERIVATIVE TEST FOR A LOCAL MAXIMUM OR MINIMUM

Let f be differentiable on an open interval containing x_0, and suppose that $f''(x_0)$ exists. Then

> **1.** if $f'(x_0) = 0$ and $f''(x_0) > 0$, f has a local minimum at x_0;
> **2.** if $f'(x_0) = 0$ and $f''(x_0) < 0$, f has a local maximum at x_0.

Remark. If $f'(x_0) = 0$ and $f''(x_0) = 0$, then f may have a local maximum, a local minimum, or neither at x_0. For example, $y = x^3$ has neither at $x = 0$, while $y = x^4$ has a minimum at 0 and $y = -x^4$ has a maximum at 0.

We now give some examples of the technique of curve sketching making use of information derived from the second derivative.

Example 4   If $f(x) = x^2 - 2x + 4$, then $f'(x) = 2x - 2$ and $f''(x) = 2$, which is greater than 0. Thus the curve is concave up for $-\infty < x < \infty$. This justifies the accuracy of the graph in Figure 1a on page 603. The function f has a local (and global) minimum at the point $(1, 3)$.

Example 5   If $f(x) = -x^2$, then $f''(x) = -2$, which is less than 0. This curve is concave down for $-\infty < x < \infty$. This justifies Figure 2a on page 604. The function f has a local (and global) maximum at the point $(0, 0)$.

Example 6   Let $f(x) = 2x^3 - 3x^2 - 12x + 5$. Sketch the curve.

Solution   We do this in several steps:

Step 1.   Compute $f'(x)$:

$$f'(x) = 6x^2 - 6x - 12 = 6(x^2 - x - 2) = 6(x - 2)(x + 1).$$

The critical points are $x = 2$ and $x = -1$. In Figure 9 we illustrate the sign of $f'(x)$. We can see that there is a local maximum at $x = -1$ $(y = 12)$ and a local minimum at $x = 2$ $(y = -15)$.

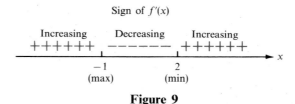

Sign of $f'(x)$

Increasing ++++++   Decreasing ------   Increasing +++++

-1 (max) 2 (min)

Figure 9

Step 2.   Compute $f''(x)$:

$$f''(x) = 12x - 6 = 6(2x - 1)$$

We have $2x - 1 < 0$ if $x < \frac{1}{2}$ and $2x - 1 > 0$ if $x > \frac{1}{2}$. In Figure 10 we illustrate the sign of $f''(x)$. Since f changes from concave down to concave up at $x = \frac{1}{2}$, we see that $x = \frac{1}{2}$ $(y = -\frac{3}{2})$ is a point of inflection.

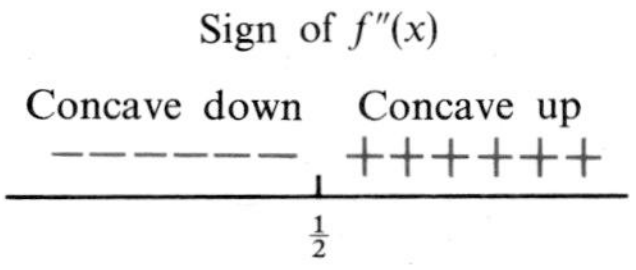

Sign of $f''(x)$

Concave down ------   Concave up +++++

$\frac{1}{2}$

Figure 10

We can also use the second derivative test to show that f has a local maximum at $x = -1$ and a local minimum at $x = 2$. When $x = 2$, $f''(x) = 6(4 - 1) = 18 > 0$, so that f has a local minimum at $(2, -15)$. When $x = -1$, $f''(x) = -18 < 0$, so that f has a local maximum at $(-1, 12)$. This information is summarized in Table 1.

TABLE 1

Critical point x_0	$f''(x_0) = 6(2x - 1)$	Sign of $f''(x_0)$	At x_0, f has a
2	18 (>0)	+	Local minimum.
-1	-18 (<0)	$-$	Local maximum.

Step 3. We plot some points:

the y-intercept $(0, 5)$;
the critical points $(-1, 12)$ and $(2, -15)$;
the point of inflection $(\frac{1}{2}, -\frac{3}{2})$.

Step 4. We draw the graph making use of the information obtained in Steps 1, 2, and 3. Note from Step 1 that f increases for $x < -1$, decreases for $-1 < x < 2$, and increases again for $x > 2$. Putting this all together, we obtain the curve in Figure 11. From this sketch we can see that the graph crosses the

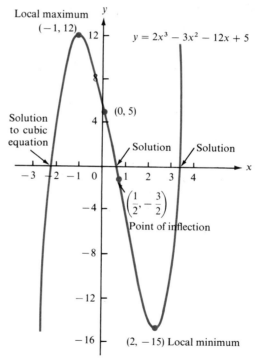

Figure 11

x-axis at three places (the three x-intercepts). This tells us that the cubic equation $2x^3 - 3x^2 - 12x + 5 = 0$ has three real solutions. Moreover, we see that two of these solutions are positive and one is negative. (One is slightly less than -2, one is between 0 and 1, and one is between 3 and 4.) This information, which comes without additional work, can often be very useful.

The next example shows that in some cases the first derivative test is more useful than the second derivative test.

Example 7   Sketch the curve $y = f(x) = x^{2/3}$.

Solution   Since $f'(x) = (2/3)x^{-1/3} = 2/3x^{1/3}$, the only critical point is at $x = 0$, since $f'(0)$ is not defined although $f(0)$ is defined. Also, $f''(0)$ is not defined (since $f'(0)$ isn't defined), so that the second derivative test won't work. However, since $f'(x) < 0$ for $x < 0$ and $f'(x) > 0$ for $x > 0$, we see that 0 is a local minimum. (It is, in fact, a global minimum.) Also, when $x \neq 0$, $f''(x) < 0$, so that the curve is concave down. It is sketched in Figure 12.

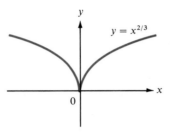

Figure 12

Example 8   Sketch the curve $y = xe^{-x}$.

Solution

$$f'(x) = \frac{d}{dx} xe^{-x} \stackrel{\overset{\text{Product rule}}{\downarrow}}{=} x\frac{d}{dx} e^{-x} + e^{-x}\frac{d}{dx}(x) = x(-e^{-x}) + e^{-x} = e^{-x}(1 - x)$$

Since $e^{-x} > 0$ for all x, we see that

$$xe^{-x} \text{ is increasing if } x < 1$$

and

$$xe^{-x} \text{ is decreasing if } x > 1.$$

Also,

$$1 \text{ is the only critical point.}$$

Now

$$f''(x) = \frac{d}{dx} e^{-x}(1 - x) \stackrel{\overset{\text{Product rule}}{\downarrow}}{=} e^{-x}(-1) + (1 - x)(-e^{-x})$$

$$= e^{-x}[-1 - (1 - x)] = e^{-x}(x - 2)$$

Thus,

$$xe^{-x} \text{ is concave up for } x > 2,$$
$$xe^{-x} \text{ is concave down for } x < 2,$$

and

2 is the only point of inflection.

The information we have obtained is summarized in Table 2.

TABLE 2

Interval	Sign of $f'(x)$	Sign of $f''(x)$	Nature of point
$x < 1$	$+$	$-$	
$x = 1$	0	$-$	Critical point (local and global maximum)
$1 < x < 2$	$-$	$-$	
$x = 2$	$-$	0	Point of inflection
$x > 2$	$-$	$+$	

It is not hard to see that

$$\text{Let } u = -x. \quad \text{See Figure 5a on page 498.}$$

$$\lim_{x \to -\infty} e^{-x} = \lim_{u \to \infty} e^{u} = \infty$$

Thus

$$\lim_{x \to -\infty} xe^{-x} = -\infty.$$

That is, as x gets large in the negative direction, so does xe^{-x}. What about $\lim_{x \to \infty} xe^{-x}$? Some sample values are given in Table 3. Evidently, $\lim_{x \to \infty} xe^{-x} = 0$. Finally, $xe^{-x} = 0$ when $x = 0$. The graph is given in Figure 13.

TABLE 3

x	e^{-x}	xe^{-x}
1	0.367879	0.367879
2	0.135335	0.27067
3	0.049787	0.149361
5	0.006738	0.03369
7	0.000912	0.006384
10	0.0000453999	0.000453999
15	0.0000003059	0.0000045885
20	0.0000000021	0.0000000412
30	9.357623×10^{-14}	$2.8072869 \times 10^{-12}$
50	1.9×10^{-22}	9.6×10^{-21}
100	3.7×10^{-44}	3.7×10^{-42}

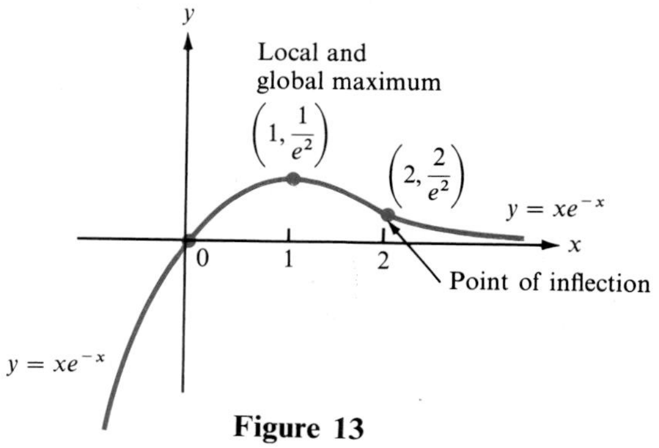

Figure 13

We now turn to a different kind of example.

Example 9   Let $f(x) = x/(1 + x)$. Graph the function.

Solution   Before doing any calculations, first note that we will have problems at $x = -1$, since the function is not defined there. We can expect that the function will "blow up" as x approaches -1 from the right or left. Now

$$f'(x) = \frac{d}{dx}\frac{x}{1 + x} = \frac{(1 + x)1 - x}{(1 + x)^2} = \frac{1}{(1 + x)^2}.$$

Therefore, f is always increasing (except at $x = -1$, where it is not defined). Also, f has *no* critical points (-1 is *not* a critical point because f is not defined at -1). Now

$$f''(x) = \frac{d}{dx}(1 + x)^{-2} = -\frac{2}{(1 + x)^3}.$$

If $x < -1$, then $1 + x < 0$, $(1 + x)^3 < 0$, and $-2/(1 + x)^3 > 0$. If $x > -1$, then $1 + x > 0$, $(1 + x)^3 > 0$, and $-2/(1 + x)^3 < 0$. Hence f is concave up if $x < -1$ and concave down if $x > -1$. Observe that

$$\lim_{x \to \infty} \frac{x}{1 + x} = \lim_{x \to \infty} \frac{1}{(1/x) + 1} = 1$$

and

$$\lim_{x \to -\infty} \frac{x}{1 + x} = \lim_{x \to -\infty} \frac{1}{(1/x) + 1} = 1.$$

Therefore, as $x \to \pm\infty$, $f(x) \to 1$. The line $y = 1$ is called a **horizontal asymptote** for the function $x/(1 + x)$. We also observe that if

$$x < -1, \quad \text{then} \quad f(x) > 1; \qquad \text{A negative over a negative with the numerator more negative than the denominator}$$

$$-1 < x < 0, \quad \text{then } f(x) < 0;$$
$$x = 0, \quad \text{then } f(x) = 0;$$
$$x > 0, \quad \text{then } 0 < f(x) < 1. \qquad \text{Denominator larger than the numerator}$$

Also, $\lim_{x \to -1^+}[x/(1 + x)] = -\infty$ (since the numerator is negative and the denominator is positive) and $\lim_{x \to -1^-}[x/(1 + x)] = \infty$.[†] Finally,

$$\lim_{x \to \pm\infty} f'(x) = \lim_{x \to \pm\infty} \frac{1}{(1 + x)^2} = 0,$$

[†] $\lim_{x \to a^+} f(x)$ is the limit as $x \to a$ from the right. Analogously, $\lim_{x \to a^-} f(x)$ is the limit as $x \to a$ from the left.

so that the tangent lines become flat (horizontal) as $x \to \infty$, and

$$\lim_{x \to -1} f'(x) = \lim_{x \to -1} \frac{1}{(1 + x)^2} = \infty,$$

so that the tangent lines become vertical as $x \to -1$ (from either side); the line $x = -1$ is called a **vertical asymptote**. We put this all together in Figure 14.

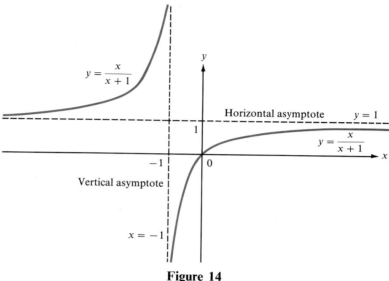

Figure 14

We now review the kinds of information that are available to help us sketch curves. The following steps should be carried out in order to obtain an accurate picture.

TO SKETCH A CURVE $y = f(x)$:

1. Calculate the derivative $dy/dx = f'$ and determine where the curve is increasing and decreasing. Find all points at which $f' = 0$ or f' does not exist and plot them.

2. Calculate the second derivative $f'' = d^2y/dx^2$ and determine where the curve is concave up and concave down.

3. Determine local maxima and minima by using either the first or the second derivative test.

4. Find all points of inflection and plot them.

5. Determine the y-intercept and (if possible) the x-intercept(s) and plot them.

6. Determine $\lim_{x \to \infty} f(x)$ and $\lim_{x \to -\infty} f(x)$. If either of these is finite, then we obtain horizontal asymptotes (as in Example 9).

7. If f is not defined at x_0, determine $\lim_{x \to x_0^+} f(x)$ and $\lim_{x \to x_0^-} f(x)$.

8. Look for vertical asymptotes. These are the lines of the form $x = x_0$, where $\lim_{x \to x_0^+} f(x) = \infty$ (or $-\infty$) or $\lim_{x \to x_0^-} f(x) = \infty$ (or $-\infty$) and $f(x_0)$ is undefined (see Example 9).

PROBLEMS 11.2   In Problems 1–30 follow the steps of this section to sketch the curve.

1. $y = x^2 + x - 30$

2. $y = 2 - x^2$

3. $y = x^2 - 5x + 3$

4. $y = 1 - x - x^2$

5. $y = x^3 + 1$

6. $y = x^3 - 3x$

7. $y = x^3 - 12x + 10$

8. $y = x^3 - 3x^2 - 45x + 25$

9. $y = 4x^3 - 3x + 2$

10. $y = x^4$

11. $y = 2x^4 - 1$

12. $y = 2x^3 - 9x^2 + 12x - 3$

13. $y = \dfrac{x^3}{3} + \dfrac{x^2}{2} - 2x - \dfrac{2}{3}$

14. $y = \sqrt{1 + x}$

15. $y = \sqrt[3]{1 + x}$

16. $y = \sqrt{1 - x^2}$

17. $y = (x - 2)^{2/3}$

18. $y = (x + 3)^{1/3}$

19. $y = \dfrac{1}{x - 3}$

20. $y = \dfrac{1}{3x + 9}$

*21. $y = \dfrac{1}{x^2 - 1}$

22. $y = \dfrac{x + 1}{x - 1}$

23. $y = \dfrac{2x + 5}{4x - 8}$

24. $y = e^{x^2}$

*25. $y = e^{1/x}$

26. $y = \ln(2 + x)$

27. $y = \ln(1 + x^2)$

28. $y = x \ln x$

29. $y = xe^x$

30. $y = x^2 e^{-x}$

31. (a) Discuss the concavity of the total cost function $C(q) = 100 + 5q - 0.01q^2$.

  (b) Find all local maxima and minima.

  (c) Sketch the curve.

32. Answer the questions of Problem 31 for the cost function $C(q) = 500 + 45q + 2.1q^2 - 0.01q^3$.

11.3 Maxima and Minima: Applications

In this section we consider the problem of finding the maximum (largest) and minimum (smallest) values of a function, $y = f(x)$, over a finite interval $[a, b]$. We then show how this theory can be used in applications.

In Section 11.1 we defined what we meant by a function having a local maximum or minimum at a point x_0. Roughly, f has a local maximum at x_0 if, among all the values of $f(x)$ in an open interval containing x_0, f takes its largest value at x_0. A similar statement can be made for a local minimum at x_0. It is true that if f has a local maximum or minimum at x_0, then x_0 is a critical point of f. That is, either $f'(x_0) = 0$ or $f(x_0)$ exists but $f'(x_0)$ does not exist. We further indicated that

if $f'(x_0) = 0$ and $f''(x_0) < 0$, then f has a local maximum at x_0

and

if $f'(x_0) = 0$ and $f''(x_0) > 0$, then f has a local minimum at x_0.

But there could be more than one local maximum or minimum in an interval. Moreover, a maximum or minimum could be reached in other ways. Look at Figure 1. The function depicted there has local maxima at x_0, x_2, and x_4 and local minima at x_1, x_3, and x_5. However, in the interval $[a, b]$ the global maximum and minimum values of f is taken at none of these critical points. The maximum value of f is taken at $x = b$, and the minimum is taken at $x = a$. Thus it is necessary to check both endpoints as well as all the critical points to find the maximum and minimum.

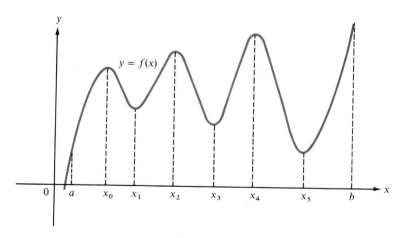

Figure 1

Another problem could arise. Consider the function

$$f(x) = x^{2/3} \quad \text{for} \quad -1 \leq x \leq 1.$$

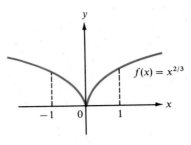

Figure 2

This is graphed in Figure 2. We see that in the interval $[-1, 1]$, f takes on its minimum value at $x = 0$, which is not an endpoint of the interval and at which $f'(0)$ is not zero. In fact, since $f'(x) = 2/3x^{1/3}$, $f'(0)$ does not even exist. Thus we have the following procedure:

If $f(x)$ is defined on the finite interval $[a, b]$, then in order to find the maximum and minimum values of f over that interval,

1. evaluate f at all critical points in (a, b);

2. evaluate $f(a)$ and $f(b)$.

Then, if M denotes the largest value of f in $[a, b]$ and m denotes the smallest value,

(a) $M =$ the largest of the values calculated in (1) and (2);

(b) $m =$ the smallest of the values calculated in (1) and (2).

Remark. In an infinite interval, the function may have neither a maximum nor a minimum. For example, $f(x) = x^2$ has a minimum (zero) but no maximum on $[0, \infty)$. The function $f(x) = x^3$ has neither a maximum nor a minimum in $(-\infty, \infty)$. However, the following is true:

If f is continuous on the closed, bounded interval $[a, b]$, then f has a maximum and a minimum in $[a, b]$.

Example 1 Find the maximum and minimum values of $f(x) = 2x^3 - 3x^2 - 12x + 5$ in the interval $[0, 4]$.

Solution $f'(x) = 6x^2 - 6x - 12 = 6(x^2 - x - 2) = 6(x - 2)(x + 1)$. The only place in the interval $(0, 4)$ where $f'(x) = 0$ is at $x = 2$. At $x = 2$, $f(x) = -15$. We then find that $f''(x) = 12x - 6 = 18$ when $x = 2$, so that f has a local minimum at $(2, -15)$ by the second derivative test, and this is the only critical point in $(0, 4)$. We also calculate $f(0) = 5$ and $f(4) = 37$. There are no points in $(0, 4)$ at which f' is not defined. Therefore, $f_{max} [0, 4] = 37$ and $f_{min} [0, 4] = -15$. This is sketched in Figure 3.

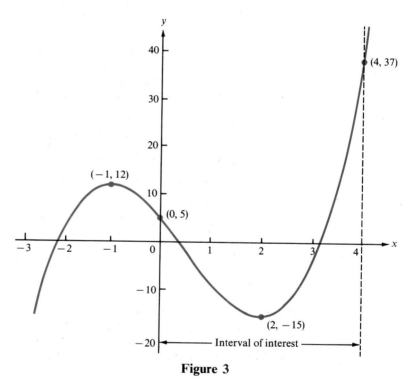

Figure 3

Example 2 Find the maximum and minimum values of $f(x) = x/(1 + x)$ in the interval $[1, 5]$.

Solution $f'(x) = 1/(1 + x)^2$, which is never zero so that there are no critical points. Since f' is defined everywhere in $[1, 5]$, the maximum and minimum values are taken at the endpoints of the interval. We have $f(1) = \frac{1}{2}$ and $f(5) = \frac{5}{6}$, so that $f_{max} [1, 5] = \frac{5}{6}$ and $f_{min} [1, 5] = \frac{1}{2}$. This is sketched in Figure 4. We emphasize that in this problem we were only interested in values of x in the interval $[1, 5]$. Note that in $[-2, 2]$, f has neither a maximum nor a minimum since $f(x) \to \infty$ as $x \to -1$ from the left and $f(x) \to -\infty$ as $x \to -1$ from the right. This doesn't contradict the boxed result above as f is not continuous in $[-2, 2]$.

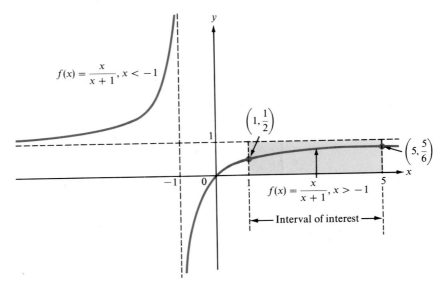

$f(x) = \dfrac{x}{x+1}, x < -1$

$\left(1, \dfrac{1}{2}\right)$

$\left(5, \dfrac{5}{6}\right)$

$f(x) = \dfrac{x}{x+1}, x > -1$

Interval of interest

Figure 4

We now turn to practical applications.

Example 3   Suppose a farmer has 1000 yards of fence that he wishes to use to fence off a rectangular plot along the bank of a river (see Figure 5). What are the dimensions of the maximum area he can enclose?

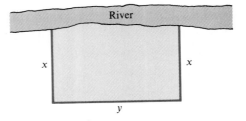

River

x x

y

Figure 5

Solution   Since one side of the rectangular plot is taken up by the river, the farmer can use the fence for the other three sides. From the figure, the area of the plot is

$$A = xy,$$

where y is the length of the side parallel to the river. Without other information, it would be impossible to solve this problem, since the area, A, is a function of the *two* variables x and y and we only know how to find the maximum and minimum values of a function of *one* variable. However, there is a way we can eliminate one of these variables. Since 1000 yards of fence will be used and the total amount of fencing (see Figure 5) is $x + x + y = 2x + y$, we have

$$2x + y = 1000$$

or, solving for y,

$$y = 1000 - 2x, \qquad \text{for } x \text{ in } (0, 500)$$

$y = 1000 - 2x \geq 0$ so $x \leq 500$.

Then

$$A = x(1000 - 2x) = 1000x - 2x^2.$$

To find the maximum value for A, we set dA/dx equal to zero to find the critical points. We have

$$\frac{dA}{dx} = 1000 - 4x = 0$$

when $x = 250$. Also, $dA/dx > 0$ if $x < 250$ and $dA/dx < 0$ if $x > 250$ so that, by the first derivative test, A is a maximum when $x = 250$. This is depicted in Figure 6. When $x = 250$ yards, then $y = 500$ yards and $A = 125{,}000$ square yards, which is the maximum area that the farmer can enclose. Note that, as is evident from Figure 6, A is positive when $0 \leq x \leq 500$, and at the endpoints $x = 0$ and $x = 500$ the area is zero, so the local maximum we have obtained is a maximum over the entire interval $[0, 500]$.

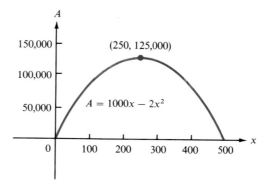

Figure 6

The last example illustrates the steps used to solve an applied maximum or minimum problem:

1. draw a picture if it makes sense to do so;

2. write all the information in the problem in mathematical terms, giving a letter name to each variable;

3. determine the variable that is to be maximized or minimized and write this as a function of the other variables in the problem;

4. using information given in the problem, eliminate all variables except one, so that the function to be maximized or minimized is written in terms of *one* of the variables of the problem;

5. determine the interval over which this one variable can be defined;

6. follow the steps of this section to maximize or minimize the function over this interval.

Example 4   Suppose that the farmer in Example 3 wishes to build his rectangular plot away from the river (so that he must use his fence for all four sides of the rectangle). How large an area can he enclose in this case?

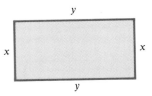

Figure 7

Solution   The situation is now as in Figure 7. The area is again given by $A = xy$, but now

$$2x + 2y = 1000,$$

or

$$2y = 1000 - 2x$$

and

$$y = 500 - x.$$

Hence, the problem is to maximize

$$A = x(500 - x) = 500x - x^2, \quad \text{for } x \text{ in } [0, 500].$$

Now

$$\frac{dA}{dx} = 500 - 2x = 0$$

when $x = 250$. Also, $d^2 A/dx^2 = -2$, so that a local maximum is achieved when $x = y = 250$ yards and the answer is $A = 62,500$ square yards. (Note that the plot is, in this case, a square.) At 0 and 500, $A = 0$, so that $x = 250$ is indeed the maximum.

Remark 1. The reasoning in Example 4 can be used to prove this statement: For a given perimeter, the rectangle containing the greatest area is a square.

Remark 2. If we do not require that the plot be rectangular, then we can enclose an even greater area. Although the proof of this fact is beyond the scope of this book, it can be shown that *for a given perimeter, the geometric shape with the largest area is a circle.* For example, if the 1000 yards of fence in Example 4 are formed in the shape of a circle, then the circle has a circumference of $2\pi r = 1000$, so that the radius of the circle $r = 1000/2\pi$ yards. Then $A = \pi r^2 = \pi(1,000,000/4\pi^2) = 1,000,000/4\pi \approx 79,577$ square yards.

Example 5 An oil importer needs to construct a number of cylindrical barrels, each of which is to hold 32π m³ (cubic meters) of oil. The cost per square meter of constructing the side of the barrel is \$2, and the cost per square meter of constructing the top and bottom is \$4. What are the dimensions of the barrel that costs the least to construct?

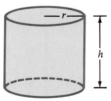

Figure 8

Solution Consider Figure 8. Let h be the height of the barrel and r be the radius of the top and bottom. Then the volume of the barrel is given by $V = \pi r^2 h = 32\pi$ m³.[†] The area of the top or the bottom is πr^2 m², and that of the side is $2\pi rh$ m² (since the circumference of the side is $2\pi r$). Thus the total cost is given by

$$C = 2\pi r^2(4) + 2\pi rh(2) = 8\pi r^2 + 4\pi rh.$$

[†] The volume of a right circular cylinder with radius r and height h is $\pi r^2 h$.

To write C as a function of one variable only (which we must do in order to solve the problem), since the volume is given as 32π, we use $32\pi = \pi r^2 h$ to obtain $h = 32\pi/\pi r^2 = 32/r^2$. Thus

$$C = 8\pi r^2 + 4\pi r \cdot \frac{32}{r^2} = 8\pi r^2 + \frac{128\pi}{r} \qquad \text{for} \quad r > 0.$$

Then,

$$\frac{dC}{dr} = 16\pi r - \frac{128\pi}{r^2}.$$

Setting this equal to zero, we obtain

$$16\pi r = \frac{128\pi}{r^2}, \quad 16r^3 = 128, \quad r^3 = 8, \quad \text{and} \quad r = 2 \text{ m}.$$

In addition, $d^2C/dr^2 = 16\pi + 256\pi/r^3$, which is >0 when $r = 2$. Hence, by the second derivative test, there is a local minimum when $r = 2$ m. When $r = 2$, $h = 32/4 = 8$ m. Note that this local minimum is a true minimum because the only endpoint of the interval occurs at $r = 0$, which makes no practical sense. In that case, the barrel can hold nothing.

Example 6   A cardboard box with a square base and an open top, is to be constructed from a square piece of cardboard 10 cm on a side by cutting out four squares of equal size at the corners and folding up the sides. What should be the dimensions of the box in order to make the volume enclosed as large as possible?

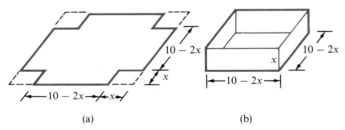

(a)   (b)

Figure 9

Solution   Refer to Figures 9a and 9b. If we cut out squares with sides of x cm, each side of the base of the box will be equal to $10 - 2x$ cm, and the height of the box will be x cm. Then the volume of the box is given by

$$V = x(10 - 2x)^2, \qquad \text{for } x \text{ in } [0, 5].$$

Now,

$$\frac{dV}{dx} = (10 - 2x)^2 + 2x(10 - 2x)(-2) = (10 - 2x)^2 - 4x(10 - 2x)$$

$$= (10 - 2x)(10 - 2x - 4x)$$

$$= (10 - 2x)(10 - 6x),$$

which is equal to zero when $10 - 2x = 0$ or $x = 5$ and $10 - 6x = 0$ or $x = \frac{5}{3}$. At $x = 5$, $V = 0$. At $x = \frac{5}{3}$, $V = \frac{2000}{27}$.

$$\text{If } 0 < x < \tfrac{5}{3}, \text{ then } \frac{dV}{dx} = (10 \overset{>0}{-} 2x)(10 \overset{>0}{-} 6x) > 0.$$

$$\text{If } \tfrac{5}{3} < x < 5, \text{ then } \frac{dV}{dx} = (10 \overset{>0}{-} 2x)(10 \overset{<0}{-} 6x) < 0.$$

Thus, by the first derivative test, the maximum volume of $\frac{2000}{27}$ cm^3 is achieved when $x = \frac{5}{3}$ cm. Note that $V = 0$ at the other endpoint ($x = 0$).

Example 7 In Example 9.2.2 on page 481 we discussed the case of the Universal Card Company, which produces greeting cards. In that example we showed that the total profit function for a certain card was given by

$$P(p) = R(p) - C(p) = -20{,}000(p^2 - 1.66p + 0.315),$$

where p is the retail price of a single card. At what price are profits maximized? What is the maximum profit?

Solution $P'(p) = -20{,}000(2p - 1.66) = 0$ when $2p = 1.66$ or $p = 0.83$. Since $P'(p) > 0$ when $p < 0.83$ and $P'(p) < 0$ when $p > 0.83$, we see, from the first derivative test, that there is a local maximum at $p = 0.83$. It is easy to see that this is a global maximum as well. Thus profit is maximized when $p = 83\cent$. The maximum profit is $-20{,}000(0.83^2 - 1.66(0.83) + 0.315) = -20{,}000(-0.3739) = \$7478$. We solved this problem in Section 9.2 by completing the square. Notice how much easier things are with a bit of calculus.

Example 8 The Kertz Leasing Company leases fleets of new cars to large corporations. The rental fee is \$2000 per car per year. However, for contracts with a fleet size of more than 10 cars, the rental fee per car is discounted by 1% for each car in the contract, up to a maximum fleet of 75 cars. How many cars leased to a single corporation in one year will produce maximum revenue and profit if each car depreciates in value \$1000 per year?

Solution Let R denote total annual revenue for a contract with a fleet size q. If 10 or fewer cars are ordered, then $R = \text{price} \cdot \text{number ordered} = 2000q$. However, if $q > 10$, then the price is reduced by 1% of \$2000 $= \frac{1}{100}(2000)$ for each car ordered (up to 75). Thus, the price *reduction* is $\frac{1}{100}2000q$ and the price per car is $2000 - \frac{1}{100}2000q$. In this case the total revenue is given by $R = \text{price} \cdot \text{number ordered} = (2000 - \frac{1}{100}2000q)q = (2000 - 20q)q$, and we have

$$R = \begin{cases} 2000q, & \text{if } 0 \le q \le 10 \\ (2000 - 20q)q, & \text{if } 10 < q \le 75 \end{cases}$$

There is also an upper bound of 75 to the maximum fleet size. So the first problem is to determine the fleet size, q, that maximizes R. For $0 \le q \le 10$, R is just a linear increasing function. For $0 \le q \le 10$, the maximum value of R is $R(10) = 20,000$. But for $x > 10$,

$$R = 2000q - 20q^2$$

and

$$\frac{dR}{dq} = 2000 - 40q$$

$$= 0 \text{ when } q = 50;$$

also,

$$\frac{d^2R}{dq^2} = -40,$$

and so $q = 50$ gives a maximum value to R on $(10, 75]$ since $R(50) = 2000 \cdot 50 - 20 \cdot 50^2 = 100,000 - 50,000 = 50,000$ is larger than $R(10) = 20,000$ and $R(75) = 37,500$. Therefore, leasing a fleet of 50 cars will yield a maximum revenue. The situation is illustrated in Figure 10.

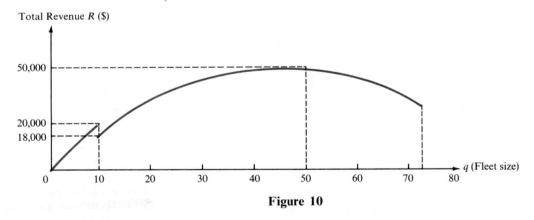

Figure 10

But this answer is not really what the company wants to know. A fleet size of 50 will maximize the revenue from one corporation, but will it maximize the company's profit? Almost certainly not. It is estimated that the value of a leased car depreciates by \$1000 over a year, so that the total loss in value of a fleet size of q cars is $1000q$. Hence, the profit made by Kertz on a fleet size of q cars is given by

$$P = \begin{cases} 2000q - 1000q, & \text{if } 0 \le q \le 10 \\ (2000 - 20q)q - 1000q, & \text{if } 10 < q \le 75 \end{cases}$$

That is,

$$P = \begin{cases} 1000q, & \text{if } 0 \le q \le 10 \\ 1000q - 20q^2, & \text{if } 10 < q \le 75 \end{cases}$$

We must look for the fleet size, q, that maximizes this profit function. In the

interval $0 \leq q \leq 10$, the profit function is linear and the maximum profit (at $q = 10$) is \$10,000.
If $q > 10$, then

$$\frac{dP}{dq} = 1000 - 40q$$

$$= 0 \text{ when } x = 25.$$

Also $d^2P/dq^2 = -20$, so $q = 25$ gives a maximum value to the profit function in the interval $10 < q \leq 75$ [note that $P(75)$ is negative]. This profit is $1000 \cdot 25 - 20 \cdot 25^2 = 25,000 - 12,500 = \$12,500$. Thus, the maximum profit is \$12,500 with 25 cars leased. The situation is illustrated in Figure 11.

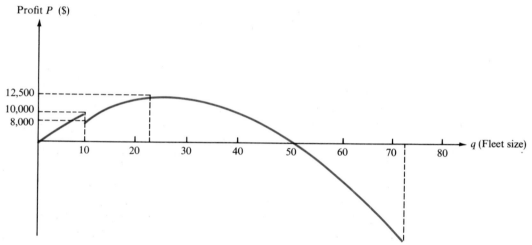

Figure 11

Example 9 **Drug Concentration in the Bloodstream** (Optional). The concentration in the bloodstream, $x = x(t)$, of a particular group of drugs approximately obeys a law of the form

$$x(t) = k(e^{-at} - e^{-bt}),$$

where k, a, and b $(a < b)$ are positive constants for a given dosage of a particular drug and where the drug has been injected into the body at time $t = 0$. We need to know the general characteristics of the drug concentration in the body for time $t > 0$. In particular, at what time t is the concentration at a maximum, and what is this maximum concentration?

Solution   First note the behavior at $t = 0$ and as $t \to \infty$. At $t = 0$, $x = 0$, and as $t \to \infty$, $x \to 0$. Also for $t > 0$, since $a < b$,

Multiplying the inequality by
$-t < 0$ reverses the inequality.
↓

$$-at > -bt$$

and so $e^{-at} > e^{-bt}$. Thus, for all $t > 0$, $x > 0$. Clearly x must have a maximum somewhere in the range $0 < t < \infty$. To find this maximum, we put $dx/dt = 0$. That is,

$$\frac{dx}{dt} = k(-ae^{-at} - (-b)e^{-bt}) = 0$$

when

$$ae^{-at} = be^{-bt}$$

or

$$ae^{(b-a)t} = be^{(b-b)t} = be^0 = b \qquad \text{We multiplied both sides by } e^{bt}.$$

or

$$e^{(b-a)t} = b/a.$$

Taking natural logarithms,

$$\ln e^{(b-a)t} = \ln \frac{b}{a}$$

or

$$(b-a)t = \ln \frac{b}{a} \qquad \ln e^u = u$$

or

$$t = \frac{1}{b-a} \ln \frac{b}{a}.$$

Also, we need to compute the sign of the second derivative when $e^{(b-a)t} = b/a$:

$$\frac{d^2x}{dt^2} = k(a^2e^{-at} - b^2e^{-bt})$$

$$= ke^{-bt}(a^2e^{(b-a)t} - b^2) \qquad \text{Factor out } e^{-bt}$$

$$= ke^{-bt}\left(a^2\frac{b}{a} - b^2\right) \qquad \text{When } e^{(b-a)t} = b/a$$

$$= ke^{-bt}b(a - b) < 0, \text{ since } b > 0, \ k > 0, \ e^{-bt} > 0, \text{ and } a < b.$$

Thus, we have a maximum value at $t = [1/(b-a)] \ln(b/a)$, and the actual maximum concentration level is given by

$$x_{max} = ke^{-bt}(e^{(b-a)t} - 1) = ke^{-bt}\left(\frac{b}{a} - 1\right) = ke^{-bt}\left(\frac{b-a}{a}\right)$$

for $t = [1/(b-a)] \ln(b/a)$.

Now, at $t = [1/(b - a)] \ln(b/a)$,

$$e^{\ln x} = x \qquad x^{-y} = \frac{1}{x^y} = \left(\frac{1}{x}\right)^y$$

$$e^{-bt} = e^{[-b/(b-a)]\ln(b/a)} = \left[e^{\ln(b/a)}\right]^{-b/(b-a)} = \left(\frac{b}{a}\right)^{-b/(b-a)} = \left(\frac{a}{b}\right)^{b/(b-a)}.$$

Thus

$$x_{\max} = k\,\frac{(b - a)}{a}\left(\frac{a}{b}\right)^{b/(b-a)}.$$

The behavior of the concentration is illustrated in Figure 12. The figure shows that, as time increases and more of the injected drug is absorbed into the bloodstream, the concentration increases. This concentration eventually reaches a maximum level, after which it decays gradually, tending to zero as $t \to \infty$.

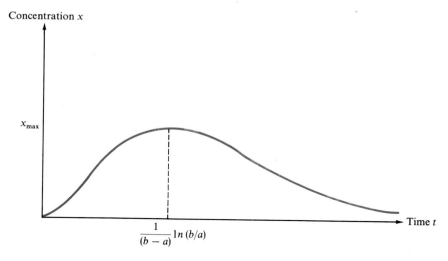

Figure 12

PROBLEMS 11.3 In Problems 1–32 find the maximum and minimum values for the given function over the indicated interval.

1. $f(x) = x^2 + x - 30; [0, 2]$

2. $f(x) = x^2 + x - 30; [-2, 0]$

3. $f(x) = x^3 - 12x + 10; [-10, 10]$

4. $f(x) = 4x^3 - 3x + 2; [-5, 5]$

5. $f(x) = x^5; (-\infty, 1]$

6. $f(x) = x^7; [-1, \infty)$

7. $f(x) = x^3 - 3x^2 - 45x + 25; [-5, 5]$ **8.** $f(x) = \dfrac{x^3}{3} + \dfrac{x^2}{2} - 2x - \dfrac{2}{3}; [-3, 3]$

9. $f(x) = x^{20}; [0, 1]$ **10.** $f(x) = x^4 - 18x^2; [-2, 2]$

11. $f(x) = (x + 1)^{1/3}; [-2, 7]$ **12.** $f(x) = (x - 2)^{2/3}; [-14, 3]$

13. $f(x) = 2x(x + 4)^3; [-1, 1]$ **14.** $f(x) = x\sqrt[3]{1 + x}; [-2, 26]$

15. $f(x) = \dfrac{1}{(x - 1)(x - 2)}; [3, 5]$ **16.** $f(x) = \dfrac{1}{(x - 1)(x - 2)}; [-3, 0]$

17. $f(x) = \dfrac{x - 1}{x - 2}; [-3, 1]$ **18.** $f(x) = \dfrac{1}{x^2 - 1}; [2, 5]$

19. $f(x) = x^{2/3}(x - 3); [-1, 1]$ **20.** $f(x) = (x - 1)^{1/5}; [-31, 33]$

21. $f(x) = 1 + x + \dfrac{1}{2x^2}; [\frac{1}{2}, \frac{3}{2}]$ **22.** $f(x) = \sqrt{x} + \dfrac{1}{\sqrt{x}}; [\frac{1}{2}, \frac{3}{2}]$

23. $f(x) = (1 + \sqrt{x} + \sqrt[3]{x})^9; [0, \infty)$ **24.** $f(x) = \dfrac{1}{x - 2}; [3, \infty)$

25. $f(x) = \begin{cases} x^2, & x \neq 3 \\ 20, & x = 3 \end{cases}; [0, 4]$ ***26.** $f(x) = \begin{cases} x^2, & x \leq 1 \\ x^3, & x \geq 1 \end{cases}; [0, 5]$

27. $f(x) = \ln(1 + x^2); [-2, 3]$ **28.** $f(x) = e^{1 + 2x}; [-1, 1]$

29. $f(x) = \dfrac{e^x}{x}; [1, 2]$ **30.** $f(x) = \dfrac{\ln^2 x}{x}; [1, 4]$

31. $f(x) = \ln(1 + e^x); [0, 1]$ ***32.** $f(x) = x \ln x; [\frac{1}{2}, 4]$

33. A rectangle has a perimeter of 300 m. What length and width will maximize its area?

34. A farmer wishes to set aside one acre of land for corn and wheat. To keep out the cows, the field is enclosed by a fence (costing 50¢ per running foot). In addition, a fence running down the middle of the field is needed, with a cost per foot of $1. Given that 1 acre = 43,560 square feet, what dimensions should the field have so as to minimize the farmer's total cost? The field is rectangular.

35. An isosceles triangle has a perimeter of 24 cm. What lengths for the sides of such a triangle maximize its area?

36. Show that, among all isosceles triangles with a given perimeter, the area enclosed is greatest when the triangle is equilateral.

37. A wire 35 cm long is cut into two pieces. One piece is bent in the shape of a square, and the other is bent in the shape of a circle. How should the wire be cut so as to minimize and maximize the total area enclosed by the pieces?

38. A cylindrical tin can is to hold 50 cm³ of tomato juice. How should the can be constructed in order to minimize the amount of material needed in its construction?

39. The can in Problem 38 costs 4¢ per cm² to construct the sides and 6¢ per cm² to construct the top and bottom. How should the can be constructed to minimize the cost of materials?

40. A rectangular box with a volume of 360 ft³ is to be constructed with a square base and top. The sides cost $1 per square foot, and the top and bottom cost $1.25 per square foot. How should the box be constructed to minimize the total cost?

41. Answer the question in Problem 40 if the top and bottom cost twice as much per square foot as the sides.

42. Find the positive number that exceeds its square by the largest amount.

43. Find the two positive numbers whose sum is 20 having the maximum product.

44. A Transylvanian submarine is traveling due east and heading straight for point P. A battleship from Luxembourg is traveling due south and heading for the same point P. Both ships are traveling at a velocity of 30 km/hr. Initially, their distances from P are 210 km for the submarine and 150 km for the battleship. The range of the submarine's torpedoes is 3 km. How close will the two vessels come? Does the submarine have a chance to torpedo the battleship?

45. Suppose that the rate of population growth of pigeons in a large city is given by

$$R = 400P^2 - \tfrac{1}{5}P^3,$$

where P is the population of pigeons. For what population level is this rate maximized?

46. A clear rectangle of glass is inserted in a colored semicircular glass window (see Figure 13). If the radius of the window is 3 feet and if the clear glass passes twice as much light as the colored glass, find the dimensions of the rectangular insert that passes the maximum light (through the entire window).

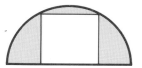

Figure 13

47. Referring to Example 8, suppose that the 1% discount when the number of cars is in excess of 10 is given only to those cars leased in excess of 10. Determine the total revenue function, and find the fleet size that yields maximum revenue.

48. In Problem 47, assume that each car depreciates by $1000 over the year. Find the profit function, and again determine the optimum fleet size.

49. A real estate office handles a large property with 200 apartments. When the rent of each apartment is $400 per month, all apartments are occupied. Experience has shown that for each $20 per month increase in rent, five apartments become vacant. Also, the cost of servicing a rented apartment is $40 a month.

    (a) What rent should the office charge in order to maximize profit?

    (b) How much is that maximum profit?

    (c) How many apartments will be rented at that maximum profit?

50. A steel company knows that if it charges $$x$ a ton it can sell $300 - x$ tons in a single order, up to a maximum of 150 tons. It costs $120 to manufacture each ton. A fixed cost of $5000 is spread equally among each of the $300 - x$ tons produced in a single batch.

    (a) How much should the company charge per ton to maximize its revenue?

    (b) What is that revenue?

    (c) How many tons are sold at the maximum revenue?

51. A manufacturer can sell q items per week at a selling price of $p(q) = 200 - 0.1q$ cents per item, and it costs $C(q) = 50q + 10,000$ cents to produce the batch of q items. How many items should be produced in a single batch to maximize the profit?

52. In a certain truck factory, the total cost of producing q trucks per week is $C(q) = q^2 + 75q + 1000$. How many trucks should be produced to maximize the profit if the number produced is limited by a production capacity of 50 per week and the sale price per truck is

$$p(q) = \begin{cases} \frac{5}{3}(125 - q), & \text{if } q \le 25, \\ \frac{500}{3}, & \text{if } 25 < q \le 50? \end{cases}$$

53. The position of a moving object is given by

$$s(t) = 4t - 6t^2 + 6, t \ge 0.$$

For what value of t is the velocity of the object a maximum?

54. In Problem 53, for what value of t is the acceleration a maximum?

55. A ball is thrown in the air with an initial vertical velocity of 64 ft/sec.

   (a) How high will the ball go (ignoring air resistance)?

   (b) After how many seconds will the ball reach its maximum height?

   [*Hint*: The height is given by the formula $s(t) = 64t - 16t^2$.]

56. The most important function of the human cough is to increase the velocity of the air going out of the windpipe (trachea). Let R_0 denote the "rest radius" of the trachea (that is, the radius when you are relaxed and not coughing; R_0 is measured in cm), let R be the contracted radius of the trachea during a cough ($R < R_0$), and let V be the average velocity of the air in the trachea when it is contracted to R cm. Under some fairly reasonable assumptions regarding the flow of air near the tracheal wall (we assume it is very slow) and the "perfect" elasticity of the wall, we can model the velocity of flow during a cough by the equation

$$V = \alpha(R_0 - R)R^2 \text{ cm/sec},$$

where α is a constant depending on the length of the trachea wall.[†]
   If you are coughing efficiently, your tracheal wall should contract in such a way as to maximize the velocity of air going out of the trachea. Show that V is maximized when the trachea is contracted by one-third of its original radius (so that $R = \frac{2}{3}R_0$). This result has been confirmed, approximately, by x-ray photographs taken during actual coughs.

57. In chemistry, a **catalyst** is defined as a substance that alters the rate of a chemical reaction without itself undergoing a change; the phenomenon is called **catalysis.** If, in a chemical reaction, the product of the reaction serves as a catalyst for the reaction, then the process is called **autocatalysis.** Suppose that in the autocatalytic process we start with an amount A of a given substance. Let x be the amount of the product (that is, the result of the process). It is reasonable to assume that the rate of reaction depends on both the amount of the product, x, and the amount of remaining substance, $A - x$. If this rate is given by

$$R = \alpha x(A - x),$$

where α is a known positive constant, for what concentration x is the rate of reaction greatest?

[†] This equation and a detailed description of this problem appear in Philip Tuchinsky's paper "The Human Cough," UMAP Project, Education Development Center, Newton, Mass., 1978.

58. Psychologists often study how long it takes a subject (animal or human) to learn a given task. Mastery of the task is described by a **learning curve,** which is a function $L = f(t)$. Here t stands for time and L stands for the percentage of the task that is mastered. For example, if $L = 75$, then the subject has mastered 75% of the task.

An electronics parts manufacturer has determined that trainees learn to assemble a given part according to the learning curve

$$L = 10t^2 - 2t^3 + 12t + 1,$$

where t is measured in days. After how many days are the trainees learning most rapidly? [*Hint*: Be careful. First determine which function is to be maximized.]

59. Find the point on the line $y = 2x + 5$ that is nearest to the origin.

60. Find the point on the line $3x + 4y = 12$ that is nearest to the point $(-1, 2)$.

11.4 Marginal Analysis and Elasticity

We saw in Chapter 10 (see Example 10.5.3 on page 552) that the derivative can be used to represent the marginal cost and marginal revenue in producing or selling a given product. The idea of margin can also be applied to other important notions in economics, such as demand, consumption, profit, and savings. The following is a summary of some of the important terms we will need in this section. Other terms will be introduced later.

(a) The **total cost function** gives the cost, C, of producing q units of a given product. A typical cost function is

$$C = aq^2 + bq + c, \qquad (1)$$

where the number c represents the **fixed cost** that will have to be incurred even if nothing is produced (for rent, depreciation, utilities, and so on). Fixed cost is often referred to as **overhead**.

(b) The **total revenue function** gives the amount, R, received for selling q units of the product. A typical revenue function is

$$R = aq + bq^2. \qquad (2)$$

Revenue can often be calculated by multiplying the price times the number of items sold.

(c) The **profit function** gives the profit, P, received when q units of the product are sold. In simple models we will have

$$P = R - C. \qquad (3)$$

(d) The **demand function** expresses the relationship between the unit price that a product can sell for and the number of units that can be sold at that price. Typically, the more units sold, the lower the price, so that if p represents the price per unit sold, dp/dq will be negative. A typical demand function is

$$p = a - bq. \qquad (4)$$

We now give some examples of how these four functions can be used.

Example 1   A toy manufacturer finds that the cost in dollars of producing q copies of a certain doll is given by

$$C = 250 + 3q + 0.01q^2.$$

The dolls can be sold for $14 each. How many should he produce to maximize his profit, assuming that he can sell all he produces?

Solution   Since the price does not vary, the revenue is $14q$ dollars. Thus

$$P = R - C = 14q - (250 + 3q + 0.01q^2)$$
$$= 11q - 0.01q^2 - 250.$$

Then the marginal profit $dP/dq = 11 - 0.02q$, which is equal to 0 when $q = 11/0.02 = 550$ units. Since $d^2P/dq^2 = -0.02$, there is a local maximum at $q = 550$. When $q = 0$, $P = -250$. Also, as q gets very large the profit becomes negative (since the term $0.01q^2$ becomes larger than the q term). Therefore, the maximum profit is $P = 11 \cdot 550 - (0.01)(550)^2 - 250 = 6050 - 3025 - 250 = \$2775$, and the answer to the problem is 550 dolls.

Example 2   A manufacturer of men's shirts figures that her exclusive "Parisian" model will cost $500 for overhead plus $9 for each shirt made. The price she can get for the shirts depends on how exclusive they are. From experience, her accountant has estimated the following demand function:

$$p = 30 - 0.2\sqrt{q},$$

where q is the number of shirts sold. How many shirts should the manufacturer produce in order to maximize her profit? (Assume that all the shirts produced will be sold.) What is the maximum profit?

Solution   From the information given, we have

$$C = 500 + 9q,$$

$$R = (\text{price}) \times (\text{number sold}) = (30 - 0.2\sqrt{q})q = 30q - 0.2q^{3/2},$$

and

$$P = R - C = 21q - 0.2q^{3/2} - 500. \tag{5}$$

Then the marginal profit $dP/dq = 21 - 0.3\sqrt{q}$, which is equal to 0 when $\sqrt{q} = 21/0.3 = 70$ or when $q = 70^2 = 4900$ shirts. Since $d^2P/dq^2 = -0.3/2\sqrt{q}$, which is less than 0, there is a local maximum at $q = 4900$. This is easily seen to be a true maximum as well and is therefore the answer to the problem. At a production level of 4900,

$$P = 21 \cdot 4900 - (0.2)(4900)^{3/2} - 500$$
$$= 102{,}900 - (0.2)(343{,}000) - 500 = \$33{,}800.$$

Note that if 4900 shirts are sold, then they will be sold at a price of $p = 30 - 0.2(70) = 30 - 14 = \$16$ each.

Example 3 A certain manufacturer has a steady demand for 50,000 refrigerators each year. The machines are not made continuously but rather in equally sized batches. Production costs are $10,000 to set up the machinery plus $100 for each refrigerator made. In addition, there is a storage (inventory) charge of $2.50 per year for each refrigerator stored.[†] If the demand is steady throughout the year, how should the manufacturer schedule his production runs so as to minimize his total costs? Assume that production is scheduled so that one new batch is completed just as the previous batch has run out.

Solution We have

$$C = \text{total cost} = \text{production cost} + \text{inventory cost} = PC + IC.$$

If batches are in lots of q refrigerators at a time, then there must be $50,000/q$ runs each year. The production cost will then be

$$PC = \left(\frac{50,000}{q}\right) \cdot 10,000 + (50,000)(100) = \frac{500,000,000}{q} + 5,000,000.$$

Now we calculate inventory costs. Because of the assumption that production is scheduled so that one new batch is completed just as the previous batch has run out, inventory starts at q units and decreases steadily (because of the steady demand) to 0 units. The average number of units in storage at any one time will then be $q/2$ units. Therefore, the inventory costs are

$$IC = 2.50\frac{q}{2} = 1.25q.$$

Then we find that

$$C = \frac{500,000,000}{q} + 5,000,000 + 1.25q.$$

The marginal cost is $dC/dq = -(500,000,000/q^2) + 1.25$. This is zero when

$$\frac{500,000,000}{q^2} = 1.25 \quad \text{or} \quad 1.25q^2 = 500,000,000$$

or when

$$q^2 = \frac{500,000,000}{1.25} = 400,000,000 \quad \text{and} \quad q = \pm 20,000.$$

Of course, the value $q = -20,000$ is meaningless. We can then verify that a minimum is indeed reached when $q = 20,000$. If $q = 20,000$, then $50,000/q = 2\frac{1}{2}$, which means that costs will be minimized when there are $2\frac{1}{2}$ runs a year, which works out (practically) to 5 runs every two years. The minimum annual cost is $5,050,000 (check this).

[†] In business applications, the total cost is often broken down into *fixed costs* and *variable costs*. Thus total cost = fixed cost + variable cost. In this problem the fixed cost is $10,000 and the variable cost is $100 per refrigerator made + $2.50 per refrigerator stored.

Often a person in business will be faced with the choice between increasing or not increasing prices. Generally the demand function will show that an increase in price will cause a drop in sales. But how much of a drop? If there were a very small drop, then revenue would increase. On the other hand, if a large drop in sales were to result, then revenue would fall. The **average price elasticity of demand** is defined as the *relative change* in quantity demanded divided by the relative change in price:

$$\text{average price elasticity of demand} = \eta_{AV} = -\frac{\Delta q/q}{\Delta p/p}. \tag{6}$$

The relative change in q is the change Δq divided by q and likewise for the relative change in p. Note that $\Delta q/q$ can also be thought of as the percentage change in q. The minus sign in equation (6) is put there so that η_{AV} will be positive, since if $\Delta p > 0$, Δq will usually be <0 (why?).

In general, if $\eta_{AV} < 1$, then the percentage decrease in demand is less than the percentage increase in price, so that an increase in price will lead to an increase in revenue. If $\eta_{AV} > 1$, then the opposite is true. If $\eta_{AV} = 1$, then the price increase will not make any difference. The loss in demand will just offset the revenue gained by the increase in price.[†]

Example 4   In Example 2 suppose that the demand function is given by

$$p(q) = 30 - 0.2\sqrt{q}. \tag{7}$$

Calculate the elasticity of demand if the price per shirt is increased from $20 to $22.

Solution   At a price of $20, we calculate from (7) that $20 = 30 - 0.2\sqrt{q}$, $\sqrt{q} = 10/0.2 = 50$, and $q = 2500$. At $p = 22$, $q = 1600$ (verify this). Therefore, we have $p = 20$, $\Delta p = 2$, $q = 2500$, $\Delta q = -900$, and

$$\eta_{AV} = -\frac{(-900)/2500}{2/20} = \frac{90}{25} = 3.6.$$

Since this number is >1, it means that the percentage loss in demand (the numerator of η_{AV}) is greater than the percentage increase in price. Therefore, the increase of $2 would result in a net *decrease* in revenue. It also would result in a net decrease in profits. To verify this, we use the profit formula (5):

$$P(2500) = 21 \cdot 2500 - 0.2(2500)^{3/2} - 500 = \$27,000$$

and

$$P(1600) = 21 \cdot 1600 - 0.2(1600)^{3/2} - 500 = \$20,300.$$

[†] In the terminology of economics, a demand curve is **elastic** if $|\eta_{AV}| > 1$, of **unit elasticity** if $|\eta_{AV}| = 1$, or **inelastic** if $|\eta_{AV}| < 1$.

There is a decrease in profits of $6,700 due to the price increase.

We return to the question whether the person in business should increase or decrease prices. Put another way, will a very small increase in price lead to more or less revenue? To answer this, let $p = p(q)$ represent the demand function. If $p(q)$ is continuous (as it is assumed to be), a small change in price, Δp, will be caused by a small change in demand, Δq. Then we define the **price elasticity of demand** when the demand is q and the price is $p(q)$ as the limit of η_{AV} as $\Delta q \to 0$. That is,

$$\eta(q) = \lim_{\Delta q \to 0} \eta_{AV} = \lim_{\Delta q \to 0} -\frac{\Delta q/q}{\Delta p/p} = \lim_{\Delta q \to 0} -\frac{p}{q\Delta p/\Delta q} = -\frac{p}{qp'}. \tag{8}$$

Example 5   The demand function for a certain electric toaster is $p(q) = 35 - 0.05\sqrt{q}$. Will a rise in price increase or decrease revenue if 10,000 toasters are in demand?

Solution   We calculate η. If $\eta > 1$, then, as before, an increase in price will cause a decrease in revenue; if $\eta < 1$, then an increase in price will cause an increase in revenue. Here

$$p'(q) = -\frac{0.05}{2\sqrt{q}} = -\frac{0.05}{2\sqrt{10,000}} = -\frac{0.05}{200} = -0.00025.$$

When $q = 10,000$, $p = 35 - 0.05\sqrt{10,000} = 30$ and

$$\eta = -\frac{p}{qp'(q)} = \frac{-30}{(10,000)(-0.00025)} = \frac{30}{2.5} = 12.$$

Since $\eta > 1$, there will be a *decrease* in revenue if the price is raised.

Example 6   Answer the question in Example 5 if 250,000 toasters are in demand.

Solution   When $q = 250,000$, $p = 35 - 0.05\sqrt{250,000} = 35 - (0.05)(500) = 10$, $p'(250,000) = -0.05/2(500) = -0.00005$, and

$$\eta = \frac{-10}{(250,000)(-0.00005)} = \frac{-10}{-12.5} = 0.8 < 1.$$

There would be an increase in revenue if prices were increased.

The notion of elasticity of demand has an interesting interpretation in global economics. Suppose that certain economists in the United States are concerned about an imbalance in the balance of payments. That is, more dollars are going out than are coming in. To offset this problem, they suggest a devaluation of the dollar. This would make dollars cheaper abroad, thereby making American goods cheaper to foreign consumers. Therefore, the foreign purchase of U.S. goods would increase, thereby leading to an increase in the number of export dollars flowing back to the United States. Or will something else happen? The economists must be careful. If the elasticity of demand for American exports is less than 1, more American products would indeed be purchased abroad, but there would be a net *decrease* in the value of the dollars paid for these goods. Things are never simple in the area of international trade.

PROBLEMS 11.4

1. The cost of producing q color televison sets is given by $C = 5000 + 250q - 0.01q^2$. The revenue received is $R = 400q - 0.02q^2$. Assuming that all sets produced will be sold, how many should be produced so as to maximize the profit?

2. What is the demand function in Problem 1?

3. In Problem 1, at a production level of 10,000 sets, will an increase in price generate an increase or decrease in revenue?

4. Bottles of whiskey cost a distiller in Scotland $2 a bottle to produce. In addition, he has fixed costs of $500. The demand function worldwide for the whiskey is given by $p = 12 - 0.001q$. How many bottles should he produce to maximize his profit?

5. The distiller of Problem 4 now sells exclusively to the United States. He must pay duty of $0.50 per bottle. The demand function does not change. How does he maximize his profit in this case?

*6. The distiller of Problem 4 shifts his sales to France, where an import duty of 20 percent of the sales price is charged. How does he now maximize his profit?

7. In Problems 4, 5, and 6, determine whether a price increase will result in an increase or decrease in revenue at a production level of 10,000 bottles.

8. Show for any problem of the type we have considered that whenever profit is maximized, the marginal cost and the marginal revenue are equal.

9. A manufacturer of kitchen sinks finds that if he produces q sinks per week, he has fixed costs of $1000, labor and materials costs of $5 per sink, and advertising costs of $100\sqrt{q}$. How many sinks should he manufacture weekly to minimize costs?

 10. The demand function for the sinks of Problem 9 is $p(q) = 25 - 0.01q$. How many sinks should be manufactured weekly to maximize profits?

11. In Problem 10, at a level of production of 2000 sinks, will an increase in price lead to an increase or decrease in revenue?

12. If the demand function for a certain manufactured good is $p(q) = 75 - 0.1\sqrt{q} - 0.002q^2$, at what level of production will it not make any difference in the revenue whether the price is increased or decreased?

13. A Detroit manufacturer has a steady annual demand for 50,000 pickup trucks. The trucks are made in batches. The costs of production include a $20,000 setup cost per batch and a cost of $2500 per truck. Inventory charges are $50 per truck per year. How is production to be scheduled so as to minimize total costs?

14. Suppose that the manufacturer in Problem 13 produces trucks continuously (so that there is only one setup cost) but that demand is not constant and is given by $q = 5000 + (300,000/\sqrt{p})$, where p is the price in dollars and q is the demand. If fewer than 10,000 trucks are sold, the manufacturer will go bankrupt. What should be the price charged for each truck so as to maximize the total profits? [*Hint*: Find the demand function by writing p as a function of q.]

15. In Problem 14 determine the elasticity function.

16. In Problem 14, at a level of production of 25,000 trucks, will it be profitable to increase prices?

17. A woman has $10,000 to invest in two companies. The return from investing q dollars in Company 1 is $4\sqrt{q}$ dollars, and the return from investing q dollars in Company 2 is $2\sqrt{q}$ dollars. How should she invest her money so as to maximize her return?

18. The Op-Pol Company does public opinion polls. Its researchers have observed that the cost of conducting a national survey of n people is

$$C(n) = 25,000 + 0.02(n - 1500)^2.$$

Of course, the more people surveyed, the better are the results (up to a point). The Op-Pol researchers have estimated that the value (in dollars, since better accuracy ensures greater profits) is given by

$$V(n) = 500,000 - 0.1(n - 8000)^2.$$

If "profit" is defined by $P(n) = V(n) - C(n)$, what is the optimal number of people to be polled (to the nearest person)?

*19. Show that, if the demand law is given as $q = q(p)$ (that is, demand as a function of price rather than vice versa), then

$$\eta(p) = -\frac{pq'(p)}{q(p)}.$$

(This expresses the elasticity in terms of price rather than in terms of demand.)

20. Show that if $q(p) = 5/p^4$, then $\eta(p) = 4$.

21. Show in general that if $q(p) = a/p^\alpha$ with $\alpha > 0$, then $\eta(p) = \alpha$.

*22. Show that if $p = p(q)$ is the demand function, then at a level of production that maximizes total revenue, the elasticity is 1.

*23. Let the cost function be $C = aq^2 + bq + c$ and the demand function $p = \alpha - \beta q$, where $a, b, c, \alpha,$ and β are positive.

(a) At what level of production is profit maximized?

(b) What is the elasticity?

(c) At what level of production does it make no difference whether the price is increased or decreased?

*24. Let $C(q)$ be the total cost function for a certain product, and let C_A denote the average cost per unit for producing q units.

(a) Explain why $C_A = C(q)/q$.

(b) Show that when the average cost is decreasing, marginal costs are less than average costs.

(c) Show that when the average cost function has a local maximum or minimum, average cost = marginal cost.

25. A dealer has an approximately constant demand rate of r per year for one of her products. She orders the product from the manufacturer in a batch size of quantity q, say, and wishes to know the optimum batch size in order to minimize her costs. Her costs arise from a set-up cost for each order of amount c_1 and a storage cost of c_2 per unit of stock per unit time. The average stock held is $q/2$, and the number of orders per year will be r/q, giving a total yearly cost of

$$C = \frac{r}{q}c_1 + c_2\frac{q}{2}.$$

Show that for minimum total costs, the optimum batch size is given by

$$q = \sqrt{2rc_1/c_2}.$$

This is known as the **economic order quantity**, or EOQ.

26. The following problem is taken from a CPA exam (November 1975): The mathematical notation for the total cost for a business is $2X^3 + 4X^2 + 3X + 5$, where X equals production volume. Which of the following is the mathematical notation for the marginal cost function for this business?

(a) $2(X^3 + 2X^2 + 1.5X + 2.5)$

(b) $6X^2 + 8X + 3$

(c) $2X^3 + 4X^2 + 3X$

(d) $3X + 5$

11.5 Related Rates of Change (Optional)

We have seen, beginning in Section 10.4, that the derivative can be interpreted as a rate of change. In many problems involving two or more variables it is necessary to calculate the rate of change of one or more of these variables with respect to time. After giving an example, we will suggest a procedure for handling problems of this type.

Example 1   A rope is attached to a pulley mounted on a 15-ft tower. The end of the rope is attached to a heavily loaded cart (see Figure 1). A worker can pull in rope at a rate of 2 ft/sec. How fast is the cart approaching the tower when it is 8 ft from the tower?

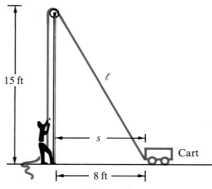

Figure 1

Solution We let s denote the horizontal distance of the cart from the tower and l the length of rope from the top of the tower to the cart, as in Figure 1. Since the speed of the cart is ds/dt, the question asks us to determine ds/dt when $s = 8$ ft. We are told that $dl/dt = 2$. To calculate ds/dt, we must first find a relationship between s and l. From the Pythagorean theorem we immediately obtain

$$15^2 + s^2 = l^2. \tag{1}$$

We now differentiate equation (1) implicitly with respect to t to find that

$$\frac{d}{dt}(15^2) + \frac{d}{dt}(s^2) = \frac{d}{dt}(l^2),$$

or

$$0 + 2s\frac{ds}{dt} = 2l\frac{dl}{dt}$$

and

$$\frac{ds}{dt} = \frac{l}{s}\frac{dl}{dt}. \tag{2}$$

When $s = 8$, $l^2 = 15^2 + 8^2 = 225 + 64 = 289$, and $l = 17$. Then inserting $s = 8$, $l = 17$, and $dl/dt = 2$ into equation (2) gives us

$$\frac{ds}{dt} = \frac{17}{8}(2) = \frac{17}{4} = 4\tfrac{1}{4} \text{ ft/sec.}$$

Thus, the cart is approaching the tower at a rate of $4\tfrac{1}{4}$ ft/sec.

The solution given above suggests that the following steps be taken to solve a problem involving the rates of change of related variables:

> **1.** If feasible, draw a picture of what is going on.
>
> **2.** Determine the important variables in the problem and find an equation relating them.
>
> **3.** Differentiate the equation obtained in step (2) with respect to t.
>
> **4.** Solve for the derivative sought.
>
> **5.** Evaluate that derivative by substituting given and calculated values of the variables in the problem.
>
> **6.** Interpret your answer in terms of the question posed in the problem.

Example 2 An oil storage tank is built in the form of an inverted right circular cone with a height of 6 m and a base radius of 2 m (see Figure 2). Oil is being pumped into the tank at a rate of 2 liters/min = 0.002 m³/min (since 1 m³ = 1000 liters). How fast is the level of the oil rising when the tank is filled to a height of 3 m?

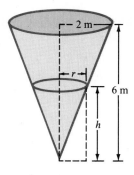

Figure 2

Solution   In mathematical terms, we are asked to calculate dh/dt when $h = 3$ m, where h denotes the height of the oil at a given time and r the radius of the cone of oil (see Figure 2). The volume of a right circular cone is $V = \frac{1}{3}\pi r^2 h$. From the data given in the problem, we see from Figure 2 (using similar triangles) that $h/r = 6/2 = 3$, or $h = 3r$ and $r = h/3$. Then

$$V = \frac{1}{3}\pi\left(\frac{h}{3}\right)^2 h = \frac{1}{27}\pi h^3 = \text{volume of oil at height } h.$$

Differentiating with respect to t and using the fact (given to us) that $dV/dt = 0.002$, we obtain

$$\frac{dV}{dt} = \frac{1}{9}\pi h^2 \frac{dh}{dt} = 0.002 \quad \text{or} \quad \frac{dh}{dt} = \frac{9(0.002)}{\pi h^2} = \frac{0.018}{\pi h^2}\text{ m/min.}$$

Then, for $h = 3$,

$$\frac{dh}{dt} = \frac{0.018}{\pi \cdot 9} = \frac{0.002}{\pi} \approx 6.37 \times 10^{-4}\text{ m/min,}$$

which is the rate at which the height of the oil is increasing.

Example 3   The cost and revenue functions for a children's watch are given by

$$C(q) = 250 + 3q + 0.01q^2$$

and

$$R(q) = 5q + 0.02q^2.$$

Here q represents monthly production. If production is increasing at a rate of 200 watches per month when production is 3000 units, find the rate of increase in profits.

Solution   We seek dP/dt, where $P = R - C$.

We have

$$\frac{dP}{dt} = \frac{dR}{dt} - \frac{dC}{dt} = \left(5\frac{dq}{dt} + 0.04q\frac{dq}{dt}\right) - \left(3\frac{dq}{dt} + 0.02q\frac{dq}{dt}\right)$$

$$= \frac{dq}{dt}(2 + 0.02q).$$

We are given that $dq/dt = 200$. Thus for $q = 3000$, we have

$$\frac{dP}{dt} = 200[2 + 0.02(3000)] = 200(2 + 60) = \$12,400/\text{month}.$$

We interpret this to mean that, if the company is producing 3000 watches and increases production by 200 watches per month, the increase in its profits will be $12,400 per month. This, of course, is under the assumption that the cost and revenue functions do not change when production is increased.

PROBLEMS 11.5 **1.** Let $xy = 6$. If $dx/dt = 5$, find dy/dt when $x = 3$.

2. Let $x/y = 2$. If $dx/dt = 4$, find dy/dt when $x = 2$.

3. A 10-ft ladder is leaning against the side of a house. As the foot of the ladder is pulled away from the house, the top of the ladder slides down along the side (see Figure 3). If the foot of the ladder is pulled away at a rate of 2 ft/sec, how fast is the ladder sliding down when the foot is 8 ft from the house?

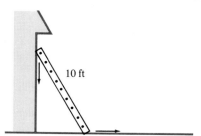

10 ft

Figure 3

4. A cylindrical water tank 6 m high with a radius of 2 m is being filled at a rate of 10 liters/min. How fast is the water rising when the water level is at a height of 0.5 m? [*Hint*: 1 liter $= 1000 \text{ cm}^3$ or $1 \text{ m}^3 = 1000$ liters.]

5. An airplane at a height of 1000 m is flying horizontally at a velocity of 500 km/hr and passes directly over a civil defense observer. How fast is the plane receding from the observer when it is 1500 m away from the observer?

6. Sand is being dropped in a conical pile at a rate of 15 m³/min. The height of the pile is always equal to its diameter. How fast is the height increasing when the pile is 7 m high?

7. When helium expands adiabatically, its pressure is related to its volume by the formula $PV^{1.67} = $ constant. At a certain time, the volume of the helium in a balloon is 18 m³ and the pressure is 0.3 kg/m². If the pressure is increasing at a rate of 0.01 kg/m²/sec, how fast is the volume changing? Is the volume increasing or decreasing?

8. A woman standing on a pier 15 ft above the water is pulling in her boat by means of a rope attached to the boat's bow. She can pull in the rope at a rate of 5 ft/min. How fast is the boat approaching the foot of the pier when the boat is 20 ft away?

9. A baseball player can run at a top speed of 25 ft/sec. A catcher can throw a ball at a speed of 120 ft/sec. The player attempts to steal third base. The catcher (who is 90 ft from third base) throws the ball toward the third baseman when the player is 30 ft from third base. What is the rate of change of the distance between the ball and the runner at the instant the ball is thrown? (See Figure 4.)

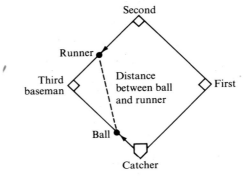

Figure 4

10. At 2 P.M. on a certain day, ship A is 100 km due north of ship B. At that moment, ship A begins to sail due east at a rate of 15 km/hr while ship B sails due north at a rate of 20 km/hr. How fast is the distance between the two ships changing at 5 P.M.? Is it increasing or decreasing?

11. A storage tank is 20 ft long, and its ends are isosceles triangles having bases and altitudes of 3 ft. Water is poured into the tank at a rate of 4 ft^3/min. How fast is the water level rising when the water in the tank is 6 in. deep?

12. Two roads intersect at right angles. A car traveling 80 km/hr reaches the intersection half an hour before a bus that is traveling on the other road at 60 km/hr. How fast is the distance between the car and the bus increasing 1 hr after the bus reaches the intersection? (See Figure 5.)

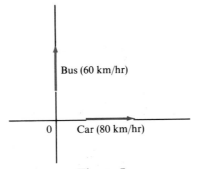

Figure 5

13. A rock is thrown into a pool of water. A circular wave leaves the point of impact and travels so that its radius increases at a rate of 25 cm/sec. How fast is the circumference of the wave increasing when the radius of the wave is 1 m?

14. The body of a snowman is in the shape of a sphere and is melting at a rate of $2 \text{ ft}^3/\text{hr}$. How fast is the radius changing when the body is 3 ft in diameter (assuming that the body stays spherical)?

15. In Problem 14, how fast is the surface area of the body changing when $d = 3$ ft?

*16. Water is leaking out of the bottom of a hemispherical tank with a radius of 6 m at a rate of $3 \text{ m}^3/\text{hr}$. If the tank was full at noon, how fast is the height of the water level in the tank changing at 3 P.M.? [*Hint:* The volume of a segment of a sphere with the radius r is $\pi h^2[r - (h/3)]$, where h is the height of the segment (see Figure 6).]

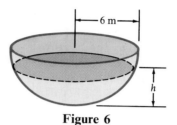

Figure 6

17. A light is affixed to the top of a 12-ft lamppost. If a 6-ft man walks away from the lamppost at a rate of 5 ft/sec, how fast is the length of his shadow increasing when he is 5 ft away?

18. Bacteria grow in circular colonies. The radius of one colony is increasing at a rate of 4 mm/day. On Wednesday, the radius of the colony is 1 mm. How fast is the area of the colony changing one week (that is, seven days) later?

19. A pill is in the shape of a right circular cylinder with a hemisphere on each end. The height of the cylinder (excluding its hemispherical ends) is half its radius. What is the rate of change of the volume of the pill with respect to the radius of the cylinder?

20. A spherical mothball is dissolving at a rate of 8π cc/hr (1 cc = 1 cubic centimeter). How fast is the radius of the mothball decreasing when the radius is 3 cm?

21. In Example 3, how is the profit changing if production is cut by 150 units per month when 3000 units are being produced?

22. The cost of producing q color television sets is given by $C = 5000 + 250q - 0.01q^2$. The revenue received is $R = 400q - 0.02q^2$. At a level of production of 200 TV sets per week, how is profit changing if production is increased by 10 sets per week?

23. Answer the question of Problem 22 if production is cut by 5 sets per week.

11.6 Differential Equations of Exponential Growth and Decay

In this section we begin to illustrate the great importance of the exponential function e^x in applications. Before citing examples, we will discuss a very basic type of mathematical model.

Let $y = f(x)$ represent some physical quantity such as the volume of a substance, the population of a certain species, the mass of a decaying radioactive substance, the number of dollars invested in bonds, and so on. Then the growth of $f(x)$ is given by its derivative dy/dx. Thus if $f(x)$ is growing at a constant rate, $dy/dx = k$ and $y = kx + C$; that is, $y = f(x)$ is a straight-line function.

It is sometimes more interesting and more appropriate to consider the **relative rate of growth,** defined by

$$\begin{array}{l} \text{relative rate} \\ \text{of growth} \end{array} = \frac{\text{actual rate of growth}}{\text{size of } f(x)} = \frac{f'(x)}{f(x)} = \frac{dy/dx}{y}.$$

The relative rate of growth indicates the percentage increase or decrease in f. For example, an increase of 100 individuals for a species with a population size of 500 would probably have a significant impact, being an increase of 20 percent. On the other hand, if the population were 1,000,000, then the addition of 100 would hardly be noticed, being an increase of only 0.01 percent.

In many applications we are told that the relative rate of growth of the given physical quantity is constant. That is,

$$\frac{dy/dx}{y} = \alpha$$

or

$$\frac{dy}{dx} = \alpha y \tag{1}$$

where α is the constant percentage increase or decrease in the quantity.

Another way to view equation (1) is that it tells us that *the function is changing at a rate proportional to itself.* If the constant of proportionality α is greater than 0, the quantity is increasing; if $\alpha < 0$, it is decreasing. Equation (1) is called a **differential equation** because it is an equation involving a derivative. Differential equations arise in a great variety of settings, as we shall soon see.

A **solution** to a differential equation is a differentiable function or set of functions that satisfies the equation. We now give you the solution of the differential equation (1):

$$y = f(x) = ce^{\alpha x} \tag{2}$$

for any real number c. Also, it may be shown that any solution to (1) is of the form (2).

Let us verify that $y = ce^{\alpha x}$ solves the equation

$$\frac{dy}{dx} = \alpha y.$$

If $y = ce^{\alpha x}$, then

See equation (10.6.5) on page 558.

See equation (10.11.5) on page 585.   Remember that $y = ce^{\alpha x}$.

$$\frac{dy}{dx} = \frac{d}{dx}\,ce^{\alpha x} = c\,\frac{d}{dx}\,e^{\alpha x} = ce^{\alpha x}\,\frac{d}{dx}\,\alpha x = ce^{\alpha x}(\alpha) = \alpha(ce^{\alpha x}) = \alpha y.$$

If $\alpha > 0$, we say that the quantity described by $f(x)$ is **growing exponentially.** If $\alpha < 0$, it is **decaying exponentially** (see Figure 1). Of course, if $\alpha = 0$, then there is no growth and $f(x)$ remains constant.

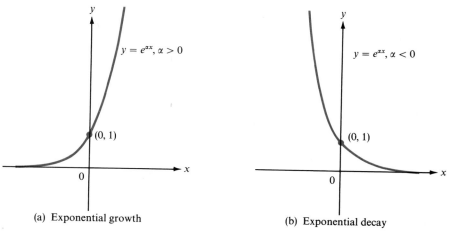

(a) Exponential growth   (b) Exponential decay

Figure 1

For a physical problem it would not make sense to have an infinite number of solutions. We can usually get around this difficulty by specifying the value of y for one particular value of x; say $y(x_0) = y_0$. This is called an **initial condition,** and it gives us a unique solution to the problem. We will see this illustrated in the examples that follow.

Example 1   (a) Find all solutions to $dy/dx = 3y$. (b) Find the solution that satisfies the initial condition $y(0) = 2$.

Solution   (a) Since $\alpha = 3$, all solutions are of the form

$$y = ce^{3x}.$$

(b) $2 = y(0) = ce^{3 \cdot 0} = c \cdot 1 = c$, so that $c = 2$ and the unique solution is $y = 2e^{3x}$.

The situation is illustrated in Figure 2. We can see that, although there are indeed an infinite number of solutions, only one passes through the point $(0, 2)$.

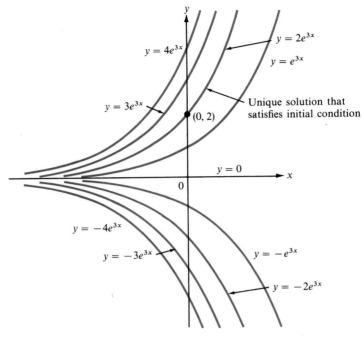

Figure 2

The problem

$$y' = \alpha y \tag{3}$$

$$y(0) = y_0 \tag{4}$$

is called an **initial value problem**. The solution to the initial value problem (3), (4) is given by

$$\boxed{y(x) = y_0 e^{\alpha x},} \tag{5}$$

since when $x = 0$,

From equation (2)   Since $e^0 = 1$
$$y(0) = ce^{\alpha 0} = ce^0 = c.$$

 Example 2   The population of a certain city grows continuously at a rate of 6% a year. If the population in 1980 was 250,000, what will the population be in 1990? in 2010?

Solution   The words "rate of 6% a year" mean that the growth of the population is equal to 6% of the population. If $P(t)$ denotes the population in t

years after the initial year 1980, then

Population growth   6%

$$\frac{dP}{dt} = 0.06P(t). \tag{6}$$

The solution to equation (6) is, from (5), given by

Population in 1980 (initial year)

$$P(t) = P(0)e^{0.6t} = 250,000e^{0.6t}.$$

Since 1990 is 10 years after 1980, we have

$$\text{population in } 1990 = P(10) = 250,000e^{(0.06)10} = 250,000e^{0.6}$$
$$= 250,000(1.8221188) = 455,530.$$

Similarly,

$$\text{population in } 2010 = P(30) = 250,000e^{(0.06)30} = 250,000e^{1.8}$$
$$= 250,000(6.049647464) = 1,512,412.$$

In both answers we rounded to the nearest integer.

Example 3   The population of a city grows continuously by a fixed percentage each year. If the population was 500,000 in 1950 and 750,000 in 1975,

(a)   what is the annual rate of growth?

(b)   what is the projected population in 1990?

Solution   (a) As in Example 2, we have

$$\frac{dP}{dt} = \alpha P(t), \tag{7}$$

where α is the (unknown) rate of growth. The solution to (7) is, with 1950 as our initial year,

$$P(t) = P(0)e^{\alpha t} = 500,000e^{\alpha t}.$$

The year 1975 corresponds to $t = 25$. Thus, we have

$$P(25) = 750,000 = 500,000e^{25\alpha}$$

or

$$e^{25\alpha} = \frac{750,000}{500,000} = 1.5,$$

and taking natural logarithms,

$\ln e = 1$

$$25\alpha \ln e = 25\alpha = \ln 1.5 = 0.4055$$

or

$$\alpha = \frac{\ln 1.5}{25} = \frac{0.4055}{25} \approx 0.0162.$$

Thus, the population is growing continuously at a rate of approximately 1.62% a year.

(b) The year 1990 corresponds to $t = 40$. Thus

$$\text{population in 1990} = P(40) \approx 500{,}000e^{(0.0162)(40)}$$
$$= 500{,}000e^{0.648} \approx 500{,}000(1.9117) = 955{,}850.$$

 Example 4 **Carbon Dating.** **Carbon dating** is a technique used by archaeologists, geologists, and others who want to estimate the ages of certain artifacts and fossils they uncover. The technique is based on certain properties of the carbon atom. In its natural state, the nucleus of the carbon atom C^{12} has 6 protons and 6 neutrons. An **isotope** of carbon C^{12} is C^{14}, which has 2 additional neutrons in its nucleus. C^{14} is **radioactive.** That is, it emits neutrons until it reaches the stable state C^{12}. We make the assumption that the ratio of C^{14} to C^{12} in the atmosphere is constant. This assumption has been shown experimentally to be valid because, although C^{14} is being constantly lost through **radioactive decay** (as this process is often termed), new C^{14} is constantly being produced by the cosmic bombardment of nitrogen in the upper atmosphere. Living plants and animals do not distinguish between C^{12} and C^{14}, so at the time of death, the ratio of C^{12} to C^{14} in an organism is the same as the ratio in the atmosphere. However, this ratio changes after death, since C^{14} is converted to C^{12} but no further C^{14} is taken in.

It has been observed that C^{14} decays at a rate proportional to its mass and that its **half-life** is approximately 5580 years.[†] That is, if a substance starts with 1 g of C^{14}, then 5580 years later it would have $\frac{1}{2}$ g of C^{14}, the other $\frac{1}{2}$ g having been converted to C^{12}.

We may now pose a question typically asked by an archaeologist: A fossil is unearthed and it is determined that the amount of C^{14} present is 40% of what it would be for a similarly sized living organism. What is the approximate age of the fossil?

Solution　Let $M(t)$ denote the mass of C^{14} present in the fossil. Since C^{14} decays at a rate proportional to its mass, we have

$$M(t) = M(0)e^{-\alpha t},$$

[†] This number was first determined in 1941 by the American chemist W. S. Libby, who based his calculations on the wood from sequoia trees, whose ages were determined by rings marking years of growth. Libby's method has come to be regarded as the archaeologist's absolute measuring scale. But in truth, this scale is flawed. Libby used the assumption that the atmosphere at all times had a constant amount of C^{14}. Recently the American chemist C. W. Ferguson of the University of Arizona deduced from his study of tree rings in 4000-year-old American giant trees that objects dated to have lived before 1500 B.C. were much older than previously considered, because Libby's "clock" allowed for a greater amount of C^{14} than actually was present. For example, a find dated at 1800 B.C. was in fact from 2500 B.C. This fact has had a considerable impact on the study of prehistoric times. For a fascinating discussion of this subject, see Gerhard Herm, *The Celts* (New York: St. Martin's Press, 1975), pp. 90–92.

where α is the constant of proportionality and $M(0)$ is the initial amount of C^{14} present. When $t = 5580$ years, $M(5580) = \frac{1}{2}M(0)$, since half the original amount of C^{14} has been converted to C^{12}. We can use this fact to solve for α, since we have

$$\tfrac{1}{2}M(0) = M(0)e^{-\alpha 5580} \qquad \text{or} \qquad e^{-5580\alpha} = \tfrac{1}{2}.$$

Taking natural logarithms,

$$-5580\alpha = \ln \tfrac{1}{2} = -\ln 2,$$

which yields

$$\alpha = \frac{\ln 2}{5580}.$$

Thus

$$M(t) = M(0)e^{-(\ln 2/5580)t}.$$

Now we are told that, after t years (from the death of the fossilized organism to the present), $M(t) = 0.4M(0)$. We are asked to determine t. Then

$$0.4M(0) = M(0)e^{-(\ln 2/5580)t} \qquad \text{or} \qquad 0.4 = e^{-(\ln 2/5580)t}.$$

Again taking natural logarithms, we obtain

$$\ln 0.4 = \frac{-\ln 2}{5580}\,t,$$

or

$$t = \frac{-5580 \ln 0.4}{\ln 2} = \frac{(-5580)(-0.9163)}{(0.6931)} \approx 7376 \text{ years.}$$

Example 5 **Newton's Law of Cooling.** **Newton's law of cooling** states that the rate of change of the temperature difference between an object and its surrounding medium is proportional to the temperature difference. If $D(t)$ denotes this temperature difference at time t and if α denotes the constant of proportionality, then

$$dD/dt = -\alpha D.$$

The minus sign indicates that this difference decreases. (If the object is cooler than the surrounding medium—usually air—it will warm up; if it is hotter, it will cool.) The solution to this differential equation is

$$D(t) = ce^{-\alpha t}.$$

If we denote the initial ($t = 0$) temperature difference by D_0, then

$$D(t) = D_0 e^{-\alpha t}$$

is the formula for the temperature difference for any $t > 0$. Notice that for t large, $e^{-\alpha t}$ is very small so that, as we have all observed, temperature differences tend to die out rather quickly.

We now may ask: In terms of the constant α, how long does it take for the temperature difference to decrease to half its original value?

Solution   The original value is D_0. We are therefore looking for a value of t for which $D(t) = \frac{1}{2}D_0$. That is, $\frac{1}{2}D_0 = D_0 e^{-\alpha t}$, or $e^{-\alpha t} = \frac{1}{2}$. Taking natural logarithms, we obtain

$$-\alpha t = \ln \tfrac{1}{2} = -\ln 2 = -0.6931 \quad \text{and} \quad t = \frac{0.6931}{\alpha}.$$

Notice that this value of t does *not* depend on the initial temperature difference D_0.

Example 6   With the air temperature equal to 30°C, an object with an initial temperature of 10°C warmed to 14°C in one hour.

(a)   What was its temperature after 2 hr?

(b)   After how many hours was its temperature 25°C?

Solution   Let $T(t)$ denote the temperature of the object. Then $D(t) = 30 - T(t) = D_0 e^{-\alpha t}$, from equation (5). But $D_0 = 30 - T(0) = 30 - 10 = 20$, so that

$$D(t) = 20e^{-\alpha t}.$$

We are given that $T(1) = 14$, so that $D(1) = 30 - T(1) = 16$ and

$$16 = D(1) = 20e^{-\alpha \cdot 1} = 20e^{-\alpha} \quad \text{or} \quad e^{-\alpha} = 0.8.$$

Taking natural logarithms,

$$-\alpha = \ln 0.8 = -0.223.$$

Thus,

$$D(t) = 30 - T(t) = 20e^{-0.223t} \quad \text{and} \quad T(t) = 30 - 20e^{-0.223t}.$$

We can now answer the two questions:

(a) $T(2) = 30 - 20e^{-(0.223)\cdot 2} = 30 - 20e^{-0.446} \approx 17.2°C.$

(b)   We need to find t such that $T(t) = 25$. That is,

$$25 = 30 - 20e^{-(0.223)t} \quad \text{or} \quad e^{-0.223t} = \tfrac{1}{4},$$

$$-0.223t = \ln \tfrac{1}{4} = -\ln 4 = -1.3863,$$

so that

$$t = \frac{1.3863}{0.223} \approx 6.2 \text{ hr} = 6 \text{ hr } 12 \text{ min.}$$

Example 7   We can now derive the compound interest formula with continuous compounding (see page 498). The value of an original principal, P, after t years with an interest rate of i compounded continuously is

$$\boxed{A(t) = Pe^{it}.}$$

Recall the simple interest formula (see page 433):

$$A(t) = Pit. \tag{8}$$

In the time period t to $t + \Delta t$, the interest earned is $A(t + \Delta t) - A(t)$. If Δt is small, then the interest paid on $A(t)$ dollars would be (from formula (8)) approximately equal to $A(t)\Delta ti$. We say "approximately" because $A(t)\Delta ti$ represents simple interest between t and $t + \Delta t$. However, the difference between this approximation and the actual interest paid is small if Δt is small. Thus

$$A(t + \Delta t) - A(t) \approx A(t)\Delta ti,$$

or dividing by Δt,

$$\frac{A(t + \Delta t) - A(t)}{\Delta t} \approx iA(t).$$

Then, taking the limit as $\Delta t \to 0$, we obtain

$$\frac{dA}{dt} = iA(t).$$

But then, from (5),

$$A(t) = A(0)e^{it}. \tag{9}$$

But $A(0) = P$, the original principal. Thus (9) becomes

$$A(t) = Pe^{it}.$$

PROBLEMS 11.6   In Problems 1–6 find all solutions to the given differential equations.

1. $\dfrac{dy}{dx} = 3y$

2. $\dfrac{dx}{dt} = 0.1x; x(0) = 2$

3. $\dfrac{dp}{dt} = -p$

4. $\dfrac{dy}{dx} = -\dfrac{1}{2}y$

5. $\dfrac{dx}{dt} = x; x(0) = 5$

6. $\dfrac{dP}{dt} = -\dfrac{P}{10}$

 The following problems require the use of a calculator.

7. The growth rate of a bacteria population is proportional to its size. Initially the population is 10,000, and after 10 days its size is 25,000. What is the population size after 20 days? after 30 days?

8. In Problem 7, suppose instead that the population after 10 days is 6000. What is the population after 20 days? after 30 days?

9. The population of a certain city grows 6% a year. If the population in 1970 was 250,000, what would be the population in 1980? in 2000?

10. The population of a certain city grows at a rate of 1.2% a year. If the population was 600,000 in 1950, what was it in 1970? What will it be in the year 2000?

11. In what year will the city in Problem 10 have a population of 1,500,000 if the 1.2% growth rate continues indefinitely?

12. The population of a certain city is declining at a rate of 3% a year. If the population was 400,000 in 1975, what will its population be in 1990? in 2000?

13. In what year will the city in Problem 12 have a population of 200,000, assuming that the rate of decline continues indefinitely?

14. In what year will the population of the city in Problem 12 fall below 100,000?

15. When the air temperature is 70°F, an object cools from 170°F to 140°F in $\frac{1}{2}$ hr.

    (a) What will the temperature be after 1 hr?

    (b) When will the temperature be 90°F? [*Hint*: Use Newton's law of cooling.]

16. A hot coal (temperature 150°C) is immersed in ice water (temperature -10°C). After 30 seconds the temperature of the coal is 60°C. Assuming that the ice water is kept at -10°C,

    (a) what is the temperature of the coal after 2 min?

    (b) when will the temperature of the coal be 0°C?

17. A fossilized leaf contains 70% of a "normal" amount of C^{14}. How old is the fossil?

18. Forty percent of a radioactive substance disappears in 100 years.

    (a) What is its half-life?

    (b) After how many years will 90% be gone?

19. Radioactive beryllium is sometimes used to date fossils found in deep-sea sediment. The decay of beryllium satisfies the equation

    $$\frac{dA}{dt} = -\alpha A,$$

    where $\alpha = 1.5 \times 10^{-7}$. What is the half-life of beryllium?

20. In a certain medical treatment a tracer dye is injected into the pancreas to measure its function rate. A normally active pancreas will secrete 4% of the dye each minute. A physician injects 0.3 g of the dye, and 30 minutes later 0.1 g remains. How much dye would remain if the pancreas were functioning normally?

21. Atmospheric pressure is a function of altitude above sea level and satisfies the differential equation $dP/da = \beta P$, where β is a constant. The pressure is measured in millibars. At sea level ($a = 0$), $P(0)$ is 1013.25 millibars (mb), which means that the atmosphere at sea level will support a column of mercury 1013.25 mm high at a standard temperature of 15°C. At an altitude of $a = 1500$ m, the pressure is 845.6 mb.

    (a) What is the pressure at $a = 4000$ m?

    (b) What is the pressure at 10 km?

(c) In California, the highest and lowest points are Mount Whitney (4418 m) and Death Valley (86 m below sea level). What is the difference in their atmospheric pressures?

(d) What is the atmospheric pressure at Mount Everest (elevation 8848 m)?

(e) At what elevation is the atmospheric pressure equal to 1 mb?

22. A bacteria population is known to grow exponentially. The data in Table 1 were collected.

TABLE 1

Number of days	Number of bacteria
5	936
10	2190
20	11,986

(a) What was the initial population?

(b) If the present growth rate were to continue, what would be the population after 60 days?

23. A bacteria population is declining exponentially. The data in Table 2 were collected.

TABLE 2

Number of hours	Number of bacteria
12	5969
24	3563
48	1269

(a) What was the initial population?

(b) How many bacteria are left after 1 week?

(c) When will there be no bacteria left (that is, $P(t) < 1$)?

11.7 Newton's Method for Solving Equations (Optional)

In this section we look at a very different kind of application of the derivative. Consider the equation

$$f(x) = 0, \tag{1}$$

where f is assumed to be differentiable in some interval (a, b). It is often important to calculate the *roots* of equation (1), that is, the values of x that satisfy the equation. For example, if $f(x)$ is a polynomial of degree 5, say, then the roots of $f(x)$ could be as in Figure 1.

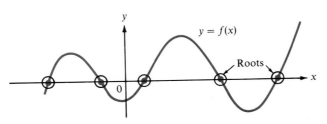

Figure 1

In the seventeenth century Newton discovered a method for estimating a solution or root by defining a sequence of numbers that become successively closer and closer to the root sought. His method is best illustrated graphically. Let $y = f(x)$, as in Figure 2. A number, x_0, is chosen arbitrarily. We then

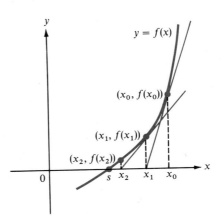

Figure 2

locate the point $(x_0, f(x_0))$ on the graph and draw the tangent line to the curve at that point. Next, we follow the tangent line down until it hits the x-axis. The point of intersection of the tangent line and the x-axis is called x_1. We then repeat the process to arrive at the next point, x_2. On the graph, we have labeled the solution to $f(x) = 0$ as s. That is, $f(s) = 0$. For our graph at least, it seems as if the points $x_0, x_1, x_2, \ldots$ are approaching the point $x = s$. In fact, this happens for quite a few functions, and the rate of approach to the solution is quite rapid.

Having briefly looked at a graphical representation of Newton's method, let us next develop a formula for giving us x_1 from x_0, x_2 from x_1, and so forth.

The slope of the tangent line at the point $(x_0, f(x_0))$ is $f'(x_0)$. Two points on this line are $(x_1, 0)$ and $(x_0, f(x_0))$. Therefore,

$$\frac{0 - f(x_0)}{x_1 - x_0} = f'(x_0). \tag{2}$$

Solving equation (2) for x_1 gives us

$$x_1 = x_0 - \frac{f(x_0)}{f'(x_0)}.$$

Similarly,

$$x_2 = x_1 - \frac{f(x_1)}{f'(x_1)}.$$

In general, we obtain

$$\boxed{x_{n+1} = x_n - \frac{f(x_n)}{f'(x_n)}.} \tag{3}$$

This last step tells us how to obtain the $(n + 1)$st point if the nth point is given, as long as $f'(x_n) \neq 0$ so that (3) is defined. Thus, if we start with a given value x_0, we can obtain $x_1, x_2, x_3, x_4, \ldots$. The formula (3) is called **Newton's formula**. The set of numbers $x_0, x_1, x_2, x_3, \ldots$ is called a **sequence**. If the numbers in the sequence get closer and closer to a certain number, s, as n gets larger and larger, then we say that the sequence **converges** to s and we write

$$\lim_{n \to \infty} x_n = s.$$

There is a theorem that tells us when Newton's method works. We will not discuss that theorem here except to say that, in most cases of interest, Newton's method works if our initial number x_0 is reasonably close to the solution of the equation. One example of the failure of Newton's method is given in Problem 12.

Example 1 Let $r > 1$. Formulate a rule for calculating the square root of r.

Solution We must find an x such that $x = \sqrt{r}$ or $x^2 = r$ or $x^2 - r = 0$. Let $f(x) = x^2 - r$. Then if $f(s) = 0$, s will be a square root of r. ($f(s) = 0$ means that $s^2 - r = 0$ or $s^2 = r$.) By Newton's formula, since $f'(x) = 2x$, we obtain the sequence $x_0, x_1, x_2, \ldots$ where x_0 is arbitrary and

$$x_{n+1} = x_n - \frac{f(x_n)}{f'(x_n)} = x_n - \frac{(x_n^2 - r)}{2x_n} = \frac{2x_n^2 - x_n^2 + r}{2x_n} = \frac{1}{2}\left(x_n + \frac{r}{x_n}\right). \tag{4}$$

Example 2 Calculate $\sqrt{2}$ by Newton's method.

Solution In formula (4), $r = 2$, so that

$$x_{n+1} = \frac{1}{2}\left(x_n + \frac{2}{x_n}\right).$$

Using a calculator, we obtain the sequence in Table 1 starting with $x_0 = 1$. We can see here the remarkable accuracy of Newton's method. An answer correct to 9 decimal places was obtained after only four steps! We were limited in accuracy only by the fact that our calculator could display only 10 digits.

TABLE 1

n	x_n	$\dfrac{2}{x_n}$	$x_n + \dfrac{2}{x_n}$	$x_{n+1} = \dfrac{1}{2}\left(x_n + \dfrac{2}{x_n}\right)$
0	1.0	2.0	3.0	1.5
1	1.5	1.333333333	2.833333333	1.416666667
2	1.416666667	1.411764706	2.828431373	1.414215686
3	1.414215686	1.414211438	2.828427125	1.414213562
4	1.414213562	1.414213562	2.828427125	1.414213562

In Example 2 we stopped when there was no change in going from x_3 to x_4. This illustrates a useful rule of thumb:

> In using Newton's method, stop when two successive iterates are the same.

Example 3   Formulate a rule for calculating the kth root of a given number r.

Solution We must find an x such that $x = r^{1/k}$ or $x^k = r$ or $f(x) = x^k - r = 0$. Then $f'(x) = kx^{k-1}$ and

$$x_{n+1} = x_n - \frac{x_n^k - r}{kx_n^{k-1}} = x_n - \frac{1}{k}\frac{x_n^k}{x_n^{k-1}} + \frac{r}{kx_n^{k-1}} = \left(1 - \frac{1}{k}\right)x_n + \frac{r}{kx_n^{k-1}}. \quad (5)$$

Example 4   Calculate $\sqrt[3]{17}$.

Solution By formula (5), with $k = 3$ and $r = 17$, we have $x_{n+1} = \frac{2}{3}x_n + 17/3x_n^2$. Values of x_n are tabulated in Table 2. The last number is correct to 9 decimal places. Again, the rapid convergence of Newton's method is illustrated. Note that we stopped because the last two iterates (x_4 and x_5) were equal.

TABLE 2

n	x_n	$\dfrac{2}{3}x_n$	x_n^2	$\dfrac{17}{3x_n^2}$	$x_{n+1} = \dfrac{2}{3}x_n + \dfrac{17}{3x_n^2}$
0	2.0	1.333333333	4.0	1.416666667	2.75
1	2.75	1.833333333	7.5625	0.7493112948	2.582644628
2	2.582644628	1.721763085	6.670053275	0.8495684267	2.571331512
3	2.571331512	1.714221008	6.611745745	0.8570605836	2.571281592
4	2.571281592	1.714187728	6.611489025	0.8570938627	2.571281591
5	2.571281591	1.714187727	6.611489020	0.8570938633	2.571281591

Example 5 Find the real roots of $p(x) = x^3 + x^2 + 7x - 3$.

Solution We differentiate to find that $p'(x) = 3x^2 + 2x + 7 = 3(x^2 + \frac{2}{3}x + \frac{7}{3}) = 3[(x + \frac{1}{3})^2 - \frac{1}{9} + \frac{7}{3}] = 3[(x + \frac{1}{3})^2 + \frac{20}{9}] > 0$, so that $p(x)$ is an increasing function. There are no critical points. Also, $p''(x) = 6x + 2 = 0$ when $x = -\frac{1}{3}$, so that $(-\frac{1}{3}, -\frac{142}{27})$ is a point of inflection. The graph of $p(x)$ is given in Figure 3. From the graph, we see that there is exactly one real root. We have

$$x_{n+1} = x_n - \frac{p(x_n)}{p'(x_n)} = x_n - \frac{x_n^3 + x_n^2 + 7x_n - 3}{3x_n^2 + 2x_n + 7}$$

$$= \frac{x_n(3x_n^2 + 2x_n + 7) - (x_n^3 + x_n^2 + 7x_n - 3)}{3x_n^2 + 2x_n + 7} = \frac{2x_n^3 + x_n^2 + 3}{3x_n^2 + 2x_n + 7}.$$

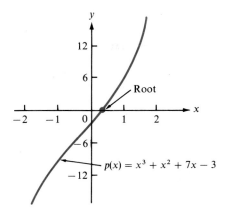

Figure 3

If we choose $x_0 = 0$, we obtain the results in Table 3. The root is $s = 0.3970992165$, correct to 10 decimal places.

TABLE 3

n	x_n	$2x_n^3 + x_n^2 + 3$	$3x_n^2 + 2x_n + 7$	$x_{n+1} = \dfrac{2x_n^3 + x_n^2 + 3}{3x_n^2 + 2x_n + 7}$
0	0	3.0	7.0	0.4285714286
1	0.4285714286	3.341107872	8.408163265	0.3973647712
2	0.3973647712	3.283385572	8.268425827	0.3970992352
3	0.3970992352	3.282923214	8.267261878	0.3970992164
4	0.3970992164	3.282923182	8.267261796	0.3970992165
5	0.3970992165	3.282923182	8.267261796	0.3970992165

The next example makes use of the material in Section 8.3.

Example 6   A new boat costs $4000. If you purchase it in 36 monthly install-
ments of $150 per month, what is the effective annual interest you are paying?

Solution   The dealer is receiving an annuity of $150 per month in exchange
for a present value of $4000. Thus, in formula (8.3.6) on page 450,

$$A_0 = 4000 = \frac{150}{i}\left[1 - \frac{1}{(1 + i)^{36}}\right].$$

Rewriting, we obtain

$$4000i = 150 - \frac{150}{(1 + i)^{36}}, \qquad 4000i(1 + i)^{36} - 150(1 + i)^{36} + 150 = 0,$$

or dividing by 50,

$$80i(1 + i)^{36} - 3(1 + i)^{36} + 3 = 0.$$

There is, of course, no formula for finding the roots of this 37th-degree
polynomial.
   However, we can estimate the root i by Newton's method. (Note that $i = 0$
is a root. We thus seek the first positive root.) Defining $P(i) = 80i(1 + i)^{36} - 3(1 + i)^{36} + 3$, we have

$$P'(i) = 80(1 + i)^{36} + (80 \cdot 36)i(1 + i)^{35} - 108(1 + i)^{35},$$

and Newton's method yields the sequence

$$i_{n+1} = i_n - \frac{P(i_n)}{P'(i_n)}$$

$$= i_n - \left[\frac{80i_n(1 + i_n)^{36} - 3(1 + i_n)^{36} + 3}{80(1 + i_n)^{36} + 2880i_n(1 + i_n)^{35} - 108(1 + i_n)^{35}}\right].$$

Dividing numerator and denominator by $(1 + i_n)^{35}$ to simplify, we obtain

$$i_{n+1} = i_n - \left[\frac{80i_n(1 + i_n) - 3(1 + i_n) + (3/(1 + i_n)^{35})}{80(1 + i_n) + 2880i_n - 108}\right]$$

$$= i_n - \left[\frac{80i_n^2 + 77i_n - 3 + (3/(1 + i_n)^{35})}{2960i_n - 28}\right].$$

Starting with $i_0 = 0.02$ (representing a monthly rate of 2%), we obtain the values in Table 4. The required interest rate is $i \approx 0.0172 = 1.72\%$ a month (correct to 4 decimal places). The annual interest is $1.72 \cdot 12 = 20.64\%$. You would be better off borrowing the money from a bank!

TABLE 4

i_n	$\dfrac{3}{(1 + i_n)^{35}}$	$\begin{array}{c}A,\\80i_n^2 + 77i_n\\-3 + \dfrac{3}{(1+i_n)^{35}}\end{array}$	$\begin{array}{c}B,\\2960i_n - 28\end{array}$	$\dfrac{A}{B}$	$i_{n+1} = i_n - \dfrac{A}{B}$
0.02	1.50008284	0.0720828401	31.2	0.0023103474	0.0176896526
0.0176896526	1.62399129	0.0111284445	24.3613717	0.0004568069	0.0172328456
0.0172328456	1.649712021	0.0003988095	23.00922298	0.0000173325	0.017215513
0.017215513	1.650696151	0.0000005635	22.9579185	0.0000000245	0.0172154884
0.0172154884	1.650697549	-0.0000000014	22.95784566	-0.0000000001	0.0172154885

PROBLEMS 11.7 In the problems below, calculate all answers to 4 decimal places of accuracy if you do not have a hand calculator. If you have one, calculate your answers to as many decimal places of accuracy as are displayed on the machine.

1. Calculate $\sqrt[4]{25}$ using Newton's method.

2. Calculate $\sqrt[5]{10}$ using Newton's method.

3. Calculate $\sqrt[6]{100}$ using Newton's method.

4. Use Newton's method to calculate the roots of $x^2 - 7x + 5 = 0$. It will be necessary to do two separate calculations using two distinct intervals. Compare this with the answers obtained by the quadratic formula.

5. Find all solutions of the equation $x^3 - 6x^2 - 15x + 4 = 0$. [*Hint*: Draw a sketch and estimate the roots to the nearest integer. Then use this estimate as your initial choice of x_0 for each of the three roots.]

6. Find all solutions of the equation $x^3 + 14x^2 + 60x + 105 = 0$.

7. Find all solutions of $x^3 - 8x^2 + 2x - 15 = 0$.

*8. Find all solutions of $x^3 + 3x^2 - 24x - 40 = 0$.

9. Using Newton's method, find a formula for finding reciprocals without dividing (the reciprocal of x is $1/x$).

10. Using the formula found in Problem 9, calculate $\frac{1}{7}$ and $\frac{1}{81}$.

11. Use Newton's method to find the unique solution to $10 - x = e^x$.

12. Find the first 10 values $x_1, x_2, \ldots, x_{10}$ for the equation $x^2 + 5x + 7 = 0$, starting with $x_0 = 0$. Explain why Newton's method must fail in this case.

*13. A new refrigerator costs $400. At 24 monthly installments of $25, what is the effective interest rate?

Review Exercises for Chapter 11

In Exercises 1–8, a function is given. (a) Find intervals over which the function is increasing and decreasing. (b) Find all critical points. (c) Find all local maxima and minima. (d) Find all points of inflection. (e) Determine intervals over which the graph is concave up and concave down. (f) Sketch the graph.

1. $y = x^2 - 3x - 4$

2. $y = x^3 - 3x^2 - 9x + 25$

3. $y = x^5 + 2$

4. $y = \sqrt[3]{x}$

5. $y = \sqrt[4]{x}$

6. $y = x(x - 1)(x - 2)(x - 3)$

7. $y = |x - 4|$

8. $y = \dfrac{1}{x + 2}$

9. A rope is attached to a pulley mounted on top of a 5-m tower. One end of the rope is attached to a heavy mass. If the rope is pulled in at a rate of $1\frac{1}{2}$ m/sec, how fast is the mass approaching the tower when it is 3 m from the base of the tower?

10. A storage tank is in the shape of an inverted right circular cone with a radius of 6 ft and a height of 14 ft. If water is pumped into the tank at a rate of 20 gal/min, how fast is the water level rising when it has reached a height of 5 ft? [*Hint:* 1 gal ≈ 0.1337 ft^3.]

In Exercises 11–13, find the maximum and minimum values for the function over the indicated interval.

11. $y = 2x^3 + 9x^2 - 24x + 3$; $[-2, 5]$

12. $y = (x - 2)^{1/3}$; $[0, 4]$

13. $y = (x + 1)/(x^2 - 4)$; $[-1, 1]$

14. What is the maximum rectangular area that can be enclosed with 800 m of wire fencing?

15. Find the point on the line $2x - 3y = 6$ that is nearest the origin.

16. A cylindrical barrel is to be constructed to hold 128π ft^3 of liquid. The cost per square foot of constructing the side of the barrel is three times that of constructing the top and bottom. What are the dimensions of the barrel that costs the least to construct?

17. The population of a certain species is given by $P(t) = 1000 + 800t + 96t^2 + 12t^3 - t^4$, where t is measured in weeks. After how many weeks is the instantaneous rate of population increase a maximum?

18. A producer of dog food finds that the cost in dollars of producing q cans of dog food is $C = 200 + 0.2q + 0.001q^2$. If the cans sell for 35¢ each, how many cans should be produced to maximize the profit?

19. If the cost function for a certain product is $C(q) = 100 + (1000/q)$ and the demand function is $p(q) = 50 - 0.02q$, at what level of production is profit maximized?

20. Calculate the price elasticity of demand for the product of Exercise 19. At what level of production will it make no difference whether prices are increased or decreased?

21. The relative annual rate of growth of a population is 15%. If the initial population is 10,000, what is the population after 5 years? after 10 years?

22. In Exercise 21, how long will it take for the population to double?

23. When a cake is taken out of the oven, its temperature is 125°C. Room temperature is 23°C. If the temperature of the cake is 80°C after 10 minutes,

    (a) what will be its temperature after 20 minutes?

    (b) how long will the cake take to cool to 25°C?

24. A fossil contains 35% of the normal amount of C^{14}. What is its approximate age?

25. What is the half-life of an exponentially decaying substance that loses 20% of its mass in one week?

26. Calculate $\sqrt[6]{135}$ to 4 decimal places using Newton's method.

27. Use Newton's method to find all roots of $x^3 - 2x^2 + 5x - 8 = 0$.

12 INTEGRATION

12.1 The Antiderivative or Indefinite Integral

In Chapters 10 and 11 we discussed the derivative and its applications. In this chapter we see how useful information can be obtained by reversing the process of differentiation. We begin with two simple examples.

Example 1 We know that $(d/dx)x^2 = 2x$. That is, x^2 is a function whose derivative is $2x$. For this reason x^2 is called an **antiderivative** of $2x$.

Example 2 We know that $(d/dx)\sqrt{x} = 1/2\sqrt{x}$. Putting this fact another way, we say that $\sqrt{x}$ is an antiderivative of $1/2\sqrt{x}$.

The Antiderivative Let f be defined on $[a, b]$. Suppose that there is a differentiable function F defined on $[a, b]$ such that

$$F'(x) = \frac{dF}{dx} = f(x), \qquad \text{for } x \text{ in } [a, b].$$

Then F is called an **antiderivative** or **indefinite integral** of f on the interval $[a, b]$, and we write

$$F = \int f(x)\, dx = \int f. \tag{1}$$

This is read "F is the integral of $f(x)$ with respect to x," or simply "F is an antiderivative of f." If such a function F exists, then f is said to be **integrable** and the process of calculating an integral is called **integration**. The variable x is called the **variable of integration**, and the function f is called the **integrand**. The dx in equation (1) indicates that the variable in the function whose antiderivative we seek is x.

Example 3   Find $\int 3x^2\, dx$.

Solution   Since $d(x^3)/dx = 3x^2$, we have

$$\int 3x^2\, dx = x^3.$$

But the derivative of any constant is zero, so that $x^3 + C$ is also an indefinite integral of $3x^2$ for any constant C. To see this, we have

$$\frac{d}{dx}(x^3 + C) = \frac{d}{dx}(x^3) + \frac{d}{dx}(C) = 3x^2 + 0 = 3x^2.$$

The last example leads to the following:

Remark.   The reason that we refer to an *indefinite* integral is that there are always an infinite number of them. If F is an indefinite integral of f, then so is $F + C$ for every constant C because

$$\frac{d}{dx}(F + C) = \frac{dF}{dx} + \frac{dC}{dx} = f + 0 = f.$$

Of course, you may ask, how do we know we have found all the antiderivatives or indefinite integrals of f? For example, are there any other functions F such that $F'(x) = 3x^2$, other than functions of the form $F(x) = x^3 + C$? The answer is no.

To see why this is so, suppose that F and G have the same derivative; that is, suppose that $F'(x) = G'(x)$. Then $(d/dx)[F(x) - G(x)] = F'(x) - G'(x) = 0$. But on page 557 we stated that the only function with a derivative equal to the zero function is a constant function. That is, $F(x) - G(x) = C$ for some constant C. We have shown that

$$\boxed{\text{if } F'(x) = G'(x), \text{ then } F(x) - G(x) = C \text{ for some constant } C.}$$

This result allows us to define the **most general antiderivative** of f as $F(x) + C$, where F is some antiderivative of f and C is an arbitrary constant. For the remainder of this section we will speak about "the integral," leaving out the word "indefinite," and will look for the most general indefinite integral in our calculations.

We now show how some integrals can be calculated. In Section 10.6 (page 556) we stated the power rule:

$$\frac{d}{dx}x^r = rx^{r-1},$$

where r is a nonzero real number. Thus,

$$\frac{d}{dx}\left(\frac{x^{r+1}}{r+1}+C\right)=\frac{(r+1)x^r}{(r+1)}+0=x^r.$$

This means that,

$$\text{if } r\neq-1, \text{ then}$$
$$\int x^r\,dx=\frac{x^{r+1}}{r+1}+C. \tag{2}$$

Note. The result does *not* hold for $r=-1$, because the denominator is $-1+1=0$ and we cannot divide by zero.

Example 4 Calculate $\int x^9\,dx$.

Solution $r=9$, so that

$$\int x^9\,dx=\frac{x^{9+1}}{9+1}+C=\frac{x^{10}}{10}+C.$$

Example 5 Calculate $\int x^{1/3}\,dx$.

Solution $r=\frac{1}{3}$, so that

$$\int x^{1/3}\,dx=\frac{x^{(1/3)+1}}{(\frac{1}{3})+1}+C=\frac{x^{4/3}}{\frac{4}{3}}+C=\frac{3}{4}x^{4/3}+C.$$

Example 6 Calculate $\int(1/\sqrt{t})\,dt$.

Solution The only difference between using x and t as the variable of integration is that t instead of x is the variable used in the antiderivative function. Then, since $1/\sqrt{t}=t^{-1/2}$,

$$\int\frac{1}{\sqrt{t}}\,dt=\int t^{-1/2}\,dt=\frac{t^{(-1/2)+1}}{(-\frac{1}{2})+1}+C=\frac{t^{1/2}}{\frac{1}{2}}+C=2\sqrt{t}+C.$$

In the last three examples it was easy to check the answer by differentiating. This should always be done, because it is always easier to differentiate than to integrate. Thus differentiation provides a method for verifying the results of calculating an antiderivative.

Equation (2) is valid if $r\neq-1$. What happens if $r=-1$? That is, what is $\int x^{-1}\,dx=\int(1/x)\,dx$? Recall, from Section 10.11, that $(d/dx)\ln x=1/x$. Thus,

$$\int\frac{1}{x}\,dx=\ln x+C \qquad \text{if} \quad x>0. \tag{3}$$

Now suppose that $x < 0$. Then $-x > 0$, $\ln(-x)$ is defined, and by the chain rule,

$$\frac{d}{dx}\ln(-x) = \frac{1}{-x}\frac{d}{dx}(-x) = -\frac{1}{x}(-1) = \frac{1}{x},$$

so that

$$\int \frac{1}{x}\,dx = \ln(-x) + C \qquad \text{if} \quad x < 0. \tag{4}$$

Since

$$|x| = \begin{cases} x, & x > 0 \\ -x, & x < 0 \end{cases}$$

we can put (3) and (4) together to obtain the important integration formula

$$\boxed{\int \frac{1}{x}\,dx = \ln|x| + C.} \tag{5}$$

Now, since $(d/dx)e^x = e^x$, we have

$$\boxed{\int e^x\,dx = e^x + C.} \tag{6}$$

We have seen that integration, the process of finding an antiderivative, is the reverse of differentiation. There is a very important distinction, however, between these two processes. Remember that a function is integrable if it has an antiderivative. It turns out that

$$\boxed{\text{every continuous function is integrable.}}$$

The problem is that, unlike differentiation, the process of finding an antiderivative may be very difficult. In fact, it may be impossible. For example, the function e^{-x^2}, which is very useful in probability theory, has an antiderivative; however, there is no way to express an antiderivative of e^{-x^2} in terms of functions with which we are familiar. We can get around this difficulty by using numerical techniques to obtain approximations to definite integrals. We shall discuss definite integrals beginning in Section 12.5 and shall discuss one numerical technique in Section 12.10.

The question remains "How do we find antiderivatives?" There are rules and techniques for finding them just as there are rules for computing derivatives. We give two rules here and discuss two important techniques of integration in Sections 12.3 and 12.4. Other techniques can be found in engineering calculus texts.

Two Integration Rules If f and g are integrable[†] and if k is any constant, then kf and $f + g$ are integrable, and we have

$$(1) \quad \int kf(x)\, dx = k \int f(x)\, dx$$

(7)

and

$$(2) \quad \int [f(x) + g(x)]\, dx = \int f(x)\, dx + \int g(x)\, dx.$$

(8)

Example 7   Calculate $\int [(3/x^2) + 6x^2]\, dx$.

Solution

$$\int \left(\frac{3}{x^2} + 6x^2 \right) dx = \int \frac{3}{x^2}\, dx + \int 6x^2\, dx = 3 \int x^{-2}\, dx + 6 \int x^2\, dx$$

$$= \frac{3x^{-2+1}}{-2+1} + \frac{6x^{2+1}}{2+1} + C = -3x^{-1} + 2x^3 + C$$

$$= -\frac{3}{x} + 2x^3 + C$$

Check.

$$\frac{d}{dx}\left(-\frac{3}{x} + 2x^3 + C \right) = -3(-x^{-2}) + 2 \cdot 3x^2 = \frac{3}{x^2} + 6x^2$$

Example 8   Calculate $\int (2x^{7/5} - 3x^{-11/9} + 17x^{17})\, dx$.

Solution

$$\int (2x^{7/5} - 3x^{-11/9} + 17x^{17})\, dx = \int 2x^{7/5}\, dx + \int -3x^{-11/9}\, dx + \int 17x^{17}\, dx$$

$$= 2 \int x^{7/5}\, dx - 3 \int x^{-11/9}\, dx + 17 \int x^{17}\, dx$$

$$= 2\frac{x^{(7/5)+1}}{(7/5)+1} - \frac{3x^{(-11/9)+1}}{(-11/9)+1} + \frac{17x^{17+1}}{17+1} + C$$

$$= \frac{2x^{12/5}}{12/5} - \frac{3x^{-2/9}}{-2/9} + \frac{17x^{18}}{18} + C$$

$$= \frac{5}{6}x^{12/5} + \frac{27}{2}x^{-2/9} + \frac{17}{18}x^{18} + C$$

[†] Remember that f is integrable if f has an antiderivative.

Check.

$$\frac{d}{dx}\left(\frac{5}{6}x^{12/5} + \frac{27}{2}x^{-2/9} + \frac{17}{18}x^{18} + C\right)$$

$$= \frac{5}{6}\cdot\frac{12}{5}x^{(12/5)-1} + \frac{27}{2}\cdot-\frac{2}{9}x^{(-2/9)-1} + \frac{17}{18}\cdot 18x^{18-1} + 0$$

$$= 2x^{7/5} - 3x^{-11/9} + 17x^{17}$$

Example 9 Compute $\int\left(\frac{1}{x} + \frac{1}{x^2} - 3e^x\right)dx$.

Solution

$$\int\left(\frac{1}{x} + \frac{1}{x^2} - 3e^x\right)dx = \int\frac{1}{x}dx + \int x^{-2}dx - 3\int e^x dx$$

$$= \ln|x| - x^{-1} - 3e^x + C$$

$$= \ln|x| - \frac{1}{x} - 3e^x + C$$

Check.

$$\frac{d}{dx}\left(\ln|x| - \frac{1}{x} - 3e^x + C\right) = \frac{d}{dx}\ln|x| - \frac{d}{dx}x^{-1} - 3\frac{d}{dx}e^x + \frac{dC}{dx}$$

$$= \frac{1}{x} + x^{-2} - 3e^x = \frac{1}{x} + \frac{1}{x^2} - 3e^x$$

Let us look more closely at the general integral of a function. Let $f(x) = 2x$. Then $\int f(x)\,dx = x^2 + C$. For every value of C, we get a different integral. But these integrals are very similar geometrically. For example, the curve $y = x^2 + 1$ is obtained by "shifting" the curve $y = x^2$ up 1 unit. More generally, the curve $y = x^2 + C$ is obtained by shifting the curve $y = x^2$ up or down $|C|$ units (up if $C > 0$ and down if $C < 0$; see Section 9.3). Some of these curves are plotted in Figure 1. These curves never intersect. To prove this, suppose that (x_0, y_0) is a point on both curves $y = x^2 + A$ and $y = x^2 + B$. Then $y_0 = x_0^2 + A = x_0^2 + B$, which implies that $A = B$. That is, if the two curves have a point in common, then the two curves are the same. Thus, if we specify one point through which the integral passes, then we know *the* integral.

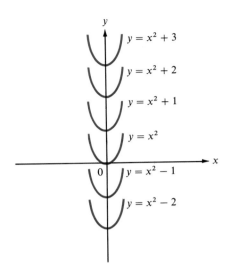

Figure 1

Example 10   Find $\int 2x \, dx$, which passes through the point $(2, 7)$.

Solution $y = \int 2x \, dx = x^2 + C$. But when $x = 2$, $y = 7$, so that $7 = 2^2 + C = 4 + C$, or $C = 3$. The solution is the function $x^2 + 3$.

Example 11   The slope of the tangent to a given curve is given by $f(x) = 6x^3 - 4x^2 + (1/x^2)$. The curve passes through the point $(1, 4)$. Find the curve.

Solution   The slope of the curve $y = F(x)$ for any x is given by $dy/dx = F'(x) = 6x^3 - 4x^2 + (1/x^2)$. Then

$$F(x) = \int \left(6x^3 - 4x^2 + \frac{1}{x^2} \right) dx = \frac{6x^4}{4} - \frac{4x^3}{3} + \frac{x^{-1}}{-1} + C$$

$$= \frac{3}{2} x^4 - \frac{4}{3} x^3 - \frac{1}{x} + C.$$

When $x = 1$, $y = 4$, so that

$$4 = \frac{3}{2}(1)^4 - \frac{4}{3}(1)^3 - \frac{1}{1} + C = \frac{3}{2} - \frac{4}{3} - 1 + C$$

$$= \frac{9 - 8 - 6}{6} + C = -\frac{5}{6} + C.$$

Then

$$C = 4 + \frac{5}{6} = \frac{29}{6} \quad \text{and} \quad F(x) = \frac{3}{2} x^4 - \frac{4}{3} x^3 - \frac{1}{x} + \frac{29}{6}.$$

PROBLEMS 12.1   In Problems 1–23 find the most general antiderivative.

1. $\displaystyle\int dx$ **2.** $\displaystyle\int x\, dx$

3. $\displaystyle\int a\, dx,$ where a is a constant **4.** $\displaystyle\int (ax + b)\, dx,$ a, b constants

5. $\displaystyle\int \tfrac{1}{2}x^2\, dx$ **6.** $\displaystyle\int 4x^3\, dx$

7. $\displaystyle\int 7x^6\, dx$ **8.** $\displaystyle\int x^{-2}\, dx$

9. $\displaystyle\int x^{1/3}\, dx$ **10.** $\displaystyle\int \frac{1}{x^5}dx$

11. $\displaystyle\int \frac{1}{x^{3/2}}\, dx$ **12.** $\displaystyle\int x^{7/9}\, dx$

13. $\displaystyle\int (1 + x + x^2 + x^3 + x^4)\, dx$ **14.** $\displaystyle\int (x^5 - 3x^3 + 1)\, dx$

15. $\displaystyle\int (x^{10} - x^8 + 14x^3 - 2x + 9)\, dx$ **16.** $\displaystyle\int \left(\sqrt{x} + \frac{1}{\sqrt{x}}\right) dx$

17. $\displaystyle\int \left(3\sqrt[3]{x} + \frac{3}{\sqrt[3]{x}}\right) dx$ **18.** $\displaystyle\int (x^{1/2} + x^{3/2} + x^{3/4})\, dx$

19. $\displaystyle\int \left(\frac{4}{3x^4} - \frac{5}{7x^5} + \frac{6}{11x^6}\right) dx$

20. $\displaystyle\int \frac{x^3 + x^2 - x}{x^{3/2}}\, dx$ [*Hint:* Divide through.]

21. $\displaystyle\int \frac{4x^{2/3} - x^{5/8} + x^{17/5}}{5x^3}\, dx$

22. $\displaystyle\int \left(-\frac{17}{x^{13/17}} + \frac{3}{x^{4/9}}\right) dx$ **23.** $\displaystyle\int \left(\frac{4}{x} - \frac{2}{x^3} + 7e^x\right) dx$

In Problems 24–29 the derivative of a function and one point on its graph are given. Find the function.

24. $\dfrac{dy}{dx} = x^3 + x^2 - 3;$ $(1, 5)$ **25.** $\dfrac{dy}{dx} = 2x(x + 1);$ $(2, 0)$

26. $\dfrac{dy}{dx} = \sqrt[3]{x} + x - \dfrac{1}{3\sqrt[3]{x}};$ $(-1, -8)$ **27.** $y' = 13x^{15/18} - 3;$ $(1, 14)$

28. $\dfrac{dy}{dx} = \dfrac{4}{x} + \dfrac{5}{x^3};$ $(2, 10)$ **29.** $y' = 3e^x - 5;$ $(0, 7)$

*30. Find all antiderivatives of $f(x) = |x|$.

12.2 Applications of the Antiderivative

In Section 10.5 we saw several applications of the derivative. In Section 11.4 we saw applications of the derivative to marginal analysis. It should not be surprising that the antiderivative can be used in similar ways. For example, if velocity is the derivative of the distance function, as we have seen, then the distance function is the integral of the velocity function. Similarly, if marginal cost is the derivative of the total cost function, then the total cost function is the integral of the marginal cost. We summarize these facts, and others, in Table 1 below.

TABLE 1

Function	Its antiderivative (indefinite integral)
Marginal cost	Total cost
Marginal revenue	Total revenue
Marginal profit	Total profit
Velocity	Distance
Acceleration	Velocity
Instantaneous population growth rate	Total population

Example 1 The marginal revenue that a manufacturer receives for his goods is given by

$$\text{MR} = 100 - 0.03q, \qquad 0 \le q \le 1000.$$

Find his total revenue function.

Solution We know that revenue is the integral of marginal revenue. Thus,

$$R(q) = \int \text{MR} = \int (100 - 0.03q)\, dq$$

$$= 100q - 0.03\frac{q^2}{2} + C = 100q - 0.015q^2 + C.$$

But the manufacturer certainly receives nothing if he sells nothing. Therefore, $R(0) = 0 = C$ and

$$R(q) = 100q - 0.015q^2 \qquad \text{for} \qquad 0 \le q \le 1000. \tag{1}$$

Example 2 A manufacturer's marginal cost for the product of Example 1 is $80 + 0.02q$. He has fixed costs of \$500. Find his total cost.

Solution The total cost, $C(q)$, is the integral of the marginal cost function.

Then

$$C(q) = \int MC = \int (80 + 0.02q)\, dq = 80q + 0.01q^2 + C. \qquad (2)$$

The fixed costs are incurred even if nothing is produced. Thus,

$$\text{fixed costs} = C(0) = 500.$$

But from equation (2), $C(0) = C$. Thus $C = 500$ and

$$C(q) = 500 + 80q + 0.01q^2. \qquad (3)$$

Example 3 In Examples 1 and 2, how much profit is made from the sale of 75 units?

Solution Profit = revenue − cost. Thus, from (1) and (3),

$$P(q) = (100q - 0.015q^2) - (500 + 80q + 0.01q^2)$$
$$= -500 + 20q - 0.025q^2$$

and

$$P(75) = -500 + 20 \cdot 75 - 0.025 \cdot 75^2 \approx \$859.38.$$

Example 4 A ball is dropped from rest from a certain height. Due to the earth's gravitational field, its velocity after t seconds is given by

$$v = 32t,$$

where v is measured in ft/sec. How far has the ball fallen after t seconds?

Solution If $s = s(t)$ represents distance, then

$$\frac{ds}{dt} = v \quad \text{or} \quad s(t) = \int v(t)\, dt = \int 32t\, dt = 16t^2 + C.$$

But $0 = s(0) = 16 \cdot 0^2 + C$, which implies that $C = 0$. Thus $s(t) = 16t^2$. For example, after 3 seconds the ball has fallen $16 \cdot 3^2 = 16(9) = 144$ feet.

Example 5 The acceleration under the pull of the earth's gravity is 9.81 m/sec². Find a formula for velocity and distance traveled by a particle under the influence of the force of gravitational attraction.

Solution $v(t) = \int a(t)\, dt = \int 9.81\, dt = 9.81t + C$. Let $v(0)$, the initial velocity of the particle being considered, be denoted by v_0. Then $v(0) = v_0 = (9.81)(0) + C$, or $C = v_0$. Hence

$$v(t) = 9.81t + v_0$$

and

$$s(t) = \int v(t)\, dt = \int (9.81t + v_0)\, dt = \frac{9.81}{2} t^2 + v_0 t + C.$$

If we let s_0 denote $s(0)$, the initial position of the particle, we obtain

$$s(0) = s_0 = \frac{9.81}{2}(0)^2 + v_0(0) + C, \quad \text{or} \quad C = s_0,$$

so that

$$\boxed{s(t) = \frac{9.81}{2}t^2 + v_0 t + s_0.}$$

This is called the **equation of motion** of the particle.

PROBLEMS 12.2   In Problems 1–5 a marginal revenue function is given. Find the total revenue function.

1. $MR = 4$

2. $MR = 4 - 0.01q$

3. $MR = 50 - 0.04q + 0.002q^2$

4. $MR = 125 - 0.02q^{3/2}$

5. $MR = 75 + 5q^{0.3} - 0.02q^{1.8}$

6. In Problem 2, what is the total revenue from sales of 200 units of the product?

 7. In Problem 5, what is the total revenue from sales of 50 units of the product?

In Problems 8–11 a marginal cost function is given, with fixed costs in parentheses. Determine the total cost function.

8. $MC = 30 - \dfrac{q}{10}$; (100)

9. $MC = 50 + 0.05q$; (600)

10. $MC = 100q^{0.1} + 0.15q^{1.3}$; (1200)

11. $MC = 750 + 0.24\sqrt{q} - 0.04q^{3/2}$; (1000)

12. In Problem 8, what is the total cost of producing 200 units of the product?

 13. In Problem 11, what is the total cost of producing 125 units of the product?

14. For a certain product, $MR = 80 - 0.04q$ and $MC = 65 - 0.02q$. Fixed costs are $350.

(a) Find the total profit function.

(b) What is the profit from producing 150 units of the product?

15. If the marginal cost function for a certain manufacturer is $MC = 25q - 0.02q^2 + 20$, and if it costs $2000 to produce 10 units of a certain product, how much does it cost to produce 100 units? 500 units?

16. If the marginal revenue function for a certain product is given by $MR = 3q - 2\sqrt{q} + 10$, how much money will the manufacturer receive if she sells 50 units? 100 units?

17. In Example 1, at what level of production will revenue begin to decrease as additional items are produced? (This is often called the **point of diminishing returns**.)

18. A particle moves with the constant acceleration of 5.8 m/sec². It starts with an initial velocity of 0.2 m/sec and an initial position of 25 m. Find the equation of motion of the particle.

19. A bullet is shot from ground level straight up into the air at an initial velocity of 2000 ft/sec. Find the function that tells us the height of the bullet after t seconds, up until the time the bullet hits the earth (assume that the only force acting on the bullet is the force of gravitational attraction, which imparts the constant acceleration $g = 32$ ft/sec²).

20. The acceleration of a moving particle starting at a position 100 m along the x-axis and moving with an initial velocity of $v_0 = 25$ m/min is given by

$$a(t) = 13\sqrt{t} \quad \text{m/min}^2.$$

Find the equation of motion of the particle.

12.3 Integration by Substitution

To this point we have only integrated the functions x^r and e^x and constant multiples and sums of these functions. In this section we discuss a technique that greatly increases the number of functions we can integrate. We begin with an example.

Example 1   Compute $\int (1 + x^2)^3 2x \, dx$.

Solution   Let $u = 1 + x^2$. Then $(du/dx) = 2x$ and $(1 + x^2)^3 2x = u^3 (du/dx)$. Now, from the chain rule,

$$\frac{d}{dx} u^4 = 4u^3 \frac{du}{dx} \qquad \text{and} \qquad \frac{d}{dx} \frac{u^4}{4} = \frac{1}{4} \frac{d}{dx} u^4 = u^3 \frac{du}{dx}.$$

This suggests that $u^4/4$ is an antiderivative of $u^3(du/dx) = (1 + x^2)^3 2x$. But $u^4/4 = (1 + x^2)^4/4$. We may differentiate this to verify that

$$\frac{d}{dx} \left[\frac{(1 + x^2)^4}{4} \right] = \frac{1}{4} \frac{d}{dx} (1 + x^2)^4 \overset{\text{Chain rule}}{=} \frac{1}{4} \cdot 4(1 + x^2)^3 \frac{d}{dx} (1 + x^2)$$

$$= (1 + x^2)^3 2x.$$

Thus, $(1 + x^2)^4/4$ is indeed an antiderivative of $(1 + x^2)^3 2x$, and we have

$$\int (1 + x^2)^3 2x \, dx = \frac{(1 + x^2)^4}{4} + C.$$

We now generate the result of Example 1. To do so, we start with the formula

$$\boxed{\int u^r \, du = \frac{u^{r+1}}{r + 1} + C, \qquad \text{if} \quad r \neq -1.} \tag{1}$$

This is equation (12.1.2) with the variable u instead of the variable x. Now, let $u = g(x)$. Then $du/dx = g'(x)$, or

$$du = g'(x)\,dx.^\dagger \qquad (2)$$

Then, using equations (1) and (2), we have the following: Let g be a differentiable function and set $u = g(x)$; then

$$\int [g(x)]^r g'(x)\,dx = \int u^r\,du = \frac{u^{r+1}}{r+1} + C = \frac{[g(x)]^{r+1}}{r+1} + C. \qquad (3)$$

Observe that formula (3) is closely related to the chain rule (see page 569). The chain rule says that

$$\frac{d}{dx}\frac{[g(x)]^{r+1}}{r+1} = \frac{1}{r+1}\frac{d}{dx}[g(x)]^{r+1}$$

Since $\dfrac{d}{dx}u^{r+1} = (r+1)u^r\dfrac{du}{dx}$

$$\downarrow$$
$$= \frac{1}{r+1}(r+1)[g(x)]^r\frac{d}{dx}[g(x)] = [g(x)]^r g'(x),$$

so that $[g(x)]^{r+1}/(r+1)$ is indeed an antiderivative of $[g(x)]^r g'(x)$.

In general, to calculate $\int f(x)\,dx$, perform the following steps:

1. Make a substitution $u = g(x)$ so that the integrand can be expressed in the form $u^r\,du$ (if possible).

2. Calculate $du = g'(x)\,dx$.

3. Write $\int f(x)\,dx$ as $\int u^r\,du$ for some number $r \neq -1$.

4. Integrate using formula (1).

5. Substitute $g(x)$ for u to obtain the answer in terms of x.

Example 2 Calculate $\int \sqrt{3+x}\,dx$.

Solution Let $u = g(x) = 3 + x$, then $du = dx$ and

$$\int \sqrt{3+x}\,dx = \int \sqrt{u}\,du = \int u^{1/2}\,du = \tfrac{2}{3}u^{3/2} + C = \tfrac{2}{3}(3+x)^{3/2} + C.$$

---

† The symbols du and dx have, for our purposes, no meaning of their own. They are written here to help you see what is going on.

Example 3   Calculate $\int (x^3 - 1)^{11/5} 3x^2 \, dx$.

Solution   Let $u = g(x) = x^3 - 1$. Then $du/dx = 3x^2$, $du = 3x^2 \, dx$, and

$$\int (x^3 - 1)^{11/5} \, 3x^2 \, dx = \int u^{11/5} \, du = \tfrac{5}{16} u^{16/5} + C = \tfrac{5}{16}(x^3 - 1)^{16/5} + C.$$

In the two examples above, the integrand was already in the form $\int u^r \, du$ for the appropriate value of r. Sometimes this is not the case, but we can salvage the problem by multiplying and dividing by an appropriate constant.

Example 4   Calculate $\int x \sqrt[3]{1 + x^2} \, dx$.

Solution   Let $u = 1 + x^2$. Then $du = 2x \, dx$. All we have is $x \, dx$. To get du, we multiply inside the integral by 2 and, to preserve the equality, divide outside the integral by 2 to obtain

$$\int x \sqrt[3]{1 + x^2} \, dx = \frac{1}{2} \int \sqrt[3]{1 + x^2} \, 2x \, dx = \frac{1}{2} \int u^{1/3} \, du$$

$$= \frac{1}{2} \cdot \frac{3}{4} u^{4/3} + C = \frac{3}{8}(1 + x^2)^{4/3} + C.$$

We were allowed to multiply and divide by the constant 2 because of the fact that $\int kf(x) \, dx = k \int f(x) \, dx$ for any constant k.

Warning.   *We cannot multiply and divide in the same way by functions that are not constants.* If we try, we get incorrect answers, such as

$$\int \sqrt{1 + x^2} \, dx = \frac{1}{2x} \int \sqrt{1 + x^2} \, 2x \, dx \qquad \text{We multiplied and divided by } 2x.$$

$$= \frac{1}{2x} \int u^{1/2} \, du \qquad \text{Where } u = 1 + x^2$$

$$= \frac{1}{2x} \cdot \frac{2}{3} u^{3/2} + C = \frac{1}{3x}(1 + x^2)^{3/2} + C.$$

But the derivative of $(1 + x^2)^{3/2}/3x$ is not even close to $\sqrt{1 + x^2}$. (Check this!) It is possible to compute (by techniques that we shall not discuss in this text) that

$$\int \sqrt{1 + x^2} \, dx = \frac{x\sqrt{x^2 + 1}}{2} + \frac{1}{2} \ln (x + \sqrt{x^2 + 1}) + C.$$

Example 5   Calculate $\int (1 + (1/t))^5/t^2 \, dt$.

Solution   Let $u = 1 + (1/t)$. Then $du = -(1/t^2) \, dt$. We then multiply and divide by -1 to obtain

$$\int \frac{[1 + (1/t)]^5}{t^2} \, dt = -\int \left(1 + \frac{1}{t}\right)^5 \left(-\frac{1}{t^2}\right) dt = -\int u^5 \, du$$

$$= -\frac{u^6}{6} + C = -\frac{[1 + (1/t)]^6}{6} + C.$$

We now turn to other functions that can be integrated by an appropriate substitution. From equation (12.1.5) on page 665, we immediately obtain

$$\boxed{\int \frac{1}{u} \, du = \ln |u| + C.} \qquad (4)$$

Let $u = g(x)$. Then $du = g'(x) \, dx$, so that (4) becomes

$$\boxed{\int \frac{g'(x)}{g(x)} \, dx = \ln |g(x)| + C.}$$

We have shown that

> the integral of a quotient in which the function in the numerator is the derivative of the function in the denominator is equal to the natural logarithm of the absolute value of the denominator.

Example 6 Calculate $\int (2x/(x^2 + 1)) \, dx$.

Solution Since $2x$ is the derivative of $x^2 + 1$,

$$\int \frac{2x}{x^2 + 1} \, dx = \int \frac{du}{u} = \ln |u| + C = \ln|x^2 + 1| + C = \ln(x^2 + 1) + C$$

(since $x^2 + 1$ is always positive).

Example 7 Calculate $\int (1/(x \ln x)) \, dx$.

Solution We write the integral as $\int [(1/x)/(\ln x)] \, dx$. If $u = \ln x$, then $du/dx = 1/x$, and we see that the numerator is the derivative of the denominator. Thus,

$$\int \frac{1/x}{\ln x} \, dx = \int \frac{du}{u} = \ln|u| + C = \ln |\ln x| + C.$$

Example 8 Calculate $\int (x^3 + 1)/(x^4 + 4x) \, dx$.

Solution Here the numerator is not quite the derivative of the denominator. The derivative of the denominator is $4x^3 + 4 = 4(x^3 + 1)$. Thus we need to multiply and divide by 4 to obtain

$$\int \frac{x^3 + 1}{x^4 + 4x}\, dx = \frac{1}{4} \int \frac{4x^3 + 4}{x^4 + 4x}\, dx = \frac{1}{4} \int \frac{du}{u} = \frac{1}{4} \ln |u| + C = \frac{1}{4} \ln |x^4 + 4x| + C.$$

Now, from equation (12.1.6) on page 665, we have

$$\boxed{\int e^u\, du = e^u + C.} \tag{5}$$

If $u = g(x)$, then $du = g'(x)\, dx$ and (5) becomes

$$\boxed{\int e^{g(x)} g'(x)\, dx = e^{g(x)} + C.}$$

Thus,

> the integral of e raised to a power times the derivative of the power is simply e to that power.

Example 9 Calculate $\int e^{x^3} \cdot 3x^2\, dx$.

Solution If $u = x^3$, then $du = 3x^2\, dx$, so that

$$\int e^{x^3} \cdot 3x^2\, dx = \int e^u\, du = e^u + C = e^{x^3} + C.$$

Example 10 Calculate $\int e^{-2x}\, dx$.

Solution If $u = -2x$, then $du = -2\, dx$, so we multiply and divide by -2 to obtain

$$\int e^{-2x}\, dx = -\frac{1}{2} \int e^{-2x}(-2)\, dx = -\frac{1}{2} \int e^u\, du = -\frac{1}{2} e^u + C = -\frac{1}{2} e^{-2x} + C.$$

Example 11 Calculate $\int (e^{\sqrt{x}}/\sqrt{x})\, dx$.

Solution If $u = \sqrt{x}$, then $du = (1/2\sqrt{x})\, dx$, and we multiply and divide by $\frac{1}{2}$ to obtain

$$\int \frac{e^{\sqrt{x}}}{\sqrt{x}}\, dx = 2 \int \frac{e^{\sqrt{x}}}{2\sqrt{x}}\, dx = 2 \int e^u\, du = 2e^u + C = 2e^{\sqrt{x}} + C.$$

Example 12 **The Gross National Product and National Debt (Optional)**   The gross national product (GNP) is defined as the sum of final products such as consumption goods and gross investment (which is the increase in inventories plus gross production of buildings and equipment). We assume that the GNP is increasing continuously at a rate of 2% per annum.

For various reasons the government has a deficit in its spending, which becomes part of the national debt. Government policy is to keep deficit spending a constant proportion, k, of the increasing GNP. It seems possible that the national debt may outstrip the increasing GNP.

Writing the national debt as D and GNP as Y, we have a pair of differential equations:

$$\frac{dY}{dt} = 0.02Y; \tag{6}$$

$$\frac{dD}{dt} = kY. \tag{7}$$

Equation (6) is a special case of the differential equation we presented in Section 11.6 (see equation (11.6.1) on page 644). The solution to (6), according to equation (11.6.5) on page 646, is

$$Y(t) = Y_0 e^{0.02t}, \tag{8}$$

where Y_0 is the GNP in year zero. We substitute (8) into (7) to obtain

$$\frac{dD}{dt} = kY_0 e^{0.02t}.$$

Thus,

$$D = \int \frac{dD}{dt}\, dt = \int kY_0 e^{0.02t}\, dt = kY_0 \int e^{0.02t}\, dt.$$

Let $u(t) = 0.02t$. Then $du = 0.02t$ and

Multiply and divide by 0.02

$$D(t) = \overset{\downarrow}{\frac{kY_0}{0.02}} \int e^{0.02t}(0.02)\, dt = \frac{kY_0}{0.02} \int e^u\, du = \frac{kY_0}{0.02} e^u + d,$$

or

$$D(t) = \frac{kY_0}{0.02} e^{0.02t} + d, \tag{9}$$

where d is an arbitrary constant.

If the initial national debt is D_0, we find, after setting $t = 0$ in equation (9) and using the fact that $e^0 = 1$, that

$$d = D_0 - \frac{kY_0}{(0.02)}$$

and

$$D(t) = D_0 + \frac{kY_0}{(0.02)} e^{0.02t} - \frac{kY_0}{0.02} = \frac{kY_0}{0.02}(e^{0.02t} - 1) + D_0.$$

Now $Y(t) = Y_0 e^{0.02t}$ and we may compute the ratio of national debt to GNP. We have

$$\frac{D(t)}{Y(t)} = \frac{\dfrac{kY_0}{0.02}(e^{0.02t-1}) + D_0}{Y_0 e^{0.02t}} \xrightarrow{\;\frac{1}{e^{0.02}} = e^{-0.02t}\;} = \frac{k}{0.02}(e^{0.02t} - 1)e^{-0.02t} + \frac{D_0}{Y_0} e^{-0.02t}$$

$$\xrightarrow{\;(e^{0.02t})(e^{-0.02t}) = 1\;} = \frac{k}{0.02}(1 - e^{-0.02t}) + \frac{D_0}{Y_0} e^{-0.02t}.$$

Now we note that as $t \to \infty$, $e^{-0.02t} \to 0$, so that

$$\frac{D(t)}{Y(t)} \to \frac{k}{0.02} \qquad \text{as } t \to \infty.$$

Thus, in the long term the ratio of national debt to GNP tends to a constant level. The rate at which it tends to this level depends on the rate of growth of GNP: The larger the rate of growth, the more quickly the ratio reaches this constant level (recall the properties of the exponential function). In many industrialized nations it is common practice for the national debt to be several times larger than the GNP. For instance, the British national debt has been between 2 and 3 times the GNP in 1946, 1923, and as early as 1818.

Several years of U.S. national debt and GNP are given in Table 1.[†]

TABLE 1

Year	National debt, D, in U.S. (billions of $)	GNP, Y, in U.S. (billions of $)	Ratio D/Y
1980	875.0	2,550.0	0.30
1976	621.8	1,663.0	0.40
1945	278.7	213.6	1.30
1939	47.6	91.1	0.50
1929	16.3	104.4	0.20
1920	24.3	88.9	0.30
1916	1.2	40.3	0.03
1868	2.6	6.8	0.40

[†] This information and further discussions of these topics can be found in Paul Samuelson, *Economics*, 11th ed. (New York: McGraw-Hill, 1980), pp. 344–346.

PROBLEMS 12.3   In Problems 1–44 carry out the indicated integration by making an appropriate substitution.

1. $\int \sqrt{9 + x}\, dx$

2. $\int \sqrt{10 + 3x}\, dx$

3. $\int \sqrt{10 - 9x}\, dx$

4. $\int (3x - 2)^4\, dx$

5. $\int (1 - x)^{10}\, dx$

6. $\int \dfrac{dx}{\sqrt{1 + x}}$

7. $\int (1 + 2x)^{3/2}\, dx$

8. $\int x(1 + x^2)^3\, dx$

9. $\int x \sqrt[3]{1 + x^2}\, dx$

10. $\int (s^4 + 1)\sqrt{s^5 + 5s}\, ds$

11. $\int \dfrac{t + 3t^2}{\sqrt{t^2 + 2t^3}}\, dt$

12. $\int (3w - 2)^{99}\, dw$

13. $\int \dfrac{dx}{\sqrt{x}(1 + \sqrt{x})^5}$

14. $\int \dfrac{w + 1}{\sqrt{w^2 + 2w - 1}}\, dw$

15. $\int \dfrac{[1 + (1/v^2)]^{5/3}}{v^3}\, dv$

16. $\int \left(\dfrac{x}{3} - 1\right)^{77}\, dx$

17. $\int (ax + b)\sqrt{ax^2 + 2bx + c}\, dx$

18. $\int (ax^2 + bx + c)\sqrt{2ax^3 + 3bx^2 + 6cx + d}\, dx$

19. $\int \dfrac{ax + b}{(ax^2 + 2bx + c)^{3/7}}\, dx$

20. $\int t \sqrt{t^2 + \alpha^2}\, dt$

21. $\int t^n \sqrt{\alpha^2 + t^{n+1}}\, dt$

22. $\int \dfrac{s^{n-1}}{\sqrt{a + bs^n}}\, ds$

23. $\int p^2 \sqrt{\alpha^3 - p^3}\, dp$

24. $\int p^5 \sqrt{\alpha^6 - p^6}\, dp$

25. $\int \dfrac{dx}{5 + x}$

26. $\int \dfrac{dx}{3 - 2x}$

27. $\int \dfrac{dx}{1 + 100x}$

28. $\int e^{2x}\, dx$

29. $\int e^{1-x}\, dx$

30. $\int e^{3x-5}\, dx$

31. $\int \dfrac{x}{1 + x^2}\, dx$

32. $\int \dfrac{x^2}{1 + x^3}\, dx$

33. $\int \dfrac{x^n}{1 + x^{n+1}}\, dx$

34. $\int \dfrac{x^3}{x^4 + 25}\, dx$

35. $\int e^{4x}\, dx$

36. $\int e^{-x}\, dx$

37. $\displaystyle\int \frac{e^{2x}}{1 + e^{2x}}\, dx$ **38.** $\displaystyle\int xe^{x^2}\, dx$

39. $\displaystyle\int x^2 e^{x^3}\, dx$ **40.** $\displaystyle\int \frac{e^{\sqrt[3]{x}}}{x^{2/3}}\, dx$

41. $\displaystyle\int \frac{e^{1/x}}{x^2}\, dx$ **42.** $\displaystyle\int \frac{\ln(1/x)}{x}\, dx$

43. $\displaystyle\int \frac{e^x}{e^x + 4}\, dx$ **44.** $\displaystyle\int \frac{1}{\ln e^x}\, dx$

45. The marginal cost incurred in the manufacture of a certain product is given by $MC = 4/\sqrt{q + 4}$. If fixed costs are $1000, find the total cost function.

46. The marginal revenue received by a manufacturer for his product is given by $MR = 100 + 0.35e^{-q/4}$.

   (a) Find the total revenue function.

   (b) What is the total revenue from sales of 50 units?

*47. The marginal cost function for a certain product is $20 + \sqrt{1 + 2q}$. What additional cost is incurred by increasing production from 24 to 60 units?

*48. A particle moves with velocity $v(t) = 1/(3\sqrt{2 + t})$ m/min after t minutes. How far does the particle travel between the times $t = 2$ and $t = 7$?

49. In Example 12, assume that the GNP grows at a rate of 3% per year.

   (a) Find the GNP in terms of the initial GNP Y_0.

   (b) Find the national debt as a function of k and D_0.

   (c) Find the long-term ratio $D(t)/Y(t)$.

12.4 Integration by Parts

In this section we discuss a technique that allows us to compute integrals in some situations when no substitution of the type discussed in Section 12.3 can be used. Our method is derived from the product rule:

$$\frac{d}{dx}(uv) = u\frac{dv}{dx} + v\frac{du}{dx}. \tag{1}$$

Integrating both sides of equation (1) with respect to x,

$$\int \frac{d}{dx}(uv)\, dx = \int u\frac{dv}{dx}\, dx + \int v\frac{du}{dx}\, dx$$

or

$$uv = \int u\, dv + \int v\, du.$$

We rearrange terms:

$$\boxed{\int u\, dv = uv - \int v\, du.} \tag{2}$$

As we shall see, the trick in using integration by parts is to rewrite an expression that is difficult to integrate in terms of an expression for which an integral is more easily obtainable.

Example 1 Calculate $\int xe^x \, dx$.

Solution We cannot integrate xe^x directly, because the x term gets in the way. However, if we set $u = x$ and $dv = e^x \, dx$, then $du = dx$, $v = \int e^x \, dx = e^x$, and

$$\int xe^x \, dx = \int u \, dv = uv - \int v \, du = xe^x - \int e^x \, dx = xe^x - e^x + C.$$

This can be checked by differentiation. Note how the x term "disappeared" in $\int v \, du$, making it easy to integrate. Example 1 is typical of the type of problem that can be solved by integration by parts.

Example 2 Calculate $\int \ln x \, dx$.

Solution There are two terms here, $\ln x$ and dx. If we are to integrate by parts, the only choices we have for u and dv are $u = \ln x$ and $dv = dx$. Then $du = (1/x) \, dx$, $v = x$, and

$$\int \ln x \, dx = \int u \, dv = uv - \int v \, du = x \ln x - \int x \cdot \frac{1}{x} \, dx$$

$$= x \ln x - \int dx = x \ln x - x + C.$$

Note. In many of the integrals involving $\ln x$, we may take $u = \ln x$ so that $du = (1/x) \, dx$ and the ln term "vanishes" in $\int v \, du$.

Example 3 Calculate $\int x^3 e^{x^2} \, dx$.

Solution Here we have several choices for u and dv. The possibilities are shown in Table 1. From the table we see that, although there are several choices for u and dv, only one works. The others fail either because (a) dv cannot be readily integrated to find v or (b) $\int v \, du$ is not any easier to integrate than the integral we started with. Hence, to complete the problem, we set $u = x^2$ and $dv = xe^{x^2} \, dx$ to obtain

$$\int x^3 e^{x^2} \, dx = \frac{x^2 e^{x^2}}{2} - \int xe^{x^2} \, dx = \frac{x^2 e^{x^2}}{2} - \frac{1}{2} \int 2xe^{x^2} \, dx$$

$$= \frac{x^2 e^{x^2}}{2} - \frac{1}{2} e^{x^2} + C = \frac{e^{x^2}}{2}(x^2 - 1) + C.$$

TABLE 1

u	dv	du	v	$uv - \int v\,du$	Comments
x^3	$e^{x^2}\,dx$	$3x^2\,dx$	$\int dv = \int e^{x^2}\,dx$ (We're stuck.)	—	Try something else.
e^{x^2}	$x^3\,dx$	$2xe^{x^2}\,dx$	$\dfrac{x^4}{4}$	$\dfrac{x^4 e^{x^2}}{4} - \dfrac{1}{2}\int x^5 e^{x^2}\,dx$	We're worse off than when we started.
xe^{x^2}	$x^2\,dx$	$e^{x^2}(1+2x^2)\,dx$	$\dfrac{x^3}{3}$	$\dfrac{x^4 e^{x^2}}{3} - \dfrac{1}{3}\int x^3 e^{x^2}(1+2x^2)\,dx$	Ditto.
x^2	$xe^{x^2}\,dx$	$2x\,dx$	$\dfrac{1}{2}e^{x^2}$	$\dfrac{x^2 e^{x^2}}{2} - \int xe^{x^2}\,dx$	We can integrate this.
$x^2 e^{x^2}$	$x\,dx$	$e^{x^2}(2x^3+2x)\,dx$	$\dfrac{x^2}{2}$	$\dfrac{x^4}{2}e^{x^2} - \int e^{x^2}(x^5+x^3)\,dx$	What a mess!
x	$x^2 e^{x^2}\,dx$	dx	$\int x^2 e^{x^2}\,dx$ (Stuck again.)	—	Try something else.
$x^3 e^{x^2}$	dx	$e^{x^2}(2x^4+3x^2)\,dx$	x	$x^4 e^{x^2} - \int e^{x^2}(2x^5+3x^3)\,dx$	This is the worst of all.

Example 3 illustrates two things to look for in choosing u and dv:

> **1**. It must be possible to evaluate $\int dv$.
>
> **2**. $\int v\,du$ should be easier to evaluate than $\int u\,dv$.

Sometimes it is necessary to integrate by parts more than once to find an antiderivative.

Example 4   Calculate $\int x^2 e^{-x}\,dx$.

Solution   We need to get rid of the x^2 term. Let $u = x^2$ and $dv = e^{-x}\,dx$. Then

$du = 2x\,dx$ and $v = \int e^{-x}\,dx = -\int e^{-x}(-1)\,dx = -e^{-x}$, so that

$$\int x^2 e^{-x}\,dx = x^2(-e^{-x}) - \int (-e^{-x}) \cdot 2x\,dx = -x^2 e^{-x} + 2\int xe^{-x}\,dx.$$

(The plus sign comes from $-\int - e^{-x}\,dx$.) We see that $\int xe^{-x}\,dx$ is simpler than $\int x^2 e^{-x}\,dx$, but it is still necessary to integrate by parts once more. Setting $u = x$ and $dv = e^{-x}\,dx$, we have $du = dx$ and $v = -e^{-x}$, so that

$$2\int xe^{-x}\,dx = -2xe^{-x} + 2\int e^{-x}\,dx = -2xe^{-x} - 2e^{-x} + C.$$

Therefore,

$$\int x^2 e^{-x}\,dx = -x^2 e^{-x} - 2xe^{-x} - 2e^{-x} + C.$$

As you have seen, integrating by parts requires a bit of ingenuity in the choice of u and dv. However, there are some guidelines that are useful. We give three of them here.

THREE GUIDELINES FOR INTEGRATING BY PARTS

1. For integrands of the form $x^n e^{ax}$ (n an integer),

set $u = x^n$ and $dv = e^{ax}\,dx$.

2. For integrands of the form $x^n \ln x$,

set $u = \ln x$ and $dv = x^n\,dx$.

3. For integrands of the form $x^n \sqrt{ax + b}$,

set $u = x^n$ and $dv = \sqrt{ax + b}\,dx$.

PROBLEMS 12.4   In Problems 1–13 compute all antiderivatives.

1. $\displaystyle\int xe^{3x}\,dx$

2. $\displaystyle\int xe^{-7x}\,dx$

3. $\displaystyle\int x^2 e^{x/4}\,dx$

4. $\displaystyle\int x \ln x\,dx$

5. $\displaystyle\int x^3 \ln x\,dx$

6. $\displaystyle\int x\sqrt{3x + 1}\,dx$

7. $\displaystyle\int x\sqrt{1 - \frac{x}{2}}\,dx$

8. $\displaystyle\int x\sqrt{1 - x}\,dx$

*9. $\displaystyle\int \frac{x + 1}{x + 2}\,dx$

*10. $\displaystyle\int \frac{x^3}{x^2 + 1}\,dx$

11. $\displaystyle\int \ln(x + 1)\,dx$

12. $\displaystyle\int \ln(2 + 3x)\,dx$

*13. $\displaystyle\int x^3 e^{-x}$

14. Find the total revenue function for a product whose marginal revenue is given by $MR = 20 + (q/2)e^{-q/4}$.

12.5 The Definite Integral

In Example 12.2.2 on page 670 we computed a total cost function when the marginal cost and fixed costs were given. Now we look at the problem in a different way.

Example 1   A manufacturer's marginal cost for a certain product is $80 + 0.02q$. What is the additional cost of increasing production from 40 to 60 units?

Solution   The total cost function is given by

$$C(q) = \int MC \, dq = \int (80 + 0.02q) \, dq = 80q + 0.01q^2 + C. \tag{1}$$

We do not know what the constant of integration, C, is, but as we shall soon see, it does not matter. Now, if $C(40)$ is the cost of producing 40 units and $C(60)$ is the cost of producing 60 units, then the additional cost of increasing production is given by

additional cost $= C(60) - C(40)$

or, from equation (1),

additional cost $= (80 \cdot 60 + 0.01(60)^2 + C) - (80 \cdot 40 + 0.01(40)^2 + C)$
$= (4836 + C) - (3216 + C) = \$1620.$

Note that in obtaining this answer we did not need to know the value C (which, as we have seen, represents fixed costs). In notation we shall introduce shortly, we have shown that

$$\int_{40}^{60} (80 + 0.02q) \, dq = 1620.$$

We now make a general definition that, as we shall see, has a great number of applications.

Definite Integral   Let f be an integrable function and let F be an antiderivative for f. Let a and b be real numbers with $b > a$. Then the **definite integral** of f over the interval $[a, b]$, written $\int_a^b f(x) \, dx$, is given by

$$\int_a^b f(x) \, dx = F(b) - F(a). \tag{2}$$

Remark. The x in (2) is called a **dummy variable** because it does not appear at the end of a computation (remember, we always end up with a number). We could just as easily write $\int_a^b f(t)\, dt$ or $\int_a^b f(u)\, du$ or, simply, $\int_a^b f$.

Example 2 Calculate $\int_0^1 x^2\, dx$.

Solution We have seen that $F(x) = x^3/3$ is an antiderivative for x^2. Thus

$$\int_0^1 x^2\, dx = F(1) - F(0) = \left(\frac{x^3}{3} \text{ evaluated at } x = 1\right) - \left(\frac{x^3}{3} \text{ evaluated at } x = 0\right)$$

$$= \frac{1}{3} - 0 = \frac{1}{3}.$$

We will use a simple notation to avoid writing the words "evaluated at" each time.

Notation. $F(x)|_a^b = F(b) - F(a)$. In Example 2 we could have written

$$\int_0^1 x^2\, dx = \frac{x^3}{3}\bigg|_0^1 = \frac{1}{3} - 0 = \frac{1}{3}.$$

Remark. It doesn't make any difference which antiderivative we choose to evaluate the definite integral. For example, if C is any constant, then

$$\int_0^1 x^2\, dx = \left(\frac{x^3}{3} + C\right)\bigg|_0^1 = \left(\frac{1}{3} + C\right) - (0 + C) = \frac{1}{3} + C - C = \frac{1}{3}.$$

The constants will always "disappear" in this manner. Thus we will use the "easiest" antiderivative in our evaluation of $\int_a^b f$, which will almost always be the one in which $C = 0$.

Example 3 Calculate $\int_0^3 x^3\, dx$.

Solution $\int x^3\, dx = x^4/4 + C$. Then

$$\int_0^3 x^3\, dx = \frac{x^4}{4}\bigg|_0^3 = \frac{3^4}{4} - 0 = \frac{81}{4}.$$

Example 4 Calculate $\int_1^2 (3x^4 - x^5)\, dx$.

Solution $\int (3x^4 - x^5)\, dx = 3x^5/5 - x^6/6 + C$. Thus

$$\int_1^2 (3x^4 - x^5)\, dx = \left(\frac{3x^5}{5} - \frac{x^6}{6}\right)\bigg|_1^2 = \left(\frac{3(2)^5}{5} - \frac{2^6}{6}\right) - \left(\frac{3(1)^5}{5} - \frac{1^6}{6}\right)$$

$$= \left(\frac{96}{5} - \frac{64}{6}\right) - \left(\frac{3}{5} - \frac{1}{6}\right) = \left(\frac{576}{30} - \frac{320}{30}\right) - \left(\frac{18}{30} - \frac{5}{30}\right)$$

$$= \frac{243}{30} = \frac{81}{10}.$$

Example 5 Compute $\int_1^3 e^x \, dx$.

Solution e^x is an antiderivative for e^x, so

$$\int_1^3 e^x \, dx = e^x \Big|_1^3 = e^3 - e^1.$$

Example 6 Compute $\int_0^2 (1/(1 + x)) \, dx$.

Solution If $u = 1 + x$, then $du = dx$ and we find that

$$\int \frac{1}{1 + x} \, dx = \int \frac{1}{u} \, du = \ln |u| = \ln |1 + x|,$$

so that

$$\int_0^2 \frac{1}{1 + x} \, dx = \ln |1 + x| \Big|_0^2 = \ln (1 + 2) - \ln (1 + 0) = \ln 3 - \ln 1 \overset{\text{Since } \ln 1 = 0}{=} \ln 3.$$

Warning. Be careful when making substitutions in definite integrals. In Example 6 it is true that if $u = 1 + x$, then

$$\int \frac{dx}{1 + x} = \int \frac{du}{u} = \ln |u| = \ln |1 + x|.$$

But it is *not* true that

$$\int_0^2 \frac{dx}{1 + x} = \int_0^2 \frac{du}{u} = \ln |u| \Big|_0^2 = \ln 2 - \ln 0.$$

In fact, $\ln 0$ is not even defined. What went wrong? In $\int_0^2 (dx/(1 + x))$, x ranges from 0 to 2. But if $u = 1 + x$, then u ranges from $1 + 0 = 1$ to $1 + 2 = 3$. Thus, the correct way to perform the substitution is as follows:

$$\int_{x=0}^{x=2} \frac{dx}{1 + x} = \int_{u=1}^{u=3} \frac{du}{u} = \ln |u| \Big|_1^3 = \ln 3 - \ln 1 = \ln 3.$$

In making a substitution in a definite integral, the limits of integration must be changed as well.

Example 7 Compute $\int_{-1}^4 x \sqrt[3]{1 + x^2} \, dx$.

Solution In Example 12.3.4 on page 675, we found the indefinite integral

$F(x) = \frac{3}{8}(1 + x^2)^{4/3}$. Thus,

$$\int_{-1}^{4} x\sqrt[3]{1 + x^2}\, dx = \frac{3}{8}(1 + x^2)^{4/3}\Big|_{-1}^{4} = \frac{3}{8}[(1 + 4^2)^{4/3} - (1 + (-1)^2)^{4/3}]$$

$$= \frac{3}{8}(17^{4/3} - 2^{4/3}).$$

Example 8   Compute $\displaystyle\int_{1}^{3} \frac{2x}{x^2 + 1}\, dx.$

Solution   Using the result of Example 12.3.6 on page 676, we have

$$\int_{1}^{3} \frac{2x}{x^2 + 1}\, dx = \ln|x^2 + 1|\Big|_{1}^{3} = \ln 10 - \ln 2 = \ln \frac{10}{2} = \ln 5.$$

The next example extends the integration by parts technique of Section 12.4 to definite integrals. We need the following formula, which extends formula 12.4.2 on page 681.

$$\int_{a}^{b} u\, dv = uv\Big|_{a}^{b} - \int_{a}^{b} v\, du.$$

Example 9   Compute $\int_0^5 xe^{-x/2}\, dx.$

Solution   We integrate by parts. If $u = x$ and $dv = e^{-x/2}\, dx$, then $du = dx$ and $v = \int e^{-x/2}\, dx = -2e^{-x/2}$. Thus,

$$\int_{0}^{5} xe^{-x/2}\, dx = uv\Big|_0^5 - \int_0^5 v\, du = -2xe^{-x/2}\Big|_0^5 + \int_0^5 2e^{-x/2}\, dx$$

$$= -10e^{-5/2} + 0 - 4e^{-x/2}\Big|_0^5 = -10e^{-5/2} - 4e^{-5/2} + 4$$
$$\uparrow$$
$$\text{Since } e^0 = 1$$

$$= 4 - 14e^{-5/2} \approx 4 - 1.149 = 2.851.$$

We close this section by citing some facts about definite integrals.

FACTS ABOUT DEFINITE INTEGRALS

1. For any constant c,

$$\int_a^b c \, dx = c(b - a). \tag{3}$$

2.
$$\int_a^b dx = b - a. \tag{4}$$

This is (3) with $c = 1$.

3. For any real number a,

$$\int_a^a f(x) \, dx = 0. \tag{5}$$

4. If $\int_a^b f(x) \, dx$ exists and $a < b$, then $\int_b^a f(x) \, dx$ is defined by

$$\int_b^a f(x) \, dx = -\int_a^b f(x) \, dx. \tag{6}$$

5. If $a < c < b$, then

$$\int_a^b f(x) \, dx = \int_a^c f(x) \, dx + \int_c^b f(x) \, dx. \tag{7}$$

6. If k is a constant, then

$$\int_a^b kf(x) \, dx = k \int_a^b f(x) \, dx. \tag{8}$$

7. $\displaystyle \int_a^b [f(x) + g(x)] \, dx = \int_a^b f(x) \, dx + \int_a^b g(x) \, dx. \tag{9}$

Example 10 Compute $\int_4^7 2 \, dx$.

Solution From equation (3), with $c = 2$, $a = 4$, and $b = 7$, we have

$$\int_4^7 2 \, dx = 2(7 - 4) = 2 \cdot 3 = 6.$$

Alternatively, since $2x$ is an antiderivative of the function $f(x) = 2$,

$$\int_4^7 2 \, dx = 2x \Big|_4^7 = 2 \cdot 7 - 2 \cdot 4 = 14 - 8 = 6.$$

Example 11 Compute $\int_5^2 x^2 \, dx$.

Solution By (6),

$$\int_5^2 x^2 \, dx = -\int_2^5 x^2 \, dx = -\frac{x^3}{3}\bigg|_2^5 = -\frac{1}{3}(5^3 - 2^3)$$

$$= -\frac{1}{3}(125 - 8) = -\frac{117}{3}.$$

Example 12 Compute $\int_1^3 17x^3 \, dx$.

Solution By (8),

$$\int_1^3 17x^3 \, dx = 17 \int_1^3 x^3 \, dx = 17\frac{x^4}{4}\bigg|_1^3 = \frac{17}{4}(3^4 - 1^4)$$

$$= \frac{17}{4}(81 - 1) = \frac{17}{4} \cdot 80 = 340.$$

Example 13 Compute $\int_0^1 (\sqrt{x} + \sqrt[3]{x}) \, dx$.

Solution From (9),

$$\int_0^1 (x^{1/2} + x^{1/3}) \, dx = \int_0^1 x^{1/2} \, dx + \int_0^1 x^{1/3} \, dx$$

$$= \frac{2}{3} x^{3/2}\bigg|_0^1 + \frac{3}{4} x^{4/3}\bigg|_0^1 = \frac{2}{3} + \frac{3}{4} = \frac{17}{12}.$$

Example 14 To illustrate (7), we verify that

$$\int_0^4 x^2 \, dx = \int_0^1 x^2 \, dx + \int_1^4 x^2 \, dx.$$

We have

$$\int_0^1 x^2 \, dx = \frac{x^3}{3}\bigg|_0^1 = \frac{1}{3},$$

$$\int_1^4 x^2 \, dx = \frac{x^3}{3}\bigg|_1^4 = \frac{64}{3} - \frac{1}{3} = \frac{63}{3},$$

$$\int_0^4 x^2 \, dx = \frac{x^3}{3}\bigg|_0^4 = \frac{64}{3}.$$

Note that

$$\frac{64}{3} = \frac{63}{3} + \frac{1}{3}.$$

We conclude this section with an application from business mathematics. It depends on the material on present value in Sections 8.2 and 8.3. In Section 8.3 we computed the present value of an annuity in which money is received at fixed intervals. Now suppose that income is received continuously. Let $a(t)$ denote a *stream of income*. We know that $\$P$ now is worth Pe^{it} dollars in t years if money is compounded continuously (see page 651). Thus the present value of $a(t)$ dollars in t years is $a(t)/e^{it} = a(t)e^{-it}$ dollars. The following can be shown.

Present Value of a Stream of Income   The present value of a stream of income received up to the time $t = T$ years, where interest is compounded continuously, is given by

$$\text{present value} = \int_0^T a(t)e^{-it}\,dt. \tag{10}$$

Example 15   An investment broker has a choice between investing $100,000 in a growing business or putting it into bonds yielding an 8% annual return, compounded continuously for 20 years. The annual income from the business is projected to be $4000(1 + e^{t/10})$ dollars after t years. Where should she put her money?

Solution   Here $a(t) = 4000(1 + e^{0.1t})$. To the banker, the money can earn 8%. Thus the present value of the investment in the business is

$$\int_0^{20} 4000(1 + e^{0.1t})e^{-0.08t}\,dt = 4000 \int_0^{20} (e^{-0.08t} + e^{0.02t})\,dt$$

$$= 4000\left\{ \frac{e^{-0.08t}}{-0.08}\Big|_0^{20} + \frac{e^{0.02t}}{0.02}\Big|_0^{20} \right\}$$

$$= 4000\left\{ \frac{1 - e^{-1.6}}{0.08} + \frac{e^{0.4} - 1}{0.02} \right\}$$

$$\approx 4000(9.9763 + 24.5912)$$

$$= (4000)(34.5675) = \$138,270.$$

Thus, if she invests in the business, she can increase the present value of her money by about 38%. Of course, this is under the assumption that there is no greater risk in the business investment.

PROBLEMS 12.5   In Problems 1–40 calculate the given definite integral.

1. $\displaystyle\int_{-1}^{2} x^4\,dx$

2. $\displaystyle\int_{2}^{5} 3s^3\,ds$

3. $\displaystyle\int_{1}^{9} \frac{\sqrt{t}}{2}\,dt$

4. $\displaystyle\int_{1}^{3} (x^2 + 3x + 5)\,dx$

5. $\int_a^b (c_1 y^2 + c_2 y + c_3)\, dy$

6. $\int_2^4 \left(\dfrac{1}{z^3} - \dfrac{1}{z^2}\right) dz$

7. $\int_1^8 \left(\dfrac{1}{\sqrt[3]{x}} + 7\sqrt[3]{x}\right) dx$

8. $\int_0^1 (1 + s^8 + s^{16} + s^{32})\, ds$

9. $\int_{-1}^1 (p^9 + p^{17})\, dp$

10. $\int_{-a}^a x^{2n+1}\, dx$, where n is a positive integer and a is a real number

11. $\int_9^{16} \dfrac{v+1}{\sqrt{v}}\, dv$ [*Hint*: Divide.]

12. $\int_1^4 \dfrac{z^2 + 2z + 5}{z^{3/2}}\, dz$

13. $\int_2^3 (y-1)(y+2)\, dy$ [*Hint*: Multiply.]

14. $\int_{-2}^2 (v^2 - 4)(v^5 + 6)\, dv$

15. $\int_0^1 (t^{3/2} - t^{2/3})(t^{4/3} - t^{3/4})\, dt$

16. $\int_0^1 (\sqrt{x} - \sqrt[3]{x})^2\, dx$

17. $\int_1^0 s^{100}\, ds$

18. $\int_{-1}^{-2} \dfrac{1}{s^{100}}\, ds$

19. $\int_0^7 \sqrt{9+x}\, dx$

20. $\int_0^3 \sqrt{10+3x}\, dx$

21. $\int_0^1 x^3 \sqrt[3]{1+3x^4}\, dx$

22. $\int_1^3 \dfrac{w+1}{\sqrt{w^2+2w-1}}\, dw$

23. $\int_1^2 \dfrac{(1+1/v^2)^{5/3}}{v^3}\, dv$

24. $\int_{-3}^3 \left(\dfrac{x}{3}-1\right)^{10} dx$

25. $\int_0^2 e^{3x}\, dx$

26. $\int_{\ln 3}^{\ln 7} e^x\, dx$

27. $\int_e^{e^2} \dfrac{1}{x}$

28. $\int_2^5 \dfrac{1}{4x}\, dx$

29. $\int_{\ln 2}^{\ln 4} xe^{x^2}\, dx$

30. $\int_2^5 \ln x\, dx$

31. $\int_0^1 \dfrac{x}{1+x^2}\, dx$

32. $\int_{-2}^1 \dfrac{x^3}{x^4+25}\, dx$

33. $\int_0^1 \dfrac{e^{2x}}{1+e^{2x}}\, dx$

34. $\int_1^4 \dfrac{e^{1/x}}{x^2}\, dx$

35. $\int_0^2 \dfrac{e^x}{e^x+4}\, dx$

36. $\int_0^2 \ln e^x\, dx$

37. $\int_0^1 xe^{3x}\, dx$

38. $\int_{1/7}^{2/7} xe^{-7x}\, dx$

39. $\int_1^e x \ln x\, dx$

40. $\int_0^5 x\sqrt{3x+1}\, dx$

In Problems 41–47 a marginal cost function is given. Find the increase or decrease in cost when the number of units purchased is increased or decreased as indicated.

41. $MC = 30 - \dfrac{q}{10}$; from 50 to 70

 42. $MC = 750 + 0.24\sqrt{q} - 0.04q^{3/2}$; from 64 to 100

 43. $MC = 750 + 0.24\sqrt{q} - 0.04q^{3/2}$; from 64 to 25

44. $MC = \dfrac{4}{\sqrt{q+4}}$; from 45 to 77

45. $MC = \dfrac{4}{\sqrt{q+4}}$; from 144 to 100

46. $MC = 20 + \sqrt{1 + 2q}$; from 40 to 60

 47. $MC = 20 + (q/2)e^{-q/4}$; from 5 to 10

48. The marginal revenue that a manufacturer receives is given by $MR = 2 - 0.02q + 0.003q^2$ dollars per additional unit sold. How much additional money does the manufacturer receive if he increases sales from 50 to 100 units?

49. The velocity of a moving particle after t seconds is given by $v(t) = t^{3/2} + 16t + 1$, where v is measured in m/sec. How far does the particle travel between $t = 4$ and $t = 9$ sec?

50. A ball is thrown down from a tower with an initial velocity of 25 ft/sec. The tower is 500 ft high.

(a) How fast is the ball traveling after 3 sec?

(b) How far does the ball travel in the first 3 sec?

*(c) How long does it take for the ball to hit the ground? [*Hint:* Use $a = 32$ ft/sec^2.]

51. The population of a species of insects is increasing at a rate of $3000/\sqrt{t}$ individuals per week. How many individuals are added to the population between the ninth and the twenty-fifth week?

52. A charged particle enters a linear accelerator. Its velocity increases with a constant acceleration from an initial velocity of 500 m/sec to a velocity of 10,500 m/sec in 1/100 sec.

(a) What is the acceleration?

(b) How far does the particle travel in 1/100 sec?

If $f(x)$ takes on the n values $x_1, x_2, \ldots, x_n$, then the *average* of these values is $(x_1 + x_2 + \cdots + x_n)/n$. That is, we add up the values the function takes and divide the sum by n, the number of points. If f is continuous over the interval $[a, b]$ then the **average value of f over $[a, b]$** is defined by

$$\text{average value} = \frac{1}{b-a} \int_a^b f(x)\, dx.$$

In Problems 53–64 find the average value of the given function over the given interval.

53. $f(x) = 3x + 5$; $[1, 2]$

54. $f(x) = C$, C a constant; $[a, b]$

55. $f(x) = x^2$; $[0, 2]$

56. $f(x) = x^3$; $[0, 1]$

57. $f(x) = x^r$; $[0, 2]$ $(r \neq -1)$

58. $f(x) = x^2 - 2x + 5$; $[-2, 2]$

59. $f(x) = \sqrt{x} + 5x^3$; $[0, 4]$

60. $f(x) = 1 + x + x^2 + x^3 + x^4$; $[0, 1]$

61. $f(x) = \dfrac{1}{x^2}$; $[1, 3]$

62. $f(x) = x^{2/3} + \dfrac{1}{\sqrt[3]{x}}$; $[1, 8]$

63. $f(x) = \ln x$; $[1, 4]$

64. $f(x) = e^{-x}$; $[0, 2]$

65. If the marginal cost of a given product is $50 - 0.05q$ dollars, what is the average cost per unit of production if 200 units are produced?

66. One day the temperature of the air t hours after noon was found to be $60 + 4t - t^2/3$ degrees (Fahrenheit). What was the average temperature between noon and 5 P.M.?

'67. A ball was dropped from rest at a height of 400 feet. What was its average velocity on the way to the ground?

68. For the marginal cost function of Problem 41, what is the average cost of producing 100 units of the product?

69. What is the present value of an annuity if the stream of income is $a(t) = 10,000(1 + e^{0.13t})$ after 10 years, assuming continuous compounding at 8%?

12.6 Area and the Definite Integral

Modern calculus has its origins in two mathematical problems of antiquity. The first of these, the problem of finding the line tangent to a given curve, was, as we have noted, not solved until the seventeenth century. Its solution (by Newton and Leibniz) gave rise to what is known as *differential calculus*. The second of these problems was to find the area enclosed by a given curve. The solution of this problem led to what is now termed *integral calculus*.

It is not known how long scientists have been concerned with finding the area bounded by a curve. In 1858 Henry Rhind, an Egyptologist from Scotland, discovered fragments of a papyrus manuscript written in approximately 1650 B.C., which came to be known as the *Rhind papyrus.*[†] The Rhind papyrus contains 85 problems and was written by the Egyptian scribe Ahmes, who wrote that he copied the problems from an earlier manuscript. In Problem 50, Ahmes assumed that the area of a circular field with a diameter of 9 units was

---

[†] For an interesting discussion of the Rhind papyrus and other similar finds, see C. B. Boyer, *A History of Mathematics*, (New York: Wiley, 1968).

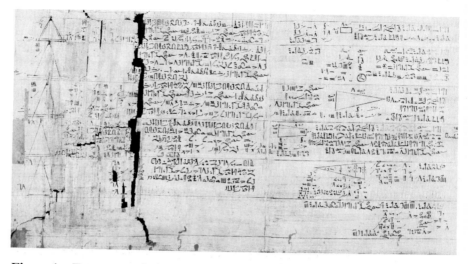

Figure 1 Fragment of the Rhind Papyrus (courtesy of the Trustees of the British Museum).

the same as the area of a square with a side of 8 units. If we compare this with the correct formula for the area of a circle, we find that

$$A = \pi r^2 = \pi \left(\frac{9}{2}\right)^2 = (\text{according to Ahmes}) \; 8^2$$

or

$$\pi \approx \frac{64}{(4.5)^2} = \frac{64}{20.25} \approx 3.16.$$

Thus we see that *before* 1650 B.C. the Egyptians could calculate the area of a circle from the formula

$$A = 3.16r^2.$$

Since $\pi \approx 3.1416$, we see that this is remarkably close, considering that the Egyptian formula dates back nearly 4000 years!

In ancient Greece there was much interest in obtaining methods for calculating the areas bounded by curves other than circles and rectangles. The problem was solved for a wide variety of curves by Archimedes of Syracuse (287–212 B.C.), who is considered by many to be the greatest mathematician who ever lived. Archimedes used what he called the *method of exhaustion* to calculate the shaded area, A, bounded by a parabola (see Figure 2).

We shall discuss Archimedes' method in the next section. It turns out that in computing an area, one is led to a new definition of the definite integral. When the definite integral is given this way, we can see how areas can easily be

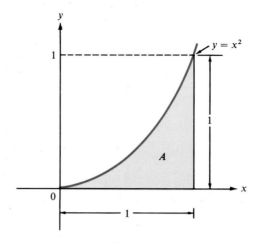

Figure 2

computed. The results we give below will be discussed in Section 12.8. We start by assuming that the function is nonnegative.

Area under a Curve I Let f be a nonnegative continuous function on the interval $[a, b]$. Then the **area** between the graph of f and the x-axis for x in the interval $[a, b]$, denoted A_a^b, is given by

$$A_a^b = \int_a^b f(x)\, dx. \qquad (1)$$

Example 1 Compute the area under the line $y = 2x$ and over the x-axis between $x = 0$ and $x = 2$.

Solution The requested area is shaded in Figure 3.

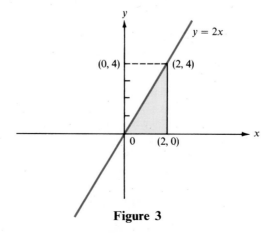

Figure 3

Then, from equation (1),

$$A_0^2 = \int_0^2 2x \, dx = x^2 \Big|_0^2 = 4.$$

We can verify this result since the area here is that of a triangle. We know that

$$\text{area of a triangle} = \tfrac{1}{2}(\text{base} \cdot \text{height}),$$

so that in our case,

$$A_0^2 = \tfrac{1}{2}(2 \cdot 4) = 4,$$

which agrees with the result we obtained with the definite integral.

Example 2 Compute the area between the curve $y = x^2$ and the x-axis for x in the interval $[0, 1]$.

Solution The requested area is sketched in Figure 4. We have

$$A_0^1 = \int_0^1 x^2 \, dx = \frac{x^3}{3} \Big|_0^1 = \frac{1}{3}.$$

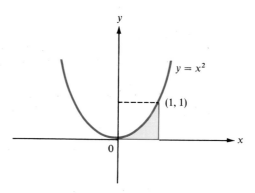

Figure 4

We now treat the case in which the function is not necessarily nonnegative.

Area under a Curve II (**Area Bounded by a Curve and the x-Axis**) Let f be a continuous function. Then

$$A_a^b = \int_a^b |f(x)| \, dx.$$

Example 3 Calculate the area bounded by the curve $y = x^3$, the x-axis, and the lines $x = -1$ and $x = 1$ (see Figure 5).

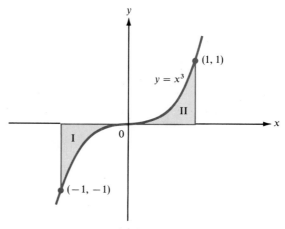

Figure 5

Solution In $[-1, 0]$, x^3 is negative, so that $|x^3| = -x^3$. In $[0, 1]$, x^3 is positive. Thus

$$A_{-1}^1 = \int_{-1}^1 f(x)\,dx, \qquad \text{where} \quad f(x) = \begin{cases} x^3, & x \geq 0 \\ -x^3, & x < 0 \end{cases}.$$

Geometrically, it appears that the area of region I in Figure 5 is equal to the area of region II.

To verify this, we have

$$\text{area of region I} = A_{-1}^0 = \int_{-1}^0 (-x^3)\,dx = \left.\frac{-x^4}{4}\right|_{-1}^0 = -\frac{1}{4}(0 - 1) = \frac{1}{4}$$

and

$$\text{area of region II} = \int_0^1 x^3\,dx = \left.\frac{x^4}{4}\right|_0^1 = \frac{1}{4}.$$

Thus,

$$A_{-1}^1 = A_{-1}^0 + A_0^1 = \tfrac{1}{4} + \tfrac{1}{4} = \tfrac{1}{2}.$$

Note that

$$\int_{-1}^1 x^3\,dx = \left.\frac{x^4}{4}\right|_{-1}^1 = \frac{1}{4}(1^4 - (-1)^4) = \frac{1}{4}(1 - 1) = 0.$$

So if the absolute value of x^3 is not taken, we get the wrong answer.

Warning. A fairly common error made by students when they first face the problem of calculating areas by integration is to employ the following *incorrect* reasoning: "If I get a negative answer when calculating an area, then all I need to do is to take the absolute value of my answer to make it right." To see why this reasoning is faulty, consider the problem of computing the area bounded by the line $y = x$ and the x-axis for x between -2 and 1. This area is

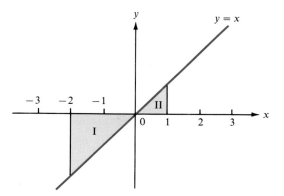

Figure 6

drawn in Figure 6. It is easy to see that the area of triangle I is 2 ($=\frac{1}{2}bh = \frac{1}{2}(2)(2)$) and the area of triangle II is $\frac{1}{2}$. Thus the total area is $\frac{5}{2}$. However, it is not difficult to show that

$$\int_{-2}^{1} x \, dx = -\frac{3}{2}.$$

This is the answer we would get if we forgot to take the absolute value in $\int_{-2}^{1} |x| \, dx$. Changing the $-\frac{3}{2}$ to $\frac{3}{2}$ will not give us the correct answer.

Example 4   Calculate the area bounded by the curve $y = x^3 - 6x^2 + 11x - 6$ and the x-axis.

Solution   We have $y = x^3 - 6x^2 + 11x - 6 = (x - 1)(x - 2)(x - 3)$. The curve is graphed in Figure 7. The desired area is the shaded part of the graph.

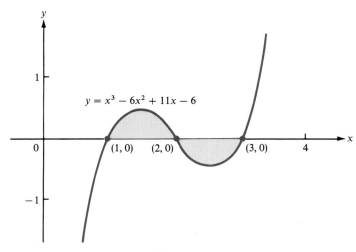

Figure 7

We know that

$$A = \int_1^3 |x^3 - 6x^2 + 11x - 6| \, dx$$

f is positive in (1, 2) so in (1, 2), |*f*| = *f*.　　　*f* is negative in (2, 3) so in (2, 3), |*f*| = −*f*.

$$= \int_1^2 (x^3 - 6x^2 + 11x - 6)\, dx + \int_2^3 -(x^3 - 6x^2 + 11x - 6)\, dx$$

$$= \left(\frac{x^4}{4} - 2x^3 + \frac{11x^2}{2} - 6x\right)\Big|_1^2 - \left(\frac{x^4}{4} - 2x^3 + \frac{11x^2}{2} - 6x\right)\Big|_2^3$$

$$= \left(4 - 16 + 22 - 12 - \frac{1}{4} + 2 - \frac{11}{2} + 6\right)$$

$$- \left(\frac{81}{4} - 54 + \frac{99}{2} - 18 - 4 + 16 - 22 + 12\right) = \frac{1}{2}.$$

Note that

$$\int_1^3 (x^3 - 6x^2 + 11x - 6)\, dx = \left(\frac{x^4}{4} - 2x^3 + \frac{11x^2}{2} - 6x\right)\Big|_1^3 = 0.$$

This illustrates why in the process of calculating area, care must be taken so that "positive" areas and "negative" areas don't cancel each other out.

Example 5   Calculate the area bounded by the curve $y = (1 + x^3)/\sqrt[4]{4x + x^4}$, the *x*-axis, and the lines $x = 1$ and $x = 3$.

Solution   It is not necessary to draw this curve to see that $(1 + x^3)/\sqrt[4]{4x + x^4} > 0$ for $1 \le x \le 3$. Thus the area is represented by the definite integral

$$\int_1^3 \frac{1 + x^3}{\sqrt[4]{4x + x^4}}\, dx.$$

To integrate, let $u = 4x + x^4$. Then $du = (4 + 4x^3)\, dx$, and we multiply and divide by 4 to obtain

$$\int \frac{1 + x^3}{\sqrt[4]{4x + x^4}}\, dx = \frac{1}{4}\int \frac{4(1 + x^3)}{\sqrt[4]{4x + x^4}}\, dx = \frac{1}{4}\int u^{-1/4}\, du$$

$$= \frac{1}{3} u^{3/4} + C = \frac{1}{3}(4x + x^4)^{3/4} + C.$$

Thus

$$\int_1^3 \frac{1 + x^3}{\sqrt[4]{x + x^4}}\, dx = \frac{1}{3}(4x + x^4)^{3/4}\Big|_1^3 = \frac{1}{3}(12 + 81)^{3/4} - \frac{1}{3}(5)^{3/4}$$

$$= \frac{1}{3}(93^{3/4} - 5^{3/4}).$$

Example 6   Compute the area between the curve $y = e^{-x}$ and the x-axis for x between 0 and 100.

Solution   The area is sketched in Figure 8.

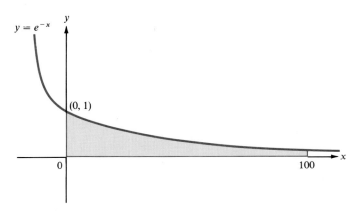

Figure 8

Then

$$A_0^{100} = \int_0^{100} e^{-x}\,dx = -e^{-x}\Big|_0^{100} = -e^{-100} + e^0 = 1 - e^{-100}.$$

On a calculator
↓

$$e^{-100} = 3.72 \times 10^{-44}$$

so, for all practical purposes, $e^{-100} = 0$ and $-e^{-x}\big|_0^{100} = -(0 - 1) = 1$.

PROBLEMS 12.6   In Problems 1–22 compute the area between the given curve and the x-axis for x in the given interval. If no interval is given, sketch the curve and determine an appropriate interval (that is, determine where the curve intersects the x-axis).

1. $y = x^3$; $[0, 2]$ **2.** $y = x^3$; $[-2, 2]$

3. $y = \sqrt{x}$; $[0, 5]$ **4.** $y = \dfrac{1}{x}$; $[1, 6]$

5. $y = \dfrac{1}{x}$; $[2, 7]$ **6.** $y = e^x$; $[0, 1]$

7. $y = x^2 - 4$ **8.** $y = x^2 + 2x - 3$; $[1, 3]$

9. $y = x^2 - 6x + 5$ **10.** $y = 9 - x^2$

11. $y = 10 + 3x - x^2$ **12.** $y = x^3 + 2x^2 - x - 2$

13. $y = (x^2 - 1)(x^2 - 4)$ **14.** $y = x^3 + 2x^2 - 13x + 10$; $[0, 3]$

15. $y = (x - a)(x - b)$, $a < b$ **16.** $y = \sqrt{x + 2}$; $[0, 7]$

17. $y = x\sqrt{x^2 + 7}$; $[0, 3]$ **18.** $y = \dfrac{x + 1}{x^3}$; $[\frac{1}{3}, \frac{1}{2}]$

19. $y = xe^{-x}$; $[0, 50]$ **20.** $y = e^{-x}$; $[0, 10^{25}]$

21. $y = xe^{-x}$; $[0, 10^6]$ **22.** $y = x^3 e^{-x}$; $[0, 10^{100}]$

23. An **even function** is a function, f, with the property that $f(-x) = f(x)$ for every real number x. (For example, 1, x^2, x^4, $|x|$, and $1/(1 + x^2)$ are all even functions.) Show that if f is even, then

$$\int_{-a}^{a} f = 2 \int_{0}^{a} f$$

for every real number a. Can you explain this fact geometrically?

24. An **odd function** f has the property that $f(-x) = -f(x)$ for every real number x. (For example, x, x^3, and $1/(x^5 + x^7)$ are all odd functions.) Show that if f is odd, then

$$\int_{-a}^{a} f = 0$$

for any real number a. Can you explain this fact geometrically?

25. Calculate the following integrals.

(a) $\displaystyle\int_{-50}^{50} (x + x^3 + x^{17})\, dx$ (b) $\displaystyle\int_{-100}^{100} (1 + x^{1/3} + x^{5/3} + x^{11/3})\, dx$

12.7 Improper Integrals (Optional)

In the examples of the last section we computed areas under curves over an interval $[a, b]$, where a and b are real numbers. However, in many applications, especially in probability and statistics, it is necessary to compute areas over infinite intervals. An integral in which one (or both) of the limits of integration is infinite is called an **improper integral**.

Example 1   Compute the area bounded by the curve $y = e^{-x}$ and the x-axis for $x \geq 0$.

Solution   The area is sketched in Figure 1. In Example 12.6.6 on page 701 we showed that

$$\int_{0}^{100} e^{-x}\, dx = 1 - 3.72 \times 10^{-44} \approx 1.$$

Now we need to compute $\int_0^\infty e^{-x}\, dx$, even though we have not yet defined what we mean by an integral with an infinite limit. However, it seems that the area under the curve in the interval $[0, \infty)$ can be computed by calculating the area in the interval $[0, N]$ and then seeing what happens as N gets large (that is, as $N \to \infty$). We compute

$$\int_{0}^{N} e^{-x}\, dx = -e^{-x} \Big|_{0}^{N} = -e^{-N} + e^0 = 1 - e^{-N}.$$

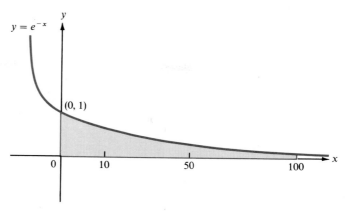

Figure 1

In Table 1 we display values of $1 - e^{-N}$ for various values of N. It is clear that as $N \to \infty$, $\int_0^N e^{-x} \, dx \to 1$. Thus we have

$$\int_0^\infty e^{-x} \, dx = 1 = \text{area under the curve } e^{-x} \text{ for } x \geq 0.$$

TABLE 1

N	e^{-N}	$\int_0^N e^{-x} \, dx = 1 - e^{-N}$
1	0.367879	0.632121
2	0.135335	0.864665
3	0.049787	0.950213
5	0.006738	0.993262
10	0.0000454	0.9999546
25	1.389×10^{-11}	$1 - 3.889 \times 10^{-11} \approx 0.99999999996$
50	1.929×10^{-22}	$1 - 1.929 \times 10^{-22}$
100	3.72×10^{-44}	$1 - 3.72 \times 10^{-44}$
200	1.384×10^{-87}	$1 - 1.384 \times 10^{-87}$

Example 1 suggests the following definitions.

Improper Integral   (a) Let f be continuous and let a be a real number. Then

$$\int_a^\infty f(x) \, dx = \lim_{N \to \infty} \int_a^N f(x) \, dx, \tag{1}$$

providing that the limit in formula (1) exists and is a finite number. In this case we say that the integral **converges**. Otherwise, the **improper integral** $\int_a^\infty f(x) \, dx$ is said to be **divergent**.

(b)
$$\int_{-\infty}^{a} f(x)\, dx = \lim_{M \to \infty} \int_{-M}^{a} f(x)\, dx, \tag{2}$$

providing that the limit in (2) exists and is a finite number. In this case the integral converges. Otherwise, the improper integral is said to be divergent.

(c)
$$\int_{-\infty}^{\infty} f(x)\, dx = \lim_{N \to \infty} \int_{0}^{N} f(x)\, dx + \lim_{M \to \infty} \int_{-M}^{0} f(x)\, dx, \tag{3}$$

providing that both of the limits in (3) exist and are finite. In this case the integral converges. If one or both of them fail to exist, then the improper integral is said to be divergent.

Example 2   Evaluate $\int_{1}^{\infty} (1/x)\, dx$.

Solution $\int_{1}^{N} (1/x)\, dx = \ln x\,|_{1}^{N} = \ln N - \ln 1 = \ln N$. But $\lim_{N \to \infty} \ln N = \infty$, so the improper integral is divergent.

Example 3   Evaluate $\int_{0}^{\infty} e^x\, dx$.

Solution $\int_{0}^{N} e^x\, dx = e^N - 1$, which approaches ∞ as $N \to \infty$, so that this improper integral is also divergent.

Example 4   Evaluate $\int_{-\infty}^{0} e^x\, dx$.

Solution $\int_{-N}^{0} e^x\, dx = 1 - e^{-N}$, which $\to 1$ as $N \to \infty$, so that $\int_{-\infty}^{0} e^x\, dx$ is convergent and is equal to 1.

Example 5   Evaluate $\int_{-\infty}^{\infty} xe^{-x^2}\, dx$.

Solution   The function xe^{-x^2} is sketched in Figure 2. $\int_{0}^{N} xe^{-x^2}\, dx = -\tfrac{1}{2}e^{-x^2}|_{0}^{N} = \tfrac{1}{2}(1 - e^{-N^2}) \to \tfrac{1}{2}$ as $N \to \infty$; and $\int_{-M}^{0} xe^{-x^2}\, dx = -\tfrac{1}{2}e^{-x^2}|_{-M}^{0} = \tfrac{1}{2}(e^{-M^2} - 1) \to$

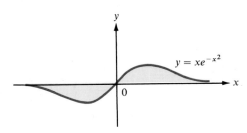

Figure 2

$-\frac{1}{2}$ as $M \to \infty$. Thus, since both limits exist,

$$\int_{-\infty}^{\infty} xe^{-x^2}\, dx = \int_{0}^{\infty} xe^{-x^2}\, dx + \int_{-\infty}^{0} xe^{-x^2}\, dx = \frac{1}{2} - \frac{1}{2} = 0.$$

Example 6 Calculate $\int_{-\infty}^{\infty} x^3\, dx$.

Solution The area to be calculated is sketched in Figure 3. Since $\lim_{N\to\infty} \int_{0}^{N} x^3\, dx = \lim_{N\to\infty} (N^4/4) = \infty$, the integral diverges. Note, however, that

$$\lim_{N\to\infty} \int_{-N}^{N} x^3\, dx = \lim_{N\to\infty} \frac{x^4}{4}\Big|_{-N}^{N} = \lim_{N\to\infty} \left(\frac{N^4}{4} - \frac{N^4}{4}\right) = \lim_{N\to\infty} 0 = 0.$$

Here $\int_{0}^{\infty} x^3\, dx = \infty$ and $\int_{-\infty}^{0} x^3\, dx = -\infty$. We simply cannot "cancel off" infinite terms. The expression $\infty - \infty$ is not defined.

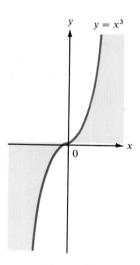

Figure 3

In computing improper integrals, the following fact is very useful. For any real number b,

$$\boxed{\lim_{N\to\infty} N^b e^{-aN} = 0 \qquad \text{if} \quad a > 0.}$$

(4)

The numbers in Table 2 indicate, as an illustration, that

$$\lim_{N\to\infty} N^3 e^{-(1/2)N} = 0.$$

TABLE 2

N	N^3	$-\frac{1}{2}N$	$e^{-(1/2)N}$	$N^3 e^{-(1/2)N}$
1	1	−0.5	0.6065	0.6065
2	8	−1	0.3679	2.943
5	125	−2.5	0.0821	10.26
10	1000	−5	0.006738	6.738
15	3375	−7.5	0.0005531	1.8667
20	8000	−10	0.0000454	0.3632
25	15,625	−12.5	0.0000037267	0.0582
35	42,875	−17.5	0.0000000251	0.00108
50	125,000	−25	1.3888×10^{-11}	0.000001736
100	1,000,000	−50	1.9287×10^{-22}	1.9287×10^{-16}

Example 7 Compute $\int_0^\infty xe^{-5x}\,dx$.

Solution We compute $\int_0^N xe^{-5x}\,dx$ by parts. Let $u = x$ and $dv = e^{-5x}\,dx$. Then $du = dx$, $v = -\frac{1}{5}e^{-5x}$, and

$$\int_0^N xe^{-5x}\,dx = x\left(-\frac{1}{5}e^{-5x}\right)\Big|_0^N - \int_0^N \left(-\frac{1}{5}e^{-5x}\right)dx$$

$$= -\frac{x}{5}e^{-5x}\Big|_0^N + \frac{1}{5}\int_0^N e^{-5x}\,dx = -\frac{N}{5}e^{-5N} - \frac{1}{25}e^{-5x}\Big|_0^N$$

$$= -\frac{N}{5}e^{-5N} - \frac{1}{25}e^{-5N} + \frac{1}{25}.$$

Now, $\lim_{N\to\infty} e^{-5N} = 0$ and, from equation (4), $\lim_{N\to\infty} Ne^{-5N} = 0$. Thus,

$$\lim_{N\to\infty}\int_0^N xe^{-5x}\,dx = \frac{1}{25} = \int_0^\infty xe^{-5x}\,dx.$$

In Section 12.5, we discussed the notion of present value, which is defined as the current worth of an investment paying a certain sum of money at some time in the future. If $a(t)$ denotes a stream of income, then the present value of the money received up to the time T years in the future is given by

$$\text{present value} = \int_0^T a(t)e^{-it}\,dt,$$

where i denotes the annual interest rate (see equation (12.5.10) on page 691). If the income is to be received indefinitely, then we would be interested in calculating the present value of the investment for an indefinitely long time in the future. This leads to the equation

$$\text{present value} = \int_0^\infty a(t)e^{-it}\,dt.$$

Example 8   A corporation makes an investment from which it expects to receive a stream of income of $2400 + 18t$ dollars, where t is the time (in years) since the investment was made. What is the present value of this investment assuming an annual interest rate of 7% (compounded continuously)?

Solution   We have $a(t) = 2400 + 18t$. Since $i = 0.07$, we have

$$\text{present value} = \int_0^\infty (2400 + 18t)e^{-0.07t}\,dt$$

$$= \lim_{N \to \infty} \int_0^N (2400 + 18t)e^{-0.07t}\,dt.$$

Then, integrating by parts, we have, with $u = 2400 + 18t$, $du = 18dt$, $dv = e^{-0.07t}\,dt$, and $v = -(1/0.07)e^{-0.07t}$,

$$\text{present value} = \lim_{N \to \infty} \left\{ \overbrace{(2400 + 18t)}^{u} \overbrace{\frac{-1}{0.07}e^{-0.07t}}^{v} \Big|_0^N - \int_0^N \overbrace{\frac{-1}{0.07}e^{-0.07t}}^{v} \overbrace{18\,dt}^{du} \right\}$$

$$= \lim_{N \to \infty} \left\{ \frac{1}{0.07}[2400 - (2400 + 18N)e^{-0.07N}] - \frac{18}{(0.07)^2}e^{-0.07t}\Big|_0^N \right\}$$

$$= \frac{2400}{0.07} + \frac{18}{(0.07)^2} \approx \$37{,}959.$$

In this calculation we used the fact that $\lim_{N \to \infty} Ne^{-0.07N} = 0$.

We close this section by giving an improper integral that is very important in statistical applications. Its computation requires techniques not discussed in this book.

$$\int_{-\infty}^{\infty} \frac{1}{\sqrt{2\pi}}e^{-x^2/2} = 1.$$

PROBLEMS 12.7   In Problems 1–24 determine whether the given integral converges or diverges. If it converges, find its value.

1. $\displaystyle\int_0^\infty e^{-2x}\,dx$ **2.** $\displaystyle\int_{-\infty}^4 e^{3x}\,dx$ **3.** $\displaystyle\int_{-\infty}^\infty e^{-0.01x}\,dx$

4. $\displaystyle\int_{-\infty}^{\infty} x^3 e^{-x^4}\, dx$ **5.** $\displaystyle\int_{-\infty}^{\infty} x^2 e^{-x^3}\, dx$ **6.** $\displaystyle\int_{16}^{\infty} \frac{dx}{\sqrt{x}}$

7. $\displaystyle\int_{1}^{\infty} \frac{1}{x^2}\, dx$ **8.** $\displaystyle\int_{1}^{\infty} \frac{dx}{x^{3/2}}$ **9.** $\displaystyle\int_{-\infty}^{-1} \frac{dx}{x^3}$

10. $\displaystyle\int_{0}^{\infty} \frac{2x}{x^2+1}\, dx$ **11.** $\displaystyle\int_{0}^{\infty} \frac{dx}{(x+1)^2}$ **12.** $\displaystyle\int_{0}^{\infty} \frac{dx}{\sqrt{x+1}}$

13. $\displaystyle\int_{0}^{\infty} \frac{dx}{x+1}$ **14.** $\displaystyle\int_{-\infty}^{\infty} \frac{x\, dx}{(x^2+1)^3}$ **15.** $\displaystyle\int_{-\infty}^{0} \frac{x}{(x^2+1)^3}\, dx$

16. $\displaystyle\int_{0}^{\infty} \frac{x^3}{(x^4+3)^3}\, dx$ **17.** $\displaystyle\int_{-\infty}^{\infty} \frac{x}{x^2+1}\, dx$ **18.** $\displaystyle\int_{-\infty}^{\infty} \frac{x}{(x^2+1)^2}\, dx$

19. $\displaystyle\int_{0}^{\infty} x e^{-2x}\, dx$ ***20.** $\displaystyle\int_{-\infty}^{\infty} x^2 e^{-x}\, dx$ **21.** $\displaystyle\int_{-\infty}^{\infty} x^3 e^{-x^4}\, dx$

22. $\displaystyle\int_{-\infty}^{0} x e^{x}\, dx$ ***23.** $\displaystyle\int_{0}^{\infty} x^2 e^{-x/10}\, dx$ ***24.** $\displaystyle\int_{0}^{\infty} x^4 e^{-x}\, dx$

In Problems 25–30 compute the area between the given curve and the x-axis for the given values of x.

25. $y = e^{-2x};\ x \geq 0$ **26.** $y = e^{-x/4};\ x \geq 0$

27. $y = e^{-bx};\ x \geq 0;\ b$ a positive number

28. $y = x e^{-bx};\ x \geq 0;\ b$ a positive number

29. $y = x e^{-4x};\ x \geq 3$ **30.** $y = x e^{-x/50};\ x \geq 0$

31. What is the present value of an annuity if the stream of income is $a(t) = 10{,}000(1 + e^{0.05t})$, paid annually over an indefinite period of time, assuming an annual interest rate of 8%?

32. What is the present value of an annuity if the stream of income is $a(t) = 20 + 2t + \frac{1}{2}t^2$ paid weekly over an indefinite period of time, assuming an annual interest rate of 6%?

*33. An investment broker has a choice between investing \$100,000 in a growing business or putting the money into bonds yielding a 12% annual return, compounded continuously for an indefinite period of time. The annual income from the business is projected to be $4000(1 + e^{t/10})$ dollars after t years. Where should she put her money? [*Hint:* See Example 12.5.15 on page 691.]

12.8 Area, Riemann Sums, and the Fundamental Theorem of Calculus

As we stated in the beginning of Section 12.6, Archimedes was the first to discover a systematic method for evaluating areas. To illustrate his **method of exhaustion**, let us estimate the area under the curve $y = x^2$ for x in $[0, 1]$. This

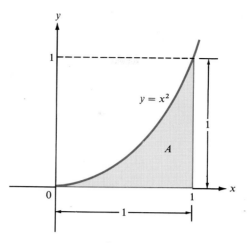

Figure 1

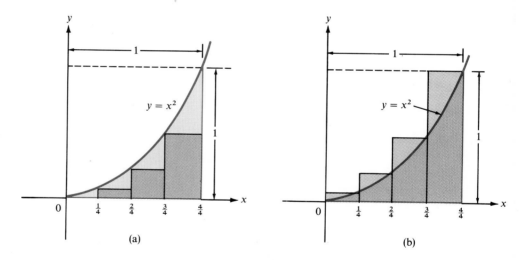

Figure 2

area is sketched in Figure 1. The shaded area A, bounded by the parabola $y = x^2$, the x-axis, and the line $x = 1$, was approximated by rectangles under the curve and over the curve (see Figure 2). If we let s denote the sum of the areas of the rectangles in Figure 2a and S the sum of the areas of the rectangles in Figure 2b, then

$$s < A < S. \tag{1}$$

Since Archimedes knew how to calculate the area of a rectangle (by multiplying its base by its height), he was able to calculate s and S exactly, thereby obtaining an estimate for the area A. Since the equation of the parabola is $y = x^2$, the height of each rectangle (which is the y-value on the curve) is the square of the x-value. Therefore, we have

$$s = \frac{1}{4}\left(\frac{1}{4}\right)^2 + \frac{1}{4}\left(\frac{2}{4}\right)^2 + \frac{1}{4}\left(\frac{3}{4}\right)^2 = \frac{1}{4}\left\{\left(\frac{1}{4}\right)^2 + \left(\frac{2}{4}\right)^2 + \left(\frac{3}{4}\right)^2\right\}$$

$$= \frac{1}{4\cdot4^2}(1^2 + 2^2 + 3^2) = \frac{1^2 + 2^2 + 3^2}{4^3} = \frac{14}{64} = \frac{7}{32}$$

and

$$S = \frac{1}{4}\left(\frac{1}{4}\right)^2 + \frac{1}{4}\left(\frac{2}{4}\right)^2 + \frac{1}{4}\left(\frac{3}{4}\right)^2 + \frac{1}{4}\left(\frac{4}{4}\right)^2 = \frac{1^2 + 2^2 + 3^2 + 4^2}{4^3}$$

$$= \frac{30}{64} = \frac{15}{32}.$$

Thus from equation (1), with $s = \frac{7}{32}$ and $S = \frac{15}{32}$ we obtain

$$0.22 \approx \frac{7}{32} < A < \frac{15}{32} \approx 0.47.$$

It was clear to Archimedes that this estimate could be improved by increasing the number of rectangles, so that the error in the estimate becomes smaller. By doubling the number of rectangles, we obtain the approximation depicted in Figure 3.

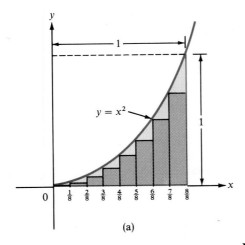

(a)

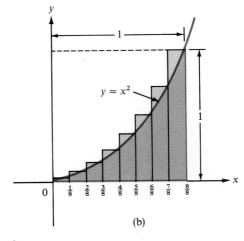

(b)

Figure 3

Here

$$s = \frac{1}{8}\left(\frac{1}{8}\right)^2 + \frac{1}{8}\left(\frac{2}{8}\right)^2 + \cdots + \frac{1}{8}\left(\frac{7}{8}\right)^2 = \frac{1^2 + 2^2 + 3^2 + 4^2 + 5^2 + 6^2 + 7^2}{8^3}$$

$$= \frac{140}{512} = \frac{35}{128}$$

and

$$S = \frac{1}{8}\left(\frac{1}{8}\right)^2 + \frac{1}{8}\left(\frac{2}{8}\right)^2 + \cdots + \frac{1}{8}\left(\frac{8}{8}\right)^2$$

$$= \frac{1^2 + 2^2 + 3^2 + 4^2 + 5^2 + 6^2 + 7^2 + 8^2}{8^3} = \frac{204}{512} = \frac{51}{128}.$$

We have shown that

$$0.27 \approx \frac{35}{128} < A < \frac{51}{128} \approx 0.40.$$

We see that the approximations have "squeezed down" somewhat. We now double the number of rectangles again to obtain

$$s = \frac{1}{16}\left(\frac{1}{16}\right)^2 + \frac{1}{16}\left(\frac{2}{16}\right)^2 + \cdots + \frac{1}{16}\left(\frac{15}{16}\right)^2$$

$$= \frac{1^2 + 2^2 + \cdots + 14^2 + 15^2}{16^3} = \frac{1240}{4096} = \frac{155}{512}$$

and

$$S = \frac{1}{16}\left(\frac{1}{16}\right)^2 + \frac{1}{16}\left(\frac{2}{16}\right)^2 + \cdots + \frac{1}{16}\left(\frac{15}{16}\right)^2 + \frac{1}{16}\left(\frac{16}{16}\right)^2$$

$$= \frac{1^2 + 2^2 + \cdots + 15^2 + 16^2}{16^3} = \frac{1496}{4096} = \frac{187}{512}.$$

Then

$$0.30 \approx \frac{155}{512} < A < \frac{187}{512} \approx 0.37$$

and the calculation of area is squeezed down further.

Using his method of exhaustion, Archimedes was able to show that, using the outer rectangles, the assumption $A > \frac{1}{3}$ led to a contradiction. Similarly, using the inner rectangles, he showed that $A < \frac{1}{3}$ led to a contradiction. From this he concluded that $A = \frac{1}{3}$.

Now, let $y = f(x)$ be a function that is positive and continuous on the interval $[a, b]$ (see Figure 4). We will show how the area bounded by this

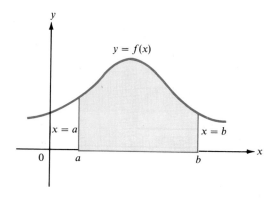

Figure 4

curve, the x-axis, and the lines $x = a$ and $x = b$ can be approximated. The method we will give here is very similar to Archimedes' method of exhaustion.

METHOD FOR COMPUTING AREA UNDER THE CURVE $y = f(x)$ OVER THE INTERVAL $[a, b]$

Step 1. Partition the interval $[a, b]$ into n parts, each having length

$$\Delta x = \frac{b - a}{n}$$

Let $a = x_0 < x_1 < x_2 < \cdots < x_n = b$ denote the endpoints of the subintervals, as illustrated in Figure 5.

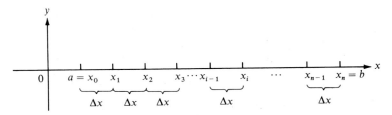

Figure 5

Step 2. Choose a point x_i^* in each interval $[x_{i-1}, x_i]$.

Step 3. Form the sum $f(x_1^*)\Delta x + f(x_2^*)\Delta x + \cdots + f(x_n^*)\Delta x$.

Suppose that the interval $[a, b]$ is partitioned into n subintervals of equal length. We denote the area bounded by the curve $y = f(x)$, the x-axis, and the lines $x = a$ and $x = b$ by A_a^b. We will approximate A_a^b by drawing rectangles whose total area is "close" to the actual area (see Figure 6). We locate the points $(x_i^*, f(x_i^*))$ on the curve for $i = 1, 2, \ldots, n$. The numbers $f(x_i^*)$ give us the heights of our n rectangles. The base of each rectangle has length x_i

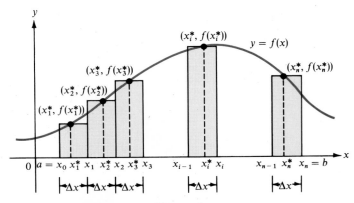

Figure 6

$- x_{i-1} = \Delta x$. This is all illustrated in Figure 6. The area, A_i, of the ith rectangle is the height of the rectangle times its length:

$$A_i = f(x_i^*)\Delta x. \tag{2}$$

Let us take a closer look at the area enclosed by these rectangles. In Figure 7, the shaded areas depict the differences between the region whose area we

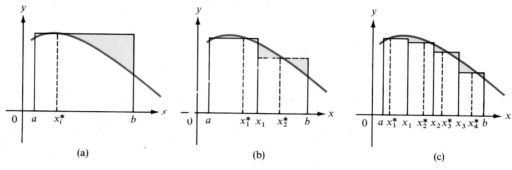

(a)   (b)   (c)

Figure 7

wish to calculate and the area enclosed by the rectangles. We see that as each rectangle becomes thinner and thinner, the area enclosed by the rectangles seems to get closer and closer to the area under the curve. But the length of the base of each rectangle is Δx, and so if Δx is reasonably small, we have the approximation

$$A_a^b \approx A_1 + A_2 + \cdots + A_i + \cdots + A_n,$$

or using equation (2),

$$\boxed{A_a^b \approx f(x_1^*)\Delta x + f(x_2^*)\Delta x + \cdots + f(x_n^*)\Delta x.} \tag{3}$$

Riemann Sum

The sum in (3) is called a **Riemann sum**.[†] As we have indicated, a Riemann sum approximates the area under a curve. As $n \to \infty$ ($\Delta x \to 0$), the approximation becomes better and better. This discussion leads to the next definition of the definite integral.

Alternative Definition of the Definite Integral[‡] Let f be continuous in $[a, b]$ and let $\Delta x = (b - a)/n$. Then

$$\int_a^b f(x)\,dx = \lim_{n \to \infty} [f(x_1^*)\Delta x + f(x_2^*)\Delta x + \cdots + f(x_n^*)\Delta x].$$

[†] G. F. B. Riemann (1826–1866) was a brilliant German mathematician.
[‡] Note that this definition, like the definition on page 685, says nothing about area. Area was discussed first to show you why this definition makes sense.

We will see shortly why the two seemingly very different definitions of the definite integral give the same number.

Example 4   Use Riemann sums to compute the area bounded by the curve $y = x^2$ and the x-axis for x in the interval $[0, 3]$.

Solution   Divide the interval into n equal subintervals, each having length $(b - a)/n = 3/n$. The partition points are

$$0 = \frac{0}{n} < \frac{3}{n} < \frac{6}{n} < \frac{9}{n} < \cdots < \frac{3(n-1)}{n} < \frac{3n}{n} = 3.$$

For convenience, we choose $x_i^* = x_i$ (the right-hand endpoint of the subinterval), so that

$$f(x_i^*) = f(x_i) = x_i^3 = \left(\frac{3i}{n}\right)^3 = \frac{27i^3}{n^3} \qquad \text{for} \quad i = 1, 2, \ldots, n$$

(see Figure 8).

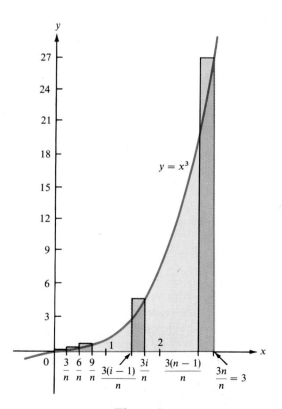

Figure 8

Then

$$A_0^3 \approx f(x_1^*)\Delta x + f(x_2^*)\Delta x + \cdots + f(x_n^*)\Delta x$$

$$= \left(\frac{27\cdot 1^3}{n^3} + \frac{27\cdot 2^3}{n^3} + \frac{27\cdot 3^3}{n^3} + \cdots + \frac{27\cdot n^3}{n^3}\right)\frac{3}{n}$$

$$= \frac{27}{n^3}\cdot\frac{3}{n}(1^3 + 2^3 + \cdots + n^3) = \frac{81}{n^4}(1^3 + 2^3 + 3^3 + \cdots + n^3). \qquad (4)$$

There is a formula for the sum of the first n cubes:

$$1^3 + 2^3 + 3^3 + \cdots + (n-1)^3 + n^3 = \left(\frac{n(n+1)}{2}\right)^2. \qquad (5)$$

You should satisfy yourself that this formula is correct by substituting some different values for n. Substituting (5) in (4), we obtain

$$A_0^3 \approx \frac{81}{n^4}\cdot\frac{n^2(n+1)^2}{4} = \frac{81}{4}\cdot\frac{(n+1)^2}{n^2}.$$

Then

Theorem 2 on page 519

$$A_0^3 = \lim_{n\to\infty}\frac{81}{4}\frac{(n+1)^2}{n^2} = \frac{81}{4}\lim_{n\to\infty}\frac{(n+1)^2}{n^2} = \frac{81}{4}\lim_{n\to\infty}\frac{n^2+2n+1}{n^2}$$

Divide top and bottom by n^2 Theorem 3 on page 519

$$= \frac{81}{4}\lim_{n\to\infty}\frac{1+\frac{2}{n}+\frac{1}{n^2}}{1} = \frac{81}{4}\left[\lim_{n\to\infty}1 + \lim_{n\to\infty}\frac{2}{n} + \lim_{n\to\infty}\frac{1}{n^2}\right]$$

$$= \frac{81}{4}(1+0+0) = \frac{81}{4}.$$

Note that this agrees with the result we get using the earlier definition of the definite integral (on p. 685). Since $x^4/4$ is an antiderivative for x^3,

$$A_0^3 = \int_0^3 x^3\,dx = \left.\frac{x^4}{4}\right|_0^3 = \frac{81}{4}.$$

The question you should be asking is "Why should the Riemann sum approximation to area have anything to do with the definite integral as defined in Section 12.5?" The answer is given by one of the most famous results in calculus.

Fundamental Theorem of Calculus   Let f be a continuous function and let F be an antiderivative for f. Let $a = x_0 < x_1 < x_2 < \cdots < x_{n-1} < x_n = b$ be a partition of $[a, b]$ into subintervals of length $\Delta x = (b - a)/n$, and let x_i^* for $i = 1, 2, \ldots, n$ be arbitrarily selected points in each subinterval. Then

$$\lim_{n \to \infty} [f(x_1^*)\Delta x + f(x_2^*)\Delta x + \cdots + f(x_n^*)\Delta x]$$

$$= \int_a^b f(x) \, dx = F(b) - F(a).$$

Remark.   This remarkable theorem tells us that if f is continuous and F is an antiderivative of f, then no matter how we choose the points x_i^* in each subinterval, the Riemann sums approach the value $F(b) - F(a)$ as the length of each subinterval approaches 0.

Let us indicate why the fundamental theorem of calculus is true. Let

$$G(x) = \int_a^x f(t) \, dt,$$

and suppose that $f(t) \geq 0$ for $a \leq t \leq b$. Then $\int_a^x f(t) \, dt$ represents the area under the curve f from a to x. That is,

$$G(x) = A_a^x.$$

Note that

$$G(a) = A_a^a = 0$$

and

$$G(b) = A_a^b = G(b) - G(a).$$

We will now indicate why G is an antiderivative for f.

Consider Figure 9, paying particular attention to the various areas under the curve $f(x)$. By definition of the derivative,

$$G'(x) = \lim_{\Delta x \to 0} \frac{G(x + \Delta x) - G(x)}{\Delta x}.$$

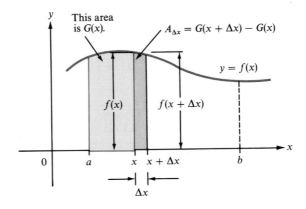

Figure 9

Since $G(x) = \int_a^x f(t)\, dt =$ area under the curve $f(x)$ between a and x, and assuming that $f > 0$ on $[a, b]$, we have

$$G(x + \Delta x) = \text{area between } a \text{ and } x + \Delta x;$$

$$G(x) = \text{area between } a \text{ and } x;$$

$$G(x + \Delta x) - G(x) = \text{area between } x \text{ and } x + \Delta x.$$

This last area is denoted $A_{\Delta x}$ in Figure 9. Now if Δx is small, then x and $x + \Delta x$ are close, and since f is continuous, $f(x + \Delta x)$ is close to $f(x)$. That is, the shaded area $A_{\Delta x}$ is approximately equal to the area of the rectangle with height $f(x)$ and base Δx. We therefore see that

$$G(x + \Delta x) - G(x) = A_{\Delta x} \approx f(x)\Delta x.$$

Then, for Δx small,

$$\frac{G(x + \Delta x) - G(x)}{\Delta x} \approx \frac{f(x)\Delta x}{\Delta x} = f(x).$$

Since this approximation gets better and better as $\Delta x \to 0$, we may assert that

$$G'(x) = \lim_{\Delta x \to 0} \frac{G(x + \Delta x) - G(x)}{\Delta x} = f(x),$$

which indicates that the derivative of G is f, as we wanted to show.

PROBLEMS 12.8   In Problems 1–10 use Riemann sums to estimate the area bounded by the given curve and the x-axis for x in the given interval. Pick x_i^* to be the right-hand endpoint of each subinterval. Then use the fundamental theorem of calculus to verify your estimate. You will need the following formulas:

(a) $1 + 2 + 3 + \cdots + n = \dfrac{n(n + 1)}{2};$

(b) $1^2 + 2^2 + 3^2 + \cdots + n^2 = \dfrac{n(n + 1)(2n + 1)}{6};$

(c) $1^3 + 2^3 + 3^3 + \cdots + n^3 = \dfrac{n^2(n + 1)^2}{4}.$

1. $y = 7x$; $[0, 4]$ 2. $y = 4x$; $[1, 2]$ 3. $y = 3x + 2$; $[0, 3]$

4. $y = \frac{1}{2}x^2$; $[0, 2]$ 5. $y = 3x^3$; $[0, 1]$ 6. $y = (1 - x)^2$; $[0, 1]$

7. $y = 1 - x^2$; $[0, 1]$ 8. $y = 1 - x^3$; $[0, 1]$ 9. $y = x^3$; $[1, 2]$

10. $y = 1 + x + x^2 + x^3$; $[0, 1]$

*11.   Let $S_n = \dfrac{1}{n}\left(\sqrt{1 - \left(\dfrac{1}{n}\right)^2} + \sqrt{1 - \left(\dfrac{2}{n}\right)^2} + \sqrt{1 - \left(\dfrac{3}{n}\right)^2} + \cdots + \sqrt{1 - \left(\dfrac{n - 1}{n}\right)^2} \right).$

(a)   Describe a region whose area is estimated by S_n.

(b)   By using general information about the area of the region described in part (a), find $\lim_{\Delta x \to 0} S_n$ where $\Delta x = 1/n$.

12.9 The Area Between Two Curves, with Applications to Economics (Optional)

In Section 12.8 we defined the definite integral as the limit of sums of areas of rectangles and showed how Archimedes used essentially this definition to find the area under a curve. In this section we show that by using similar methods, the area between two curves can be computed. We then apply these methods to the analysis of certain problems in economics.

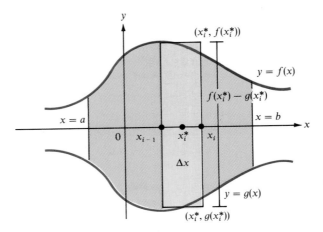

Figure 1

Consider the two functions f and g as graphed in Figure 1. We shall calculate the area between the curves $y = f(x)$ and $y = g(x)$ and the lines $x = a$ and $x = b$ by adding up the areas of a large number of rectangles. Note that $f(x) \geq g(x)$ for $a \leq x \leq b$. We partition $[a, b]$ into n subintervals of length $\Delta x = (b - a)/n$. Then a typical rectangle (see Figure 1) has the area

$$[f(x_i^*) - g(x_i^*)] \, \Delta x$$

where x_i^* is a point in the interval $[x_{i-1}, x_i]$ and $\Delta x = (b - a)/n = x_i - x_{i-1}$. Then, proceeding exactly as we have proceeded before, we find that

$$\text{area} = \lim_{n \to \infty} \sum_{i=1}^{n} [f(x_i^*) - g(x_i^*)] \Delta x = \int_a^b [f(x) - g(x)] \, dx.$$

This formula is valid as long as $f(x) \geq g(x)$ in $[a, b]$.

Example 1 Find the area bounded by $y = x^2$ and $y = 4x$, for x between 0 and 1.

Solution In any problem of this type, it is helpful to draw a graph. In Figure

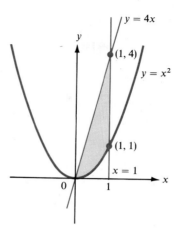

Figure 2

2, the required area is shaded. We have

$$A = \int_0^1 (4x - x^2)\, dx = \left(2x^2 - \frac{x^3}{3}\right)\Big|_0^1 = 2 - \frac{1}{3} = \frac{5}{3}.$$

Example 2   Find the area bounded by the curves $y = x^3$ and $y = \sqrt{x}$.

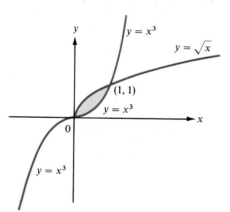

Figure 3

Solution   This problem makes sense only if the curves intersect at two points or more, as in Figure 3. (Otherwise, we would have to be given other bounding lines.) We need to find the points of intersection of these two curves. To find them, we set the two functions equal. If $x^3 = \sqrt{x}$, then $x^6 = x$ or $x^6 - x = x(x^5 - 1) = 0$. This occurs when $x = 0$ and $x = 1$. Then

$$A = \int_0^1 (\sqrt{x} - x^3)\, dx = \left(\frac{2x^{3/2}}{3} - \frac{x^4}{4}\right)\Big|_0^1 = \frac{2}{3} - \frac{1}{4} = \frac{5}{12}.$$

In these last two examples, note that we have had no trouble deciding which function came first in the expression $f(x) - g(x)$. We always put the larger function first. Thus in $[0, 1]$, $\sqrt{x} \geq x^3$, so that $\sqrt{x}$ comes first.

Example 3 Find the area bounded by the two curves $y = x^2 + 3x + 5$ and $y = -x^2 + 5x + 9$.

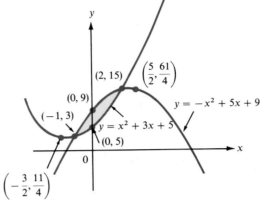

Figure 4

Solution We first sketch the two curves (see Figure 4). To find the points of intersection, we set $x^2 + 3x + 5 = -x^2 + 5x + 9$. This leads to the equation $2x^2 - 2x - 4 = 0$, which has roots $x = -1$ and $x = 2$. Thus

$$A = \int_{-1}^{2} [(-x^2 + 5x + 9) - (x^2 + 3x + 5)]\, dx$$

$$= \int_{-1}^{2} (-2x^2 + 2x + 4)\, dx = \left(-\frac{2x^3}{3} + x^2 + 4x\right)\Bigg|_{-1}^{2} = 9.$$

Example 4 Find the area bounded by the two curves $y = x^2 + 3x + 5$ and $y = -x^2 + 5x + 9$ and the line $x = 4$.

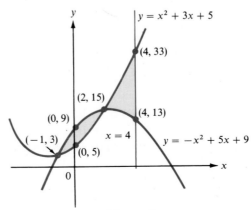

Figure 5

Solution This is more complicated than Example 3. See Figure 5. There are now two areas to be added together. In the first (calculated in Example 3),

$$-x^2 + 5x + 9 \geq x^2 + 3x + 5.$$

In the second,

$$x^2 + 3x + 5 \geq -x^2 + 5x + 9.$$

We therefore break the calculation into two parts:

$$A = \int_{-1}^{2} [(-x^2 + 5x + 9) - (x^2 + 3x + 5)]\, dx$$

$$+ \int_{2}^{4} [(x^2 + 3x + 5) - (-x^2 + 5x + 9)]\, dx$$

$$= 9 + \int_{2}^{4} (2x^2 - 2x - 4)\, dx = 9 + \left(\frac{2x^3}{3} - x^2 - 4x \right)\Big|_{2}^{4} = 9 + \frac{52}{3} = \frac{79}{3}.$$

Note that $(-x^2 + 5x + 9) - (x^2 + 3x + 5)$ in $[-1, 2]$ and $(x^2 + 3x + 5) - (-x^2 + 5x + 9)$ in $[2, 4]$ can be written as $|(x^2 + 3x + 5) - (-x^2 + 5x + 9)|$ in $[-1, 4]$. (Why?)

We can generalize the result of the last example to obtain the following rule:

> *The area between the curves* $y = f(x)$ *and* $y = g(x)$ *between* $x = a$ *and* $x = b$ $(a < b)$ *is given by*
>
> $$A = \int_{a}^{b} |f(x) - g(x)|\, dx.$$

This rule forces the integrand to be positive, so that we cannot run into the problem of adding negative areas (although it does not make the calculation any easier).

We now show how these ideas can be applied to a problem in economics. For the first application we use the terms discussed in Section 11.4 (on page 631).

Let $p = p(q)$ be the demand function for a certain product. A typical demand function is like the function graphed in Figure 6. There is a point q_{max} at which the demand is exhausted so that no more of the product can even be given away. Likewise, there is a price p_{max} at which no items can be sold.

In a typical market situation, prices do not fluctuate instantly with demand but rather are held constant over a certain period of time (after which, in today's marketplace, they will usually rise—even if demand falls). Let this fixed price be denoted by $\bar{p}$ (see Figure 7). Choose a small subinterval, $[q_{i-1}, q_i]$, as shown. If the demand is q_i, then the correct price should be $p(q_i)$ (according to the demand function). Since the price is only $\bar{p}$, the consumer is saving the

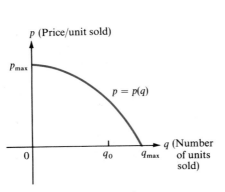

Figure 6

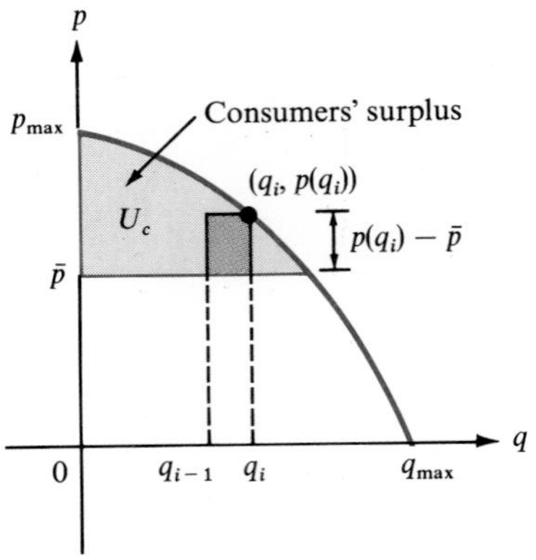

Figure 7

amount $p(q_i) - \bar{p}$ for each unit purchased. If Δq units are purchased, the amount saved is $[p(q_i) - \bar{p}]\Delta q$. If we add up these amounts, we obtain

$$\text{amount saved} \approx [p(q_1) - \bar{p}]\Delta q + [p(q_2) - \bar{p}]\Delta q + \cdots + [p(q_n) - \bar{p}]\Delta q. \quad (1)$$

We see that the sum in equation (1) approximates the area bounded by the curve $p(q)$, the line $q = 0$ (the p-axis), and the line $p = \bar{p}$.

Consumers' Surplus   If a product has the demand function $p(q)$ and is sold for the price p, then the **consumers' surplus** is the area of the region bounded by the demand curve, the line $p = \bar{p}$, and the line $q = 0$. This is the "triangular" region in Figure 7. It is denoted U_c.

Example 5   Let the demand function for a product be given by $p = 30 - 0.2\sqrt{q}$. If the price per unit is fixed at $10, calculate the consumers' surplus.

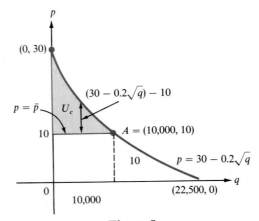

Figure 8

Solution   Look at Figure 8. The consumers' surplus is the shaded area of the graph. To calculate this area, we need to find the value of q that corresponds to $p = 10$ (the point A). We set $30 - 0.2\sqrt{q} = 10$ and find that $0.2\sqrt{q} = 20$, or $\sqrt{q} = 100$ and $q = 10,000$. Now we compute

$$\begin{array}{l}\text{area under demand curve} \\ \text{from } q = 0 \text{ to } q = 10,000\end{array} = \int_0^{10,000} (30 - 0.2\sqrt{q})\, dq = (30q - \tfrac{2}{15}q^{3/2})|_0^{10,000}$$

$$= 30 \cdot 10,000 - \tfrac{2}{15}10,000^{3/2} = 166,666.67.$$

This is the total area under the curve. To find the consumers' surplus, we must subtract the area of the unshaded rectangle under the demand curve. We have

$$\text{area of rectangle} = \text{base} \cdot \text{height} = 10,000 \cdot 10 = 100,000.$$

Thus,

$$U_c = 166,666.67 - 100,000 = \$66,666.67.$$

Suppose now that we are given a **supply function** $s = s(q)$. This function gives the relationship between the expected price of a product and the number of units the manufacturer will produce. It is reasonable to assume that as the expected price increases, the number of units the manufacturer will produce also increases, so that $s(q)$ is an increasing function. One supply function is sketched in Figure 9. The amount s_{min} is the minimum price that must be paid before the manufacturer will produce anything. It is related to the manufacturer's fixed cost (or overhead). If items are sold for $\$\bar{s}$ per unit and if $\bar{s} > s_{min}$, then the manufacturer will earn more money initially than if he had sold at his expected price per unit. This gain (at least on paper) is called the **producers' surplus**. It is represented, graphically, by the shaded area of Figure 9 and is denoted U_p.

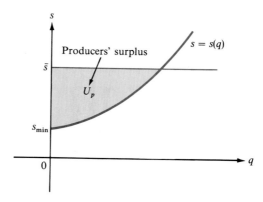

Figure 9

Producers' Surplus   If a product has the supply function $s(q)$ and is sold for $\bar{s}$, then the **producers' surplus** is the area of the region bounded by the supply curve, the line $s = \bar{s}$, and the line $q = 0$.

Example 6 Let $s(q) = 250 + 3q + 0.01q^2$. If items are sold for \$425 each, calculate the producers' surplus.

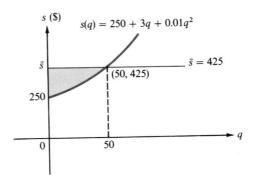

Figure 10

Solution The producers' surplus is the shaded region of Figure 10. When $\bar{s} = \$425$, $425 = 250 + 3q + 0.01q^2$. Then $0.01q^2 + 3q - 175 = 0$ and

$$q = \frac{-3 \pm \sqrt{9 + 4(0.01)(175)}}{0.02}$$

$$= \frac{-3 \pm \sqrt{16}}{0.02} = 50(-3 \pm 4).$$

The only positive solution is $q = 50$.

The area of the rectangle in Figure 10 is $50 \cdot 425 = 21{,}250$. The area under the curve $s = s(q)$ is

$$\text{area under supply curve} = \int_0^{50} (250 + 3q + 0.01q^2)\, dq$$

$$= 250q + \frac{3}{2}q^2 + \frac{0.01}{3}q^3 \bigg|_0^{50}$$

$$= 250 \cdot 50 + \frac{3}{2} \cdot 50^2 + \frac{0.01}{3} \cdot 50^3 = 16{,}666.67.$$

Then the producers' surplus is given by

$$U_p = \text{area under rectangle} - \text{area under supply curve}$$
$$= 21{,}250 - 16{,}666.67 = \$4583.33.$$

If we put these two ideas together, we may define **pure competition** as the situation that obtains when the price is set so that supply equals demand. Typically, we have a picture like the one in Figure 11. We can see, graphically, that in a pure market competitive system a certain level of production guarantees "profit" to the consumer as well as the producer.

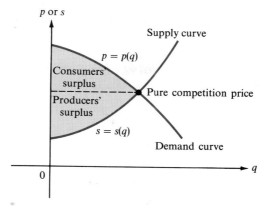

Figure 11

PROBLEMS 12.9   In Problems 1–14 calculate the area bounded by the given curves and lines.

1. $y = x^2$, $y = x$

2. $y = x^2$, $y = x^3$

3. $y = x^2$, $y = x^3$, $x = 3$

4. $y = 2x^2 + 3x + 5$, $y = x^2 + 3x + 6$

5. $y = 3x^2 + 6x + 8$, $y = 2x^2 + 9x + 18$

6. $y = x^2 - 7x + 3$, $y = -x^2 - 4x + 5$

*7. $y = 2x$, $y = \sqrt{4x - 24}$, $y = 0$, $y = 10$

8. $y = x^3$, $y = x^3 + x^2 + 6x + 5$

9. $y = x^4 + x - 81$, $y = x$

10. $xy^2 = 1$, $x = 5$, $y = 5$

11. $\sqrt{x} + \sqrt{y} = 4$, $x = 0$, $y = 0$

12. $x + y^2 = 8$, $x + y = 2$

13. $x = y^2$, $x^2 = 6 - 5y$

*14. $xy^2 = 1$, $y = 3 - 2\sqrt{x}$

15. Find the area of the triangle with vertices at (1, 6), (2, 4), and (−3, 7). [*Hint*: Find the equations of the straight lines forming the sides and draw a sketch.]

16. Find the area of the triangle with vertices at (2, 0), (3, 2), and (6, 7).

17. If the demand function is given by $p = 10 - 0.01q$, find the consumers' surplus if the product is sold for $5.

18. If the demand function is $p(q) = 5 + (180/\sqrt{1 + q})$ and the product is sold for $10, find the consumers' surplus.

19. If the supply function for a certain product is $s(q) = 400 + 2q + 0.02q^2$ and the product sells for $462.50, find the producers' surplus.

20. Given the demand function $p(q) = 175 - q$ and the supply function $s(q) = 50 + q + 0.01q^2$ for a certain product,

(a) calculate the pure competition price;

(b) calculate the consumers' and producers' surplus at that price.

21. Answer the questions of Problem 20 for the demand function $p = 30 - \sqrt{q}$ and the supply function $s = 10 + 3\sqrt{q}$.

▦ 12.10 Numerical Integration: The Trapezoidal Rule (Optional)

Consider the problem of evaluating

$$\int_0^1 \sqrt{1 + x^3}\, dx \qquad \text{or} \qquad \int_0^1 e^{x^2}\, dx.$$

Since both $\sqrt{1 + x^3}$ and e^{x^2} are continuous in $[0, 1]$, we know that both the definite integrals given here exist. They represent the areas under the curves $y = \sqrt{1 + x^3}$ and $y = e^{x^2}$ for x between 0 and 1. The problem is that none of the methods we have studied (or any other method, for that matter) will enable us to find the antiderivative of $\sqrt{1 + x^3}$ or e^{x^2}. This is because neither antiderivative can be expressed in terms of the functions we know.

In fact, there are a great number of continuous functions for which an antiderivative cannot be expressed in terms of functions we know. In those cases we cannot use the fundamental theorem of calculus to evaluate a definite integral. Nevertheless, it may be very important to approximate the value of such an integral. For that reason, many methods have been devised to approximate the value of a definite integral to as many decimal places as are deemed necessary. All these techniques come under the heading of **numerical integration**. We will not discuss this vast subject in great generality here. Rather we will introduce a reasonably effective method for estimating a definite integral: the **trapezoidal rule**. For a more complete discussion of numerical integration, you are referred to a book on numerical analysis.[†]

Consider the problem of calculating

$$\int_a^b f(x)\, dx.$$

By the results of Section 12.8,

$$\int_a^b f(x)\, dx = \lim_{\Delta x \to 0} [f(x_1^*)\Delta x + f(x_2^*)\Delta x + \cdots + f(x_n^*)\Delta x]. \qquad (1)$$

In other words, when the lengths of the subintervals in a partition of $[a, b]$ are small, the sum in the right-hand side of equation (1) gives us a crude approximation to the integral. Here, if f is nonnegative on $[a, b]$, the area is approximated by a sum of areas of rectangles. We saw some examples of this type of approximation in Sections 12.8 and 12.9. We now develop a more efficient way to approximate the integral.

Let f be as in Figure 1 and let us partition the interval $[a, b]$ by the equally spaced points

$$a = x_0 < x_1 < x_2 < \cdots < x_{i-1} < x_i < \cdots < x_n = b,$$

where $x_i - x_{i-1} = \Delta x = (b - a)/n$. In Figure 1, we have indicated that the area

---

[†] One reasonably elementary book in this area is by Conte and deBoor, *Elementary Numerical Analysis: An Algorithmic Approach*, 3rd ed. (New York: McGraw-Hill, 1980).

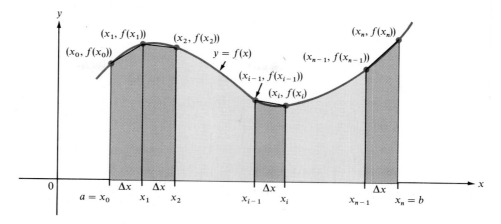

Figure 1

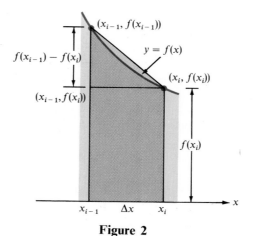

Figure 2

under the curve can be approximated by the sum of the areas of n trapezoids. One typical trapezoid is sketched in Figure 2. The area of the trapezoid is the area of the rectangle plus the area of the triangle. But the area of the rectangle is $f(x_i)\Delta x$, and the area of the triangle is $\frac{1}{2}[f(x_{i-1}) - f(x_i)]\Delta x$, so that

$$\text{area of trapezoid} = f(x_i)\Delta x + \tfrac{1}{2}[f(x_{i-1}) - f(x_i)]\Delta x$$
$$= \tfrac{1}{2}[f(x_{i-1}) + f(x_i)]\Delta x.^{\dagger}$$

[†] Note that this is the same as the average of the area of the rectangle R_{i-1}, whose height is $f(x_{i-1})$ (the left-hand endpoint), and the area of the rectangle R_i, whose height is $f(x_i)$ (the right-hand endpoint). That is, $\frac{1}{2}[(\text{area of } R_{i-1}) + (\text{area of } R_i)] = \frac{1}{2}[f(x_{i-1})\Delta x + f(x_i)\Delta x] = \frac{1}{2}[f(x_{i-1}) + f(x_i)]\Delta x$.

Then

$$\int_a^b f(x)\, dx \approx \text{sum of the areas of the trapezoids}$$

$$= \tfrac{1}{2}[f(x_0) + f(x_1)]\Delta x + \tfrac{1}{2}[f(x_1) + f(x_2)]\Delta x + \cdots$$
$$+ \tfrac{1}{2}[f(x_{n-2}) + f(x_{n-1})]\Delta x + \tfrac{1}{2}[f(x_{n-1}) + f(x_n)]\Delta x,$$

or

$$\int_a^b f(x)\, dx \approx \tfrac{1}{2}\Delta x[f(x_0) + 2f(x_1) + 2f(x_2) + \cdots + 2f(x_{n-1}) + f(x_n)]. \qquad (2)$$

The approximation formula (2) is called the **trapezoidal rule** for numerical integration. Note that since $\Delta x = (b - a)/n$, we can write (2) as

$$\int_a^b f(x)dx \approx \frac{b - a}{2n}[f(x_0) + 2f(x_1) + 2f(x_2) + \cdots + 2f(x_{n-1}) + f(x_n)].$$

Example 1 Estimate $\int_1^2 (1/x)\, dx$ by using the trapezoidal rule, with $n = 5$ and $n = 10$.

Solution (a) Here $n = 5$ and

$$\Delta x = \frac{b - a}{n} = \frac{2 - 1}{5} = \frac{1}{5} = 0.2.$$

Then $x_0 = 1$, $x_1 = 1.2$, $x_2 = 1.4$, $x_3 = 1.6$, $x_4 = 1.8$, and $x_5 = 2$. From (2),

$$\int_1^2 \frac{1}{x}\, dx \approx \frac{1}{2}\Delta x[f(x_0) + 2f(x_1) + 2f(x_2) + 2f(x_3) + 2f(x_4) + f(x_5)]$$

$$= \frac{0.2}{2}\left(\frac{1}{1} + \frac{2}{1.2} + \frac{2}{1.4} + \frac{2}{1.6} + \frac{2}{1.8} + \frac{1}{2}\right)$$

$$\approx 0.1(1 + 1.6667 + 1.4286 + 1.25 + 1.1111 + 0.5)$$

$$= 0.1(6.9564) = 0.6956.$$

(b) Now $n = 10$ and $\Delta x = 1/10 = 0.1$, so that $x_0 = 1$, $x_1 = 1.1, \ldots, x_9 = 1.9$, and $x_{10} = 2$. Thus

$$\int_1^2 \frac{1}{x}\, dx \approx \frac{1}{2}(0.1)\left[1 + \frac{2}{1.1} + \frac{2}{1.2} + \frac{2}{1.3} + \frac{2}{1.4} + \frac{2}{1.5} + \frac{2}{1.6} + \frac{2}{1.7} + \frac{2}{1.8} + \frac{2}{1.9} + \frac{1}{2}\right]$$

$$\approx 0.05[1 + 1.8182 + 1.6667 + 1.5385 + 1.4286 + 1.3333 + 1.25$$

$$+ 1.1765 + 1.1111 + 1.0526 + 0.5]$$

$$= 0.05[13.8755] = 0.6938.$$

We can check our calculations by integrating:

$$\int_1^2 \frac{1}{x}\,dx = \ln x \Big|_1^2 = \ln 2 - \ln 1 = \ln 2 \approx 0.6931.$$

You can see that by increasing the number of intervals, we increase the accuracy of our answer. This, of course, is not surprising. However, we are naturally led to ask what kind of accuracy we can expect by using the trapezoidal rule. In general, two kinds of errors are encountered when using a numerical method to integrate. The first kind we have already encountered. This is the error obtained by approximating the curve between the points $(x_{i-1}, f(x_{i-1}))$ and $(x_i, f(x_i))$ by the straight line joining those points. Since we now consider the function at a finite or *discrete* number of points, the error incurred by this approximation is called **discretization error**. However, we will always encounter another kind of error. As you saw in Example 1, we rounded our calculations to four decimal places. Each such "rounding" led to an error in our calculation. The accumulated effect of this rounding is called **round-off error**. Note that, as we increase the number of intervals in our calculation, we improve the accuracy of our approximation to the area under the curve. This, evidently, has the effect of reducing the discretization error. On the other hand, an increase in the number of subintervals leads to an increase in the number of computations, which in turn leads to an increase in the accumulated round-off error. In fact, there is a delicate balance between these two types of errors, and often there is an "optimal" number of intervals to be chosen so as to minimize the total error. Round-off error depends on the type of device used for the computations (pencil and paper, hand calculator, computer, and so on), and will not be discussed further here. However, we can give a formula for estimating the discretization error incurred in using the trapezoidal rule.

Let the sum in (2) be denoted by T and let ε_n^T denote the discretization error:

$$\boxed{\varepsilon_n^T = \int_a^b f(x)\,dx - T}$$

when n subintervals are used. It is then possible to prove the following:

Error Formula for Trapezoidal Rule Let f, f', and f'' be continuous on $[a, b]$. If $|f''(x)| < M$ for all x in $[a, b]$, then

$$\boxed{|\varepsilon_n^T| \le M \frac{(b-a)^3}{12n^2}.} \tag{3}$$

 Example 2   Find a bound on the discretization error incurred when estimating $\int_1^2 (1/x)\,dx$ using the trapezoidal rule with n subintervals.

Solution $f(x) = 1/x$, $f'(x) = -1/x^2$, and $f''(x) = 2/x^3$. Hence $f''(x)$ is bounded above by 2 for x in $[1, 2]$. Then, from (3),

$$|\varepsilon_n^T| \le \frac{2(2 - 1)^3}{12n^2} = \frac{1}{6n^2}.$$

For example, for $n = 5$ we calculated $\int_1^2 (1/x)\, dx \approx 0.6956$. Then the actual error is

$$\varepsilon_n^T \approx 0.6931 - 0.6956 = -0.0025.$$

This compares with a maximum possible error of $1/6n^2 = 1/(6 \cdot 25) = 1/150 \approx 0.0067$. For $n = 10$, the actual error is

$$\varepsilon_{10}^T \approx 0.6931 - 0.6938 = -0.0007.$$

This compares with a maximum possible error of $1/6n^2 = 1/600 \approx 0.0017$. Hence we see, in this example at least, that the error bound (3) is a crude estimate of the actual error. Nevertheless, even this crude bound allows us to estimate the accuracy of our calculation in the cases where we *cannot* check our answer by integrating. Of course, these are the only cases of interest, since we would not use a numerical technique if we could calculate the answer exactly.

Example 3   Use the trapezoidal rule to estimate $\int_0^2 e^{x^2}\, dx$ with a maximum error of 1.

Solution   We must choose n large enough so that $|\varepsilon_n^T| \le 1$. For $f(x) = e^{x^2}$, we have $f'(x) = 2xe^{x^2}$ and $f''(x) = (2 + 4x^2)e^{x^2}$. Since this is an increasing function, its maximum over the interval $[0, 2]$ occurs at 2. Then $M = f''(2) = 18e^4 \approx 983$. Hence, from (3),

$$|\varepsilon_n^T| \le \frac{M(b - a)^3}{12n^2} \le \frac{(983)2^3}{12n^2} \approx \frac{655}{n^2}.$$

We need $655/n^2 \le 1$ or $n^2 \ge 655$ or $n \ge \sqrt{655}$. The smallest n that meets this requirement is $n = 26$. Hence, we use the trapezoidal rule with $n = 26$ and $\Delta x = (b - a)/n = 2/26 = 1/13$. We have $x_0 = 0$, $x_1 = 1/13$, $x_2 = 2/13, \ldots, x_{25} = 25/13$, and $x_{26} = 26/13 = 2$. Then

$$\int_0^2 e^{x^2}\, dx \approx \frac{1}{2} \cdot \frac{1}{13} \left[e^0 + 2e^{(1/13)^2} + 2e^{(2/13)^2} + \cdots + 2e^{(25/13)^2} + e^{(26/13)^2} \right]$$

$$\approx \frac{1}{26}(1 + 2.012 + 2.048 + 2.109 + 2.199 + 2.319 + 2.475 + 2.673$$

$$+ 2.921 + 3.230 + 3.614 + 4.092 + 4.689 + 5.437 + 6.378 + 7.572$$

$$+ 9.097 + 11.059 + 13.603 + 16.933 + 21.328 + 27.184 + 35.060$$

$$+ 45.756 + 60.427 + 80.751 + 54.598)$$

$$= \frac{1}{26}(430.564) \approx 16.560.$$

This answer is correct to within 1 unit.[†]

PROBLEMS 12.10 In Problems 1–10, (a) estimate the given definite integral using the trapezoidal rule over the given number of intervals; (b) use formula (3) to obtain a bound for the error of the trapezoidal approximation; (c) calculate the integral exactly; and (d) compare the actual error in your computation with the maximum possible error found in part (b).

1. $\int_0^1 x \, dx$; 4 intervals $(n = 4)$

2. $\int_{-2}^2 x \, dx$; 6 intervals $(n = 6)$

3. $\int_0^1 x^2 \, dx$; 4 intervals $(n = 4)$

4. $\int_0^1 e^x \, dx$; 4 intervals $(n = 4)$

5. $\int_0^2 e^x \, dx$; 6 intervals $(n = 6)$

6. $\int_1^2 \frac{1}{x^2} \, dx$; 6 intervals $(n = 6)$

7. $\int_1^2 \sqrt{x} \, dx$; 8 intervals $(n = 8)$

8. $\int_0^3 \frac{1}{\sqrt{1 + x}} \, dx$; 6 intervals $(n = 6)$

9. $\int_2^5 \frac{x}{\sqrt{x^2 + 1}} \, dx$; 6 intervals $(n = 6)$

10. $\int_1^e \ln x \, dx$; 6 intervals $(n = 6)$

In Problems 11–20 approximate the given integral using the trapezoidal rule.

11. $\int_0^1 \sqrt{x + x^2} \, dx$; 4 intervals

12. $\int_0^1 e^{\sqrt{x}} \, dx$; 6 intervals

13. $\int_0^1 e^{x^3} \, dx$; 8 intervals

14. $\int_1^2 \sqrt{\ln x} \, dx$; 10 intervals

15. $\int_{-1}^1 e^{-x^2} \, dx$; 10 intervals

16. $\int_0^1 \sqrt{1 + x^3} \, dx$; 10 intervals

17. $\int_0^1 \frac{dx}{\sqrt{1 + x^3}}$; 10 intervals

18. $\int_0^1 x e^{x^3} \, dx$; 10 intervals

19. $\int_0^1 \ln(1 + e^x) \, dx$; 8 intervals

20. $\int_1^3 \frac{x^2}{\sqrt[3]{1 + x}} \, dx$; 10 intervals

In Problems 21–26 find a bound on the discretization error using the trapezoidal rule.

21. Integral of Problem 13

22. Integral of Problem 12

23. Integral of Problem 15

24. Integral of Problem 16

25. Integral of Problem 17

26. Integral of Problem 18

[†] Values of the function $\int_0^x e^{t^2} \, dt$ have been tabulated. To 6 decimal places, the correct value of $\int_0^2 e^{t^2} \, dt$ is 16.452627. Thus our answer is actually correct to within 0.11.

27. The integral $(1/\sqrt{2\pi}) \int_{-a}^{a} e^{-x^2/2}\, dx$ is very important in probability theory. (It is the density function of the unit normal distribution.) Using the trapezoidal rule, estimate $(1/\sqrt{2\pi}) \int_{-1}^{1} e^{-x^2/2}\, dx$ with an error of less than 0.01. [*Hint*: Show that $\int_{-a}^{a} e^{-x^2/2}\, dx = 2\int_{0}^{a} e^{-x^2/2}\, dx$.]

28. Estimate $(1/\sqrt{2\pi}) \int_{-5}^{5} e^{-x^2/2}\, dx$ with an error of less than 0.01.

*29. (a) Estimate $(1/\sqrt{2\pi}) \int_{-50}^{50} e^{-x^2/2}\, dx$ with an error of less than 0.1.

    (b) Can you guess what happens to $(1/\sqrt{2\pi}) \int_{-N}^{N} e^{-x^2/2}\, dx$ as N grows without bound?

Review Exercises for Chapter 12

In Exercises 1–20, compute the given definite or indefinite integral.

1. $\displaystyle\int x^5\, dx$

2. $\displaystyle\int 3x^{7/3}\, dx$

3. $\displaystyle\int_{0}^{1} (x - \sqrt[3]{x})\, dx$

4. $\displaystyle\int_{2}^{5} \frac{3}{1+x}\, dx$

5. $\displaystyle\int \frac{2x^2}{1+x^3}\, dx$

6. $\displaystyle\int_{0}^{2} e^{-4x}\, dx$

7. $\displaystyle\int_{0}^{2} (t^3 + 3t + 5)\, dt$

8. $\displaystyle\int_{1}^{8} \frac{ds}{\sqrt[3]{s}}$

9. $\displaystyle\int \frac{du}{(u+3)^3}$

10. $\displaystyle\int_{0}^{1} x\sqrt{x^2+1}\, dx$

11. $\displaystyle\int_{0}^{1} \frac{x}{x^2+1}\, dx$

12. $\displaystyle\int_{1}^{\sqrt{2}} \frac{[1 - 1/t^2]^4}{t^3}\, dt$

13. $\displaystyle\int_{2}^{3} x^2(1-x^3)^5\, dx$

14. $\displaystyle\int_{0}^{1} xe^{-x}\, dx$

15. $\displaystyle\int \frac{dx}{3x\ln x}$

16. $\displaystyle\int_{2}^{4} \frac{\ln x}{x}\, dx$

17. $\displaystyle\int_{2}^{4} \frac{\ln x}{2}\, dx$

18. $\displaystyle\int \frac{e^{-1/x^2}}{x^3}\, dx$

19. $\displaystyle\int_{0}^{1} (1 + x + x^2 + x^3 + x^4 + x^5)\, dx$

20. $\displaystyle\int_{0}^{1} (x^{1/2} + x^{1/3} + x^{1/4} + x^{1/5})\, dx$

21. If the marginal cost of a certain product is $20 - (q/5)$ and fixed costs are $300, find the total cost function.

22. If the marginal revenue of a certain product is $200 - 0.03q^{4/3}$, find the total revenue function.

23. If the marginal cost to produce a certain product is $200 + 0.65q + 0.001q^2$, what is the cost of increasing production from 30 to 50 units?

In Exercises 24–28, find the area between the given curve and the x-axis for x in the given interval. If no interval is given, sketch the curve and determine an appropriate interval.

24. $y = 3x - 7$; $[-2, 5]$

25. $y = \sqrt{x+1}$; $[0, 15]$

26. $y = x^3 - 7x^2 + 7x + 15$

27. $y = -x^2 - x + 2$

28. $y = \dfrac{1}{1+x}$; $[0, 5]$

In Exercises 29–32, determine whether the given integral converges. If so, find its value.

29. $\displaystyle\int_0^\infty e^{-x/3}\,dx$

30. $\displaystyle\int_0^\infty e^{x/3}\,dx$

31. $\displaystyle\int_0^\infty xe^{-4x}\,dx$

32. $\displaystyle\int_1^\infty \frac{x}{(x^2+4)^2}\,dx$

33. Estimate $\int_0^1 (x^2/2)\,dx$ by using Riemann sums.

34. A particle is moving with the velocity $v(t) = t + 1/\sqrt{1+t}$ m/sec.

   (a) How far does the particle move in the first 15 seconds?

   (b) What is the average velocity of the particle?

35. The demand function for a certain product is $p(q) = 150 - q$. Find the consumers' surplus if the product is sold for $100.

36. The supply function for the product of Exercise 35 is $s(q) = 75 + q + 0.01q^2$. Find the producers' surplus if the product is sold for $125.

37. (a) Find the pure competition price for the product of Exercises 35 and 36.

   (b) Calculate the consumers' and producers' surplus at that price.

38. Estimate $\int_0^1 e^{x^3}\,dx$ using the trapezoidal rule with 4 subintervals.

39. Estimate $\int_0^1 (dx/\sqrt{1+x^4})$ using the trapezoidal rule with 6 subintervals.

13 INTRODUCTION TO MULTIVARIABLE CALCULUS

13.1 Functions of Two or More Variables

For most of the functions you have so far encountered in this book, we have been able to write $y = f(x)$. This means that we could write the variable y explicitly in terms of the single variable x. However, in a great variety of applications it is necessary to write the quantity of interest in terms of two or more variables.

Example 1 The compound interest formula (see formula (8.1.3) on page 434) is given by

$$A(t) = P(1 + i)^t,$$

where $A(t)$, the amount of money in an investment after t years, is a function of the three variables:

$$P = \text{the initial amount invested};$$

$$i = \text{the annual interest rate};$$

$$t = \text{the number of years}.$$

We should write this function as

$$A(P, i, t) = P(1 + i)^t$$

to indicate that A depends on three variables. For example, if $P = 1000$, $i = 0.06$, and $t = 5$, then, as we found in Example 8.1.3 on page 434, $A(1000, 0.06, 5) = 1000(1.06)^5 = \$1338.23$.

Example 2 In Example 11.4.1 on page 632, we discussed the total cost function

$$C(q) = 250 + 3q + 0.01q^2.$$

That is, $C(q)$ is the cost of producing q units of a certain product. Suppose, instead, that the manufacturer produces two items. Let q_1 represent the number of units of the first item produced, and let q_2 represent the number of units of the second item produced. Then the cost function is a function of the *two* variables q_1 and q_2. A typical cost function is

$$C(q_1, q_2) = aq_1^2 + bq_2^2 + cq_1 + dq_2 + k,$$

where a, b, c, d, and k are constants.

Example 3 A manufacturer produces portable radios and portable cassette tape players. He estimates that the cost of producing q_1 radios and q_2 tape players is given by

$$C(q_1, q_2) = 300 + 40q_1 + 60q_2 + 0.01q_1^2 + 0.02q_2^2.$$

Find

(a) the cost of producing 50 radios and 100 tape players.

(b) the cost of producing 75 of each.

Solution

(a) $C(50,100) = 300 + 40 \cdot 50 + 60 \cdot 100 + (0.01)50^2 + (0.02)100^2$
$\qquad\qquad = 300 + 2000 + 6000 + 25 + 200 = \$8525$

(b) $C(75,75) = 300 + 40 \cdot 75 + 60 \cdot 75 + (0.01)75^2 + (0.02)75^2$
$\qquad\qquad = \$7968.75$

Example 4 The Cobb–Douglas Production Function In a manufacturing process, costs are typically divided between labor costs and capital costs. The total sum spent on production is usually fixed, but often a manufacturer has some choice in allocating money between capital and labor. For example, if part of the process is automated, more money will be spent on capital and less will be spent on labor.

Suppose that L units of labor and K units of capital are used in production. How many units will be produced? Economists have determined that, in some cases, the answer is given by the **Cobb–Douglas production function**:

$$\text{number of units produced} = F(L, K) = cL^a K^{1-a},$$

where c and a are constants that depend on the particular manufacturing process.

In the manufacture of a certain type of die, the Cobb–Douglas production function is given by

$$F(L, K) = 200L^{2/5} K^{3/5}.$$

(a) How many units are produced if 100 units of labor and 300 units of capital are used?

(b) If the number of units of labor and capital are both doubled, what is the change in the number of units produced?

Solution

(a)
$$F(100, 300) = 200(100)^{0.4}(300)^{0.6}$$
$$\approx 200(6.31)(30.64) \approx 38{,}668 \text{ units}$$

(b)   We are asked to determine what happens to F if L becomes $2L$ and K becomes $2K$. We have

$$F(2L, 2K) = 200(2L)^{2/5}(2K)^{3/5} = 200 \cdot 2^{2/5}L^{2/5}2^{3/5}K^{3/5}$$
$$= \underbrace{2^{2/5}2^{3/5}}_{2^{5/5}\,=\,2^{1}\,=\,2}\underbrace{[200L^{2/5}K^{3/5}]}_{F(L,K)} = 2F(L, K).$$

We have shown that production is doubled if both labor and production costs are doubled.

Example 5   According to **Poiseuille's law** (see Problem 10.7.38 on page 568), the resistance, R, of a blood vessel of length l and radius r is given by

$$R(l, r) = \frac{\alpha l}{r^4},$$

where α is a constant of proportionality. Find

(a) $R(5, 2)$.   (b) $R(7.5, 1.5)$.

Solution

(a) $R(5, 2) = \dfrac{\alpha \cdot 5}{2^4} = \dfrac{5\alpha}{16} = 0.3125\alpha$

(b) $R(7.5, 1.5) = \dfrac{\alpha \cdot 7.5}{(1.5)^4} = \dfrac{7.5}{(1.5)^4}\alpha = \dfrac{7.5}{5.0625}\alpha \approx 1.48\alpha$

Example 6   The volume of a right circular cone is given by $V(r, h) = \frac{1}{3}\pi r^2 h$ (see Figure 1). Find

(a)   the volume of a right circular cone of radius 2 in and height 3.5 in.

(b)   the volume of a right circular cone of radius 3.25 in and height 7.25 in.

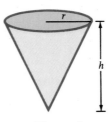

Figure 1

Solution

(a) $V(2, 3.5) = \frac{1}{3}\pi 2^2(3.5) = \frac{14}{3}\pi \approx 14.66 \text{ in}^3$.

(b) $V(3.25, 7.25) = \frac{1}{3}\pi(3.25)^2(7.25) \approx 80.19 \text{ in}^3$.

In general, we have the following definition:

Function of Two Variables Let D be a subset of $\mathbb{R}^2$. Then a **function of two variables**, f, is a rule that assigns to each ordered pair (x, y) in D a unique real number, which we denote $f(x, y)$. The set D is called the **domain** of f. For each (x, y) in D, we usually write the function as

$$z = f(x, y).$$

Example 7 Let $f(x, y) = x^2 + y^3$. Then the domain of $f = \mathbb{R}^2$, since $x^2 + y^3$ makes sense for all real numbers x and y. Find

(a) $f(2, 3)$. (b) $f(5, -1)$.

Solution

(a) $f(2, 3) = 2^2 + 3^3 = 4 + 27 = 31$

(b) $f(5, -1) = 5^2 + (-1)^3 = 25 - 1 = 24$

Example 8 Let $z = f(x, y) = \sqrt{4 - x^2 - y^2}$. Find

(a) domain of f. (b) $f(0, 1)$. (c) $f(-1, 1)$.

(d) $f(1.2, 1.3)$. (e) $f(0, -2)$.

Solution

(a) The square root of a number is defined only if the number is nonnegative. Thus, the domain of $f = \{(x, y): 4 - x^2 - y^2 \geq 0\} = \{(x, y): x^2 + y^2 \leq 4\}$. This is the set of points on and inside the circle of radius 2 centered at $(0, 0)$. The domain of f is sketched in Figure 2.

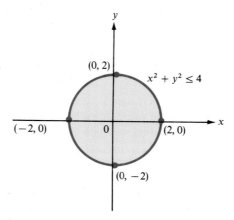

Figure 2

(b) $f(0, 1) = \sqrt{4 - 0^2 - 1^2} = \sqrt{3}$.

(c) $f(-1, 1) = \sqrt{4 - (-1)^2 - 1^2} = \sqrt{4 - 1 - 1} = \sqrt{2}.$

(d) $f(1.2, 1.3) = \sqrt{4 - (1.2)^2 - (1.3)^2} = \sqrt{4 - 1.44 - 1.69} = \sqrt{0.87} \approx 0.93.$

(e) $f(0, -2) = \sqrt{4 - 0 - (-2)^2} = \sqrt{0} = 0.$

Example 9 Let $z = f(x, y) = \ln(2x - y + 1)$. Find

(a) domain of f. (b) $f(3, 2)$. (c) $f(5, -7)$.

Solution

(a) $\ln x$ is defined only for $x > 0$. Thus, domain of $f = \{(x, y): 2x - y + 1 > 0\} = \{(x, y): y < 2x + 1\}$. This is the equation of the half-plane "below" (but not including) the line $y = 2x + 1$. It is sketched in Figure 3.

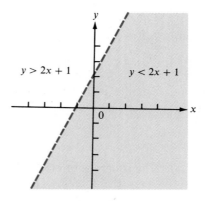

Figure 3

(b) $f(3, 2) = \ln(2 \cdot 3 - 2 + 1) = \ln 5 \approx 1.61.$

(c) $f(5, -7) = \ln(2 \cdot 5 - (-7) + 1) = \ln 18 \approx 2.89.$

Caution. In Figures 1 and 2 we sketched the *domains* of two different functions. We did *not* sketch their graphs. The graph of a function of two or more variables is more complicated and will be discussed shortly.

In economics, a useful model is the model of *a monopolist producing two commodities*. In this model a firm produces or sells two goods, and the sales of these two goods are related. The goods could be complementary (so that sales of one increases sales of the other) or substitutes (so that sales of one reduces sales of another). Rather than discussing this topic in general, we provide an example.[†]

[†] A complete discussion of this topic can be found in the excellent book by R. W. Quincey and F. Neal, *Using Mathematics in Economics* (London: Butterworths, 1973), pp. 153–156.

Example 10 A retail store sells electric broilers and toasters. Each broiler costs the store $12, and each toaster costs the store $8. If there are no fixed costs, then the total cost function is

$$C(q_1, q_2) = 12q_1 + 8q_2,$$

where q_1 and q_2 are the numbers of broilers and toasters bought, respectively. The company has determined that the quantities of the two items sold depend on the prices of both items and has obtained the following demand functions (see Section 11.4 for a discussion of demand functions):

$$q_1 = 150 - 3p_1 + p_2 \qquad (1)$$

and

$$q_2 = 270 + p_1 - 2p_2, \qquad (2)$$

where p_1 is the retail price of a broiler and p_2 is the retail price of a toaster.

(a) Write the total profit as a function of the prices p_1 and p_2.

(b) What is the profit when broilers are sold for $15 and toasters are sold for $10?

Solution

(a) We have profit (P) = revenue (R) − cost (C). The revenue for selling q_1 broilers at price p_1 is $p_1 q_1$. Similarly, the revenue from selling q_2 toasters at price p_2 is $p_2 q_2$. Thus

$$R = p_1 q_1 + p_2 q_2$$

and

$$P(p_1, p_2) = R - C = p_1 q_1 + p_2 q_2 - (12q_1 + 8q_2)$$

Using (1) and (2)
$\downarrow$

$$\begin{aligned}
&= p_1(150 - 3p_1 + p_2) + p_2(270 + p_1 - 2p_2) \\
&\quad - [12(150 - 3p_1 + p_2) + 8(270 + p_1 - 2p_2)] \\
&= (150p_1 - 3p_1^2 + p_1 p_2) + (270p_2 + p_1 p_2 - 2p_2^2) \\
&\quad - [(1800 - 36p_1 + 12p_2) + (2160 + 8p_1 - 16p_2)],
\end{aligned}$$

or

$$P(p_1, p_2) = -3960 + 178p_1 + 274p_2 + 2p_1 p_2 - 3p_1^2 - 2p_2^2.$$

(b) $P(15, 10) = -3960 + 178 \cdot 15 + 274 \cdot 10 + 2 \cdot 15 \cdot 10 - 3 \cdot 15^2 - 2 \cdot 10^2$
$= \$875.$

The *graph* of a function $y = f(x)$ is a set of points in the xy-plane. In order to draw the graph of a function $z = f(x, y)$, we need *three* dimensions. In Section 2.1 we showed how any point in a plane can be represented as an ordered pair of real numbers. It is not surprising, then, that any point in space can be represented by an **ordered triple** of real numbers:

$$(a, b, c), \tag{3}$$

where a, b, and c are real numbers.

Three-Dimensional Space   The set of ordered triples of the form (3) is called **real three-dimensional space** and is denoted $\mathbb{R}^3$. There are many ways to represent a point in $\mathbb{R}^3$. However, the most common representation, given by (3), is very similar to the representation of a point in the plane by its x- and y-coordinates. We begin, as before, by choosing a point in $\mathbb{R}^3$ and calling it the **origin**, denoted by 0. Then we draw three mutually perpendicular axes, called the **coordinate axes**, which we label the **x-axis**, the **y-axis**, and the **z-axis**. These axes can be selected in a variety of ways, but the most common selection has the x- and y-axes drawn horizontally, with the z-axis vertical. On each axis, we choose a positive direction and measure distance along each axis as the number of units in this positive direction measured from the origin.

The two basic systems of drawing these axes are depicted in Figure 4. If the axes are placed as in Figure 4a, then the system is called a **right-handed system**; if they are placed as in Figure 4b, the system is a **left-handed system**.

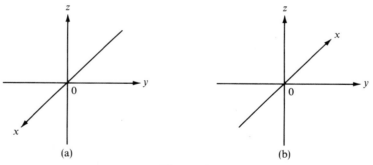

(a)   (b)

Figure 4

In the figures, the arrows indicate the positive directions on the axes. The reason for this choice of terms is as follows: In a right-handed system, if you place your right hand so that your index finger points in the positive direction of the x-axis while your middle finger points in the positive direction of the y-axis, then your thumb will point in the positive direction of the z-axis. This is illustrated in Figure 5. For a left-handed system, the same rule will work for your left hand. For the remainder of this book, we will follow common practice and depict the coordinate axes using a right-handed system.

If you have trouble visualizing the placement of these axes, do the following. Face any uncluttered corner (on the floor) of the room in which you are sitting. Call the corner the origin. Then the x-axis lies along the floor, along the wall, and to your left; the y-axis lies along the floor, along the wall, and to your right; and the z-axis lies along the vertical intersection of the two perpendicular walls. This is illustrated in Figure 6.

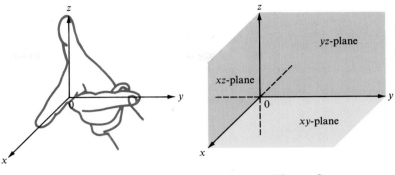

Figure 5   Figure 6

The three axes in our system determine three **coordinate planes**, which we will call the xy-plane, the xz-plane, and the yz-plane. The xy-plane contains the x- and y-axes and is simply the plane with which we have been dealing in most of this book. The xz- and yz-planes can be thought of in a similar way.

Having built our structure of coordinate axes and planes, we can describe any point P in space in a unique way:

$$P = (x, y, z),$$

where the first coordinate, x, is the distance from the yz-plane to P (measured in the positive direction of the x-axis), the second coordinate, y, is the distance from the xz-plane to P (measured in the positive direction of the y-axis), and the third coordinate, z, is the distance from the xy-plane to P (measured in the positive direction of the z-axis). Thus, for example, any point in the xy-plane has z-coordinate 0; any point in the xz-plane has y-coordinate 0; and any point in the yz-plane has x-coordinate 0. Some representative points are sketched in Figure 7.

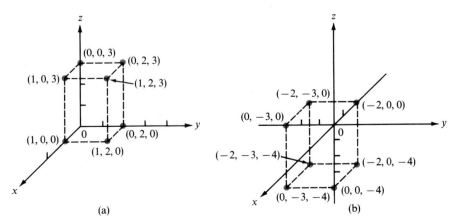

(a)   (b)

Figure 7

In this system, the three coordinate planes divide $\mathbb{R}^3$ into eight **octants**, just as in $\mathbb{R}^2$ the two coordinate axes divide the plane into four quadrants. The first octant is always chosen to be the one in which the three coordinates are positive. The coordinate system we have just established is often referred to as the **rectangular coordinate system** or the **Cartesian coordinate system**.

Once we know about three-dimensional space, we can sketch the graphs of functions of two variables. These are given by

$$\text{\textbf{graph} of } f = \{(x, y, z): z = f(x, y)\}.$$

The graph of a function $z = f(x, y)$ in $\mathbb{R}^3$ is often called a surface in $\mathbb{R}^3$. More generally, a **surface** in space is the set of points (x, y, z) in $\mathbb{R}^3$ that satisfy an equation of the form $F(x, y, z) = 0$.

However, a graph or surface in $\mathbb{R}^3$ is much more difficult to obtain than a graph in $\mathbb{R}^2$. To show you how complicated things can be, we provide in Figure 8 a computer-drawn sketch of the surface $z = x^3 + y^3 - x^2 + y + 2$. In the next section we provide sketches of some common surfaces in $\mathbb{R}^3$.

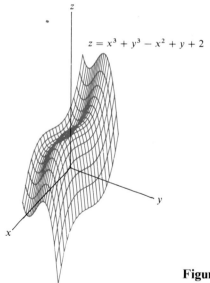

$$z = x^3 + y^3 - x^2 + y + 2$$

Figure 8

We have defined a function of two variables. The definition of a function of three or more variables is similar. Example 1 illustrated a function of three variables. We will give no further example here except to note that the only essential difference between functions of two variables and functions of three variables is that it is impossible to draw the graph of a function of three variables (we would need four dimensions to do so).

PROBLEMS 13.1   In Problems 1–10, (a) find the domain of the given function and (b) evaluate the function at the given point.

1. $f(x, y) = x^5 + y^3$; $(3, -2)$ **2.** $f(x, y) = 2xy$; $(4, 5)$

3. $f(x, y) = \sqrt{9 - x^2 - y^2}$; (1, 2)

4. $f(x, y) = \dfrac{x + y}{x - y}$; (−3, 1)

5. $f(x, y) = \sqrt[3]{1 - x^2 - y^2}$; (2, 5)

6. $f(x, y) = \sqrt[3]{1 - x^2 - y^2}$; $(\frac{1}{2}, \frac{1}{3})$

7. $f(x, y) = e^{x + 2y}$; (−2, 1)

8. $f(x, y) = \dfrac{x^2 - y}{y^2 - x}$; (4, 0)

9. $f(x, y) = \dfrac{x - y + 2 + y^3}{4 + x^2 - y}$; (7, −2)

10. $f(x, y) = \ln(x^2 - y^2)$; (−3, 1)

In Problems 11–20 sketch the point in three-dimensional space.

11. (3, 0, 0) **12.** (0, −5, 0) **13.** (0, 0, 7) **14.** (1, 2, 0)

15. (0, 1, −2) **16.** (−1, 0, 3) **17.** (2, −1, 5) **18.** (4, 1, 6)

19. (2, 2, 2) **20.** (−1, −1, −1)

21. The cost of producing q_1 units of product A and q_2 units of product B is given by

$$C(q_1, q_2) = 250 + 3q_1 + 2.5q_2 - (0.003)q_1^2 - (0.007)q_2^2.$$

Find the total cost of producing

(a) 200 of A and 150 of B.

(b) 300 of A and 250 of B.

 22. The Cobb–Douglas production function for a given product is

$$F(L, K) = 250L^{0.7}K^{0.3}.$$

(a) Compute $F(50, 80)$.

(b) Show that if labor and capital costs both triple, then the total output (number of units produced) triples as well.

 *23. The Cobb–Douglas production function for a given product is

$$F(L, K) = 500L^{1/3}K^{2/3}.$$

(a) Compute $F(250, 150)$.

(b) If labor costs double while capital costs are halved, what is the change in the total output? (Give this change as a percentage increase or decrease.)

(c) Answer the question in part (b) if labor costs are halved while capital costs double.

24. In Example 5 assume that $\alpha = 0.3$. Find R if

(a) $l = 6$ and $r = 3$.

(b) $l = 5.32$ and $r = 1.79$.

25. The volume of a box is given by

$$V(l, h, w) = lhw,$$

where l is its length, h is its height, and w is its width. Find the volume of the box whose dimensions are

(a) $l = 3$ cm, $h = 4$ cm, $w = \frac{1}{2}$ cm.

(b) $l = 6$ in, $h = 3$ in, $w = 2$ in.

(c) $l = w = h = 5$ in.

26. The temperature T at any point on an object in space is given by
$$T(x, y, z) = 20 + 3x^2 + 4y^2 + 2z^2.$$
Find the temperature at the points
   (a) $(2, 1, 4)$.
   (b) $(3, 2, 6)$.

27. In Example 10, what is the profit if broilers and toasters are sold for $38 and $52, respectively.

28. A jeweler sells ordinary and digital watches. Each ordinary watch costs her $8, and each digital watch costs her $25. The demand functions for the two watches are
$$q_1 = 80 - 2.5p_1 + 0.8p_2 \quad \text{for the ordinary watches}$$
and
$$q_2 = 120 + p_1 - 1.8p_2 \quad \text{for the digital watches.}$$
   (a) Find the total profit as a function of the prices of the two watches.
   (b) What is the profit when ordinary watches are sold for $24 and digital watches are sold for $30?

29. In Problem 28, what is the profit if ordinary and digital watches are sold for $30 and $40, respectively?

13.2 Partial Derivatives

In this section we show one of the ways a function of several variables can be differentiated. The idea is simple. Let $z = f(x, y)$. If we keep one of the variables fixed, say y, then f can be treated as a function of x only and we can calculate the derivative (if it exists) of f with respect to x. This new function is called the *partial derivative of f with respect to x* and is denoted $\partial f/\partial x$.[†] Before giving a more formal definition, we give an example.

Example 1 Let $= f(x, y) = x^2 y + y/x$. Calculate $\partial f/\partial x$.

Solution Treating y as if it were constant, we have
$$\frac{\partial f}{\partial x} = \frac{\partial}{\partial x}\left(x^2 y + \frac{y}{x}\right) = \frac{\partial}{\partial x}(x^2 y) + \frac{\partial}{\partial x}\left(\frac{y}{x}\right) = 2xy - \frac{y}{x^2}.$$

[†] The symbol ∂ of partial derivatives is not a letter from any alphabet but an invented mathematical symbol that may be read "partial." Historically, the difference between an ordinary and partial derivative was not recognized at first, and the same symbol d was used for both. The symbol ∂ was introduced in the eighteenth century by the mathematicians Alexis Fontaine des Bertins (1705–1771), Leonhard Euler (1707–1783), Alexis-Claude Clairaut (1713–1765), and Jean Le Rond d'Alembert (1717–1783) and was used in their development of the theory of partial differentiation.

Partial Derivative   Let $z = f(x, y)$. Then

(a)   the **partial derivative of f with respect to x** is the function

$$\frac{\partial z}{\partial x} = \frac{\partial f}{\partial x} = \lim_{\Delta x \to 0} \frac{f(x + \Delta x, y) - f(x, y)}{\Delta x}.$$ (1)

$\partial f / \partial x$ is defined at every point (x, y) in the domain of f such that the limit (1) exists.

(b)   the **partial derivative of f with respect to y** is the function

$$\frac{\partial z}{\partial y} = \frac{\partial f}{\partial y} = \lim_{\Delta y \to 0} \frac{f(x, y + \Delta y) - f(x, y)}{\Delta y}.$$ (2)

$\partial f / \partial y$ is defined at every point (x, y) in the domain of f such that the limit (2) exists.

Remark 1.   This definition allows us to calculate partial derivatives in the same way we calculate ordinary derivatives: by allowing only one of the variables to vary.

Remark 2.   The partial derivatives $\partial f / \partial x$ and $\partial f / \partial y$ give us the rate of change of f as each of the variables x and y changes with the other one held fixed. They do *not* tell us how f changes when x and y change simultaneously.

Remark 3.   It should be emphasized that, although the functions $\partial f / \partial x$ and $\partial f / \partial y$ are computed with one of the variables held constant, each is a function of both variables.

Example 2   Let $f(x, y) = \sqrt{x + y^2}$. Calculate $\partial f / \partial x$ and $\partial f / \partial y$.

Solution   We have $f(x, y) = (x + y^2)^{1/2}$. Then

$$\frac{\partial f}{\partial x} = \frac{1}{2\sqrt{x + y^2}} \frac{\partial}{\partial x} (x + y^2) = \frac{1}{2\sqrt{x + y^2}} (1 + 0) = \frac{1}{2\sqrt{x + y^2}},$$

since we are treating y as a constant. Also,

$$\frac{\partial f}{\partial y} = \frac{1}{2\sqrt{x + y^2}} \frac{\partial}{\partial y} (x + y^2) = \frac{1}{2\sqrt{x + y^2}} (0 + 2y) = \frac{y}{\sqrt{x + y^2}},$$

since we are treating x as a constant.

Example 3   Let $z = e^{2x + 3y}$. Calculate $\partial z / \partial x$ and $\partial z / \partial y$.

Solution   Let $u = 2x + 3y$. Then, treating y as a constant, we see that $\partial u / \partial x = 2$, so that by the chain rule,

$$\frac{\partial z}{\partial x} = e^u \frac{\partial u}{\partial x} = 2e^{2x + 3y}.$$

Similarly,

$$\frac{\partial z}{\partial y} = 3e^{2x+3y}.$$

Example 4   Let $f(x, y) = (1 + x^2 + y^5)^{4/3}$. Calculate $\partial f/\partial x$ and $\partial f/\partial y$ at the point $(3, 1)$.

Solution

$$\frac{\partial f}{\partial x} = \frac{4}{3}(1 + x^2 + y^5)^{1/3}\frac{\partial}{\partial x}(1 + x^2 + y^5)$$

$$= \frac{4}{3}(1 + x^2 + y^5)^{1/3} \cdot 2x = \frac{8x}{3}(1 + x^2 + y^5)^{1/3}.$$

At $(3, 1)$,

$$\frac{\partial f}{\partial x} = \frac{(8)(3)}{3}(1 + 3^2 + 1^5)^{1/3} = 8\sqrt[3]{11};$$

$$\frac{\partial f}{\partial y} = \frac{4}{3}(1 + x^2 + y^5)^{1/3}\frac{\partial}{\partial y}(1 + x^2 + y^5)$$

$$= \frac{4}{3}(1 + x^2 + y^5)^{1/3} \cdot 5y^4 = \frac{20y^4}{3}(1 + x^2 + y^5)^{1/3}.$$

At $(3, 1)$,

$$\frac{\partial f}{\partial y} = \frac{20}{3}\sqrt[3]{11}.$$

 Example 5   In Example 13.1.4 on page 735, we discussed the Cobb–Douglas production function:

$$F(L, K) = 200L^{2/5}K^{3/5}.$$

(a)   Compute $\partial F/\partial L$ and $\partial F/\partial K$.
(b)   Evaluate these partial derivatives at $L = 100$, $K = 300$.

Solution

(a) $$\frac{\partial F}{\partial L} = 200\left(\frac{\partial}{\partial L}L^{2/5}\right)K^{3/5} = 200\left(\frac{2}{5}L^{-3/5}\right)K^{3/5} = 80L^{-3/5}K^{3/5};$$

$$\frac{\partial F}{\partial K} = 200L^{2/5}\left(\frac{\partial}{\partial K}K^{3/5}\right) = 200L^{2/5}\left(\frac{3}{5}K^{-2/5}\right) = 120L^{2/5}K^{-2/5}.$$

(b) $$\frac{\partial F}{\partial L}(100, 300) = 200(100^{-3/5})(300^{3/5}) \approx 200(0.063)(30.639) = 386.0514;$$

$$\frac{\partial F}{\partial K}(100, 300) = 200(100^{2/5})(300^{-2/5}) \approx 200(6.31)(0.102) = 128.724.$$

Note. The quantity $\partial F/\partial L$ is called the **marginal productivity of labor**, and $\partial F/\partial K$ is called the **marginal productivity of capital**.

We now obtain a geometric interpretation of the partial derivative. Let $z = f(x, y)$. As we saw in Section 13.1, this is the equation of a surface in $\mathbb{R}^3$. To obtain $\partial z/\partial x$, we hold y fixed at some constant value y_0. The equation $y = y_0$ is a plane in space parallel to the xz-plane (whose equation is $y = 0$). Thus, if y is constant, $\partial z/\partial x$ is the rate of change of f with respect to x as x changes along the curve C, which is at the intersection of the surface $z = f(x, y)$ and the plane $y = y_0$. This is indicated in Figure 1. To be more precise, if (x_0, y_0, z_0) is a point on the surface $z = f(x, y)$, then $\partial z/\partial x$ evaluated at (x_0, y_0) is the slope of the line tangent to the surface at the point (x_0, y_0, z_0), which lies in the plane $y = y_0$. Analogously, $\partial z/\partial y$ evaluated at (x_0, y_0) is the slope of the line tangent to the surface at the point (x_0, y_0, z_0), which lies in the plane $x = x_0$ (since x is held fixed in order to calculate $\partial z/\partial y$). This is illustrated in Figure 2.

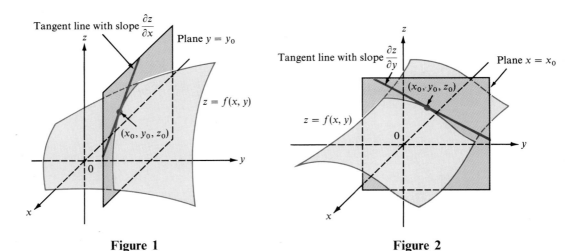

Figure 1   **Figure 2**

There are other ways to denote partial derivatives. We will often write

$$z_x = f_x = \frac{\partial f}{\partial x} \quad \text{and} \quad z_y = f_y = \frac{\partial f}{\partial y}.$$

If f is a function of other variables, say s and t, then we may write $\partial f/\partial s = f_s$ and $\partial f/\partial t = f_t$.

Example 6   The volume of a cone of radius r and height h is given by $V = \frac{1}{3}\pi r^2 h$. Then the rate of change of V with respect to r (with h fixed) is given by

$$\frac{\partial V}{\partial r} = V_r = \frac{\partial}{\partial r}(\tfrac{1}{3}\pi r^2 h) = \tfrac{2}{3}\pi rh,$$

and the rate of change of V with respect to h (with r fixed) is given by

$$\frac{\partial V}{\partial h} = V_h = \frac{\partial}{\partial h}(\tfrac{1}{3}\pi r^2 h) = \tfrac{1}{3}\pi r^2.$$

Partial derivatives of functions of three or more variables are computed in the same way.

Example 7 Let $w = f(x, y, z) = xz + e^{y^2 z} + \sqrt{xy^2 z^3}$. Calculate $\partial w/\partial x$, $\partial w/\partial y$, and $\partial w/\partial z$.

Solution To calculate $\partial w/\partial x$, we keep y and z fixed. Then

$$\frac{\partial w}{\partial x} = \frac{\partial f}{\partial x} = f_x = \frac{\partial}{\partial x}xz + \frac{\partial}{\partial x}e^{y^2 z} + \frac{1}{2\sqrt{xy^2 z^3}}\frac{\partial}{\partial x}(xy^2 z^3)$$

$$= z + 0 + \frac{y^2 z^3}{2\sqrt{xy^2 z^3}} = z + \frac{y^2 z^3}{2\sqrt{xy^2 z^3}}.$$

To calculate $\partial w/\partial y$, we keep x and z fixed. Then

$$\frac{\partial w}{\partial y} = \frac{\partial f}{\partial y} = f_y = \frac{\partial}{\partial y}xz + e^{y^2 z}\frac{\partial}{\partial y}(y^2 z) + \frac{1}{2\sqrt{xy^2 z^3}}\frac{\partial}{\partial y}(xy^2 z^3)$$

$$= 0 + 2yze^{y^2 z} + \frac{2xyz^3}{2\sqrt{xy^2 z^3}} = 2yze^{y^2 z} + \frac{xyz^3}{\sqrt{xy^2 z^3}}.$$

To calculate $\partial w/\partial z$, we keep x and y fixed. Then

$$\frac{\partial w}{\partial z} = \frac{\partial f}{\partial z} = f_z = x + e^{y^2 z}\frac{\partial}{\partial z}y^2 z + \frac{1}{2\sqrt{xy^2 z^3}}\frac{\partial}{\partial z}(xy^2 z^3)$$

$$= x + y^2 e^{y^2 z} + \frac{3xy^2 z^2}{2\sqrt{xy^2 z^3}}.$$

Example 8 Let C denote the oxygen consumption (per unit weight) of a fur-bearing animal, let T_b denote its internal body temperature (in °C), let T_f denote the outside temperature of its fur, and let w denote its weight (in kg). It has been experimentally determined that a reasonable model for the oxygen consumption of the animal is given by

$$C = \frac{5(T_b - T_f)}{2w^{2/3}}.$$

Calculate

(a) C_{T_b}.   (b) C_{T_f}.   (c) C_w.

Solution

(a) $C_{T_b} = \dfrac{\partial C}{\partial T_b} = \dfrac{\partial}{\partial T_b}\left(\dfrac{5T_b}{2w^{2/3}} - \dfrac{5T_f}{2w^{2/3}}\right) = \dfrac{5}{2w^{2/3}}.$

(b) $C_{T_f} = -\dfrac{5}{2w^{2/3}}.$

(c) $C_w = \dfrac{\partial}{\partial w}\dfrac{5}{2}(T_b - T_f)w^{-2/3} = -\dfrac{5}{3}(T_b - T_f)w^{-5/3}.$

Note that (a) and (b) imply that, since $w > 0$, an increase in internal body temperature leads to an increase in oxygen consumption (if T_f and w do not change), whereas an increase in fur temperature leads to a decrease in oxygen consumption (with T_b and w held constant). Does this make sense intuitively? Furthermore, if T_b and T_f are held constant, then assuming that $T_b > T_f$, an increase in the animal's weight will lead to a decrease in its oxygen consumption per unit weight.

Example 9   A tire company manufactures two types of truck snow tires: regular and radial. If it produces q_1 regular tires and q_2 snow tires, its total cost function is given by

$$C(q_1, q_2) = 1200 + 45q_1 + 70q_2 - 0.01q_1^2 - 0.02q_2^2.$$

Find the marginal cost function for each type of tire.

Solution   The marginal cost is, as we have seen, a measure of how a change in the number of units produced of one item affects the total cost. To compute the marginal cost of producing regular tires, we must determine how a change in the number of regular tires produced affects a change in the total cost when the number of radial tires produced stays constant. Thus, we compute

$$\text{marginal cost function for regular tires} = \frac{\partial C}{\partial q_1} = 45 - 0.02q_1$$

and

$$\text{marginal cost function for radial tires} = \frac{\partial C}{\partial q_2} = 70 - 0.04q_2.$$

For example, at a level of production of $(150, 200)$, an increase of 1 unit in regular tire production will increase total costs by approximately $45 - 0.02 \times (150) = 45 - 3 = \$42$. Similarly, an increase of 1 unit in radial tire production will increase costs by approximately $70 - 0.04(200) = 70 - 8 = \$62$.

Example 10   A retail store sells electric broilers and toasters. Each broiler costs the store $12, and each toaster costs the store $8. As in Example 13.1.10 on page 739, the demand functions for the two items are

$$q_1 = 150 - 3p_1 + p_2$$

and

$$q_2 = 270 + p_1 - 2p_2.$$

If broilers and toasters are initially selling for $15 and $10, respectively, what is the marginal profit of the broilers?

Solution   In Example 13.1.10 we obtained the profit as a function of the retail prices:

$$P(p_1, p_2) = -3960 + 178p_1 + 274p_2 + 2p_1p_2 - 3p_1^2 - 2p_2^2.$$

If the price of the broiler is changing and the price of the toaster is fixed, then the profit is changing by an amount $\partial P/\partial p_1$. This is the marginal profit for broilers. But

$$\frac{\partial P}{\partial p_1} = 178 + 2p_2 - 6p_1.$$

When $p_1 = 15$ and $p_2 = 10$, we obtain

$$\frac{\partial P}{\partial p_1} = 178 + 2\cdot 10 - 6\cdot 15 = 108.$$

That is, if $p_1 = \$15$ and $p_2 = \$10$, then a $1 increase in the price of a broiler will result in approximately a $108 increase in profits. Remember, this is true if p_2, the price of the toaster, does not change.

Second Partial Derivatives   We have seen that if $y = f(x)$, then

$$y' = \frac{df}{dx} \quad \text{and} \quad y'' = \frac{d^2f}{dx^2} = \frac{d}{dx}\left(\frac{df}{dx}\right).$$

That is, the second derivative of f is the derivative of the first derivative of f. Analogously, if $z = f(x, y)$, then we can differentiate each of the two "first" partial derivatives $\partial f/\partial x$ and $\partial f/\partial y$ with respect to both x and y to obtain four **second partial derivatives**, as follows.

(a)   Differentiate twice with respect to x:

$$\frac{\partial^2 z}{\partial x^2} = \frac{\partial^2 f}{\partial x^2} = f_{xx} = \frac{\partial}{\partial x}\left(\frac{\partial f}{\partial x}\right).$$

(b)   Differentiate first with respect to x and then with respect to y:

$$\frac{\partial^2 z}{\partial y\, \partial x} = \frac{\partial^2 f}{\partial y\, \partial x} = f_{xy} = \frac{\partial}{\partial y}\left(\frac{\partial f}{\partial x}\right).$$

(c)   Differentiate first with respect to y and then with respect to x:

$$\frac{\partial^2 z}{\partial x\, \partial y} = \frac{\partial^2 f}{\partial x\, \partial y} = f_{yx} = \frac{\partial}{\partial x}\left(\frac{\partial f}{\partial y}\right).$$

(d)   Differentiate twice with respect to y:

$$\boxed{\frac{\partial^2 z}{\partial y^2} = \frac{\partial^2 f}{\partial y^2} = f_{yy} = \frac{\partial}{\partial y}\left(\frac{\partial f}{\partial y}\right).}$$

Remark 1.   The derivatives $\partial^2 f/\partial x\, \partial y$ and $\partial^2 f/\partial y\, \partial x$ are called the **mixed second partials**.

Remark 2.   It is much easier to denote the second partials by f_{xx}, f_{xy}, f_{yx}, and f_{yy}. We will therefore use this notation for the remainder of this section. Note that the symbol f_{xy} indicates that we differentiate first with respect to x and then with respect to y.

Example 11   Let $z = f(x, y) = x^3 y^2 - xy^5$. Calculate the four second partial derivatives.

Solution   We have $f_x = 3x^2 y^2 - y^5$ and $f_y = 2x^3 y - 5xy^4$. Then

$$f_{xx} = \frac{\partial}{\partial x}(f_x) = 6xy^2;$$

$$f_{xy} = \frac{\partial}{\partial y}(f_x) = 6x^2 y - 5y^4;$$

$$f_{yx} = \frac{\partial}{\partial x}(f_y) = 6x^2 y - 5y^4;$$

$$f_{yy} = \frac{\partial}{\partial y}(f_y) = 2x^3 - 20xy^3.$$

Example 12   Let $z = f(x, y) = e^{2x+3y}$. Calculate the four second partial derivatives of f.

Solution   We have $f_x = 2e^{2x+3y}$ and $f_y = 3e^{2x+3y}$ from Example 3. Then

$$f_{xx} = 4e^{2x+3y};$$
$$f_{xy} = 6e^{2x+3y};$$
$$f_{yx} = 6e^{2x+3y};$$
$$f_{yy} = 9e^{2x+3y}.$$

In the last two examples we saw that $f_{xy} = f_{yx}$. This is no accident.

Equality of Mixed Partials   Suppose that f, f_x, f_y, f_{xy}, and f_{yx} are all continuous at (x_0, y_0). Then

$$\boxed{f_{xy}(x_0, y_0) = f_{yx}(x_0, y_0).}$$

The definition of second partial derivatives and the equality of mixed partials is easily extended to functions of three variables. If $w = f(x, y, z)$, then we have nine second partial derivatives (assuming that they exist):

$$\frac{\partial^2 f}{\partial x^2} = f_{xx}, \qquad \frac{\partial^2 f}{\partial y\, \partial x} = f_{xy}, \qquad \frac{\partial^2 f}{\partial z\, \partial x} = f_{xz},$$

$$\frac{\partial^2 f}{\partial x\, \partial y} = f_{yx}, \qquad \frac{\partial^2 f}{\partial y^2} = f_{yy}, \qquad \frac{\partial^2 f}{\partial z\, \partial y} = f_{yz},$$

$$\frac{\partial^2 f}{\partial x\, \partial z} = f_{zx}, \qquad \frac{\partial^2 f}{\partial y\, \partial z} = f_{zy}, \qquad \frac{\partial^2 f}{\partial z^2} = f_{zz}.$$

Example 13   Let $f(x, y, z) = xy^3 - zx^5 + x^2yz$. Calculate all nine second partial derivatives and show that all three pairs of mixed partials are equal.

Solution   We have

$$f_x = y^3 - 5zx^4 + 2xyz,$$
$$f_y = 3xy^2 + x^2z,$$

and

$$f_z = -x^5 + x^2y.$$

Then

$$f_{xx} = -20zx^3 + 2yz, \qquad f_{yy} = 6xy, \qquad f_{zz} = 0,$$

$$f_{xy} = \frac{\partial}{\partial y}(y^3 - 5zx^4 + 2xyz) = 3y^2 + 2xz,$$

$$f_{yx} = \frac{\partial}{\partial x}(3xy^2 + x^2z) = 3y^2 + 2xz,$$

$$f_{xz} = \frac{\partial}{\partial z}(y^3 - 5zx^4 + 2xyz) = -5x^4 + 2xy,$$

$$f_{zx} = \frac{\partial}{\partial x}(-x^5 + x^2y) = -5x^4 + 2xy,$$

$$f_{yz} = \frac{\partial}{\partial z}(3xy^2 + x^2z) = x^2,$$

$$f_{zy} = \frac{\partial}{\partial y}(-x^5 + x^2y) = x^2.$$

PROBLEMS 13.2   In Problems 1–18 calculate $\partial z/\partial x$ and $\partial z/\partial y$.

1. $z = x^2y$ **2.** $z = \dfrac{x}{y}$

3. $z = x^3 + \sqrt{y}$

4. $z = x^3 y^5$

5. $z = x^2 + 7y^2$

6. $z = 9y^2 - 2x^4$

7. $z = 4xy + 9y^5$

8. $z = 3x^2 y^4 - x^3 y^9$

9. $z = 17xy^4 - 3x^{20}$

10. $z = x^{100} + y^{200} + 2x^2 y^2$

11. $z = \dfrac{1+x}{1-y}$

12. $z = \sqrt{2x + 3y}$

13. $z = \dfrac{4x}{y^5}$

14. $z = \dfrac{x+y}{x-y}$

15. $z = \ln(2x - 5y)$

16. $z = e^{xy^3}$

17. $z = (2x + \ln y)^{3/2}$

18. $z = e^{\ln(x^2 y^4)}$

In Problems 19–22 evaluate the given partial derivative at the given point.

19. $f(x, y) = x^3 - y^4; f_x(1, -1)$

20. $f(x, y) = \ln(x^2 + y^4); f_y(3, 1)$

21. $f(x, y) = 2xy^6; f_y(4, 1)$

22. $f(x, y) = e^{3x - 2y}; f_x(-4, 3)$

In Problems 23–29 calculate $\partial w/\partial x$, $\partial w/\partial y$, $\partial w/\partial z$.

23. $w = xyz$

24. $w = x^2 + y^2 + z^2$

25. $w = \sqrt{x + y + z}$

26. $w = \dfrac{x+y}{z}$

27. $w = e^{x + 2y + 3z}$

28. $w = \ln(x^3 + y^2 + z)$

29. $w = e^{xy/z}$

30. Find the equation of the line tangent to the surface $z = x^3 - 4y^3$ at the point $(1, -1, 5)$ that

(a) lies in the plane $x = 1$;

(b) lies in the plane $y = -1$.

31. Find the equation of the line tangent to the surface $x^2 + 4y^2 + 4z^2 = 9$ that lies on the plane $y = 1$ at the point $(1, 1, 1)$.

32. A fur-bearing animal weighing 10 kg has a constant internal body temperature of 23°C. Using the model of Example 8, if the outside temperature is dropping, how is the oxygen consumption of the animal changing when the outside temperature of its fur is 5°C?

33. The cost to a manufacturer of producing q_1 units of product A and q_2 units of product B is given (in dollars) by

$$C(q_1, q_2) = 250 + 3q_1 + 2.5q_2 - (0.003)q_1^2 - 0.007q_2^2.$$

Calculate the marginal cost function of each of the two products.

34. In Problem 33, what is the marginal cost of each item when 200 units of item A and 150 units of item B are produced?

35. The revenue received from the manufacturer of Problem 33 is given by

$$R(q_1, q_2) = \ln(1 + 50q_1 + 75q_2) + \sqrt{1 + 40q_1 + 125q_2}.$$

Calculate the marginal revenue from each of the two products.

36. If a particle is falling in a fluid, then according to **Stokes' law** the velocity of the particle is given by

$$V = \frac{2g}{9}(\rho_P - \rho_f)\frac{r^2}{v},$$

where g is the acceleration due to gravity, ρ_P is the density of the particle, ρ_f is the density of the fluid, r is the radius of the particle (in cm), and v is the absolute viscosity of the liquid. Calculate $V\rho_P$, $V\rho_f$, V_r, and V_v.

37. In Example 10, what is the marginal profit function of the toasters?

38. In Example 10, what is the marginal profit of a broiler if broilers and toasters are selling for \$12 each?

39. In Problem 13.1.28, on page 744, what is the marginal profit function of ordinary watches?

40. In Problem 13.1.28, what is the marginal profit function of digital watches?

 41. The Cobb–Douglas production function for a given product is

$$F(L, K) = 500L^{1/3}K^{2/3}.$$

(a) Compute $\partial F/\partial L$ and $\partial F/\partial K$.

(b) Evaluate these partial derivatives at $L = 200$ and $K = 350$.

 42. Answer the questions of Problem 41 for the function

$$F(L, K) = 250L^{0.7}K^{0.3}.$$

43. Compute $\partial F/\partial L$ and $\partial F/\partial K$ for the general Cobb–Douglas production function

$$F(L, K) = cL^a K^{1-a}.$$

In Problems 44–53 calculate all second partial derivatives and show that all pairs of mixed partials are equal.

44. $f(x, y) = x^2 y$

45. $f(x, y) = xy^2$

46. $f(x, y) = \dfrac{x}{y}$

47. $f(x, y) = \ln(3x - 4y)$

48. $f(x, y) = e^{3x - 4y}$

49. $f(x, y) = \dfrac{x + y}{x - y}$

50. $f(x, y, z) = xyz$

51. $f(x, y, z) = \dfrac{xy}{z}$

52. $f(x, y, z) = x^2 y^3 z^4$

53. $f(x, y, z) = \ln(xy + z)$

13.3 Maxima and Minima for a Function of Two Variables

In Section 11.3 we discussed methods for obtaining maximum and minimum values for a function of one variable. We defined a critical point to be a number x_0 at which $f'(x_0) = 0$ or for which $f(x_0)$ exists but $f'(x_0)$ does not. Then we indicated that local maxima and minima occurred at critical points and gave conditions on first and second derivatives that ensured a critical point was a local maximum or minimum.

The theory of maxima and minima for functions of two or more variables is more complicated, but some of the basic ideas are the same. To simplify things we shall assume that for all the functions under consideration, the second partial derivatives exist.

Neighborhood A **neighborhood** N of a point (x_0, y_0) in $\mathbb{R}^2$ is the inside of a circle centered at (x_0, y_0). This is illustrated in Figure 1.

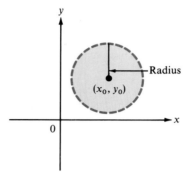

Figure 1

Local Maxima and Minima Suppose that f is defined at all points in a neighborhood N of (x_0, y_0). Then f has

(a) a **local maximum** at (x_0, y_0) if $f(x, y) \leq f(x_0, y_0)$ for all points in N.

(b) a **local minimum** at (x_0, y_0) if $f(x, y) \geq f(x_0, y_0)$ for all points in N.

A rough sketch of a function with several local maxima is given in Figure 2.

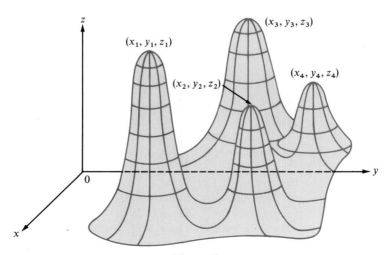

Figure 2

Critical Point   The point (x_0, y_0) is a **critical point** of f if

$$\frac{\partial f}{\partial x}(x_0, y_0) = 0 \quad \text{and} \quad \frac{\partial f}{\partial y}(x_0, y_0) = 0.$$

Necessary Condition for a Local Maximum or Minimum   If the partial derivatives exist and if f has a local maximum or minimum at (x_0, y_0) then (x_0, y_0) is a critical point.

So far, the theory is similar to the theory for a function of one variable. We know that if $f'(x_0) = 0$, then f may have a local maximum or minimum at x_0, or it may have neither. The same is true for a function of two variables, as the following three examples illustrate.

Example 1   Let $f(x, y) = 1 + x^2 + 3y^2$. Now $\partial f/\partial x = 2x$, $\partial f/\partial y = 6y$, and these are zero only when $x = y = 0$. That is, $(0, 0)$ is the only critical point. Since $x^2 \geq 0$ and $y^2 \geq 0$, it is evident that the minimum value of f occurs at $(0, 0)$. That is, $(0, 0)$ is a local (and global) minimum. The graph of f is sketched in Figure 3.

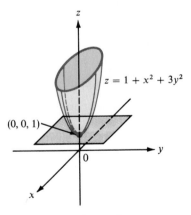

Figure 3

Example 2   Let $f(x, y) = 1 - x^2 - 3y^2$. Now $\partial f/\partial x = -2x$ and $\partial f/\partial y = -6y$, so that $(0, 0)$ is, as in Example 1, the only critical point. Since $-x^2 \leq 0$ and $-3y^2 \leq 0$, f has a local maximum at the critical point $(0, 0)$. The graph of f is sketched in Figure 4.

Example 3   Let $f(x, y) = y^2 - x^2$. Then $\partial f/\partial x = -2x$ and $\partial f/\partial y = 2y$, so that $(0, 0)$ is the only critical point. But $(0, 0)$ is *neither* a local maximum nor a local minimum for f. To see this, we simply note that f can take positive and negative values in any neighborhood of $(0, 0)$, because $f(x, y) > 0$ if $|x| < |y|$ and $f(x, y) < 0$ if $|x| > |y|$. This is illustrated in Figure 5. The surface sketched in Figure 5 is called, for obvious reasons, a **saddle surface**.

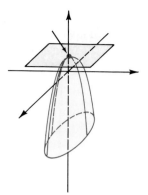

Figure 4

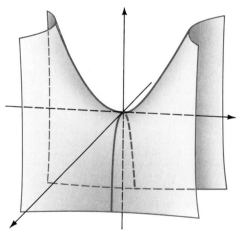

Figure 5

Saddle Point   If (x_0, y_0) is a critical point of f but f does not have a local maximum or local minimum at (x_0, y_0), then (x_0, y_0) is called a **saddle point** of f.

Examples 1, 2, and 3 indicate that more is needed to determine whether a critical point is a local maximum, a local minimum, or a saddle point (in most cases, it will not be at all obvious). As with functions of one variable, the answer has something to do with the signs of the second partial derivatives of f. However, now the situation is more complicated. The proof of the following result is beyond the scope of this text.

Second Derivatives Test   Let f and all its partial derivatives exist in a neighborhood of the critical point (x_0, y_0). Let

$$D(x, y) = f_{xx}(x, y)f_{yy}(x, y) - [f_{xy}(x, y)]^2,$$

and let D denote $D(x_0, y_0)$.

---

1. If $D > 0$ and $f_{xx}(x_0, y_0) > 0$, then f has a local minimum at (x_0, y_0).

2. If $D > 0$ and $f_{xx}(x_0, y_0) < 0$, then f has a local maximum at (x_0, y_0).

3. If $D < 0$, then (x_0, y_0) is a saddle point of f.

4. If $D = 0$, then any of the preceding situations is possible.

Example 4 Let $f(x, y) = 1 + x^2 + 3y^2$. Then, as we saw in Example 1, $(0, 0)$ is the only critical point of f. But $f_{xx} = 2$, $f_{yy} = 6$, and $f_{xy} = 0$, so that $D(0, 0) = 2 \cdot 6 - 0^2 = 12$ and $f_{xx} > 0$, which *proves* that f has a local minimum at $(0, 0)$.

Example 5 Let $f(x, y) = -x^2 - y^2 + 2x + 4y + 5$. Determine the nature of the critical points of f.

Solution $\partial f/\partial x = -2x + 2$, and $\partial f/\partial y = -2y + 4$. Setting $\partial f/\partial x = \partial f/\partial y = 0$, we obtain

$$-2x + 2 = 0,$$
$$-2y + 4 = 0;$$

or

$$2x = 2,$$
$$2y = 4;$$

or

$$x = 1, \; y = 2.$$

Thus $(1, 2)$ is the only critical point. But $f_{xx} = -2$, $f_{yy} = -2$, and $f_{xy} = 0$, so that $D = (-2)(-2) = 4$. Because $D(1, 2) = 4 > 0$ and $f_{xx} < 0$, there is a local maximum at $(1, 2)$. At $(1, 2)$, $f(1, 2) = 10$.

Example 6 Let $f(x, y) = 2x^3 - 24xy + 16y^3$. Determine the nature of the critical points of f.

Solution $\partial f/\partial x = 6x^2 - 24y$, and $\partial f/\partial y = -24x + 48y^2$. At a critical point, we have

$$6(x^2 - 4y) = 0 \quad \text{and} \quad -24(x - 2y^2) = 0.$$

To obtain the critical points, we must solve the simultaneous equations

$$x^2 - 4y = 0,$$
$$x - 2y^2 = 0.$$

The second equation tells us that $x = 2y^2$. Substituting this into the first equation yields

$$4y^4 - 4y = 0 = 4y(y - 1),$$

which has solutions $y = 0$ and $y = 1$. When $y = 0$, $x = 0$, and when $y = 1$, $x = 2 \cdot 1^2 = 2$. Thus the critical points are $(0, 0)$ and $(2, 1)$. Now $f_{xx} = 12x$, $f_{yy} = 96y$, and $f_{xy} = -24$, so that

$$D(x, y) = (12x)(96y) - 24^2 = 1152xy - 576.$$

Since $D(0, 0) = -576 < 0$, $(0, 0)$ is a saddle point. Since $D(2, 1) = 1728 > 0$ and $f_{xx}(2, 1) = 24 > 0$, $(2, 1)$ is a local minimum.

Example 7   Refer to Example 13.1.10 on page 739. How should the retail store price its broilers and toasters in order to maximize profit?

Solution   In Example 13.1.10, we obtained the profit function

$$P(p_1, p_2) = -3960 + 178p_1 + 274p_2 + 2p_1p_2 - 3p_1^2 - 2p_2^2,$$

where p_1 is the retail price of a broiler and p_2 is the retail price of a toaster. Then

$$\frac{\partial P}{\partial p_1} = 178 + 2p_2 - 6p_1$$

and

$$\frac{\partial P}{\partial p_2} = 274 + 2p_1 - 4p_2.$$

Setting $\partial P/\partial p_1 = 0$ and $\partial P/\partial p_2 = 0$, we obtain the simultaneous equations

$$6p_1 - 2p_2 = 178,$$
$$-2p_1 + 4p_2 = 274.$$

Multiplying the first equation by 2 and adding the second equation,

$$12p_1 - 4p_2 = 356$$
$$\underline{-2p_1 + 4p_2 = 274}$$
$$10p_1 \qquad = 630,$$

and

$$p_1 = 63.$$

Since $2p_2 = -178 + 6p_1 = -178 + 378 = 200$, we find that

$$p_2 = 100.$$

The only critical point is, therefore, $(63, 100)$. Now

$$P_{p_1p_1} = -6, \qquad P_{p_2p_2} = -4 \qquad \text{and} \qquad P_{p_1p_2} = 2.$$

Thus, $D = (-6)(-4) - 2^2 = 20 > 0$ and $P_{p_1p_1}(63, 100) < 0$. Thus P has a local maximum at $(63, 100)$. At these prices, the total profit is

$$P(63, 100) = -3960 + 178 \cdot 63 + 274 \cdot 100 + 2 \cdot 63 \cdot 100 - 3 \cdot 63^2 - 2 \cdot 100^2$$
$$= \$15,347.$$

Example 8   A rectangular wooden box with an open top is to contain 50 cm^3. Ignoring the thickness of the wood, how is the box to be constructed so as to use the smallest amount of wood?

Solution   If the dimensions of the box are x, y, and z, then

$$V = xyz = 50.$$

We must minimize the area of the sides of the box, where the area is given by $A = xy + 2xz + 2yz$ (see Figure 6). To use the techniques of this section, we need to express A in terms of two variables only. We see that $z = 50/xy$, so the problem becomes one of minimizing $A(x, y)$ where

$$A(x, y) = xy + 2x\frac{50}{xy} + 2y\frac{50}{xy} = xy + \frac{100}{y} + \frac{100}{x}.$$

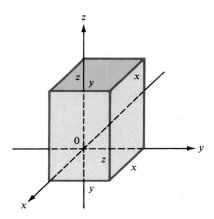

Figure 6

Now

$$\frac{\partial A}{\partial x} = y - \frac{100}{x^2} \quad \text{and} \quad \frac{\partial A}{\partial y} = x - \frac{100}{y^2}.$$

These are zero when

$$y = \frac{100}{x^2} \quad \text{and} \quad x = \frac{100}{y^2}.$$

From the first equation, we have

$$yx^2 = 100.$$

Substituting $x = 100/y^2$ from the second equation yields

$$100 = yx^2 = y\left(\frac{100}{y^2}\right)^2 = y\left(\frac{100^2}{y^4}\right) = \frac{100^2}{y^3},$$

$$1 = \frac{100}{y^3}, \qquad \text{We divided the left and right sides by 100.}$$

$$y^3 = 100, \quad \text{and} \quad y = \sqrt[3]{100} = 100^{1/3}.$$

Then

$$x = \frac{100}{y^2} = \frac{100}{(100^{1/3})^2} = \frac{100}{100^{2/3}} = 100^1(100^{-2/3}) = 100^{1/3} = \sqrt[3]{100}.$$

Thus, $(\sqrt[3]{100}, \sqrt[3]{100})$ is the only critical point.

$$A_{xx} = \frac{200}{x^3}, \qquad A_{yy} = \frac{200}{y^3}, \qquad \text{and} \qquad A_{xy} = 1.$$

If $x = \sqrt[3]{100}$, then $x^3 = 100$. Similarly, $y^3 = 100$ at the critical point. Thus $A_{xx}(\sqrt[3]{100}, \sqrt[3]{100}) = 200/100 = 2$, $A_{yy}(\sqrt[3]{100}, \sqrt[3]{100}) = 2$, and $A_{xy}(\sqrt[3]{100}, \sqrt[3]{100}) = 1$, so that

$$D(\sqrt[3]{100}, \sqrt[3]{100}) = A_{xx}A_{yy} - (A_{xy})^2 = 2 \cdot 2 - 1 = 3 > 0$$

and

$$A_{xx} = 2 > 0.$$

This means that A has a local minimum at $(\sqrt[3]{100}, \sqrt[3]{100})$. Finally, for $x = y = \sqrt[3]{100}$,

$$z = \frac{50}{xy} = \frac{50}{x^2} = \frac{50}{(\sqrt[3]{100})^2} = \frac{50\sqrt[3]{100}}{(\sqrt[3]{100})^3} = \frac{50\sqrt[3]{100}}{100} = \frac{\sqrt[3]{100}}{2}.$$

That is, $z = x/2 = y/2$. We conclude that we can minimize the amount of wood needed by building a box with a square base and a height equal to half its length (or width).

We now provide an example from statistics.

Example 9   **Regression Lines (Optional)**   We can use the theory of this section in an interesting way to derive a result that is very useful for statistical analysis and, in fact, any analysis involving the use of a great deal of data. Suppose n data points $(x_1, y_1), (x_2, y_2), \ldots, (x_n, y_n)$ are collected. For example, the x's may represent average tree growth and the y's average daily temperature in a given year in a certain forest. Or x may represent a week's sales and y a week's profit for a certain business. The question arises as to whether we can "fit" these data points to a straight line. That is, is there a straight line that runs "more or less" through the points? If so, then we can write y as a linear function of x, with obvious computational advantages.

The problem is to find the "best" straight line, $y = mx + b$, passing through or near these points. Look at Figure 7. If (x_i, y_i) is one of our n points, then, on the line $y = mx + b$, corresponding to x_i we obtain $y_i = mx_i + b$. The "error," ε_i, between the y value of our actual point and the "approximating" value on the line is given by

$$\varepsilon_i = y - mx_i - b.$$

One way to choose the approximating line is to use the line that minimizes the

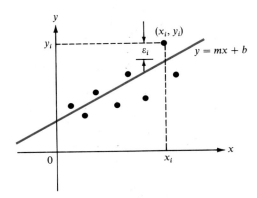

Figure 7

sum of the squares of the errors. This is called the **least-squares** criterion for choosing the line.

Before continuing, we remind you of the notation used in Chapter 6 (see page 326). We denote the sum $x_1 + x_2 + \cdots + x_n$ by $\sum x_i$, where the Greek letter sigma Σ stands for the word *sum*.

Now we want to choose m and b such that the function

$$f(m, b) = \varepsilon_1^2 + \varepsilon_2^2 + \cdots + \varepsilon_n^2 = \sum \varepsilon_i^2 = \sum (y_i - mx_i - b)^2$$

is a minimum. To do this, we calculate

$$\frac{\partial}{\partial m} (y_i - mx_i - b)^2 = -2x_i(y_i - mx_i - b)$$

and

$$\frac{\partial}{\partial b} (y_i - mx_i - b)^2 = -2(y_i - mx_i - b).$$

Hence,

$$\frac{\partial f}{\partial m} = -2 \sum x_i(y_i - mx_i - b)$$

and

$$\frac{\partial f}{\partial b} = -2 \sum (y_i - mx_i - b).$$

Setting $\partial f/\partial m = 0$ and $\partial f/\partial b = 0$ and rearranging terms, we obtain

$$\sum (x_i y_i - mx_i^2 - bx_i) = 0$$
$$\sum (y_i - mx_i - b) = 0.$$

This leads to the system of two equations in the unknowns m and b:

$$(\sum x_i^2)m + (\sum x_i)b = \sum x_i y_i \tag{1}$$

and

$$\left(\sum x_i\right)m + nb = \sum y_i. \tag{2}$$

Here we have used the fact that $\sum b = nb$. The system (1) and (2) is not hard to solve for m and b. To do so, we multiply both sides of (1) by n and both sides of (2) by $\sum x_i$ and then subtract to finally obtain

$$m = \frac{n \sum x_i y_i - [\sum x_i][\sum y_i]}{n \sum x_i^2 - [\sum x_i]^2} \tag{3}$$

and

$$b = \frac{[\sum x_i^2][\sum y_i] - [\sum x_i][\sum x_i y_i]}{n \sum x_i^2 - [\sum x_i]^2}. \tag{4}$$

We will leave it to you to check that the numbers m and b given in (3) and (4) do indeed provide a minimum. The line $y = mx + b$ given by (3) and (4) is called the **regression line** for the n points.

Remark.   Equations (3) and (4) make sense only if

$$n \sum x_i^2 - \left(\sum x_i\right)^2 \neq 0.$$

But in fact,

$$n \sum x_i^2 - \left(\sum x_i\right)^2 \geq 0$$

and is equal to zero only when all the x_i's are equal (in which case the regression line is the vertical line $x = x_i$). We will not prove this fact.

Example 10   Find the regression line for the points $(1, 2)$, $(2, 4)$, and $(5, 5)$.

Solution   We tabulate some appropriate values in Table 1. Then, from (3) and (4),

$$m = \frac{3(35) - (8)(11)}{3(30) - 8^2} = \frac{17}{26} \approx 0.654$$

and

$$b = \frac{(30)(11) - 8(35)}{26} = \frac{50}{26} \approx 1.923.$$

Thus the regression line is

$$y = 0.654x + 1.923.$$

This is all illustrated in Figure 8.

TABLE 1

i	x_i	y_i	x_i^2	$x_i y_i$
1	1	2	1	2
2	2	4	4	8
3	5	5	25	25
Σ	8	11	30	35

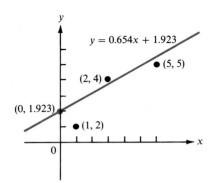

Figure 8

PROBLEMS 13.3

In Problems 1–16 determine the nature of the critical points of the given function.

1. $f(x, y) = 7x^2 - 8xy + 3y^2 + 1$

2. $f(x, y) = x^2 + y^3 - 3xy$

3. $f(x, y) = x^2 + 3y^2 + 4x - 6y + 3$

4. $f(x, y) = x^2 + y^2 + 4xy + 6y - 3$

5. $f(x, y) = x^2 + y^2 + 4x - 2y + 3$

6. $f(x, y) = xy^2 + x^2 y - 3xy$

7. $f(x, y) = x^3 + 3xy^2 + 3y^2 - 15x + 2$

8. $f(x, y) = x^3 + y^3 - 3xy$

9. $f(x, y) = \dfrac{1}{y} - \dfrac{1}{x} - 4x + y$

10. $f(x, y) = \dfrac{1}{x} + \dfrac{2}{y} + 2x + y + 1$

11. $f(x, y) = x^2 - xy + y^2 + 2x + 2y$

12. $f(x, y) = xy + \dfrac{8}{x} + \dfrac{1}{y}$

13. $f(x, y) = (4 - x - y)xy$

14. $f(x, y) = 2x^2 + y^2 + 2/x^2 y$

15. $f(x, y) = 4x^2 + 12xy + 9y^2 + 25$

16. $f(x, y) = x^{25} - y^{25}$

17. Find three positive numbers whose sum is 50 and whose product is a maximum.

18. Find three positive numbers, x, y, and z, whose sum is 50 and such that the product $xy^2 z^3$ is a maximum.

19. Find three numbers whose sum is 50 and the sum of whose squares is a minimum.

20. What is the maximum volume of an open-top rectangular box that can be built from 8 m^2 of wood?

21. In Problem 13.1.28 on page 744, how should the jeweler price her ordinary and digital watches in order to maximize her profit?

22. A company uses two types of raw materials, I and II, for its product. If it uses q_1 units of I and q_2 units of II, it can produce U units of the finished item where

$$U(q_1, q_2) = 8q_1 q_2 + 32q_1 + 40q_2 - 4q_1^2 - 6q_2^2.$$

Each unit of I costs $10, and each unit of II costs $4. Each unit of the product can be sold for $40. How can the company maximize its profits?

23. A packing crate holding 480 cubic feet is to be constructed from three different materials. The cost of constructing the bottom of the crate is $10 per square foot, the cost of material for the sides is $8 per square foot, and the cost of building the top is $5 per square foot.

(a) What are the dimensions of the crate that will minimize the total construction cost?

(b) What is the minimum cost?

24. A major oil company sells both oil and natural gas. Each unit of oil costs the company $25, and each unit of gas costs $15. The revenue (in dollars) received from selling i units of oil and g units of gas is given by

$$R(i, g) = 60i + 50g - 0.02ig - 0.3i^2 - 0.2g^2.$$

(a) How many units of each should the company sell in order to maximize profits?

(b) What is the maximum profit?

25. A manufacturer is developing a new electronic garage-door opener. She hopes to build the device for $100 and sell it for $250. After considerable market research, it is determined that the number of units that can be sold at that price depends on the amount of money spent on advertising (a) and the amount spent on product development (d). The relationship is given by

$$N(a, d) = \text{number of units sold} = \frac{300a}{a + 3} + \frac{160d}{d + 5}.$$

(a) Write down a function that gives profit P as a function of a and d.

(b) Explain why there is no expenditure on advertising and product development that will maximize profit. Is this a realistic model?

26. The prices for two products, denoted by p_1 and p_2, are related to the quantities sold of the two products, x_1 and x_2, by the relations

$$x_1 = 32 - 2p_1,$$

$$x_2 = 22 - p_2.$$

Furthermore, the total cost of producing and selling the products is related to the quantities sold by the function

$$C(x_1, x_2) = \tfrac{1}{2}x_1^2 + 2x_1x_2 + x_2^2 + 73.$$

(a) Develop a mathematical model that shows profit as a function of the quantities produced.

(b) Determine the prices and quantities that maximize profit.

27. A shoe store has determined that its earnings in thousands of dollars, $E(x_1, x_2)$, can be roughly represented as a function of x_1, its investment in inventory in thousands of dollars, and x_2, its expenditure on advertising in thousands of dollars:

$$E(x_1, x_2) = -3x_1^2 + 2x_1x_2 - 6x_2^2 + 30x_1 + 24x_2 - 86.$$

Find the maximum earnings and the amount of advertising expenditure and inventory investment that yields this maximum.

28. Find the regression line for the points $(1, 1)$, $(2, 3)$, and $(3, 6)$. Sketch the line and the points.

29. Find the regression line for the points $(-1, 3)$, $(1, 2)$, $(2, 0)$, and $(4, -2)$. Sketch the line and the points.

30. Find the regression line for the points $(1, 4)$, $(3, -2)$, $(5, 8)$, and $(7, 3)$. Sketch the line and the points.

31. Show that $\sqrt{x^2 + y^2}$ has a local minimum at $(0, 0)$ but that $(0, 0)$ is *not* a critical point.

*32. Let $f(x, y) = 4x^2 - 4xy + y^2 + 5$.

   (a) Show that f has an infinite number of critical points.

   (b) Show that f has a local minimum at each critical point.

   (c) Show that $D = 0$ at each critical point.

*33. For $f(x, y) = -4x^2 + 4xy - y^2 - 5$, answer the questions of Problem 32 with the word *maximum* replacing the word *minimum*.

34. (a) Show that $(0, 0)$ is the only critical point of $f(x, y) = x^3 - y^3$.

   (b) Show that $(0, 0)$ is a saddle point.

   (c) Show that $D = 0$ at $(0, 0)$.

Note. Problems 32, 33, and 34 illustrate the fact that when $D = 0$ at a critical point, f can have a local maximum, a local minimum, or a saddle point.

13.4 Constrained Maxima and Minima—Lagrange Multipliers (Optional)

In the last section we saw how to find the maximum and minimum of a function of two variables by taking partial derivatives and applying a second derivative test. It often happens that side conditions (or *constraints*) are attached to a problem. For example, if we are asked to find the shortest distance from a point (x_0, y_0) to a line $y = mx + b$, we could write this problem as

$$\text{minimize the function:} \quad z = \sqrt{(x - x_0)^2 + (y - y_0)^2}$$

$$\text{subject to the constraint:} \quad y - mx - b = 0.$$

As another example, suppose that the retail store of Example 13.1.10 on page 739 or 13.2.10 on page 749 has, for marketing purposes, determined that the price of the broiler must always be 50% greater than the price of the toaster. Then, to maximize profit, the problem becomes

$$\text{maximize the function:} \quad P(p_1, p_2) = -3960 + 178p_1 + 274p_2$$
$$+ 2p_1p_2 - 3p_1^2 - 2p_2^2$$

$$\text{subject to the constraint: } p_1 = 1.5p_2 \text{ or } p_1 - 1.5p_2 = 0.$$

We now generalize these two examples. Let f and g be functions of two variables. Then we can formulate a **constrained maximization** (or **minimization**) problem as follows:

$$\boxed{\begin{aligned} &\text{maximize (or minimize):} \quad z = f(x, y) \\ &\text{subject to the constraint:} \quad g(x, y) = 0. \end{aligned}}$$

A method for solving problems of this type was developed by the great French mathematician Joseph-Louis Lagrange (1736–1813). Using facts about the geometry of surfaces in three-dimensional space, Lagrange made the following remarkable discovery.

Lagrange Multiplier Method

If, subject to the constraint $g(x, y) = 0$, f takes its maximum or minimum value at a point (x_0, y_0), then, if all partial derivatives exist, there is a number λ such that

$$\frac{\partial f}{\partial x}(x_0, y_0) = \lambda \frac{\partial g}{\partial x}(x_0, y_0) \tag{1}$$

and

$$\frac{\partial f}{\partial y}(x_0, y_0) = \lambda \frac{\partial g}{\partial y}(x_0, y_0). \tag{2}$$

The number λ is called a **Lagrange multiplier**. We shall illustrate the Lagrange multiplier technique with a number of examples.

Example 1   Maximize: $f(x, y) = x^2 + y^2$
subject to: $2x + 5y = 10$.

Solution   Since $2x + 5y = 10$, we have $2x + 5y - 10 = 0$. We then define $g(x, y) = 2x + 5y - 10$ and the problem becomes

maximize: $f(x, y) = x^2 + y^2$
subject to: $g(x, y) = 2x + 5y - 10 = 0$.

Then we compute

$$\frac{\partial f}{\partial x} = 2x, \qquad \frac{\partial g}{\partial x} = 2,$$

$$\frac{\partial f}{\partial y} = 2y, \qquad \frac{\partial g}{\partial y} = 5.$$

At a maximizing point we have, from (1) and (2),

$$\frac{\partial f}{\partial x} = \lambda \frac{\partial g}{\partial x} \qquad \text{and} \qquad \frac{\partial f}{\partial y} = \lambda \frac{\partial g}{\partial y},$$

$$2x = 2\lambda \qquad \text{and} \qquad 2y = 5\lambda.$$

Solving each equation for λ, we obtain

$$\lambda = x \qquad \text{and} \qquad \lambda = \frac{2y}{5}, \qquad \text{or} \qquad x = \frac{2y}{5}.$$

But substituting this into the constraint equation, we obtain

$$2x + 5y = 10,$$

$$2\left(\frac{2y}{5}\right) + 5y = 10,$$

$$\frac{4y}{5} + 5y = 10,$$

$$\frac{29}{5}y = 10,$$

$$y = \frac{5}{29} \cdot 10 = \frac{50}{29}.$$

Then $x = 2y/5 = \frac{2}{5} \cdot \frac{50}{29} = \frac{20}{29}$. Hence f is maximized at $x = \frac{20}{29}$, $y = \frac{50}{29}$ and the maximum value is

$$f\left(\frac{20}{29}, \frac{50}{29}\right) = \left(\frac{20}{29}\right)^2 + \left(\frac{50}{29}\right)^2 = \frac{2900}{841} \approx 3.45.$$

Example 2   Find the maximum and minimum values of $f(x, y) = xy^2$ subject to the condition $x^2 + y^2 = 1$.

Solution   Since $x^2 + y^2 = 1$, we have $x^2 + y^2 - 1 = 0 = g(x, y)$. Then

$$\frac{\partial f}{\partial x} = y^2, \qquad \frac{\partial g}{\partial x} = 2x,$$

$$\frac{\partial f}{\partial y} = 2xy, \qquad \frac{\partial g}{\partial y} = 2y.$$

At a maximizing or minimizing point, we have, from (1) and (2),

$$y^2 = 2x\lambda,$$

$$2xy = 2y\lambda.$$

Multiplying the first equation by y and the second by x, we obtain

$$y^3 = 2xy\lambda,$$

$$2x^2y = 2xy\lambda$$

or

$$y^3 = 2x^2y.$$

But $x^2 = 1 - y^2$ because $g(x, y) = x^2 + y^2 - 1 = 0$, so that

$$y^3 = 2(1 - y^2)y = 2y - 2y^3$$

or

$$3y^3 = 2y.$$

The solutions to this last equation are $y = 0$ and $y = \pm\sqrt{\frac{2}{3}}$. This leads to the six points

$$(1, 0), (-1, 0), \left(\frac{1}{\sqrt{3}}, \sqrt{\frac{2}{3}}\right), \left(-\frac{1}{\sqrt{3}}, \sqrt{\frac{2}{3}}\right), \left(\frac{1}{\sqrt{3}}, -\sqrt{\frac{2}{3}}\right), \left(-\frac{1}{\sqrt{3}}, -\sqrt{\frac{2}{3}}\right).$$

Evaluating $f(x, y) = xy^2$ at these points, we have

$$f(1, 0) = f(-1, 0) = 0, f\left(\frac{1}{\sqrt{3}}, \sqrt{\frac{2}{3}}\right) = f\left(\frac{1}{\sqrt{3}}, -\sqrt{\frac{2}{3}}\right) = \frac{2}{3\sqrt{3}},$$

and

$$f\left(-\frac{1}{\sqrt{3}}, \sqrt{\frac{2}{3}}\right) = f\left(-\frac{1}{\sqrt{3}}, -\sqrt{\frac{2}{3}}\right) = -\frac{2}{3\sqrt{3}}.$$

Thus the maximum value of f is $2/(3\sqrt{3})$, and the minimum value of f is $-2/(3\sqrt{3})$. Note that there is neither a maximum nor a minimum at $(1, 0)$ and $(-1, 0)$, even though conditions (1) and (2) hold at these two points.

Lagrange multipliers can be used for functions of three or more variables. Rather than describe this theory in detail, we give an example.

Example 3   In Example 13.3.8 on page 760, we found a way to minimize the amount of wood needed to construct an open-top rectangular box with a fixed volume of 50 cm³. The problem was to minimize $A(x, y, z) = xy + 2xz + 2yz$ subject to the constraint $xyz = 50$, or $g(x, y, z) = xyz - 50 = 0$. Solve this problem using Lagrange multipliers.

Solution   We have

$$\frac{\partial A}{\partial x} = y + 2z, \qquad \frac{\partial g}{\partial x} = yz,$$

$$\frac{\partial A}{\partial y} = x + 2z, \qquad \frac{\partial g}{\partial y} = xz,$$

$$\frac{\partial A}{\partial z} = 2x + 2y, \qquad \frac{\partial g}{\partial z} = xy.$$

Lagrange's method works equally well for functions of three variables. We have, at a minimizing point,

$$\frac{\partial A}{\partial x} = \lambda \frac{\partial g}{\partial x},$$

$$\frac{\partial A}{\partial y} = \lambda \frac{\partial g}{\partial y},$$

$$\frac{\partial A}{\partial z} = \lambda \frac{\partial g}{\partial z}.$$

These lead to the equations

$$y + 2z = \lambda yz,$$

$$x + 2z = \lambda xz,$$

$$2x + 2y = \lambda xy.$$

To solve, we multiply the three equations by x, y, and z, respectively, to obtain

$$xy + 2xz = \lambda xyz,$$

$$xy + 2yz = \lambda xyz,$$

$$2xz + 2yz = \lambda xyz.$$

Thus

$$xy + 2xz = xy + 2yz = 2xz + 2yz.$$

The first equality indicates that

$$2xz = 2yz$$

and, since $z \neq 0$ (since otherwise xyz would equal 0), we have

$$x = y.$$

Then, the second equality tells us that

$$xy = 2xz,$$

or

$$x^2 = 2xz.$$

And since $x \neq 0$,

$$x = 2z.$$

Thus we have $x = y = 2z$ and, since $xyz = 50$,

$$xyz = (2z)(2z)z = 50 \qquad \text{or} \qquad 4z^3 = 50,$$

so that

$$z = \sqrt[3]{\frac{50}{4}} = \sqrt[3]{\frac{100}{8}} = \frac{\sqrt[3]{100}}{2}$$

and

$$x = y = \sqrt[3]{100}.$$

This is the answer we obtained earlier.

Example 4 In Example 13.1.10 on page 739, how can profit be maximized if the price of a broiler must be 50% greater than the price of a toaster?

Solution As you saw in the beginning of this section, the problem is

maximize: $P(p_1, p_2) = -3960 + 178p_1 + 274p_2 + 2p_1p_2 - 3p_1^2 - 2p_2^2$

subject to the constraint: $g(p_1, p_2) = p_1 - 1.5p_2 = 0$.

We have

$$\frac{\partial P}{\partial p_1} = 178 + 2p_2 - 6p_1, \qquad \frac{\partial g}{\partial p_1} = 1,$$

$$\frac{\partial P}{\partial p_2} = 274 + 2p_1 - 4p_2, \qquad \frac{\partial g}{\partial p_2} = -1.5.$$

Then, setting $\frac{\partial P}{\partial p_1} = \lambda \frac{\partial g}{\partial p_1}$ and $\frac{\partial P}{\partial p_2} = \lambda \frac{\partial P}{\partial p_2}$, we obtain the equations

$$178 + 2p_2 - 6p_1 = \lambda,$$

$$274 + 2p_1 - 4p_2 = -1.5\lambda.$$

If we multiply the first equation by 1.5, we obtain the system

$$267 - 9p_1 + 3p_2 = 1.5\lambda$$

$$274 + 2p_1 - 4p_2 = -1.5\lambda$$

We add:

$$541 - 7p_1 - p_2 = 0$$

or

$$7p_1 + p_2 = 541.$$

But $p_1 = 1.5p_2$, so $7p_1 = 10.5p_2$, and we obtain

$$10.5p_2 + p_2 = 541$$

or

$$11.5p_2 = 541.$$

Then

$$p_2 \approx \$47.04$$

and

$$p_1 = 1.5p_2 \approx 1.5(47.04) = \$70.56.$$

Thus, subject to the new constraint, profits are maximized when broilers sell for \$70.56 and toasters sell for \$47.04. The total profit is then

$$P(70.56, 47.04) = -3960 + 178(70.56) + 274(47.04)$$
$$+ 2(70.56)(47.04) - 3(70.56)^2 - 2(47.04)^2$$
$$\approx \$8765.26.$$

Note that this is considerably smaller than the unconstrained profit of \$15,347 found in Example 13.3.7 on page 759.

 Example 5   In Example 13.1.4 on page 735, we discussed the Cobb-Douglas production function. In particular, we discussed the function

$$F(L, K) = 200L^{2/5}K^{3/5},$$

which represents the number of units of a certain type of die produced if L units of labor and K units of capital are used in production. Suppose that each unit of labor costs \$400 and each unit of capital costs \$500. If \$50,000 is available for production costs, how many units of labor and capital should be used in order to maximize output? How many units will be produced?

Solution   The total cost is

$$400L + 500K.$$

But the total cost is fixed at \$50,000, so we have

$$400L + 500K = 50,000$$

or

$$400L + 500K - 50,000 = 0.$$

Thus, our problem is

maximize: $F(L, K) = 200L^{2/5}K^{3/5}$

subject to: $G(L, K) = 400L + 500K - 50,000 = 0.$

Now

$$\frac{\partial F}{\partial L} = 80L^{-3/5}K^{3/5},$$

$$\frac{\partial F}{\partial K} = 120L^{2/5}K^{-2/5},$$

$$\frac{\partial G}{\partial L} = 400,$$

$$\frac{\partial G}{\partial K} = 500.$$

At a maximizing point,

$$\frac{\partial F}{\partial L} = \lambda \frac{\partial G}{\partial L} \quad \text{and} \quad \frac{\partial F}{\partial K} = \lambda \frac{\partial G}{\partial K},$$

so that

$$80L^{-3/5}K^{3/5} = 400\lambda \quad \text{and} \quad 120L^{2/5}K^{-2/5} = 500\lambda.$$

Thus

$$\lambda = \frac{80L^{-3/5}K^{3/5}}{400} = \frac{120L^{2/5}K^{-2/5}}{500}$$

and

$$\tfrac{1}{5}L^{-3/5}K^{3/5} = \tfrac{6}{25}L^{2/5}K^{-2/5}.$$

We multiply both sides by $L^{3/5}K^{2/5}$:

$$\tfrac{1}{5}(\overbrace{L^{-3/5}L^{3/5}}^{I}\overbrace{K^{3/5}K^{2/5}}^{K}) = \tfrac{6}{25}(\overbrace{L^{2/5}L^{3/5}}^{L}\overbrace{K^{-2/5}K^{2/5}}^{I}),$$

$$\tfrac{1}{5}K = \tfrac{6}{25}L,$$

$$K = \tfrac{30}{25}L = \tfrac{6}{5}L.$$

But

$$400L + 500K = 50,000,$$

$$400L + 500(\tfrac{6}{5})L = 50,000,$$

$$400L + 600L = 50,000,$$

$$1000L = 50,000,$$

$$L = 50,$$

$$K = \tfrac{6}{5}L = 60.$$

Thus 50 units of labor and 60 units of capital should be used, and the maximum number of units produced is

$$F(50, 60) = 200(50)^{2/5}(60)^{3/5} \approx 11,156 \text{ units}.$$

We can use the result of the last example to make an interesting observation. In Example 13.2.5 on page 746, we said that $\partial F/\partial L$ is called the marginal productivity of labor and $\partial F/\partial K$ is called the marginal productivity of capital. At our maximizing values $L = 50$, $K = 60$, we have

$$\frac{\text{marginal productivity of labor}}{\text{marginal productivity of capital}} = \frac{\partial F/\partial L}{\partial F/\partial K} = \frac{80(50)^{-3/5}(60)^{3/5}}{120(50)^{2/5}(60)^{-2/5}}$$

$$= \frac{2}{3}\left[\frac{(60)^{3/5}(60)^{2/5}}{(50)^{2/5}(50)^{3/5}}\right]$$

$$= \frac{2}{3}\cdot\frac{60}{50} = \frac{2}{3}\cdot\frac{6}{5} = \frac{4}{5}.$$

In addition,

$$\frac{\text{unit cost of labor}}{\text{unit cost of capital}} = \frac{400}{500} = \frac{4}{5}.$$

This is no coincidence! It is a general law of economics that

> when labor and capital investments are such as to maximize production, the ratio of marginal productivity of labor to marginal productivity of capital is equal to the ratio of the unit cost of labor to the unit cost of capital.

We conclude this section with two observations. First, while the outlined steps make the method of Lagrange multipliers seem easy, it should be noted that solving three nonlinear equations in three unknowns or four such equations in four unknowns often entails very involved algebraic manipulations. Second, no method is given for determining whether a solution found actually yields a maximum, a minimum, or neither. Fortunately, in many practical applications the existence of a maximum or a minimum can readily be inferred from the nature of the particular problem.

PROBLEMS 13.4

1. Find the maximum and minimum values of $5x + 2xy + 7y$ subject to the condition $x - y = 5$.

2. Find the maximum and minimum values of $x^2 + 5xy + 2y^2$ subject to the condition $2x + 3y = 4$.

*3. Find the maximum and minimum values of $x^2 + 2y^2$ subject to the condition $x + 2y = 6$.

4. Find the maximum and minimum values of $2x^2 + xy + y^2 - 2y$ subject to the condition $y = 2x - 1$.

5. Find the maximum and minimum values of $x^2 + y^2 + z^2$ subject to the condition $z^2 = x^2 - 1$.

6. Find the maximum and minimum values of $x^3 + y^3 + z^3$ if (x, y, z) lies on the sphere $x^2 + y^2 + z^2 = 4$.

7. Find the maximum and minimum values of $x + y + z$ if (x, y, z) lies on the sphere $x^2 + y^2 + z^2 = 1$.

8. Use Lagrange multipliers to find the minimum distance from the point $(1, 2)$ to the line $2x + 3y = 5$.

9. Use Lagrange multipliers to find the minimum distance from the point $(3, -2)$ to the line $y = 2 - x$.

10. Using Lagrange multipliers, show that among all rectangles with the same perimeter, the square encloses the greatest area.

**11. Show that among all triangles having the same perimeter, the equilateral triangle has the greatest area.

12. The base of an open-top rectangular box costs $3 per square meter to construct, but the sides cost only $1 per square meter. Find the dimensions of the box of greatest volume that can be constructed for $36.

13. In Example 4, how can the profit be maximized if the price of a broiler is twice that of a toaster?

14. In Problem 13.1.28 on page 744, how can the jeweler maximize her profit if digital watches sell for 25% more than ordinary watches?

15. In Problem 13.1.28, how can the jeweler maximize her profit if ordinary watches sell for 10% more than digital watches?

16. A manufacturing company has three plants I, II, and III, which produce x, y, and z units, respectively, of a certain product. The annual revenue from this production is given by

$$R(x, y, z) = 6xyz^2 - 400{,}000x - 400{,}000y - 400{,}000z.$$

If the company is to produce 1000 units annually, how should it allocate production so as to maximize profits?

17. A firm has $250,000 to spend on labor and raw materials. The output of the firm is αxy, where α is a constant and x and y are, respectively, the quantity of labor and raw materials consumed. If the unit price of hiring labor is $5000 and the unit price of raw materials is $2500, find the ratio of x to y that maximizes output.

18. The Cobb–Douglas production function for a certain product is

$$F(L, K) = 250L^{0.7}K^{0.3}.$$

Suppose that each unit of labor costs $200, that each unit of capital costs $350, and that $25,000 is available for production costs.

(a) How many units of labor and capital should be used to maximize output?

(b) How many units will be produced?

(c) Compute the ratio of marginal productivity of labor to marginal productivity of capital at levels of labor and capital costs that maximize output.

19. A product has the Cobb–Douglas production function

$$F(L, K) = 500L^{1/3}K^{2/3}.$$

Answer the questions of Problem 18 assuming that each unit of labor costs $1000, each unit of capital costs $1600, and a total of $250,000 is available for production.

20. The temperature of a point (x, y, z) on the unit sphere is given by $T(x, y, z) = xy + yz$. What is the hottest point on the sphere? [*Hint*: The unit sphere has the equation $x^2 + y^2 + z^2 = 1$.]

21. A can of dog food is advertised to contain 80 units of protein. Two types of meat are used in making up the food. Each unit of liver costs 30¢, and each unit of horsemeat costs 16¢. If l units of liver and h units of horsemeat are in the can, then the number of units of protein is

$$N(l, h) = 4l^2 + 2.5h^2.$$

(a) How many units of each meat should be put in a can of dog food to meet the advertised claim at minimum cost?

(b) What is the minimum cost?

22. Bellingham Health Care (BHC) is a nonprofit foundation providing medical treatment to emotionally distressed children. BHC has hired you as a business consultant to aid the foundation in the development of a hiring policy that would be consistent with its overall goal of providing the most meaningful patient service possible given scarce foundation resources. In your initial analysis, you have determined that *service* can be described as a function of medical (M) and social services (S) staff input, as follows:

$$\text{service} = M + 0.5S + 0.5MS - S^2.$$

BHC's staff budget for the coming year is $600,000. Annual employment costs total $15,000 for each social service staff member and $30,000 for each medical staff member.

(a) Construct the function you might use to determine the optimal (*service*-maximizing) social service–medical employment combination.

(b) Determine the optimal combination of social service and medical staff for BHC.

23. A field representative for a major pharmaceutical firm has just received the following information from a marketing research consultant who has been analyzing his recent sales performance. The consultant estimates that time spent in the two major metropolitan areas that compose his sales territory will result in monthly sales as indicated by the equation

$$\text{sales} = 500A - 20A^2 + 300B - 10B^2.$$

Here A and B represent the number of days spent in each metropolitan area respectively. Assuming that a working month is composed of twenty business days, what is the optimal number of days the salesperson should spend in each city?

24. Mary Moore, the marketing manager for a large midwestern department store chain, is trying to decide on the optimal allocation of her advertising budget for the next quarter. Mary has data indicating that television and newspaper advertising have this impact on sales:

$$\text{total revenue} = 20T + 5N + 20TN - T^2,$$

where T = units of television advertising and N = units of newspaper advertising. Each television ad unit costs \$10, and a unit of newspaper advertising costs \$5. Assuming a \$1000 budget constraint, what are the revenue-maximizing units of television and newspaper advertising?

25. Chet Hammond, operations manager for Northern States Nursery, is attempting to determine his equipment requirements for the cultivation operation in the nursery's conifer plantation. The equipment required for this work is already owned by the firm, so Hammond's decision at this point is merely one of allocating machines to this particular area. Relevant costs for the decision include \$50 in variable expenses per week for each cultivator employed and \$150 per week for each employee working in the plantation. From historical records Hammond has estimated that the number of acres that can be cultivated each week is given by the equation

$$A = 2E^2 + EC + 15C - (2/3)C^2,$$

where A represents acres cultivated, E is the number of employees, and C is the number of cultivators used. If the budget for the cultivation activities is \$2550 per week and Hammond wants to maximize the area cultivated each week, what combination of employees and equipment should he use?

26. Heller Manufacturing has two production facilities that manufacture baseball gloves. Production costs at the two facilities differ because of varying labor rates, local property taxes, type of equipment, capacity, and so on. The Dayton plant has weekly production costs that can be represented as a function of the number of gloves produced; that is,

$$C_1(x_1) = x_1^2 - x_1 + 5,$$

where x_1 is the weekly production volume in thousands of units and $C_1(x_1)$ is the cost in thousands of dollars. The Hamilton plant's weekly production costs are given by

$$C_2(x_2) = x_2^2 + 2x_2 + 3,$$

where x_2 is the weekly production volume in thousands of units and $C_2(x_2)$ is the cost in thousands of dollars. Heller Manufacturing would like to produce 8000 gloves per week at the lowest possible cost.

(a) Formulate a mathematical model that can be used to determine the number of gloves to produce each week at each facility.

(b) Find the solution to your mathematical model to determine the optimal number of gloves to produce at each facility.

*27. For a simple lens of focal length f, the object distance d and the image distance i are related by the formula

$$\frac{1}{d} + \frac{1}{i} = \frac{1}{f}.$$

A given lens has a focal length of 50 cm.

(a) What is the minimum value of the object-image distance $d + i$?

(b) For what values of d and i is this minimum achieved?

Review Exercises for Chapter 13

In Exercises 1–6 (a) find the domain of the given function and (b) evaluate the function at the given point.

1. $f(x, y) = x^3 y^5$; (2, 1)

2. $f(x, y) = \sqrt{x^2 - y^2}$; (−5, 4)

3. $f(x, y) = \dfrac{1}{\sqrt{x^2 + y^2}}$; (−7, −3)

4. $f(x, y) = \ln(2y - 3x)$; (−2, 1)

5. $f(x, y, z) = \sqrt{1 - x^2 - y^2 - z^2}$; $(\frac{1}{2}, \frac{1}{3}, \frac{1}{4})$

6. $f(x, y, z) = \dfrac{1}{\sqrt{x^2 + y^2 + z^2 - 1}}$; (2, −1, 3)

7. Sketch the point (−3, 1, 2) in three-dimensional space.

8. Sketch the point (2, 0, −4) in three-dimensional space.

In Exercises 9–14 calculate all first partial derivatives.

9. $f(x, y) = 2y + 3x$

10. $f(x, y) = \dfrac{y}{x}$

11. $f(x, y) = 4x^3 y^7$

12. $f(x, y) = \dfrac{1}{\sqrt{x^2 - y^3}}$

13. $f(x, y, z) = \ln(x - y + 4z)$

14. $f(x, y, z) = \dfrac{1}{\sqrt{x^2 + y^2 + z^2}}$

In Exercises 15–20 calculate all second partial derivatives and show that all pairs of mixed partials are equal.

15. $f(x, y) = xy^3$

16. $f(x, y) = \sqrt{x^2 - y^2}$

17. $f(x, y) = 17x + 205y$

18. $f(x, y) = \dfrac{x + y}{x - y}$

19. $f(x, y, z) = x^3yz^4$

20. $f(x, y, z) = \dfrac{x - y}{z}$

In Exercises 21–25 determine the nature of the critical points of the given function.

21. $f(x, y) = 6x^2 + 14y^2 - 16xy + 2$

22. $f(x, y) = x^5 - y^5$

23. $f(x, y) = \dfrac{1}{y} + \dfrac{2}{x} + 2y + x + 4$

24. $f(x, y) = 49 - 16x^2 + 24xy - 9y^2$

25. $f(x, y) = x^2 + y^2 + \dfrac{2}{xy^2}$

26. Find the minimum distance from the point $(3, -2)$ to the line $y = 2x + 3$.

27. What is the maximum volume of an open-top rectangular box that can be built from 10 m^2 of wood?

28. What is the smallest amount of wood needed to build an open-top rectangular box enclosing a volume of 25 m^3?

29. Find the regression line through the points $(1, 4)$, $(-2, 3)$, $(1, 1)$, and $(2, 6)$. Sketch the line and the points.

30. A bicycle shop owner sells three-speed and ten-speed bicycles. Each three-speed bicycle costs her \$90, and each ten-speed bicycle costs her \$125. The demand functions for the bicycles are

$$q_1 = 100 - 1.5p_1 + 0.6p_2 \quad \text{for the three-speeds}$$

and

$$q_2 = 160 + 0.8p_1 - 2p_2 \quad \text{for the ten-speeds.}$$

(a) Find the total profit as a function of the prices of the two types of bicycles.

(b) What is the profit when three-speed bicycles are sold for \$110 and ten-speed bicycles are sold for \$150?

31. In Exercise 30, what prices will bring maximum profits to the shop owner?

32. In Exercise 30, what prices will bring maximum profits to the shop owner if ten-speed bicycles must be sold for 25% more than three-speed bicycles?

A1 REVIEW OF THREE TOPICS IN ALGEBRA

A1.1 Polynomials

You will encounter polynomials often in this text. A **polynomial** is an expression of the form

$$p(x) = a_n x^n + a_{n-1} x^{n-1} + \cdots + a_2 x^2 + a_1 x + a_0. \qquad (1)$$

Here $a_0, a_1, \ldots, a_n$ are real numbers. They are called the **coefficients** of the polynomial p. In equation (1) the number a_n is called the **leading coefficient** of p and a_0 is called the **constant term**.

We assume in (1) that $a_n \neq 0$. Then p is a polynomial of **degree n**. We denote the degree of p by deg p. A polynomial of *degree zero* is a constant.

Example 1

(a) $x^2 - 2x + 4$ is a polynomial of degree 2, or a *quadratic* polynomial.

(b) $2x^3 - 4x^2 + 5x - 7$ is a polynomial of degree 3, or a *cubic* polynomial.

(c) $5x - 6$ is a first-degree or *linear* polynomial.

(d) $5x^7 - 6x^4 + 3x - 2$ is a polynomial of degree 7.

In this section we will discuss the addition, subtraction, multiplication, and factoring of polynomials.

We begin by giving a general rule for adding or subtracting two polynomials.

ADDITION RULE FOR POLYNOMIALS

To add (or subtract) two polynomials, add (or subtract) the coefficients of corresponding terms.

Example 2 Let $p(x) = 2x^3 - 7x^2 + 4x - 5$ and $q(x) = 5x^2 - 2x + 6$. Find

(a) $p(x) + q(x)$.

(b) $p(x) - q(x)$.

Solution

(a) $p(x) + q(x) = (2x^3 - 7x^2 + 4x - 5) + (5x^2 - 2x + 6)$
$= 2x^3 - 7x^2 + 4x - 5 + 5x^2 - 2x + 6$
$= 2x^3 - 2x^2 + 2x + 1.$

(b) $p(x) - q(x) = (2x^3 - 7x^2 + 4x - 5) - (5x^2 - 2x + 6)$
$= 2x^3 - 7x^2 + 4x - 5 - 5x^2 + 2x - 6$
$= 2x^3 - 12x^2 + 6x - 11.$

RULE FOR MULTIPLYING A POLYNOMIAL BY A CONSTANT

To multiply a polynomial by a constant, multiply each term in the polynomial by that constant.

Example 3 Let $p(x) = 5x^4 - 2x^3 + 3x^2 - x + 7$. Find $4p(x)$.

Solution By the rule above,

$$4p(x) = 4 \cdot 5x^4 + 4 \cdot (-2x^3) + 4 \cdot 3x^2 - 4 \cdot (-1x) + 4 \cdot 7 \cdot 1$$
$$= 20x^4 - 8x^3 + 12x^2 - 4x + 28.$$

Example 4 Let $p(x) = 5x^4 - 2x^3 + 3x^2 - x + 7$ and $q(x) = x^5 - 2x^4 + 3x^2 - 5$. Compute $4p(x) - 3q(x)$.

Solution We computed $4p(x)$ in Example 3. Then $3q(x) = 3x^5 - 6x^4 + 9x^2 - 15$, so that

$$4p(x) - 3q(x) = (0 - 3)x^5 + (20 - (-6))x^4 + (-8 - 0)x^3$$
$$+ (12 - 9)x^2 + (-4 - 0)x + (28 - (-15))1$$
$$= -3x^5 + 26x^4 - 8x^3 + 3x^2 - 4x + 43.$$

To multiply two polynomials together, use the following rule:

DISTRIBUTIVE PROPERTY

Let a, b, and c denote polynomials, then

$$a(b + c) = ab + ac \qquad (2)$$

and

$$(a + b)c = ac + bc. \qquad (3)$$

Example 5   Using equation (2), $3x(5x^2 + 2) = (3x)(5x^2) + (3x)(2)$
$$= 15x^3 + 6x.$$

Example 6   Let $p(x) = 2x^2 - 3x + 4$ and $q(x) = x^2 + 4x - 5$. Using the distributive property, find $p(x)q(x)$.

Solution

$$
\begin{aligned}
p(x)q(x) &= (2x^2 - 3x + 4)(x^2 + 4x - 5)\\
&= (2x^2 - 3x + 4)x^2 + (2x^2 - 3x + 4)(4x)\\
&\quad + (2x^2 - 3x + 4)(-5) \qquad \text{From (2)}\\
&= 2x^4 - 3x^3 + 4x^2 + 8x^3 - 12x^2 + 16x - 10x^2\\
&\quad + 15x - 20 \qquad \text{From (3)}\\
&= 2x^4 + 5x^3 - 18x^2 + 31x - 20
\end{aligned}
$$

Remark. The preceding example illustrates three things that happen when we multiply two polynomials together.

If deg $p = m$ and deg $q = n$, then deg $pq = m + n$.

The leading coefficient of pq is the product of the leading coefficients of p and q,

The constant term of pq is the product of the constant terms of p and q.

Thus, in Example 6, deg $pq = 4 = 2 + 2$, the leading coefficient of $pq = 2 = 2 \cdot 1$, and the constant term of $pq = -20 = 4 \cdot (-5)$.

It is often important to be able to *factor* a given polynomial, that is, to write the polynomial as the product of two or more polynomials of smaller degree. We will give some examples of factored polynomials and then suggest methods for factoring a general polynomial.

Example 7 From Example 6,

$$2x^4 + 5x^3 - 18x^2 + 31x - 20 = (2x^2 - 3x + 4)(x^2 + 4x - 5).$$

Example 8 $x^2 - 3x - 10 = (x - 5)(x + 2)$

Example 9 $x^3 - x^2 - 14x + 24 = (x - 2)(x^2 + x - 12)$
$$= (x - 2)(x + 4)(x - 3)$$

Example 10 $x^3 + x^2 + 2x + 2 = (x + 1)(x^2 + 2)$

For all practical purposes, the only polynomials that can be factored with any ease are *quadratic polynomials*. Thus we will start by indicating how these can be factored.

First, we make the obvious observation that either a quadratic polynomial can be factored into the product of two linear polynomials or it cannot. If it cannot, it is called **irreducible**. In Example 8, we factored $x^2 - 3x - 10$ into the product of the linear terms $(x - 5)$ and $(x + 2)$. In Example 10, we obtained the quadratic term $x^2 + 2$, which is irreducible. We will soon see why this is so.

The preceding discussion can help us find a method for factoring a quadratic polynomial. Consider the equation

$$ax^2 + bx + c = 0, \qquad (4)$$

which is a **quadratic equation**. Since $ax^2 + bx + c = a[x^2 + (b/a)x + (c/a)]$, we can factor $ax^2 + bx + c$ if we can factor $x^2 + (b/a)x + (c/a)$. Thus we can assume that the leading coefficient of $ax^2 + bx + c$ is 1. That is, we assume that $a = 1$. Then the quadratic equation (4) becomes

$$x^2 + bx + c = 0. \qquad (5)$$

Suppose that

$$x^2 + bx + c = (x - r)(x - s) \qquad (6)$$

for some real numbers r and s. It also means that r and s are *roots* or *solutions* of the quadratic equation (5), since if $x = r$, for example, then

$$x^2 + bx + c = r^2 + br + c = (r - r)(r - s) = 0(r - s) = 0.$$

Thus we have shown the following important fact:

> $x^2 + bx + c$ can be factored if the quadratic equation
> (5) has one or two real solutions.

We include the case of one real solution because, in (6), r and s could be equal.

It is now easy to see why $x^2 + 2$ must be irreducible. If not, then there is a real number r such that $r^2 + 2 = 0$. But for every real number r, $r^2 \geq 0$, so that $r^2 + 2 \geq 2$ and, therefore, $x^2 + 2 = 0$ has no real roots.

Example 11 The numbers 2 and -5 are solutions of a quadratic equation with leading coefficient 1. Find the quadratic.

Solution From (6), we have

$$x^2 + bx + c = (x - 2)(x - (-5)) = (x - 2)(x + 5) = x^2 + 3x - 10.$$

In the next section we will give three methods for finding all real solutions of a quadratic equation. Here, we simply make some observations to help us factor the polynomial $x^2 + bx + c$. Multiplying through in (6), we have

$$x^2 + bx + c = x^2 - (r + s)x + rs,$$

so that

$$-(r + s) = b \qquad \text{and} \qquad rs = c. \tag{7}$$

Thus, if b and c are integers and we want to factor, as in (6), with r and s also integers, we seek numbers r and s that satisfy (7).

Example 12 Factor $x^2 + 4x - 21$ into two terms with integer coefficients.

Solution From (7) we seek integers r and s such that

$$r + s = -4 \qquad \text{and} \qquad rs = -21.$$

The only choices (if we insist on integers) for r and s that have a product of -21 are 3 and -7, -3 and 7, 21 and -1, or -21 and 1. But $3 + (-7) = -4$, $-3 + 7 = 4$, $21 + (-1) = 20$, and $-21 + 1 = -20$, so the choice is 3 and -7 and we have

$$x^2 + 4x - 21 = (x - 3)[x - (-7)] = (x - 3)(x + 7).$$

We can also write this as $(x + 7)(x - 3)$.

Example 13 Factor $3x^2 - 18x + 27$ into terms with integer coefficients.

Solution We first factor out the 3 to make the leading coefficient 1:

$$3x^2 - 18x + 27 = 3(x^2 - 6x + 9).$$

Now we seek integers r and s such that

$$r + s = 6 \qquad \text{and} \qquad rs = 9.$$

Four choices of r and s give a product of 9. These are 1, 9; -1, -9; 3, 3; and -3, -3. Clearly, only the choice 3, 3 gives a sum of 6. Thus we may write

$$3x^2 - 18x + 27 = 3(x - 3)(x - 3) = 3(x - 3)^2.$$

Some terms occur fairly frequently. We list four of them here. Let a be a real number, then

1. $x^2 - 2ax + a^2 = (x - a)(x - a) = (x - a)^2.$	(8)
2. $x^2 + 2ax + a^2 = (x + a)(x + a) = (x + a)^2.$	(9)
3. $x^2 - a^2 = (x - a)(x + a).$	(10)
4. $x^2 + a^2$ is irreducible if $a \neq 0.$	(11)

Facts 1, 2, and 3 are easily verified by multiplication; $x^2 + a^2$ with $a \neq 0$ is irreducible for the same reason that $x^2 + 2$ is irreducible.

In the next section we shall provide a method to determine when a quadratic is irreducible. If it can be factored, we shall show how this can always be done.

We now turn to the factoring of higher-order polynomials. In general, this is much more difficult. Although we shall not attempt to do so here, it can be shown that any polynomial in the form (1) can be factored as a product of linear and irreducible quadratic factors. To find a factoring, we use the same basic idea we used in factoring a quadratic. Suppose, for example, that deg $p = n$ and

$$p(x) = (x - r)q(x),$$

where q is also a polynomial with deg $q = n - 1$. That is, suppose that a factoring of p contains the linear term $x - r$. Then r is a root of the equation $p(x) = 0$. Thus, *we can find linear factors of $p(x)$ by finding roots of $p(x) = 0$.* But how do we find such roots? Alas, this is, in general, an extremely difficult problem if deg $p > 2$. The only method we will use here is trial and error. This leads to a general method for factoring a polynomial.

TO FACTOR A POLYNOMIAL IN THE FORM (1)

Step 1. Factor out a_n so the leading coefficient of the polynomial to be factored is 1.

Step 2. If $p(x) = a_n p_1(x) = 0$, find (if possible) a root, r, of $p_1(x) = 0$. Use trial and error by trying the divisors (factors) of the constant term in $p_1(x)$.

Step 3. (a) Write $p(x) = a_n(x - r)q(x)$, where $q(x)$ is obtained by long division. If $q(x)$ is a quadratic, then factor $q(x)$ if possible; if $q(x)$ is irreducible, you are done.

(b) If $q(x)$ is not a quadratic, then seek a root, s, of $q(x) = 0$. If one is found, then write

$$p(x) = a_n(x - r)(x - s)q_1(x),$$

where deg $q_1 = $ deg $p - 2$.

Step 4. Try to factor $q_1(x)$ as in Step 3a or 3b.

Step 5. Continue the process until either you can find no more roots or until only linear and irreducible quadratic factors appear.

Example 14 Factor $p(x) = x^3 + 5x^2 - 17x - 21$.

Solution First, look for a root of $p(x) = 0$. The most obvious first tries are ± 1. Here

$$p(1) = 1^3 + 5 \cdot 1^2 - 17 \cdot 1 - 21 = 1 + 5 - 17 - 21$$
$$= -32 \neq 0$$

and

$$p(-1) = (-1)^3 + 5(-1)^2 - 17(-1) - 21 = -1 + 5 + 17 - 21 = 0.$$

Success! We have "discovered" that -1 is a root, so we know that $x - (-1) = x + 1$ is a factor, and we can write

$$p(x) = x^3 + 5x^2 - 17x - 21 = (x + 1)q(x).$$

Thus, $q(x) = (x^3 + 5x^2 - 17x - 21)/(x + 1)$, and we can find $q(x)$ by long division. This is essentially the same long division you learned in grade school:

$$
\begin{array}{r}
x^2 + 4x\ \ - 21 \\
x + 1\overline{\smash{\big)}\ x^3 + 5x^2 - 17x - 21.} \\
\underline{x^3 +\ \ x^2} \\
4x^2 - 17x \\
\underline{4x^2 +\ \ 4x} \\
-21x - 21 \\
\underline{-21x - 21} \\
0
\end{array}
$$

Thus $q(x) = x^2 + 4x - 21$ and, using the result of Example 12,

$$x^3 + 5x^2 - 17x - 21 = (x + 1)(x^2 + 4x - 21) = (x + 1)(x - 3)(x + 7).$$

Example 15   Factor $p(x) = -2x^3 + 6x^2 - 4x + 12$.

Solution   First factor out the -2:

$$p(x) = -2(x^3 - 3x^2 + 2x - 6)$$

and seek a root of $p_1(x) = x^3 - 3x^2 + 2x - 6 = 0$. We try the factors of -6. After trial and error, we find that $p_1(3) = 3^3 - 3(3^2) + 2 \cdot 3 - 6 = 0$, so that

$$p(x) = -2(x - 3)q(x),$$

where $q(x) = (x^3 - 3x^2 + 2x - 6)/(x - 3)$. Now the long division looks like this:

$$
\begin{array}{r}
x^2 + 2 \\
x - 3 \overline{\smash{)}\, x^3 - 3x^2 + 2x - 6.} \\
\underline{x^3 - 3x^2 } \\
2x - 6 \\
\underline{2x - 6} \\
0
\end{array}
$$

Thus,

$$p(x) = -2(x - 3)(x^2 + 2),$$

and this is as far as we can go because $x^2 + 2$ is irreducible.

We close this section by noting that our method will not always result in a factoring. For example,

$$p(x) = x^4 + 3x^2 + 2 = (x^2 + 1)(x^2 + 2)$$

has *no* linear factors, so any attempt to find a root of $x^4 + 3x^2 + 2 = 0$ will be fruitless. However, in this case the factoring could be obtained by treating the fourth-degree polynomial as a quadratic in the variable x^2.

Techniques like the one just mentioned sometimes work, but in general it is extremely difficult to factor a polynomial.

PROBLEMS A1.1   In Problems 1–6 determine the degree of the given polynomial.

1. $x^4 - 1$ **2.** $x^2 - 3x + 2$

3. 6 **4.** $5x - 6$

5. $x^8 - 2x + 3$ **6.** $-8x^7 - 5x^3 + 2x - 4$

In Problems 7–15 let $p(x) = 2x^2 - 3x + 4$ and $q(x) = 3x^3 - x^2 + 5x - 3$. Compute the following:

7. $2p(x)$ **8.** $3q(x)$ **9.** $p(x) + q(x)$

10. $p(x) - q(x)$ **11.** $q(x) - p(x)$ **12.** $-2p(x) + 3q(x)$

13. $4p(x) - 7q(x)$ **14.** the degree of pq **15.** $p(x)q(x)$

In Problems 16–22 let $p(x) = x^4 - 3$ and $q(x) = x^7 - 2x + 3$. Compute the following:

16. $p(x) + q(x)$ **17.** $3q(x)$ **18.** $-8p(x)$

19. $q(x) - p(x)$ **20.** $p(x) - q(x)$ **21.** degree pq

22. $p(x)q(x)$

In Problems 23–42 perform the multiplication and determine the degree of the product.

23. $(x + 2)(x + 4)$ **24.** $x(x - 6)$

25. $(x + 5)(x - 5)$ **26.** $(2x - 3)(x + 2)$

27. $(3x - 5)(-5x + 2)$ **28.** $(x - a)(x + a)$

29. $(x^2 - 1)(x^2 + 1)$ **30.** $(x^2 - 4x + 3)(2x^2 - x + 2)$

31. $(-3x^2 - 4x + 2)(6x^2 - 3x + 2)$ **32.** $x^2(x^3 - 1)$

33. $(x^3 - 1)(x^2 + 1)$ **34.** $(x^3 - 1)(x^3 + 1)$

35. $(3x^3 - 3x + 2)(4x^2 - 5)$ **36.** $(ax^2 + bx + c)(dx^2 + ex + f)$

37. $(ax^3 + bx)(cx^3 + dx + e)$ **38.** $(x^4 + 1)(x^5 - 1)$

39. $x^{10}(x^{20} - 2)$ **40.** $(x^{10} + x^5 - 2)(x^{10} - x^5 - 2)$

41. $(x^4 - 2x^2 + 3)(x^4 + 2x^2 - 3)$ **42.** $(3x^8 - 2x^4 + 3x + 1)(6x^7 - 5x^2 + 6)$

In Problems 43–64 factor the given quadratic polynomial into linear factors with integer coefficients.

43. $x^2 - 4x + 3$ **44.** $x^2 + 4x + 3$

45. $x^2 + 2x + 1$ **46.** $x^2 - 2x + 1$

47. $x^2 + 6x + 5$ **48.** $x^2 + x - 6$

49. $x^2 - x - 42$ **50.** $x^2 - 13x + 42$

51. $x^2 + 13x + 42$ **52.** $x^2 + x - 42$

53. $x^2 - 6x - 16$ **54.** $x^2 - 5x$

55. $x^2 - ax$ **56.** $x^2 + ax$

57. $bx^2 - 3cx$ **58.** $3x^2 - 12x + 12$

59. $-5x^2 + 65x - 210$ **60.** $2x^2 - 6x - 36$

61. $-4x^2 - 36x - 72$ **62.** $x^2 + (a - b)x - ab$

63. $2x^2 - 2(a + b)x + 2ab$ **64.** $3x^2 + 3(b - a)x - 3ab$

*65. Show that $x^2 + 2x + 2$ is irreducible.

*66. Show that $x^2 - 4x + 7$ is irreducible.

In Problems 67–76, all the roots of a polynomial with leading coefficient 1 are given. Find the polynomial.

67. $1, 4$ **68.** $-1, -4$ **69.** $-1, 4$

70. $1, -4$ **71.** $0, 3$ **72.** $0, 1, 6$

73. $-1, 2, 4$ **74.** $-1, 1, 2$ **75.** $2, 2, 5$

76. $2, -3, 3, 7$

In Problems 77–93 factor the given polynomial into linear and irreducible quadratic terms.

77. $x^3 + 2x^2 - x - 2$ **78.** $x^3 + 2x^2 + x + 2$

79. $x^3 - 5x^2 + 7x - 3$ **80.** $2x^3 + 2x^2 - 40x$

81. $-3x^3 + 3x^2 + 24x - 36$ **82.** $x^3 - 12x^2 + 29x + 42$

83. $4x^3 - 20x^2 + 28x - 12$ ***84.** $x^3 + 5x^2 + 8x + 6$

85. $x^4 + x^2$ **86.** $x^4 - 16$

87. $x^4 - 5x^2 + 4$ **88.** $x^4 + x^2 - 6$

89. $x^4 - x^2 - 42$ **90.** $x^4 - 3x^3 - 28x^2 + 132x - 144$

91. $x^5 + x^3 + x$ ***92.** $x^5 + 3x^3 - x^2 - 3$

93. $x^8 - 1$

94. Show that, for any real number a,

$$x^3 - a^3 = (x - a)(x^2 + ax + a^2).$$

95. Show that, for any real number a,

$$x^3 + a^3 = (x + a)(x^2 - ax + a^2).$$

96. Show that, for any real number a,

$$x^4 - a^4 = (x - a)(x^3 + ax^2 + a^2x + a^3).$$

*97. Show that, for any positive real number a,

$$x^4 + a^4 = (x^2 + \sqrt{2a}x + a)(x^2 - \sqrt{2a}x + a),$$

where each quadratic term is irreducible.

In Problems 98–113 use formulas (8), (9), (10), and (11), and the results of Problems 94–97 to factor the given polynomial, if possible.

98. $x^2 - \frac{1}{4}$ **99.** $x^2 + \frac{1}{4}$ **100.** $x^2 + \frac{2}{3}x + \frac{1}{9}$

101. $x^2 - \frac{2}{3}x + \frac{1}{9}$ **102.** $25x^2 + 9$ **103.** $25x^2 - 9$

104. $36x^2 - 81$ **105.** $49x^2 - 14x + 1$ **106.** $x^3 - 8$

107. $x^3 + 8$ **108.** $x^3 + \frac{1}{8}$ **109.** $x^3 - \frac{1}{8}$

110. $x^4 - 81$ ***111.** $x^4 + 81$ **112.** $x^4 - \frac{1}{16}$

113. $625x^4 - 1$

114. Let $p(x)$ be given by equation (1). Suppose that $r_1, r_2, \ldots, r_n$ are n real roots of $p(x) = 0$. Show that

$$r_1 \cdot r_2 \cdot \ldots \cdot r_n = a_0.$$

A1.2 Quadratic Equations

As you saw in Section A1.1, a **quadratic equation** is an equation of the form

$$ax^2 + bx + c = 0, \qquad (1)$$

where a, b, and c are real numbers with $a \neq 0$. Since $a \neq 0$, we can divide both sides of equation (1) by a to obtain

$$x^2 + \frac{b}{a}x + \frac{c}{a} = 0.$$

Thus, we can assume that the leading coefficient in (1) is 1.

In this section we shall show that we can always either solve equation (1) (that is, find all of its solutions) or show that it has no real solutions. There are three methods for solving quadratic equations.

Method 1 Factoring This method is useful only when the quadratic polynomial is easily factored. More often, some other technique is required.

To use factoring to solve a quadratic equation, we need the following rule:

> If a and b are real numbers and $ab = 0$, then either $a = 0$ or $b = 0$.

Example 1   Find all solutions to the quadratic equation $x^2 + 4x - 21 = 0$.

Solution   We see that

$$0 = x^2 + 4x - 21 = (x - 3)(x + 7).$$

From the rule above, we must have

$$x - 3 = 0 \quad \text{or} \quad x + 7 = 0.$$

Thus we obtain the two solutions $x = 3$ and $x = -7$. Note that if $x \neq 3$ and $x \neq -7$, then $x - 3 \neq 0$ and $x + 7 \neq 0$, so that $(x - 3)(x + 7) = x^2 + 4x - 21 \neq 0$. That is, 3 and -7 are the only solutions to the equation.

In general, we have the following rule:

> The quadratic equation $ax^2 + bx + c$ has at most two real roots.

Example 2 Solve the quadratic equation $3x^2 - 18x + 27 = 0$.

Solution From Example 13 in Appendix A1.1,

$$3x^2 - 18x + 27 = 3(x - 3)^2 = 0$$

only when $x = 3$. This is the only root. It is called a *double root* of the equation.

Example 3 The quadratic equation $x^2 + 2 = 0$ has no real solutions, as we have already observed.

Method 2 Completing the Square This method relies on the following algebraic facts:

$$x^2 + bx + c = x^2 + 2\left(\frac{b}{2}\right)x + \frac{b^2}{4} + \left(c - \frac{b^2}{4}\right) = \left(x + \frac{b}{2}\right)^2 + \left(c^2 - \frac{b^2}{4}\right). \quad (2)$$

Note that in obtaining (2) we added and subtracted the term $b^2/4$. The term $\left(x + \frac{b}{2}\right)^2$ is a square, and this gives the method its name.

Example 4 Solve the quadratic equation $x^2 + 6x - 40 = 0$ by completing the square.

Solution $0 = x^2 + 6x - 40 = x^2 + 2(3)x + 3^2 - 40 - 3^2 = (x + 3)^2 - 49 = 0$. Thus,

$$(x + 3)^2 = 49 \quad \text{and} \quad (x + 3) = \pm 7.$$

If $x + 3 = 7$, then $x = 4$, and if $x + 3 = -7$, then $x = -10$. Thus the roots are 4 and -10, as is easily verified. Note that we could have obtained this answer by factoring.

Example 5 Solve the quadratic equation $x^2 + 3x - 7 = 0$.

Solution

$$0 = x^2 + 3x - 7 = x^2 + 2\left(\frac{3}{2}\right)x + \left(\frac{3}{2}\right)^2 - 7 - \left(\frac{3}{2}\right)^2$$

$$= \left(x + \frac{3}{2}\right)^2 - 7 - \frac{9}{4} = \left(x + \frac{3}{2}\right)^2 - \frac{37}{4}.$$

Thus,

$$\left(x + \frac{3}{2}\right)^2 = \frac{37}{4} \quad \text{and} \quad x + \frac{3}{2} = \frac{\pm\sqrt{37}}{2}.$$

If $x + \frac{3}{2} = \sqrt{37}/2$, then $x = (\sqrt{37} - 3)/2$. If $x + \frac{3}{2} = -\sqrt{37}/2$, then $x = (-\sqrt{37} - 3)/2$. Thus, the two roots are $(\sqrt{37} - 3)/2$ and $(-\sqrt{37} - 3)/2$. Note that we could *not* have obtained this answer by factoring.

Example 6 Solve the quadratic equation $x^2 + 2x + 2 = 0$.

Solution $x^2 + 2x + 2 = x^2 + 2x + 1 + 1 = (x + 1)^2 + 1 = 0$. Thus, $(x + 1)^2 = -1$, which is impossible if x is a real number. Hence, the quadratic equation has no real roots.

Method 3 The Quadratic Formula We now obtain a formula for solving any quadratic equation in the form (1). We solve the equation $ax^2 + bx + c = 0$ by completing the squares:

$$ax^2 + bx + c = a\left(x^2 + \frac{b}{a} + \frac{c}{a}\right) = a\left[x^2 + 2\left(\frac{b}{2a}\right)x + \left(\frac{b}{2a}\right)^2 + \left(\frac{c}{a} - \left(\frac{b}{2a}\right)^2\right)\right]$$

$$= a\left[\left(x + \frac{b}{2a}\right)^2 + \left(\frac{c}{a} - \frac{b^2}{4a^2}\right)\right] = a\left[\left(x + \frac{b}{2a}\right)^2 + \frac{4ac - b^2}{4a^2}\right] = 0.$$

$$(3)$$

The last step followed from the fact that

$$\frac{c}{a} = \frac{(4a)c}{(4a)a} = \frac{4ac}{4a^2} \quad \text{and} \quad \frac{4ac}{4a^2} - \frac{b^2}{4a^2} = \frac{4ac - b^2}{4a^2}.$$

(We will say more about this kind of manipulation in the next section.) Then, after dividing both sides of (3) by a, we obtain

$$\left(x + \frac{b}{2a}\right)^2 = -\left(\frac{4ac - b^2}{4a^2}\right) = \frac{b^2 - 4ac}{4a^2}$$

and

$$x + \frac{b}{2a} = \pm\sqrt{\frac{b^2 - 4ac}{4a^2}} = \pm\frac{\sqrt{b^2 - 4ac}}{2a}.$$

Thus, if $b^2 - 4ac > 0$,

$$x = \frac{-b}{2a} \pm \frac{\sqrt{b^2 - 4ac}}{2a} = \frac{-b \pm \sqrt{b^2 - 4ac}}{2a}$$

are the two roots of (1). If $b^2 - 4ac < 0$, then (1) has no real roots. The expression $b^2 - 4ac$ is called the **discriminant** of the quadratic equation (1). In sum,

Let $ax^2 + bx + c = 0$ be a quadratic equation with discriminant $b^2 - 4ac$. Then

1. if $b^2 - 4ac > 0$, the equation has two solutions given by

$$x = \frac{-b + \sqrt{b^2 - 4ac}}{2a} \quad \text{and} \quad x = \frac{-b - \sqrt{b^2 - 4ac}}{2a}.$$

2. if $b^2 - 4ac = 0$, the equation has the unique solution

$$x = \frac{-b}{2a}.$$

3. if $b^2 - 4ac < 0$, there are no real solutions.

Example 7   Solve the quadratic equation $3x^2 - 8x + 2 = 0$.

Solution $a = 3$, $b = -8$, and $c = 2$. Then $b^2 - 4ac = (-8)^2 - 4(3)(2) = 64 - 24 = 40 > 0$, so that there are two solutions. We have

$$x = \frac{-(-8) \pm \sqrt{40}}{2 \cdot 3} = \frac{8 \pm \sqrt{40}}{6} = \frac{8 \pm \sqrt{10 \cdot 4}}{6} = \frac{8 \pm 2\sqrt{10}}{6} = \frac{4 \pm \sqrt{10}}{3}.$$

Thus, the solutions are $x = (4 + \sqrt{10})/3$ and $x = (4 - \sqrt{10})/3$.

Example 8   Solve the quadratic equation $3x^2 - 8x + 9 = 0$.

Solution   Here $b^2 - 4ac = (-8)^2 - 4(3)(9) = 64 - 108 = -44 < 0$, so that there are no real solutions.

Example 9   Solve the quadratic equation $9x^2 - 24x + 16 = 0$.

Solution $b^2 - 4ac = (-24)^2 - 4(9)(16) = 576 - 576 = 0$. Thus the unique solution is $x = -b/2a = -(-24)/(2 \cdot 9) = \frac{24}{18} = \frac{4}{3}$.

In economics, the **demand function** expresses the relationship between the unit price, p, a product can sell for and the number of units, q, that can be sold at that price. Generally, the more units sold, the lower the unit price.

A **supply function** gives the relationship between the expected price, s, of a product and the number of units, q, the manufacturer will produce. It is reasonable to assume that as the expected price increases, the number of units the manufacturer will produce also increases, so that q increases as s increases. It is often the case that supply and demand functions are quadratic.

Example 10 **Market Equilibrium** If the supply function for a commodity is given by $s = q^2 + 100$ and the demand function is given by $p = -20q + 2500$, find the point of **market equilibrium**; that is, find the number of units, q, where supply equals demand.

Solution At market equilibrium, both equations will have the same value. That is,

$$s = p$$

or

$$q^2 + 100 = -20q + 2500$$

or

$$q^2 + 20q - 2400 = 0$$

or

$$(q - 40)(q + 60) = 0$$

or

$$q = 40 \text{ or } q = -60.$$

But $q = -60$ has no meaning in this problem (you cannot buy or sell a negative number of items). Thus the equilibrium point occurs when 40 units are sold. Then $s = p = \$1700$.

Example 11 **Break-even Analysis** The total cost per week of producing Ace electric shavers is $C = 360 + 10q + 0.2q^2$. The limitations of the plant permit only 80 shavers to be produced each week. If the price per unit sold is $50 - 0.2q$, at what level(s) of production will the break-even point occur?

Solution The break-even point occurs when total cost, C, equals total revenue, R. But R = total revenue = (price per unit)·(number of units sold). In this problem $R = (50 - 0.2q)q = 50q - 0.2q^2$. Thus, setting $C = R$, we have

$$360 + 10q + 0.2q^2 = 50q - 0.2q^2$$

or

$$0.4q^2 - 40q + 360 = 0.$$

We can simplify this equation by dividing through by 0.4;

$$q^2 - 100q + 900 = 0$$

or

$$(q - 10)(q - 90) = 0.$$

But $q = 90$ is not meaningful in this problem, because the formula for C is valid only for values of $q \leq 80$. Thus $q = 10$ and $C = R = \$480$.

PROBLEMS A1.2 In Problems 1–36 solve the quadratic equations by one of the three methods presented in this section.

1. $x^2 - 4x + 3 = 0$
2. $x^2 + 4x + 3 = 0$
3. $x^2 + 4x + 5 = 0$
4. $x^2 + 4x + 4 = 0$
5. $x^2 + 2x + 1 = 0$
6. $x^2 + 2x + 3 = 0$
7. $x^2 - 2x + 1 = 0$
8. $x^2 - 2x - 3 = 0$
9. $x^2 + 6x + 5 = 0$
10. $x^2 + 6x + 9 = 0$
11. $x^2 + 6x + 10 = 0$
12. $x^2 - 13x + 42 = 0$
13. $x^2 + 13x + 42 = 0$
14. $x^2 - 5x = 0$
15. $x^2 + 5x = 0$
16. $x^2 - ax = 0$
17. $x^2 + ax = 0$
18. $x^2 - 7x + 3 = 0$
19. $x^2 - 3x + 7 = 0$
20. $3x^2 - 12x + 12 = 0$
21. $-5x^2 + 65x - 210 = 0$
22. $-4x^2 - 36x - 72 = 0$
23. $3x^2 + 4x + 5 = 0$
24. $3x^2 + 4x - 5 = 0$
25. $2x^2 - 8x + 15 = 0$
26. $3x^2 + 5x - 2 = 0$
27. $5x^2 - 3x + 2 = 0$
28. $6x^2 - x - 1 = 0$
29. $4x^2 - 8x - 5 = 0$
30. $8x^2 + 4x + 5 = 0$
31. $x^2 - x + 1 = 0$
32. $x^2 + x + 1 = 0$
33. $x^2 - x - 1 = 0$
34. $x^2 + x - 1 = 0$
35. $\dfrac{x^2}{4} + \dfrac{x}{3} - \dfrac{1}{6} = 0$
*36. $x - \dfrac{3}{x} = 4, x \neq 0$

37. If the demand function for a commodity is given by $p = 216 - 2q$ and the supply function is $s = q^2 + 8q + 16$, find the equilibrium quantity and the equilibrium price.

38. The monthly total cost for a commodity is given by $C = 360 + 40q + 0.1q^2$. If the manufacturer receives \$60 for each item sold, find the break-even point(s).

39. If the supply and demand functions for a commodity are $s = q - 10$ and $p = 1200/q$, what is the equilibrium price and what is the corresponding number of units supplied and demanded?

 40. If $C = 400 + 25q + 0.13q^2$ and the price per unit sold is $45 - 0.17q$, at what level(s) of production will the break-even point occur?

A1.3 Simplifying Rational Expressions

In this book we will, at various times, encounter rational expressions. A **rational expression** is a sum of terms, each taking the form $p(x)/q(x)$, where p and q are polynomials.

Example 1 The following are rational expressions:

(a) $\dfrac{1}{x}$.

(b) $\dfrac{x^2 - 1}{x^3 + x^2 - 2}$.

(c) $x + \dfrac{3}{x^2}$.

(d) $x^3 + \dfrac{5x^5 - 1}{2x + 3} - \dfrac{x}{x + 2}$.

In this section we shall show how rational expressions can be simplified. To do so, we use several rules. These rules derive from the following five rules about real numbers:

Let x, y, z, and w be real numbers with $y \neq 0$ and $w \neq 0$. Then

1. $\dfrac{x}{y} = \dfrac{xw}{yw}$

2. $\dfrac{x}{y} + \dfrac{z}{y} = \dfrac{x + z}{y}$

3. $\dfrac{x}{y} - \dfrac{z}{y} = \dfrac{x - z}{y}$

4. $\dfrac{x}{y} \cdot \dfrac{z}{w} = \dfrac{xz}{yw}$

5. $\dfrac{x/y}{z/w} = \dfrac{x}{y} \cdot \dfrac{w}{z}$

Now, let $a(x)$, $b(x)$, $c(x)$, and $d(x)$ be rational expressions with $b(x) \neq 0$ and $d(x) \neq 0$.

Rule 1. $\dfrac{a(x)}{b(x)} = \dfrac{a(x)d(x)}{b(x)d(x)}$ Multiplying numerator and denominator by a nonzero expression leaves the value of the fraction unchanged.

Rule 2. $\dfrac{a(x)}{b(x)} + \dfrac{c(x)}{b(x)} = \dfrac{a(x) + c(x)}{b(x)}$ Addition property

Rule 3. $\dfrac{a(x)}{b(x)} - \dfrac{c(x)}{b(x)} = \dfrac{a(x) - c(x)}{b(x)}$ Subtraction property

Rule 4. $\dfrac{a(x)}{b(x)} \cdot \dfrac{c(x)}{d(x)} = \dfrac{a(x)c(x)}{b(x)d(x)}$ Multiplication property

Rule 5. $\dfrac{a(x)/b(x)}{c(x)/d(x)} = \dfrac{a(x)}{b(x)} \cdot \dfrac{d(x)}{c(x)}$ Division property

---

The division property can be restated as: *To divide one rational expression by another, invert the divisor and multiply.*

Example 2 Simplify the expression $15x^2/20x$.

Solution

By Rule 1

$$\frac{15x^2}{20x} = \frac{(5x)(3x)}{(5x)(4)} \overset{\downarrow}{=} \frac{3x}{4}.$$

Example 3 Simplify the expression $[(2x+1)/4x^2] - [(3x^3-4)/4x^2]$.

Solution By the subtraction property,

$$\frac{2x+1}{4x^2} - \frac{3x^3-4}{4x^2} = \frac{2x+1-(3x^3-4)}{4x^2} = \frac{2x+1-3x^3+4}{4x^2} = \frac{2x-3x^3+5}{4x^2}.$$

Example 4 Simplify the expression $(x^2-5x+6)/(x^2+6x-16)$.

Solution

By Rule 1

$$\frac{x^2-5x+6}{x^2+6x-16} = \frac{(x-2)(x-3)}{(x-2)(x+8)} \overset{\downarrow}{=} \frac{x-3}{x+8}$$

Example 5 Simplify the expression $(1/x)/(3/x^2)$.

Solution By using first the division property, then the multiplication property, and last Rule 1, we have

$$\frac{1/x}{3/x^2} = \frac{1}{x} \cdot \frac{x^2}{3} = \frac{x^2}{3x} = \frac{x \cdot x}{x \cdot 3} = \frac{x}{3}.$$

Example 6 Write the expression $x + (1/x)$ as a rational expression.

Solution First we need to write each term with the same *common denominator*. Since the only denominator in sight is x, we multiply the first term by x/x (by Rule 1) and then use the addition rule to obtain

$$x + \frac{1}{x} = \frac{x \cdot x}{x} + \frac{1}{x} = \frac{x^2}{x} + \frac{1}{x} = \frac{x^2+1}{x}.$$

In this book, when we ask you to simplify a rational expression that is the sum of two or more terms, we will require that the answer be given as a rational function in *lowest terms*, that is, a rational function $r(x) = p(x)/q(x)$ for which $p(x)$ and $q(x)$ have no common factors.

Example 7 Simplify the expression

$$\frac{y}{y^2 + 1} + \frac{2y^3 - 3}{y + 3}.$$

Solution We seek first to write both terms with the same denominator. The easiest way to find this common denominator is to multiply together the denominators already present. In this example, the common denominator is $(y^2 + 1)(y + 2)$. Then

$$\frac{y}{y^2 + 1} + \frac{2y^3 - 3}{y + 2} = \frac{y(y + 2)}{(y^2 + 1)(y + 2)} + \frac{(2y^3 - 3)(y^2 + 1)}{(y^2 + 1)(y + 2)}$$

$$= \frac{y(y + 2) + (2y^3 - 3)(y^2 + 1)}{(y^2 + 1)(y + 2)}$$

$$= \frac{y^2 + 2y + 2y^5 + 2y^3 - 3y^2 - 3}{(y^2 + 1)(y + 2)}$$

$$= \frac{2y^5 + 2y^3 - 2y^2 + 2y - 3}{(y^2 + 1)(y + 2)}$$

$$= \frac{2y^5 + 2y^3 - 2y^2 + 2y - 3}{y^3 + 2y^2 + y + 2}.$$

↑

This last step is optional.

Since the numerator and denominator have no common factors (you should check that $(y^2 + 1)$ and $(y + 2)$ are not factors of the numerator), we are done.

Example 8 Simplify the expression $(1/x) + (1/x^2) + (1/x^3)$.

Solution If we multiply the denominators to obtain a common divisor, we end up with x^6. Then

$$\frac{1}{x} + \frac{1}{x^2} + \frac{1}{x^3} = \frac{x^5}{x \cdot x^5} + \frac{x^4}{x^2 \cdot x^4} + \frac{x^3}{x^3 \cdot x^3} = \frac{x^5 + x^4 + x^3}{x^6}$$

$$= \frac{x^3(x^2 + x + 1)}{x^3 \cdot x^3} = \frac{x^2 + x + 1}{x^3}.$$

We could have avoided the intermediate step by noting that it is possible, at the beginning, to write everything with the denominator x^3. Then the computation is

$$\frac{1}{x} + \frac{1}{x^2} + \frac{1}{x^3} = \frac{x^2}{x \cdot x^2} + \frac{x}{x^2 \cdot x} + \frac{1}{x^3} = \frac{x^2 + x + 1}{x^3}.$$

PROBLEMS A1.3 In the following problems simplify the given rational expression.

1. $\dfrac{6x}{3}$

2. $\dfrac{5}{10x}$

3. $\dfrac{2x+4}{6x^2+8}$

4. $\dfrac{3z}{6z^3}$

5. $\dfrac{12y^4}{24y^7}$

6. $\dfrac{4xz}{2xz}$

7. $\dfrac{x^2y^2}{x^3y^3}$

8. $\dfrac{4s+3}{12s^2+9s}$

9. $\dfrac{3}{x}\cdot\dfrac{x}{6}$

10. $\dfrac{y}{16}\cdot\dfrac{4}{3y^2}$

11. $\dfrac{25}{z^3}\cdot\dfrac{z^5}{100}$

12. $\dfrac{3}{s^2}\cdot\dfrac{s^3+1}{6}$

13. $\dfrac{x}{x+1}\cdot\dfrac{(x+1)^2}{x^4}$

14. $\dfrac{a^n}{b^n}\cdot\dfrac{b^{n-1}}{a^{n-1}} \quad a,b\neq 0$

15. $\dfrac{x/(x+1)}{x/(x+2)}$

16. $\dfrac{(x+1)/x}{(x+1)^2/(x+2)}$

17. $\dfrac{z/(z+1)}{(z+1)/z}$

18. $\dfrac{(w+2)/(w+3)}{(w+3)/(w+4)}$

19. $\dfrac{(w^2+1)/w}{w+1/w}$

20. $\dfrac{x^2-1}{(x-1)^2}$

21. $\dfrac{x^2+1}{(x+1)^2}$

22. $\dfrac{x^2-3x+2}{x^2-6x+8}$

23. $\dfrac{x^2+4x+3}{x^2-6x-7}$

24. $\dfrac{y^2+3y-18}{y^2-6y+9}$

25. $\dfrac{z^2+2z-8}{z^2-5z-36}$

26. $\dfrac{w^2+4w-21}{w^2+8w+7}$

27. $\dfrac{x^2-\frac{5}{6}x+\frac{1}{6}}{x^2-x-\frac{1}{4}}$

28. $\dfrac{y^2-\frac{1}{3}y-\frac{2}{9}}{y^2+\frac{7}{12}y+\frac{1}{12}}$

29. $\dfrac{z^2-\frac{2}{3}z-\frac{5}{36}}{z^2-2z+\frac{35}{36}}$

30. $\dfrac{x^2-y^2}{(x-y)^2}$

31. $\dfrac{x^3-y^3}{(x-y)^3}$

32. $\dfrac{x^3+y^3}{(x+y)^3}$

33. $\dfrac{(4z-12)/(z^2-4z+3)}{(5z+25)/(z^2+2z-15)}$

34. $\dfrac{(z^2+2z+1)/(z^2+4z+3)}{(z^2-6z-7)/(z^2+9z+8)}$

35. $\dfrac{(x^2+1)/(x^3-2x^2+x-2)}{(x^4+2x^2+1)/(x^3-x^2+x+1)}$

36. $\dfrac{(x^4-y^4)/(x^4-2x^2y^2+y^4)}{(x^3-xy^2-yx^2+y^3)/(x^2-y^2)}$

37. $2 + \dfrac{y}{2}$

38. $6 - \dfrac{5}{z}$

39. $\frac{1}{2} + 2x$

40. $\dfrac{3}{s} + \dfrac{7}{s}$

41. $\dfrac{x - 3}{x^2} + \dfrac{2x - 5}{x^2}$

42. $\dfrac{5}{x^3} - \dfrac{7x^2 + 5}{x^3}$

43. $\dfrac{\frac{1}{2}}{x} - \dfrac{\frac{1}{2}}{x - 1}$

44. $\dfrac{1}{x} + \dfrac{2}{x^2}$

45. $\dfrac{3}{y} - \dfrac{y}{2}$

46. $\dfrac{1}{z^2} - \dfrac{1}{z - 1}$

47. $\dfrac{2}{3s} - \dfrac{4}{5s}$

48. $\dfrac{3}{2x} + \dfrac{x}{5x^2}$

49. $\dfrac{7}{2y^2} + \dfrac{8}{3y^2}$

50. $\dfrac{1}{x - 1} - \dfrac{1}{x - 2}$

51. $\dfrac{3}{x - a} + \dfrac{4}{x - b}$

52. $\dfrac{1}{x - 1} + \dfrac{1}{x - 2} + \dfrac{1}{x - 3}$

53. $\dfrac{3}{x + 5} - \dfrac{6}{x - 7}$

54. $\dfrac{1}{x^2 - 1} + \dfrac{3}{x - 1}$

55. $\dfrac{2x}{x^2 - 4} + \dfrac{5}{x - 2} - \dfrac{3}{x + 2}$

56. $\dfrac{6}{x^2 - 5x + 4} - \dfrac{3}{x - 1}$

57. $\dfrac{x}{x^2 - 1} + \dfrac{3x - 2}{x^2 + 1}$

58. $\dfrac{3}{x^2} - \dfrac{5}{x^4} + \dfrac{2x}{x^6}$

59. $\dfrac{6}{x - 1} - \dfrac{2x}{(x - 1)^2}$

60. $\dfrac{1}{x + 2} + \dfrac{5x + 3}{(x + 2)^2} - \dfrac{6x^2 + 3x - 2}{(x + 2)^3}$

61. $\dfrac{3}{y - 3} - \dfrac{7}{y + 6} + \dfrac{2y - 3}{y^2 + 3y - 18}$

62. $\dfrac{-2}{z + 4} - \dfrac{3}{z - 2} + \dfrac{7z - 5}{z^2 + 2z - 8}$

63. $\dfrac{1 - \dfrac{1}{s}}{1 + \dfrac{1}{s}}$

64. $\dfrac{\dfrac{1}{x} - \dfrac{3}{x - 1}}{\dfrac{2}{x - 1} + \dfrac{4}{x}}$

65. $\dfrac{\dfrac{3}{x} - \dfrac{5}{y}}{\dfrac{6}{y} + \dfrac{2}{x}}$

66. $\dfrac{\dfrac{4}{x - 2} + \dfrac{3}{x + 5}}{(2x - 5)/(x^2 + 3x - 10)}$

67. $\dfrac{\dfrac{1}{x} - \dfrac{3}{x^2} + \dfrac{7x}{x^3}}{\dfrac{-4}{x} + \dfrac{3x - 2}{x^2} + \dfrac{3x^2 - 5x + 2}{x^3}}$

TABLES

TABLE 1 Compound Interest—$(1 + i)^t$

t \ i	½% 0.005	1% 0.010	1½% 0.015	2% 0.020	2½% 0.025	3% 0.030	3½% 0.035	4% 0.040
1	1.005000	1.010000	1.015000	1.020000	1.025000	1.030000	1.035000	1.040000
2	1.010025	1.020100	1.030225	1.040400	1.050625	1.060900	1.071225	1.081600
3	1.015075	1.030301	1.045678	1.061208	1.076891	1.092727	1.108718	1.124864
4	1.020151	1.040604	1.061364	1.082432	1.103813	1.125509	1.147523	1.169859
5	1.025251	1.051010	1.077284	1.104081	1.131408	1.159274	1.187686	1.216653
6	1.030378	1.061520	1.093443	1.126162	1.159693	1.194052	1.229255	1.265319
7	1.035529	1.072135	1.109845	1.148686	1.188686	1.229874	1.272279	1.315932
8	1.040707	1.082857	1.126493	1.171659	1.218403	1.266770	1.316809	1.368569
9	1.045911	1.093685	1.143390	1.195093	1.248863	1.304773	1.362897	1.423312
10	1.051140	1.104622	1.160541	1.218994	1.280085	1.343916	1.410599	1.480244
11	1.056396	1.115668	1.177949	1.243374	1.312087	1.384234	1.459970	1.539454
12	1.061678	1.126825	1.195618	1.268242	1.344889	1.425761	1.511069	1.601032
13	1.066986	1.138093	1.213552	1.293607	1.378511	1.468534	1.563956	1.665074
14	1.072321	1.149474	1.231756	1.319479	1.412974	1.512590	1.618695	1.731676
15	1.077683	1.160969	1.250232	1.345868	1.448298	1.557967	1.675349	1.800944
16	1.083071	1.172579	1.268986	1.372786	1.484506	1.604706	1.733986	1.872981
17	1.088487	1.184304	1.288020	1.400241	1.521618	1.652848	1.794676	1.947900
18	1.093929	1.196147	1.307341	1.428246	1.559659	1.702433	1.857489	2.025817
19	1.099399	1.208109	1.326951	1.456811	1.598650	1.753506	1.922501	2.106849
20	1.104896	1.220190	1.346855	1.485947	1.638616	1.806111	1.989789	2.191123
21	1.110420	1.232392	1.367058	1.515666	1.679582	1.860295	2.059431	2.278768
22	1.115972	1.244716	1.387564	1.545980	1.721571	1.916103	2.131512	2.369919
23	1.121552	1.257163	1.408377	1.576899	1.764611	1.973587	2.206114	2.464716
24	1.127160	1.269735	1.429503	1.608437	1.808726	2.032794	2.283328	2.563304
25	1.132796	1.282432	1.450945	1.640606	1.853944	2.093778	2.363245	2.665836
26	1.138460	1.295256	1.472710	1.673418	1.900293	2.156591	2.445959	2.772470
27	1.144152	1.308209	1.494800	1.706886	1.947800	2.221289	2.531567	2.883369
28	1.149873	1.321291	1.517222	1.741024	1.996495	2.287928	2.620172	2.998703
29	1.155622	1.334504	1.539981	1.775845	2.046407	2.356566	2.711878	3.118651
30	1.161400	1.347849	1.563080	1.811362	2.097568	2.427262	2.806794	3.243398
31	1.167207	1.361327	1.586526	1.847589	2.150007	2.500080	2.905031	3.373133
32	1.173043	1.374941	1.610324	1.884541	2.203757	2.575083	3.006708	3.508059
33	1.178908	1.388690	1.634479	1.922231	2.258851	2.652335	3.111942	3.648381
34	1.184803	1.402577	1.658996	1.960676	2.315322	2.731905	3.220860	3.794316
35	1.190727	1.416603	1.683881	1.999890	2.373205	2.813862	3.333590	3.946089
36	1.196681	1.430769	1.709140	2.039887	2.432535	2.898278	3.450266	4.103933
37	1.202664	1.445076	1.734777	2.080685	2.493349	2.985227	3.571025	4.268090
38	1.208677	1.459527	1.760798	2.122299	2.555682	3.074783	3.696011	4.438813
39	1.214721	1.474123	1.787210	2.164745	2.619574	3.167027	3.825372	4.616366
40	1.220794	1.488864	1.814018	2.208040	2.685064	3.262038	3.959260	4.801021
41	1.226898	1.503752	1.841229	2.252200	2.752190	3.359899	4.097834	4.993061
42	1.233033	1.518790	1.868847	2.297244	2.820995	3.460696	4.241258	5.192784
43	1.239198	1.533978	1.896880	2.343189	2.891520	3.564517	4.389702	5.400495
44	1.245394	1.549318	1.925333	2.390053	2.963808	3.671452	4.543342	5.616515
45	1.251621	1.564811	1.954213	2.437854	3.037903	3.781596	4.702359	5.841176
46	1.257879	1.580459	1.983526	2.486611	3.113851	3.895044	4.866941	6.074823
47	1.264168	1.596263	2.013279	2.536344	3.191697	4.011895	5.037284	6.317816
48	1.270489	1.612226	2.043478	2.587070	3.271490	4.132252	5.213589	6.570528
49	1.276842	1.628348	2.074130	2.638812	3.353277	4.256219	5.396065	6.833349
50	1.283226	1.644632	2.105242	2.691588	3.437109	4.383906	5.584927	7.106683

TABLE 1 Compound Interest (continued)

t \ i	5% 0.050	6% 0.060	7% 0.070	8% 0.080	9% 0.090	10% 0.100	12% 0.120	15% 0.150	18% 0.180
1	1.050000	1.060000	1.070000	1.080000	1.090000	1.100000	1.120000	1.150000	1.180000
2	1.102500	1.123600	1.144900	1.166400	1.188100	1.210000	1.254400	1.322500	1.392400
3	1.157625	1.191016	1.225043	1.259712	1.295029	1.331000	1.404928	1.520875	1.643032
4	1.215506	1.262477	1.310796	1.360489	1.411582	1.464100	1.573519	1.749006	1.938778
5	1.276282	1.338226	1.402552	1.469328	1.538624	1.610510	1.762342	2.011357	2.287758
6	1.340096	1.418519	1.500730	1.586874	1.677100	1.771561	1.973823	2.313061	2.699554
7	1.407100	1.503630	1.605781	1.713824	1.828039	1.948717	2.210681	2.660020	3.185474
8	1.477455	1.593848	1.718186	1.850930	1.992563	2.143589	2.475963	3.059023	3.758859
9	1.551328	1.689479	1.838459	1.999005	2.171893	2.357948	2.773079	3.517876	4.435454
10	1.628895	1.790848	1.967151	2.158925	2.367364	2.593742	3.105848	4.045558	5.233836
11	1.710339	1.898299	2.104852	2.331639	2.580426	2.853117	3.478550	4.652391	6.175926
12	1.795856	2.012196	2.252192	2.518170	2.812665	3.138428	3.895976	5.350250	7.287593
13	1.885649	2.132928	2.409845	2.719624	3.065805	3.452271	4.363493	6.152788	8.599359
14	1.979932	2.260904	2.578534	2.937194	3.341727	3.797498	4.887112	7.075706	10.147244
15	2.078928	2.396558	2.759032	3.172169	3.642482	4.177248	5.473566	8.137062	11.973748
16	2.182875	2.540352	2.952164	3.425943	3.970306	4.594973	6.130394	9.357621	14.129023
17	2.292018	2.692773	3.158815	3.700018	4.327633	5.054470	6.866041	10.761264	16.672247
18	2.406619	2.854339	3.379932	3.996019	4.717120	5.559917	7.689966	12.375454	19.673251
19	2.526950	3.025600	3.616528	4.315701	5.141661	6.115909	8.612762	14.231772	23.214436
20	2.653298	3.207135	3.869684	4.660957	5.604411	6.727500	9.646293	16.366537	27.393035
21	2.785963	3.399564	4.140562	5.033834	6.108808	7.400250	10.803848	18.821518	32.323781
22	2.925261	3.603537	4.430402	5.436540	6.658600	8.140275	12.100310	21.644746	38.142061
23	3.071524	3.819750	4.740530	5.871464	7.257874	8.954302	13.552347	24.891458	45.007632
24	3.225100	4.048935	5.072367	6.341181	7.911083	9.849733	15.178629	28.625176	53.109006
25	3.386355	4.291871	5.427433	6.848475	8.623081	10.834706	17.000064	32.918953	62.668627
26	3.555673	4.549383	5.807353	7.396353	9.399158	11.918177	19.040072	37.856796	73.948980
27	3.733456	4.822346	6.213868	7.988061	10.245082	13.109994	21.324881	43.535315	87.259797
28	3.920129	5.111687	6.648838	8.627106	11.167140	14.420994	23.883866	50.065612	102.966560
29	4.116136	5.418388	7.114257	9.317275	12.172182	15.863093	26.749930	57.575454	121.500541
30	4.321942	5.743491	7.612255	10.062657	13.267678	17.449402	29.959922	66.211772	143.370638
31	4.538039	6.088101	8.145113	10.867669	14.461770	19.194342	33.555113	76.143538	169.177355
32	4.764941	6.453387	8.715271	11.737083	15.763329	21.113777	37.581726	87.565068	199.629278
33	5.003189	6.840590	9.325340	12.676050	17.182028	23.225154	42.091533	100.699829	235.562548
34	5.253348	7.251025	9.978114	13.690134	18.728411	25.547670	47.142517	115.804803	277.963802
35	5.516015	7.686087	10.676581	14.785344	20.413968	28.102437	52.799620	133.175524	327.997292
36	5.791816	8.147252	11.423942	15.968172	22.251225	30.912681	59.135574	153.151852	387.036804
37	6.081407	8.636087	12.223618	17.245626	24.253835	34.003949	66.231843	176.124630	456.703426
38	6.385477	9.154252	13.079271	18.625276	26.436680	37.404343	74.179664	202.543324	538.910049
39	6.704751	9.703507	13.994820	20.115298	28.815982	41.144778	83.081224	232.924824	635.913857
40	7.039989	10.285718	14.974458	21.724521	31.409420	45.259256	93.050970	267.863544	750.378342
41	7.391988	10.902861	16.022670	23.462483	34.236268	49.785181	104.217087	308.043079	885.446449
42	7.761588	11.557033	17.144257	25.339482	37.317532	54.763699	116.723137	354.249538	1044.826813
43	8.149667	12.250455	18.344355	27.366640	40.676110	60.240069	130.729914	407.386971	1232.895630
44	8.557150	12.985482	19.628460	29.555972	44.336960	66.264076	146.417503	468.495014	1454.816849
45	8.985008	13.764611	21.002452	31.920449	48.327286	72.890484	163.987604	538.769272	1716.683868
46	9.434258	14.590487	22.472623	34.474085	52.676742	80.179532	183.666117	619.584656	2025.686981
47	9.905971	15.465917	24.045707	37.232012	57.417649	88.197485	205.706051	712.522362	2390.310638
48	10.401270	16.393872	25.728907	40.210573	62.585237	97.017234	230.390776	819.400711	2820.566559
49	10.921333	17.377504	27.529930	43.427419	68.217908	106.718957	258.037670	942.310814	3328.268524
50	11.467400	18.420154	29.457025	46.901613	74.357520	117.390853	289.002193	1083.657440	3927.356873

TABLE 2 Present Value—$(1 + i)^{-t}$

t \ i	$\frac{1}{2}$% 0.005	1% 0.010	$1\frac{1}{2}$% 0.015	2% 0.020	$2\frac{1}{2}$% 0.025	3% 0.030	$3\frac{1}{2}$% 0.035	4% 0.040
1	0.995025	0.990099	0.985222	0.980392	0.975610	0.970874	0.966184	0.961538
2	0.990075	0.980296	0.970662	0.961169	0.951814	0.942596	0.933511	0.924556
3	0.985149	0.970590	0.956317	0.942322	0.928599	0.915142	0.901943	0.888996
4	0.980248	0.960980	0.942184	0.923845	0.905951	0.888487	0.871442	0.854804
5	0.975371	0.951466	0.928260	0.905731	0.883854	0.862609	0.841973	0.821927
6	0.970518	0.942045	0.914542	0.887971	0.862297	0.837484	0.813501	0.790315
7	0.965690	0.932718	0.901027	0.870560	0.841265	0.813092	0.785991	0.759918
8	0.960885	0.923483	0.887711	0.853490	0.820747	0.789409	0.759412	0.730690
9	0.956105	0.914340	0.874592	0.836755	0.800728	0.766417	0.733731	0.702587
10	0.951348	0.905287	0.861667	0.820348	0.781198	0.744094	0.708919	0.675564
11	0.946615	0.896324	0.848933	0.804263	0.762145	0.722421	0.684946	0.649581
12	0.941905	0.887449	0.836387	0.788493	0.743556	0.701380	0.661783	0.624597
13	0.937219	0.878663	0.824027	0.773033	0.725420	0.680951	0.639440	0.600574
14	0.932556	0.869963	0.811849	0.757875	0.707727	0.661118	0.617782	0.577475
15	0.927917	0.861349	0.799852	0.743015	0.690466	0.641862	0.596891	0.555265
16	0.923300	0.852821	0.788031	0.728446	0.673625	0.623167	0.576706	0.533908
17	0.918707	0.844377	0.776385	0.714163	0.657195	0.605016	0.557204	0.513373
18	0.914136	0.836017	0.764912	0.700159	0.641166	0.587395	0.538361	0.493628
19	0.909588	0.827740	0.753607	0.686431	0.625528	0.570286	0.520156	0.474642
20	0.905063	0.819544	0.742470	0.672971	0.610271	0.553676	0.502566	0.456387
21	0.900560	0.811430	0.731498	0.659776	0.595386	0.537549	0.485571	0.438834
22	0.896080	0.803396	0.720688	0.646839	0.580865	0.521893	0.469151	0.421955
23	0.891622	0.795442	0.710037	0.634156	0.566697	0.506692	0.453286	0.405726
24	0.887186	0.787566	0.699544	0.621721	0.552875	0.491934	0.437957	0.390121
25	0.882772	0.779768	0.689206	0.609531	0.539391	0.477606	0.423147	0.375117
26	0.878380	0.772048	0.679021	0.597579	0.526235	0.463695	0.408838	0.360689
27	0.874010	0.764404	0.668986	0.585862	0.513400	0.450189	0.395012	0.346817
28	0.869662	0.756836	0.659099	0.574375	0.500878	0.437077	0.381654	0.333477
29	0.865335	0.749342	0.649359	0.563112	0.488661	0.424346	0.368748	0.320651
30	0.861030	0.741923	0.639762	0.552071	0.476743	0.411987	0.356278	0.308319
31	0.856746	0.734577	0.630308	0.541246	0.465115	0.399987	0.344230	0.296460
32	0.852484	0.727304	0.620993	0.530633	0.453771	0.388337	0.332590	0.285058
33	0.848242	0.720103	0.611816	0.520229	0.442703	0.377026	0.321343	0.274094
34	0.844022	0.712973	0.602774	0.510028	0.431905	0.366045	0.310476	0.263552
35	0.839823	0.705914	0.593866	0.500028	0.421371	0.355383	0.299977	0.253415
36	0.835645	0.698925	0.585090	0.490223	0.411094	0.345032	0.289833	0.243669
37	0.831487	0.692005	0.576443	0.480611	0.401067	0.334983	0.280032	0.234297
38	0.827351	0.685153	0.567924	0.471187	0.391285	0.325226	0.270562	0.225285
39	0.823235	0.678370	0.559531	0.461948	0.381741	0.315754	0.261413	0.216621
40	0.819139	0.671653	0.551262	0.452890	0.372431	0.306557	0.252572	0.208289
41	0.815064	0.665003	0.543116	0.444010	0.363347	0.297628	0.244031	0.200278
42	0.811009	0.658419	0.535089	0.435304	0.354485	0.288959	0.235779	0.192575
43	0.806974	0.651900	0.527182	0.426769	0.345839	0.280543	0.227806	0.185168
44	0.802959	0.645445	0.519391	0.418401	0.337404	0.272372	0.220102	0.178046
45	0.798964	0.639055	0.511715	0.410197	0.329174	0.264439	0.212659	0.171198
46	0.794989	0.632728	0.504153	0.402154	0.321146	0.256737	0.205468	0.164614
47	0.791034	0.626463	0.496702	0.394268	0.313313	0.249259	0.198520	0.158283
48	0.787098	0.620260	0.489362	0.386538	0.305671	0.241999	0.191806	0.152195
49	0.783182	0.614119	0.482130	0.378958	0.298216	0.234950	0.185320	0.146341
50	0.779286	0.608039	0.475005	0.371528	0.290942	0.228107	0.179053	0.140713

TABLE 2 Present Value (continued)

t \ i	5% 0.050	6% 0.060	7% 0.070	8% 0.080	9% 0.090	10% 0.100	12% 0.120	15% 0.150	18% 0.180
1	0.952381	0.943396	0.934579	0.925926	0.917431	0.909091	0.892857	0.869565	0.847458
2	0.907029	0.889996	0.873439	0.857339	0.841680	0.826446	0.797194	0.756144	0.718184
3	0.863838	0.839619	0.816298	0.793832	0.772183	0.751315	0.711780	0.657516	0.608631
4	0.822702	0.792094	0.762895	0.735030	0.708425	0.683013	0.635518	0.571753	0.515789
5	0.783526	0.747258	0.712986	0.680583	0.649931	0.620921	0.567427	0.497177	0.437109
6	0.746215	0.704961	0.666342	0.630170	0.596267	0.564474	0.506631	0.432328	0.370432
7	0.710681	0.665057	0.622750	0.583490	0.547034	0.513158	0.452349	0.375937	0.313925
8	0.676839	0.627412	0.582009	0.540269	0.501866	0.466507	0.403883	0.326902	0.266038
9	0.644609	0.591898	0.543934	0.500249	0.460428	0.424098	0.360610	0.284262	0.225456
10	0.613913	0.558395	0.508349	0.463193	0.422411	0.385543	0.321973	0.247185	0.191064
11	0.584679	0.526788	0.475093	0.428883	0.387533	0.350494	0.287476	0.214943	0.161919
12	0.556837	0.496969	0.444012	0.397114	0.355535	0.318631	0.256675	0.186907	0.137220
13	0.530321	0.468839	0.414964	0.367698	0.326179	0.289664	0.229174	0.162528	0.116288
14	0.505068	0.442301	0.387817	0.340461	0.299246	0.263331	0.204620	0.141329	0.098549
15	0.481017	0.417265	0.362446	0.315242	0.274538	0.239392	0.182696	0.122894	0.083516
16	0.458112	0.393646	0.338735	0.291890	0.251870	0.217629	0.163122	0.106865	0.070776
17	0.436297	0.371364	0.316574	0.270269	0.231073	0.197845	0.145644	0.092926	0.059980
18	0.415521	0.350344	0.295864	0.250249	0.211994	0.179859	0.130040	0.080805	0.050830
19	0.395734	0.330513	0.276508	0.231712	0.194490	0.163508	0.116107	0.070265	0.043077
20	0.376889	0.311805	0.258419	0.214548	0.178431	0.148644	0.103667	0.061100	0.036506
21	0.358942	0.294155	0.241513	0.198656	0.163698	0.135131	0.092560	0.053131	0.030937
22	0.341850	0.277505	0.225713	0.183941	0.150182	0.122846	0.082643	0.046201	0.026218
23	0.325571	0.261797	0.210947	0.170315	0.137781	0.111678	0.073788	0.040174	0.022218
24	0.310068	0.246979	0.197147	0.157699	0.126405	0.101526	0.065882	0.034934	0.018829
25	0.295303	0.232999	0.184249	0.146018	0.115968	0.092296	0.058823	0.030378	0.015957
26	0.281241	0.219810	0.172195	0.135202	0.106393	0.083905	0.052521	0.026415	0.013523
27	0.267848	0.207368	0.160930	0.125187	0.097608	0.076278	0.046894	0.022970	0.011460
28	0.255094	0.195630	0.150402	0.115914	0.089548	0.069343	0.041869	0.019974	0.009712
29	0.242946	0.184557	0.140563	0.107328	0.082155	0.063039	0.037383	0.017369	0.008230
30	0.231377	0.174110	0.131367	0.099377	0.075371	0.057309	0.033378	0.015103	0.006975
31	0.220359	0.164255	0.122773	0.092016	0.069148	0.052099	0.029802	0.013133	0.005911
32	0.209866	0.154957	0.114741	0.085200	0.063438	0.047362	0.026609	0.011420	0.005009
33	0.199873	0.146186	0.107235	0.078889	0.058200	0.043057	0.023758	0.009931	0.004245
34	0.190355	0.137912	0.100219	0.073045	0.053395	0.039143	0.021212	0.008635	0.003598
35	0.181290	0.130105	0.093663	0.067635	0.048986	0.035584	0.018940	0.007509	0.003049
36	0.172657	0.122741	0.087535	0.062625	0.044941	0.032349	0.016910	0.006529	0.002584
37	0.164436	0.115793	0.081809	0.057986	0.041231	0.029408	0.015098	0.005678	0.002190
38	0.156605	0.109239	0.076457	0.053690	0.037826	0.026735	0.013481	0.004937	0.001856
39	0.149148	0.103056	0.071455	0.049713	0.034703	0.024304	0.012036	0.004293	0.001573
40	0.142046	0.097222	0.066780	0.046031	0.031838	0.022095	0.010747	0.003733	0.001333
41	0.135282	0.091719	0.062412	0.042621	0.029209	0.020086	0.009595	0.003246	0.001129
42	0.128840	0.086527	0.058329	0.039464	0.026797	0.018260	0.008567	0.002823	0.000957
43	0.122704	0.081630	0.054513	0.036541	0.024584	0.016600	0.007649	0.002455	0.000811
44	0.116861	0.077009	0.050946	0.033834	0.022555	0.015091	0.006830	0.002134	0.000687
45	0.111297	0.072650	0.047613	0.031328	0.020692	0.013719	0.006098	0.001856	0.000583
46	0.105997	0.068538	0.044499	0.029007	0.018984	0.012472	0.005445	0.001614	0.000494
47	0.100949	0.064658	0.041587	0.026859	0.017416	0.011338	0.004861	0.001403	0.000418
48	0.096142	0.060998	0.038867	0.024869	0.015978	0.010307	0.004340	0.001220	0.000355
49	0.091564	0.057546	0.036324	0.023027	0.014659	0.009370	0.003875	0.001061	0.000300
50	0.087204	0.054288	0.033948	0.021321	0.013449	0.008519	0.003460	0.000923	0.000255

TABLE 3 Future Value of an Annuity: $s_{\overline{n}|i} = \dfrac{(1+i)^n - 1}{i}$

$\dfrac{i}{n}$	$\frac{1}{2}\%$ 0.005	1% 0.010	$1\frac{1}{2}\%$ 0.015	2% 0.020	$2\frac{1}{2}\%$ 0.025	3% 0.030	$3\frac{1}{2}\%$ 0.035	4% 0.040
1	1.000000	1.000000	1.000000	1.000000	1.000000	1.000000	1.000000	1.000000
2	2.005000	2.010000	2.015000	2.020000	2.025000	2.030000	2.035000	2.040000
3	3.015025	3.030100	3.045225	3.060400	3.075625	3.090900	3.106225	3.121600
4	4.030100	4.060401	4.090903	4.121608	4.152516	4.183627	4.214943	4.246464
5	5.050251	5.101005	5.152267	5.204040	5.256329	5.309136	5.362466	5.416323
6	6.075502	6.152015	6.229551	6.308121	6.387737	6.468410	6.550152	6.632975
7	7.105879	7.213535	7.322994	7.434283	7.547430	7.662462	7.779408	7.898294
8	8.141409	8.285671	8.432839	8.582969	8.736116	8.892336	9.051687	9.214226
9	9.182116	9.368527	9.559332	9.754628	9.954519	10.159106	10.368496	10.582795
10	10.228026	10.462213	10.702722	10.949721	11.203382	11.463879	11.731393	12.006107
11	11.279167	11.566835	11.863262	12.168715	12.483466	12.807796	13.141992	13.486351
12	12.335562	12.682503	13.041211	13.412090	13.795553	14.192030	14.601962	15.025805
13	13.397240	13.809328	14.236830	14.680332	15.140442	15.617790	16.113030	16.626838
14	14.464226	14.947421	15.450382	15.973938	16.518953	17.086324	17.676986	18.291911
15	15.536548	16.096896	16.682138	17.293417	17.931927	18.598914	19.295681	20.023588
16	16.614230	17.257864	17.932370	18.639285	19.380225	20.156881	20.971030	21.824531
17	17.697301	18.430443	19.201355	20.012071	20.864730	21.761588	22.705016	23.697512
18	18.785788	19.614748	20.489376	21.412312	22.386349	23.414435	24.499691	25.645413
19	19.879717	20.810895	21.796716	22.840559	23.946007	25.116868	26.357180	27.671229
20	20.979115	22.019004	23.123667	24.297370	25.544658	26.870374	28.279682	29.778079
21	22.084011	23.239194	24.470522	25.783317	27.183274	28.676486	30.269471	31.969202
22	23.194431	24.471586	25.837580	27.298984	28.862856	30.536780	32.328902	34.247970
23	24.310403	25.716302	27.225144	28.844963	30.584427	32.452884	34.460414	36.617889
24	25.431955	26.973465	28.633521	30.421862	32.349038	34.426470	36.666528	39.082604
25	26.559115	28.243200	30.063024	32.030300	34.157764	36.459264	38.949857	41.645908
26	27.691911	29.525631	31.513969	33.670906	36.011708	38.553042	41.313102	44.311745
27	28.830370	30.820888	32.986678	35.344324	37.912001	40.709634	43.759060	47.084214
28	29.974522	32.129097	34.481479	37.051210	39.859801	42.930923	46.290627	49.967583
29	31.124395	33.450388	35.998701	38.792235	41.856296	45.218850	48.910799	52.966286
30	32.280017	34.784892	37.538681	40.568079	43.902703	47.575416	51.622677	56.084938
31	33.441417	36.132740	39.101762	42.379441	46.000271	50.002678	54.429471	59.328335
32	34.608624	37.494068	40.688288	44.227030	48.150278	52.502759	57.334502	62.701469
33	35.781667	38.869009	42.298612	46.111570	50.354034	55.077841	60.341210	66.209527
34	36.960575	40.257699	43.933092	48.033802	52.612885	57.730177	63.453152	69.857909
35	38.145378	41.660276	45.592088	49.994478	54.928207	60.462082	66.674013	73.652225
36	39.336105	43.076878	47.275969	51.994367	57.301413	62.275944	70.007603	77.598314
37	40.532785	44.507647	48.985109	54.034255	59.733948	66.174223	73.457869	81.702246
38	41.735449	45.952724	50.719885	56.114940	62.227297	69.159449	77.028895	85.970336
39	42.944127	47.412251	52.480684	58.237238	64.782979	72.234233	80.724906	90.409150
40	44.158847	48.886373	54.267894	60.401983	67.402554	75.401260	84.550278	95.025516
41	45.379642	50.375237	56.081912	62.610023	70.087617	78.663298	88.509537	99.826536
42	46.606540	51.878989	57.923141	64.862223	72.839808	82.023196	92.607371	104.819598
43	47.839572	53.397779	59.791988	67.159468	75.660803	85.483892	96.848629	110.012382
44	49.078770	54.931757	61.688868	69.502657	78.552323	89.048409	101.238331	115.412877
45	50.324164	56.481075	63.614201	71.892710	81.516131	92.719861	105.781673	121.029392
46	51.575785	58.045885	65.568414	74.330564	84.554034	96.501457	110.484031	126.870568
47	52.833664	59.626344	67.551940	76.817176	87.667885	100.396501	115.350973	132.945391
48	54.097832	61.222608	69.565219	79.353519	90.859582	104.408396	120.388257	139.263206
49	55.368321	62.834834	71.608698	81.940590	94.131072	108.540648	125.601846	145.833735
50	56.645163	64.463182	73.682828	84.579401	97.484349	112.796867	130.997910	152.667084

TABLE 3 Future Value of an Annuity (continued)

i n	5% 0.050	6% 0.060	7% 0.070	8% 0.080	9% 0.090	10% 0.100	12% 0.120	15% 0.150	18% 0.180
1	1.000000	1.000000	1.000000	1.000000	1.000000	1.000000	1.000000	1.000000	1.000000
2	2.050000	2.060000	2.070000	2.080000	2.090000	2.100000	2.120000	2.150000	2.180000
3	3.152500	3.183600	3.214900	3.246400	3.278100	3.310000	3.374400	3.472500	3.572400
4	4.310125	4.374616	4.439943	4.506112	4.573129	4.641000	4.779328	4.993375	5.215432
5	5.525631	5.637093	5.750739	5.866601	5.984711	6.105100	6.352847	6.742381	7.154210
6	6.801913	6.975319	7.153291	7.335929	7.523335	7.715610	8.115189	8.753738	9.441968
7	8.142008	8.393838	8.654021	8.922803	9.200435	9.487171	10.089012	11.066799	12.141522
8	9.549109	9.897468	10.259803	10.636628	11.028474	11.435888	12.299693	13.726819	15.326996
9	11.026564	11.491316	11.977989	12.487558	13.021036	13.579477	14.775656	16.785842	19.085855
10	12.577893	13.180795	13.816448	14.486562	15.192930	15.937425	17.548735	20.303718	23.521309
11	14.206787	14.971643	15.783599	16.645487	17.560293	18.531167	20.654583	24.349276	28.755144
12	15.917127	16.869941	17.888451	18.977126	20.140720	21.384284	24.133133	29.001667	34.931070
13	17.712983	18.882138	20.140643	21.495297	22.953385	24.522712	28.029109	34.351917	42.218663
14	19.598632	21.015066	22.550488	24.214920	26.019189	27.974983	32.392602	40.504705	50.818022
15	21.578564	23.275970	25.129022	27.152114	29.360916	31.772482	37.279715	47.580411	60.965266
16	23.657492	25.672528	27.888054	30.324283	33.003399	35.949730	42.753280	55.717472	72.939014
17	25.840366	28.212880	30.840217	33.750226	36.973705	40.544703	48.883674	65.075093	87.068036
18	28.132385	30.905653	33.999033	37.450244	41.301338	45.599173	55.749715	75.836357	103.740283
19	30.539004	33.759992	37.378965	41.446263	46.018458	51.159090	63.439681	88.211811	123.413534
20	33.065954	36.785591	40.995492	45.761964	51.160120	57.274999	72.052442	102.443583	146.627970
21	35.719252	39.992727	44.865177	50.422921	56.764530	64.002499	81.698736	118.810120	174.021006
22	38.505214	43.392290	49.005739	55.456755	62.873338	71.402749	92.502584	137.631638	206.344786
23	41.430475	46.995828	53.436141	60.893296	69.531939	79.543024	104.602894	159.276384	244.486849
24	44.501999	50.815577	58.176671	66.764759	76.789813	88.497327	118.155241	184.167843	289.494480
25	47.727099	54.864512	63.249038	73.105940	84.700896	98.347059	133.333870	212.793018	342.603489
26	51.113454	59.156383	68.676470	79.954415	93.323977	109.181765	150.333935	245.711969	405.272114
27	54.669126	63.705766	74.483823	87.350768	102.723135	121.099942	169.374008	283.568768	479.221092
28	58.402583	68.528112	80.697691	95.338830	112.968217	134.209936	190.698889	327.104080	566.480888
29	62.322712	73.639798	87.346529	103.965936	124.135356	148.630930	214.582754	377.169693	669.447449
30	66.438848	79.058186	94.460786	113.283211	136.307541	164.494024	241.332684	434.745148	790.947990
31	70.760790	84.801677	102.073041	123.345868	149.575218	181.943426	271.292606	500.956921	934.318634
32	75.298829	90.889778	110.218154	134.213537	164.036987	201.137768	304.847721	577.100456	1103.495987
33	80.063771	97.343165	118.933425	145.950621	179.800316	222.251545	342.429447	664.665527	1303.125259
34	85.066959	104.183755	128.258764	158.626671	196.982344	245.476700	384.520981	765.365349	1538.687805
35	90.320307	111.434780	138.236877	172.316805	215.710756	271.024368	431.663494	881.170158	1816.651611
36	95.836323	119.120867	148.913460	187.102148	236.124723	299.126804	484.463116	1014.345680	2144.648895
37	101.628139	127.268119	160.337402	203.070320	258.375950	330.039490	543.598686	1167.497543	2531.685699
38	107.709546	135.904520	172.561020	220.315947	282.629784	364.043434	609.830536	1343.622162	2988.389130
39	114.095023	145.058458	185.640291	238.941221	309.066463	401.447781	684.010201	1546.165482	3527.299164
40	120.799774	154.761967	199.635113	259.056519	337.882442	442.592556	767.091423	1779.090302	4163.213013
41	127.839763	165.047684	214.609570	280.781040	369.291862	487.851810	860.142395	2046.953857	4913.591370
42	135.231752	175.950546	230.632240	304.243523	403.528133	537.636993	964.359482	2354.996918	5799.037842
43	142.993340	187.507578	247.776497	329.583004	440.845665	592.400696	1081.082611	2709.246460	6843.864624
44	151.143005	199.758032	266.120853	356.949650	481.521774	652.640762	1211.812531	3116.633453	8076.760254
45	159.700155	212.743513	285.749313	386.505615	525.858734	718.904839	1358.230026	3585.128448	9531.577148
46	168.685163	226.508125	306.751762	418.426067	574.186028	791.795319	1522.217636	4123.897705	11248.260986
47	178.119423	241.098612	329.224388	452.900150	626.862762	871.974854	1705.883759	4743.482361	13273.947876
48	188.025394	256.564529	353.270092	490.132164	684.280411	960.172333	1911.589813	5456.004761	15664.258545
49	198.426664	272.958401	378.999001	530.342735	746.865646	1057.189575	2141.980591	6275.405457	18484.825195
50	209.347996	290.335903	406.528927	573.770157	815.083557	1163.908524	2400.018250	7217.716309	21813.093750

TABLE 4 Present Value of an Ordinary Annuity: $a_{\overline{n}|i} = \dfrac{1 - (1 + i)^{-n}}{i}$

n \ i	$\frac{1}{2}\%$ 0.005	1% 0.010	$1\frac{1}{2}\%$ 0.015	2% 0.020	$2\frac{1}{2}\%$ 0.025	3% 0.030	$3\frac{1}{2}\%$ 0.035	4% 0.040
1	0.995025	0.990099	0.985222	0.980392	0.975610	0.970874	0.966184	0.961538
2	1.985099	1.970395	1.955883	1.941561	1.927424	1.913470	1.899694	1.886095
3	2.970248	2.940985	2.912200	2.883883	2.856024	2.828611	2.801637	2.775091
4	3.950496	3.901966	3.854385	3.807729	3.761974	3.717098	3.673079	3.629895
5	4.925866	4.853431	4.782645	4.713460	4.645828	4.579707	4.515052	4.451822
6	5.896384	5.795476	5.697187	5.601431	5.508125	5.417191	5.328553	5.242137
7	6.862074	6.728195	6.598214	6.471991	6.349391	6.230283	6.114544	6.002055
8	7.822959	7.651678	7.485925	7.325481	7.170137	7.019692	6.873956	6.732745
9	8.779064	8.566018	8.360517	8.162237	7.970866	7.786109	7.607687	7.435332
10	9.730412	9.471305	9.222185	8.982585	8.752064	8.530203	8.316605	8.110896
11	10.677027	10.367628	10.071118	9.786848	9.514209	9.252624	9.001551	8.760477
12	11.618932	11.255077	10.907505	10.575341	10.257765	9.954004	9.663334	9.385074
13	12.556151	12.133740	11.731532	11.348374	10.983185	10.634955	10.302738	9.985648
14	13.488708	13.003703	12.543382	12.106249	11.690912	11.296073	10.920520	10.563123
15	14.416625	13.865053	13.343233	12.849264	12.381378	11.937935	11.517411	11.118387
16	15.339925	14.717874	14.131264	13.577709	13.055003	12.561102	12.094117	11.652296
17	16.258632	15.562251	14.907649	14.291872	13.712198	13.166118	12.651321	12.165669
18	17.172768	16.398269	15.672561	14.992031	14.353364	13.753513	13.189682	12.659297
19	18.082356	17.226008	16.426168	15.678462	14.978891	14.323799	13.709837	13.133939
20	18.987419	18.045553	17.168639	16.351433	15.589162	14.877475	14.212403	13.590326
21	19.887979	18.856983	17.900137	17.011209	16.184549	15.415024	14.697974	14.029160
22	20.784059	19.660379	18.620824	17.658048	16.765413	15.936917	15.167125	14.451115
23	21.675681	20.455821	19.330861	18.292204	17.332110	16.443608	15.620410	14.856842
24	22.562866	21.243387	20.030405	18.913926	17.884986	16.935542	16.058368	15.246963
25	23.445638	22.023156	20.719611	19.523456	18.424376	17.413148	16.481515	15.622080
26	24.324018	22.795204	21.398632	20.121036	18.950611	17.876842	16.890352	15.982769
27	25.198028	23.559608	22.067617	20.706898	19.464011	18.327031	17.285365	16.329586
28	26.067689	24.316443	22.726717	21.281272	19.964889	18.764108	17.667019	16.663063
29	26.933024	25.065785	23.376076	21.844385	20.453550	19.188455	18.035767	16.983715
30	27.794054	25.807708	24.015838	22.396456	20.930293	19.600441	18.392045	17.292033
31	28.650800	26.542285	24.646146	22.937702	21.395407	20.000428	18.736276	17.588494
32	29.503284	27.269589	25.267139	23.468335	21.849178	20.388766	19.068865	17.873551
33	30.351526	27.989693	25.878954	23.988564	22.291881	20.765792	19.390208	18.147646
34	31.195548	28.702666	26.481728	24.498592	22.723786	21.131837	19.700684	18.411198
35	32.035371	29.408580	27.075595	24.998619	23.145157	21.487220	20.000661	18.664613
36	32.871016	30.107505	27.660684	25.488842	23.556251	21.832252	20.290494	18.908282
37	33.702504	30.799510	28.237127	25.969453	23.957318	22.167235	20.570525	19.142579
38	34.529854	31.484663	28.805052	26.440641	24.348603	22.492462	20.841087	19.367864
39	35.353089	32.163033	29.364583	26.902589	24.730344	22.808215	21.102500	19.584485
40	36.172228	32.834686	29.915845	27.355479	25.102775	23.114772	21.355072	19.792774
41	36.987291	33.499689	30.458961	27.799489	25.466122	23.412400	21.599104	19.993052
42	37.798300	34.158108	30.994050	28.234794	25.820607	23.701359	21.834883	20.185627
43	38.605274	34.810008	31.521232	28.661562	26.166446	23.981902	22.062689	20.370795
44	39.408232	35.455454	32.040622	29.079963	26.503849	24.254274	22.282791	20.548841
45	40.207196	36.094508	32.552337	29.490160	26.833024	24.518713	22.495450	20.720040
46	41.002185	36.727236	33.056490	29.892314	27.154170	24.775449	22.700918	20.884654
47	41.793219	37.353699	33.553192	30.286582	27.467483	25.024708	22.899438	21.042936
48	42.580318	37.973959	34.042554	30.673120	27.773154	25.266707	23.091244	21.195131
49	43.363500	38.588079	34.524683	31.052078	28.071369	25.501657	23.276564	21.341472
50	44.142786	39.196118	34.999688	31.423606	28.362312	25.729764	23.455618	21.482185

TABLE 4   Present Value of an Ordinary Annuity (continued)

n \ i	5% 0.050	6% 0.060	7% 0.070	8% 0.080	9% 0.090	10% 0.100	12% 0.120	15% 0.150	18% 0.180
1	0.952381	0.943396	0.934579	0.925926	0.917431	0.909091	0.892857	0.869565	0.847458
2	1.859410	1.833393	1.808018	1.783265	1.759111	1.735537	1.690051	1.625709	1.565642
3	2.723248	2.673012	2.624316	2.577097	2.531295	2.486852	2.401831	2.283225	2.174273
4	3.545951	3.465106	3.387211	3.312127	3.239720	3.169865	3.037349	2.854978	2.690062
5	4.329477	4.212364	4.100197	3.992710	3.889651	3.790787	3.604776	3.352155	3.127171
6	5.075692	4.917324	4.766540	4.622880	4.485919	4.355261	4.111407	3.784483	3.497603
7	5.786373	5.582381	5.389289	5.206370	5.032953	4.868419	4.563757	4.160420	3.811528
8	6.463213	6.209794	5.971299	5.746639	5.534819	5.334926	4.967640	4.487322	4.077566
9	7.107822	6.801692	6.515232	6.246888	5.995247	5.759024	5.328250	4.771584	4.303022
10	7.721735	7.360087	7.023582	6.710081	6.417658	6.144567	5.650223	5.018769	4.494086
11	8.306414	7.886875	7.498674	7.138964	6.805191	6.495061	5.937699	5.233712	4.656005
12	8.863252	8.383844	7.942686	7.536078	7.160725	6.813692	6.194374	5.420619	4.793225
13	9.393573	8.852683	8.357651	7.903776	7.486904	7.103356	6.423548	5.583147	4.909513
14	9.898641	9.294984	8.745468	8.244237	7.786150	7.366687	6.628168	5.724476	5.008062
15	10.379658	9.712249	9.107914	8.559479	8.060688	7.606080	6.810864	5.847370	5.091578
16	10.837770	10.105895	9.446649	8.851369	8.312558	7.823709	6.973986	5.954235	5.162354
17	11.274066	10.477260	9.763223	9.121638	8.543631	8.021553	7.119630	6.047161	5.222334
18	11.689587	10.827603	10.059087	9.371887	8.755625	8.201412	7.249670	6.127966	5.273164
19	12.085321	11.158116	10.335595	9.603599	8.950115	8.364920	7.365777	6.198231	5.316241
20	12.462210	11.469921	10.594014	9.818147	9.128546	8.513564	7.469444	6.259331	5.352746
21	12.821153	11.764077	10.835527	10.016803	9.292244	8.648694	7.562003	6.312462	5.383683
22	13.163003	12.041582	11.061240	10.200744	9.442425	8.771540	7.644646	6.358663	5.409901
23	13.488574	12.303379	11.272187	10.371059	9.580207	8.883218	7.718434	6.398837	5.432120
24	13.798642	12.550358	11.469334	10.528758	9.706612	8.984744	7.784316	6.433771	5.450949
25	14.093945	12.783356	11.653583	10.674776	9.822580	9.077040	7.843139	6.464149	5.466906
26	14.375185	13.003166	11.825779	10.809978	9.928972	9.160945	7.895660	6.490564	5.480429
27	14.643034	13.210534	11.986709	10.935165	10.026580	9.237223	7.942554	6.513534	5.491889
28	14.898127	13.406164	12.137111	11.051078	10.116128	9.306567	7.984423	6.533508	5.501601
29	15.141074	13.590721	12.277674	11.158406	10.198283	9.369606	8.021806	6.550877	5.509831
30	15.372451	13.764831	12.409041	11.257783	10.273654	9.426914	8.055184	6.565980	5.516806
31	15.592811	13.929086	12.531814	11.349799	10.342802	9.479013	8.084986	6.579113	5.522717
32	15.802677	14.084043	12.646555	11.434999	10.406240	9.526376	8.111594	6.590533	5.527726
33	16.002549	14.230230	12.753790	11.513888	10.464441	9.569432	8.135352	6.600463	5.531971
34	16.192904	14.368141	12.854009	11.586934	10.517835	9.608575	8.156564	6.609099	5.535569
35	16.374194	14.498246	12.947672	11.654568	10.566821	9.644159	8.175504	6.616607	5.538618
36	16.546852	14.620987	13.035208	11.717193	10.611763	9.676508	8.192414	6.623137	5.541201
37	16.711287	14.736780	13.117017	11.775179	10.652993	9.705917	8.207513	6.628815	5.543391
38	16.867893	14.846019	13.193473	11.828869	10.690820	9.732651	8.220993	6.633752	5.545247
39	17.017041	14.949075	13.264928	11.878582	10.725523	9.756956	8.233030	6.638045	5.546819
40	17.159086	15.046297	13.331709	11.924613	10.757360	9.779051	8.243777	6.641778	5.548152
41	17.294368	15.138016	13.394120	11.967235	10.786569	9.799137	8.253372	6.645025	5.549281
42	17.423208	15.224543	13.452449	12.006699	10.813366	9.817397	8.261939	6.647848	5.550238
43	17.545912	15.306173	13.506962	12.043240	10.837950	9.833998	8.269589	6.650302	5.551049
44	17.662773	15.383182	13.557908	12.077074	10.860505	9.849089	8.276418	6.652437	5.551737
45	17.774070	15.455832	13.605522	12.108402	10.881197	9.862808	8.282516	6.654293	5.552319
46	17.880066	15.524370	13.650020	12.137409	10.900181	9.875280	8.287961	6.655907	5.552813
47	17.981016	15.589028	13.691608	12.164267	10.917597	9.886618	8.292822	6.657310	5.553231
48	18.077158	15.650027	13.730474	12.189136	10.933575	9.896926	8.297163	6.658531	5.553586
49	18.168722	15.707572	13.766799	12.212163	10.948234	9.906296	8.301038	6.659592	5.553886
50	18.255925	15.761861	13.800746	12.233485	10.961683	9.914814	8.304498	6.660515	5.554141

A-32 TABLES

TABLE 5 The Unit Normal Distribution

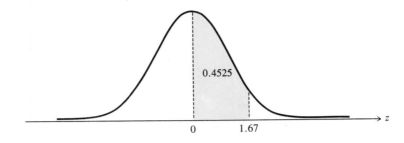

z	0.00	0.01	0.02	0.03	0.04	0.05	0.06	0.07	0.08	0.09
0.0	0.0000	0.0040	0.0080	0.0120	0.0160	0.0199	0.0239	0.0279	0.0319	0.0359
0.1	0.0398	0.0438	0.0478	0.0517	0.0557	0.0596	0.0636	0.0675	0.0714	0.0753
0.2	0.0793	0.0832	0.0871	0.0910	0.0948	0.0987	0.1026	0.1064	0.1103	0.1141
0.3	0.1179	0.1217	0.1255	0.1293	0.1331	0.1368	0.1406	0.1443	0.1480	0.1517
0.4	0.1554	0.1591	0.1628	0.1664	0.1700	0.1736	0.1772	0.1808	0.1844	0.1879
0.5	0.1915	0.1950	0.1985	0.2019	0.2054	0.2088	0.2123	0.2157	0.2190	0.2224
0.6	0.2257	0.2291	0.2324	0.2357	0.2389	0.2422	0.2454	0.2486	0.2517	0.2549
0.7	0.2580	0.2611	0.2642	0.2673	0.2704	0.2734	0.2764	0.2794	0.2823	0.2852
0.8	0.2881	0.2910	0.2939	0.2967	0.2995	0.3023	0.3051	0.3078	0.3106	0.3133
0.9	0.3159	0.3186	0.3212	0.3238	0.3264	0.3289	0.3315	0.3340	0.3365	0.3389
1.0	0.3413	0.3438	0.3461	0.3485	0.3508	0.3531	0.3554	0.3577	0.3599	0.3621
1.1	0.3643	0.3665	0.3686	0.3708	0.3729	0.3749	0.3770	0.3790	0.3810	0.3830
1.2	0.3849	0.3869	0.3888	0.3907	0.3925	0.3944	0.3962	0.3980	0.3997	0.4015
1.3	0.4032	0.4049	0.4066	0.4082	0.4099	0.4115	0.4131	0.4147	0.4162	0.4177
1.4	0.4192	0.4207	0.4222	0.4236	0.4251	0.4265	0.4279	0.4292	0.4306	0.4319
1.5	0.4332	0.4345	0.4357	0.4370	0.4382	0.4394	0.4406	0.4418	0.4429	0.4441
1.6	0.4452	0.4463	0.4474	0.4484	0.4495	0.4505	0.4515	0.4525	0.4535	0.4545
1.7	0.4554	0.4564	0.4573	0.4852	0.4591	0.4599	0.4608	0.4616	0.4625	0.4633
1.8	0.4641	0.4649	0.4656	0.4664	0.4671	0.4678	0.4686	0.4692	0.4699	0.4706
1.9	0.4713	0.4719	0.4726	0.4732	0.4738	0.4744	0.4750	0.4756	0.4761	0.4767
2.0	0.4772	0.4778	0.4783	0.4788	0.4793	0.4798	0.4803	0.4808	0.4812	0.4817
2.1	0.4821	0.4826	0.4830	0.4838	0.4838	0.4842	0.4846	0.4850	0.4854	0.4857
2.2	0.4861	0.4864	0.4868	0.4871	0.4875	0.4878	0.4881	0.4884	0.4887	0.4890
2.3	0.4893	0.4896	0.4898	0.4901	0.4904	0.4906	0.4909	0.4911	0.4913	0.4916
2.4	0.4918	0.4920	0.4922	0.4925	0.4927	0.4929	0.4931	0.4932	0.4934	0.4936
2.5	0.4938	0.4940	0.4941	0.4943	0.4945	0.4946	0.4948	0.4949	0.4951	0.4952
2.6	0.4953	0.4955	0.4956	0.4957	0.4959	0.4960	0.4961	0.4962	0.4963	0.4964
2.7	0.4965	0.4966	0.4967	0.4968	0.4969	0.4970	0.4971	0.4972	0.4973	0.4974
2.8	0.4974	0.4975	0.4976	0.4977	0.4977	0.4978	0.4979	0.4979	0.4980	0.4981
2.9	0.4981	0.4982	0.4982	0.4983	0.4984	0.4984	0.4985	0.4985	0.4986	0.4986
3.0	0.4987	0.4987	0.4987	0.4988	0.4988	0.4989	0.4989	0.4989	0.4990	0.4990

TABLE 6 The Exponential Function e^x

x	e^x	e^{-x}	x	e^x	e^{-x}
0.00	1.0000	1.0000	3.0	20.086	0.0498
0.05	1.0513	0.9512	3.1	22.198	0.0450
0.10	1.1052	0.9048	3.2	24.533	0.0408
0.15	1.1618	0.8607	3.3	27.113	0.0369
0.20	1.2214	0.8187	3.4	29.964	0.0334
0.25	1.2840	0.7788	3.5	33.115	0.0302
0.30	1.3499	0.7408	3.6	36.598	0.0273
0.35	1.4191	0.7047	3.7	40.447	0.0247
0.40	1.4918	0.6703	3.8	44.701	0.0224
0.45	1.5683	0.6376	3.9	49.402	0.0202
0.50	1.6487	0.6065	4.0	54.598	0.0183
0.55	1.7333	0.5769	4.1	60.340	0.0166
0.60	1.8221	0.5488	4.2	66.686	0.0150
0.65	1.9155	0.5220	4.3	73.700	0.0136
0.70	2.0138	0.4966	4.4	81.451	0.0123
0.75	2.1170	0.4724	4.5	90.017	0.0111
0.80	2.2255	0.4493	4.6	99.484	0.0101
0.85	2.3396	0.4274	4.7	109.95	0.0091
0.90	2.4596	0.4066	4.8	121.51	0.0082
0.95	2.5857	0.3867	4.9	134.29	0.0074
1.0	2.7183	0.3679	5.0	148.41	0.0067
1.1	3.0042	0.3329	5.1	164.02	0.0061
1.2	3.3201	0.3012	5.2	181.27	0.0055
1.3	3.6693	0.2725	5.3	200.34	0.0050
1.4	4.0552	0.2466	5.4	221.41	0.0045
1.5	4.4817	0.2231	5.5	244.69	0.0041
1.6	4.9530	0.2019	5.6	270.43	0.0037
1.7	5.4739	0.1827	5.7	298.87	0.0033
1.8	6.0496	0.1653	5.8	330.30	0.0030
1.9	6.6859	0.1496	5.9	365.04	0.0027
2.0	7.3891	0.1353	6.0	403.43	0.0025
2.1	8.1662	0.1225	6.5	665.14	0.0015
2.2	9.0250	0.1108	7.0	1096.6	0.0009
2.3	9.9742	0.1003	7.5	1808.0	0.0006
2.4	11.023	0.0907	8.0	2981.0	0.0003
2.5	12.182	0.0821	8.5	4914.8	0.0002
2.6	13.464	0.0743	9.0	8103.1	0.0001
2.7	14.880	0.0672	9.5	13,360	0.00007
2.8	16.445	0.0608	10.0	22,026	0.00004
2.9	18.174	0.0550			

TABLE 7 The Natural Logarithm Function
$\ln x = \log_e x$

x	ln x	x	ln x	x	ln x
0.0	—	4.5	1.5041	9.0	2.1972
0.1	−2.3026	4.6	1.5261	9.1	2.2083
0.2	−1.6094	4.7	1.5476	9.2	2.2192
0.3	−1.2040	4.8	1.5686	9.3	2.2300
0.4	−0.9163	4.9	1.5892	9.4	2.2407
0.5	−0.6931	5.0	1.6094	9.5	2.2513
0.6	−0.5108	5.1	1.6292	9.6	2.2618
0.7	−0.3567	5.2	1.6487	9.7	2.2721
0.8	−0.2231	5.3	1.6677	9.8	2.2824
0.9	−0.1054	5.4	1.6864	9.9	2.2925
1.0	0.0000	5.5	1.7047	10	2.3026
1.1	0.0953	5.6	1.7228	11	2.3979
1.2	0.1823	5.7	1.7405	12	2.4849
1.3	0.2624	5.8	1.7579	13	2.5649
1.4	0.3365	5.9	1.7750	14	2.6391
1.5	0.4055	6.0	1.7918	15	2.7081
1.6	0.4700	6.1	1.8083	16	2.7726
1.7	0.5306	6.2	1.8245	17	2.8332
1.8	0.5878	6.3	1.8405	18	2.8904
1.9	0.6419	6.4	1.8563	19	2.9444
2.0	0.6931	6.5	1.8718	20	2.9957
2.1	0.7419	6.6	1.8871	25	3.2189
2.2	0.7885	6.7	1.9021	30	3.4012
2.3	0.8329	6.8	1.9169	35	3.5553
2.4	0.8755	6.9	1.9315	40	3.6889
2.5	0.9163	7.0	1.9459	45	3.8067
2.6	0.9555	7.1	1.9601	50	3.9120
2.7	0.9933	7.2	1.9741	55	4.0073
2.8	1.0296	7.3	1.9879	60	4.0943
2.9	1.0647	7.4	2.0015	65	4.1744
3.0	1.0986	7.5	2.0149	70	4.2485
3.1	1.1314	7.6	2.0281	75	4.3175
3.2	1.1632	7.7	2.0142	80	4.3820
3.3	1.1939	7.8	2.0541	85	4.4427
3.4	1.2238	7.9	2.0669	90	4.4998
3.5	1.2528	8.0	2.0794	95	4.5539
3.6	1.2809	8.1	2.0919	100	4.6052
3.7	1.3083	8.2	2.1041	200	5.2983
3.8	1.3350	8.3	2.1163	300	5.7038
3.9	1.3610	8.4	2.1282	400	5.9915
4.0	1.3863	8.5	2.1401	500	6.2146
4.1	1.4110	8.6	2.1518	600	6.3069
4.2	1.4351	8.7	2.1633	700	6.5511
4.3	1.4586	8.8	2.1748	800	6.6846
4.4	1.4816	8.9	2.1861	900	6.8024

ANSWERS TO ODD-NUMBERED PROBLEMS

Chapter 1

Section 1.1, page 7

1. Yes   **3.** Yes   **5.** No   **7.** Yes   **9.** Yes   **11.** (a), (c), (f), (g), and (i) are true.

13. (a) Female or executive   (b) Male employees   (c) Nonexecutives   (d) Females with vested interest
(e) Executives with no vested interest   (f) Males with vested interest   (g) Female or executive or one with a vested interest
(h) Executives with no vested interest   (i) Females or executives with a vested interest   (j) Female executives with a vested interest
(k) Female executives or one with a vested interest   (l) Females who are either executives or have a vested interest

17. (a) 20   (b) 50   (c) 0

19. (a)

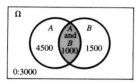

(b) $A \cup B$: People with either antigen; $A \cap B$: People with both antigens; $A \cap 0$: Empty set (ϕ); $(A \cup B) \cap 0$: Empty set (ϕ);
A^c: People without antigen A; $A^c \cap B$: People with antigen B only

21. (a) 875, (b) 314, (c) 413, (d) 66, (e) 0, (f) 459, (g) 551, (h) 528, (i) 797

23. ϕ represents the empty set; $\{\phi\}$ is a set with one element, namely the symbol ϕ.

Section 1.2, page 13

1. $\{-3, 0, 1, 2, 4, 7\}$ **3.** $\{-8, 1, 5\}$ **5.** $\{-8, -3, 0, 1, 2, 4, 5, 7\}$ **7.** $\{1\}$ **9.** $\frac{2}{9}$ **11.** $\frac{1}{99,000}$ **13.** $\frac{205}{999}$ **15.** $[-7, 5]$ **17.** ϕ

19. ϕ **21.** ϕ **23.** $[1, 8)$ **25.** $\{8\}$ **27.** $\{x \mid x \le 5 \text{ or } x > 8\} = (-\infty, 5] \cup (8, \infty)$ **31.** $(-\infty, b) \cap (a, \infty)$ **33.** $(-\infty, 7) \cap (-3, \infty)$

35. $(-\infty, 6] \cap (-1, \infty)$ **37.** $(-\infty, 83] \cap (4, 8)$

Section 1.3, page 21

1. $x = 2$ **3.** $y = \dfrac{5}{4}$ **5.** $x = -\dfrac{3}{2}$ **7.** $x = 2$ **9.** $y = \dfrac{1}{7}$ **11.** $x = \dfrac{c - a}{b}$ **13.** $x = \dfrac{c + a}{b}$ **15.** $x \le 7$; $(-\infty, 7]$ **17.** $x \ge 7$; $[7, \infty)$

19. $x < 9$; $(-\infty, 9)$ **21.** $x < -2$; $(-\infty, -2)$ **23.** $x > \dfrac{5}{3}$; $\left(\dfrac{5}{3}, \infty\right)$ **25.** $x \ge -\dfrac{6}{7}$; $\left[-\dfrac{6}{7}, \infty\right)$ **27.** $-1 < x \le 4$; $(-1, 4]$

29. $-\dfrac{1}{2} \le x \le 1$; $\left[-\dfrac{1}{2}, 1\right]$ **31.** $\dfrac{1}{2} < x < 3$; $\left(\dfrac{1}{2}, 3\right)$ **33.** $1 \le x \le \dfrac{10}{7}$; $\left[1, \dfrac{10}{7}\right]$ **35.** $-\dfrac{2}{3} < x \le 1$; $\left(-\dfrac{2}{3}, 1\right]$ **37.** $-3 \le x < 12$; $[-3, 12)$

39. $0 \le x < \dfrac{2}{3}$; $\left[0, \dfrac{2}{3}\right)$ **41.** $\dfrac{ad - c}{b} \le x < \dfrac{ed - c}{b}$; $\left[\dfrac{ad - c}{b}, \dfrac{ed - c}{b}\right)$ **43.** (a) $C = 1650 + 35x$ (b) \$9175

Section 1.4, page 27

1. 512   **3.** 2   **5.** $-\dfrac{1}{512}$ **7.** $-\dfrac{1}{2}$ **9.** 100   **11.** $\dfrac{1}{100}$ **13.** 1024   **15.** $\dfrac{1}{1024}$ **17.** -1 **19.** -1 **21.** $\dfrac{5}{3}$ **23.** $\dfrac{26}{81}$ **25.** $\dfrac{b}{a}$

27. $\dfrac{1}{81}$ **29.** 2   **31.** $\dfrac{1}{36}$ **33.** 9   **35.** 140.2961   **37.** $\dfrac{1}{10}$ **39.** $\dfrac{1}{10}$ **41.** $\dfrac{1}{100}$ **43.** $\dfrac{1}{8}$ **45.** 256   **47.** $\dfrac{1}{4}$ **49.** $-\dfrac{1}{512}$

51. Does not exist   **53.** $\dfrac{1}{2}$ **55.** $\dfrac{1}{32}$ **57.** -12.1257 **59.** $0.0000000001 = 10^{-10}$ **61.** 4651.6787   **63.** 23.131   **65.** $\dfrac{1}{16}$

67. $\dfrac{1}{64}$ **69.** 1   **71.** $\dfrac{1}{8}$ **73.** $\dfrac{1}{27}$ **75.** $\dfrac{1}{(-4)^7} = -\dfrac{1}{16,384}$ **77.** $-|a^{m/n}|$ or $|a^{m/n}|$

79. We need first to find an x such that $x^n = a$; but n is even. So $x^n \ge 0$ for all x and, if $a < 0$, then $x^n \ne a$ for all x.   **83.** $n = 14$
85. $n = 38$ **87.** 4^5 **89.** 1000^{1001} **93.** 68

Section 1.5, page 31

1. 1   **3.** -1 **5.** 2   **7.** 0
9. $(-\infty, -5] \cup [5, \infty)$

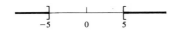

11. $(-\infty, \infty)$

13. $(-\infty, \infty)$

15. $(1, 3)$

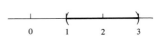

17. $(-\infty; -7] \cup [1, \infty)$

19. $\left(-\dfrac{7}{2}, -\dfrac{1}{2}\right)$

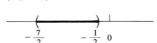

21. $(-\infty, 4] \cup [6, \infty)$

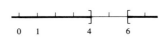

23. $(-\infty, -2) \cup \left(-\dfrac{2}{3}, \infty\right)$

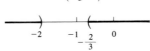

25. $\left(-\infty, \dfrac{4}{3}\right] \cup \left[\dfrac{8}{5}, \infty\right)$

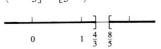

27. $\left(-\infty, -\dfrac{53}{3}\right) \cup \left(\dfrac{19}{3}, \infty\right)$

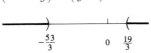

29. $\left(-\infty, \dfrac{c-b}{a}\right] \cup \left[\dfrac{-c-b}{a}, \infty\right)$

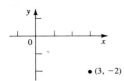

$\dfrac{c-b}{a}$ $\dfrac{-c-b}{a}$

31. $\left(\dfrac{5}{4}, \infty\right)$

0   $\dfrac{5}{4}$

39. (a) $|x-3|<7$ (b) $|x|\geq 3$ (c) $|x-8|>3$ (d) $\left|x-\dfrac{11}{2}\right|\geq\dfrac{7}{2}$ (e) $|x+3|\leq 7$

Review Exercises for Chapter 1, page 33

1. Yes   **3.** No   **5.** (a) All workers   (b) Empty set   (c) Female executives   (d) Males or clerical workers   (e) Males
(f) Nonexecutives   (g) Male executives   (h) Males who are not clerical workers
(i) Male executives, or clerical workers of either sex   (j) Workers who are not male clerical workers   (k) Female clerical workers
(l) Executives or males who are not clerical workers   (m) Males who are neither clerical workers nor executives
7. $[-8, 9)$ **9.** $(-\infty, -8) \cup [4, \infty)$ **11.** $(6, 12]$ **13.** $[9, 12]$ **15.** $(-\infty, -2] \cup [9, \infty)$ **17.** $[2, 4)$ **19.** $\dfrac{134}{999}$ **21.** $-\dfrac{5}{3}$ **23.** 5

25. $(-\infty, 7); x < 7$ **27.** $\left(-\infty, \dfrac{5}{2}\right]; x \leq \dfrac{5}{2}$ **29.** $[-2, 5); -2 \leq x < 5$ **31.** $\left(\dfrac{5}{4}, \dfrac{13}{4}\right]; \dfrac{5}{4} < x \leq \dfrac{13}{4}$

33. $(9, \infty) \cup (-\infty, -1); x > 9$ or $x < -1$ **35.** $(1, 4); 1 < x < 4$ **37.** ϕ **39.** $\left(\dfrac{2}{3}, 2\right); \dfrac{2}{3} < x < 2$

41. $\left(\dfrac{26}{5}, \infty\right) \cup (-\infty, -2); x > \dfrac{26}{5}$ or $x < -2$ **43.** 3   **45.** $\dfrac{1}{3}$ **47.** $\dfrac{1}{8}$ **49.** 1.6   **51.** -1 **53.** $\dfrac{27}{8}$ **55.** $\dfrac{1}{2}$ **57.** $\dfrac{1}{2}$ **59.** 2.716   **61.** $\dfrac{3}{7}$

63. 2.754   **65.** $\dfrac{1}{3}$ **67.** 856

Chapter 2

Section 2.1, page 38

1. IV

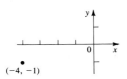

• $(3, -2)$

3. On x-axis

(2, 0)

5. III

$(-4, -1)$

7. I

$\left(\dfrac{1}{2}, \dfrac{1}{3}\right)$

9. On y-axis

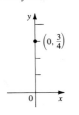

$\left(0, \dfrac{3}{4}\right)$

11. 5   **13.** $\sqrt{101}$ **15.** $\sqrt{29}$ **17.** $\sqrt{a^2 + b^2}$

Section 2.2, page 44

1. 6   **3.** −2   **5.** −$\frac{1}{5}$ **7.** −1   **9.** −$\frac{6}{5}$ **11.** $\frac{58}{9}$ **13.** (a) x-intercept = 0; y-intercept = 0   (b)

15. (a) x-intercept = 1; y-intercept = 1
  (b)

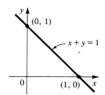

17. (a) x-intercept = $\frac{4}{3}$; y-intercept = −4
  (b)

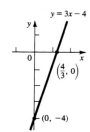

19. (a) x-intercept = −6; y-intercept = 3
  (b)

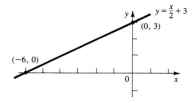

21. (a) x-intercept = 4; y-intercept = −8
  (b)

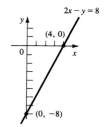

23. (a) x-intercept = 8; y-intercept = 4
  (b)

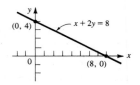

25. (a) x-intercept = −4; y-intercept = −8
  (b)

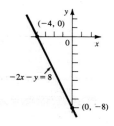

27. (a) x-intercept $= -4$; y-intercept $= 3$

(b)

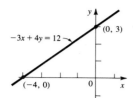

$-3x + 4y = 12$

$(0, 3)$

$(-4, 0)$

29. $R = 0.5q$

q	R
0	0
1	0.5
2	1
4	2

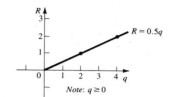

$R = 0.5q$

Note: $q \geq 0$

31. (a) $C = 6 + 0.07(k - 30) = 0.07k + 3.9$

(b)

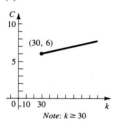

$(30, 6)$

Note: $k \geq 30$

(c) \$9.15

33. (a) $T = 7434 + 0.44(x - 28{,}800) = 0.44x - 5238$

(b)

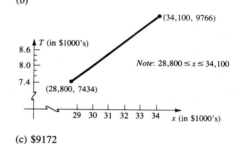

$(34{,}100, 9766)$

T (in \$1000's)

Note: $28{,}800 \leq x \leq 34{,}100$

$(28{,}800, 7434)$

x (in \$1000's)

(c) \$9172

Section 2.3, page 55

1. -2 **3.** 1   **5.** $-\dfrac{7}{13}$ **7.** 0   **9.** -1 **11.** $\dfrac{d-b}{c-a}$ **13.** Neither   **15.** Parallel   **17.** Neither   **19.** Perpendicular

23. $P - S$: $y + 1 = -\dfrac{2}{3}(x - 4)$ or $y - 3 = -\dfrac{2}{3}(x + 2)$

$S - I$: $y = -\dfrac{2}{3}x + \dfrac{5}{3}$

Std: $2x + 3y = 5$

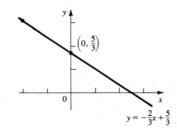

$\left(0, \dfrac{5}{3}\right)$

$y = -\dfrac{2}{3}x + \dfrac{5}{3}$

25. $P - S$: $y + 7 = 0(x - 4)$

$S - I$: $y = 0x - 7$

Std: $y = -7$

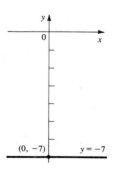

$(0, -7)$ $y = -7$

27. $P - S$: $y + \dfrac{1}{2} = -\dfrac{3}{16}(x - 3)$ or $y = -\dfrac{3}{16}\left(x - \dfrac{1}{3}\right)$

$S - I$: $y = -\dfrac{3}{16}x + \dfrac{1}{16}$

Std: $3x + 16y = 1$

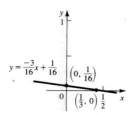

$y = \dfrac{-3}{16}x + \dfrac{1}{16}$ $\left(0, \dfrac{1}{16}\right)$

$\left(\dfrac{1}{3}, 0\right)$ $\dfrac{1}{2}$

29. $P - S$: $y + 1 = 1(x - 5)$ or $y - 2 = 1(x - 8)$

$S - I$: $y = x - 6$

Std: $x - y = 6$

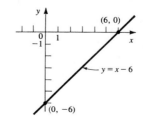

$(6, 0)$

$y = x - 6$

$(0, -6)$

31. $P - S$: $y - 1 = \frac{3}{7}(x + 5)$

$S - I$: $y = \frac{3}{7}x + \frac{22}{7}$

Std: $-3x + 7y = 22$

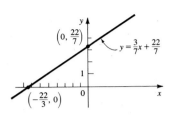

33. $P - S$: $y - b = c(x - a)$; $S - I$: $y = cx + (b - ac)$; Std: $cx - y = ac - b$; Graph depends on values of a, b, and c.

35. $y \mp \frac{5}{7}x + \frac{25}{7}$ **37.** $y = -\frac{3}{4}x + \frac{13}{4}$ **39.** $\left(\frac{22}{5}, -\frac{13}{5}\right)$ **41.** No intersection   **43.** $\left(\frac{1}{4}, \frac{13}{4}\right)$

45. (a) $C = 200 + 0.2x$; $R = 0.4x$ (b)

47. 100 kwh   **49.** 0.63

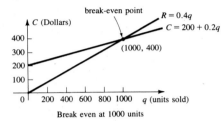

Break even at 1000 units

Section 2.4, page 62

1. $\left(-\frac{13}{5} \quad -\frac{11}{5}\right)$ **3.** No solution   **5.** $\left(\frac{11}{2} \quad -30\right)$ **7.** Infinite number of solutions of form $\left(x \quad \frac{2}{3}x\right)$ **9.** $(-1 \quad 2)$

11. $\left(\frac{c}{b + a} \quad \frac{c}{b + a}\right)$ if $b + a \neq 0$ **13.** $ab \neq 0$ **15.** $a = 0$, $b = 0$ and either c or d is nonzero.   **17.** None

19. Infinite number of solutions of form $\left(x \quad \frac{2x - 5}{3}\right)$ **21.** $\left(\frac{67}{45} \quad \frac{2}{15}\right)$ **23.** $A = 3,500,000$; $B = 500,000$

25. No solution will use all materials.   **27.** 50 acres of corn; 450 acres of soybeans

Section 2.5, page 76

1. $(2 \quad -3 \quad 1)$ **3.** $(3 \quad 0 \quad 0)$ **5.** $(-9 \quad 30 \quad 14)$ **7.** No solution   **9.** $\left(-\frac{4}{5}z \quad \frac{9}{5}z \quad z\right)$ for $z \in \mathbb{R}$ **11.** $\left(-1 \quad \frac{5}{2} + \frac{1}{2}z \quad z\right)$ for $z \in \mathbb{R}$

13. No solution   **15.** $\left(\frac{20}{13} - \frac{4}{13}w \quad \frac{-28}{13} + \frac{3}{13}w \quad \frac{-45}{13} + \frac{9}{13}w \quad w\right)$ for $w \in \mathbb{R}$

17. $\left(18 - 4w \quad \frac{-15}{2} + 2w \quad -31 + 7w \quad w\right)$ for $w \in \mathbb{R}$ **19.** No solution   **21.** No   **23.** Yes   **25.** No   **27.** Yes   **29.** No

31. $\begin{pmatrix} 1 & 0 \\ 0 & 1 \end{pmatrix}$ **33.** $\begin{pmatrix} 1 & 0 & 0 \\ 0 & 1 & 0 \\ 0 & 0 & 1 \end{pmatrix}$ **35.** $\begin{pmatrix} 1 & 0 \\ 0 & 1 \\ 0 & 0 \end{pmatrix}$ **37.** $(1100 \quad 1450 \quad 840)$ **39.** 6 days in England; 4 days in France; 4 days in Spain

41. No unique solution (2 equations in 3 unknowns); if 200 shares of McDonald's, then 100 shares of Hilton and 300 shares of Eastern

43. $p_1 = \frac{1}{4}$; $p_2 = \frac{1}{8}$; $p_3 = \frac{1}{8}$; $p_4 = \frac{1}{8}$; $p_5 = \frac{3}{8}$ **45.** Graze $\frac{72}{7} - \frac{2}{7}R$, $R \geq 6$; Move $\frac{96}{7} - \frac{5}{7}R$, $R \geq 6$; Rest at least 6 hours   **47.** $2a - c = b$

49. $(1.90081 \quad 4.19411 \quad -11.34852)$ **51.** $(0 \quad 0)$ **53.** $(0 \quad 0 \quad 0)$ **55.** $\left(\frac{1}{6}z \quad \frac{5}{6}z \quad z\right)$ **57.** $(0 \quad 0)$ **59.** $(-4w \quad 2w \quad 7w \quad w)$

61. $(0 \quad 0)$ **63.** $(0 \quad 0 \quad 0)$ **67.** $a_{11}a_{22}a_{33} + a_{12}a_{23}a_{31} + a_{13}a_{21}a_{32} - a_{11}a_{23}a_{32} - a_{12}a_{21}a_{33} - a_{13}a_{22}a_{31} \neq 0$

Section 2.6, page 84

1. $x_1 = 800$; $x_2 = 600$; $x_3 = 100$; $x_4 = 0$; $x_5 = 400$; $x_6 = 500$; $x_7 = 0$

Review Exercises for Chapter 2, page 85

1. (a) IV   (b) III   (c) I   (d) II

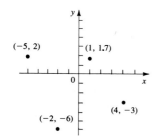

3. (a) -10 (b) x-intercept $= 0$; y-intercept $= 0$

(c)

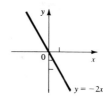

5. (a) 10   (b) x-intercept $= \dfrac{4}{7}$; y-intercept $= -4$

(c)

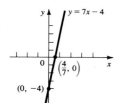

7. (a) 4   (b) No x-intercept; y-intercept $= 4$

(c)

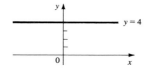

9. $S-I$: $y = 3x + 10$; $P - S$: $y - 4 = 3(x + 2)$; Std: $-3x + y = 10$ **11.** $S - I$: $y = 2x + 6$; $P - S$: $y - 4 = 2(x + 1)$; Std: $-2x + y = 6$

13. $S - I$: $y = 0(x - 8)$; $P - S$: $(y + 8) = 0(x - 3)$ or $(y + 8) = 0(x + 8)$; Std: $y = -8$ **15.** $-2x + 3y = 13$ or $y = \dfrac{2}{3}x + \dfrac{13}{3}$

17. Infinite number of solutions of form $\left(x \quad \dfrac{3 - x}{2} \right)$ **19.** $\left(-\dfrac{1}{3} \quad 0 \quad \dfrac{7}{3} \right)$ **21.** $(0 \quad 0 \quad 2)$ **23.** $(0 \quad 0 \quad 0)$ **25.** $(0 \quad 0)$

27. $(0 \quad 2 \quad -1 \quad 3)$ **29.** $(-3w \quad -2w \quad 4w \quad w)$ for $w \in \mathbb{R}$ **31.** Yes   **33.** No   **35.** No   **37.** $\begin{pmatrix} 1 & 0 & 0 & -\dfrac{1}{5} \\ 0 & 1 & 0 & \dfrac{7}{5} \\ 0 & 0 & 1 & \dfrac{14}{5} \end{pmatrix}$

Chapter 3

Section 3.1, page 94

1. 2×2; square   **3.** 2×2; square   **5.** 3×2 **7.** 2×4 **9.** 3×1 **11.** 3×3; square   **13.** 1×4 **15.** No   **17.** No   **19.** Yes

21. $\begin{pmatrix} 6 & 12 \\ 21 & -6 \end{pmatrix}$ **23.** $\begin{pmatrix} -1 & -2 & -5 \\ -7 & 2 & 0 \end{pmatrix}$ **25.** Not possible   **27.** $\begin{pmatrix} 4 & 8 & 45 \\ 18 & 11 & 23 \\ -15 & 25 & 29 \end{pmatrix}$ **29.** $\begin{pmatrix} 5 & 6 & 2 \\ 3 & 4 & 1 \\ 0 & -7 & 2 \end{pmatrix}$ **31.** $\begin{pmatrix} 3 & 9 \\ 6 & 15 \\ -3 & 6 \end{pmatrix}$

33. $\begin{pmatrix} 2 & 2 \\ -2 & -1 \\ 6 & -1 \end{pmatrix}$ **35.** $\begin{pmatrix} 0 & 0 \\ 0 & 0 \\ 0 & 0 \end{pmatrix}$ **37.** $\begin{pmatrix} -2 & 4 \\ 7 & 15 \\ -15 & 10 \end{pmatrix}$ **39.** $\begin{pmatrix} 4 & 10 \\ 17 & 22 \\ -9 & 1 \end{pmatrix}$ **41.** $\begin{pmatrix} 0 & 6 \\ 5 & 14 \\ -9 & 9 \end{pmatrix}$ **43.** $\begin{pmatrix} 1 & -5 & 0 \\ -3 & 4 & -5 \\ -14 & 13 & -1 \end{pmatrix}$

45. $\begin{pmatrix} 1 & 1 & 5 \\ 9 & 5 & 10 \\ 7 & -7 & 3 \end{pmatrix}$ **47.** $\begin{pmatrix} -1 & -1 & -1 \\ -3 & -3 & -10 \\ -7 & 3 & 5 \end{pmatrix}$ **49.** $\begin{pmatrix} -1 & -1 & -5 \\ -9 & -5 & -10 \\ -7 & 7 & -3 \end{pmatrix}$ **55.** $\begin{pmatrix} 1 & 1 & 1 & 0 \\ 1 & 1 & 1 & 0 \\ 1 & 1 & 1 & 1 \\ 0 & 0 & 1 & 1 \end{pmatrix}$

57. (a) 1.5 (b) 6 (c) $7C = \begin{pmatrix} 7 & 3.5 & 21 & 56 & 1.75 \\ 10.5 & 14 & 42 & 42 & 2.1 \\ 14 & 10.5 & 28 & 63 & 4.2 \end{pmatrix}$ **59.** $0.6S = \begin{pmatrix} 3.6 & 11.4 & 8.4 & 27.6 \\ 4.8 & 16.8 & 7.2 & 24 \\ 2.4 & 15.6 & 10.2 & 33 \end{pmatrix}$

61. (a) $\begin{pmatrix} 415 & 91 & 6 & 77 \\ 65 & 281 & 4 & 63 \\ 31 & 19 & 8 & 29 \end{pmatrix}$ (b) $\begin{pmatrix} 77 \\ 63 \\ 29 \end{pmatrix}$ (c) $(65 \quad 281 \quad 4 \quad 63)$ (d) 63 (e) There is no 4,2 component since there is no 4th row.

Section 3.2, page 107

1. 2 **3.** 32 **5.** 32 **9.** \$1535 **11.** $\begin{pmatrix} 8 & 20 \\ -4 & 11 \end{pmatrix}$ **13.** $\begin{pmatrix} -3 & -3 \\ 1 & 3 \end{pmatrix}$ **15.** $\begin{pmatrix} 13 & 35 & 18 \\ 20 & 26 & 20 \end{pmatrix}$ **17.** $\begin{pmatrix} 19 & -17 & 34 \\ 8 & -12 & 20 \\ -8 & -11 & 7 \end{pmatrix}$

19. $\begin{pmatrix} 18 & 15 & 35 \\ 9 & 21 & 13 \\ 10 & 9 & 9 \end{pmatrix}$ **21.** $(7 \quad 16)$ **23.** $\begin{pmatrix} 3 & -2 & 1 \\ 4 & 0 & 6 \\ 5 & 1 & 9 \end{pmatrix}$ **25.** $\begin{pmatrix} a & b & c \\ d & e & f \\ g & h & j \end{pmatrix}$

27. $b_{11} = \dfrac{a_{22}}{a_{11}a_{22} - a_{12}a_{21}}; \ b_{12} = \dfrac{-a_{12}}{a_{11}a_{22} - a_{12}a_{21}}; \ b_{21} = \dfrac{-a_{21}}{a_{11}a_{22} - a_{12}a_{21}}; \ b_{22} = \dfrac{a_{11}}{a_{11}a_{22} - a_{12}a_{21}}$

29. (a) 5 (b) 4 (c) $\begin{pmatrix} 1 & 3 & 1 & 2 \\ 0 & 1 & 2 & 2 \end{pmatrix}$ **31.** (a) $(100 \quad 200 \quad 400 \quad 100)$ (b) $\begin{pmatrix} 46 \\ 34 \\ 15 \\ 10 \end{pmatrix}$ (c) \$18,400

33. (a) $\begin{pmatrix} 80,000 & 45,000 & 40,000 \\ 50 & 20 & 10 \end{pmatrix}$ (b) $\begin{pmatrix} 1 \\ 3 \\ 1 \end{pmatrix}$ (c) Money: \$255,000; Shares: 120

35. (a) $\begin{pmatrix} \frac{1}{5} & \frac{2}{5} & \frac{1}{5} \\ 0 & 1 & \frac{1}{2} \\ \frac{1}{10} & 0 & 0 \\ \frac{1}{8} & \frac{1}{8} & \frac{1}{16} \end{pmatrix}$ (b) $\begin{pmatrix} 2 \\ 1 \\ 3 \end{pmatrix}$ (c) Bread: $\frac{7}{5}$ loaves; Milk: $\frac{5}{2}$ quarts; Coffee: $\frac{1}{5}$ pound; Cheese: $\frac{9}{16}$ pound

37. $\begin{pmatrix} 0 & -8 \\ 32 & 32 \end{pmatrix}$ **39.** $A^2 = \begin{pmatrix} 0 & 0 & 1 & 0 \\ 0 & 0 & 0 & 1 \\ 0 & 0 & 0 & 0 \\ 0 & 0 & 0 & 0 \end{pmatrix}; A^3 = \begin{pmatrix} 0 & 0 & 0 & 1 \\ 0 & 0 & 0 & 0 \\ 0 & 0 & 0 & 0 \\ 0 & 0 & 0 & 0 \end{pmatrix}; A^4 = A^5 = \begin{pmatrix} 0 & 0 & 0 & 0 \\ 0 & 0 & 0 & 0 \\ 0 & 0 & 0 & 0 \\ 0 & 0 & 0 & 0 \end{pmatrix}$

43. Sums of rows of P^2 still yield 1.

45. (a) 2, 4, 1, 3; Player 2 is first. (b) Score is 1 point for each victory, plus $\frac{1}{2}$ point for each victory of a defeated opponent.

47. $A(B + C) = AB + AC = \begin{pmatrix} 45 & 35 \\ 1 & 16 \end{pmatrix}$

Section 3.3, page 115

1. $\begin{pmatrix} 2 & -1 \\ 4 & 5 \end{pmatrix}\begin{pmatrix} x \\ y \end{pmatrix} = \begin{pmatrix} 3 \\ 7 \end{pmatrix}$ **3.** $\begin{pmatrix} 3 & 6 & -7 \\ 2 & -1 & 3 \end{pmatrix}\begin{pmatrix} x \\ y \\ z \end{pmatrix} = \begin{pmatrix} 0 \\ 1 \end{pmatrix}$ **5.** $\begin{pmatrix} 0 & 1 & -1 \\ 1 & 0 & 1 \\ 3 & 2 & 0 \end{pmatrix}\begin{pmatrix} x \\ y \\ z \end{pmatrix} = \begin{pmatrix} 7 \\ 2 \\ -5 \end{pmatrix}$ **7.** $\begin{aligned} x + y - z &= 7 \\ 4x - y + 5z &= 4 \\ 6x + y + 3z &= 20 \end{aligned}$ **9.** $\begin{aligned} 2x \quad\;\; + z &= 2 \\ -3x + 4y \quad &= 3 \\ 5y + 6z &= 5 \end{aligned}$

11. $\begin{aligned} x \quad &= 2 \\ y \quad &= 3 \\ z &= -5 \\ w &= 6 \end{aligned}$ **13.** $\begin{aligned} 6x + 2y + z &= 2 \\ -2x + 3y + z &= 4 \\ 0x + 0y + 0z &= 2 \end{aligned}$ **15.** $\begin{aligned} 7x + 2y &= 1 \\ 3x + y &= 2 \\ 6x + 9y &= 3 \end{aligned}$ **17.** $(4 - 2y + 4z \quad y \quad z)$ for $y, z \in \mathbb{R}$ **19.** $(0 \quad 0 \quad 0)$

21. $\left(1 \quad -\frac{1}{3} \quad \frac{1}{2} \quad 4\right)$ **23.** 20 units of each type of fertilizer

Section 3.4, page 128

1. $\begin{pmatrix} 2 & -1 \\ -3 & 2 \end{pmatrix}$ **3.** $\begin{pmatrix} 0 & 1 \\ 1 & 0 \end{pmatrix}$ **5.** No inverse **7.** $\frac{1}{6}\begin{pmatrix} 2 & -2 & -2 \\ 0 & 3 & 6 \\ 0 & 0 & -6 \end{pmatrix} = \begin{pmatrix} \frac{1}{3} & -\frac{1}{3} & -\frac{1}{3} \\ 0 & \frac{1}{2} & 1 \\ 0 & 0 & -1 \end{pmatrix}$ **9.** No inverse **11.** No inverse

13. $\dfrac{1}{9}\begin{pmatrix} 21 & -3 & -3 & -6 \\ 4 & -1 & -4 & 1 \\ -1 & -2 & 1 & 2 \\ -15 & 6 & 6 & 3 \end{pmatrix} = \begin{pmatrix} \frac{7}{3} & -\frac{1}{3} & -\frac{1}{3} & -\frac{2}{3} \\ \frac{4}{9} & -\frac{1}{9} & -\frac{4}{9} & \frac{1}{9} \\ -\frac{1}{9} & -\frac{2}{9} & \frac{1}{9} & \frac{2}{9} \\ -\frac{5}{3} & \frac{2}{3} & \frac{2}{3} & \frac{1}{3} \end{pmatrix}$ **15.** $\begin{pmatrix} 0 & 1 & 0 & 2 \\ 1 & -1 & -2 & 2 \\ 0 & 1 & 3 & -3 \\ -2 & 2 & 3 & -2 \end{pmatrix}$ **19.** 20 units of each grade of fertilizer

21. 575 acres of soybeans; $191\frac{2}{3}$ acres of corn; $233\frac{1}{3}$ acres of wheat **23.** 3 chairs and 2 tables **25.** 4 units of A and 5 units of B

27. 5 vice-presidents; 10 division managers; 10 assistant division managers **29.** $\begin{pmatrix} 1 & 0 \\ 0 & 1 \end{pmatrix}$; Yes

31. $\begin{pmatrix} 1 & 0 & 0 \\ 0 & 1 & 0 \\ 0 & 0 & 1 \end{pmatrix}$; Yes **33.** $\begin{pmatrix} 1 & 0 & -5 \\ 0 & 1 & -14 \\ 0 & 0 & 0 \end{pmatrix}$; No **35.** $\begin{pmatrix} 1 & 0 & 0 & \frac{1}{7} \\ 0 & 1 & 0 & \frac{29}{7} \\ 0 & 0 & 1 & \frac{10}{7} \\ 0 & 0 & 0 & 0 \end{pmatrix}$ **39.** $\begin{pmatrix} \frac{1}{2} & -\frac{1}{6} & \frac{7}{30} \\ 0 & \frac{1}{3} & -\frac{4}{15} \\ 0 & 0 & \frac{1}{5} \end{pmatrix}$

Section 3.5, page 140

1. $\begin{pmatrix} 72.65306 \\ 55.10204 \\ 65.30612 \end{pmatrix}$

3. For example, $a_{21} + a_{22} + a_{23} = 0.7 + 0.3 + 0.8 = 1.8$. This means that the demand for industry 2's output exceeds the output of industry 2.

5. $\begin{pmatrix} 57.14286 \\ 178.57143 \\ 73.80952 \end{pmatrix}$ **7.** $A = \begin{pmatrix} \frac{1}{6} & \frac{5}{18} & \frac{5}{12} \\ \frac{9}{40} & \frac{1}{4} & \frac{1}{2} \\ \frac{3}{10} & \frac{2}{9} & \frac{1}{3} \end{pmatrix}$; $I - A = \begin{pmatrix} \frac{5}{6} & -\frac{5}{18} & -\frac{5}{12} \\ -\frac{9}{40} & \frac{3}{4} & -\frac{1}{2} \\ -\frac{3}{10} & -\frac{2}{9} & \frac{2}{3} \end{pmatrix}$ **9.** $\begin{pmatrix} 896.14679 \\ 1017.02752 \\ 922.27523 \end{pmatrix}$ **11.** $\begin{pmatrix} 500.21739 \\ 478.95652 \\ 529.56522 \end{pmatrix}$

13. (a) $A = \begin{pmatrix} 0.293 & 0 & 0 \\ 0.014 & 0.207 & 0.017 \\ 0.044 & 0.010 & 0.216 \end{pmatrix}$; $I - A = \begin{pmatrix} 0.707 & 0 & 0 \\ -0.014 & 0.793 & -0.017 \\ -0.044 & -0.010 & 0.784 \end{pmatrix}$ (b) $\begin{pmatrix} 195492.2207 \\ 25932.85859 \\ 13580.33966 \end{pmatrix}$ **15.** $\begin{pmatrix} 16179.30153 \\ 9661.53840 \\ 2498.87179 \\ 18108.3929 \end{pmatrix}$

Section 3.6, page 146

1. 166 97 187 107 226 132 96 57 183 107 172 99 221 129 67 40 204 120 **3.** FROGS CAN JUMP

Review Exercises for Chapter 3, page 147

1. $\begin{pmatrix} -6 & 3 \\ 0 & 12 \\ 6 & 9 \end{pmatrix}$ **3.** $\begin{pmatrix} 16 & 2 & 3 \\ -20 & 10 & -1 \\ -36 & 8 & 16 \end{pmatrix}$ **5.** $\begin{pmatrix} 17 & 39 & 41 \\ 14 & 20 & 42 \end{pmatrix}$ **7.** $\begin{pmatrix} 9 & 10 \\ 30 & 32 \end{pmatrix}$ **9.** $(AB)C = A(BC) = \begin{pmatrix} 25 & 74 \\ 132 & 222 \end{pmatrix}$

11. $\begin{pmatrix} 1 & -2 \\ 0 & 0 \end{pmatrix}$; Not invertible **13.** $\begin{pmatrix} 1 & 0 & -\frac{2}{3} \\ 0 & 1 & -\frac{1}{3} \\ 0 & 0 & 0 \end{pmatrix}$; Not invertible **15.** $\begin{pmatrix} 1 & -3 \\ 2 & 5 \end{pmatrix}\begin{pmatrix} x \\ y \end{pmatrix} = \begin{pmatrix} 4 \\ 7 \end{pmatrix}$; Solution is $\begin{pmatrix} \frac{41}{11} \\ -\frac{1}{11} \end{pmatrix}$

17. $\begin{pmatrix} 2 & 0 & 4 \\ -1 & 3 & 1 \\ 0 & 1 & 2 \end{pmatrix}\begin{pmatrix} x \\ y \\ z \end{pmatrix} = \begin{pmatrix} 7 \\ -4 \\ 5 \end{pmatrix}$; Solution is $\begin{pmatrix} -\frac{41}{6} \\ -\frac{16}{3} \\ \frac{31}{6} \end{pmatrix}$

19. (a) $A = \begin{pmatrix} 0.245 & 0.102 & 0.051 \\ 0.099 & 0.291 & 0.279 \\ 0.433 & 0.372 & 0.011 \end{pmatrix}$; $I - A = \begin{pmatrix} 0.755 & -0.102 & -0.051 \\ -0.099 & 0.709 & -0.279 \\ -0.433 & -0.372 & 0.989 \end{pmatrix}$ (b) $\begin{pmatrix} 18.20792 \\ 73.16603 \\ 66.74600 \end{pmatrix}$ (in billions of dollars)

21. THIS IS THE END OF CHAPTER THREE

Chapter 4

Section 4.1, page 157

1.

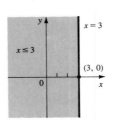

3.

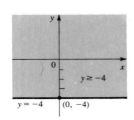

5.

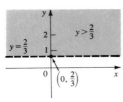

7.

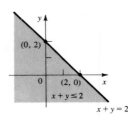

9.

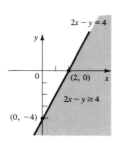

11.

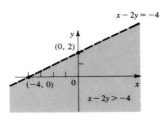

13.

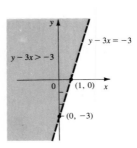

15.

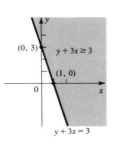

17.

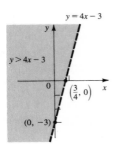

19.

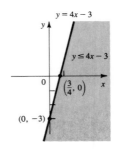

21.

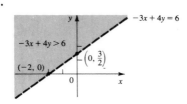

23.

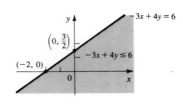

25.

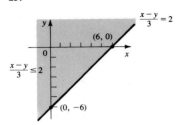

$$\frac{x-y}{3} = 2$$

(6, 0)

0

$$\frac{x-y}{3} \le 2$$

(0, −6)

27.

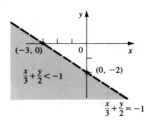

(−3, 0)

0

$$\frac{x}{3} + \frac{y}{2} < -1$$

(0, −2)

$$\frac{x}{3} + \frac{y}{2} = -1$$

29.

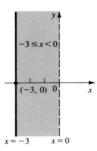

$$-3 \le x < 0$$

(−3, 0) 0

$x = -3$ $x = 0$

31.

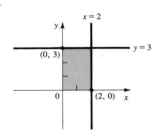

$x = 2$

(0, 3) $y = 3$

0   (2, 0)   x

33.

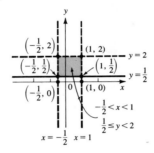

$\left(-\frac{1}{2}, 2\right)$ (1, 2)

$y = 2$

$\left(-\frac{1}{2}, \frac{1}{2}\right)$ $\left(1, \frac{1}{2}\right)$

$y = \frac{1}{2}$

$\left(-\frac{1}{2}, 0\right)$ (1, 0)

$-\frac{1}{2} < x < 1$

$\frac{1}{2} \le y < 2$

$x = -\frac{1}{2}$ $x = 1$

35.

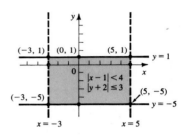

(−3, 1)   (0, 1)   (5, 1)

$y = 1$

$|x - 1| < 4$

$|y + 2| \le 3$

(−3, −5)   (5, −5)

$y = -5$

$x = -3$ $x = 5$

37.

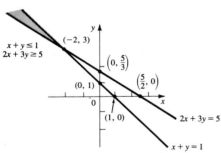

$x + y \le 1$

$2x + 3y \ge 5$

(−2, 3)

$\left(0, \frac{5}{3}\right)$

(0, 1) $\left(\frac{5}{2}, 0\right)$

0

(1, 0) $2x + 3y = 5$

$x + y = 1$

39.

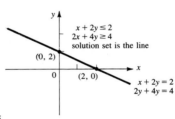

$x + 2y \le 2$

$2x + 4y \ge 4$

solution set is the line

(0, 2)

0   (2, 0)

$x + 2y = 2$

$2y + 4y = 4$

41.

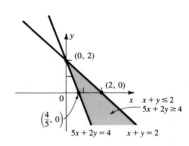

(0, 2)

(2, 0)

0    $x + y \le 2$

$5x + 2y \ge 4$

$\left(\frac{4}{5}, 0\right)$

$5x + 2y = 4$ $x + y = 2$

43.

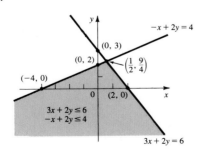

$-x + 2y = 4$

(0, 3)

(0, 2) $\left(\frac{1}{2}, \frac{9}{4}\right)$

(−4, 0)

0   (2, 0)

$3x + 2y \le 6$

$-x + 2y \le 4$

$3x + 2y = 6$

45.

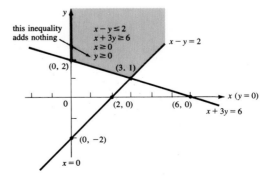

this inequality adds nothing

$x - y \le 2$

$x + 3y \ge 6$

$x \ge 0$

$y \ge 0$

$x - y = 2$

(0, 2)   (3, 1)

0   (2, 0)   (6, 0) x $(y = 0)$

$x + 3y = 6$

(0, −2)

$x = 0$

Section 4.2, page 170

1. 65 chairs; 40 tables; Profit = $525 **3.** 380 chairs; no tables; Profit = $1900 **5.** 380 chairs; no tables; Profit = $3040

7. 5 mgd from each; Cost = $4000 per day **9.** (1, 3); $f = 13$

11. Any point on the line $2x + 3y = 10$ between $\left(0, \dfrac{10}{3}\right)$ and (2, 2) will yield the maximum value $f = 10$. **13.** $\left(\dfrac{10}{11}, \dfrac{10}{11}\right); f = \dfrac{80}{11}$

15. (0, 1); $f = 12$ **17.** (2, 1); $g = 13$ **19.** (1, 0); $g = 3$

21. Any point on the line $4s_1 + 2s_2 = 1200$ will yield 1200 lb. Choose s_1 of species S_1 with s_1 an integer between 234 and 300, inclusive. Then $s_2 = 600 - 2s_1$.

23. $s_1 = 200$; $s_2 = 400$; $f = 1600$ **25.** $s_1 = 260$; $s_2 = 120$; $f = 380$ fish **27.** $s_1 = 200$; $s_2 = 400$; $f = 600$ lbs

29. Food I = $\dfrac{2}{3}$ lb; Food II = $\dfrac{7}{3}$ lb; Cost per lb = $\$\dfrac{8}{9} \approx 89¢$ **31.** $s_1 = 1$; $s_2 = 5$; Energy = 13 units **33.** b **35.** c **37.** a **39.** c **41.** c

43. a **47.** (1 4 10); (0 4 11); $\left(\dfrac{13}{3} \quad \dfrac{32}{3} \quad 0\right)$; (0 15 0); (2 0 11); (0 0 13); $\left(\dfrac{43}{5} \quad 0 \quad 0\right)$; (0 0 0)

49. (a) Minimize $C = 2x_1 + 2.5x_2 + 0.8x_3$ subject to:
$$\begin{aligned} x_1 + x_2 + 10x_3 &\geq 1 \quad x_1 \geq 0 \\ 100x_1 + 10x_2 + 10x_3 &\geq 50; \ x_2 \geq 0 \\ 10x_1 + 100x_2 + 10x_3 &\geq 10 \quad x_3 \geq 0 \end{aligned}$$

(b) (0 0 0); $\left(0 \quad 0 \quad \dfrac{1}{10}\right)$; (0 0 5); (0 0 1); (0 1 0); (0 5 0); $\left(0 \quad \dfrac{1}{10} \quad 0\right)$; (1 0 0); $\left(\dfrac{1}{2} \quad 0 \quad 0\right)$; (1 0 0);

$\left(0 \quad \dfrac{49}{9} \quad -\dfrac{4}{9}\right)$; $\left(0 \quad \dfrac{1}{11} \quad \dfrac{1}{11}\right)$; $\left(0 \quad -\dfrac{4}{9} \quad \dfrac{49}{9}\right)$; $\left(\dfrac{49}{99} \quad 0 \quad \dfrac{5}{99}\right)$; (1 0 0); $\left(\dfrac{4}{9} \quad 0 \quad \dfrac{5}{9}\right)$; $\left(\dfrac{4}{9} \quad \dfrac{5}{9} \quad 0\right)$; (1 0 0);

$\left(\dfrac{49}{99} \quad \dfrac{5}{99} \quad 0\right)$; $\left(\dfrac{53}{108} \quad \dfrac{5}{108} \quad \dfrac{5}{108}\right)$

(c) (0 0 5); (0 5 0); (1 0 0); (1 0 0); (1 0 0); $\left(\dfrac{4}{9} \quad 0 \quad \dfrac{5}{9}\right)$; $\left(\dfrac{4}{9} \quad \dfrac{5}{9} \quad 0\right)$; (1 0 0); $\left(\dfrac{53}{108} \quad \dfrac{5}{108} \quad \dfrac{5}{108}\right)$

(d)

Feasible solution	Cost (in dollars)
(0 0 5)	4.00
(0 5 0)	12.50
(1 0 0)	2.00
$\left(\dfrac{4}{9} \quad 0 \quad \dfrac{5}{9}\right)$	1.33
$\left(\dfrac{4}{9} \quad \dfrac{5}{9} \quad 0\right)$	2.28
$\left(\dfrac{53}{108} \quad \dfrac{5}{108} \quad \dfrac{5}{108}\right)$	1.13

(e) Cost is $\$\dfrac{245}{216} \approx \$1.13$ when $x_1 = \dfrac{53}{108} \approx 0.49$ gallon of milk, $x_2 = \dfrac{5}{108} \approx 0.046$ pound of beef, and $x_3 = \dfrac{5}{108} \approx 0.046$ dozen eggs are consumed daily. Note: $\dfrac{5}{108} \cdot 12 = 0.555$, so consume slightly more than $\dfrac{1}{2}$ of an egg daily.

51. 8,000 cans of Mix 1; 18,000 cans of Mix 2; 4,000 cans of Mix 3; Sales: $1160

Section 4.3, page 183

1. $\begin{aligned} x_1 + x_2 + s_1 &= 3 \\ 2x_1 + x_2 + s_2 &= 7 \end{aligned}$; $s_1, s_2 \geq 0$ **3.** $\begin{aligned} 2x_1 + x_2 + s_1 &= 10 \\ 3x_1 + 2x_2 + s_2 &= 30; \ s_1, s_2, s_3 \geq 0 \\ 4x_1 + 7x_2 + s_3 &= 20 \end{aligned}$

5. $\begin{aligned} 7x_1 + x_2 + 3x_3 + x_4 + s_1 &= 8 \\ 3x_1 + 2x_2 + 5x_3 + 12x_4 + s_2 &= 12; \ s_1, s_2, s_3 \geq 0 \\ 2x_1 + 5x_2 + 8x_3 + 2x_4 + s_3 &= 9 \end{aligned}$ **7.** $\begin{aligned} x_1 &= \dfrac{20}{3} - \dfrac{7}{3}x_2 - \dfrac{1}{3}s_2 \\ s_1 &= -\dfrac{5}{3} + \dfrac{1}{3}x_2 + \dfrac{1}{3}s_2 \end{aligned}$

9. (a) $\begin{aligned} 2x_1 + 5x_2 + s_1 &= 12 \\ 4x_1 + 9x_2 + s_2 &= 20 \end{aligned}; s_1, s_2 \geq 0$ **(b)** $\begin{aligned} x_1 &= -4 + \frac{9}{2}s_1 - \frac{5}{2}s_2 \\ x_2 &= 4 - 2s_1 + s_2 \end{aligned}$

11. (a) $\begin{aligned} x_1 + 2x_2 + x_3 + s_1 &= 8 \\ 2x_1 + 5x_2 + 5x_3 + s_2 &= 35 \end{aligned}; s_1, s_2 \geq 0$ **(b)** $\begin{aligned} x_1 &= -30 - 5s_1 + 2s_2 + 5x_3 \\ x_2 &= 19 + 2s_1 - s_2 - 3x_3 \end{aligned}$

13. (a) $\begin{aligned} x_1 + 2x_2 + x_3 + s_1 &= 8 \\ 2x_1 + 5x_2 + 5x_3 + s_2 &= 35 \end{aligned}; s_1, s_2 \geq 0$ **(b)** $\begin{aligned} x_2 &= 1 - \frac{3}{5}x_1 - s_1 + \frac{1}{5}s_2 \\ x_3 &= 6 + \frac{1}{5}x_1 + s_1 - \frac{2}{5}s_2 \end{aligned}$

15. (a) $\begin{aligned} x_1 + 3x_2 + s_1 &= 5 \\ 2x_1 + 7x_2 + s_2 &= 20 \\ 3x_1 + 8x_2 + s_3 &= 40 \end{aligned}; s_1, s_2, s_3 \geq 0$ **(b)** $\begin{aligned} x_1 &= 24 + \frac{8}{5}s_2 - \frac{7}{5}s_3 \\ x_2 &= -4 - \frac{3}{5}s_2 + \frac{2}{5}s_3 \\ s_1 &= -7 + \frac{1}{5}s_2 + \frac{1}{5}s_3 \end{aligned}$

17. (a) $\begin{aligned} 2x_1 + 4x_2 + 8x_3 + s_1 &= 12 \\ 2x_1 + 5x_2 + 12x_3 + s_2 &= 25 \\ 3x_1 + 6x_2 + 13x_3 + s_3 &= 60 \end{aligned}; s_1, s_2, s_3 \geq 0$ **(b)** $\begin{aligned} s_1 &= 12 - 2x_1 - 4x_2 - 8x_3 \\ s_2 &= 25 - 2x_1 - 5x_2 - 12x_3 \\ s_3 &= 60 - 3x_1 - 6x_2 - 13x_3 \end{aligned}$

19. (a) Same as 17(a) **(b)** $\begin{aligned} x_1 &= \frac{1}{2}[25 - 5x_2 - 12x_3 - s_2] \\ s_1 &= -13 + x_2 + 4x_3 + s_2 \\ s_3 &= \frac{45}{2} + \frac{3}{2}x_2 + 5x_3 + \frac{3}{2}s_2 \end{aligned}$ **21.** b

Section 4.4, page 200

1.

②	−1	②	1	0		1
−2	0	3	0	1		2
1	1	1	0	0		f

3.

1	2	3	1	0	0	5
2	3	1	0	1	0	3
③	1	②	0	0	1	1
2	−1	3	0	0	0	f

5.

2	3	1	0		7
⑤	8	0	1		4
1	−1	0	0		f

7.

1	①	1	1	0	5
②	1	③	0	1	6
3	2	4	0	0	f

9.

1	①	1	1	0	0	5
1	−2	−2	0	1	0	6
②	−1	1	0	0	1	4
1	1	−3	0	0	0	f

11.

x_1	x_2	x_3	s_1	s_2	
$\frac{2}{3}$	1	$\frac{2}{3}$	$\frac{1}{3}$	0	$\frac{2}{3}$	x_2
−1	0	4	0	1	2	s_2
$-\frac{1}{3}$	0	$-\frac{4}{3}$	$-\frac{2}{3}$	0	$f - \frac{4}{3}$

13.

x_1	x_2	x_3	s_1	s_2	
1	1	1	1	0	1	x_3
−2	−1	0	−1	1	1	s_2
−1	−2	0	−3	0	$f - 3$

15. $\left(\frac{4}{5} \; 0\right); f = \frac{4}{5}$ **17.** $\left(\frac{3}{5} \; \frac{8}{5}\right); f = \frac{52}{5}$

19. $(0 \; 5 \; 0); f = 5$; or $(3 \; 2 \; 0)$ or . . . . Actually, there are an infinite number of points in the constraint set that yield $f = 5$.

21. $(2 \; 1); f = 18$ **23.** $\left(\frac{5}{4} \; \frac{25}{8} \; 0\right); f = \frac{15}{2}$ **25.** $\left(\frac{13}{4} \; 0 \; \frac{1}{2}\right); f = \frac{15}{4}$ **27.** A = 137.5; B = 25; X = 0; Profit: $8500

29. 0 gallons chocolate; 300 gallons vanilla; 75 gallons banana; Profit: $341.25

31. 2.5 oz of syrup; 1.5 oz of cream; 4 oz of soda water; 4 oz of ice cream; Calories: 422

33. 4 rings; 10 pr. of earrings; 0 pins; 3 necklaces; Profit: $1600; Note: Jeweler can make 2 pins instead of a ring, with the same profit, so there are more solutions.

35. (a) 160 of category I; 0 of category II; 0 of category III; Revenue: $16,000   (b) 160 of category I; 0 of category II; 0 of category III; Number: 160

37. 1 of species I; 5 of species II; Energy: 13 units

Section 4.5, page 213

1. $\begin{pmatrix} -1 & 6 \\ 4 & 5 \end{pmatrix}$ **3.** $\begin{pmatrix} 2 & -1 & 1 \\ 3 & 2 & 4 \end{pmatrix}$ **5.** $\begin{pmatrix} 1 & -1 & 1 \\ 2 & 0 & 5 \\ 3 & 4 & 5 \end{pmatrix}$ **7.** $\begin{pmatrix} 1 & 0 \\ 0 & 1 \\ 1 & 0 \\ 0 & 1 \end{pmatrix}$ **9.** $\begin{pmatrix} a & d & g \\ b & e & h \\ c & f & j \end{pmatrix}$

11. Minimize $g = 5y_1 + 7y_2 + y_3$
Subject to: $y_1 + 3y_2 + y_3 \geq 2$; $y_1 \geq 0$
$2y_1 + 2y_2 + y_3 \geq 5$ $y_2 \geq 0$
$y_3 \geq 0$

13. Maximize $f = x_1 + x_2$
Subject to: $2x_1 + x_2 \leq 2$; $x_1 \geq 0$
$x_1 + 2x_2 \leq 3$ $x_2 \geq 0$

15. Minimize $g = 5y_1 + 6y_2$
Subject to: $y_1 + 2y_2 \geq 1$; $y_1 \geq 0$
$y_1 + y_2 \geq 1$ $y_2 \geq 0$
$y_1 + 3y_2 \geq 1$

17. Maximize $f = 13x_1 + 21x_2 + 11x_3$
Subject to: $x_1 + 4x_2 - 3x_3 \leq 2$; $x_1 \geq 0$
$2x_1 + x_2 - x_3 \leq 5$ $x_2 \geq 0$
$x_1 + 2x_2 + 4x_3 \leq 3$ $x_3 \geq 0$

19. Minimize $g = 12y$
Subject to: $y \geq 1$
$2y \geq 2$
$3y \geq -1$
$4y \geq 5$
$y \geq 0$

21. $\left(\frac{1}{3} \quad \frac{1}{3}\right)$; $g = \frac{5}{3}$ **23.** $\left(\frac{41}{7} \quad 0 \quad \frac{50}{7}\right)$; $g = \frac{232}{7}$

25. $\frac{2}{3}$ lb of food I; $\frac{7}{3}$ lb of food II; Cost $\$\frac{8}{3}$; Cost per pound $= \$\frac{8}{9} \approx 89¢$ **27.** (20 40 40); Cost $= \$12$

Review Exercises for Chapter 4, page 215

1.

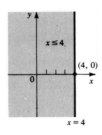

3.

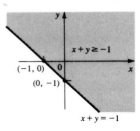

5.

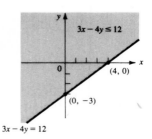

7.

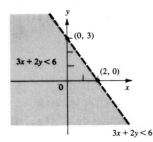

9.

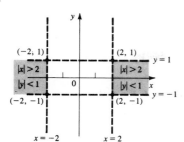

11.

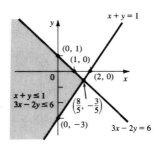

13. $\left(\frac{4}{5}, \frac{12}{5}\right); f = \frac{68}{5}$ **15.** $\left(\frac{16}{9}, \frac{10}{9}\right); g = \frac{68}{9}$ **17.** Unbounded (No maximum) **19.** Minimize $g = 5y_1 + 6y_2$
Subject to: $-y_1 + 2y_2 \geq 2; \ y_1 \geq 0$
$y_1 - 3y_2 \geq 3 \ y_2 \geq 0$ **21.** $\begin{pmatrix} 2 & 1 \\ 3 & 7 \end{pmatrix}$

23. $\begin{pmatrix} 1 & 4 & 3 \\ 3 & 1 & 0 \\ 2 & -6 & 5 \end{pmatrix}$ **25.**

x_1	x_2	x_3	s_1	s_2	s_3		
1	2	1	1	0	0	13	s_1
3	4	-2	0	1	0	6	s_2
-4	6	3	0	0	1	11	s_3
2	1	4	0	0	0	f	

27. Minimize $g = 13y_1 + 6y_2 + 11y_3$
Subject to: $y_1 + 3y_2 - 4y_3 \geq 2; \ y_1 \geq 0$
$2y_1 + 4y_2 + 6y_3 \geq 1 \ y_2 \geq 0$
$y_1 - 2y_2 + 3y_3 \geq 4 \ y_3 \geq 0$

29. (#25) (4 0 9); $f = 44$ (#27—dual) $\left(\frac{22}{7} \ 0 \ \frac{2}{7}\right); g = 44$ **31.**

x_1	x_2	x_3	s_1	s_2	s_3		
3	-6	9	1	0	0	3	s_1
2	8	-10	0	1	0	-1	s_2
1	3	5	0	0	1	4	s_3
4	12	8	0	0	0	f	

33. $\left(\frac{1}{13} \ \frac{15}{26} \ \frac{34}{13}\right); g = \frac{263}{26}$ **35.** (4 1 0); $g = 12$ **37.** (2 0 1); $f = 3$ **39.** $\left(0 \ \frac{15}{8} \ \frac{3}{8}\right); g = 3$

41. $\frac{80}{3}$ cakes; $\frac{400}{3}$ batches of cookies; Total cost: \$98; Problem is to minimize $1.425x + 0.45y$ where $x = $ # of cakes and $y = $ # of batches of cookies.

Chapter 5

Section 5.2, page 224

1. $2 \cdot 2 = 4$ **3.** $6 \cdot 6 = 36$ **5.** $2 \cdot 6 = 12$ **7.** $3 \cdot 2 \cdot 4 = 24$ **9.** $7 \cdot 6 \cdot 5 = 210$ **11.** $2 \cdot 2 \cdot 2 = 8$
13. No; Three letters can be chosen in $26^3 = 17,576$ ways. Four letters are needed which yield $26^4 = 456,976$ possible classifications.
15. $4^3 = 64$ **17.** $6^3 5^2 = 5400$ **19.** $26 \cdot 9^4 = 170,586$ **21.** $8 \cdot 7 \cdot 6 = 336$ **23.** $47 \cdot 51 \cdot 54 \cdot 55 = 7,119,090$

Section 5.3, page 235

1. $7! = 5040$ **3.** $6! = 720$ **5.** 60 **7.** 720 **9.** 19,958,400 **11.** 40,320 **13.** 1 **15.** $n!$ **17.** $\frac{n!}{2}$ **19.** $P_{7,4} = 840$

21. $50! \approx 3.0414 \times 10^{64}$ **23.** $\frac{50!}{40!} \approx 3.7276 \times 10^{16}$ **25.** $\frac{5!}{3!} = 20$ **27.** 10 **29.** 10 **31.** 120 **33.** 126 **35.** 495 **37.** 13 **39.** 8

41. 1 **43.** 1 **45.** n **47.** $\frac{n(n-1)}{2}$ **49.** $\binom{10}{4} = 420$ **51.** $\binom{52}{13} = 635,013,559,680$ **53.** $\binom{9}{6} = 84$ **55.** $\binom{10}{3}\binom{8}{2} = 3360$

57. $\binom{13}{1}\binom{4}{3}\binom{12}{1}\binom{4}{2} = 3744$ **59.** $\binom{10}{4} = 210$ **61.** 6 **63.** $\binom{10}{1}\left[\binom{4}{1}\right]^5 = 10,240$ (Including straight flushes)

65. $\binom{13}{2}\binom{4}{2}\binom{4}{2}\binom{11}{1}\binom{4}{1} = 123,552$ **67.** 15 **69.** $\binom{8}{5} = 56$ **71.** $P_{20,2} = 380$ **73.** $4^{20} \approx 1.0995 \times 10^{12}$

75. $\binom{100}{37}\binom{63}{51} \approx 9.1261 \times 10^{39}$ **77.** $P_{9,2}P_{7,2} = 3024$ **79.** $\binom{15}{4}\binom{11}{3}\binom{8}{5} = 12,612,600$

81. $\binom{12}{8} + \binom{12}{9} + \binom{12}{10} + \binom{12}{11} + \binom{12}{12} = 794$ **83.** $5 \cdot 2 \cdot 4! = 240$

85. $9\left[\binom{3}{2} \cdot 8 \cdot 8 \cdot 7 \cdot 6 \cdot 5 + \binom{3}{1} \cdot 8 \cdot 7 \cdot \binom{4}{1} \cdot 7 \cdot 6 \cdot 5 + 8 \cdot 7 \cdot 6 \cdot \binom{4}{2} \cdot 6 \cdot 5\right] + \left[9 \cdot 8 \cdot 7 \cdot \binom{4}{2} \cdot 6 \cdot 5\right] = 2,268,000$ **87.** $9 \cdot 10^4 = 90,000$

89. (a) $4^5 = 1024$ (b) $4 \cdot 3 \cdot 2 \cdot 1 \cdot 4 \cdot 5 = 480$ **91.** (a) 32 (b) 8 (c) 4

Section 5.4, page 246

1. (a) $S = \{A \text{ clubs, 2 clubs, } \ldots, J \text{ spades, } Q \text{ spades, } K \text{ spades}\} = \{\text{Cards in a standard deck}\}$ (b) $\dfrac{1}{52}$

3. (a) $S = \{\text{All groups of 5 persons}\}$ (b) $\dfrac{1}{\binom{10}{5}} = \dfrac{1}{252}$

5. (a) $S = \{AB: A, B \text{ are different persons, in order, from the committee}\}$ (b) $\dfrac{1}{P_{6,2}} = \dfrac{1}{30}$

7. (a) $\dfrac{1}{12!} = \dfrac{1}{479,001,600} \approx 0.000000002$ (b) $1 - \dfrac{1}{12!} \approx 0.999999998$ (c) 0 **9.** (a) $\dfrac{\binom{20}{6}}{\binom{28}{6}} \approx 0.1029$ (b) $\dfrac{\binom{20}{3}\binom{8}{3}}{\binom{28}{6}} \approx 0.1695$

11. $\dfrac{\binom{13}{1}\binom{4}{3}\binom{12}{1}\binom{4}{2}}{\binom{52}{5}} = \dfrac{3744}{2,598,960} \approx 0.00144 \approx \dfrac{1}{694}$ **13.** $\dfrac{\binom{13}{1}\binom{4}{3}\binom{12}{2}\binom{4}{1}\binom{11}{1}\binom{4}{1}}{\binom{52}{5}} \approx 0.0423 \approx \dfrac{1}{24}$

15. $\dfrac{\binom{13}{1}\binom{4}{2}\binom{12}{3}\binom{4}{1}\binom{4}{1}\binom{4}{1}}{\binom{52}{5}} \approx 0.423 \approx \dfrac{1}{2.4}$ **17.** $\dfrac{1}{4^5} = \dfrac{1}{1024}$

19. $1 - \underset{P(\text{All match})}{\dfrac{186}{560}} - \underset{P(\text{None match})}{\dfrac{58}{560}} = 1 - \dfrac{244}{560} = \dfrac{316}{560} \approx 0.564$

21. (a) $\dfrac{2}{16} = \dfrac{1}{8}$ (b) $\dfrac{5}{16}$ **23.** (a) $S = \{\text{Any group of 4 people from among the 15}\}$ (b) $\dfrac{1}{\binom{15}{4}} = \dfrac{1}{1365}$ (c) $\dfrac{1}{1365}$

25. (a) $A \cup B = \{\text{Oldest 2 children are girls or youngest 3 children are girls}\}$; $A \cap B = \{G\ G\ G\ G\ G\}$; $A - B = \{\text{Oldest 2 are girls and at least one boy}\}$; $B - A = \{\text{Youngest 3 are girls and at least one boy}\}$
(b) $P(A) = \dfrac{1}{4}$; $P(B) = \dfrac{1}{8}$; $P(A \cap B) = \dfrac{1}{32}$; $P(A \cup B) = \dfrac{11}{32}$; $P(A - B) = \dfrac{7}{32}$; $P(B - A) = \dfrac{3}{32}$

Section 5.5, page 257

1. $P(2) = \dfrac{1}{36} = P(12)$; $P(3) = \dfrac{2}{36} = \dfrac{1}{18} = P(11)$; $P(4) = \dfrac{3}{36} = \dfrac{1}{12} = P(10)$; $P(5) = \dfrac{4}{36} = \dfrac{1}{9} = P(9)$; $P(6) = \dfrac{5}{36} = P(8)$; $P(7) = \dfrac{6}{36} = \dfrac{1}{6}$

3. $P(8) + P(9) + P(10) + P(11) + P(12) = \dfrac{15}{36} = \dfrac{5}{12}$ **5.** $P(4 \le x \le 9) = \dfrac{27}{36} = \dfrac{3}{4}$ **7.** $P(\le 9) = \dfrac{30}{36} = \dfrac{5}{6}$

9. (a) 0.67 (b) 0.39 (c) 0.15 (d) 0.48 (e) 0.72 **11.** (a) 0.16 (b) 0.39 (c) 0.12 (d) 0.45 (e) 0.72

13. No; 2 is in both events. **15.** No; Straight flushes are in both sets. **17.** No; Infestation can occur with below average growth.

19. 0.78 **21.** (a) 0.22 (b) 0.48 (c) 0.85 (d) 0.07 **23.** (a) $\dfrac{1}{\binom{12}{3}} = \dfrac{1}{220}$ (b) $\dfrac{\binom{8}{3}}{\binom{12}{3}} = \dfrac{56}{220} = \dfrac{14}{55}$ (c) $\dfrac{164}{220} = \dfrac{41}{55}$ (d) $\dfrac{216}{220} = \dfrac{54}{55}$

25. $P(4 \text{ girls}) = \frac{1}{16} = 0.0625$, so $P(\text{at least 1 boy}) = 1 - P(4 \text{ girls}) > 0.9$ **27.** $\frac{194,038}{1,434,640} \approx 0.1353$

31. In favor: 3 to 5; against: 5 to 3 **33.** In favor: 1 to 5; against: 5 to 1 **35.** $\frac{5}{17}$ **37.** 40%

Section 5.6, page 268

1. $P(A|B) = \frac{2}{3}$; $P(B|A) = \frac{1}{3}$ **3.** $P(A|B) = P(B|A) = 0$ **5.** $\frac{2}{11}$ **7.** $\frac{3}{7}$

9. (a) $A|B =$ three oldest are of the same sex given that three youngest are of the same sex; $B|A =$ three youngest are of the same

sex given that three oldest are (b) $P(A) = \frac{1}{4}$; $P(B) = \frac{1}{4}$; $P(A|B) = P(B|A) = \frac{\frac{1}{16}}{\frac{1}{4}} = \frac{1}{4}$

11. (a) $H \cap C =$ smokers with heart disease; $H - C =$ those with heart disease who don't smoke; $C - H =$ smokers without heart disease; $H^c \cap C^c =$ nonsmokers without heart disease (b) $P(H \cap C) = 0.12$; $P(H - C) = 0.08$; $P(C - H) = 0.38$; $P(H^c \cap C^c) = 0.42$

13. $\frac{3}{7}$ **15.** $\frac{13}{21}$ **17.** $6\frac{2}{3}\%$ **19.** $\dfrac{\binom{4}{3}\binom{12}{1}\binom{4}{2}}{\binom{4}{3}\binom{48}{2} + \binom{4}{4}\binom{48}{1}} = \frac{288}{4560} \approx 0.0632$ **21.** $\frac{1}{106}$ **23.** $\frac{3992}{40,000} \approx 0.0998$

25. 0.205 **27.** $\binom{10}{2} \cdot \frac{1}{365} \cdot \frac{364}{365} \cdot \frac{363}{365} \cdot \frac{362}{365} \cdot \frac{361}{365} \cdot \frac{360}{365} \cdot \frac{359}{365} \cdot \frac{358}{365} \cdot \frac{357}{365} \approx 0.11162$ **29.** $\frac{0.122}{0.34} \approx 0.3588$ **31.** 0.52

33. (a) $\frac{1}{9}$ (b) $\frac{2}{17}$ **35.** (a) 0.381 (b) $(1 - 0.3^3)\left(\frac{1}{8}\right) + (1 - 0.3^2)\left(\frac{3}{8}\right) + (1 - 0.3)\left(\frac{3}{8}\right) = 0.725375$

Section 5.7, page 273

1. No **3.** Yes **5.** Yes **7.** Yes **9.** No **11.** (a) 0.1 (b) 0.45 (c) 0.3 **13.** No **15.** No

19. $P(A \cap B \cap C) = 0 \ne P(A)P(B)P(C) = \frac{1}{8}$ **21.** (a) $P(A) = P(B) = \frac{1}{2}$; $P(A \cap B) = \frac{1}{6}$ (b) $P(A|B) = P(B|A) = \frac{1}{3}$ (c) No (d) No

23. No

Section 5.8, page 282

1. $\frac{0.027}{0.221} \approx 0.122$ **3.** $\frac{0.506}{0.9533} \approx 0.531$ **5.** $\frac{0.27}{0.55} \approx 0.491$ **7.** $\frac{0.225}{0.285} \approx 0.789$ **9.** $\frac{0.3}{0.8} \approx 0.375$

11. $P(\text{1st task completed}|\text{Received pellet}) = 0.4$; $P(\text{2nd task completed}|\text{Received pellet}) = 0.3$

13. $P(\text{Strain I}|\text{Infection}) = \frac{1}{4}$; $P(\text{Strain II}|\text{Infection}) = \frac{3}{16}$; $P(\text{Strain III}|\text{Infection}) = \frac{9}{16}$

15. $\dfrac{(0.3)(0.05)x}{(0.3)(0.05)x + 0.7x} = \frac{0.015}{0.715} \approx 0.021$; Here x is the probability that an unvaccinated person contracts rubella.

17. $\frac{0.5}{1.1} = \frac{5}{11}$ **19.** $\frac{0.45}{0.7} \approx 0.643$ **21.** (a) 0.7 (b) 0.226 **23.** $\frac{20}{719} \approx 0.028$

Section 5.9, page 288

1. $\frac{15}{64} \approx 0.234$ **3.** $\frac{120 \cdot 2^7}{3^{10}} \approx 0.260$ **5.** $(0.4)^{12} \approx 0.0000167$ **7.** $56(0.12)^5(0.88)^3 \approx 0.000946$ **9.** $161,700(0.01)^3(0.99)^{97} \approx 0.061$

11. (a) $B(8, 3, 0.8) \approx 0.0092$ (b) $B(8, 6, 0.8) + B(8, 7, 0.8) + B(8, 8, 0.8) \approx 0.797$ (c) $1 - B(8, 8, 0.8) \approx 0.832$

13. (a) $B(7, 2, 0.15) \approx 0.210$ (b) $1 - B(7, 0, 0.15) \approx 0.679$ (c) $\frac{0.210}{0.679} \approx 0.309$

15. 230 $(B(230, 0, 0.01) \approx 0.0991$; $B(229, 0, 0.01) \approx 0.100106)$ **17.** $\frac{(0.8)^{15}}{(0.8)^{10}} \approx 0.328$ **19.** $\frac{1}{32} = 0.03125$

21. (a) $(0.8)(0.04) + (0.85)(0.96) = 0.848$ (b) $B(5, 3, 0.2) = 0.0512$ (c) $\dfrac{(0.0512)(0.04)}{(0.0512)(0.04) + (0.138178)(0.96)} \approx 0.0152$

23. (a) $(0.999)^{500} \approx 0.606$ (b) $1 - [(0.999)^{500} + 500(0.999)^{499}(0.001)] \approx 0.0901$

25. (a) 7 (P(At least one of 7 children is a surviving male) ≈ 0.962747) (b) 11 **27.** $\dbinom{20}{4}(0.15)^4(0.85)^{16}(0.55)^4 \approx 0.0167$

Review Exercises for Chapter 5, page 290

1. 16 **3.** 240 **5.** 504 **7.** 2520 **9.** (a) 126 (b) 21 (c) 9 (d) 9 (e) 210 **11.** $12! = 479{,}001{,}600$ **13.** $\dfrac{7!}{3!2!} = 420$

15. 5148 (Including straight flushes) **17.** $\dfrac{\dbinom{13}{5}\dbinom{39}{0}}{\dbinom{13}{2}\dbinom{39}{3} + \dbinom{13}{3}\dbinom{39}{2} + \dbinom{13}{4}\dbinom{39}{1} + \dbinom{13}{5}\dbinom{39}{0}} \approx 0.001349$ **19.** $\dfrac{1}{12}$

21. $\dbinom{50}{18}\dbinom{32}{13} \approx 6.27 \times 10^{21}$

23. (a) $\dfrac{1}{1024}$ (b) $\dfrac{120}{1024} \approx 0.117$ (c) $1 - \dfrac{1}{1024} = \dfrac{1023}{1024} \approx 0.99902$ (d) $1 - \dfrac{10}{1024} - \dfrac{1}{1024} \approx 0.98926$

25. (a) $4^4 = 256$ (b) $4! = 24$ **27.** (a) $\dfrac{1}{8}$ (b) $\dfrac{5}{16}$ (c) $\dfrac{1}{3}$ **29.** (a) $\dfrac{1}{\dbinom{11}{5}} = \dfrac{1}{462} \approx 0.00216$ (b) $\dfrac{7}{\dbinom{11}{5}} \approx 0.0152$

31. (a) $\dfrac{80{,}000 + 336{,}000 + 16{,}000}{972{,}000} = \dfrac{432{,}000}{972{,}000} = \dfrac{4}{9}$ (b) $\dfrac{16{,}000}{432{,}000} = \dfrac{1}{27}$ **33.** (a) $\dfrac{5}{36}$ (b) $\dfrac{2}{11}$ (c) $\dfrac{2}{5}$

35. (a) 0.02 (b) 0.46 (c) $\dfrac{0.02}{0.48} \approx 0.417$ **37.** 0.65 **39.** $\dfrac{1}{P_{18,5}} = \dfrac{1}{1{,}028{,}160} \approx 0.000000973$

41. No; $P(A \cap B) = 0 \neq 0.06 = P(A)P(B)$ **43.** $A = \{\text{Even}\}$; $B = \{\text{Odd}\}$ or $A = \{x\}$; $B = \{y\}$ where $x \neq y$ are integers between 1 and 6

45. $1 - B(10, 0, 0.03) = 1 - (0.97)^{10} \approx 0.263$ **47.** (a) 0.4325 (b) $\dfrac{0.18}{0.4325} \approx 0.416$ (c) $\dfrac{0.0875}{0.5675} \approx 0.154$

Chapter 6

Section 6.1, page 300

1. (a) $\{0, 1, 2, 3\}$ (b) $p(0) = \dfrac{1}{8}$; $p(1) = \dfrac{3}{8}$; $p(2) = \dfrac{3}{8}$; $p(3) = \dfrac{1}{8}$ (c) $F(x) = \begin{cases} 0, & x < 0 \\ \dfrac{1}{8}, & 0 \leq x < 1 \\ \dfrac{1}{2}, & 1 \leq x < 2 \\ \dfrac{7}{8}, & 2 \leq x < 3 \\ 1, & x \geq 3 \end{cases}$ (d) 1 (e) 0 (f) $\dfrac{3}{8}$

(g)

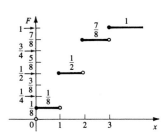

3. (a) {1, 2, 3, 4, 5, 6, 7, 8, 9, 10} **(b)** {0, 1, 2} **(c)** $p(0) = \frac{2}{5}; p(1) = \frac{2}{5}; p(2) = \frac{1}{5}; F(x) = \begin{cases} 0, & x < 0 \\ \frac{2}{5}, & 0 \le x < 1 \\ \frac{4}{5}, & 1 \le x < 2 \\ 1, & x \ge 2 \end{cases}$ **(d)** 1 **(e)** 1

(f)

5. (a) {0, 1, 2, 3, 4, 5} **(b)** $p(0) = \left(\frac{5}{6}\right)^5 \approx 0.4019; p(1) = 5\left(\frac{1}{6}\right)\left(\frac{5}{6}\right)^4 \approx 0.4019; p(2) = 10\left(\frac{1}{6}\right)^2\left(\frac{5}{6}\right)^3 \approx 0.1608;$

$p(3) = 10\left(\frac{1}{6}\right)^3\left(\frac{5}{6}\right)^2 \approx 0.0322; p(4) = 5\left(\frac{1}{6}\right)^4\left(\frac{5}{6}\right) \approx 0.0032; p(5) = \left(\frac{1}{6}\right)^5 \approx 0.0001; F(x) = \begin{cases} 0, & x < 0 \\ 0.4019, & 0 \le x < 1 \\ 0.8038, & 1 \le x < 2 \\ 0.9645, & 2 \le x < 3 \\ 0.9967, & 3 \le x < 4 \\ 0.9999, & 4 \le x < 5 \\ 1, & x \ge 5 \end{cases}$ **(c)** 0.9967 **(d)** 1

(e)

7. Range = {0, 1, 2, 3, 4, 5}; $p(0) \approx 0.237$; $p(1) \approx 0.396$; $p(2) \approx 0.264$; $p(3) \approx 0.088$; $p(4) \approx 0.015$; $p(5) \approx 0.001$;

$F(x) = \begin{cases} 0, & x < 0 \\ 0.237, & 0 \le x < 1 \\ 0.633, & 1 \le x < 2 \\ 0.897, & 2 \le x < 3 \\ 0.985, & 3 \le x < 4 \\ 0.999, & 4 \le x < 5 \\ 1, & x \ge 5 \end{cases}$

9. (a) This is a binomial random variable with $n = 20$ and $p = 0.1$. Thus $p(k) = \binom{20}{k}(0.1)^k(0.9)^{20-k}$ for $k = 0, 1, 2, \ldots, 20$.
(b) $p(0) + p(1) + p(2) + p(3) + p(4) \approx 0.9568$ **(c)** $1 - p(0) - p(1) - p(2) - p(3) \approx 0.1330$
11. (a) Range of X = {1, 2, 3, ...}; Range of Y = {0, 1, 2, ...} **(b)** $P(X = 2) = (1-p)^2 p + (1-p)^3 p; P(Y = 1) = (1-p)p + (1-p)^2 p;$
$P(Y > 2) = 1 - [p + (1-p)p + (1-p)^2 p + (1-p)^3 p + (1-p)^4 p]$
13. (a) $0.6^4 = 0.1296$ **(b)** Range of X = {4, 5, 6, ...}; $p(k) = \binom{k-1}{3}(0.6)^4(0.4)^{k-4}$ **(c)** $p(7) \approx 0.1659$;
$P(X > 5) = 1 - p(4) - p(5) \approx 0.663$
15. (a) {0, 1, 2, ..., 10} **(b)** $p(k) = \binom{10}{k}\left(\frac{1}{2}\right)^{10}$ if $k = 0, 1, 2, \ldots, 10$ **(c)** $B(10, 6, 0.6) \approx 0.251$

17. Range $= \{0, 1, 2, 3\}$; $p(0) = \dfrac{1}{11}$; $p(1) = \dfrac{9}{22}$; $p(2) = \dfrac{9}{22}$; $p(3) = \dfrac{1}{11}$; $F(x) = \begin{cases} 0, & x < 0 \\ \dfrac{1}{11}, & 0 \le x < 1 \\ \dfrac{1}{2}, & 1 \le x < 2 \\ \dfrac{10}{11}, & 2 \le x < 3 \\ 1, & 3 \le x \end{cases}$

19. Range $= \{-3, -1, 1, 3\}$; $p(-3) = \dfrac{1}{27}$; $p(-1) = \dfrac{6}{27} = \dfrac{2}{9}$; $p(1) = \dfrac{12}{27} = \dfrac{4}{9}$; $p(3) = \dfrac{8}{27}$

21. Range $= \{0, 1, 2, 3, 4\}$; This is a binomial random variable with $n = 4$ and $p = 0.4$.

23. (a) $\{1, 2, 3 \ldots\}$ (b) $p(1) = 0.3$; $p(3) = 0.7^2(0.3) = 0.147$; $p(6) = 0.7^5(0.3) \approx 0.0504$; $p(n) = (0.7)^{n-1}(0.3)$
(c) $F(1) = 0.3$; $F(3) = 0.3 + 0.21 + 0.147 = 0.657$; $F(6) \approx 0.8824$ (d) $F(x) = \begin{cases} 0, & x < 1 \\ 1 - 0.7^n, & n \le x < n + 1 \end{cases}$

Section 6.2, page 309

1. $\dfrac{3}{2}$ **3.** 0.8 **5.** $\dfrac{5}{6}$ **7.** 1.25 **9.** 2 **11.** $E(X) = \dfrac{1}{3}\left[1 + 2\left(\dfrac{2}{3}\right) + 3\left(\dfrac{2}{3}\right)^2 + \cdots + 25\left(\dfrac{2}{3}\right)^{24}\right] + 25\left(\dfrac{2}{3}\right)^{25} \approx 2.999881194$ **13.** 0 **15.** $-\dfrac{1}{4}$

17. 4 **19.** $E(X) = p[1 + (1 - p) + 2(1 - p)^2 + 2(1 - p)^3 + 3(1 - p)^4 + 3(1 - p)^5 + 4(1 - p)^6 + 4(1 - p)^7 + \cdots] = \dfrac{1}{p(2 - p)}$;

$E(Y) = p[(1 - p) + (1 - p)^2 + 2(1 - p)^3 + 2(1 - p)^4 + 3(1 - p)^5 + 3(1 - p)^6 + 4(1 - p)^7 + 4(1 - p)^8 + \cdots] = \dfrac{1 - p}{p(2 - p)}$;

Note: The sums were obtained, for example, by computing $(1 - p)E(x)$, subtracting it from $E(X)$, and using the fact that
$1 + a + a^2 + a^3 + \cdots = \dfrac{1}{1 - a}$ if $|a| < 1$. **21.** 250,000 **23.** 2.8 **25.** \$2250

27. $E(k) = 0.2k + 30$ if $100 \le k \le 150$ and $E(k) = 97.5 - 0.25k$ if $150 \le k \le 200$ programs are ordered. Maximum profit is \$60 when $k = 150$.

29. 100 gallons (Similar reasoning as in Problem 27) **31.** 2 **33.** $\dfrac{5}{2}$ **35.** 4.5 **37.** a **39.** a

Section 6.3, page 323

1. (a)

Interval	Count	Frequency	Cumulative Frequency
209.5–219.5	1	1	1
219.5–229.5	0	0	1
229.5–239.5	0	0	1
239.5–249.5	‖	2	3
249.5–259.5	‖	2	5
259.5–269.5	‖	2	7
269.5–279.5	‖	2	9
279.5–289.5	ＴＨＬ	5	14
289.5–299.5	‖	2	16
299.5–309.5	ＴＨＬ ‖	7	23
309.5–319.5	ＴＨＬ ＴＨＬ	10	33
319.5–329.5	‖‖	4	37
329.5–339.5	ＴＨＬ ‖‖	8	45
339.5–349.5	ＴＨＬ ‖	7	52
349.5–359.5	ＴＨＬ ‖‖‖	9	61
359.5–369.5	‖	2	63
369.5–379.5	ＴＨＬ ‖	6	69
379.5–389.5	‖	2	71
389.5–399.5	‖	1	72
399.5–409.5	‖‖	3	75
409.5–419.5	‖	2	77
419.5–429.5	‖	1	78
429.5–439.5	‖	1	79
439.5–449.5	‖	1	80

(b)

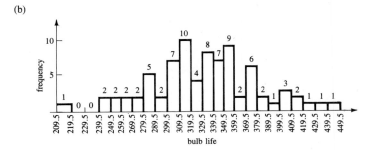

(c)

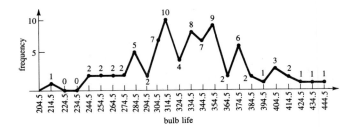

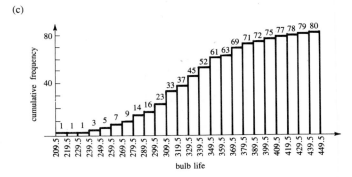

3. (a)

Outcome	Count	Frequency
1	☰ l	6
2	☰ ☰ lll	13
3	☰ ll	7
4	☰	5
5	☰ ☰ llll	14
6	☰	5

(b)

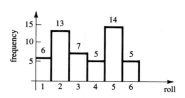

(c)

(d) Probably not; too many 2s and 5s   (e) $\frac{26}{50} = 0.52$ (f) $\frac{31}{50} = 0.62$

5. (a)

Height	Count	Frequency	Cumulative Frequency
90.5–100.5	||	2	2
100.5–110.5	⊪ ||	7	9
110.5–120.5	⊪ ⊪ |	11	20
120.5–130.5	⊪ ⊪ ⊪ |||	18	38
130.5–140.5	⊪ ⊪ ||||	14	52
140.5–150.5	⊪ ||	7	59
150.5–160.5	|	1	60

(b)

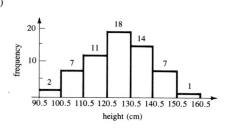

(c)

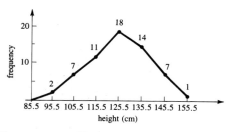

(d) $\frac{43}{60} \approx 0.7167$ (e) $\frac{38}{60} \approx 0.6333$

7. (a)

Interval	Frequency	Cumulative Frequency
0.5–10.5	1	1
10.5–20.5	1	2
20.5–30.5	3	5
30.5–40.5	4	9
40.5–50.5	10	19
50.5–60.5	13	32
60.5–70.5	15	47
70.5–80.5	12	59
80.5–90.5	6	65
90.5–100.5	5	70

(b)

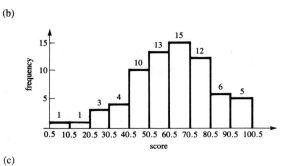 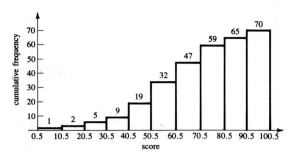

(c)

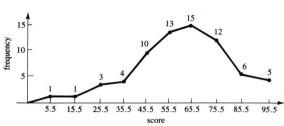

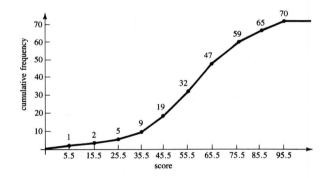

(d) $\dfrac{27}{70} \approx 0.3857$ (e) $\dfrac{47}{70} \approx 0.6714$

9. (a)

Interval	Count	Frequency	Cumulative Frequency
0.5–5.5		0	0
5.5–10.5	I	1	1
10.5–15.5		0	1
15.5–20.5	I	1	2
20.5–25.5	I	1	3
25.5–30.5	II	2	5
30.5–35.5		0	5
35.5–40.5	IIII	4	9
40.5–45.5	IIII	4	13
45.5–50.5	⊬HL I	6	19
50.5–55.5	⊬HL I	6	25
55.5–60.5	⊬HL II	7	32
60.5–65.5	⊬HL II	7	39
65.5–70.5	⊬HL III	8	47
70.5–75.5	⊬HL I	6	53
75.5–80.5	⊬HL I	6	59
80.5–85.5	IIII	4	63
85.5–90.5	II	2	65
90.5–95.5	III	3	68
95.5–100.5	II	2	70

(b)

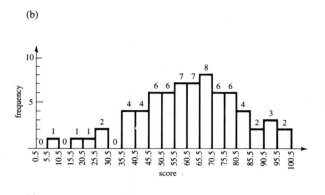

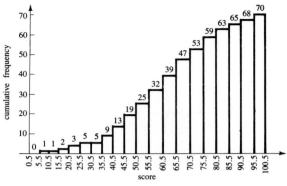

(c)

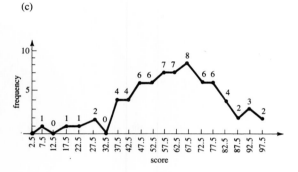

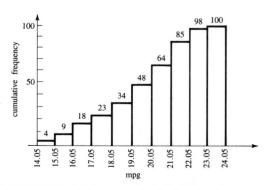

11. (a)

Interval	Frequency	Cumulative Frequency
14.05–15.05	4	4
15.05–16.05	5	9
16.05–17.05	9	18
17.05–18.05	5	23
18.05–19.05	11	34
19.05–20.05	14	48
20.05–21.05	16	64
21.05–22.05	21	85
22.05–23.05	13	98
23.05–24.05	2	100

(b)

(c)

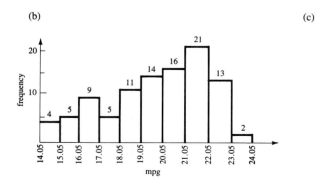

13. The histogram of Problem 11 gives the best indication. Too fine a division (#12) leads to too much fluctuation. We see the trees but not the forest. Too coarse a division (#10) fails to give enough information.

Section 6.4, page 330

1. $\bar{x} = 6.86$; Median $= 7$ **3.** $\bar{x} = 1.57$; Median $= 1.6$ **5.** $\bar{x} = 113.11$; Median $= 4$

7. $\bar{x} = 1.655$; Median $= 1.635$ **9.** $\bar{x} = 50$; Median $= 50$ **11.** $\bar{x} = 55.56$; Median $= 100$

13. $\bar{x} = 124.6$; Median $= 124$ **15.** $\bar{x} = \dfrac{7495}{60} \approx 124.92$ **17.** $\bar{x} = \dfrac{4295}{70} \approx 61.36$ **19.**

(6.3.10) $\bar{x} = 19.75$;
(6.3.11) $\bar{x} = 19.72$;
(6.3.12) $\bar{x} = 19.715$

21. $\bar{x} = \$1932.97$ millions; Median $= \$945.7$ millions **23.** $\bar{x} = 5.19$; Median $= 4.9$

25. $\bar{x} = \$181.39$; Median $= \$179.72$ **27.** $\bar{x} = 1{,}125{,}935.69$; Median $= 1{,}215{,}680$

29. $\bar{x} = 0.6445$; Median $= 0.636$ **31.** 778 **33.** 100 **35.** 92

Section 6.5, page 341

1. $s^2 = 7.81$; $s = 2.79$ **3.** $s^2 = 0.756$; $s = 0.869$ **5.** $s^2 = 106{,}905.61$; $s = 326.96$ **7.** $s^2 = 0.03226$; $s = 0.1796$

9. $s^2 = 2857.14$; $s = 53.45$ **13.** $s^2 = 183.05$; $s = 13.53$ **15.** $s^2 = 180.58$; $s = 13.44$ **17.** $s^2 = 386.94$; $s = 19.67$

19. $s^2 = 6{,}398{,}911.7$; $s = 2529.6$ **21.** $s^2 = 2.544$; $s = 1.59$ **23.** $s^2 = 883.89$; $s = 29.73$

25. $s^2 = 5134831.7 \times 10^4 \approx 5.13 \times 10^{10}$; $s = 226{,}601.67$ **27.** $s^2 = 0.00336$; $s = 0.058$ **29.** $\sigma^2 = \dfrac{2}{3}$; $\sigma = \sqrt{\dfrac{2}{3}} \approx 0.816$

31. $\sigma^2 = \dfrac{1}{2}$; $\sigma = \sqrt{\dfrac{1}{2}} \approx 0.707$ **33.** $\sigma^2 = 0.8875$; $\sigma = 0.942$ **35.** $\sigma^2 = 1.28$; $\sigma = 1.13$ **37.** $\sigma^2 = 2$; $\sigma = \sqrt{2}$

39. $\sigma^2 = \dfrac{8}{3}$; $\sigma = \sqrt{\dfrac{8}{3}} \approx 1.633$ **41.** $\sigma^2 = 2.16$; $\sigma = 1.47$

Section 6.6, page 359

1. 0.3849 **3.** 0.4484 **5.** 0.0526 **7.** 0.8849 **9.** 0.1151 **11.** 0.9099 **13.** 0.0862 **15.** 0.7071 **17.** 0.0062 **19.** 0.7066

21. 0.3085 **23.** 0.0228 **25.** 0.3085 **27.** 0.3721 **29.** 0.8944

31. (a) $Z = \dfrac{X - 5}{2}$ (b) $P(3 \leq X) = 0.8413$; $P(X \leq 6) = 0.6915$; $P(2 \leq X \leq 8) = 0.8664$ (c) $a = 8.3$; $b = 9.66$

33. 0.3085 **35.** (a) 0.2743 (b) 0.3446 (c) 0.3674 (d) 0.6149 **37.** (a) 0.9772 (b) 0.9987 **39.** 0.9544

41. (a) $\dfrac{0.1587}{0.5} = 0.3174$ (b) 0.3174 **43.** 0.1587 **45.** (a) $\dfrac{0.0228}{0.5} = 0.0456$ (b) $\dfrac{0.0228}{0.1587} \approx 0.1437$

47. (a) 0.6915 (b) $\dfrac{0.5}{0.6915} \approx 0.7231$ **49.** (a) 0.4649 (b) 0.6915 **51.** (76.4, 119.6) **53.** 25.96 in.

57. 95%: (26.08, 33.92); 99%: (24.84, 35.16)

Review Exercises for Chapter 6, page 362

1. (a) $\{0, 1, 2, 3, 4\}$ (b) $p(0) = \dfrac{1}{16}$; $p(1) = \dfrac{1}{4}$; $p(2) = \dfrac{3}{8}$; $p(3) = \dfrac{1}{4}$; $p(4) = \dfrac{1}{16}$; $F(x) = \begin{cases} 0, & x < 0 \\[4pt] \dfrac{1}{16}, & 0 \leq x < 1 \\[4pt] \dfrac{5}{16}, & 1 \leq x < 2 \\[4pt] \dfrac{11}{16}, & 2 \leq x < 3 \\[4pt] \dfrac{15}{16}, & 3 \leq x < 4 \\[4pt] 1, & x \geq 4 \end{cases}$ (c) 1 (d) $\dfrac{11}{16}$

(e)

(f) 2 (g) $\sigma^2 = 1$, $\sigma = 1$

3. (a) {0, 1, 2, 3, 4} (b) Binomial with $n = 4$ and $p = \frac{1}{12}$ (c) $\frac{1}{3}$ (d) $\sigma^2 = \frac{11}{36} \approx 0.306$; $\sigma \approx 0.553$ **5.** 2.34

7. (a)

Roll on die	Count	Frequency
2	||	2
3	||	2
4	⊦⊦⊦	4
5	⊦⊦⊦ |	6
6	⊦⊦⊦ |	6
7	⊦⊦⊦ ||||	9
8	||||	5
9	|||	3
10	|||	3
11	||	2
12	|||	3

(b)

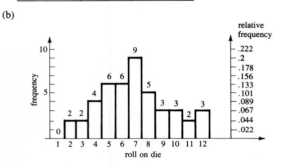

(c)

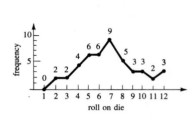

(d) $\bar{x} \approx 6.89$ (e) $s^2 \approx 6.78$; $s \approx 2.60$

9. $\bar{x} = 6.286$; $s^2 = 5.238$; $s = 2.287$; Median = 7   **11.** $\bar{x} = 32.75$; $s^2 = 135.929$; $s = 11.659$; Median = 31

13. $\bar{x} = 0.29$; $s^2 = 0.00929$; $s = 0.0964$; Median = 0.306   **15.** 0.4484   **17.** 0.2743   **19.** 0.5221   **21.** 0.2266

23. 0.8301   **25.** (a) 0.3446 (b) 0.1151 (c) 0.5138   **27.** (a) 0.0060 (b) 0.0222 (c) 0.6826

Chapter 7

Section 7.1, page 379

1. Yes   **3.** No   **5.** No   **7.** Yes   **9.** Yes   **11.** Yes   **13.** Yes   **15.** No   **17.** Yes   **19.** Yes

21. $\mathbf{p}_1 = \left(\frac{5}{8} \quad \frac{3}{8}\right)$; $\mathbf{p}_2 = \left(\frac{19}{32} \quad \frac{13}{32}\right)$; $\mathbf{p}_3 = \left(\frac{77}{128} \quad \frac{51}{128}\right)$ **23.** $\mathbf{p}_1 = \left(\frac{2}{3} \quad \frac{1}{3}\right)$; $\mathbf{p}_2 = \left(\frac{22}{72} \quad \frac{50}{72}\right)$; $\mathbf{p}_3 = \left(\frac{433}{864} \quad \frac{431}{864}\right)$

25. $\mathbf{p}_1 = \left(\frac{11}{48} \quad \frac{13}{24} \quad \frac{11}{48}\right)$; $\mathbf{p}_2 = \left(\frac{25}{72} \quad \frac{11}{36} \quad \frac{25}{72}\right)$; $\mathbf{p}_3 = \left(\frac{29}{108} \quad \frac{25}{54} \quad \frac{29}{108}\right)$

27. $\mathbf{p}_1 = (0.2214 \quad 0.4 \quad 0.3786)$; $\mathbf{p}_2 = (0.3034 \quad 0.4418 \quad 0.2548)$; $\mathbf{p}_3 = (0.3098 \quad 0.4994 \quad 0.1908)$

29. $\mathbf{p}_1 = \mathbf{p}_2 = \mathbf{p}_3 = \left(\frac{1}{3} \quad \frac{1}{3} \quad \frac{1}{3}\right)$ **31.** Regular; $\left(\frac{2}{5} \quad \frac{3}{5}\right)$ **33.** Not regular   **35.** Regular; $\left(\frac{b}{1-a+b} \quad \frac{1-a}{1-a+b}\right)$ **37.** Not regular

39. Regular; $(0.2843 \quad 0.3768 \quad 0.3390)$ **41.** (a) $(0 \quad 0 \quad 1)$ (b) Yes   **43.** $T = \begin{pmatrix} 0.98 & 0.02 \\ 0.3 & 0.7 \end{pmatrix}$ **45.** 93.75%

47. 50 correct, 50 incorrect = 50%   **49.** $\begin{pmatrix} 0.6 & 0.1 & 0.1 & 0.1 & 0.1 \\ 0.1 & 0.6 & 0.1 & 0.1 & 0.1 \\ 0.1 & 0.1 & 0.6 & 0.1 & 0.1 \\ 0.1 & 0.1 & 0.1 & 0.6 & 0.1 \\ 0.1 & 0.1 & 0.1 & 0.1 & 0.6 \end{pmatrix}$ **51.** $T = \begin{pmatrix} 0.6 & 0.3 & 0.1 \\ 0.2 & 0.5 & 0.3 \\ 0.1 & 0.2 & 0.7 \end{pmatrix}$ **53.** $\begin{pmatrix} 0 & \frac{7}{10} & 0 & \frac{3}{10} \\ \frac{1}{3} & 0 & \frac{1}{3} & \frac{1}{3} \\ 0 & 1 & 0 & 0 \\ \frac{1}{3} & \frac{1}{3} & \frac{1}{3} & 0 \end{pmatrix}$

55. $679.15 **57.** Democrat: 54.64%;
Republican: 35.05%;
Independent: 10.31%

Section 7.2, page 390

1. Yes; 1 **3.** Yes; 2 **5.** No **7.** Yes; 2 **9.** Yes; 2

11. $T = \begin{pmatrix} \frac{1}{3} & \frac{2}{3} \\ 0 & 1 \end{pmatrix}$; $T' = \begin{pmatrix} \frac{1}{3} & \frac{2}{3} \end{pmatrix}$; $R = \begin{pmatrix} \frac{1}{3} \end{pmatrix}$; $S = \begin{pmatrix} \frac{2}{3} \end{pmatrix}$; $Q = \begin{pmatrix} \frac{3}{2} \end{pmatrix}$; $A = (1)$

13. $T' = \begin{pmatrix} \frac{1}{3} & \frac{1}{3} & \frac{1}{3} \\ \frac{1}{2} & 0 & \frac{1}{2} \end{pmatrix}$; $R = \begin{pmatrix} \frac{1}{3} & \frac{1}{3} \\ \frac{1}{2} & 0 \end{pmatrix}$; $S = \begin{pmatrix} \frac{1}{3} \\ \frac{1}{2} \end{pmatrix}$; $Q = \begin{pmatrix} 2 & \frac{2}{3} \\ 1 & \frac{4}{3} \end{pmatrix}$; $A = \begin{pmatrix} 1 \\ 1 \end{pmatrix}$

15. $T' = (0.4 \quad 0.4 \quad 0.2)$; $R = (0.4)$; $S = (0.4 \quad 0.2)$; $Q = \begin{pmatrix} \frac{5}{3} \end{pmatrix}$; $A = \begin{pmatrix} \frac{2}{3} & \frac{1}{3} \end{pmatrix}$

17. $T' = \begin{pmatrix} \frac{1}{3} & \frac{1}{3} & \frac{1}{6} & \frac{1}{6} \\ \frac{1}{2} & 0 & 0 & \frac{1}{2} \end{pmatrix}$; $R = \begin{pmatrix} \frac{1}{3} & \frac{1}{6} \\ 0 & 0 \end{pmatrix}$; $S = \begin{pmatrix} \frac{1}{3} & \frac{1}{6} \\ \frac{1}{2} & \frac{1}{2} \end{pmatrix}$; $Q = \begin{pmatrix} \frac{3}{2} & \frac{1}{4} \\ 0 & 1 \end{pmatrix}$; $A = \begin{pmatrix} \frac{5}{8} & \frac{3}{8} \\ \frac{1}{2} & \frac{1}{2} \end{pmatrix}$

19. $T' = \begin{pmatrix} \frac{1}{8} & \frac{1}{4} & \frac{1}{8} & \frac{1}{8} & \frac{3}{8} \\ \frac{1}{7} & \frac{2}{7} & \frac{1}{7} & \frac{2}{7} & \frac{1}{7} \\ \frac{1}{4} & \frac{1}{2} & 0 & \frac{1}{8} & \frac{1}{8} \end{pmatrix}$; $S = \begin{pmatrix} \frac{1}{8} & \frac{3}{8} \\ \frac{2}{7} & \frac{1}{7} \\ \frac{1}{8} & \frac{1}{8} \end{pmatrix}$; $R = \begin{pmatrix} \frac{1}{8} & \frac{1}{4} & \frac{1}{8} \\ \frac{1}{7} & \frac{2}{7} & \frac{1}{7} \\ \frac{1}{4} & \frac{1}{2} & 0 \end{pmatrix}$; $Q = \begin{pmatrix} 1.321 & 0.642 & 0.257 \\ 0.367 & 1.734 & 0.294 \\ 0.514 & 1.028 & 1.211 \end{pmatrix}$; $A = \begin{pmatrix} 0.381 & 0.619 \\ 0.578 & 0.422 \\ 0.509 & 0.491 \end{pmatrix}$

21. $\begin{pmatrix} 0.2 & 0.8 & 0 & 0 & 0 \\ 0 & 0.2 & 0.8 & 0 & 0 \\ 0 & 0 & 0.2 & 0.8 & 0 \\ 0 & 0 & 0 & 0.2 & 0.8 \\ 0 & 0 & 0 & 0 & 1 \end{pmatrix}$

23. (a) $E_0, E_1, E_2, \ldots, E_8$ where E_i is "G_2 has i dollars." E_0 and E_8 are absorbing.

(b) $\mathbf{p}_1 = \begin{pmatrix} \frac{3}{7} & 0 & \frac{4}{7} & 0 & 0 & 0 & 0 \end{pmatrix}$; $\mathbf{p}_2 = \begin{pmatrix} \frac{3}{7} & \frac{12}{49} & 0 & \frac{16}{49} & 0 & 0 & 0 \end{pmatrix}$ (c) 0.723

25. 2 **27.** (a) $T = \begin{pmatrix} 0 & 0.35 & 0 & 0.65 & 0 \\ 0 & 0 & 0.4 & 0.6 & 0 \\ 0 & 0 & 0 & 0.3 & 0.7 \\ 0 & 0 & 0 & 1 & 0 \\ 0 & 0 & 0 & 0 & 1 \end{pmatrix}$ (b) 231 **29.** $\frac{15}{28}(2000) \approx 1071$

Section 7.3, page 403

1. (a) $T = \begin{pmatrix} 0.8 & 0.2 & 0 & 0 & 0 \\ 0.4 & 0.5 & 0.1 & 0 & 0 \\ 0 & 0.4 & 0.5 & 0.1 & 0 \\ 0 & 0 & 0.4 & 0.5 & 0.1 \\ 0 & 0 & 0 & 0.5 & 0.5 \end{pmatrix}$ (b) T^3 has no zeros.

(c) $\begin{pmatrix} \frac{80}{133} & \frac{40}{133} & \frac{10}{133} & \frac{5}{266} & \frac{1}{266} \end{pmatrix} \approx (0.6015 \quad 0.3008 \quad 0.0752 \quad 0.0188 \quad 0.0036)$

3. (a) 8.9% (b) 0.2% (c) 21.5% **5.** 0.657 **7.** $\frac{10}{3}$ **9.** $(0.1)(2.1) + (0.9)(3.1) = 3$ **11.** $\begin{matrix} & G & K \\ G & \begin{pmatrix} 0.6 & 0.4 \\ K & 0 & 1 \end{pmatrix} \end{matrix}$ **13.** 6

15. $\begin{pmatrix} 0.15 & 0.45 & 0.4 \\ 0.15 & 0.45 & 0.4 \\ 0 & 0 & 1 \end{pmatrix}$ **17.** $\frac{5}{8}$

Section 7.4, page 414

1. Strictly determined; $\mathbf{p} = (0 \quad 1 \quad 0)$; $\mathbf{q} = \begin{pmatrix} 1 \\ 0 \\ 0 \end{pmatrix}$ **3.** Not strictly determined   **5.** Strictly determined; $\mathbf{p} = (0 \quad 0 \quad 1)$; $\mathbf{q} = \begin{pmatrix} 1 \\ 0 \end{pmatrix}$

7. Not strictly determined   **9.** Strictly determined; $\mathbf{p} = (0 \quad 1 \quad 0 \quad 0)$; $\mathbf{q} = \begin{pmatrix} 0 \\ 0 \\ 1 \\ 0 \end{pmatrix}$ **11.** $\begin{pmatrix} -2 & 3 \\ 3 & -4 \end{pmatrix}$; Not strictly determined

13. $\begin{pmatrix} -2 & 3 & -4 \\ 3 & -4 & 5 \\ -4 & 5 & -6 \end{pmatrix}$; Not strictly determined   **15.** $\begin{array}{c} \\ I \\ N \\ D \end{array} \begin{array}{c} A\backslash B \end{array} \begin{pmatrix} \begin{array}{ccc} I & N & D \\ -1 & -3 & -11 \\ 4 & 0 & -5 \\ 9 & 3 & -1 \end{array} \end{pmatrix}$; A and B should both decrease prices.

17. $\begin{array}{c} D\backslash R \\ 0 \\ 1 \\ 2 \end{array} \begin{pmatrix} \begin{array}{ccc} 0 & 1 & 2 \\ 0 & -3 & -7 \\ 3 & 0 & -3 \\ 7 & 3 & 0 \end{array} \end{pmatrix}$; Both should campaign for 2 days in region with 60,000 voters.

19. $\begin{array}{c} \text{P}\backslash\text{R} \\ \text{Station} \\ \text{Person} \end{array} \begin{pmatrix} \begin{array}{cc} \text{Home} & \text{Not home} \\ 1 & -2 \\ -1.5 & 0 \end{array} \end{pmatrix}$; Not strictly determined

21. The company should be logical; the union should seek a legal solution.

Section 7.5, page 426

1. 4   **3.** $\frac{3}{2}$ **5.** 2   **7.** $\frac{8}{3}$ **9.** $\frac{149}{40} = 3.725$ **11.** $\mathbf{p}_0 = \begin{pmatrix} \frac{1}{2} & \frac{1}{2} \end{pmatrix}$; $\mathbf{q}_0 = \begin{pmatrix} \frac{1}{2} \\ \frac{1}{2} \end{pmatrix}$; $v = \frac{1}{2}$

13. $\mathbf{p}_0 = \begin{pmatrix} \frac{4}{5} & \frac{1}{5} \end{pmatrix}$; $\mathbf{q}_0 = \begin{pmatrix} \frac{2}{5} \\ \frac{3}{5} \end{pmatrix}$; $v = \frac{2}{5}$ **15.** $\mathbf{p}_0 = (1 \quad 0)$; $\mathbf{q}_0 = \begin{pmatrix} 1 \\ 0 \end{pmatrix}$; $v = 0$; Fair

17. $\mathbf{p}_0 = (1 \quad 0)$; $\mathbf{q}_0 = \begin{pmatrix} 0 \\ 1 \end{pmatrix}$; $v = -1$ **19.** $\mathbf{p}_0 = (1 \quad 0)$; $\mathbf{q}_0 = \begin{pmatrix} 1 \\ 0 \end{pmatrix}$; $v = 3$ **21.** $\mathbf{p}_0 = \begin{pmatrix} \frac{3}{5} & \frac{2}{5} \end{pmatrix}$; $\mathbf{q}_0 = \begin{pmatrix} 0 \\ \frac{4}{5} \\ \frac{1}{5} \end{pmatrix}$; $v = \frac{18}{5}$

23. $\mathbf{p}_0 = \begin{pmatrix} \frac{5}{6} & 0 & \frac{1}{6} & 0 \end{pmatrix}$; $\mathbf{q}_0 = \begin{pmatrix} \frac{5}{6} \\ 0 \\ 1 \\ \frac{1}{6} \end{pmatrix}$; $v = \frac{29}{6}$ **27.** $\mathbf{p}_0 = \begin{pmatrix} \frac{19}{36} & \frac{17}{36} \end{pmatrix}$; $\mathbf{q}_0 = \begin{pmatrix} \frac{19}{36} \\ \frac{17}{36} \end{pmatrix}$; Unfair since $v = \frac{1}{36} \neq 0$ **29.** Any strategy is optimal.

31. Harvest early.   **33.** Plant 1500 acres of crop II only.   **37.** $\mathbf{p}_0 = (1 \quad 0)$; $\mathbf{q}_0 = \begin{pmatrix} 0 \\ 1 \end{pmatrix}$; $v_A = 2$; $v_B = 4$

39. (a) If $t = 0$, then $\mathbf{p}_0 = (1 \quad 0)$ and $\mathbf{q}_0 = \begin{pmatrix} 1 \\ 0 \end{pmatrix}$; If $0 < t < 1$, then $\mathbf{p}_0 = (1 - t \quad t)$ and $\mathbf{q}_0 = \begin{pmatrix} 1 - t \\ t \end{pmatrix}$; If $t = 1$, then $\mathbf{p}_0 = (0 \quad 1)$ and

$\mathbf{q}_0 = \begin{pmatrix} 0 \\ 1 \end{pmatrix}$ (b) If $0 \leq t < \frac{1}{2}$, then $\mathbf{p}_0 = (1 \quad 0)$ and $\mathbf{q}_0 = \begin{pmatrix} 0 \\ 1 \end{pmatrix}$; If $\frac{1}{2} \leq t \leq 1$, then $\mathbf{p}_0 = (1 \quad 0)$ and $\mathbf{q}_0 = \begin{pmatrix} 1 \\ 0 \end{pmatrix}$ (c) If $0 \leq t < \frac{1}{2}$, then

$\mathbf{p}_0 = (0 \quad 1)$ and $\mathbf{q}_0 = \begin{pmatrix} 0 \\ 1 \end{pmatrix}$; If $\frac{1}{2} < t \leq 1$, then $\mathbf{p}_0 = (1 \quad 0)$ and $\mathbf{q}_0 = \begin{pmatrix} 0 \\ 1 \end{pmatrix}$; If $t = \frac{1}{2}$, then $\mathbf{p}_0$ can be any strategy

and $\mathbf{q}_0 = \begin{pmatrix} 0 \\ 1 \end{pmatrix}$

41. (b) $v = \dfrac{2t - 1}{4t - 2} = \dfrac{1}{2}$

Review Exercises for Chapter 7, page 429

1. No   **3.** Yes   **5.** Yes   **7.** Yes   **9.** No

11. (a) $\mathbf{p}_1 = \left(\frac{1}{16} \quad \frac{15}{16}\right)$; $\mathbf{p}_2 = \left(\frac{1}{128} \quad \frac{127}{128}\right)$; $\mathbf{p}_3 = \left(\frac{1}{1024} \quad \frac{1023}{1024}\right)$ (b) Not regular

13. (a) $\mathbf{p}_1 = \left(\frac{1}{4} \quad \frac{1}{3} \quad \frac{5}{12}\right)$; $\mathbf{p}_2 = \left(\frac{1}{3} \quad \frac{3}{8} \quad \frac{7}{24}\right)$; $\mathbf{p}_3 = \left(\frac{5}{16} \quad \frac{1}{3} \quad \frac{17}{48}\right)$ (b) Regular (T^2 has no 0 components.)   (c) $\mathbf{t} = \left(\frac{1}{3} \quad \frac{1}{3} \quad \frac{1}{3}\right)$

15. (All answers rounded to 3 decimal places) (a) $\mathbf{p}_1 = (0.242 \quad 0.475 \quad 0.284)$; $\mathbf{p}_2 = (0.250 \quad 0.477 \quad 0.273)$;
$\mathbf{p}_3 = (0.251 \quad 0.478 \quad 0.272)$ (b) Yes   (c) $\mathbf{t} = (0.250 \quad 0.478 \quad 0.272)$

17. (a) 1   (b) $T' = \left(\frac{1}{2} \quad \frac{1}{2}\right)$; $R = \left(\frac{1}{2}\right)$; $Q = (2)$; $S = \left(\frac{1}{2}\right)$; $A = (1)$

19. (a) 2   (b) $T' = \left(\frac{1}{5} \quad \frac{2}{5} \quad \frac{2}{5}\right)$; $R = \left(\frac{2}{5}\right)$; $S = \left(\frac{1}{5} \quad \frac{2}{5}\right)$; $Q = \left(\frac{5}{3}\right)$; $A = \left(\frac{1}{3} \quad \frac{2}{3}\right)$

21. (a) 2   (b) $T' = \begin{pmatrix} \frac{1}{3} & \frac{1}{3} & \frac{1}{6} & \frac{1}{6} \\ \frac{1}{2} & 0 & \frac{1}{4} & \frac{1}{4} \end{pmatrix}$; $R = \begin{pmatrix} \frac{1}{3} & \frac{1}{6} \\ \frac{1}{2} & \frac{1}{4} \end{pmatrix}$; $S = \begin{pmatrix} \frac{1}{3} & \frac{1}{6} \\ 0 & \frac{1}{4} \end{pmatrix}$; $Q = \begin{pmatrix} \frac{9}{5} & \frac{2}{5} \\ \frac{6}{5} & \frac{8}{5} \end{pmatrix}$; $A = \begin{pmatrix} \frac{3}{5} & \frac{2}{5} \\ \frac{2}{5} & \frac{3}{5} \end{pmatrix}$

23.

From \ To	G	K
G	0.75	0.25
K	0	1

25. 17   **27.** $\begin{pmatrix} 0.0625 & 0.6875 & 0.25 \\ 0.0625 & 0.6875 & 0.25 \\ 0 & 0 & 1 \end{pmatrix}$ **29.** $\frac{1}{3}$ **31.** Yes; $\mathbf{p}_0 = (0 \quad 1)$; $\mathbf{q}_0 = \begin{pmatrix} 1 \\ 0 \end{pmatrix}$

33. Yes; $\mathbf{p}_0 = (0 \quad 1)$; $\mathbf{q}_0 = \begin{pmatrix} 1 \\ 0 \end{pmatrix}$ **35.** Yes; $\mathbf{p}_0 = (0 \quad 0 \quad 1)$; $\mathbf{q}_0 = \begin{pmatrix} 0 \\ 0 \\ 1 \end{pmatrix}$ **37.** $\frac{33}{8} = 4.125$ **39.** $\frac{1}{10}$ **41.** $\mathbf{p}_0 = (1 \quad 0)$; $\mathbf{q}_0 = \begin{pmatrix} 1 \\ 0 \end{pmatrix}$; $v = 3$

43. $\mathbf{p}_0 = \left(\frac{3}{7} \quad \frac{4}{7}\right)$; $\mathbf{q}_0 = \begin{pmatrix} \frac{1}{7} \\ \frac{6}{7} \end{pmatrix}$; $v = \frac{17}{7}$ **45.** $\mathbf{p}_0 = \left(\frac{1}{2} \quad 0 \quad \frac{1}{2}\right)$, $\mathbf{q}_0 = \begin{pmatrix} \frac{1}{8} \\ \frac{7}{8} \\ 0 \end{pmatrix}$; $v = -\frac{1}{2}$

Chapter 8

Note: All calculator answers in this chapter were obtained on a Texas Instruments calculator displaying 10 digits.

Section 8.1, page 440

1. $200   **3.** $225   **5.** $17,205   **7.** $6381.41; $1381.41   **9.** $6416.79; $1416.79   **11.** $12,144.56; $4144.56
13. $12,396.78; $4396.78   **15.** $12,421.64; $4421.64   **17.** $21,023.49; $11,023.49   **19.** $21,168.37; $11,168.37   **21.** $1800

23. Annually: 79.08%; Quarterly: 81.40%   **25.** $5\frac{1}{8}$% computed semiannually   **27.** 3.61%   **29.** 7.46%

31. (a) 12.36%   (b) 12.55%   (c) 12.68%   (d) 12.75%   **33.** (a) 55.07 yr   (b) 22.11 yr   (c) 13.87 yr   (d) 11.12 yr   (e) 7.46 yr
35. (a) 55.48 yr   (b) 22.52 yr   (c) 14.27 yr   (d) 11.53 yr   (e) 7.86 yr   **37.** 13.84%   **39.** 2.71%   **41.** 141.5

Section 8.2, page 445

1. $3331.71   **3.** $3289.25   **5.** $2554.00   **7.** $2484.71   **9.** $2018.04   **11.** $2007.31   **13.** $293,318.28   **15.** $290,104.97
17. 498,343   **19.** $8162.98   **21.** 9.6278%   **23.** $180,533.60

Section 8.3, page 454

1. $5474.86   **3.** $7968.71   **5.** $137,161.28   **7.** $166,820.80   **9.** $7957.13   **11.** $4491.29   **13.** $3072.28   **15.** $31,958.39
17. $123,160.96   **19.** $3577.57   **21.** $12.58   **23.** $30,060.77   **25.** $1072.96   **27.** $1074.81   **29.** $4915.69   **31.** $943.40

33. $1256.11 **35.** $37,333.83 **37.** Buy it on time (Present value ≈ 800 + 2930 = $3730) **39.** $17.60 **41.** $10,713.01
43. $2618.97 **45.** $6004.25 **47.** $2258.34 **49.** c (Total present value = $14,426.80 minus original investment = $1000 net)

Section 8.4, page 460

1. $758.68 **3.** $569.84 **5.** $219.20 **7.** $86.94 **9.** $1641.33 **11.** $1358.68 **13.** $1769.84 **15.** $719.20 **17.** $286.94
19. $3703.83 **21.** $3720.81 **23.** (a) $2504.56 (b) $4466.29 **25.** (a) $1212.88 (b) $3334.16 **27.** $1008.24
29. (a) $4287.37 (b) $24,808.30 **31.** $116,873.36

Section 8.5, page 465

1. $1800 **3.** $22,500 **5.** 7.5% **7.** 10.53% **9.** 2.61% **11.** $928.94 **13.** $951.68 **15.** $1231.15 **17.** $933.14 **19.** $1049.34
21. $10,824.15 **23.** $6785.83 **25.** $23,693.75

Review Exercises for Chapter 8, page 466

1. $4200 **3.** $7813.56 **5.** $12,915.78 **7.** $16,376.79 **9.** $13,661.16 **11.** 8.24%
13. (a) 23.19 yr (b) 10.75 yr (c) 8.75 yr (d) 5.86 yr **15.** (a) 23.11 yr (b) 10.66 yr (c) 8.67 yr (d) 5.78 yr **17.** $6719.84
19. $2121.55 **21.** $2055.30 **23.** 9.76% **25.** $8892.34 **27.** $1507.69 **29.** $784,085.21 **31.** $7019.69 **33.** $1053.76
35. $61,775.48 **37.** $4068.83 **39.** $1202.10 **41.** $1056.93 **43.** $88.84 **45.** $1506.93 **47.** $151.84 **49.** $1571.86
51. (a) $162.01 (b) $11,563.03 **53.** $6375 **55.** 8% **57.** $2729.34

Chapter 9

Section 9.1, page 477

1. $f(0) = 1, f(1) = \frac{1}{2}, f(-2) = -1, f(-5) = -\frac{1}{4}$ **3.** $f(0) = 1, f(-3) = 28, f(2) = 13, f(10) = 301$
5. $f(0) = 0, f(2) = 16, f(-2) = 16, f(\sqrt{5}) = 25$ **7.** $g(0) = 1, g(-1) = 0, g(3) = 2, g(7) = \sqrt{8}$
9. $h(0) = 1, h(2) = 7, h(\frac{1}{3}) = \frac{13}{9}, h(-\frac{1}{2}) = \frac{3}{4}$ **11.** yes **13.** yes **15.** no **17.** yes **19.** yes **21.** yes
25. Domain: $\mathbb{R}$; range: $\mathbb{R}$ **27.** Domain: $(-\infty, 0) \cup (0, \infty)$; range: $(0, \infty)$
29. Domain: $(-\infty, -1) \cup (-1, \infty)$; range: $(-\infty, 0) \cup (0, \infty)$ **31.** Domain: $[1, \infty)$; range: $[0, \infty)$
33. Domain: $(-\infty, 0) \cup (0, \infty)$; range: $(0, \infty)$ **35.** Domain: $\mathbb{R}$; range: $[0, \infty)$ **37.** Domain: $\mathbb{R}$; range: $[0, \infty)$

39.

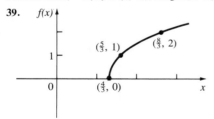

41.

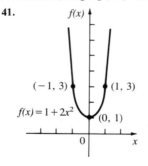

43.

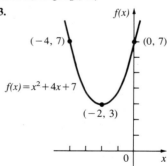

45. $f(x + \Delta x) = x^2 + 2x(\Delta x) + (\Delta x)^2$; $\dfrac{f(x + \Delta x) - f(x)}{\Delta x} = 2x + \Delta x$ **47.** Domain: $(-\infty, 0) \cup (0, \infty)$; range: $\{-1, 1\}$
49. $A(W) = 25W - W^2$; domain: $(0, 25)$; range: $(0, 12.5^2)$ **51.**
$$d(t) \begin{cases} \sqrt{30^2(3 - t)^2 + 90^2}, & 0 \le t < 3 \\ 180 - 30t, & 3 \le t < 6 \\ 30t - 180, & 6 \le t < 9 \\ \sqrt{90^2 + 30^2(t - 9)^2}, & 9 \le t \le 12 \end{cases}$$

53. To Canadian $= x + 0.12x$; back to U.S. $= (x + 0.12x) - 0.12(x + 0.12x) = 0.9856x$; double conversion, $f(x) = 0.9856x$

55. $P(x) = 0.3x - 40$ **57.** $A(1) = 1183$, $A(8) = 1187$, $A(30) = 1203$, $A(60) = 1248$, $A(88) = 1204$

Section 9.2, page 484

1. $(f + g)(x) = -2x - 5$, domain $= \mathbb{R}$; $(f - g)(x) = 6x - 5$, domain $= \mathbb{R}$; $(f \cdot g)(x) = -8x^2 + 20x$, domain $= \mathbb{R}$;

$(f/g)(x) = \dfrac{2x - 5}{-4x}$, domain $=$ all $\mathbb{R}$ except $\{0\}$

3. $(f + g)(x) = \sqrt{x + 2} + \sqrt{2 - x}$, domain $= [-2, 2]$; $(f - g)(x) = \sqrt{x + 2} - \sqrt{2 - x}$, domain $= [-2, 2]$;

$(f \cdot g)(x) = \sqrt{4 - x^2}$, domain $= [-2, 2]$; $(f/g)(x) = \sqrt{\dfrac{x + 2}{2 - x}}$, domain $= [-2, 2)$

5. $(f + g)(x) = 2 - |x| + x^5$, domain $= \mathbb{R}$; $(f - g)(x) = |x| + x^5$, domain $= \mathbb{R}$; $(f \cdot g)(x) = 1 - |x| + x^5 - |x|x^5$,

domain $= \mathbb{R}$; $(f/g)(x) = \dfrac{1 + x^5}{1 - |x|}$, domain $= \mathbb{R}$ excluding $\{-1, 1\}$

7. $(f + g)(x) = \sqrt[5]{x + 2} + \sqrt[4]{x - 3}$, domain $[3, \infty)$; $(f - g)(x) = \sqrt[5]{x + 2} - \sqrt[4]{x - 3}$, domain $[3, \infty)$;

$(f \cdot g)(x) = \sqrt[5]{x + 2} \cdot \sqrt[4]{x - 3}$, domain $[3, \infty)$; $(f/g)(x) = \dfrac{\sqrt[5]{x + 2}}{\sqrt[4]{x - 3}}$, domain $(3, \infty)$

9. $(f \circ g)(x) = 2x + 1$, domain: $\mathbb{R}$; $(g \circ f)(x) = 2x + 2$, domain: $\mathbb{R}$

11. $(f \circ g)(x) = 15x + 11$, domain $\mathbb{R}$; $(g \circ f)(x) = 15x + 27$, domain $\mathbb{R}$

13. $(f \circ g)(x) = \dfrac{x - 1}{3x - 1}$, domain $\mathbb{R}$ excluding $\{0, \frac{1}{3}\}$; $(g \circ f)(x) = \dfrac{-2}{x}$, domain $\mathbb{R}$ excluding $\{0, -2\}$

15. $(f \circ g)(x) = \sqrt{1 - \sqrt{x - 1}}$, domain $[1, 2]$; $(g \circ f)(x) = \sqrt{\sqrt{1 - x} - 1}$, domain $(-\infty, 0]$

19. $g(x) = x - 5$ and $g(x) = 5 - x$ **21.** Domain k: $[0, \infty)$; **23.** $ad + b = bc + d$

$f(x) = x^{5/7}$;
$g(x) = 1 + x$;
$h(x) = \sqrt{x}$

25. (a) $c(P) = 168000 - 3200p$ (b) $R(P) = 2000p - 400p^2$ (c) $P(p) = 5200p - 400p^2 - 168,000$

(d) Profit maximized at $p = -\frac{13}{2}$; maximum profit $= -\$168,000$ (This is a minimum loss of $\$168,000$.)

Section 9.3, page 491

1. (a)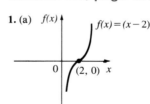

(b)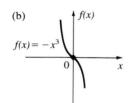

(c)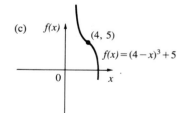

3. (a) $(x - 2)^2 + 3$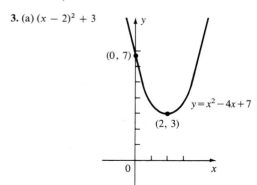

(b) $(x + 4)^2 - 14$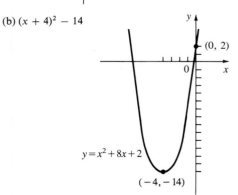

3. (c) $(x + \frac{3}{2})^2 + \frac{7}{4}$

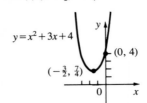

$y = x^2 + 3x + 4$

$(0, 4)$

$(-\frac{3}{2}, \frac{7}{4})$

(d) $y = -(x - 1)^2 - 2$

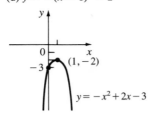

$(1, -2)$

-3

$y = -x^2 + 2x - 3$

(e) $y = -(x + \frac{5}{2})^2 + \frac{57}{4}$

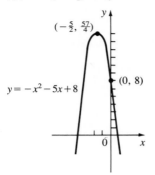

$(-\frac{5}{2}, \frac{57}{4})$

$(0, 8)$

$y = -x^2 - 5x + 8$

5.

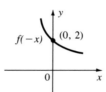

$f(x - 2)$

$(2, 2)$

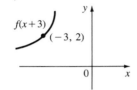

$f(x + 3)$

$(-3, 2)$

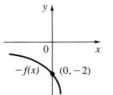

$-f(x)$

$(0, -2)$

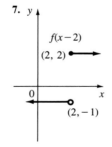

$f(-x)$

$(0, 2)$

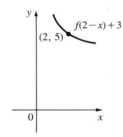

$f(2 - x) + 3$

$(2, 5)$

7.

$f(x - 2)$

$(2, 2)$

$(2, -1)$

$f(x + 3)$

$(-3, 2)$

$(-3, -1)$

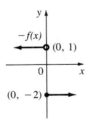

$-f(x)$

$(0, 1)$

$(0, -2)$

$f(-x)$

$(0, 2)$

$(0, -1)$

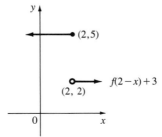

$(2, 5)$

$f(2 - x) + 3$

$(2, 2)$

9.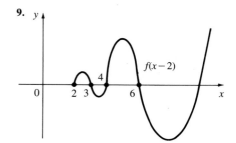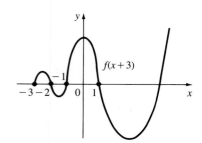

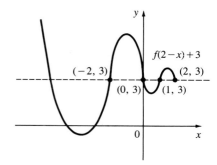 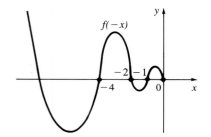

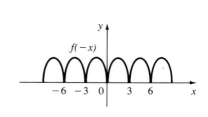

11.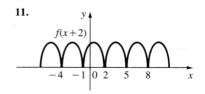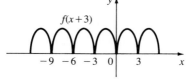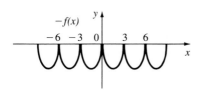

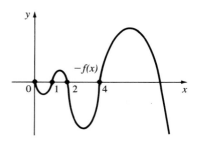 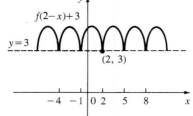

Section 9.4, page 499

1.
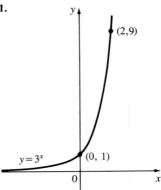

3.
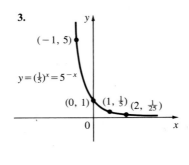

5.
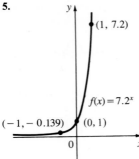

7.
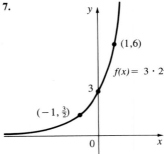

9.
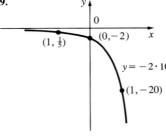

11.
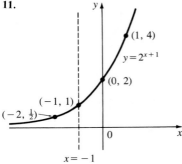

13.
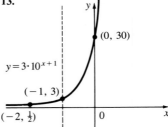

15. $e^{2.5} \approx 12.18249396$ **17.** $e^{-0.6} \approx 0.548811636$ **19.** $3^{\sqrt{2}} \approx 4.728804386$ **21.** 8753.36

23. Semiannual compounding when money is withdrawn at the end of compounding period

25. (a) \$11,274.97 (b) \$16,160.74 (c) \$33201.17

27. (a) 1.138828376 (b) 0.690730947 (c) 62.177918 (d) 0.0720789 (e) 123.965095, 0.236924

Section 9.5, page 506

1. 4 **3.** 8 **5.** 1 **7.** 3 **9.** −2 **11.** 2 **13.** −3 **15.** −4π **17.** 5 **19.** $-\frac{1}{2}$ **21.** −1 **23.** $\pm\sqrt{6}$ **25.** $\sqrt{2}$ **27.** e^{π} **29.** 4
31. $\frac{1}{2}$ **33.** ln 4 **35.** $x = (\ln 2 - 1)/2$ **37.** $\frac{1}{2}e^{1/3}$ **39.** $^{-3/2}$ **41.** (ln 8)/2 **43.** $x = 2$ **45.** $e^2/(e^2 - 1)$ **47.** 8 **49.** e^{12}
51. (a) $10^{3/2.5}$ (b) $2.5 \log 5 \approx 1.75$ (c) $10^{20.1/2.5}$ (d) $2.5 \log 45,000 \approx 11.63$ (e) 1.6

53. (a) $\ln 0.8 \approx -0.2231435$ (accurate to 6 places); $\ln 1.2 \approx 0.182321542$ (accurate to 7 places)
(b) $\ln (2) = \ln (\frac{3}{2}) + \ln (\frac{4}{3}) \approx 0.693143053$ (accurate to 5 places)   (c) $\ln 3 = \ln 2 + \ln \frac{3}{2} \approx 1.098608$ (accurate to 5 places);
$\ln 8 = 3 \ln 2 = 2.079429$ (accurate to 4 places)

55. 20.3%   **57.** 10.14 years   **59.** 10.517%

Review Exercises for Chapter 9, page 509

1. yes, domain f = range f = $\mathbb{R}$ **3.** yes, domain f = range f = $\mathbb{R} - \{0\}$ **5.** yes, domain $f = [-2, \infty)$, range $f = [0, \infty)$

7. yes, domain $f = \mathbb{R}$, range $f = [-\frac{1}{2}, \frac{1}{2}]$ **9.** yes, domain $f = (-\infty, -\sqrt{6}] \cup [\sqrt{6}, \infty)$, range $f = [0, \infty)$

15.

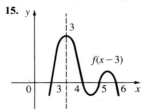

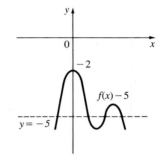

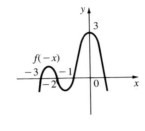

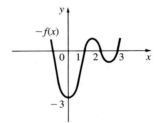

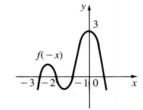

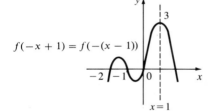

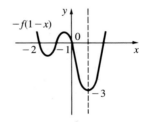

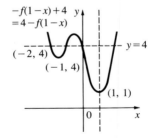

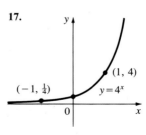

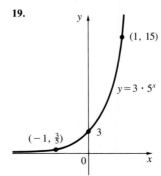

17.

19.

21. $e^{1.7} \approx 5.4739474$ **23.** $(\frac{1}{3})^{2.3} \approx 0.799137$ **25.** 2 **27.** $x = \frac{1}{2}\ln 4 = \ln 2$ **29.** $x = (3e+1)/(e-1)$ **31.** $x = 2$
33. $x = e/10$ **35.** \$16,160.74

Chapter 10

Section 10.2, page 527

1. (a)

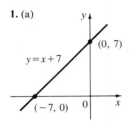

(b) $f(3) = 10, f(1) = 8, f(2.5) = 9.5, f(1.5) = 8.5, f(2.1) = 9.1, f(1.9) = 8.9, f(2.01) = 9.01, f(1.99) = 8.99$ (c) 9
3. (a)

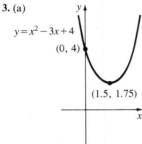

(b) $f(-0.5) = 5.75, f(-1.5) = 10.75, f(-0.9) = 7.51, f(-1.1) = 8.51, f(-0.99) = 7.9501, f(-1.01) = 8.0501$ (c) 8
5. (a) $f(2)$ is not defined since division by zero is not allowed. (b) 12 **7.** 45 **9.** $\frac{1}{2}$ **11.** 0 **13.** 729 **15.** 2 **17.** 2 **19.** $\frac{1}{2}$
21. $\frac{1}{2}$ **23.** 12
25. (a) $f(3) = 0.5238, f(1) = -0.2727, f(2.5) = 0.1488, f(1.5) = -0.2843, f(2.1) = -0.0863, f(1.9) = -0.1751,$
$f(2.01) = -0.1289, f(1.99) = 0.1377, f(2.001) = -0.1329, f(1.999) = -0.1388$ (b) -0.13 seems good from the table.
(c) $f(2) = -0.1333$
27. (a), (b), (c)

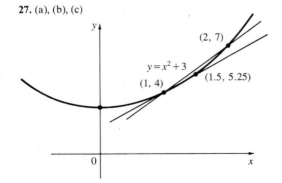

(d) slope of the line between $(1, 4)$ and $(1 + \Delta x, (1 + \Delta x)^2 + 3)$
(e) 2

29. (a)

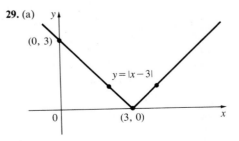

(b) 0

31. Yes; $\lim\limits_{x \to 2} f(x) = \lim\limits_{x \to 2} g(x) = 11$ **33.** 5 **35.** 0 **37.** 20 **39.** 24 **41.** -4 **43.** 128 **45.** $-\frac{1}{5}$ **47.** $-\frac{75}{19}$ **49.** $\frac{3}{5}$

51. Does not exist **53.** 243 **55.** 0 **57.** 5 **59.** Does not exist **61.** ∞ **63.** ∞ **65.** Does not exist **67.** ∞ **69.** ∞ **71.** 1

73. 0 **75.** -1 **77.** $\frac{2}{3}$ **79.** 0 **81.** 0 **83.** (b) 5

Section 10.3, page 536

1. Continuous on $(-\infty, \infty)$ **3.** Discontinuous at $x = 4$, continuous on $(-\infty, 4)$ and $(4, \infty)$

5. Discontinuous at $x = -1$, continuous on $(-\infty, -1)$ and $(-1, \infty)$ **7.** Continuous on $(-\infty, \infty)$ **9.** $(0, \infty)$

11. Discontinuities: $x = 1$ and $x = -1$; continuous on $(-\infty, -1)$, $(-1, 1)$, and $(1, \infty)$

13. Discontinuous at $x = 2$, continuous on $(-\infty, 2)$ and $(2, \infty)$ **15.** $\alpha = 2$ and $\alpha = -2$

19.

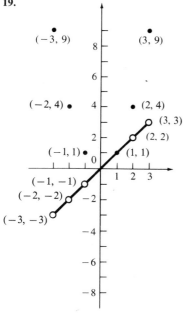

f is continuous at $x = 1$ and $x = 0$.

Section 10.4, page 549

1. 280 **3.** $\begin{cases} 16, & 0 \le q < 500 \\ 14, & q > 500 \end{cases}$

5. (a) and (b)

Δx	$f(2 + \Delta x)$	$(f(2 + \Delta x) - f(2))/\Delta x$
0.5	18.75	13.5
0.1	13.23	12.3
0.01	12.1203	12.03
0.001	12.012003	12.003
-0.01	11.8803	11.97
-0.001	11.988003	11.997

(c) $f'(x) = 6x$; $f'(2) = 12$ (d) $y = 12x - 12$

7. (a) and (b)

$f(1 + \Delta x)$	$(f(1 + \Delta x) - f(1))/\Delta x$
1.5	2.2475
1.1	2.2404
1.01	2.4938
1.001	2.4994
0.99	2.5063
0.999	2.5006

(c) $f'(x) = 5/2\sqrt{x}$; $f'(1) = \frac{5}{2}$ (d) $y = \frac{5}{2}x + \frac{5}{2}$

9. $f'(x) = -4$, $y = -4x + 6$ **11.** $f'(x) = 3x^2$, $y = 12x - 16$ **13.** $f'(x) = 2x$, $y = 2x$

15. $f'(x) = 2x - 1$, $y = x + 1$ **17.** $f'(x) = -\dfrac{1}{x^2}$, $y = -9x + 6$

Section 10.5, page 554

1. (a)

t	h	t	h	t	h	t	h
0.0	0.0	0.6	39.24	1.2	66.96	1.8	83.16
0.1	7.34	0.7	44.66	1.3	70.46	1.9	84.74
0.2	14.36	0.8	49.76	1.4	73.64	2.0	86.00
0.3	21.06	0.9	54.54	1.5	76.50		
0.4	27.44	1.0	59	1.6	79.04		
0.5	33.50	1.1	63.14	1.7	81.26		

(b) 28.6 ft/sec   **(c)** 25.4 ft/sec   **(d)** 27 ft/sec, averaging parts (b) and (c)   **(e)** $26.84 \leq v \leq 27.16$ **(f)** 27   **3.** 9   **5.** $\frac{1}{4}$ **7.** 40
9. $v(3) = 420$; $v(10) = 1400$ **11. (a)** $C'(q) = 6 - 0.02q + 0.03q^2$ **(b)** $q = \frac{1}{3}$ is where marginal cost is lowest.
13. $R'(q) = 20q - 0.0006q^2$

Section 10.6, page 560

1. $5x^4$ **3.** $8x^3$ **5.** $-\frac{2}{3}x^{-5/3}$ **7.** $\frac{3}{4}t^{-1/4}$ **9.** $-4z^{-5}$ **11.** $144r^{11}$ **13.** $6x + 19$ **15.** $5t^4 + (\frac{1}{2})t^{-1/2}$ **17.** $100z^{99} + 1000z^9$
19. $24r^7 - 48r^5 - 28r^3 + 4r$ **21.** $\frac{3}{4}x^{-1/4} - \frac{7}{8}x^{-1/8}$ **23.** $-t^{-2} + 4t^{-7/3}$ **25.** $-r^{-2} - 4r^{-3} - 9r^{-4}$ **27.** $y = 6x + 7$
29. $y = 32x - 26$ **31.** $y = 10x - 7$ **33.** $y = -\frac{1}{2}x + \frac{3}{2}$ **37.** $2 - 0.02q$ **39.** $0.01 - 0.0006q^2$ **41.** $2000 - 0.125q^{3/2}$
43. 36 m/sec   **45.** 11 ft/sec   **47. (a)** 1.5925   **(b)** 2.25   **(c)** 1.4   **49.** $x = 1024$ **51.** $y = 2x$ and $y = -6x$
53. (a) $30t^2 - 30t - 20\sqrt{t} + 3000$ **(b)** $60t - 30 - \dfrac{10}{\sqrt{t}}$ units/hr   **(c)** $P'_T(4) = 205$ units/hr; $P'_T(16) = 927$ units/hr

Section 10.7, page 566

1. $6x^2 + 2$ **3.** $\dfrac{-1}{(5x - 3)^2}$ **5.** $\dfrac{3t^2 + \frac{5}{2}t^{5/2}}{(1 + \sqrt{t})^2}$ **7.** $(1 + x + x^5)(-1 + 6x^5) + (2 - x + x^6)(1 + 5x^4)$
9. $\dfrac{-3 - 10x^4 + 10x^5 + 5x^6 + x^{10}}{(1 + x + x^5)^2}$ **11.** $\dfrac{1}{\sqrt{t}(1 - \sqrt{t})^2}$ **13.** $5v^4 - \frac{5}{2}v^{3/2} + 7v^{5/2} - 2$ **15.** $\dfrac{-2v^{5/2} - 3 - 10v}{2v^{-3/2}(v^3 - \sqrt{v})^2}$ **17.** 0
19. $-\frac{1}{2}r^{-3/2}$ **21.** $\dfrac{-9t^4 - 2}{2t\sqrt{t}(t^4 + 2)^2}$ **23.** $-6x^{-7}$ **25.** $-\frac{15}{7}x^{-4}$ **27.** $y = 28x - 20$ **29.** $y = -\frac{7}{2}u + \frac{11}{2}$
33. $(x^2 + 1)(x^3 + 2)(4x^3) + (x^2 + 1)(x^4 + 3)(3x^2) + (x^3 + 2)(x^4 + 3)(2x)$ **35.** $\dfrac{f^2\dfrac{dg}{dx} + g^2\dfrac{df}{dx}}{(f + g)^2}$
37. (a) $p'(q) = \dfrac{-20 - 1.2q - 0.02q^2}{(10 + q + 0.02q^2)^2}$ **(b)** $p'(10) = \dfrac{-34}{484}$; $p'(100) = \dfrac{-340}{96100}$ **39.** (b, e) and (c, d)

Section 10.8, page 572

1. $3(x + 1)^2$ **3.** $\dfrac{2(\sqrt{x} + 2)^3}{\sqrt{x}}$ **5.** $36x^5(1 + x^6)^5$ **7.** $5(x^2 - 4x + 1)^4(2x - 4)$ **9.** $-\frac{6}{5}\left(\dfrac{t + 1}{t - 1}\right)^{-2/5}\left(\dfrac{1}{(t - 1)^2}\right)$
11. $2(u^5 + u^4 + u^3 + u^2 + u + 1)(5u^4 + 4u^3 + 3u^2 + 2u + 1)$ **13.** $-8y(y^2 - 3)^{-5}$
15. $12x^3(x^2 + 2)^5(x^4 + 3)^2 + 10x(x^4 + 3)^3(x^2 + 2)^4$ **17.** $\dfrac{-3t^2 + 2t - 4}{(t + 2)^5\sqrt{t^2 + 1}}$
19. $\dfrac{(u^2 + 1)^2(u^2 - 1)[19u^4 - 40u^3 - 4u^2 + 8u + 1]}{2(u - 2)^{3/2}}$ **21.** $\frac{8}{9}x^{1/3}(1 + x^{4/3})^{-1/3}$

23. $\dfrac{1}{2\sqrt{x + \sqrt{1 + \sqrt{x}}}}\left[1 + \dfrac{1}{2\sqrt{1 + \sqrt{x}}}\dfrac{1}{2\sqrt{x}}\right]$ **25.** $-5(y^{-2} + y^{-3} + y^{-7})^{-6}(-2y^{-3} - 3y^{-4} - 7y^{-8})$

27. (a) $C^1(q) = 1.65(30 + 1.5q)^{0.1}$ (b) $C^1(100) = 2.7734$ **31.** (a) $y = \sqrt{r^2 - x^2}$ (b) $y = \dfrac{-x_0}{\sqrt{r^2 - x_0^2}}(x - x_0) + y_0$

33. (a) $R'(20) = 2.778$ (b) $R'(70) = -12.245$ (c) $q = 40$ (d) $R(40) = \$937.50$

Section 10.9, page 576

1. $d^2y/dx^2 = 0$; $d^3y/dx^3 = 0$ **3.** $d^2y/dx^2 = 8$; $d^3y/dx^3 = 0$ **5.** $d^2y/dx^2 = -\frac{1}{4}x^{-3/2}$; $d^3y/dx^3 = \frac{3}{8}x^{-5/2}$

7. $d^2y/dx^2 = \frac{2}{9}(x + 1)^{-1/3}$; $d^3y/dx^3 = \dfrac{-2}{27}(x + 1)^{-4/3}$

9. $d^2y/dx^2 = -x^2(1 - x^2)^{-3/2} - (1 - x^2)^{-1/2}$; $d^3y/dx^3 = -3x^3(1 - x^2)^{-5/2} - 3x(1 - x^2)^{-3/2}$

11. $d^2y/dx^2 = r(r - 1)x^{r-2}$; $d^3y/dx^3 = r(r - 1)(r - 2)x^{r-3}$ **13.** $d^2y/dx^2 = 2a$; $d^3y/dx^3 = 0$

15. $d^2y/dx^2 = 30(x + 1)^{-7}$; $d^3y/dx^3 = -210(x + 1)^{-8}$ **19.** (a) $s(0) = 3$ (b) $s'(0) = 2$ (c) $a(0) = s''(0) = -8$ (e) $t = \frac{2}{3}$

Section 10.10, page 582

1. $-x^2/y^2$ **3.** $-\sqrt{y}/\sqrt{x}$ **5.** $-y^2/x^2$ **7.** $\dfrac{(2x/3)(x^2 + y)^{-2/3} - \frac{1}{2}(x + y)^{-1/2}}{\frac{1}{2}(x + y)^{-1/2} - \frac{1}{3}(x^2 + y)^{-2/3}}$ **9.** $-x/y$ **11.** $\dfrac{2}{15(3xy + 1)^4} - \dfrac{y}{x}$ **13.** x/y

15. $\dfrac{3x}{5y}$ **17.** $-y/x$ **19.** $-y/x$ **21.** $-(y/x)^{15/8}$ **23.** $-y/x$ **25.** $\dfrac{y - 2xy^3 - 3x^2y^2}{3x^2y^2 + 2x^3y - x}$ **27.** $y = \frac{2}{3}x$

29. Tangent vertical at $(0, 1)$; tangent horizontal at $(1, 0)$ **31.** No vertical tangent; no horizontal tangent

33. Tangent vertical at $(a, 0)$ and $(-a, 0)$; tangent horizontal at $(0, b)$ and $(0, -b)$ **35.** $\dfrac{m}{n}x^{(m/n - 1)}$

Section 10.11, page 589

1. $\dfrac{1}{1 + x}$ **3.** $-e^{-x}$ **5.** $1/x$ **7.** $\dfrac{5}{1 + 5x}$ **9.** $-\dfrac{1}{x^2}e^{1/x}$ **11.** $\dfrac{1}{x \ln x}$ **13.** 1 **15.** $\dfrac{1}{x - 1} - \dfrac{1}{x + 1}$ **17.** $\dfrac{4}{x}(1 + \ln x)^3$ **19.** $\dfrac{\ln x - 1}{(\ln x)^2}$

21. $-xe^{-x} + e^{-x}$ **23.** $\dfrac{1}{(1 - x)^2}e^{1/(1 - x)}$ **25.** $x^2e^x + 2xe^x$ **27.** $\frac{9}{20}e^{-1/10}$ **29.** $s''(10) \approx 0.01$ **31.** $y'' = -\dfrac{1}{(1 + x)^2}$

33. $y'' = x^{-4}e^{1/x} + 2x^{-3}e^{1/x}$ **37.** $\dfrac{dy}{dx} = \frac{4}{3}\left(\dfrac{xe^x}{x^5 + 1}\right)^{4/3}\left(\dfrac{1}{x} + 1 - \dfrac{5x^4}{x^5 + 1}\right)$

39. $\dfrac{dy}{dx} = x^{2x}(2 \ln x + 2)$

Review Exercises for Chapter 10, page 590

1.   **3.** $1^3 - 3(1) + 2 = 0$ **5.** $\frac{76}{31}$ **7.** 2 **9.** 1 **11.** -1 **13.** Does not exist   **15.** $\frac{25}{7}$ **17.** -6 **19.** $\frac{1}{3}$ **21.** -1

x	$f(x)$
3	6
1	4
2.5	4.75
1.5	3.75
2.1	4.11
1.9	3.91
2.01	4.0101
1.99	3.9901

$\lim_{x \to 2} f(x) = 4$

23. 0 **25.** (a) yes   (b) no   **27.** Continuous on $(0, \infty)$ **29.** Continuous on $(-\infty, 6)$ and $(6, \infty)$

31. Continuous on $(-\infty, -2)$, $(-2, 2)$, and $(2, \infty)$ **33.** Continuous on $(-\infty, -3)$ and $(-3, \infty)$

35. Continuous on $(-\infty, -3), (-3, 3)$ and $(3, \infty)$ **37.** $C'(q) = 8 - 0.04q$; yes **39.** $6x - 6$ **41.** $\frac{5}{7}x^{-2/7}$ **43.** $350x^{349}$

45. $\frac{8}{3}(4x)^{-1/3}$ **47.** $\frac{1}{2}x^{-1/2} - 2x - \frac{5}{2}x^{3/2}$ **49.** $3e^{3x}$ **51.** $\dfrac{4}{1 + 4x}$ **53.** $2e^x/(1 - e^x)^2$ **55.** $\frac{3}{4}(1 + x + x^5)^{-1/4}(1 + 5x^4)$

57. $\dfrac{(1 + x + x^3)(1 + 2x) - (1 + x + x^2)(1 + 3x^2)}{(1 + x + x^3)^2}$ **59.** $-\frac{10}{7}x(1 + x)^4(1 - x^2)^{-2/7} + 4(1 + x)^3(1 - x^2)^{5/7}$ **61.** $\dfrac{9y}{2y^{-1/3} - 9x}$

63. $-y/x$ **65.** $x^{x^2}(x + 2x \ln x)$ **67.** $y = -\frac{8}{25}x + \frac{89}{50}$ **69.** $y = x$ **71.** $y'' = -\frac{1}{4}(1 + x)^{-3/2}; y''' = \frac{3}{8}(1 + x)^{-5/2}$

73. $y'' = 4(x - 1)^{-3}; y''' = -12(x - 1)^{-4}$

75. $y'' = -\frac{1}{8}(x + x^{3/2})^{-3/2}(1 + \frac{3}{2}x^{1/2}); y''' = -\frac{1}{8}(x + x^{3/2})^{-3/2}(\frac{3}{4}x^{-1/2}) - \frac{1}{8}[-\frac{3}{2}(x + x^{3/2})^{-5/2}(1 + \frac{3}{2}x^{1/2})^2]$

77. $y'' = \dfrac{2 - 2x^2}{(1 + x^2)^2}; y''' = \dfrac{4x(x^2 - 3)}{(1 + x^2)^3}$

Chapter 11

Section 11.1, page 602

1. (a) Increasing; $x > -\frac{1}{2}$; decreasing: $x < -\frac{1}{2}$ (b) $x = -\frac{1}{2}$ (c) y intercept: -30; x intercepts: $5, -6$
(d) Minimum: $-30\frac{1}{4}$ at $x = -\frac{1}{2}$ (e)

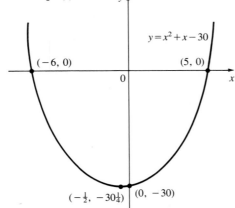

3. (a) Increasing: $x > \frac{5}{2}$; decreasing: $x < \frac{5}{2}$ (b) $x = \frac{5}{2}$ (c) y-intercept: 3; x-intercepts: $\dfrac{5 \pm \sqrt{13}}{2}$
(d) Minimum: $-\frac{13}{4}$ at $x = \frac{5}{2}$ (e)

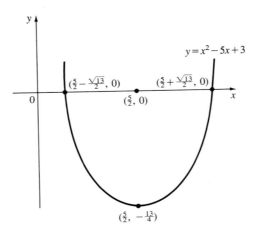

5. (a) Increasing: $x < -\frac{1}{2}$; decreasing: $x > -\frac{1}{2}$ (b) $x = -\frac{1}{2}$ (c) y-intercept: 1; x-intercepts: $\dfrac{-1 \pm \sqrt{5}}{2}$

(d) Maximum: $\frac{5}{4}$ at $x = -\frac{1}{2}$ (e)

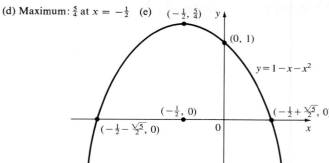

7. (a) Increasing: $x < \frac{1}{2}$; decreasing: $x > \frac{1}{2}$ (b) $x = \frac{1}{2}$ (c) y-intercept: 0; x-intercepts: 0, 1 (d) Maximum: $\frac{1}{4}$ at $x = \frac{1}{2}$
(e)

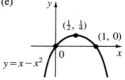

9. (a) Increasing: $x < -1$ and $x > 1$; decreasing: $-1 < x < 1$ (b) $x = -1$, $x = 1$
(c) x intercepts: $-\sqrt{3}$, 0, $\sqrt{3}$; y intercept: 0 (d) Maximum: 2 at $x = -1$; minimum: -2 at $x = 1$
(e)

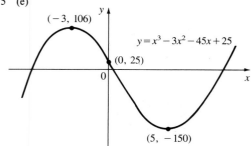

11. (a) Increasing: $x < -3$ or $x > 5$; decreasing: $-3 < x < 5$ (b) $x = -3$, $x = 5$ (c) y intercept: 25
(d) Maximum: 106 at $x = -3$; minimum: -150 at $x = 5$ (e)

13. (a) Increasing: $x > 0$, decreasing: $x < 0$ (b) $x = 0$
(c) y-intercept = x-intercept = $(0, 0)$
(d)   Minimum: 0 at $x = 0$
(e)

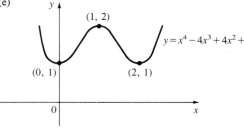

15. (a) Increasing: $x < 0$; decreasing: $x > 0$ (b) $x = 0$
(c) y-intercept: 1, x-intercepts -1 and 1
(d) Maximum: 1 at $x = 0$
(e)

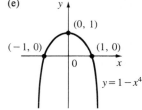

17. (a) Decreasing: $(-\infty, 0)$ and $(1, 2)$; increasing: $(0, 1)$ and $(2, \infty)$ (b) $x = 0$, $x = 1$, and $x = 2$
(c) y-intercept of 1, no x-intercept   (d) Minimum: 1 at $x = 0$; minimum: 1 at $x = 2$, maximum: 2 at $x = 1$.
(e)

$y = x^4 - 4x^3 + 4x^2 + 1$

19. (a) Decreasing: $(-\infty, \infty)$ (b) None
(c) x-intercept = y-intercept = $(0, 0)$ (d) None
(e)

$y = 1 - e^x$

21. (a) Increasing: $(0, \infty)$, decreasing: $(-\infty, 0)$ (b) $x = 0$
(c) y-intercept: 1   (d) Minimum: 1 at $x = 0$
(e)

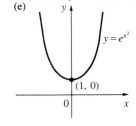

25.

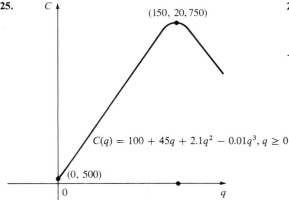

$C(q) = 100 + 45q + 2.1q^2 - 0.01q^3,\ q \geq 0$

27.

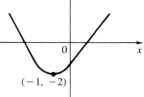

29.

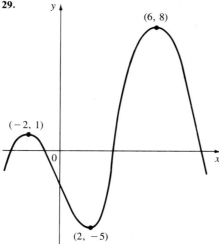

(6, 8)

(−2, 1)

0

(2, −5)

Section 11.2, page 614

1. Same as Problem 1, Section 11.1   **3.** Same as Problem 3, Section 11.1

5.

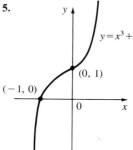

$y = x^3 + 1$

(0, 1)

(−1, 0)

0

7.

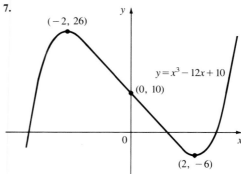

(−2, 26)

$y = x^3 - 12x + 10$

(0, 10)

0

(2, −6)

9.

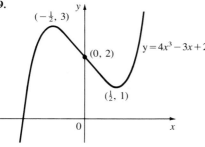

$(-\frac{1}{2}, 3)$

(0, 2)

$y = 4x^3 - 3x + 2$

$(\frac{1}{2}, 1)$

0

11.

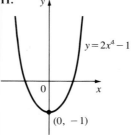

$y = 2x^4 - 1$

0

(0, −1)

13.

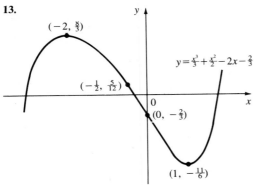

$(-2, \frac{8}{3})$

$y = \frac{x^3}{3} + \frac{x^2}{2} - 2x - \frac{2}{3}$

$(-\frac{1}{2}, \frac{5}{12})$

0

$(0, -\frac{2}{3})$

$(1, -\frac{11}{6})$

15.

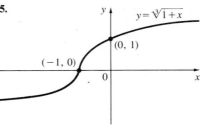

$y = \sqrt[3]{1+x}$

$(0, 1)$

$(-1, 0)$

0

17.

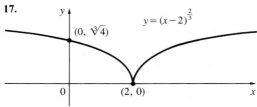

$y = (x-2)^{\frac{2}{3}}$

$(0, \sqrt[3]{4})$

0 $(2, 0)$

19.

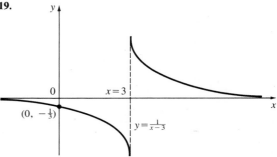

0 $x = 3$

$(0, -\frac{1}{3})$

$y = \frac{1}{x-3}$

21.

$y = \frac{1}{x^2-1}$

$x = -1$ $x = 1$

0

$(0, -1)$

23.

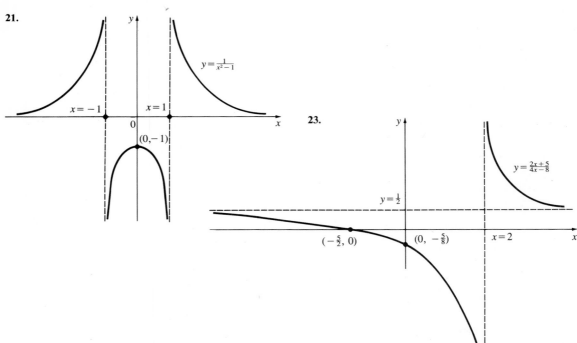

$y = \frac{2x+5}{4x-8}$

$y = \frac{1}{2}$

$(-\frac{5}{2}, 0)$ $(0, -\frac{5}{8})$ $x = 2$

25.

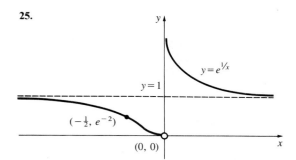

27.

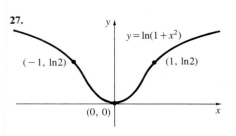

29.

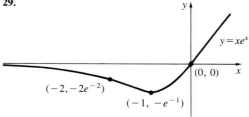

31. (a) Graph is concave down everywhere.   (b) Maximum: 725 at $q = 250$.

Section 11.3, page 627

1. Maximum: -24; minimum: $-30\frac{1}{4}$ **3.** Maximum: 890; minimum: -870 **5.** Maximum: 1; no minimum
7. Maximum: 106; minimum: -150 **9.** Maximum: 1; minimum: 0   **11.** Maximum: 2; minimum: -1
13. Maximum: 250; minimum: -54 **15.** Maximum: $\frac{1}{2}$; minimum: $\frac{1}{12}$ **17.** Maximum: $\frac{4}{5}$; minimum: 0
19. Maximum: 0; minimum: -4 **21.** Maximum: $\frac{7}{2}$; minimum: $\frac{5}{2}$ **23.** Minimum: 1; no maximum
25. Maximum: 20; minimum: 0   **27.** Maximum: ln 10; minimum: 0   **29.** Maximum: $e^2/2$; minimum: $e/2$
31. Maximum: $\ln(1 + e)$; minimum: ln 2   **33.** $l = w = 75$ cm   **35.** All sides $= 8$ cm   **37.** All wire in a circle
39. $r = \sqrt[3]{50/3\pi}$; $h = \sqrt[3]{450/\pi}$ **41.** $\sqrt[3]{180}$ by $\sqrt[3]{180}$ by $360/(180)^{2/3}$ **43.** 10 and 10   **45.** 1333 pigeons
47. Fleet size: 60; revenue: \$70,000   **49.** (a) \$620   (b) \$841,000   (c) 145   **51.** 750   **53.** $t = 0$ **55.** (a) 64 ft   (b) $t = 2$
57. $x = A/2$ **59.** $(-2, 1)$

Section 11.4, page 636

1. 7500   **3.** No change   **5.** Export 4750 bottles   **7.** Increased revenue (all problems)   **9.** $q = 0$ **11.** Increase
13. Make 8 batches   **15.** $\eta(q) = (q - 5000)/2q$ **17.** \$8000 in company 1 and \$2000 in company 2
23. (a) $q = \dfrac{\alpha - b}{2(\beta + a)}$ (b) $\eta(q) = \dfrac{\alpha}{\beta q} - 1$ (c) $q = \dfrac{\alpha}{2\beta}$

Section 11.5, page 641

1. $-\frac{10}{3}$ **3.** $\frac{3}{2}$ ft/sec   **5.** 372.678 km/hr   **7.** -0.359 m³/sec; decreasing   **9.** -121.75 ft/sec   **11.** 0.4 ft/sec   **13.** 50π cm/sec
15. $-\frac{8}{3}$ ft³/hr   **17.** 5 ft/sec   **19.** $\frac{11}{2}\pi r^2$ **21.** $-\$9300/$mo.   **23.** $-\$730/$wk.

Section 11.6, page 651

1. $y = ce^{3x}$ **3.** $p = ce^{-t}$ **5.** $x = 5e^t$ **7.** After 20 days: 62,500; after 30 days: 156,250

9. In 1980: 455,530; in 2000: 1,512,412 **11.** 2026 **13.** 1998 **15.** (a) 119°F (b) 2.256 hr **17.** 2871.3 yr **19.** 4.62×10^6 yr
21. (a) 625.53 mb (b) 303.42 mb (c) −429.03 mb (d) 348.63 mb (e) 57,396.3 m
23. (a) 10,001 (b) $7.28 \approx 7$ (c) 8.92 days

Section 11.7, page 659

1. $x_1 = 2.5$, $x_5 = 2.236069$ **3.** $x_1 = 2.5$, $x_5 = 2.154435$
5. Root 1: $x_1 = -2$, $x_4 = -2.090521$; root 2: $x_1 = 1$, $x_5 = 0.2438485$; root 3: $x_1 = 7$, $x_5 = 7.846672$
7. One root: $x_1 = 7$, $x_{16} = 10.61573$ **9.** For reciprocal of r, $x_{n+1} = 2x_n - rx_n^2$ **11.** $x_1 = 3$, $x_5 = 2.07058$
13. Annual interest: 42.416 %

Review Exercises for Chapter 11, page 660

1. (a) Decreasing: $(-\infty, \frac{3}{2})$; increasing: $(\frac{3}{2}, \infty)$ (b) $x = \frac{3}{2}$
(c) Minimum: $-\frac{25}{4}$ at $x = \frac{3}{2}$ (d) None
(e) Always concave up
(f)

3. (a) Increasing: $(-\infty, \infty)$ (b) Critical at $x = 0$
(c) No maximum or minimum (d) (0, 2)
(e) Concave up on $(0, \infty)$, concave down on $(-\infty, 0)$
(f)

5. (a) Increasing: $(0, \infty)$ (b) Critical at $x = 0$
(c) No maximum; minimum: 0 at $x = 0$
(d) None (e) Always concave down
(f)

7. (a) Increasing: $(4, \infty)$; decreasing: $(-\infty, 4)$ (b) $x = 4$
(c) Minimum: 0 at $x = 4$ (d) None (e) No concavity
(f)

9. −2.915 m/sec **11.** Maximum: 358; minimum: −10 **13.** Maximum: 0; minimum: $-\frac{2}{3}$ **15.** $(\frac{12}{13}, -\frac{18}{13})$ **17.** $t = 8$
19. $q = 1250$ **21.** $P(5) = 21,170$, $P(10) = 44,817$ **23.** (a) 54.853°C (b) 67.557 min **25.** 3.106 wk
27. One root; $x_1 = 1$, $x_5 = 1.752172$

Chapter 12

Section 12.1, page 669

1. $x + C$ **3.** $ax + C$ **5.** $\dfrac{x^3}{6} + C$ **7.** $x^7 + C$ **9.** $\frac{3}{4}x^{4/3} + C$ **11.** $-2/\sqrt{x} + C$ **13.** $x + \dfrac{x^2}{2} + \dfrac{x^3}{3} + \dfrac{x^4}{4} + \dfrac{x^5}{5} + C$

15. $\dfrac{x^{11}}{11} - \dfrac{x^9}{9} + \dfrac{7x^4}{2} - x^2 + 9x + C$ **17.** $\frac{9}{4}x^{4/3} + \frac{9}{2}x^{2/3} + C$ **19.** $-\frac{4}{9}x^{-3} + \dfrac{5x^{-4}}{28} - \dfrac{6x^{-5}}{55} + C$

21. $-\frac{3}{5}x^{-4/3} + \frac{8}{55}x^{-11/8} + \frac{1}{7}x^{7/5} + C$ **23.** $4 \ln|x| + x^{-2} + 7e^x + C$ **25.** $y = \frac{2}{3}x^3 + x^2 - \frac{28}{3}$ **27.** $y = \frac{78}{11}x^{11/6} - 3x + \frac{109}{11}$

29. $y = 3e^x - 5x + 4$

Section 12.2, page 672

1. $4q$ **3.** $R(q) = 50q - 0.02q^2 + 0.002\dfrac{q^3}{3}$ **5.** $R(q) = 75q + \frac{50}{13}q^{1.3} - \dfrac{q^{2.8}}{140}$ **7.** \$3963.54   **9.** $C(q) = 50q + 0.025q^2 + 600$

11. $750q + 0.16q^{3/2} - \frac{2}{125}q^{5/2} + 1000$ **13.** \$92,178.52   **15.** $C(100) = \$120,890$, $C(500) = \$2,302,223.33$ **17.** 3333 units

19. $H(t) = -16t^2 + 2000t$

Section 12.3, page 680

1. $\frac{2}{3}(9 + x)^{3/2} + C$ **3.** $-\frac{2}{27}(10 - 9x)^{3/2} + C$ **5.** $-\dfrac{(1 - x)^{11}}{11} + C$ **7.** $\frac{1}{5}(1 + 2x)^{5/2} + C$ **9.** $\frac{3}{8}(1 + x^2)^{4/3} + C$

11. $(t^2 + 2t^3)^{1/2} + C$ **13.** $-\frac{1}{2}(1 + \sqrt{x})^{-4} + C$ **15.** $-\frac{3}{16}\left(1 + \dfrac{1}{v^2}\right)^{8/3} + C$ **17.** $\frac{1}{3}(ax^2 + 2bx + C)^{3/2} + C$

19. $\frac{7}{8}(ax^2 + 2bx + c)^{4/7} + C$ **21.** $\dfrac{2}{3(n + 1)}(\alpha^2 + t^{n+1})^{3/2} + C$ **23.** $-\frac{2}{9}(\alpha^3 - p^3)^{3/2} + C$ **25.** $\ln|x + 5| + C$

27. $\frac{1}{100} \ln|1 + 100x| + C$ **29.** $-e^{1-x} + C$ **31.** $\frac{1}{2} \ln(1 + x^2) + C$ **33.** $\dfrac{1}{n + 1} \ln|1 + x^{n+1}| + C$ **35.** $\frac{1}{4}e^{4x} + C$

37. $\frac{1}{2} \ln(1 + e^{2x}) + C$ **39.** $\frac{1}{3}e^{x^3} + C$ **41.** $-e^{1/x} + C$ **43.** $\ln(e^x + 4) + C$ **45.** $C(q) = 8\sqrt{q + 4} + 984$ **47.** \$1049.33

49. (a) $Y_0 e^{0.03t}$ (b) $\dfrac{kY_0}{0.03}(e^{0.03t} - 1) + D_0$ (c) $\dfrac{k}{0.03}(1 - e^{0.03t}) + \dfrac{D_0}{Y_0}e^{-0.03t}$

Section 12.4, page 684

1. $\dfrac{x}{3}e^{3x} - \frac{1}{9}e^{3x} + C$ **3.** $4x^2e^{x/4} - 32xe^{x/4} + 128e^{x/4} + C$ **5.** $\dfrac{x^4(4 \ln x - 1)}{16} + C$ **7.** $-\frac{4}{3}x\left(1 - \dfrac{x}{2}\right)^{3/2} - \frac{16}{15}\left(1 - \dfrac{x}{2}\right)^{5/2} + C$

9. $x - \ln|x + 2| + C$ **11.** $x \ln(x + 1) + \ln(x + 1) - x + C$ **13.** $-x^3e^{-x} - 3x^2e^{-x} - 6xe^{-x} - 6e^{-x} + C$

Section 12.5, page 691

1. $\frac{33}{5}$ **3.** $\frac{26}{3}$ **5.** $\dfrac{c_1}{3}(b^3 - a^3) + \dfrac{c_2}{2}(b^2 - a^2) + c_3(b - a)$ **7.** $\frac{3}{2}(8^{2/3} - 1^{2/3}) + \frac{21}{4}(8^{4/3} - 1) = 83\frac{1}{4}$ **9.** 0   **11.** $\frac{80}{3}$ **13.** $\frac{41}{6}$

15. $\dfrac{875}{26,013}$ **17.** $-\frac{1}{101}$ **19.** $\frac{74}{3}$ **21.** $\frac{1}{16}(4^{4/3} - 1)$ **23.** $-\frac{3}{16}((\frac{5}{4})^{8/3} - (2)^{8/3})$ **25.** $\frac{1}{6}(e^6 - 1)$ **27.** 1   **29.** $\frac{1}{2}(4^{\ln 2} - 2^{\ln 2})$ **31.** $\ln\sqrt{2}$

33. $\frac{1}{2}\ln\left(\dfrac{1 + e^2}{2}\right)$ **35.** $\ln\left(\dfrac{e^2 + 4}{5}\right)$ **37.** $\dfrac{2e^3 + 1}{9}$ **39.** $\dfrac{e^2 + 1}{4}$ **41.** \$480   **43.** $-\$28,837.70$ **45.** $8(\sqrt{104} - \sqrt{148}) \approx -\$15.74$

47. \$102.86   **49.** 609.4 m   **51.** 12,000   **53.** $\frac{19}{2}$ **55.** $\frac{4}{3}$ **57.** $(2^{r+1} - 1)/(2r + 2)$ **59.** $\frac{244}{3}$ **61.** $\frac{1}{3}$ **63.** $\frac{4}{3}\ln 4 - 1$

65. \$45 per unit   **67.** -80 ft/sec   **69.** \$198,578.13

Section 12.6, page 701

1. 4   **3.** $\frac{10}{3}\sqrt{5}$ **5.** $\ln(\frac{7}{2})$ **7.** $\frac{32}{3}$ **9.** $\frac{32}{3}$ **11.** $\frac{343}{6}$ **13.** 8   **15.** $(b - a)^3/6$ **17.** $\frac{1}{3}(64 - 7\sqrt{7})$ **19.** $1 - 51e^{-50}$

21. $-(10^6 + 1)e^{-(10)^6} + 1$ **25.** (a) 0   (b) 200

Section 12.7, page 707

1. Converges to $\frac{1}{2}$ **3.** Diverges   **5.** Diverges   **7.** Converges to 1   **9.** Converges to $-\frac{1}{2}$ **11.** Converges to 1   **13.** Diverges
15. Converges to $-\frac{1}{4}$ **17.** Diverges   **19.** Converges to $\frac{1}{4}$ **21.** Converges to 0   **23.** Converges to 2000   **25.** $\frac{1}{2}$ **27.** $1/b$
29. $\frac{3}{4}e^{-12} + \frac{1}{16}e^{-12}$ **31.** \$458,333.33   **33.** Invest in the business!

Section 12.8, page 717

1. 56   **3.** $\frac{39}{2}$ **5.** $\frac{3}{4}$ **7.** $\frac{2}{3}$ **9.** $\frac{15}{4}$ **11.** (a) Quarter of circle of radius 1   (b) $\pi/4$

Section 12.9, page 725

1. $\frac{1}{6}$ **3.** $\frac{137}{12}$ **5.** $\frac{343}{6}$ **7.** $\frac{355}{3}$ **9.** $388\frac{4}{5}$ **11.** $42\frac{2}{3}$ **13.** $\frac{27}{5}$ **15.** 3.5   **17.** \$1250   **19.** \$833.33   **21.** (a) 25   (b) \$41.67   (c) \$125

Section 12.10, page 731

1. (a) $\frac{1}{2}$ (b) 0   (c) $\frac{1}{2}$ (d) Estimate is exact   **3.** (a) $\frac{11}{32}$ (b) 0.01042   (c) $\frac{1}{3}$ (d) $\frac{1}{96}$
5. (a) 6.448104763   (b) 0.136834   (c) 6.3890561   (d) 0.059048   **7.** (a) 1.218760835   (b) 0.0003255   (c) 1.21895   (d) 0.00019
9. (a) 2.861259   (b) 0.0067   (c) 2.8629515   (d) 0.00169   **11.** 0.8194483   **13.** 1.352446   **15.** 1.488737   **17.** 0.9091617
19. 0.9841199   **21.** 0.05309   **23.** 0.01333   **25.** 0.000802   **27.** 0.999936   **29.** (a) 1   (b) It approaches 1

Review Exercises for Chapter 12, page 732

1. $\dfrac{x^6}{6} + C$ **3.** $-\frac{1}{4}$ **5.** $\frac{2}{3}\ln|1 + x^3| + C$ **7.** 20   **9.** $-\dfrac{1}{2(u + 3)^2} + C$ **11.** $\frac{1}{2}\ln 2$ **13.** 17,155,451.5   **15.** $\frac{1}{3}\ln|\ln x| + C$

17. $3\ln 2 - 1$ **19.** $\frac{49}{20}$ **21.** $C(q) = 20q - \dfrac{q^2}{10} + 300$ **23.** \$4552.67   **25.** 42   **27.** $\frac{9}{2}$ **29.** 3   **31.** $\frac{1}{16}$ **33.** $\frac{1}{6}$ **35.** \$1250

37. (a) \$117.71   (b) \$521.24 consumers' surplus; \$745.64 producers' surplus   **39.** 0.9253959

Chapter 13

Section 13.1, page 742

1. (a) $\{(x, y): x \in \mathbb{R}, y \in \mathbb{R}\}$ (b) 235   **3.** (a) $\{(x, y): x^2 + y^2 \le 9\}$ (b) 2   **5.** (a) $\{(x, y): x \in \mathbb{R}, y \in \mathbb{R}\}$ (b) -3.0366
7. (a) $\{(x, y): x \in \mathbb{R}, y \in \mathbb{R}\}$ (b) 1   **9.** (a) $\{(x, y): y \ne x^2 + 4\}$ (b) $\frac{3}{55}$
11.

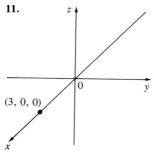

13.

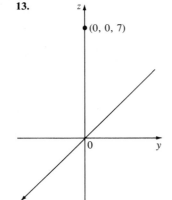

15.

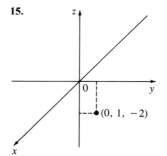

17.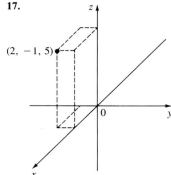

$(2, -1, 5)$

19.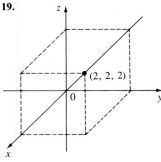

$(2, 2, 2)$

21. (a) \$947.50 (b) \$1067.50 **23.** (a) $25,000.\sqrt[3]{45}$ (b) 21% decrease (c) 26% increase
25. (a) 6 cm^3 (b) 36 in^3 (c) 125 in^3 **27.** \$11,264 **29.** \$1984

Section 13.2, page 752

1. $\dfrac{\partial z}{\partial x} = 2xy, \dfrac{\partial z}{\partial y} = x^2$ **3.** $\dfrac{\partial z}{\partial x} = 3x^2, \dfrac{\partial z}{\partial y} = \dfrac{1}{2\sqrt{y}}$ **5.** $\dfrac{\partial z}{\partial x} = 2x, \dfrac{\partial z}{\partial y} = 14y$ **7.** $\dfrac{\partial z}{\partial x} = 4y, \dfrac{\partial z}{\partial y} = 4x + 45y^4$

9. $\dfrac{\partial z}{\partial y} = 17y^4 - 60x^{19}, \dfrac{\partial z}{\partial y} = 68xy^3$ **11.** $\dfrac{\partial z}{\partial x} = \dfrac{1}{1-y}, \dfrac{\partial z}{\partial y} = \dfrac{1+x}{(1-y)^2}$ **13.** $\dfrac{\partial z}{\partial x} = \dfrac{4}{y^5}, \dfrac{\partial z}{\partial y} = -\dfrac{20x}{y^6}$

15. $\dfrac{\partial z}{\partial x} = \dfrac{2}{2x - 5y}, \dfrac{\partial z}{\partial y} = \dfrac{-5}{2x - 5y}$ **17.** $\dfrac{\partial z}{\partial x} = 3(2x + \ln y)^{1/2}, \dfrac{\partial z}{\partial y} = \dfrac{3(2x + \ln y)^{1/2}}{2y}$ **19.** 3 **21.** 48

23. $\dfrac{\partial w}{\partial x} = yz, \dfrac{\partial w}{\partial y} = xz, \dfrac{\partial w}{\partial z} = xy$ **25.** $\dfrac{\partial w}{\partial x} = \dfrac{\partial w}{\partial y} = \dfrac{\partial w}{\partial z} = \dfrac{1}{2\sqrt{x + y + z}}$ **27.** $\dfrac{\partial w}{\partial x} = e^{x+2y+3z}, \dfrac{\partial w}{\partial y} = 2e^{x+2y+3z}, \dfrac{\partial w}{\partial z} 3e^{x+2y+3z}$

29. $\dfrac{\partial w}{\partial x} = \dfrac{y}{z} e^{xy/z}, \dfrac{\partial w}{\partial y} = \dfrac{x}{z} e^{xy/z}, \dfrac{\partial w}{\partial z} = -\dfrac{xy}{z^2} e^{xy/z}$ **31.** $z = -\frac{1}{4}x + \frac{5}{4}; y = 1$

33. Product A: $\dfrac{\partial C}{\partial q_1} = 3 - 0.006q_1$; product B: $\dfrac{\partial C}{\partial q_2} = 2.5 - 0.014q_2$

35. $\dfrac{\partial R}{\partial q_1} = \dfrac{50}{1 + 50q_1 + 75q_2} + \dfrac{20}{\sqrt{1 + 40q_1 + 125q_2}}; \dfrac{\partial R}{\partial q_2} = \dfrac{75}{1 + 50q_1 + 75q_2} + \dfrac{125}{2\sqrt{1 + 40q_1 + 125q_2}}$

37. $\dfrac{\partial p}{\partial p_2} = 274 + 2p_1 - 4p_2$ **39.** $\dfrac{\partial p}{\partial p_1} = 75 + 1.8p_2 - 5p_1$

41. (a) $\dfrac{\partial F}{\partial L} = \dfrac{500}{3}\left(\dfrac{K}{L}\right)^{2/3}, \dfrac{\partial F}{\partial K} = \dfrac{1000}{3}\left(\dfrac{L}{K}\right)^{1/3}$ (b) $\dfrac{\partial F}{\partial L} = 242.03, \dfrac{\partial F}{\partial K} = 276.61$ **43.** $\dfrac{\partial F}{\partial L} = ca\left(\dfrac{K}{L}\right)^{1-a}, \dfrac{\partial F}{\partial K} = c(1 - a)\left(\dfrac{L}{K}\right)^a$

45. $f_{xy} = 2y, f_{yx} = 2y, f_{xx} = 0, f_{yy} = 2x$ **47.** $f_{xx} = \dfrac{-9}{(3x - 4y)^2}, f_{yx} = \dfrac{12}{(3x - 4y)^2}, f_{xy} = \dfrac{12}{(3x - 4y)^2}, f_{yy} = \dfrac{-16}{(3x - 4y)^2}$

49. $f_{xy} = \dfrac{-2(x + y)}{(x - y)^3}, f_{xx} = \dfrac{4y}{(x - y)^3}, f_{yx} = \dfrac{-2(x + y)}{(x - y)^3}, f_{yy} = \dfrac{4x}{(x - y)^3}$

51. $f_{xx} = 0, f_{xy} = f_{yx} = z^{-1}, f_{xz} = f_{zx} = -yz^{-2}, f_{yy} = 0, f_{yz} = f_{zy} = -xz^{-2}, f_{zz} = 2xyz^{-3}$

53. $f_{xx} = \dfrac{-y^2}{(xy + z)^2}, f_{yy} = \dfrac{-x^2}{(xy + z)^2}, f_{xy} = f_{yx} = \dfrac{z}{(xy + z)^2}, f_{xz} = f_{zx} = \dfrac{-y}{(xy + z)^2}, f_{zz} = \dfrac{-1}{(xy + z)^2}, f_{yz} = f_{zy} = \dfrac{-x}{(xy + z)^2}$

Section 13.3, page 764

1. $(0, 0)$, local minimum **3.** $(-2, 1)$, local minimum **5.** $(-2, 1)$, local minimum
7. $(\sqrt{5}, 0)$, local minimum; $(-\sqrt{5}, 0)$, local maximum; $(-1, 2)$, saddle point; $(-1, -2)$, saddle point

9. $(\frac{1}{2}, 1)$, saddle point; $(-\frac{1}{2}, 1)$, local minimum; $(\frac{1}{2}, -1)$, local maximum; $(-\frac{1}{2}, -1)$, saddle point
11. $(-2, -2)$, local minimum   **13.** $(0, 0)$, saddle point; $(0, 4)$, saddle point; $(4, 0)$, saddle point; $(\frac{4}{3}, \frac{4}{3})$, maximum
15. All points (x, y) such that $2x + 3y = 0$ are critical; local minima   **17.** $\frac{50}{3}, \frac{50}{3}, \frac{50}{3}$ **19.** $\frac{50}{3}, \frac{50}{3}, \frac{50}{3}$
21. $p_1 = \$37.63$, $p_2 = \$62.87$ **23.** (a) 8 ft $\times$ 8 ft $\times$ 7.5 ft   (b) \$2880
25. (a) $P(a, d) = 150(N(a, d))$ **27.** $(x_1, x_2) = (6, 3)$; yields a maximum of \$40,000   **29.** $y = -\frac{27}{26}x + \frac{30}{13}$

Section 13.4, page 774

1. local minimum of $-\frac{71}{2}$ at $(-\frac{1}{2}, -\frac{11}{2})$; no maximum   **3.** local maximum of 12 at $(2, 2)$; no minimum
5. local minimum of 1 at $(1, 0, 0)$ and $(-1, 0, 0)$; no maximum
7. local maximum of $\sqrt{3}$ at $(1/\sqrt{3}, 1/\sqrt{3}, 1/\sqrt{3})$; local minimum of $-\sqrt{3}$ at $(-1/\sqrt{3}, -1/\sqrt{3}, -1/\sqrt{3})$ **9.** $d = \sqrt{2}/2$
11. Hint: area of a triangle with sides a, b, c is $\sqrt{s(s-a)(s-b)(s-c)}$ where $s = \dfrac{a+b+c}{2}$ **13.** $p_1 = \$63$, $p_2 = \$31.50$

15. $p_1 = \$46.61$, $p_2 = \$42.37$ **17.** $\dfrac{x}{y} = \frac{1}{2}$ **19.** (a) $K = \frac{625}{6}$, $L = \frac{250}{3}$ (b) 48,350   (c) $\frac{1000}{1600} = \frac{5}{8}$

21. (a) $l = 3.71$, $h = 3.16$ (b) \$1.62   **23.** $A = 10, B = 10$ **25.** $C = 6, E = 15$
27. (a) minimum $i + d = 200$ (b) $i = d = 100$

Review Exercises for Chapter 13, page 777

1. (a) $\{(x, y) | x \in \mathbb{R}, y \in \mathbb{R}\}$ (b) 8   **3.** (a) $\mathbb{R} \times \mathbb{R} - \{(0, 0)\}$ (b) $\sqrt{58}/58$ **5.** (a) $\{(x, y, z) | x^2 + y^2 + z^2 \le 1\}$ (b) $\sqrt{83}/12$
7.

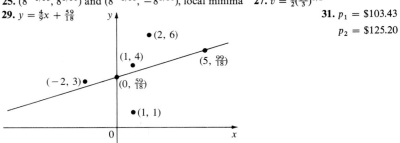

9. $f_x = 3, f_y = 2$ **11.** $f_x = 12x^2 y^7, f_y = 28x^3 y^6$ **13.** $f_x = \dfrac{1}{x - y + 4z}, f_y = \dfrac{-1}{x - y + 4z}, f_z = \dfrac{4}{x - y + 4z}$
15. $f_{xx} = 0, f_{xy} = f_{yx} = 3y^2, f_{yy} = 6xy$ **17.** $f_{xx} = f_{xy} = f_{yx} = f_{yy} = 0$
19. $f_{xx} = 6xyz^4, f_{xy} = f_{yx} = 3x^2 z^4, f_{yy} = 0, f_{yz} = f_{zy} = 4x^3 z^3, f_{zz} = 12x^3 yz^2, f_{xz} = f_{zx} = 12x^2 yz^3$ **21.** $(0, 0)$, local minimum
23. $(-\sqrt{2}, -\sqrt{2}/2)$, local maximum; $(\sqrt{2}, -\sqrt{2}/2)$, saddle point; $(\sqrt{2}, \sqrt{2}/2)$, local minimum $(-\sqrt{2}, \sqrt{2}/2)$, saddle point
25. $(8^{-1/15}, 8^{1/10})$ and $(8^{-1/15}, -8^{1/10})$, local minima   **27.** $v = \frac{1}{2}(\frac{10}{3})^{3/2}$
29. $y = \frac{4}{9}x + \frac{59}{18}$ **31.** $p_1 = \$103.43$ $p_2 = \$125.20$

Appendix A.1

Section A1.1, page A-8

1. 4   **3.** 0   **5.** 8   **7.** $4x^2 - 6x + 8$ **9.** $3x^3 + x^2 + 2x + 1$ **11.** $3x^3 - 3x^2 + 8x - 7$ **13.** $-21x^3 + 15x^2 - 47x + 37$
15. $6x^5 - 11x^4 + 25x^3 - 25x^2 + 29x - 12$ **17.** $3x^7 - 6x + 9$ **19.** $x^7 - x^4 - 2x + 6$ **21.** $4 + 7 = 11$
23. $x^2 + 6x + 8$, degree $= 2$ **25.** $x^2 - 25$, degree $= 2$ **27.** $-15x^2 + 31x - 10$, degree $= 2$ **29.** $x^4 - 1$, degree $= 4$
31. $-18x^4 - 15x^3 + 18x^2 - 14x + 4$, degree $= 4$ **33.** $x^5 + x^3 - x^2 - 1$, degree $= 5$
35. $12x^5 - 27x^3 + 8x^2 + 15x - 10$, degree $= 5$ **37.** $acx^6 + (ad + bc)x^4 + aex^3 + bdx^2 + ebx$, degree $= 6$
39. $x^{30} - 2x^{10}$, degree $= 30$ **41.** $x^8 - 4x^4 + 12x^2 - 9$, degree $= 8$ **43.** $(x - 1)(x - 3)$ **45.** $(x + 1)(x + 1)$
47. $(x + 5)(x + 1)$ **49.** $(x - 7)(x + 6)$ **51.** $(x + 7)(x + 6)$ **53.** $(x - 8)(x + 2)$ **55.** $x(x - a)$ **57.** $x(bx - 3c)$
59. $-5(x - 6)(x - 7)$ **61.** $-4(x + 6)(x + 3)$ **63.** $2(x - a)(x - b)$ **67.** $x^2 - 5x + 4$ **69.** $x^2 - 3x - 4$ **71.** $x^2 - 3x$
73. $x^3 - 5x^2 + 2x + 8$ **75.** $x^3 - 9x^2 + 24x - 20$ **77.** $(x - 1)(x + 1)(x + 2)$ **79.** $(x - 1)(x - 1)(x - 3)$
81. $-3(x - 2)^2(x + 3)$ **83.** $4(x - 1)(x - 1)(x - 3)$ **85.** $x^2(x^2 + 1)$ **87.** $(x - 1)(x + 1)(x - 2)(x + 2)$
89. $(x^2 - 7)(x^2 + 6)$ **91.** $x(x^2 + x + 1)(x^2 - x + 1)$ **93.** $(x^2 + x\sqrt{2} + 1)(x^2 - x\sqrt{2} + 1)(x + 1)(x - 1)$ **95.** $x^3 + a^3$
99. Irreducible   **101.** $(x - \frac{1}{3})(x - \frac{1}{3})$ **103.** $(5x - 3)(5x + 3)$ **105.** $(7x - 1)(7x - 1)$ **107.** $(x + 2)(x^2 - 2x + 4)$
109. $(x - \frac{1}{2})(x^2 + \frac{1}{2}x + \frac{1}{4})$ **111.** $(x^2 + \sqrt{6}x + 3)(x^2 - \sqrt{6}x + 3)$ **113.** $(25x^2 + 1)(5x - 1)(5x + 1)$

Section A1.2, page A-16

1. $x = 3$ or $x = 1$ **3.** No real roots   **5.** $x = -1$ **7.** $x = 1$ **9.** $x = -1$ or $x = -5$ **11.** No real roots
13. $x = -6$ or $x = -7$ **15.** $x = 0$ or $x = -5$ **17.** $x = 0$ or $x = -a$ **19.** No real roots   **21.** $x = 6$ or $x = 7$
23. No real roots   **25.** No real roots   **27.** No real roots   **29.** $x = \frac{5}{2}$ or $x = -\frac{1}{2}$ **31.** No real roots   **33.** $\frac{1}{2} \pm \dfrac{\sqrt{5}}{2}$

35. $-\frac{2}{3} \pm \dfrac{\sqrt{10}}{3}$ **37.** Equilibrium quantity is 10 units; equilibrium price is \$196.00.

39. Units supplied and demanded $= 40$; equilibrium price is \$30.00.

Section A1.3, page A-20

1. $2x$ **3.** $\dfrac{x + 2}{3x^2 + 4}$ **5.** $\dfrac{1}{2y^3}$ **7.** $\dfrac{1}{xy}$ **9.** $\frac{1}{2}$ **11.** $\dfrac{z^2}{4}$ **13.** $\dfrac{x + 1}{x^3}$ **15.** $\dfrac{x + 2}{x + 1}$ **17.** $\dfrac{z^2}{(z + 1)^2}$ **19.** 1   **21.** Cannot be simplified

23. $\dfrac{x + 3}{x - 7}$ **25.** $\dfrac{z - 2}{z - 9}$ **27.** $\dfrac{2(6x^2 - 5x + 1)}{3(4x^2 - 4x - 1)}$ **29.** $\dfrac{6z + 1}{6z - 7}$ **31.** $\dfrac{x^2 + xy + y^2}{x^2 - 2xy + y^2}$ **33.** $\dfrac{4(z - 3)}{5(z - 1)}$ **35.** $\dfrac{x^3 - x^2 + x + 1}{(x - 2)(x^2 + 1)^2}$

37. $\dfrac{4 + y}{2}$ **39.** $\dfrac{1 + 4x}{2}$ **41.** $\dfrac{3x - 8}{x^2}$ **43.** $\dfrac{-1}{2x(1 - x)}$ **45.** $\dfrac{6 - y^2}{2y}$ **47.** $-\dfrac{2}{15s}$ **49.** $\dfrac{37}{6y^2}$ **51.** $\dfrac{7x - 3b - 4a}{(x - a)(x - b)}$ **53.** $\dfrac{-3x - 51}{(x + 5)(x - 7)}$

55. $\dfrac{4x + 16}{x^2 - 4}$ **57.** $\dfrac{4x^3 - 2x^2 - 2x + 2}{x^4 - 1}$ **59.** $\dfrac{4x - 6}{(x - 1)^2}$ **61.** $\dfrac{-2y + 36}{(y - 3)(y + 6)}$ **63.** $\dfrac{s - 1}{s + 1}$ **65.** $\dfrac{3y - 5x}{6x + 2y}$ **67.** $\dfrac{x(x + 4)}{2x^2 - 7x + 2}$

INDEX